AF595094

Cell Biology of Viral Infections

Cell Biology of Viral Infections

Editor

Pierre-Yves Lozach

MDPI • Basel • Beijing • Wuhan • Barcelona • Belgrade • Manchester • Tokyo • Cluj • Tianjin

Editor
Pierre-Yves Lozach
INRAE—Universitätsklinikum Heidelberg
Germany

Editorial Office
MDPI
St. Alban-Anlage 66
4052 Basel, Switzerland

This is a reprint of articles from the Special Issue published online in the open access journal *Cells* (ISSN 2073-4409) (available at: https://www.mdpi.com/journal/cells/special_issues/cell_viral_infection).

For citation purposes, cite each article independently as indicated on the article page online and as indicated below:

LastName, A.A.; LastName, B.B.; LastName, C.C. Article Title. *Journal Name* **Year**, *Volume Number*, Page Range.

ISBN 978-3-0365-0146-8 (Hbk)
ISBN 978-3-0365-0147-5 (PDF)

Cover image courtesy of Pierre-Yves Lozach.

Contents

About the Editor

Pierre-Yves Lozach obtained his PhD degree in Virology at the Pasteur Institute. In 2007, he joined the lab of Ari Helenius (ETH Zurich, Switzerland), first as a Marie Curie postdoctoral fellow and then as a group leader. In 2012, he was appointed as tenure track Assistant- Professor at INRS-Institut Armand-Frappier in Canada, and in 2013, granted the CellNetworks Chair in Virology at the University of Heidelberg. He is now group leader, heading the laboratory 'Cell Biology of Arboviral Infections' with dual affiliation to both the University Hospital Heidelberg (Heidelberg, Germany) and the National Institute of Agricultural Research and Environment (Lyon, France).

Editorial

Cell Biology of Viral Infections

Pierre-Yves Lozach

CellNetworks–Cluster of Excellence and Center for Integrative Infectious Diseases Research (CIID), Virology, University Hospital Heidelberg, 69120 Heidelberg, Germany; pierre-yves.lozach@med.uni-heidelberg.de

Received: 3 November 2020; Accepted: 5 November 2020; Published: 7 November 2020

Abstract: Viruses exhibit an elegant simplicity, as they are so basic, but so frightening. Although only a few are life threatening, they have substantial implications for human health and the economy, as exemplified by the ongoing coronavirus pandemic. Viruses are rather small infectious agents found in all types of life forms, from animals and plants to prokaryotes and archaebacteria. They are obligate intracellular parasites, and as such, subvert many molecular and cellular processes of the host cell to ensure their own replication, amplification, and subsequent spread. This special issue addresses the cell biology of viral infections based on a collection of original research articles, communications, opinions, and reviews on various aspects of virus-host cell interactions. Together, these articles not only provide a glance into the latest research on the cell biology of viral infections, but also include novel technological developments.

Keywords: 3D cell culture; cell death; host-virus interactions; inclusion bodies; intracellular trafficking; organoid; pluripotent stem cells; signaling pathways; cell stress responses; virus

---

Viruses are particles made up of nucleic acids and proteins, and sometimes, of lipids and glycans. Viral particles differ considerably from one isolate to another in terms of their structural, molecular, and genomic organization. Their sizes range from a few tens of nanometers to micrometers, and viral genomes can encode anywhere from a few genes to several hundreds. As obligate intracellular parasites, viruses exploit host cells to replicate, amplify, and subsequently spread from cell to cell and from host to host. Thousands of viruses have been sequenced thus far, but only a few are known to cause fatal diseases. However, viral infections have a colossal impact on public health, agricultural productivity, and the economy. International trade, urbanization, deforestation, and climate change, and human activity, in brief, are all factors that promote not only the emergence and re-emergence of viruses, but also their global spread. Recent illustrations are the Zika pandemic and the lasting problem created by the infamous coronavirus SARS-CoV-2 worldwide. Understanding the infection process at the molecular and cellular level is an obvious prerequisite for the development of diagnostic tools, preventive approaches, and therapeutic strategies against viruses.

Research on the cell biology of viral infections remains limited to a relatively small number of viruses, thus posing a serious challenge in our preparedness against future emerging viral infectious diseases. However, many breakthroughs in the field have occurred over the past decades, shedding light on the cellular life cycles of important pathogens such as human immunodeficiency (HIV) and hepatitis viruses. On the other hand, many of these investigations have led to major advances in molecular and cellular biology. To cite only a few, exciting discoveries include protein G of the vesicular stomatitis virus, which contributed to defining the exocytotic machinery, the many viruses employed as functional cargo to decipher endocytic pathways, viral fusion proteins, and developmental biology, and the retroviral reverse transcriptase, currently used in various biotechnology applications. The list is not limited to cell biology and can be easily expanded to fields as diverse as immunology, neurobiology, and molecular medicine. Viruses somehow represent interesting molecular and cellular bridges between all these scientific disciplines. This is exactly what this special issue aims to illustrate, bringing together original results and discussions on various aspects of viral infection cell biology and their implications for other topics in biology.

Infection starts when viruses attach to the host cell surface, which invariably implies binding to one or more cellular receptors, including proteins, carbohydrates, and lipids. The expression of these molecules is often tissue-specific. Although other factors and processes contribute to the infection program, the identification of receptors and coreceptors helps to define the virus tropism, and sometimes, to explain the virus-induced pathogenesis. Wielgat and colleagues discuss a possible link among coronaviruses, sialic acids as attachment factors, immune escape, and disease severity [1]. Ideally, preventing infection requires approaches targeting early virus–host cell interactions, before the release of the viral genome into the cytosol. In this sense, Luteijn et al. identified a peptide with broad-spectrum antiviral activity that targets phosphatidylserine in the viral particle envelope and prevents infection by poxviruses, HIV, hepatitis B virus, and Rift Valley fever virus [2]. This finding is consistent with the fact that viruses often use phosphatidylserine and apoptotic mimicry as a mechanism of entry into host cells [3].

After attachment to the cell surface and prior to viral replication, viruses must gain access to the cytosol, and most viruses are sorted into the endocytic machinery [4]. To release their genomic material into the cytosol, enveloped viruses fuse their membrane with that of endosomal vesicles, and nonenveloped viruses usually inflict damage in the endosomal membrane. Daussy and colleagues reviewed the different strategies by which nonenveloped viruses enter the cytosol and provided the latest knowledge on how these viruses control and prevent cellular stress responses following membrane injuries such as inflammation and autophagy [5]. After delivery into the cytosol, the virus journey is not necessarily over, as they very often must reach a specific subcellular location to initiate viral replication, a point illustrated by Grikscheit et al. [6]. By combining live cell microscopy and computer-based image analysis, the researchers provided evidence that the Ebola virus (EBOV) finely regulates actin polymerization to ensure the directed long-distance transport of its nucleocapsid.

After penetration into the cytosol, viral replication usually takes place in viral factories wherein nucleic acids and specific viral and cellular proteins accumulate. Two studies by the Hoenen laboratory reported that EBOV recruits carbamoyl-phosphate synthetase 2, aspartate transcarbamylase, dihydroorotase (CAD) and the nuclear RNA export factor NXF1 within inclusion bodies in infected cells [7,8]. CAD facilitates the replication and transcription of the EBOV genome, while NXF1 contributes to the expression of EBOV proteins. Van Huizen and McInerney reviewed how Old World alphaviruses manipulate the phosphatidylinositol-3-kinase (PI3K)-AKT signaling pathway and discuss the benefit for viral replication [9]. Of note, viral factories can be either devoid of any lipidic layer, such as Negri bodies, or delimited by cellular membranes [10]. In the case of hantaviruses, Davies et al. established that the Tula virus induces a dramatic reorganization of the Golgi to replicate [11]. Once viral replication begins, cells are infected.

The fight between host cell defenses and viruses is permanent at each stage of the viral life cycle. The work by the Grandvaux team points out the existence of a noncanonical pathway leading to an antiviral, immunoregulatory response [12]. The data support the view that the two factors STAT2 and IRF9 regulate distinct pools of genes with functions related to antiviral and immunoregulatory responses when the levels of both interferon β and tumor necrosis factor are elevated. Another example is that bovine herpesvirus 1 infection results in the proteasomal degradation of the transporter associated with antigen processing (TAP), a key player in major histocompatibility complex class I (MHC-I)-restricted antigen presentation. Wachalska et al. found that the process depends on the host cell factor p97, which is involved in the complex ER-associated degradation pathway [13]. Saulle and colleagues observed that ERAP2/Iso3 expression is induced by infection with different viruses, including SARS-CoV-2 [14]. ERAP2/Iso3 is an alternatively spliced isoform of rs2248374-G ERAP2 that, unlike the full transcript, is incapable of trimming peptides to be loaded onto MHC-I molecules. These results suggest that viruses trigger ERAP2/Iso3 expression to escape the adaptive immune response, which could contribute to the severity of virus-induced disease. In summary, cells have acquired many molecular weapons to control viral infections, but viruses have developed equally as many strategies to evade cell defenses.

To infect and replicate in host cells, viruses display not only the capacity to escape immune responses, but also to control cellular stress responses such as degradative pathways, autophagy, and cell death. Vector-borne viruses typically persist in their vector, but not in other hosts [15,16]. In this regard, much more work is needed to improve our understanding of the vector-borne virus life cycle and transmission to humans and other vertebrates.

Viruses often assemble within viral factories, but they sometimes depend on the endoplasmic reticulum (ER), Golgi, and subsequent exocytic pathways for the maturation and folding of their own surface and fusion proteins. This is the case, for instance, for viruses that bud at the plasma membrane such as influenza virus, HIV, and coronaviruses. The henipavirus fusion protein also requires endocytic uptake for cleavage by endosomal proteases, followed by recycling to the cell surface for incorporation into budding viral particles. In this regard, Fischer and colleagues highlight the interdependence of the exocytic and endocytic machinery to produce active viral fusion proteins and infectious Cedar particles, a middle pathogenic henipavirus [17].

The pathways by which viruses leave host cells are complex and difficult to investigate, but it is obvious that they are just as diverse as viral entry pathways, as various exocytic host factors and processes have been observed. Regarding this issue, the laboratory of Urtzi Garaigorta identified Erlin-1, a cholesterol-binding protein in the ER, as a proviral host cell factor participating in the production of infectious hepatitis C virus particles [18]. Chen and colleagues exemplified the central role of ion flow in the trafficking of intracellular vesicles, addressing the importance of calcium channels and pumps in virus penetration, assembly, and exit [19]. The review by Oechslin et al. nicely illustrates all these molecular and cellular processes in the context of hepatitis E virus (HEV), presenting the most recent information on the HEV life cycle, from virus entry and replication to assembly and release of new viral progenies [20].

Viral particles are engaged to transfer the viral genome from an infected cell to a noninfected cell. However, an increasing number of studies on cell-associated viral transmission have indicated the existence of alternative routes for the spread of viral material. Sid Ahmed et al. analyzed how the cellular environment may impact HIV spread in vivo [21]. This review developed from a recent paper by the same group showing that a 3D collagen environment restricts cell-free HIV infection, but promotes cell-associated viral propagation [22]. Other viruses interfere with cell and tissue differentiation as they spread. In this respect, Bilz and colleagues established that rubella virus hampers the formation of embryoid bodies and other 3D cell aggregates derived from human-induced pluripotent stem cells [23]. Whether this blockage is due to viral replication or required for viral spread remains to be clarified. This study suggests that the clinical signs of congenital rubella syndrome are due to an impairment of human development.

In this special issue, I intend to illustrate the possible viral life cycle stages, addressing virology issues from the perspective of cell biology, and from other angles. I strongly encouraged researchers to engage in discussions on the latest knowledge, technical issues, and technological challenges as well as prospects in the fields of viral infections, cell biology, and beyond. Based on 6 reviews, 1 opinion, 12 original research articles, and 1 communication reporting on a wide variety of viruses, this special issue can be considered a success. One of these reviews also demonstrates the interest in using viruses to develop new therapeutic approaches against noninfectious diseases. Nguyen et al. examined the potential of oncolytic viruses genetically modified to encode antitumor interleukine-12 in cancer therapy, with a special focus on herpes simplex virus [24]. Two research articles documenting new methods to study and analyze viral infections include the one by Tekes and colleagues reporting a protocol for culturing organoids derived from the ileum and colon and showing their usefulness in studying feline coronavirus infection [25], and the one by Potratz et al. describing a novel quantitative approach, based on computer analysis of microscopic images to analyze the cell tropism of neurotropic rabies virus [26].

I, however, do not want to give the impression that this special issue provides an exhaustive picture of the cell biology of viral infections. Some discussions are missing, due to a lack of experts in some cases, but the omissions are not deliberate. Virus receptors, endocytic pathways, and further processing of genomic material are important topics. The role of exosomes in virus dissemination and that of protein aggregation in virus-induced diseases are fascinating [27,28]. I also would have liked to provide a greater emphasis on technology, for instance, by including works on nanobodies, electron microscopy techniques, and OMICS. I only hope that this special issue provides the field of viral infection cell biology future research directions and stimulates research in some understudied and nascent areas. Without the fantastic contributions of authors, and the great help of Dolly Wang and other editorial assistants from Cells, this special issue would not have been possible. I would therefore like to thank all these great people.

Lastly, it is difficult in these pandemic times to edit a special issue on the cell biology of viral infections without further mentioning the recent viral zoonotic pandemic and the new coronavirus. In a blueprint list, the World Health Organization (WHO) identified pathogens for which there is an urgent need to develop diagnostics, therapies, and research. The list includes, among others, coronaviruses, and a dozen other zoonotic viruses. The successive epidemics and pandemics of SARS-CoV, Chikungunya virus, Middle East respiratory syndrome coronavirus, EBOV, and Zika virus are all signals over the past 20 years that should have alerted us. Nevertheless, emerging viruses continue to appear of interest only during epidemics and pandemics, which has not contributed to sustained funding, and therefore, to promoting long-term research on emerging zoonotic viruses. Many reasons, too many to discuss herein, are responsible for this situation. We can, however, legitimately speculate about how much time would have been saved in the fight against SARS-CoV version 2.0, if only a few of the financial losses due to global lockdown had been invested in research against emerging viruses, notably coronaviruses. The ten or so viruses on the WHO blueprint list account for less than 5% of the total virology research work referenced by PubMed.gov (Figure 1). Although incomplete, this approach provides a good idea of the collective effort that has been made to date, and as a result, of the important research effort, which remains to be made in the battle against these pathogens and future pandemics.

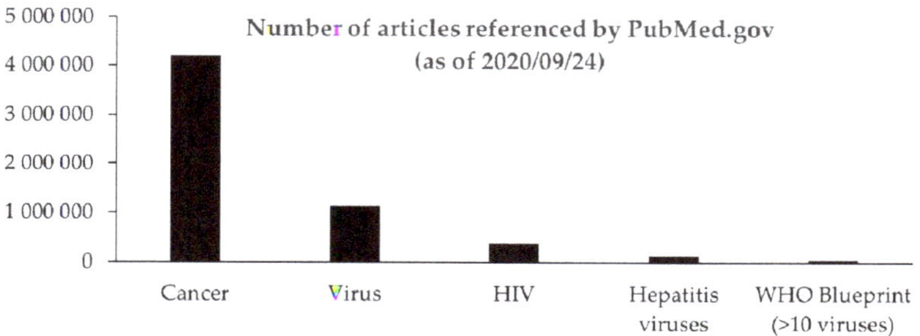

Figure 1. Number of articles on emerging viruses referenced by PuMed.gov. The keywords cancer, virus, HIV, and hepatitis viruses (A to E) were inputted into the search engine. Similarly, the WHO blueprint list represents the sum of results obtained for the next 10 keywords: Crimean–Congo hemorrhagic fever, Ebola, Marburg, Lassa fever, Middle East and severe acute respiratory syndrome coronavirus, Nipah, henipa, Rift Valley fever, and Zika.

Funding: This research by the Lozach group was funded by CellNetworks Heidelberg, the Deutsche Forschungsgemeinschaft (DFG, grant numbers LO-2338/1-1 and LO-2338/3-1).

Conflicts of Interest: The author declares no conflict of interest.

References

1. Wielgat, P.; Rogowski, K.; Godlewska, K.; Car, H. Coronaviruses: Is Sialic Acid a Gate to the Eye of Cytokine Storm? From the Entry to the Effects. *Cells* **2020**, *9*, 1963. [CrossRef]
2. Luteijn, R.D.; Praest, P.; Thiele, F.; Sadasivam, S.M.; Singethan, K.; Drijfhout, J.W.; Bach, C.; de Boer, S.M.; Lebbink, R.J.; Tao, S.; et al. A Broad-Spectrum Antiviral Peptide Blocks Infection of Viruses by Binding to Phosphatidylserine in the Viral Envelope. *Cells* **2020**, *9*, 1989. [CrossRef]
3. Amara, A.; Mercer, J. Viral apoptotic mimicry. *Nat. Rev. Microbiol.* **2015**, *13*, 461–469. [CrossRef]
4. Boulant, S.; Stanifer, M.; Lozach, P.Y. Dynamics of virus-receptor interactions in virus binding, signaling, and endocytosis. *Viruses* **2015**, *7*, 2794–2815. [CrossRef]
5. Daussy, C.F.; Wodrich, H. "Repair Me if You Can": Membrane Damage, Response, and Control from the Viral Perspective. *Cells* **2020**, *9*, 2042. [CrossRef]
6. Grikscheit, K.; Dolnik, O.; Takamatsu, Y.; Pereira, A.R.; Becker, S. Ebola Virus Nucleocapsid-Like Structures Utilize Arp2/3 Signaling for Intracellular Long-Distance Transport. *Cells* **2020**, *9*, 1728. [CrossRef]
7. Brandt, J.; Wendt, L.; Bodmer, B.S.; Mettenleiter, T.C.; Hoenen, T. The Cellular Protein CAD is Recruited into Ebola Virus Inclusion Bodies by the Nucleoprotein NP to Facilitate Genome Replication and Transcription. *Cells* **2020**, *9*, 1126. [CrossRef] [PubMed]
8. Wendt, L.; Brandt, J.; Bodmer, B.S.; Reiche, S.; Schmidt, M.L.; Traeger, S.; Hoenen, T. The Ebola Virus Nucleoprotein Recruits the Nuclear RNA Export Factor NXF1 into Inclusion Bodies to Facilitate Viral Protein Expression. *Cells* **2020**, *9*, 187. [CrossRef]
9. Van Huizen, E.; McInerney, G.M. Activation of the PI3K-AKT Pathway by Old World Alphaviruses. *Cells* **2020**, *9*, 970. [CrossRef]
10. Nevers, Q.; Albertini, A.A.; Lagaudriere-Gesbert, C.; Gaudin, Y. Negri bodies and other virus membrane-less replication compartments. *Biochim. Biophys. Acta Mol. Cell Res.* **2020**, *1867*, 118831. [CrossRef] [PubMed]
11. Davies, K.A.; Chadwick, B.; Hewson, R.; Fontana, J.; Mankouri, J.; Barr, J.N. The RNA Replication Site of Tula Orthohantavirus Resides within a Remodelled Golgi Network. *Cells* **2020**, *9*, 1569. [CrossRef]
12. Mariani, M.K.; Dasmeh, P.; Fortin, A.; Caron, E.; Kalamujic, M.; Harrison, A.N.; Hotea, D.I.; Kasumba, D.M.; Cervantes-Ortiz, S.L.; Mukawera, E.; et al. The Combination of IFN β and TNF Induces an Antiviral and Immunoregulatory Program via Non-Canonical Pathways Involving STAT2 and IRF9. *Cells* **2019**, *8*, 919. [CrossRef]
13. Wąchalska, M.; Graul, M.; Praest, P.; Luteijn, R.D.; Babnis, A.W.; Wiertz, E.J.H.J.; Bieńkowska-Szewczyk, K.; Lipińska, A.D. Fluorescent TAP as a Platform for Virus-Induced Degradation of the Antigenic Peptide Transporter. *Cells* **2019**, *8*, 1590. [CrossRef]
14. Saulle, I.; Vanetti, C.; Goglia, S.; Vicentini, C.; Tombetti, E.; Garziano, M.; Clerici, M.; Biasin, M. A New ERAP2/Iso3 Isoform Expression Is Triggered by Different Microbial Stimuli in Human Cells. Could It Play a Role in the Modulation of SARS-CoV-2 Infection? *Cells* **2020**, *9*, 1951. [CrossRef]
15. Léger, P.; Lozach, P.-Y. Bunyaviruses: From transmission by arthropods to virus entry into the mammalian host first-target cells. *Future Virol.* **2015**, *10*, 859–881. [CrossRef]
16. Mazelier, M.; Rouxel, R.N.; Zumstein, M.; Mancini, R.; Bell-Sakyi, L.; Lozach, P.Y. Uukuniemi Virus as a Tick-Borne Virus Model. *J. Virol.* **2016**, *90*, 6784–6798. [CrossRef]
17. Fischer, K.; Groschup, M.H.; Diederich, S. Importance of Endocytosis for the Biological Activity of Cedar Virus Fusion Protein. *Cells* **2020**, *9*, 2054. [CrossRef] [PubMed]
18. Whitten-Bauer, C.; Chung, J.; Gómez-Moreno, A.; Gomollón-Zueco, P.; Huber, M.D.; Gerace, L.; Garaigorta, U. The Host Factor Erlin-1 is Required for Efficient Hepatitis C Virus Infection. *Cells* **2019**, *8*, 1555. [CrossRef] [PubMed]
19. Chen, X.; Cao, R.; Zhong, W. Host Calcium Channels and Pumps in Viral Infections. *Cells* **2020**, *9*, 94. [CrossRef]
20. Oechslin, N.; Moradpour, D.; Gouttenoire, J. On the Host Side of the Hepatitis E Virus Life Cycle. *Cells* **2020**, *9*, 1294. [CrossRef]
21. Sid Ahmed, S.; Bundgaard, N.; Graw, F.; Fackler, O.T. Environmental Restrictions: A New Concept Governing HIV-1 Spread Emerging from Integrated Experimental-Computational Analysis of Tissue-Like 3D Cultures. *Cells* **2020**, *9*, 1112. [CrossRef]

22. Imle, A.; Kumberger, P.; Schnellbacher, N.D.; Fehr, J.; Carrillo-Bustamante, P.; Ales, J.; Schmidt, P.; Ritter, C.; Godinez, W.J.; Muller, B.; et al. Experimental and computational analyses reveal that environmental restrictions shape HIV-1 spread in 3D cultures. *Nat. Commun.* **2019**, *10*, 2144. [CrossRef]
23. Bilz, N.C.; Willscher, E.; Binder, H.; Böhnke, J.; Stanifer, M.L.; Hübner, D.; Boulant, S.; Liebert, U.G.; Claus, C. Teratogenic Rubella Virus Alters the Endodermal Differentiation Capacity of Human Induced Pluripotent Stem Cells. *Cells* **2019**, *8*, 870. [CrossRef]
24. Nguyen, H.-M.; Guz-Montgomery, K.; Saha, D. Oncolytic Virus Encoding a Master Pro-Inflammatory Cytokine Interleukin 12 in Cancer Immunotherapy. *Cells* **2020**, *9*, 400. [CrossRef]
25. Tekes, G.; Ehmann, R.; Boulant, S.; Stanifer, M.L. Development of Feline Ileum- and Colon-Derived Organoids and Their Potential Use to Support Feline Coronavirus Infection. *Cells* **2020**, *9*, 2085. [CrossRef]
26. Potratz, M.; Zaeck, L.; Christen, M.; te Kamp, V.; Klein, A.; Nolden, T.; Freuling, C.M.; Müller, T.; Finke, S. Astrocyte Infection during Rabies Encephalitis Depends on the Virus Strain and Infection Route as Demonstrated by Novel Quantitative 3D Analysis of Cell Tropism. *Cells* **2020**, *9*, 412. [CrossRef]
27. Leger, P.; Nachman, E.; Richter, K.; Tamietti, C.; Koch, J.; Burk, R.; Kummer, S.; Xin, Q.; Stanifer, M.; Bouloy, M.; et al. NSs amyloid formation is associated with the virulence of Rift Valley fever virus in mice. *Nat. Commun.* **2020**, *11*, 3281. [CrossRef] [PubMed]
28. Ezzat, K.; Pernemalm, M.; Palsson, S.; Roberts, T.C.; Jarver, P.; Dondalska, A.; Bestas, B.; Sobkowiak, M.J.; Levanen, B.; Skold, M.; et al. The viral protein corona directs viral pathogenesis and amyloid aggregation. *Nat. Commun.* **2019**, *10*, 2331. [CrossRef]

Publisher's Note: MDPI stays neutral with regard to jurisdictional claims in published maps and institutional affiliations.

Opinion

Coronaviruses: Is Sialic Acid a Gate to the Eye of Cytokine Storm? From the Entry to the Effects

Przemyslaw Wielgat [1,*], Karol Rogowski [2], Katarzyna Godlewska [3] and Halina Car [1,2]

1 Department of Clinical Pharmacology, Medical University of Bialystok, Waszyngtona 15A, 15274 Bialystok, Poland; hcar@umb.edu.pl

2 Department of Experimental Pharmacology, Medical University of Bialystok, Szpitalna 37, 15295 Bialystok, Poland; karolrogovsky@gmail.com

3 Department of Haematology, Medical University of Bialystok, M. Sklodowskiej-Curie 24A, 15276 Bialystok, Poland; kasia89@piasta.pl

* Correspondence: przemyslaw.wielgat@umb.edu.pl; Tel.: +48-85-7450-647

Received: 11 August 2020; Accepted: 24 August 2020; Published: 25 August 2020

Abstract: Coronaviruses (CoVs) are a diverse family of the enveloped human and animal viruses reported as causative agents for respiratory and intestinal infections. The high pathogenic potential of human CoVs, including SARS-CoV, MERS-CoV and SARS-CoV-2, is closely related to the invasion mechanisms underlying the attachment and entry of viral particles to the host cells. There is increasing evidence that sialylated compounds of cellular glycocalyx can serve as an important factor in the mechanism of CoVs infection. Additionally, the sialic acid-mediated cross-reactivity with the host immune lectins is known to exert the immune response of different intensity in selected pathological stages. Here, we focus on the last findings in the field of glycobiology in the context of the role of sialic acid in tissue tropism, viral entry kinetics and immune regulation in the CoVs infections.

Keywords: coronavirus; MERS-CoV; SARS-CoV; SARS-CoV-2; sialic acid; Siglec

1. Introduction

According to the International Committee of Taxonomy of Viruses, the subfamily of *Coronavirinae* comprises four genera of coronaviruses (CoV), characterized by high biodiversity and infectious potential [1]. The epidemiological observation in the last decades focused on the invasion of CoVs revealed that α-, β-, γ- and δ-CoVs, cause in mammals and birds a respiratory and intestinal infections that are mainly mild and asymptomatic, but severe and fatal courses were also observed [2]. In the last years, the extensive studies on the highly pathogenic human coronaviruses, especially on SARS- and MERS-CoVs helped to understand their biology but has also driven the discovery of new therapeutic strategies in case of epidemics [3,4]. However, a novel CoV named SARS-CoV-2 was described as the causative agent of the severe acute respiratory syndrome with high invasiveness and mortality, and disease it causes, called COVID-19, was officially declared a pandemic by the World Health Organization (WHO) [5–7]. In the face of the massive SARS-CoV-2 invasion, the state of knowledge of CoVs turned out to be insufficient in the field of their structure, infection mechanisms, serology, epidemiology and, finally, effective therapy.

The CoVs belong to the enveloped viruses with single-stranded large RNA genome sizing about 30,000 nucleotides [8]. The viral particle contains several classes of proteins that play structural and non-structural functions. The non-structural proteins belong to the family of replicases that are processed by proteases into 16 proteins functionally involved in genome transcription and replication [9]. Among four structural proteins, nucleocapsid (N) protein form complexes with RNA, whereas spike (S), membrane (M) and envelope (E) proteins participate in viral envelope formation, mediates attachment to the host cell surface receptors and may suppress the host immune response [10,11]. Given the invasive potential, the S protein play pivotal role in the fusion between the viral and host cell membranes, and in some CoVs interferes with interaction between infected and surrounding uninfected cells. The mechanisms underlying the attachment of virus via S protein recruit molecular systems which are specific feature of each CoV genera and a potential candidates in targeting therapy that may prevent a virus from being able to enter a cell. The N-terminal domain (NTD) and C-domain, called receptor binding domain (RBD), of S protein recognize and bind host membrane proteins [12]. As shown previously, the infection with morbid and deadly human CoVs, MERS- and SARS-CoV, is mediated by the high-affinity interaction between viral RBD and respectively, dipeptidyl peptidase 4 (DPP4) and angiotensin-converting enzyme 2 (ACE2) in host cells [13]. However, there is evidence that the main viral entry mechanisms may be supported by sialic acid-based machinery resulting in high diffusion speed [14–16]. Given that the sialoglycans are a widespread component of cellular glycocalyx in various tissues, their possible regulatory function in CoVs can open new view on predictive diagnostics and future therapy. It is particularly important in the context of SARS-CoV-2 infections characterized by an overzealous and unexpected immune response contributing to severe acute respiratory syndrome and concomitant highly limited and not fully established drug therapy. This paper briefly focuses on the engagement of sialic acids in tropism and infectivity of CoVs, and their role in the regulation of overreaction of the body's immune system.

2. Sialic Acids in CoVs Infections

2.1. Sialic Acids–Structural and Functional Players in the Epidemiological Events

Sialic acids are the nine carbon keto-sugars covalently attached to the terminal end of glycoconjugates on the cell surface and secreted glycoproteins and glycolipids. The biological and physical properties of sialoglycans, including the negative charge and high hydrophilicity of cell membranes, regulate cell-cell and cell-extracellular matrix interactions [17–20]. Due to sialic acid glycotopes ubiquity it is biologically important in biological recognition and immune responses [21–28]. According to Läubli and Varki sialoglycans expressed in the mammalian glycocalyx can be considered as self-associated molecular patterns (SAMPs) that can interact with the individual immune receptors presented on the same cell membranes and orchestrate inflammatory reactions within damaged tissues [29]. In contrast to SAMPs, the pathogen-associated molecular patterns (PAMPs) express the ability to recognize several immune receptors that underlie the mechanisms of chronic inflammation and neurodegeneration, as well as impaired immune surveillance in pathogen infections. The recruitment of the interplay between sialic acids and sialic acid-binding immunoglobulin-like lectins (Siglecs) underlie the proinflammatory signaling in response to some pathogens eg. *Haemophilus influenzae* and *Escherichia coli* [30–32]. In the field of virology, sialic acids were the first identified virus receptors that facilitate post-attachment events in the entry process [14]. The invasion by recognizing and binding to cell surface sialic acids, and in consequence fusion with the cellular membrane, is one of the best-established mechanisms of viral entry in host cells, as demonstrated with highly pathogenic influenza A, -B, and -C strains in vivo and in vitro studies [33–35]. Depending on the serotype of envelope glycoproteins that interact selectively with sialoglycans of the host, viruses can recognize and bind α2-3-, α2-6- and/or α2-8-linked sialic acid moieties, however, these preferences are additionally limited by natural host origin. For example, the receptor-binding specificity of human influenza viruses is focused preferably on α2-3- and α2-6-linked sialic acids which predominates on epithelial cells in the nasal mucosa, pharynx, larynx, trachea and ocular cells and strongly correlate with clinical symptoms [36–39]. The binding of α2-3- and/or α2-6-linked sialic acids as cellular receptors recruited in the host cell entry processes was described in several viral families isolated from humans, including *Adenoviridae*, *Picornaviridae*, *Reoviridae*, *Paramyxoviridae* and *Polyomaviridae* [40–44]. In the context of the virus-host interaction, some viruses genera employ the sialic acid- and Siglec-based mechanisms that avoid or deactivates host defence after pathogen invasion and underlie in the pathogenesis of acquired immune deficiency syndrome (AIDS) [45–48].

2.2. Coronavirinae-Obviousnesses and Surprises

Individual coronaviruses use specific cell receptors, which was used in differentiation and taxonomic classification to one of four genera. The ability to recognize host cell receptors by specific types of viruses is not synonymous with determining their pathogenic properties. It has been shown that NL63-CoV, one of seven known CoVs to infect humans, exerts lower affinity for human ACE2 (hACE2) than SARS-CoVs. These differences are related to the primary and secondary structure of S protein and determinate the clinical course of infection [49]. The infection of the upper respiratory tract with the low pathogenic CoVs including 229E-CoV, OC43-CoV, NL63-CoV, and HKU-CoV is manifested by the mild and cold-like symptoms. In contrast, highly pathogenic MERS-CoV, SARS-CoV and SARS-CoV-2 cause atypical pneumonia with the high rate of mortality [2]. The pathological changes in lungs following SARS- and MERS-CoV infection are accompanied by an excessive immune response and described as "cytokine storm" [50–52]. The clinical and experimental observation showed that individuals with severe or lethal SARS present very low serum level of anti-inflammatory IL-10 but high levels of proinflammatory cytokines and chemokines, e.g., IL-1, IL-6, IL-12, CCL2, CXCL10 and CXCL9, in comparison to patients with mild disease [53]. These changes were correlated with increased populations of neutrophils and monocytes in peripheral blood and lungs of infected individuals. In line, the in vitro studies with cultured monocytic THP-1 cell line and human peripheral blood monocytes infected with MERS-CoV confirmed the up-regulated production of pro-inflammatory factors. In animal models employed to study SARS, young mice developed lethal disease featured by pulmonary oedema, diffuse alveolar damage and excessive influx of monocytes/macrophages accompanied by delayed interferone-α/β (IFN-α/β) response, whereas targeting IFN signaling showed protective effects and better prognositic outcomes [4].

One of the greatest difficulties in the management of SARS-Cov-2 infection that underlie COVID-19 pandemic is the not fully understood hypersensitivity or relative resistance in some individuals in the context of the immune response [54]. These differences are genetically dependent and closely related to the structure and function of particular immune receptors. It has been shown that the polymorphisms of Toll-like receptor 1 (TLR1) and TLR4, linked to single nucleotide polymorphism, predispose individuals to "cytokine storm" in response to bacterial products [55,56]. The level of cytokine and chemokine expression results from the interplay between PAMPs and the pathogen recognition receptors (PRRs) of the host. The PAMPs-PRRs axis is featured by high specificity of recognition and interaction capacity between a pathogenic molecule (microbial, protozoan or viral) and engaging immune receptor. In result, the enhanced immune signalling pathways lead to a stage of hyperresponsiveness to PRRs agonists [57]. It has been shown that respiratory syncytial virus (RSV)-infected airway epithelial cells become hypersensitive to endotoxin by enhancing TLR4 expression with production of IL-8 and tumor necrosis factor α (TNFα) as well as increased mitogen-activated protein (MAP) kinase activity [58].

As suggested above, sialic acid-based pattern recognition by Siglec receptors can be considered the axis of SAMPs/PAMPs/PRRs in self-nonself discrimination by the innate immune system. The sialic acid-Siglec immune checkpoint has been described as part of the molecular mimicry system in various bacterial, viral and malignant invasions, but their involvement in CoVs infection and related pathologies is not fully understood [28]. However, extensive research on the COVID-19 pandemic has shed new light on CoVs biology, in particular on infection mechanisms and their importance in prognosis and therapy.

In the following we focus on the involvement of sialoglycans in the human CoVs biology, including the highly invasive species of the genus Betacoronavirus, in the field of mechanisms of host cell infection and the control of the immune response.

The sialic acid binding capacity is one of many determinants of biological diversity in the family of CoVs [14]. Several types of CoVs infect the host cells in a sialic acid dependent manner, however the mechanisms of virus entry are different than those described in influenza viruses A and B. In contrast to influenza A and B viral HA and NA, the corresponding mechanisms presented by CoVs are based on enzymatic activities of hemagglutinin esterase (HE) and hemagglutinin esterase fusion proteins (HEF) that differ from each other with the ligand bound in the opposite orientation [59]. Among CoVs, the lineage A of the beta-CoVs, such as human OC43-CoV and HKU1-CoV, use 9-*O*-acetylated sialic acid (9-*O*-Ac-sialic acid) as a substrate for HE. The detachment of O-acetyl residues promotes the release of viruses from the host infected cell and prevents the formation of virus aggregates [60]. Interestingly, the transmissible gastroenteritis virus (TGEV) enter their host cells via double mechanism based on binding of sialic acid and aminopeptidase N (APN), whereas its respiratory variant, the porcine respiratory coronavirus (PR-CoV) lacks HA activity and sialic acid binding capacity [61,62]. Moreover, several members of CoVs genera, such as human respiratory 229E-CoV and NL62-CoV, enter their host cells in sialic acid independent manner and recruit only the APN [63] (Figure 1).

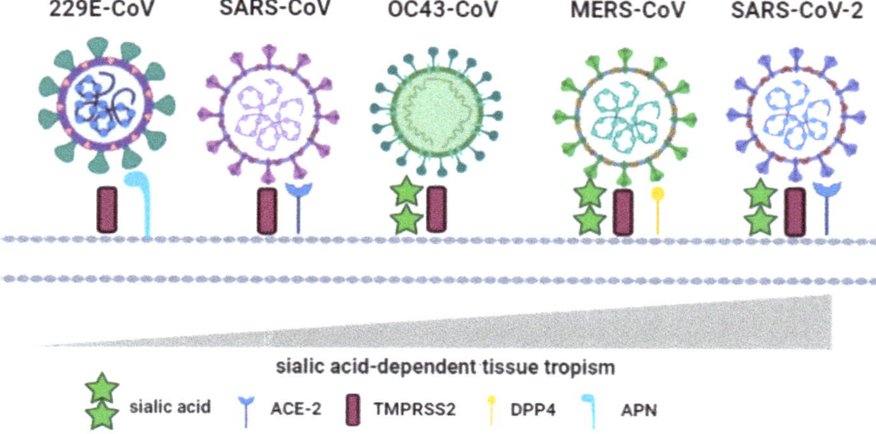

Figure 1. Receptor recognition pattern during the CoVs infections. The CoVs invade host cells through the attachment, binding and entry mechanism based on the sialic acid and protein receptors. After sialic acid-mediated virus attachment and its spike protein activation by transmembrane serine protease 2 (TMPRSS2), the entry event is associated with the binding of specific protein receptor: ACE-2 (angiotensin converting enzyme type 2); DPP4 (dipeptidyl peptidase 4); APN (aminopeptidase N). The ability to recognize sialoglycans determines virus tissue tropism and clinical manifestations among infected organs.

2.3. *MERS-CoV versus SARS-CoV—One Genus and Two Scenarios*

Despite some similarities confirmed by genomic and phylogenetic analyses, MERS- and SARS-CoV differ from each other in several structural features that are crucial for infective potential [64]. As highly pathogenic human β-CoVs, they cause a wide spectrum of clinical manifestations in respiratory tracts, however, the rate of clinical changes seem to be unfavorable in MERS-CoV infections [4]. While the clinical course of SARS resulted in acute respiratory illness presents in three distinct phases, the MERS clinical manifestations in respiratory tracts rapidly progress to pneumonia with a 35% overall mortality rate [65]. Comparative studies from human autopsies and animal models of SARS and MERS strongly suggest a pathogenic role for "cytokine storm" derived from inflammatory monocyte-macrophages (IMMs) and neutrophils. In line with this, enhanced serum cytokine and chemokine levels during MERS-CoV infection strongly correlate with increased neutrophil and monocyte numbers in the lungs and the peripheral blood, indicating a possible role for these cells in described pathology [4,66].

As mentioned above, the attachment and fusion with the cell membrane of the host is critical step of viral infection. In the consequence of the encounter of the target cell, the S protein is divided between S1 and S2 subunits responsible for receptor binding and membrane fusion, respectively. In the most CoVs, the C-terminal domain of S1 subunit recognizes specific host receptors, whereas N-terminal domain, depending on the virus type, recognizes specific sugar molecules, and thereby facilitate the initial binding of viral particles to the cell surface. In SARS-CoV, the S protein is cleaved into S1 and S2 subunits by host transmembrane protease serine 2 (TMPRSS2) [67–69]. To enter host cells, virus utilizes human ACE2 as the functional receptor recognized and linked via viral RBD of S1 subunit. This mechanism was confirmed in multiple in vitro studies with naturally ACE-2-expressing colonic epithelial LoVo cells and alveolar epithelial A549 cells [70,71]. Whereas the structure of RBD is systematically studied, the NTD of SARS-CoV is not fully characterized. However, the current state of structural analyses suggests that SARS-Cov, unlike other β-Covs, does not recognize glycans, including sialic acids, via NTD in S1 subunit [72]. For MERS-CoV, human lung epithelial cells susceptibility correlates with DPP4 expression in these cells. Analysis of MERS-CoV S1 subunit structure by cryoelectron microscopy showed four individually folded domains (S1A-S1D) implicated in receptor binding. The engagement of binding domain closely depends on virus species [73]. Among β-CoVs, low invasive bovine CoV (BCoV) and human OC43-CoV recruit S1A domain for binding of sialic acid-based entry receptors [74]. Binding of MERS-CoV to the cell surface entry receptor DPP4 occurs via S1B whereas sialic acid-binding activity was assigned to S subdomain S1A. In addition, the analysis with 60-metric lumazine synthase nanoparticle presenting S1A domains revealed high preference for α2.3-linked over α2.6-linked sialic acids. The importance of sialic acid in the mechanism of MERS-CoV infection was presented by Li et al. [73]. They showed that lower sialylated Vero cells exert lower susceptibility to MERS-CoV infection compared to highly sialic acid expressing human airway Calu-3 epithelial cells. Moreover, enzymatic depletion of sialic acids by neuraminidase from the cell surface or their modification by 9-*O*-acetylation or 5-*N*-glycolylation hindered or dampened these interactions and thereby viral entry [73]. As demonstrated in influenza viruses infections, the sialic acid-binding preference and distribution of sialoglycans is a critical regulator of viral tropism. Therefore, site of MERS-CoV replication colocalize with α2,3-linked sialic acid in vivo. According to Li et al., the differential distribution of α2,3-linked sialic acid in humans and camels correlates with predominant sites of MERS-CoV replication in the lower and upper respiratory tracts, respectively [73]. Finally, the identified MERS-CoV interaction with the host sialosides in the dual-receptors binding mechanism, has been characterized by fast kinetics and was required for viral entry [75].

2.4. SARS-CoV-2 versus SARS-CoV—Are Sialic Acids Really Important?

The comparing analysis of genetic sequences revealed that SARS-CoV-2 shared about 79% of its genome with SARS-CoV [76,77]. In addition, the overall sequence similarity for the whole S protein was 78% and 76% for RBD [78]. The close identity between S1 subdomains indicate that SARS-Cov-2 also utilizes ACE2 as the receptor. Despite similarities, SARS-CoV-2 show several structural differences that may influence infection kinetics and clinical symptoms in individuals. First, quantification of the kinetics of ligand-receptor interaction by surface plasmon resonance showed that SARS-CoV-2 binding to ACE2 is about 20-fold higher compared to SARS-CoV [79]. Second, clinical course of COVID-19 in some patients range from mild symptoms to deadly "cytokine storm", however, asymptomatic individuals, similar to those in MERS, may pose a significant public health problem [80]. To better understand the mechanisms underlying these phenomenon, the total genomic sequence of human CoVs was compared with the library of viruses. Studies on the complementarity of SARS-CoV-2 S protein identified another possible binding region responsible for the interaction with the host receptors. Despite the SARS-CoV-2 is less genetically similar to MERS-CoV (about 50%) than SARS-CoV (about 79%), it has been hypothesized that infection with both SARS-CoV-2 and MERS-CoV depends on similar mechanism of the attachment to the host cells that facilitates virus entry and mediates the immune response [81–83]. The structural properties of several SARS-CoV-2 and MERS-CoV strains were analysed using CROSS*align*, a tool for prediction of the structural similarities of two RNA profiles of different length. In the preliminary studies, group of Vandelli have found that the structural region between nucleotides 23,000 and 24,000 of S region that corresponds to amino acids 330–500 of S protein domain, is highly conserved across SARS-CoV-2 strains and responsible for ACE2 binding. However, the structural profile in SARS-CoV-2 at nucleotides 22,500–23,000 corresponds to amino acids 243–302 that share identity in MERS-CoV, but not in SARS-CoV, and bind to sialic acids during virus attachment to the host cells [82]. This observation was confirmed by Milanetti and coworkers in the preliminary analysis using quantitative assessment of the geometrical shape complementarity between interacting proteins. They used experimental structures of SARS-CoV-2, MERS-CoV and SARS-CoV S protein and sialic acid complexes, extracted their binding regions and described through Zernicke descriptors. The selected region in the analyzed viruses expressed the highest similarity when SARS-CoV-2 and MERS-CoV were compared. In addition, the MERS-CoV S protein region interacting with sialic acids and its analogue in SARS-CoV-2 showed electrostatic similarity whereas the compared region in SARS-CoV present not compatible charge that disturb the interaction with sialylated receptors. Milanetti suggest that the sialic acid binding capacity in SARS-CoV-2 seem to be a part of complex cell entry mechanism that explain highly effective infection of the upper and its progession in low airways [83]. This hypothesis seem to be confirmed by the findings of Fantini and coworkers [81]. They identified the specific area of NTD that interacts with oligosaccharide part of ganglioside GM1. Molecular and structural studies revealed that amnio acid residues 100–175 of the NTD, especially Phe-135, Asn-137 and Arg-158 are critical for the interaction with the host GM1 molecule.

Beside the upper respiratory tract, the eye exposure to infectious fluids may be associated with increased risk for SARS-CoV-2 transmission. According to goup of Craeger the conjunctival and corneal epithelia and the nasolacrimal system contain both α2-3-linked and α2-6-linked sialic acid receptors preferentially interacting with numerous human viruses [84]. The eye contamination with CoVs was confirmed by Loon and coworkers using PCR analysis of tears samples collected from probably positive patients [85]. Recently, the clinical observation by Colavita et al. have revealed that SARS-CoV-2 RNA was detected in ocular swab whereas it was undetectable in nasal swab form patient diagnosed for COVID-19 [86].

As mentioned above, changes in sialylation pattern are of particular importance in glycan-dependent biological recognition and contribute to immune disturbances and related pathologies. According to WHO reports, severe symptoms and worse prognosis among patients with COVID-19 are statistically more frequent among individuals with cardiovascular and chronic respiratory diseases than those with no pre-existing chronic medical disfunctions [87]. The clinical-related advances in glycobiology are focused on the prognostic value of sialylated epitopes as markers of pathology. For example, it has been shown that serum sialic acid forecast both coronary heart disease and stroke fatality and reflects the progress of an atherosclerotic process [88–90]. The clinical observation and experimental models described an enhanced level of total serum sialic acid and sialoglycans on the human vein endothelial cells as a marker of pathology and significant risk factor of coronary artery disease (CAD) and angiopathy [91,92]. Moreover, the glycome analysis in asthma and COPD showed that the alterations in sialoglycans correspond to the progress of disease [93–95]. According to Guo, the lung pathology inducing factors, e.g., tobacco, can be considered as the crucial players in the promotion of SARS-CoV-2 infection and associated with the risk of severe COVID-19 [95]. Given the harmful effect of cigarette smoke (CS) on respiratory and circulatory parameters, heavy smokers are at higher risk of developing severe COVID-19 symptoms [96,97]. Beside the clinical consequences, CS induces multiple molecular changes in the lung cells. Our previous studies revealed that human alveolar epithelial A549 cells present enhanced α2-3- and α-2-8-, but not α2-6-sialylation pattern when exposed to CS [98]. Additionally, Leung et al. found that patients with chronic obstructive pulmonary disease (COPD) and individuals who are still smoking presents higher levels of ACE-2 in their airways [99]. Therefore, it is reasonable to hypothesize that dual-mechanism based on ACE2 and sialic acid might put them at an increased risk of developing severe COVID-19 infections.

2.5. Does Siglec-Sialic Acid Axis Participates in CoVs Infections?

The level of immune system activity is the major implication for the protection of body systems against pathogens. As multiple investigations indicate, the response in several individuals for exposure to the same pathogen is different ranging from asymptomatic course to "cytokine storm" and death. The most coronaviruses-related worse prognosis is due to the immune system becomes hypersensitive in its response, not direct virus-caused damage. It is well known that a difference in the way populations respond to infection is largely controlled by genetics that shapes immune profiles, however, the immune mechanisms engaged in CoVs infection, in particular SARS-CoV-2, are still not fully understood [100,101].

In the context of participation of sialic acid in the mechanisms of CoVs entry, we discussed above scenario showed cellular sialoglycans as receptors for recognizing proteins expressed in the viral envelope. On the other hand, several pathogens express sialoglycan epitopes that act as a ligand recognized by cell membrane receptors [102] (Figure 2).

The virus-associated enzymes can exert modulatory effect on the cell membrane glycocalyx and unmask Siglecs by disruption of their *cis* ligands. It has been shown that unmasking of Siglec-1 (sialoadhesin, Sn) facilitate the CoV uptake into macrophages [103–105]. Another mechanism of potential importance includes interactions between sialic acids and Siglec receptors resulting in divergent immune system regulation. The role of CoV-originated sialoglycans in immune modulation was not extensively studied and the poor data have rather speculative value. Is it reasonable to consider the interplay between Siglecs and sialic acids as a controlling mechanism in CoVs invasion machinery? Varki and Angata hypothesize that expression of sialic acids by the envelope of CoV can affect Siglec receptors biology in the hosts and thereby regulate the reactivity of innate immune cells [102]. The inhibitory receptors, such as Siglec-5, -7 and -9, are implicated in molecular mimicry mechanism that allow several pathogens to avoid immunesurveillance as shown in the hepatitis B virus (HBV) infections [106,107]. In the context of Siglec-mediated immune regulation, the clinical differences in the CoVs infections can be interpreted by the variety of human phenotypes. Among the Siglec family members, the paired receptors express extremely similar amino acid sequence of extracellular domain and comparable tissue distribution resulting in the same ligand recognition capacity whereas the diverse intracellular signalling pathways trigger the opposite effects. The paired Siglec-5 and Siglec-14 proteins, gene fusion products, are widely expressed on myeloid-derived cells residing and infiltrating human airways, however their distribution in individuals is closely related to the frequencies of wild-type and SIGLEC-14 null alleles in various populations [108,109]. Therefore, the expression of activating Siglec-14 receptor predominates in the European population whereas inhibitory Siglec-5 is widely expressed in Asians. Clinically, the expression of Siglec-5 is closely linked to reduced bactericidal and virucidal abilities during infections with *Streptococcus*, *Neisseria*, *Pseudomonas*, *Campylobacter* and *HIV* [110]. Moreover, expression of Siglec-14 predispose to the potentiated inflammatory response in bacterial *(Haemophilus influenzae)* and viral infections (influenza virus), which are the cause of chronic respiratory diseases. It was confirmed by elevated expression of IL-1β in primary macrophages from SIGLEC14$^{+/+}$, but not SIGLEC14$^{-/-}$ individuals [111]. Interestingly, the expression and function of paired Siglec-5/14 receptors undergo multiple modification in response to several pathology promoting factors. In the experimental model of exposure to cigarette smoke (CS), an increase in the expression of paired Siglec-5/14 receptors in THP-1 cells was observed [98]. It was accompanied by simultaneous alterations in immune activity with an enhanced intracellular interleukin IL-1β and interleukin 10 (IL-10) expression, and reduced phagocytic capacity. Given the CS-induced functional alterations in the immune response, it is reasonable to speculate that sialic acid-Siglec axis can be an important mechanism of exacerbations in respiratory diseases and the immune system performance in viral infections in smokers.

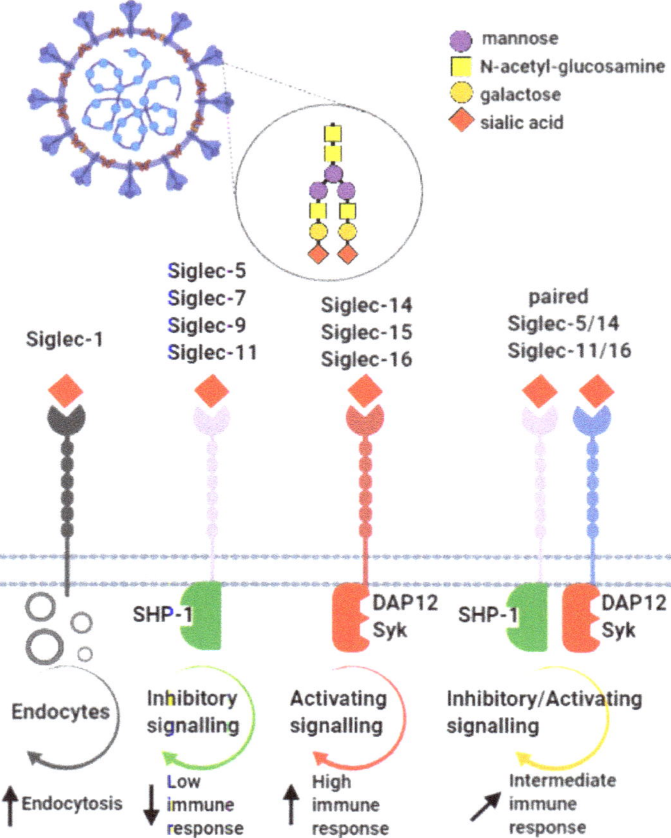

Figure 2. The possible effects of the CoVs targeting to Siglecs. The viral envelope glycans interact with sialic acid-binding lectins expressed in the host cell-dependent manner. Depending on the individual Siglec expression profile, the interactions with the viral sialic acid ligands exerts the immune response of different intensity. Abbreviations: DAP12-DNAX activation protein of 12 kDa; Syk-spleen tyrosine kinase; SHP-1-Src homology region 2 domain-containing phosphatase-1; SHP-2-Src homology region 2 domain-containing phosphatase-2.

Additionally, previous studies have revealed that conventional therapies may influence Siglec-mediated immune regulation. During the SARS epidemic in 2003, the early anti-inflammatory treatment of corticosteroids enhanced plasma viral titre leading to exacerbated disease [112]. The strong suppressive action of corticosteroids may explain these effect, however, Siglec-based mechanism can be also involved. The cytometric analysis of cell population isolated from patients with chronic obstructive pulmonary disease (COPD) has demonstrated that inhaled corticosteroids increase the expression of Siglec-5/14 in CD14+ cells [113]. Similarly, the enhancement of Siglec-9 in neutrophils was observed after management with dexamethasone (Dex) [114]. Therefore, the clinical assessment of Siglecs profiles can be considered before the therapeutic intervention with corticosteroids in individuals with respiratory comorbidities in CoVs infections.

Interestingly, the running clinical trials on the potential therapeutic management of COVID-19 showed that 4-aminoquinoline derivatives, chloroquine (CLQ) and hydroxychloroquine (CLQ-OH), prevent the virus from binding to ACE2 receptor [115]. Based on the in vitro studies, it is believed that chloroquine inhibits terminal glycosylation and thereby prevent SARS-CoV-2 from binding with gangliosides [116]. As structural and molecular modelling approaches have shown, this effect seems to be linked with direct interaction of chloroquine (CLQ) and hydroxychloroquine (CLQ-OH) with structural motif including amino acid residues 111–158 of the NTD that prevent of binding the sialylated GM1 with the high affinity [81]. From the molecular point of view, it may suggest the therapeutic potential of CLQ and CLQ-OH, however, the recent randomized trials of CLQ-OH as postexposure prophylaxis of COVID-19 did not fully confirm this action, therefore additional clinical and molecular observation are needed.

3. Conclusions and Perspectives

Although the biological importance of sialic acid is well described, its engagement in the pathogenesis of CoVs infections remains not fully understood. Recent findings in the field of virus attachment and entry to the host cells indicate that the recognizing capacity of defined sialoglycan epitopes becomes a distinctive feature among the coronaviruses family. In the most cases, the binding of sialic acid is a vulnerable part of dual mechanism of viral invasion, that could be targeted for the future development of therapeutics. On the other hand, the in vivo and in vitro studies on the role of sialic acid-Siglec axis can help the understanding of the immune-based pathogenesis of viral infection and its role in the divergent clinical course among the population, including the burden of asymptomatic individuals. The state of knowledge in this area is highly progressive, however, there are several questions in the field of glycobiology remain still unanswered: (i) Could sialic acid act as a host cellular marker to regulate the risk of CoVs attachment? (ii) Do several CoVs express sialic acids epitopes to regulate Siglec function? (iii) Is sialic acid-Siglecs interplay in CoVs infections relevant to cell function and activation? (iv) Are conventional drugs potential modulators of cellular sialome in the CoVs attachment process? (v) Do comorbidities and pathological factors associated with increased sialylation pattern predispose to SARS-CoV-2 infections and/or their severe course?

The prevention of viruses attachment and entry is the main challenge of modern virology. In the light of current SARS-CoV-2 pandemic, the novel immunotherapeutic approaches to COVID-19 should be developed in accordance to known and well established molecular mechanisms. Based on the role of sialoglycans in cell biology mentioned in this review, the targeting the patients' sialome-modulating machinery may inhibit the viral invasion through multiple potential mechanisms of action including modifications of the host receptors, modulation of endocytosis, prevention of virus-host cell interaction, and immunomodulation.

Author Contributions: Conceptualization: P.W. and K.R.; Writing—Original Draft Preparation: P.W. and K.R.; Writing—Review and Editing: K.G.; Supervision: H.C. All authors have read and agreed to the published version of the manuscript.

Funding: This work was supported by the grants from Medical University of Bialystok: SUB/1/DN/19/001/1166 (P.W.) and SUB/1/DN/20/002/1166 (P.W.).

Conflicts of Interest: The authors declare no conflict of interest.

References

1. International Committee on Taxonomy of Viruses Executive Committee. The new scope of virus taxonomy: Partitioning the virosphere into 15 hierarchical ranks. *Nat. Microbiol.* **2020**, *5*, 668–674. [CrossRef]
2. Pillaiyar, T.; Meenakshisundaram, S.; Manickam, M. Recent discovery and development of inhibitors targeting coronaviruses. *Drug. Discov. Today* **2020**, *25*, 668–688. [CrossRef]
3. Tao, X.; Hill, T.E.; Morimoto, C.; Peters, C.J.; Ksiazek, T.G.; Tseng, C.T. Bilateral entry and release of Middle East respiratory syndrome coronavirus induces profound apoptosis of human bronchial epithelial cells. *J. Virol.* **2013**, *87*, 9953–9958. [CrossRef]

4. Channappanavar, R.; Perlman, S. Pathogenic human coronavirus infections: Causes and consequences of cytokine storm and immunopathology. *Semin. Immunopathol.* **2017**, *39*, 529–539. [CrossRef]
5. Huang, C.; Wang, Y.; Li, X.; Ren, L.; Zhao, J.; Hu, Y.; Zhang, L.; Fan, G.; Xu, J.; Gu, X.; et al. Clinical features of patients infected with 2019 novel coronavirus in Wuhan, China. *Lancet* **2020**, *395*, 497–506. [CrossRef]
6. Zhu, N.; Zhang, D.; Wang, W.; Li, X.; Yang, B.; Song, J.; Zhao, X.; Huang, B.; Shi, W.; Lu, R.; et al. A novel coronavirus from patients with pneumonia in China, 2019. *N. Engl. J. Med.* **2020**, *382*, 727–733. [CrossRef]
7. Walls, A.C.; Park, Y.J.; Tortorici, M.A.; Wall, A.; McGuire, A.T.; Veesler, D. Structure, function, and antigenicity of the SARS-CoV-2 spike glycoprotein. *Cell* **2020**, *181*, 281–292. [CrossRef]
8. Schwegmann-Wessels, C.; Herrler, G. Sialic acids as receptor determinants for coronaviruses. *Glycoconj. J.* **2006**, *23*, 51–58. [CrossRef]
9. Harcourt, B.H.; Jukneliene, D.; Kanjanahaluethai, A.; Bechill, J.; Severson, K.M.; Smith, C.M.; Rota, P.A.; Baker, S.C. Identification of severe acute respiratory syndrome coronavirus replicase products and characterization of papain-like protease activity. *J. Virol.* **2004**, *78*, 13600–13612. [CrossRef]
10. McBride, R.; van Zyl, M.; Fielding, B.C. The coronavirus nucleocapsid is a multifunctional protein. *Viruses* **2014**, *6*, 2991–3018. [CrossRef]
11. Arndt, A.L.; Larson, B.J.; Hogue, B.G. A conserved domain in the coronavirus membrane protein tail is important for virus assembly. *J. Virol.* **2010**, *84*, 11418–11428. [CrossRef]
12. Zeng, F.; Hon, C.C.; Yip, C.W.; Law, K.M.; Yeung, Y.S.; Chan, K.H.; Peiris, J.S.M.; Leung, F.C.C. Quantitative comparison of the efficiency of antibodies against S1 and S2 subunit of SARS coronavirus spike protein in virus neutralization and blocking of receptor binding: Implications for the functional roles of S2 subunit. *FEBS Lett.* **2006**, *580*, 5612–5620. [CrossRef]
13. Wan, Y.; Shang, J.; Sun, S.; Tai, W.; Chen, J.; Geng, Q.; He, L.; Chen, Y.; Wu, J.; Shi, Z.; et al. Molecular mechanism for antibody-dependent enhancement of coronavirus entry. *J. Virol.* **2020**, *94*, e02015–e02019. [CrossRef]
14. Matrosovich, M.; Herrler, G.; Klenk, H.D. Sialic acid receptors of viruses. *Top. Curr. Chem.* **2015**, *367*, 1–28. [CrossRef]
15. Tortorici, M.A.; Walls, A.C.; Lang, Y.; Wang, C.; Li, Z.; Koerhuis, D.; Boons, G.-J.; Bosch, B.-J.; Rey, F.A.; de Groot, R.J.; et al. Structural basis for human coronavirus attachment to sialic acid receptors. *Nat. Struct. Mol. Biol.* **2019**, *26*, 481–489. [CrossRef]
16. Qing, E.; Hantak, M.; Perlman, S.; Gallagher, T. Distinct roles for sialoside and protein receptors in coronavirus infection. *mBio* **2020**, *11*, e02764-19. [CrossRef]
17. Schnaar, R.L. The biology of gangliosides. *Adv. Carbohydr. Chem. Biochem.* **2019**, *76*, 113–148. [CrossRef]
18. Trinchera, M.; Aronica, A.; Dall'Olio, F. Selectin ligands sialyl-Lewis a and sialyl-Lewis x in gastrointestinal cancers. *Biology* **2017**, *6*, 16. [CrossRef]
19. Bork, K.; Horstkorte, R.; Weidemann, W. Increasing the sialylation of therapeutic glycoproteins: The potential of the sialic acid biosynthetic pathway. *J. Pharm. Sci.* **2009**, *98*, 3499–3508. [CrossRef]
20. Bas, M.; Terrier, A.; Jacque, E.; Dehenne, A.; Pochet-Béghin, V.; Beghin, C.; Dezetter, A.-S.; Dupont, G.; Engrand, A.; Beaufils, B.; et al. Fc sialylation prolongs serum half-life of therapeutic antibodies. *J. Immunol.* **2019**, *202*, 1582–1594. [CrossRef]
21. Varki, A. Sialic acids in human health and disease. *Trends Mol. Med.* **2008**, *14*, 351–360. [CrossRef]
22. Miyagi, T.; Takahashi, K.; Hata, K.; Shiozaki, K.; Yamaguchi, K. Sialidase significance for cancer progression. *Glycoconj. J.* **2012**, *29*, 567–577. [CrossRef]
23. Vajaria, B.N.; Patel, K.R.; Begum, R.; Patel, P.S. Sialylation: An avenue to target cancer cells. *Pathol. Oncol. Res.* **2016**, *22*, 443–447. [CrossRef]
24. Pearce, O.M.; Läubli, H. Sialic acids in cancer biology and immunity. *Glycobiology* **2016**, *26*, 111–128. [CrossRef]
25. Chiodelli, P.; Urbinati, C.; Paiardi, G.; Monti, E.; Rusnati, M. Sialic acid as a target for the development of novel antiangiogenic strategies. *Future Med. Chem.* **2018**, *10*, 2835–2854. [CrossRef]
26. McMillan, S.J.; Crocker, P.R. CD33-related sialic-acid-binding immunoglobulin-like lectins in health and disease. *Carbohydr. Res.* **2008**, *343*, 2050–2056. [CrossRef]
27. Cagnoni, A.J.; Pérez Sáez, J.M.; Rabinovich, G.A.; Mariño, K.V. Turning-off signaling by siglecs, selectins, and galectins: Chemical inhibition of glycan-dependent interactions in cancer. *Front. Oncol.* **2016**, *6*, 109. [CrossRef]

28. Bärenwaldt, A.; Läubli, H. The sialoglycan-Siglec glyco-immune checkpoint—A target for improving innate and adaptive anti-cancer immunity. *Expert Opin. Targets* **2019**, *23*, 839–853. [CrossRef]
29. Läubli, H.; Varki, A. Sialic acid-binding immunoglobulin-like lectins (Siglecs) detect self-associated molecular patterns to regulate immune responses. *Cell. Mol. Life Sci.* **2020**, *77*, 593–605. [CrossRef]
30. Sarkar, M.; Reneer, D.V.; Carlyon, J.A. Sialyl-Lewis x-independent infection of human myeloid cells by Anaplasma phagocytophilum strains HZ and HGE1. *Infect. Immun.* **2007**, *75*, 5720–5725. [CrossRef]
31. Seidman, D.; Hebert, K.S.; Truchan, H.K.; Miller, D.P.; Tegels, B.K.; Marconi, R.T.; Carlyon, J.A. Essential domains of Anaplasma phagocytophilum invasins utilized to infect mammalian host cells. *PLoS Pathog.* **2015**, *11*, e1004669. [CrossRef]
32. Maginnis, M.S. Virus-receptor interactions: The key to cellular invasion. *J. Mol. Biol.* **2018**, *430*, 2590–2611. [CrossRef]
33. Kumlin, U.; Olofsson, S.; Dimock, K.; Arnberg, N. Sialic acid tissue distribution and influenza virus tropism. *Influenza Other Respir. Viruses* **2008**, *2*, 147–154. [CrossRef]
34. Connor, R.J.; Kawaoka, Y.; Webster, R.G.; Paulson, J.C. Receptor specificity in human, avian, and equine H2 and H3 influenza virus isolates. *Virology* **1994**, *205*, 17–23. [CrossRef]
35. Matrosovich, M.; Tuzikov, A.; Bovin, N.; Gambaryan, A.; Klimov, A.; Castrucci, M.R.; Donatelli, I.; Kawaoka, Y. Early alterations of the receptor-binding properties of H1, H2, and H3 avian influenza virus hemagglutinins after their introduction into mammals. *J. Virol.* **2000**, *74*, 8502–8512. [CrossRef]
36. Shinya, K.; Ebina, M.; Yamada, S.; Ono, M.; Kasai, N.; Kawaoka, Y. Avian flu: Influenza virus receptors in the human airway. *Nature* **2006**, *440*, 435–436. [CrossRef]
37. Ibricevic, A.; Pekosz, A.; Walter, M.J.; Newby, C.; Battaile, J.T.; Brown, E.G.; Holtzman, M.J.; Brody, S.L. Influenza virus receptor specificity and cell tropism in mouse and human airway epithelial cells. *J. Virol.* **2006**, *80*, 7469–7480. [CrossRef]
38. Nicholls, J.M.; Chan, M.C.; Chan, W.Y.; Wong, H.K.; Cheung, C.Y.; Kwong, D.L.W.; Wong, M.P.; Chui, W.H.; Poon, L.L.M.; Tsao, S.W.; et al. Tropism of avian influenza A (H5N1) in the upper and lower respiratory tract. *Nat. Med.* **2007**, *13*, 147–149. [CrossRef]
39. França, M.; Stallknecht, D.E.; Howerth, E.W. Expression and distribution of sialic acid influenza virus receptors in wild birds. *Avian Pathol.* **2013**, *42*, 60–71. [CrossRef]
40. Connolly, J.L.; Barton, E.S.; Dermody, T.S. Reovirus binding to cell surface sialic acid potentiates virus-induced apoptosis. *J. Virol.* **2001**, *75*, 4029–4039. [CrossRef]
41. Villar, E.; Barroso, I.M. Role of sialic acid-containing molecules in paramyxovirus entry into the host cell: A minireview. *Glycoconj. J.* **2006**, *23*, 5–17. [CrossRef]
42. Gee, G.V.; Dugan, A.S.; Tsomaia, N.; Mierke, D.F.; Atwood, W.J. The role of sialic acid in human polyomavirus infections. *Glycoconj. J.* **2006**, *23*, 19–26. [CrossRef]
43. Zocher, G.; Mistry, N.; Frank, M.; Hähnlein-Schick, I.; Ekström, J.-O.; Arnberg, N.; Stehle, T. A sialic acid binding site in a human picornavirus. *PLoS Pathog.* **2014**, *10*, e1004401. [CrossRef]
44. Chandra, N.; Liu, Y.; Liu, J.X.; Frängsmyr, L.; Wu, N.; Silva, L.M.; Lindström, M.; Chai, W.; Domellöf, F.P.; Feizi, T.; et al. Sulfated glycosaminoglycans as viral decoy receptors for human adenovirus type 37. *Viruses* **2019**, *11*, 247. [CrossRef]
45. Xie, J.; Christiaens, I.; Yang, B.; Trus, I.; Devriendt, B.; Cui, T.; Wei, R.; Nauwynck, H.J. Preferential use of Siglec-1 or Siglec-10 by type 1 and type 2 PRRSV strains to infect PK15S1-CD163 and PK15S10-CD163 cells. *Vet. Res.* **2018**, *49*, 67. [CrossRef]
46. Perez-Zsolt, D.; Martinez-Picado, J.; Izquierdo-Useros, N. When dendritic cells go viral: The role of Siglec-1 in host defense and dissemination of enveloped viruses. *Viruses* **2019**, *12*, 8. [CrossRef]
47. Mikulak, J.; Di Vito, C.; Zaghi, E.; Mavilio, D. Host immune responses in HIV-1 infection: The emerging pathogenic role of Siglecs and their clinical correlates. *Front. Immunol.* **2017**, *8*, 314. [CrossRef]
48. Arokiasamy, S.; Balderstone, M.J.M.; De Rossi, G.; Whiteford, J.R. Syndecan-3 in inflammation and angiogenesis. *Front. Immunol.* **2020**, *10*, 3031. [CrossRef]
49. Mathewson, A.C.; Bishop, A.; Yao, Y.; Kemp, F.; Ren, J.; Chen, H.; Xu, X.; Berkhout, B.; van der Hoek, L.; Jones, I.M.; et al. Interaction of severe acute respiratory syndrome-coronavirus and NL63 coronavirus spike proteins with angiotensin converting enzyme-2. *J. Gen. Virol.* **2008**, *89*, 2741–2745. [CrossRef]

50. Lau, S.K.P.; Lau, C.C.Y.; Chan, K.H.; Clara, P.Y.L.; Chen, H.; Jin, D.-Y.; Chan, J.F.W.; Woo, P.C.Y.; Yuen, K.-Y. Delayed induction of proinflammatory cytokines and suppression of innate antiviral response by the novel Middle East respiratory syndrome coronavirus: Implications for pathogenesis and treatment. *J. Gen. Virol.* **2013**, *94*, 2679–2690. [CrossRef]
51. Li, Y.; Chen, M.; Cao, H.; Zhu, Y.; Zheng, J.; Zhou, H. Extraordinary GU-rich single-strand RNA identified from SARS coronavirus contributes an excessive innate immune response. *Microbes Infect.* **2013**, *15*, 88–95. [CrossRef]
52. Sun, X.; Wang, T.; Cai, D.; Hu, Z.; Chen, J.; Liao, H.; Zhi, L.; Wei, H.; Zhang, Z.; Qiu, Y.; et al. Cytokine storm intervention in the early stages of COVID-19 pneumonia. *Cytokine Growth Factor. Rev.* **2020**, *53*, 38–42. [CrossRef]
53. Tisoncik, J.R.; Korth, M.; Simmons, C.P.; Farrar, J.; Martin, T.R.; Katze, M.G. Into the eye of the cytokine storm. *Microbiol. Mol. Biol. Rev.* **2012**, *76*, 16–32. [CrossRef]
54. Li, X.; Geng, M.; Peng, Y.; Meng, L.; Lu, S. Molecular immune pathogenesis and diagnosis of COVID-19. *J. Pharm. Anal.* **2020**, *10*, 102–108. [CrossRef]
55. Molteni, M.; Gemma, S.; Rossetti, C. The role of Toll-like receptor 4 in infectious and noninfectious inflammation. *Mediat. Inflamm.* **2016**, *2016*, 6978936. [CrossRef]
56. Perrin-Cocon, L.; Aublin-Gex, A.; Sestito, S.E.; Shirey, K.A.; Patel, M.C.; André, P.; Blanco, J.C.; Vogel, S.N.; Peri, F.; Lotteau, V. TLR4 antagonist FP7 inhibits LPS-induced cytokine production and glycolytic reprogramming in dendritic cells, and protects mice from lethal influenza infection. *Sci. Rep.* **2017**, *7*, 40791. [CrossRef]
57. Thompson, M.R.; Kaminski, J.J.; Kurt-Jones, E.A.; Fitzgerald, K.A. Pattern recognition receptors and the innate immune response to viral infection. *Viruses* **2011**, *3*, 920–940. [CrossRef]
58. Hu, T.; Yu, H.; Lu, M.; Yuan, X. TLR4 and nucleolin influence cell injury, apoptosis and inflammatory factor expression in respiratory syncytial virus-infected N2a neuronal cells. *J. Cell. Biochem.* **2019**, *120*, 16206–16218. [CrossRef]
59. Zeng, Q.; Langereis, M.A.; van Vliet, A.L.; Huizinga, E.G.; de Groot, R.J. Structure of coronavirus hemagglutinin-esterase offers insight into corona and influenza virus evolution. *Proc. Natl. Acad. Sci. USA* **2008**, *105*, 9065–9069. [CrossRef]
60. Hulswit, R.J.G.; Lang, Y.; Bakkers, M.J.G.; Li, W. Human coronaviruses OC43 and HKU1 bind to 9-O-acetylated sialic acids via a conserved receptor-binding site in spike protein domain A. *Proc. Natl. Acad. Sci. USA* **2019**, *116*, 2681–2690. [CrossRef]
61. Schultze, B.; Krempl, C.; Ballesteros, M.L.; Shaw, L.; Schauer, R.; Enjuanes, L.; Herrler, G. Transmissible gastroenteritis coronavirus, but not the related porcine respiratory coronavirus, has a sialic acid (N-glycolylneuraminic acid) binding activity. *J. Virol.* **1996**, *70*, 5634–5637. [CrossRef]
62. Cui, T.; Theuns, S.; Xie, J.; Van den Broeck, W.; Nauwynck, H.J. Role of porcine aminopeptidase N and sialic acids in porcine coronavirus infections in primary porcine enterocytes. *Viruses* **2020**, *12*, 402. [CrossRef]
63. Wong, A.H.M.; Tomlinson, A.C.A.; Zhou, D.; Satkunarajah, M.; Chen, K.; Sharon, C.; Desforges, M.; Talbot, P.J.; Rini, J.M. Receptor-binding loops in alphacoronavirus adaptation and evolution. *Nat. Commun.* **2017**, *8*, 1735. [CrossRef]
64. Eckerle, I.; Müller, M.A.; Kallies, S.; Gotthardt, D.N.; Drosten, C. In-vitro renal epithelial cell infection reveals a viral kidney tropism as a potential mechanism for acute renal failure during Middle East Respiratory Syndrome (MERS) Coronavirus infection. *Virol. J.* **2013**, *10*, 359. [CrossRef]
65. Alsolamy, S.; Arabi, Y.M. Infection with Middle East respiratory syndrome coronavirus. *Can. J. Respir.* **2015**, *51*, 102.
66. Gu, J.; Gong, E.; Zhang, B.; Zheng, J.; Gao, Z.; Zhong, Y.; Zou, W.; Zhan, J.; Wang, S.; Xie, Z.; et al. Multiple organ infection and the pathogenesis of SARS. *J. Exp. Med.* **2005**, *202*, 415–424. [CrossRef]
67. Belouzard, S.; Chu, V.C.; Whittaker, G.R. Activation of the SARS coronavirus spike protein via sequential proteolytic cleavage at two distinct sites. *Proc. Natl. Acad. Sci. USA* **2009**, *106*, 5871–5876. [CrossRef]
68. Shirato, K.; Kawase, M.; Matsuyama, S. Middle East respiratory syndrome coronavirus infection mediated by the transmembrane serine protease TMPRSS2. *J. Virol.* **2013**, *87*, 12552–12561. [CrossRef]
69. Hoffmann, M.; Kleine-Weber, H.; Pöhlmann, S. A multibasic cleavage site in the spike protein of SARS-CoV-2 is essential for infection of human lung cell. *Mol. Cell.* **2020**, *78*, 779–784. [CrossRef]

70. Chan, P.K.; To, K.F.; Lo, A.W.; Cheung, J.L.K.; Chu, I.; Au, F.W.L.; Tong, J.H.M.; Tam, J.S.; Sung, J.J.Y.; Ng, H.-K. Persistent infection of SARS coronavirus in colonic cells in vitro. *J. Med. Virol.* **2004**, *74*, 1–7. [CrossRef]
71. Ma, D.; Chen, C.B.; Jhanji, V.; Yuan, X.-L.; Liang, J.-J.; Huang, Y.; Cen, L.-P.; Ng, T.K. Expression of SARS-CoV-2 receptor ACE2 and TMPRSS2 in human primary conjunctival and pterygium cell lines and in mouse cornea. *Eye* **2020**, *34*, 1–8. [CrossRef]
72. Ou, X.; Liu, Y.; Lei, X.; Li, P.; Mi, D.; Ren, L.; Guo, L.; Guo, R.; Chen, T.; Hu, J.; et al. Characterization of spike glycoprotein of SARS-CoV-2 on virus entry and its immune cross-reactivity with SARS-CoV. *Nat. Commun.* **2020**, *11*, 1620. [CrossRef]
73. Li, W.; Hulswit, R.J.G.; Widjaja, I.; Stalin Raj, V.; McBride, R.; Peng, W.; Widagdo, W.; Alejandra Tortorici, M.; van Dieren, B.; Lang, Y.; et al. Identification of sialic acid-binding function for the Middle East respiratory syndrome coronavirus spike glycoprotein. *Proc. Natl. Acad. Sci. USA* **2017**, *114*, E8508–E8517. [CrossRef]
74. Huang, X.; Dong, W.; Milewska, A.; Golda, A.; Qi, Y.; Zhu, Q.K.; Marasco, W.A.; Baric, R.S.; Sims, A.C.; Pyrc, K.; et al. Human coronavirus HKU1 spike protein uses O-acetylated sialic acid as an attachment receptor determinant and employs hemagglutinin-esterase protein as a receptor-destroying enzyme. *J. Virol.* **2015**, *89*, 7202–7213. [CrossRef]
75. Petrosillo, N.; Viceconte, G.; Ergonul, O.; Ippolito, G.; Petersen, E. COVID-19, SARS and MERS: Are they closely related? *Clin. Microbiol. Infect.* **2020**, *26*, 729–734. [CrossRef]
76. Ceccarelli, M.; Berretta, M.; Venanzi Rullo, E.; Nunnari, G.; Cacopardo, B. Differences and similarities between Severe Acute Respiratory Syndrome (SARS)-CoronaVirus (CoV) and SARS-CoV-2. Would a rose by another name smell as sweet? *Eur. Rev. Med. Pharm. Sci.* **2020**, *24*, 2781–2783. [CrossRef]
77. Wang, H.; Li, X.; Li, T.; Zhang, S.; Wang, L.; Wu, X.; Liu, J. The genetic sequence, origin, and diagnosis of SARS-CoV-2. *Eur. J. Clin. Microbiol. Infect. Dis.* **2020**, *39*, 1629–1635. [CrossRef]
78. Wan, Y.; Shang, J.; Graham, R.; Baric, R.S.; Li, F. Receptor recognition by the novel coronavirus from Wuhan: An analysis based on decade-long structural studies of SARS coronavirus. *J. Virol.* **2020**, *94*, e00127-20. [CrossRef]
79. Xia, S.; Liu, M.; Wang, C.; Xu, W.; Lan, Q.; Feng, S.; Qi, F.; Bao, L.; Du, L.; Liu, S.; et al. Inhibition of SARS-CoV-2 (previously 2019-nCoV) infection by a highly potent pan-coronavirus fusion inhibitor targeting its spike protein that harbors a high capacity to mediate membrane fusion. *Cell. Res.* **2020**, *30*, 343–355. [CrossRef]
80. Al-Tawfiq, J.A. Asymptomatic coronavirus infection: MERS-CoV and SARS-CoV-2 (COVID-19). *Travel. Med. Infect. Dis.* **2020**, *35*, 101608. [CrossRef]
81. Fantini, J.; Di Scala, C.; Chahinian, H.; Yahi, N. Structural and molecular modelling studies reveal a new mechanism of action of chloroquine and hydroxychloroquine against SARS-CoV-2 infection. *Int. J. Antimicrob. Agents* **2020**, *55*, 105960. [CrossRef]
82. Vandelli, A.; Monti, M.; Milanetti, E.; Rupert, J.; Zacco, E.; Bechara, E.; Ponti, R.D.; Tartaglia, G.G. Structural analysis of SARS-CoV-2 and predictions of the human interactome. *bioRxiv* **2020**. [CrossRef]
83. Milanetti, E.; Miotto, M.; Di Rienzo, L.; Monti, M.; Gosti, G.; Ruocco, G. In-Silico evidence for two receptors based strategy of SARS-CoV-2. *bioRxiv* **2020**. [CrossRef]
84. Creager, H.M.; Kumar, A.; Zeng, H.; Maines, T.R.; Tumpey, T.M.; Belser, J.A. Infection and replication of influenza virus at the ocular surface. *J. Virol.* **2018**, *92*, e02192-17. [CrossRef]
85. Loon, S.C.; Teoh, S.C.; Oon, L.L.; Se-Thoe, S.-Y.; Ling, A.-E.; Leo, Y.-S.; Leong, H.-N. The severe acute respiratory syndrome coronavirus in tears. *Br. J. Ophthalmol.* **2004**, *88*, 861–863. [CrossRef]
86. Colavita, F.; Lapa, D.; Carletti, F.; Lalle, E.; Bordi, L.; Marsella, P.; Nicastri, E.; Bevilacqua, N.; Giancola, M.L.; Corpolongo, A.; et al. SARS-CoV-2 isolation from ocular secretions of a patient with COVID-19 in Italy with prolonged viral RNA detection. *Ann. Intern. Med.* **2020**, *173*, 242–243. [CrossRef]
87. World Health Organization. *Report of the WHO-China Joint Mission on Coronavirus Disease 2019 (COVID-19)*; 14–20 February 2020; WHO: Geneva, Switzerland, 2020.
88. Chappey, B.; Beyssen, B.; Foos, E.; Ledru, F.; Guermonprez, J.L.; Gaux, J.C.; Myara, I. Sialic acid content of LDL in coronary artery disease: No evidence of desialylation in subjects with coronary stenosis and increased levels in subjects with extensive atherosclerosis and acute myocardial infarction: Relation between desialylation and in vitro peroxidation. *Arter. Thromb. Vasc. Biol.* **1998**, *18*, 876–883.
89. Nigam, P.K.; Narain, V.S.; Kumar, A. Sialic acid in cardiovascular diseases. *Indian J. Clin. Biochem.* **2006**, *21*, 54–61. [CrossRef]

90. Gopaul, K.P.; Crook, M.A. Sialic acid: A novel marker of cardiovascular disease? *Clin. Biochem.* **2006**, *39*, 667–681. [CrossRef]
91. Zhang, C.; Chen, J.; Liu, Y.; Xu, D. Sialic acid metabolism as a potential therapeutic target of atherosclerosis. *Lipids Health Dis.* **2019**, *18*, 173. [CrossRef]
92. Zhang, Y.; Zheng, Y.; Li, J.; Nie, L.; Hu, Y.; Wang, F.; Liu, H.; Fernandes, S.M.; Zhong, Q.; Li, X.; et al. Immunoregulatory Siglec ligands are abundant in human and mouse aorta and are up-regulated by high glucose. *Life Sci.* **2019**, *216*, 189–199. [CrossRef]
93. Kirkeby, S.; Jensen, N.E.; Mandel, U.; Poulsen, S.S. Asthma induction in mice leads to appearance of alpha2-3- and alpha2-6-linked sialic acid residues in respiratory goblet-like cells. *Virchows Arch.* **2008**, *453*, 283–290. [CrossRef]
94. Rathod, S.; Shori, T.; Sarda, T.S.; Raj, A.; Jadhav, P. Comparative analysis of salivary sialic acid levels in patients with chronic obstructive pulmonary disease and chronic periodontitis patients: A biochemical study. *Indian J. Dent. Res.* **2018**, *29*, 22–25. [CrossRef]
95. Guo, F.R. Active smoking is associated with severity of coronavirus disease 2019 (COVID-19): An update of a meta-analysis. *Tob. Induc. Dis.* **2020**, *18*, 37. [CrossRef]
96. Vardavas, C.I.; Nikitara, K. COVID-19 and smoking: A systematic review of the evidence. *Tob. Induc. Dis.* **2020**, *18*, 20. [CrossRef]
97. Liu, W.; Tao, Z.W.; Wang, L.; Yuan, M.-L.; Liu, K.; Zhou, L.; Wei, S.; Deng, Y.; Liu, J.; Liu, H.-G.; et al. Analysis of factors associated with disease outcomes in hospitalized patients with 2019 novel coronavirus disease. *Chin. Med. J.* **2020**, *133*, 1032–1038. [CrossRef]
98. Wielgat, P.; Trofimiuk, E.; Czarnomysy, R.; Holownia, A.; Braszko, J.J. Sialylation pattern in lung epithelial cell line and Siglecs expression in monocytic THP-1 cells as cellular indicators of cigarette smoke-induced pathology in vitro. *Exp. Lung Res.* **2018**, *44*, 167–177. [CrossRef]
99. Leung, J.M.; Yang, C.X.; Tam, A.; Shaipanich, T.; Hackett, T.-L.; Singhera, G.K.; Dorscheid, D.R.; Sin, D.D. ACE-2 expression in the small airway epithelia of smokers and COPD patients: Implications for COVID-19. *Eur. Respir. J.* **2020**, *55*, 2000688. [CrossRef]
100. Reghunathan, R.; Jayapal, M.; Hsu, L.Y.; Chng, H.-H.; Tai, D.; Leung, B.P.; Melendez, A.J. Expression profile of immune response genes in patients with Severe Acute Respiratory Syndrome. *BMC Immunol.* **2005**, *6*, 2. [CrossRef]
101. Frieman, M.; Heise, M.; Baric, R. SARS coronavirus and innate immunity. *Virus Res.* **2008**, *133*, 101–112. [CrossRef]
102. Varki, A.; Angata, T. Siglecs—The major subfamily of I-type lectins. *Glycobiology* **2006**, *16*, 1R–27R. [CrossRef]
103. Teuton, J.R.; Brandt, C.R. Sialic acid on herpes simplex virus type 1 envelope glycoproteins is required for efficient infection of cells. *J. Virol.* **2007**, *81*, 3731–3739. [CrossRef]
104. Vanderheijden, N.; Delputte, P.L.; Favoreel, H.W.; Vandekerckhove, J.; Van Damme, J.; van Woensel, P.A.; Nauwynck, H.J. Involvement of sialoadhesin in entry of porcine reproductive and respiratory syndrome virus into porcine alveolar macrophages. *J. Virol.* **2003**, *77*, 8207–8215. [CrossRef]
105. Delputte, P.L.; Nauwynck, H.J. Porcine arterivirus infection of alveolar macrophages is mediated by sialic acid on the virus. *J. Virol.* **2004**, *78*, 8094–8101. [CrossRef]
106. Zheng, Y.; Ma, X.; Su, D.; Zhang, Y.; Yu, L.; Jiang, F.; Zhou, X.; Feng, Y.; Ma, F. The roles of Siglec7 and Siglec9 on natural killer cells in virus infection and tumour progression. *J. Immunol. Res.* **2020**, *2020*, 6243819. [CrossRef]
107. Zhao, D.; Jiang, X.; Xu, Y.; Yang, H.; Gao, D.; Li, X.; Gao, L.; Ma, C.; Liang, X. Decreased Siglec-9 expression on natural killer cell subset associated with persistent HBV replication. *Front. Immunol.* **2018**, *9*, 1124. [CrossRef]
108. Yamanaka, M.; Kato, Y.; Angata, T.; Narimatsu, H. Deletion polymorphism of SIGLEC14 and its functional implications. *Glycobiology* **2009**, *19*, 841–846. [CrossRef]
109. Tsai, C.M.; Riestra, A.M.; Ali, S.R.; Fong, J.J.; Liu, J.Z.; Hughes, G.; Varki, A.; Nizet, V. Siglec-14 enhances NLRP3-inflammasome activation in macrophages. *J. Innate. Immun.* **2019**, 1–11. [CrossRef]
110. Chang, Y.C.; Nizet, V. The interplay between Siglecs and sialylated pathogens. *Glycobiology* **2014**, *24*, 818–825. [CrossRef]
111. Ali, S.R.; Fong, J.J.; Carlin, A.F.; Busch, T.D.; Linden, R.; Angata, T.; Areschoung, T.; Parast, M.; Varki, N.M.; Murray, J.C.; et al. Siglec-5 and Siglec-14 are polymorphic paired receptors that modulate neutrophil and amnion signaling responses to group B Streptococcus. *J. Exp. Med.* **2014**, *211*, 1231–1242. [CrossRef]

112. Auyeung, T.W.; Lee, J.S.; Lai, W.K.; Choi, C.H.; Lee, H.K.; Lee, J.S.; Li, P.C.; Lok, K.H.; Ng, Y.Y.; Ming, W.W.; et al. The use of corticosteroid as treatment in SARS was associated with adverse outcomes: A retrospective cohort study. *J. Infect.* **2005**, *51*, 98–102. [CrossRef]
113. Wielgat, P.; Mroz, R.M.; Stasiak-Barmuta, A.; Szepiel, P.; Chyczewska, E.; Braszko, J.J.; Holownia, A. Inhaled corticosteroids increase siglec-5/14 expression in sputum cells of COPD patients. *Adv. Exp. Med. Biol.* **2015**, *839*, 1–5. [CrossRef]
114. Zeng, Z.; Li, M.; Wang, M.; Wu, X.; Li, Q.; Ning, Q.; Zhao, J.; Xu, Y.; Xie, J. Increased expression of Siglec-9 in chronic obstructive pulmonary disease. *Sci. Rep.* **2017**, *7*, 10116. [CrossRef]
115. Singh, A.K.; Singh, A.; Shaikh, A.; Singh, R.; Misra, A. Chloroquine and hydroxychloroquine in the treatment of COVID-19 with or without diabetes: A systematic search and a narrative review with a special reference to India and other developing countries. *Diabetes Metab. Syndr.* **2020**, *14*, 241–246. [CrossRef]
116. Liu, J.; Cao, R.; Xu, M.; Wang, X.; Zhang, H.; Hu, H.; Li, Y.; Hu, Z.; Zhong, W.; Wang, M. Hydroxychloroquine, a less toxic derivative of chloroquine, is effective in inhibiting SARS-CoV-2 infection in vitro. *Cell. Discov.* **2020**, *6*, 16. [CrossRef]

Article

A Broad-Spectrum Antiviral Peptide Blocks Infection of Viruses by Binding to Phosphatidylserine in the Viral Envelope

Rutger D. Luteijn [1,†,‡], Patrique Praest [1,†], Frank Thiele [2,†], Saravanan Manikam Sadasivam [3], Katrin Singethan [2], Jan W. Drijfhout [4], Christian Bach [2], Steffen Matthijn de Boer [5], Robert J. Lebbink [1], Sha Tao [6], Markus Helfer [2], Nina C. Bach [7], Ulrike Protzer [2], Ana I. Costa [1], J. Antoinette Killian [3], Ingo Drexler [6,§] and Emmanuel J. H. J. Wiertz [1,*,§]

1 Department of Medical Microbiology, University Medical Center Utrecht, 3508 GA Utrecht, The Netherlands; rdluteijn@gmail.com (R.D.L.); p.praest-2@umcutrecht.nl (P.P.); R.J.Lebbink-2@umcutrecht.nl (R.J.L.); A.I.CorreiadeAlmeidaCosta@umcutrecht.nl (A.I.C.)

2 Institute of Virology, Technical University and Helmholtzzentrum Munich, 81675 Munich, Germany; frank.thiele@helmholtz-muenchen.de (F.T.); katrin.singethan@tum.de (K.S.); ch-bach@gmx.de (C.B.); helfer.markus@gmail.com (M.H.); protzer@tum.de (U.P.)

3 Membrane Biochemistry and Biophysics, Bijvoet Centre for Biomolecular Research, Utrecht University, 3584 CH Utrecht, The Netherlands; mssaravananbt@gmail.com (S.M.S.); J.A.Killian@uu.nl (J.A.K.)

4 Department of Immunohematology and Blood Transfusion, Leiden University Medical Center, 2333 ZA Leiden, The Netherlands; j.w.drijfhout@lumc.nl

5 Virology Division, Department of Infectious Diseases & Immunology, Faculty of Veterinary Medicine, Utrecht University, 3584 CL Utrecht, The Netherlands; matthijn.de.boer@intravacc.nl

6 Institute for Virology, University Hospital Düsseldorf, Heinrich-Heine-University, 40225 Düsseldorf, Germany; sha.tao@med.uni-duesseldorf.de (S.T.); Ingo.Drexler@med.uni-duesseldorf.de (I.D.)

7 Institute of Chemistry, Chair of Organic Chemistry II, TU Munich, 85748 Garching, Germany; nina.bach@tum.de

* Correspondence: E.Wiertz@umcutrecht.nl; Tel.: +31-88-7550862

† These authors contributed equally to this paper.

‡ Present address: Virology Division, Department of Infectious Diseases and Immunology, Faculty of Veterinary Medicine, Utrecht University, 3584 CL Utrecht, The Netherlands.

§ These authors contributed equally to this paper.

Received: 11 August 2020; Accepted: 27 August 2020; Published: 29 August 2020

Abstract: The ongoing threat of viral infections and the emergence of antiviral drug resistance warrants a ceaseless search for new antiviral compounds. Broadly-inhibiting compounds that act on elements shared by many viruses are promising antiviral candidates. Here, we identify a peptide derived from the cowpox virus protein CPXV012 as a broad-spectrum antiviral peptide. We found that CPXV012 peptide hampers infection by a multitude of clinically and economically important enveloped viruses, including poxviruses, herpes simplex virus-1, hepatitis B virus, HIV-1, and Rift Valley fever virus. Infections with non-enveloped viruses such as Coxsackie B3 virus and adenovirus are not affected. The results furthermore suggest that viral particles are neutralized by direct interactions with CPXV012 peptide and that this cationic peptide may specifically bind to and disrupt membranes composed of the anionic phospholipid phosphatidylserine, an important component of many viral membranes. The combined results strongly suggest that CPXV012 peptide inhibits virus infections by direct interactions with phosphatidylserine in the viral envelope. These results reiterate the potential of cationic peptides as broadly-acting virus inhibitors.

Keywords: antiviral peptide; enveloped viruses; membrane phosphatidylserine; envelope disruption

1. Introduction

The continuous threat of conventional viruses and emergence of new virus species requires an ongoing search for new antiviral compounds. Broad-spectrum antiviral inhibitors have become promising therapeutic candidates [1]. These compounds act on common elements shared by many groups of viruses, including the synthesis of viral RNA and DNA, viral proteases, and glycosylation of viral proteins (reviewed by [2]). An important class of broad-spectrum antivirals targets the lipid envelope of viruses [3–7]. These include not only antibodies and chemical compounds, but also cationic peptides present as defense peptides in the human host and other organisms. In essence, the applicability of such peptides as antiviral compounds targeting the envelope of viruses will depend on their ability to discriminate viral from human membranes.

Enveloped viruses, in contrast to non-enveloped ones, have their nucleocapsid surrounded by a lipid membrane. Depending on the virus, this lipid bilayer is derived from the plasma membrane or intracellular organelles. Although the membrane is host cell-derived, it may differ functionally and structurally from the membrane of origin in several aspects, including the lipid composition. Compared to host cell membranes, the envelope of many viruses is enriched for the phospholipid phosphatidylserine (PS). PS is the most abundant anionic lipid in eukaryotic cells. In resting cells, PS is mainly contained within the cell, with only very limited exposure to the extracellular environment. PS is enriched in some intracellular organelles, but the majority of PS is asymmetrically distributed on the inner leaflet of the plasma membrane [8]. This organization results from constantly flipping PS from the outer membrane leaflet to the inner membrane leaflet by ATP-dependent flippases. As a result, the outer exposed membrane leaflet of the cellular plasma membrane is devoid of PS. In apoptotic cells, this organization is lost due to the inactivation of flippases and the activation of scramblases. The resulting PS exposure on apoptotic cells and cell debris induces uptake by surrounding cells through PS receptors. Many viruses take advantage of this uptake mechanism by exposing PS on their envelope, thereby facilitating virus entry. The increase in PS on the virus membrane may be due to the lack of an active flippase. Alternatively, some viruses actively accumulate PS at sites of virus budding, or bud from subcellular microdomains enriched in PS [9].

Another important aspect of the host membrane is its property of self-renewal upon injury. This fast-acting mechanism involves several events, including the detection of the damaged site, exocytosis of nearby cytosolic vesicles, cytoskeletal remodeling, endocytosis of the damaged membrane, and reconstitution of membrane homeostasis [10,11]. The lipid membrane loses its self-renewal capacity once surrounding the viral nucleocapsid, thereby becoming particularly prone to membrane damage [5]. The structural and functional differences between virus and host cell membranes make viral membranes ideal targets for antiviral therapy.

Here, we identify a peptide with broad-spectrum antiviral activity, whose sequence is derived from the cowpox virus protein CPXV012. Within virus-infected cells, this protein helps to evade the immune system by inhibiting the transporter associated with antigen processing (TAP), thereby interfering with MHC I-dependent antigen presentation [12–16]. We pinpointed the segment of the CPXV012 protein responsible for blocking TAP: a peptide comprising the C-terminal domain of CPXV012. It should be noted that we have no indication that this peptide is produced upon CPXV infection. Interestingly, when doing functional assays with this peptide, upon infection we noticed that the percentage of infected cells decreased significantly. This finding was the basis for the current study. Here, we show that this peptide inhibits infection of many enveloped viruses by interacting with virus particles. Variants of the CPXV012 peptide revealed that basic residues within the peptide are important for this inhibitory effect. Furthermore, CPXV012 strongly interacts with lipid monolayers and membranes enriched for the anionic phospholipid PS. These results suggest that this CPXV012 peptide disturbs the integrity of viral membranes enriched for PS, and thus may be explored as an antiviral agent against a broad range of enveloped viruses.

2. Materials and Methods

2.1. Cells and Viruses

MelJuSo (MJS) and BHK21 cells were cultivated with RPMI 1640 containing 10% FCS, 100 U/mL penicillin, 100 µg/mL streptomycin, and 2 mM L-glutamine (complete medium). HEK293T, HeLa, Huh7.5, and Vero cells were grown in DMEM supplemented with 10% FCS (5% for Vero cells), 100 U/mL penicillin, 100 µg/mL streptomycin, 1% nonessential amino acids, 2 mM L-glutamine, and 1% sodium pyruvate (complete DMEM). HepRG cells were differentiated in Williams E medium supplemented with 10% FCS (Fetalclone II from Hyclone, Thermo Scientific; Breda, The Netherlands), 100 U/mL penicillin, 100 µg/mL streptomycin, glutaMax, 0.023 IE/mL human insulin (Sanofi Aventis, Amsterdam, Netherlands), 4.7 µg/mL hydrocortisone (Pfizer, Capelle a/d IJssel, The Netherlands), 80 µg/mL Gentamicin, and 1.8% DMSO. HepG2.2.15 cells were cultivated in Williams E medium with 10% FCS, 100 U/mL penicillin, 100 µg/mL streptomycin, and 1% nonessential amino acids. LC5-RIC cells (EASY-HIT assay) were maintained as described previously [17].

Recombinant modified vaccinia virus Ankara (MVA) expressing eGFP under the natural early/late promoter p7.5 was used in this study (MVA-eGFP) [18]. MVA-eGFP was propagated and titrated in chicken embryonic fibroblasts (CEFs) according to standard methodology [19]. Vaccinia virus expressing eGFP (VACV-eGFP) was a generous gift from Dr. Jon Yewdell (NIH, Bethesda, USA). VACV-eGFP was propagated and titrated on CV-1 cells. Cowpox virus strain Brighton Red (CPXV-BR) expressing RFP/eGFP virus was originally obtained from Dr. Karsten Tischer (FU Berlin, Germany). HSV-1 expressing the VP16-eGFP was kindly provided by Dr. Peter O'Hare (Imperial College London, UK). Stocks were prepared and titrated on Vero cells.

HBV particles were concentrated from the supernatant of HepG2.2.15 cells as described previously [20]. Infectious HIV-1 stocks were prepared as described before [17]. Measles virus expressing eGFP (MV-eGFP) was generated as previously described [21]. Vesicular stomatitis virus expressing Luciferase (VSV-deltaG (Luc)) was a kind gift from Dr. Gert Zimmer (Institute of Virology and Immunology, Mittelhäusern, Switzerland). Recombinant adenovirus expressing GFP and ovalbumin (AdGOva) was kindly provided by Dr. Percy Knolle (Institutes of Molecular Medicine and Experimental Immunology, Bonn, Germany) and virus stocks were prepared as previously described [22]. Non-spreading Rift Valley fever virus (RVFV) replicon particles were produced and titrated as described previously [23]. Coxsackie B3 virus expressing Renilla Luciferase (RLuc-CVB3) was produced and titrated as described previously [24].

2.2. Peptides

The inhibitory peptide used in this study comprises amino acids 36 to 69 of the CPXV012 protein (QEGISRFKICPYHWYKQHMSLLFRRYYHKLDSII). The control peptide is derived from the pseudorabies virus TAP inhibitor UL49.5 and comprises amino acid residues 28 to 59 of this protein (sequence: STEGPLPLLREESRINFWNAAUAARGVPVDQP). U in amino acid sequence refers to alpha-amino-n-butyric acid, see below.

Synthetic peptides were prepared by normal Fmoc chemistry using preloaded Tentagel resins, using PyBop/N-methylmorpholine for in situ activation and 20% piperidine in N-methylpyrrolidinone for Fmoc removal [25]. Couplings were performed for 75 min. After final Fmoc removal, peptides were cleaved with trifluoroacetic acid/H2O 19:1 *v/v* containing additional scavengers when C or W residues were present in the peptide sequence. Peptides were isolated by ether/pentane precipitation. Peptides were checked for purity using reversed-phase HPLC–mass spectrometry and for integrity using MALDI-TOF mass spectrometry, showing the calculated molecular masses. All peptides were synthesized with an N-terminal biotin moiety and a C-terminal amide. Cysteines were replaced by the isosteric alpha-amino-n-butyric acid. Lyophilized peptides were dissolved in DMSO. Peptide concentrations were confirmed by the NanoDrop spectrophotometer (Thermo Scientific, Amsterdam, The Netherlands) using the absorbance of Trp at 280 nm. Peptides were aliquoted in Eppendorf

LoBind microcentrifuge tubes (Sigma-Aldrich, Zwijndrecht, The Netherlands) and kept at −20 °C for short-term storage.

2.3. Cell Viability Assays

Cells were incubated with CPXV012 peptide, control peptide, or DMSO at the indicated concentrations for at 37 °C for 20 to 24 h. Thereafter, cell viability was measured using different assays. DAPI staining was performed to discriminate between live and dead cells using flow cytometry. Cell Titer-Blue Cell Viability Assay (Promega, Leiden, The Netherlands) was performed according to the manufacturer's instructions. Neutral Red uptake (Applichem, Darmstadt, Germany) was performed as described previously [26]. WST-1 salt conversion assay (Roche, Woerden, The Netherlands) was performed according to the manufacturer's instructions at 7 or 20 h post exposure to the peptides. No difference was observed between 7 and 20 h post exposure.

2.4. Virus Inhibition Assays

2.4.1. MVA, VACV, and Cowpox

MJS, BHK21, HeLa, and HEK-293T cells (1×10^5) were seeded in 24-well plates in a total volume of 400 µL and incubated overnight at 37 °C. Cells were infected with MVA-eGFP, VACV-eGFP, or CPXV-RFP/eGFP at 37 °C for 18 to 20 h in medium containing 0.1% DMSO and the indicated concentrations of either CPXV012 peptide or control peptide UL49.5 or 0.1% DMSO only as vehicle control. Flow cytometric analysis was performed to quantify the percentage of infected cells as indicated by eGFP or RFP expression.

Where mentioned, MVA-eGFP was pretreated with 50, 100, or 150 µg/mL CPXV012 peptide, control peptide, or DMSO vehicle at 37 °C for 1 h. Thereafter, the virus-CPXV012 peptide mixture was diluted 1:10 by volume and used to infect MJS cells (corresponding to an MOI of 10) resulting in a final concentration of 5, 10, or 15 µg/mL peptide in the culture medium. As a control, MJS cells were incubated with the same concentrations of CPXV012 peptide (5, 10, and 15 µg/mL) during infection with MVA-eGFP that was not pretreated with peptides. After 20 h, the cytometric analysis was performed to quantify the percentage of infected cells indicated by eGFP-expression.

For qPCR analysis (Figure 1B), MJS cells were infected with MVA-eGFP (MOI 10) in the presence of 100 µg/mL of either CPXV012 peptide or control peptide in medium with 0.1% DMSO or in 0.1% DMSO only as vehicle control at 37 °C for 20 h. RNA was isolated using the RNeasy Mini Kit (Qiagen, Venlo, The Netherlands). A total of 3 µg of RNA was digested with 10 U of DNase I (Roche, Woerden, The Netherlands) and cDNA synthesis was performed by using 200 U Superscript III RNase H reverse transcriptase (Invitrogen, Amsterdam, The Netherlands), 7.5 pmol oligo$(dT)_{12–18}$ (Invitrogen), 20 U of RNasin (Promega), and 10 mM each deoxynucleoside triphosphate (Qiagen). For semiquantitative analysis of viral mRNAs, cDNA was used as the template for a PCR with the indicated primers: B8R (fwd ATCCGCATTTCCAAAGAATG, rev ACATGTCACCGCGTTTGTAA), H3L (fwd GTCTTGAAGGCAATGCATGA, rev TCCCGATGATAGACCTCCAG) and G8R (fwd ATC GAT AAA CTG CGC CAA AT), and rev CTC CGC GGT AGA ACA CTG AT). Quantitative RT-PCR analysis was performed by using the LightCycler DNA Master SYBR green I kit (Roche) and LightCycler 1.5 (Roche). Gene expression was determined using the $2^{-\Delta\Delta CT}$ method. The results are quantified relative to the housekeeping gene GAPDH [27,28].

2.4.2. HSV-1

MJS cells (10^5/well) were seeded in a 24-well plate and cultured overnight at 37 °C. HSV-1-eGFP was mixed with CPXV012 peptide, control peptide UL49.5, or DMSO only in complete medium. DMSO concentration was 0.1% with or without peptide. The medium of cells was replaced with virus/peptide inoculum (MOI 0.1) and cells were incubated for 16 h at 37 °C. Cells were harvested, fixed using 1% formaldehyde, and the number of eGFP-positive cells was determined using flow cytometry.

2.4.3. HBV

Confluent HepRG cells cultured in 48-well plates were differentiated for two weeks. Medium was removed and cells were left untreated, or CPXV012 peptide in medium containing 0.1% DMSO was added at the concentrations indicated. Medium containing 0.1% DMSO was used as vehicle control. Cells were infected with HBV (MOI 200) and incubated for 16 h. Medium was removed, cells were washed three times with PBS, fresh medium was added, and cultures were incubated for another 12 days with a medium change every three days. Thereafter, the supernatant was collected after centrifugation to remove cellular debris (5 min, 350× *g*) and HBeAg was quantified using commercial immunoassays as described [20]. HBV DNA was extracted from the supernatant using the BioSprint 96 One-For-All Vet Kit (Qiagen) and was quantified by qPCR.

2.4.4. HIV-1

The EASY-HIT assay was performed as described previously [17]. Briefly, 1×10^4 LC5-RIC cells were plated in 96-well plates and incubated at 37 °C for 24 h. Thereafter, cells were treated with medium containing 0.1% DMSO and the indicated concentrations of CPXV012 peptide or 0.1% DMSO only as vehicle control for 1 h prior to infection with HIV inoculum. After 48 h, cultures were assayed for cellular reporter gene expression by quantification of total fluorescence intensity of each culture using a fluorescence microplate reader (step 1). To assess titers of infectious virus in the culture supernatant (virion production in primary infected cells), 20 μL of medium was added to fresh LC5-RIC cells and incubated for another 72 h before reporter gene expression was quantified (step 2).

2.4.5. Measles and VSV

Vero cells (1×10^5 cells/well) were plated in 12-well plates and left untreated or treated with medium containing 0.1% DMSO and the indicated concentrations of CPXV012 peptide or 0.1% DMSO only as vehicle control. Cells were infected with MV-eGFP (MOI 0.1). Once maximum giant cell formation was observed at 48 h post infection, microscopic fluorescence images were taken by using an inverted microscope CKX41 (Olympus) with an LCachN/10×/0.40 Phc/1/FN22 UIS objective. Thereafter, the medium was removed, and infection was measured based on the expression of eGFP. Fluorescence was detected using an Infinite 200 PRO Tecan reader (fluorescence bottom reading for cell-based assays). For VSV infections, Huh7.5 cells (2 × 105 cells/well) were seeded in 12-well plates, and peptides were added at the concentrations indicated. DMSO was used as vehicle control. Cells were infected with VSVdeltaG (Luc) (MOI 0.6). At 18 h post infection, luciferase was measured using the luciferase assay system (Promega) and normalized to the protein content of the individual sample as determined by a Bradford assay.

2.4.6. Adenovirus

HEK-293 cells (4×10^5 cells/well) were seeded in 12-well plates in a total volume of 1 mL and incubated overnight at 37 °C. Cells were treated with medium containing 0.1% DMSO and the indicated concentrations of CPXV012 peptide or 0.1% DMSO only as vehicle control at 37 °C for 1 h. Afterward, cells were infected with an MOI of 10 with AdGOva for 24 h. Photometric analysis was performed with the Infinite 200 PRO Tecan to quantify the percentage of infected cells based on GFP-expression.

2.4.7. Rift Valley Fever Virus

MJS cells (2×10^4 cells/well) were seeded in a 96-well plate and incubated for two days at 37 °C. Cells were treated with medium containing 0.1% DMSO and the indicated concentrations of either CPXV012 peptide or control peptide UL49.5 or 0.1% DMSO only as vehicle control. After 15 min, RVFV-eGFP was added to the cells. Twenty-four hours after infection, cells were harvested and the percentage of eGFP-positive cells was determined using flow cytometry.

2.4.8. Coxsackievirus B3

MJS cells (10^4 cells/well) were seeded in a 96-well plate. After overnight incubation at 37 °C, cells were infected with RLuc-CVB3 in the presence of medium containing 0.1% DMSO and the indicated concentrations of CPXV012 peptide or 0.1% DMSO only as vehicle control. After 7 h of infection, cells were lysed and renilla luciferase expression levels were quantified using the Renilla Luciferase Assay System kit (Promega).

2.5. Preparation of Large Unilamellar Vesicles (LUVs)

Calcein-encapsulated LUVs were prepared using 1,2-dioleoyl-*sn*-glycero-3-phosphocholine (DOPC) or a mixture of DOPC and 1,2-dioleoyl-sn-glycero-3-phospho-L-serine (DOPS) (Avanti Polar Lipids) in a 7:3 molar ratio. A stock solution of DOPC and DOPS in chloroform (10 mM) were mixed in a glass tube. The solvent was evaporated with dry nitrogen gas yielding a lipid film that was subsequently kept in a vacuum desiccator for 20 min. Lipid films were hydrated for 30 min in buffer containing 10 mM Tris, 50 mM NaCl at pH 7.4 resulting in a total lipid concentration of 10 mM. For calcein-encapsulated LUVs, 50 mM of calcein was added during hydration. The lipid suspension was freeze-thawed ten cycles, at temperatures of −80 and +40 °C, respectively, and eventually extruded 10 times through 0.2 μM-pore size filters (Anotop 10, Whatman, UK). For the preparation of calcein-encapsulated LUVs, free calcein was separated from calcein-filled LUVs using size exclusion column chromatography (Sephadex G-50 fine) and eluted with 10 mM Tris-HCl, 150 mM NaCl buffer at pH 7.4. Finally, the phospholipid content of lipid stock solutions and vesicle preparations was determined as inorganic phosphate according to Rouser [29].

2.6. Circular Dichroism

Circular dichroism (CD) experiments were performed as described previously (15). Briefly, the CD spectra of 625 μM LUVs and/or 100 μg/mL peptides diluted in 10 mM MES buffer (pH 6.2) were recorded on a Jasco 810 spectropolarimeter (Jasco, Easton, MD) over a wavelength range of 200 to 250 nm. Each reported spectrum is the average of five independent scans recorded every 1 nm at a scan rate of 20 nm/min at room temperature in cuvettes with a path length of 1.0 mm.

2.7. Langmuir Monolayers

Peptide-induced changes in the surface pressure of a monomolecular layer (monolayer) of phospholipids at a constant surface area were measured using a Langmuir Microtrough XL device (Kibron, Helsinki, Finland). A Teflon trough was filled with 16 mL PBS (pH 7.4) and lipid monolayers of DOPC or DOPC/DOPS (ratio 7:3) were spread from a 0.5 mM stock solution in chloroform at the air–buffer interface. The buffer below the lipid monolayer (subphase) was continuously stirred using a magnetic stirrer. Upon stabilization of the initial surface pressure to 25 mN/m, a freshly prepared stock of peptide in DMSO was injected into the subphase, resulting in a final peptide concentration of 0.25 μM.

2.8. Membrane Permeability Assay

Membrane permeability was measured in standard 96-well transparent microtiter plates in a plate reader (Spectrafluor, Tecan, Salzburg, Austria). Peptides (5 μL of 0.2 mM in DMSO) were added to calcein-containing LUVs (195 μL of lipid vesicles (50 μM) in 10 mM Tris–HCl, 100 mM NaCl buffer (pH 7.4)). As positive control human islet amyloid polypeptide (hIAPP; Bachem) 5 μL of a 0.2 mM in DMSO) was added to calcein-containing LUVs. As a negative control, murine IAPP ((mIAPP; Bachem) 5 μL of 0.2 mM in DMSO) was added to calcein-containing LUVs. For blank only, 5 μL DMSO was added to calcein-containing LUVs. Directly after the addition of all components, the microtiter plate was shaken for 10 s. Fluorescence was measured from the top, every 5 min, using a 485 nm excitation filter and a 535 nm emission filter at 25 °C. The maximum leakage at the end of each measurement was determined by adding 1 µL of 10% Triton X-100 to a final concentration of 0.05% (*v/v*). The release of fluorescent dye was calculated according to Equation (1):

$$L(t) = (Ft - F0)/(F100 - F0) \quad (1)$$

L(t) is the fraction of dye released (normalized to membrane leakage), Ft is the measured fluorescence intensity, and F0 and F100 are the fluorescence intensities at times t = 0 and after addition of Triton X-100, respectively. All membrane leakage assays were performed two times, each in duplicate, on different days, using different IAPP stock solutions.

2.9. Statistical Analysis

Statistical significance was analyzed by one-way ANOVA testing, followed by Dunnett's multiple comparisons test (the mean of each column was compared to that of the DMSO control), using GraphPad Prism 8.0.1. In Figure 1D, statistical analysis was performed on the plotted data transformed as follows: Y = Log(Y). *p*-values of <0.05 were considered significant (* $p < 0.05$; ** $p < 0.01$, *** $p < 0.001$).

3. Results

3.1. CPXV012 Peptide Prevents Infection by Different Viruses in Cell Culture

We tested a number of peptides for their ability to inhibit TAP and subsequent MHC-I antigen presentation. These peptide sequences were based on the active domains of viral TAP-inhibiting proteins CPXV012 (from cowpox virus) and UL49.5 (from pseudorabies virus) [30]. While testing their ability to block MHC I antigen presentation during virus infection, we found that the CPXV012 peptide was uniquely capable of inhibiting virus reporter expression. To assess the effect of CPXV012 peptide on viral infection, different infection inhibition assays were performed using a variety of viruses and cell types. Cells were treated with CPXV012 peptide (34 amino acids, QEGISRFKICPYHWYKQHMSLLFRRYYHKLDSII) or a similar-length control peptide derived from the pseudorabies virus protein UL49.5. The viral inoculum was mixed with the peptide at different concentrations and immediately added to cells. After incubation for the indicated times, infection was determined based on viral gene expression, or viral DNA or mRNA synthesis.

Infection of the human melanoma cell line MelJuSo (MJS) with recombinant modified vaccinia virus Ankara (MVA) expressing eGFP (MVA-eGFP) was confirmed via microscopy (Figure 1A). The cell monolayer showed the typical cytopathic effect upon infection (in the presence of DMSO, peptide vehicle control; bright field) and eGFP fluorescence could be detected. A similar pattern was verified in the presence of the control peptide. In contrast, upon infection and incubation with the CPXV012 peptide (100 µg/mL), the monolayer was microscopically indiscernible from that of the uninfected control. MVA-eGFP infection was inhibited in a concentration-dependent manner (Figure 1B). Inhibition starts at 50 µg/mL and is stronger upon increasing peptide concentrations, reaching 98.3% (± 0.3) at 150 µg/mL CPXV012 peptide (Figure 1B). To test potential cytotoxicity of the peptides, the amount of DAPI-negative (live, i.e., cells whose membrane was not compromised/permeable) cells was quantified. We did not observe any decrease in live cells up to the highest concentrations (200 µg/mL) (Figure 1C). To further test any effect on cell viability, other assays were performed. The WST-1 assay measures the net metabolic activity of the cells (it is based on the enzymatic conversion of the WST-1 salt into the colored dye formazan in viable cells), and the Neutral Red uptake assay relies on the staining of lysosomes in viable cells upon active transport of the cationic dye. MJS cell viability was not severely impaired upon addition of up to 200 µg/mL CPXV012 peptide using the WST-1 assay (Figure S1A), and the Neutral Red assay (Figure S1B). Cell titer blue assays were performed to test the viability of other cell lines (Vero and Huh7.5 cells) used in this study in the setting of other viral infections. Again, no significant effect was observed at the highest peptide concentrations (Figure S1C).

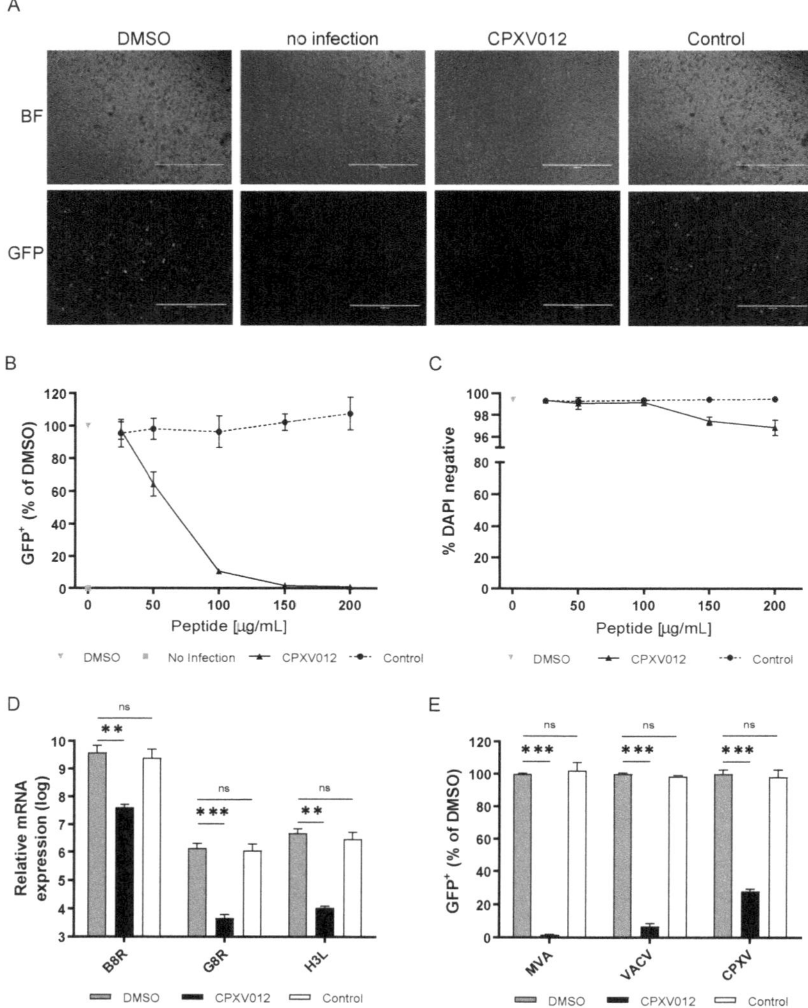

Figure 1. CPXV012 peptide inhibits poxvirus infection. (**A**) MJS cells were infected with MVA-eGFP (MOI 10) under different conditions (0.1% DMSO control, CPXV012 and control peptide UL49.5 in 0.1% DMSO). At 20 h.p.i., infection was assessed by fluorescence microscopy (bar size 1 mm). Results are representative of at least three independent experiments. BF: Bright field. GFP: Green fluorescent protein (MVA infection). (**B**) MJS cells were infected with MVA-eGFP (MOI 10) in the presence of 0.1% DMSO vehicle control, or CPXV012 peptide, or control peptide, both in 0.1% DMSO. At 18 to 20 h post infection, cells were harvested and analyzed for viral gene expression. The number of eGFP-positive cells was quantified using flow cytometry. Mean ± S.E.M. of three independent experiments is shown. (**C**) MJS cells were treated for 20 h with the indicated concentrations of CPXV012 or control peptide. Subsequently, cells were harvested and stained with DAPI for flow cytometric analysis. S.E.M. of three independent experiments is shown. (**D**) MJS cells infected with MVA-eGFP (MOI 10) in the presence of 100 µg/mL CPXV012 or UL49.5 (control) peptide in 0.1% DMSO or 0.1% DMSO only were lysed 20 h

post infection. RNA was isolated and qPCR was performed for expression of viral genes B8R, H3L, and G8R. Mean ± S.E.M. of three independent experiments is shown. Log-transformed data were analyzed with one-way ANOVA followed by multiple comparisons Dunnett's test (the mean of each column was compared to that of the DMSO control). (**E**) CPXV012 peptide inhibits infection with the poxviruses MVA-eGFP, VACV-eGFP, and CPXV-RFP/eGFP. MJS cells were infected with VACV-eGFP (MOI 10) or CPXV-RFP/eGFP (MOI 10) in the presence of 150 µg/mL CPXV012 or UL49.5 control peptide in 0.1% DMSO or 0.1% DMSO only. After 18 to 20 h, infection was quantified by cytometric analysis of fluorescent cells. Mean ± S.E.M. of three independent experiments is shown using different batches of peptide. Data were analyzed with one-way ANOVA followed by multiple comparisons Dunnett's test (the mean of each column was compared to that of the DMSO control). (** $p < 0.01$; *** $p < 0.001$).

To confirm the inhibitory effect of CPXV012 peptide on infection of MJS with MVA-eGFP, qPCR analysis for expression of viral genes was performed (Figure 1D). CPXV012 peptide decreased the expression of the early-expressed gene B8R up to two logs, relative to its expression upon infection in the presence of DMSO vehicle only. The decrease in expression of the intermediate-expressed gene G8R and late expressed gene H3L was more pronounced (more than two logs). No effect was seen if cells were treated with the control peptide compared to the DMSO sample (Figure 1D). To assess whether the inhibitory effect on MVA-eGFP infection also occurs in other cell types, HEK-293T, BHK21, and HeLa cells were used (Figure S2). In all cell lines, a strong decrease in infection (87.4 to 94.2%) was observed.

To investigate whether other poxviruses besides MVA were inhibited by CPXV012 peptide, we tested the effect on Vaccinia virus (VACV) strain WR and cowpox virus strain BR (CPXV). Again, infection of MJS with both viruses was inhibited (93 ± 1.7% for VACV and 72 ± 1.1% for CPXV) by CPXV012 peptide (150 µg/mL) while no effect was seen with the control peptide in comparison to the DMSO sample (Figure 1E). Thus, CPXV012 peptide inhibits infection of different members of the poxvirus family.

To further characterize the inhibitory effect of CPXV012, in vitro infection inhibition assays were performed with a collection of viruses that differ in genome content (RNA or DNA) and structural composition (enveloped or non-enveloped). These viruses include herpes simplex virus-1 (HSV-1), Hepatitis B virus (HBV), HIV-1, adenovirus, measles virus, vesicular stomatitis virus (VSV), coxsackievirus B3 (CVB3), and Rift Valley fever virus (RVFV) (Table S1, Figures 2 and 3).

The inhibitory effect of CPXV012 peptide on HSV-1-eGFP, a large dsDNA virus from the herpesvirus family, was evaluated upon infection of MJS cells. The cells were infected with HSV-1-eGFP in the presence of peptide or DMSO vehicle control. After 16 h, the amount of eGFP-positive cells was determined using flow cytometry (Figure 2A). As observed for the poxviruses, CPXV012 peptide showed a dose-dependent inhibition of HSV-1 infection (Figure 2A; 75.9 ± 5.7% when using 150 µg/mL peptide).

For HBV, an enveloped dsDNA virus of the hepadnavirus family, infectivity was monitored by quantifying the amount of viral envelope protein (HBeAg) and viral DNA in the supernatant of HepRG-infected cells (Figure 2B). Both HBeAg and viral DNA were decreased in the supernatant of cells treated with CPXV012 peptide in a concentration-dependent manner. At a peptide concentration of 100 µg/mL, a decrease of 84.0 ± 3.0% of HBeAg and 73.6 ± 2.3% of viral DNA was observed.

The infectivity of HIV-1, an enveloped ssRNA virus, was monitored using the EASY-HIT assay on LC5-RIC reporter cells. In this essay, both early and late phases of HIV replication are assessed: in step 1, the levels of a fluorescent reporter protein induced during the early phase of HIV replication are quantified, and in step 2 the production of infectious virions in primary infected cells is determined. HIV infection was reduced in the presence of CPXV012 peptide, but not in the presence of DMSO vehicle control (Figure 2C). Both primary viral infection (step 1) and viral replication (step 2) were dose-dependently affected, with an inhibition of 82.7 ± 4.9% (step 1) and 62.2 ± 2.4% (step 2) at the highest peptide concentration (100 µg/mL).

For RVFV, an enveloped ssRNA virus of the Bunyavirus family, infection was monitored by viral eGFP expression 24 h post infection (Figure 2D). Infection was reduced by CPXV012 peptide in a concentration-dependent manner (65.5 ± 2.3% at 160 μg/mL).

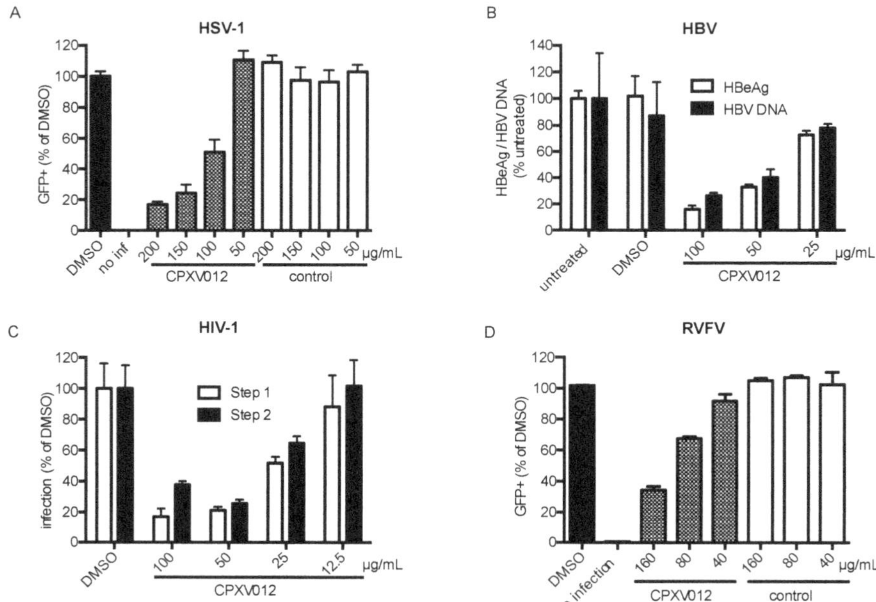

Figure 2. CPXV012 peptide inhibits infection with HSV-1, HBV, HIV, and RVFV. (**A**) CPXV012 peptide inhibits HSV-1 infection. MJS cells were left uninfected or infected with HSV-1 (MOI 0.1) in the presence of CPXV012 or UL49.5 as control at the indicated peptide concentrations in 0.1% DMSO or 0.1% DMSO only. At 16 h after infection, cells were harvested, and the amount of eGFP-positive cells was quantified using flow cytometry. Mean ± S.E.M. of three independent experiments is shown. (**B**) Effect of CPXV012 peptide on HBV infection. Differentiated HepRG cells were left untreated or CPXV012 peptide in 0.1% DMSO was added using the concentrations indicated or 0.1% DMSO only as control. Cells were infected with HBV (MOI 200) for 16 h. Subsequently, cells were washed and fresh medium was added. After 12 days, the supernatant was analyzed for HBeAg and HBV DNA. Mean ± S.E.M. of three independent experiments is shown. (**C**) CPXV012 peptide inhibits HIV infection. LC5-RIC cells were treated with the indicated concentrations of CPXV012 peptide in 0.1% DMSO or 0.1% DMSO only as control for 1 h and infected with HIV for 48 h according to the EASY-HIT assay system (17). Cellular reporter expression was quantified using a fluorescence microplate reader to assess the ability of HIV to infect LC5-RIC cells (Step 1). Next, 20 μL of culture supernatant was added to fresh LC5-RIC cells and incubated for another 72 h before fluorescence detection was performed to assess virion production from the first round of infection (Step 2). Data are shown as the percentage of infection referred to the respective DMSO control which was set to 100%. Mean ± S.E.M. of three independent experiments is shown. (**D**) CPXV012 peptide inhibits infection with RVFV. MJS cells were left uninfected or infected with RVFV-eGFP in the presence of CPXV012 or control peptide (UL49.5) at the indicated peptide concentrations in 0.1% DMSO or 0.1% DMSO only. After 24 h, cells were harvested and the amount of eGFP-positive cells was quantified using flow cytometry. Mean ± S.E.M. of three independent experiments is shown.

In contrast, CPXV012 peptide showed no effect on infection by the measles virus, an enveloped ssRNA virus of the paramyxovirus family (Figure 3A). The formation of measles virus-induced syncytia was unchanged (Figure S3). Replication of measles virus and cell-to-cell spread was assessed by quantification of viral titers present in infected cultures using plaque assays. No difference was found between CPXV012 peptide-treated samples, the DMSO vehicle control, or the control peptide (data not shown). For VSV, an enveloped ssRNA virus of the rhabdovirus family, virus-encoded Renilla luciferase activity was measured 18 h after infection (Figure 3A). Luciferase activity was unchanged in the presence of CPXV012 peptide.

Next, the effect of CPXV012 on the infection of two non-enveloped viruses was tested. The infectivity of adenovirus, a dsDNA virus, was not altered in the presence of CPXV012 peptide, as measured by viral eGFP expression (Figure 3B). For the coxsackie B3 virus, a non-enveloped ssRNA virus, infection was monitored by the activity of the virus-encoded Renilla luciferase. No change in Renilla luciferase activity was observed in the presence of CPXV012 or control peptide (Figure 3B).

These results identify the CPXV012 peptide as an antiviral peptide acting on enveloped viruses (summarized in Table S1).

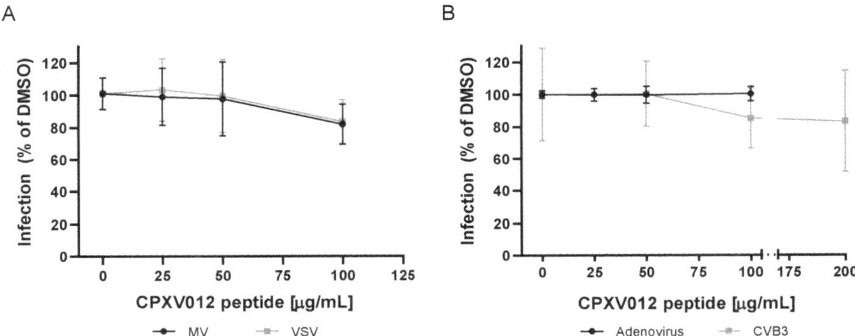

Figure 3. CPXV012 peptide does not inhibit infection with Measles virus, VSV, Adenovirus, or Coxsackie B3 virus and no difference was observed with the control peptide UL49.5 (data not shown). (**A**). Measles virus infection: Vero cells were treated with the indicated concentrations of CPXV012 peptide or DMSO as control, and infected with eGFP-expressing measles virus (MV-eGFP) (MOI 0.1). Once maximum giant cell formation was observed (approximately 48 h post infection), infection was quantified by detection of eGFP using a microplate reader. VSV infection: Huh7.5 cells were treated with the indicated concentrations of CPXV012 peptide in 0.1% DMSO or 0.1% DMSO only, and infected with luciferase-expressing VSV-deltaG (Luc) (MOI 0.6). At 18 h post infection, luciferase activity was measured to assess VSV infection of the culture. Data are shown as percentage of infection referred to the respective DMSO control which was set to 100%. Mean ± S.E.M. of each three independent experiments is shown. (**B**) Adenovirus infection: HEK-293 cells were treated with the indicated concentrations of CPXV012 peptide or DMSO for 1 h and infected with eGFP-expressing adenovirus (AdGOva) (MOI 10) for 24 h. Infection was quantified by cytometric analysis of eGFP expression. Coxsackievirus B3 infection: MJS cells were infected with RLuc-CVB3 in the presence of the indicated concentration of peptide or DMSO as vehicle control. After 7 h, cells were lysed and Renilla Luciferase expression was quantified. Data are shown as percentage of infection referred to the respective DMSO control which was set to 100%. Mean ± S.E.M. of each three independent experiments are shown.

3.2. CPXV012 May Bind to Viral Particles

To find out whether CPXV012 peptide mediates its inhibitory effect by interacting directly with viral particles, we performed a modified inhibition assay using MJS cells and MVA-eGFP. Instead of simultaneously adding CPXV012 peptide and virus to the cells, viral particles were pretreated with CPXV012 peptide for 1 h at 37 °C. After this preincubation, the virus-CPXV012 peptide mixture was diluted ten times and used to infect cells. Thus, the final peptide concentration during infection was 5 to 15 μg/mL (Figure 4A). Pretreating the virus with a high concentration of CPXV012 peptide before infection affected infectivity to a similar level as the presence of a high concentration of CPXV012 peptide during infection (Figure 4B). This effect was not due to the remaining CPXV012 peptide in the culture medium, as the diluted peptide alone was not sufficient to block infection when added directly to the cells. These data suggest that the peptide can directly act on viral particles.

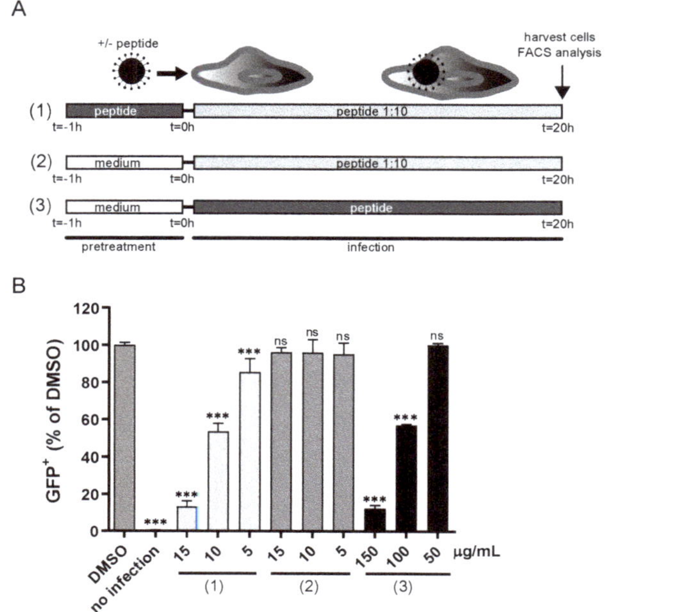

Figure 4. CPXV012 peptide inhibits infection by binding to viral particles. (**A**) Timeline of infection experiment as shown in (**B**). (1) MVA-eGFP was pretreated with 50, 100, or 150 µg/mL peptide or DMSO in culture medium at 37 °C for 1 h. Subsequently, the virus-peptide pre-incubation mixture was diluted ten times and used to infect MJS cells (corresponding to an MOI of 10), resulting in a final concentration of 5, 10, and 15 µg/mL peptide in the culture medium. (2–3) MVA-eGFP was incubated in culture medium only at 37 °C for 1 h. The virus mixture was added to MJS cells in the presence of 5, 10, or 15 µg/mL peptide (2) or 50, 100, or 150 µg/mL peptide (3) in the culture medium. (**B**) After 18 to 20 h of infection as shown in (**A**), the amount of infected eGFP-positive cells was quantified using flow cytometry. Mean ± S.E.M. of three independent experiments is shown using different batches of peptide. Data were analyzed with one-way ANOVA followed by multiple comparisons Dunnett's test (the mean of each column was compared to that of the DMSO control). (*** $p < 0.001$).

3.3. CPXV012 Peptide Variants Differentially Affect Virus Infection

To determine the amino acid residues within the CPXV012 peptide crucial for virus inhibition, alanine substitution variants of the CPXV012 peptide were synthesized. These variants had small stretches of amino acid residues replaced by alanine residues (Figure 5A). To test the inhibitory capacity of these variants, MJS cells were infected with MVA-eGFP in the presence of 100 µg/mL of each of these peptides (Figure 5B). CPXV012 variants with the N-terminal five amino acid residues (Ala1) or the C-terminal four amino acid residues substituted by alanine (Ala7) inhibited MVA infection to similar levels as wild-type CPXV012 peptide. Substituting amino acid residues 6 to 20 for alanine (CPXV012-Ala2/Ala3/Ala4) significantly affected the inhibitory capacity of the peptides. Substituting amino acid residues 20 to 30 for alanine (Ala5 and Ala6) completely abolished the inhibitory effect of CPXV012 peptide. Interestingly, the CPXV012 peptide variants with lower or no inhibitory capacity also had a lower net positive charge at pH 7.4 compared to the wild type peptide, due to the substitution of the charged amino acid residues lysine, arginine, or histidine, the latter being only partially charged.

To test the role of these amino acid residues in virus inhibition, a peptide was synthesized with the charged amino acid residues replaced for alanine (Ala8). Indeed, this CPXV012 peptide variant lost its inhibitory capacity, thus confirming the role of basic amino acid residues in inhibiting virus infection.

A

```
WT:   QEGISRFKICPYHWYKQHMSLLFRRYYHKLDSII
Ala1: AAAAARFKICPYHWYKQHMSLLFRRYYHKLDSII
Ala2: QEGISAAAAAPYHWYKQHMSLLFRRYYHKLDSII
Ala3: QEGISRFKICAAAAAKQHMSLLFRRYYHKLDSII
Ala4: QEGISRFKICPYHWYAAAAALLFRRYYHKLDSII
Ala5: QEGISRFKICPYHWYKQHMSAAAAAYYHKLDSII
Ala6: QEGISRFKICPYHWYKQHMSLLFRRAAAAADSII
Ala7: QEGISRFKICPYHWYKQHMSLLFRRYYHKLAAAA
Ala8: QEGISAFAICPYAWYAQAMSLLFAAYYAALDSII
```

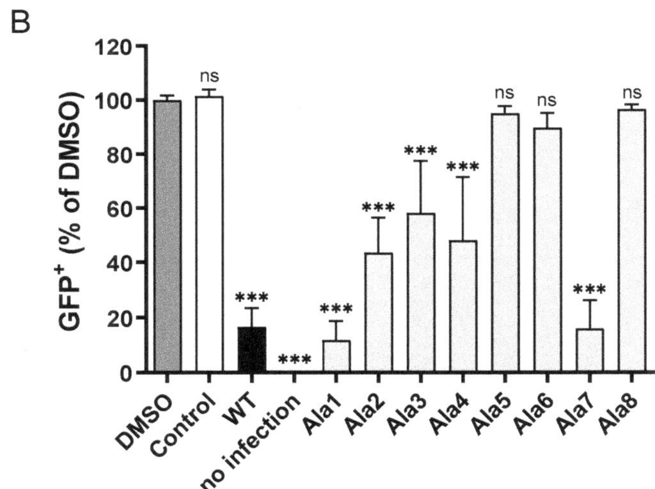

Figure 5. CPXV012 peptide variants differentially affect virus infection. (**A**) Amino acid sequence of CPXV012 peptide variants used in the experiment (**B**). (**B**) MJS cells were infected with MVA-eGFP at an MOI of 20 in the presence of 100 µg/mL of the peptide indicated or DMSO as vehicle control. After 20 h, cells were harvested, and the amount of eGFP-positive cells was quantified using flow cytometry. Mean ± S.E.M. of three independent experiments is shown using different batches of peptide. Data were analyzed with one-way ANOVA followed by multiple comparisons Dunnett's test (the mean of each column was compared to that of the DMSO control). (*** $p < 0.001$).

3.4. CPXV012 Peptide Interacts with Charged Phospholipids

The preferential inhibition of enveloped viruses suggests that CPXV012 peptide interacts with a common structure within these viral particles. A major constituent of these virions is the phospholipids forming the lipid bilayer of the viral envelope. To test the interaction between CPXV012 peptide and phospholipids, Langmuir monolayers were formed using the zwitterionic phospholipid phosphatidylcholine (PC) and the anionic phosphatidylserine (PS) on an aqueous buffer. Upon stabilization of the monolayer at an initial surface pressure of 25 mN/m, CPXV012 peptide was injected into the monolayer aqueous subphase and the surface pressure was measured for ~25 min (Figure 6A).

A change in surface pressure is interpreted as the integration of CPXV012 peptide into the lipid monolayer. Although the surface pressure of monolayers formed by the zwitterionic PC changed rapidly upon the addition of CPXV012 peptide, the shift in surface pressure was much higher for monolayers formed by a 7:3 mixture of PC and the negatively charged PS (Figure 6B). These data suggest a preferred interaction between CPXV012 and the anionic PS. This is further supported by the circular dichroism (CD) experiments used to determine the secondary structure of CPXV012 peptide (Figure 6C). The CD spectrum of CPXV012 was measured in the presence of large unilamellar vesicles (LUVs), lipid vesicles consisting of PC only, or of PC and PS. In aqueous MES buffer or in the presence of PC, CD spectra of CPXV012 peptide were rather similar and becoming more negative at a shorter wavelength, suggesting the presence of a significant amount of random coil structure. However, upon addition of PS to PC lipids using a similar ratio as for the monolayer experiments, the CPXV012 peptide acquired a different CD spectrum with minima around 208 and 222 nm, typical for an alpha-helical structure. Thus, CPXV012 peptide adopts different secondary structures, depending on the presence of PS.

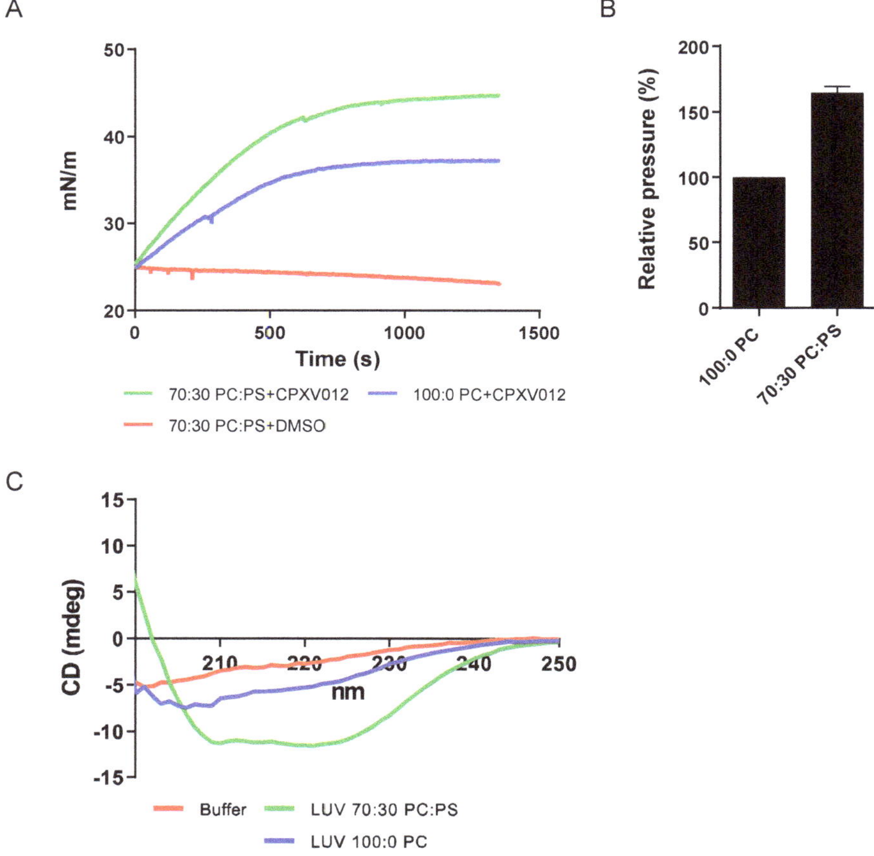

Figure 6. CPXV012 peptide interacts with PS. (**A**) CPXV012 peptide preferentially integrates into lipid monolayers composed of PC:PS (70:30). Langmuir monolayers of PC or PC:PS (70:30) with an initial

surface pressure of 25 mN/m were formed over an aqueous subphase. Peptide was injected into the subphase at t = 0 s. As vehicle control, DMSO was injected in the subphase of PC:PS (70:30) monolayers. Results are representative of three independent experiments. (**B**) Quantification of CPXV012 peptide integration into lipid monolayers composed of PC:PS (70:30), as compared to monolayers composed of PC only (set at 100%). Mean ± S.E.M. of three independent experiments is shown. (**C**) CPXV012 peptide adopts different secondary structures depending on the presence of PS. CD spectrum of CPXV012 peptide was determined in the presence of aqueous buffer or LUVs composed of PC or PC:PS (70:30). Results are representative of three independent experiments.

To gain further insight into the interaction between CPXV012 peptide and lipid membranes, the effect of the peptide on the integrity of lipid membrane vesicles was evaluated by membrane leakage assays using LUVs. These LUVs were composed of PC:PS and contained the self-quenching fluorophore calcein in their lumen. As soon as vesicle integrity is compromised, calcein leakage induces an increase in fluorescence, which can be measured in the supernatant using a fluorometer. As described previously, addition of human IAPP, but not murine IAPP, induced leakage of LUVs composed of PC and PS, whereas LUVs composed of PC alone were more resistant to leakage by human IAPP (Figure 7A,B) [31]. The addition of 25 μg/mL CPXV012 peptide disrupted the LUV membranes composed of both PC and PS, whereas membranes containing PC alone were unaffected. As described for human IAPP [32], membrane leakage by CPXV012 peptide is not instant but is preceded by a lag phase of several hours.

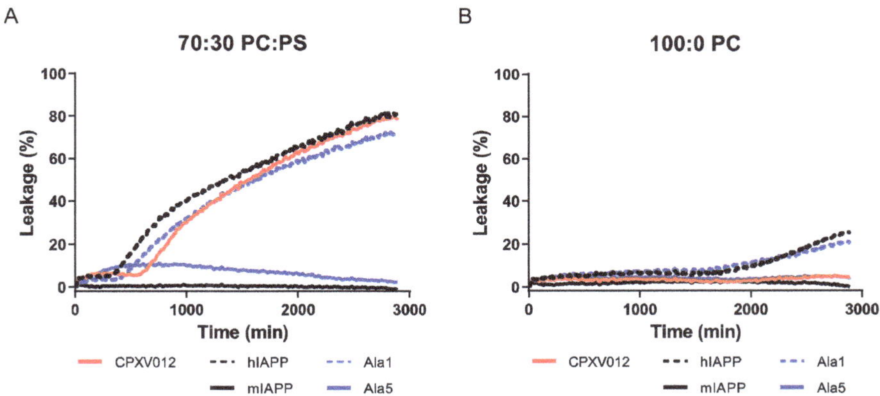

Figure 7. CPXV012 peptide disrupts membranes composed of PS. Calcein-containing PC:PS (7:3) LUVs (**A**) or PC LUVs (**B**) were incubated with human IAPP (hIAPP) as the positive control, murine IAPP (mIAPP) as the negative control, CPXV012 peptide, CPXV012 peptide variants Ala1 or Ala5 (see Figure 5A for amino acid sequences). Calcein leakage from the vesicle lumen was quantified using a microplate reader. Membrane leakage was compared to fully disintegrated detergent-treated LUVs (set at 100%). Results are representative of three independent experiments.

Two alanine substitution peptides, Ala1 and Ala5 (see Figure 5), were also tested. CPXV012 peptide variant Ala1, which inhibited virus infection, induced membrane leakage of PC:PS LUVs to similar levels and kinetics as wt CPXV012 peptide. In contrast to CPXV012, this variant was able to induce slight leakage of LUVs composed of PC only. Ala5, which had no effect on virus infection, did not compromise the integrity of either one of the LUV membranes (Figure 7B).

In conclusion, the capacity of CPXV012 peptide to inhibit virus infections correlates with its ability to disrupt lipid membranes that contain PS.

4. Discussion

In this study, we identified the CPXV012 peptide as a broadly-acting antiviral agent that probably interacts with the anionic phospholipid PS. The antiviral potency of the CPXV012 peptide can be considered modest, yet we believe that this peptide may be a candidate for biochemical improvement. The interaction between the CPXV012 peptide and PS is likely mediated by opposing electrostatic forces determined by cationic residues within the peptide and the anionic head groups of PS. In addition, hydrophobic and aromatic amino acid residues (which make up more than 30% of CPXV012 peptide) may facilitate penetration of the peptide into the lipid phase of the membrane. As shown for other peptides, such membrane interactions may promote the formation of an alpha-helical secondary structure of otherwise unstructured peptides [33]. Due to the distribution of charged and hydrophobic residues over the length of the CPXV012 peptide, the alpha-helix can be expected to have somewhat of an amphipathic character. This may allow it to bind with an orientation that is roughly parallel to the membrane surface [34] or to assemble into oligomeric transmembrane pores [35]. These biophysical properties of CPXV012 peptide are shared with many antimicrobial host defense peptides [36], that form a crucial innate immune barrier by targeting the negatively charged surfaces of bacteria and certain fungi. Like CPXV012 peptide, these small peptides (up to 50 residues) have a net positive charge and high affinity for anionic phospholipids. The structural rearrangements occurring upon membrane binding are crucial for the antimicrobial activity and can trigger pore formation, membrane depolarization, and disruption of the bacterial membrane [37,38].

Although less studied, antimicrobial peptides may also affect virus infection by different mechanisms. These mechanisms include interference with viral entry (binding to receptors, fusion), inhibition of viral replication, reduction/suppression of viral gene expression, immunomodulation, viral aggregation, and disruption of the integrity of the viral membrane (reviewed in [6,7]). While a list of peptides is reported to disrupt viral envelopes, their mode of action has not been formally tested. Still, a number of them were shown to directly affect viral membranes, although the specific interacting partners remain elusive. Examples include the human cathelicidin LL-37 and its murine equivalent mCRAMP, which disrupt the membranes of RSV, influenza A virus and VACV, and can prevent virus infection in vivo [39–41]. Although their effect on the infectivity of other viruses is not known, the structural similarities between these cationic peptides and CPXV012 peptide may suggest that these and other antimicrobial peptides possess a broader antiviral activity. Electron microscopy has also provided indications of envelope disruption with the cationic peptides Tachyplesin (VSV) [7], Temporin B (HSV-1) [42], and more recently Urumin (H1-bearing influenza A viruses) [43].

Membrane leakage induced by CPXV012 peptide was preceded by a lag phase, as was also observed for human IAPP. For human IAPP, this lag phase originates from the formation of transient oligomeric intermediates prior to the assembly of membrane-disruptive aggregates [32]. Likewise, disruption of membranes by CPXV012 peptides may be linked to the formation of (transient) oligomeric structures. The shorter lag time in the presence of PS is possibly due to more efficient binding of the peptide to the membrane: upon favorable electrostatic interactions, CPXV012 peptides achieve a higher local concentration and therefore the assembly process is supported. The lag phase of several hours appears to be inconsistent with the almost immediate effect observed in virus infection assays. There can be several reasons for this apparent discrepancy. First, the local peptide concentration at the membrane surface in the virus assays is likely to be much higher than in the leakage experiments, facilitating the oligomerization process. Second, antiviral activity may not require the formation of pores that are large enough to allow passage of calcein. Instead, binding to the viral membrane and/or insertion of hydrophobic amino acid side chains between the lipid acyl chains by itself already may be sufficient for an antiviral effect. Third, other factors may be involved in the virus infection assays, such as interactions with viral (glycol)proteins in addition to PS, that help accelerate peptide oligomerization on the viral surface.

Our data indicate that the CPXV012 peptide neutralizes the infectivity of viruses from diverse families. In vitro assays indicate that the CPXV012 peptide directly interacts with membranes.

The disruptive behavior of the CPXV012 peptide on artificial membrane vesicles suggests that the CPXV012 peptide may similarly affect viral membranes. As viral particles do not have a membrane repair mechanism, this damaging effect may particularly destabilize and disrupt viral particles. Extensive membrane damage may also hamper virus fusion to the host cell membrane by impacting the fluidity and curvatures of lipid membranes. Lipids including PS play an essential role in the induction of these membrane curvatures and subsequent membrane fusion [44,45]. Even in the absence of membrane damage, the interaction between CPXV012 peptide and PS in the viral envelope may interfere with virus infectivity by sterically blocking the interaction between viral PS and host cell PS receptors. A similar inhibitory mechanism was observed using the PS-specific antibody Bavituximab [3,46], or the PS-binding protein Annexin V [47,48]. CPXV012 peptide may also indirectly affect virus infection by binding to PS on the host cell membrane. Specific microdomains of the host cell plasma membrane are enriched for PS, including lipid rafts [49]. These sites are used by certain viruses as an entry point [50]. Although PS typically would reside in the inner leaflet of the plasma membrane, there are indications that some viruses induce redistribution of PS to the outer membrane to promote/facilitate entry [50]. In addition, some viruses use lipid rafts as budding sites when exiting the host cell [51].

The effect of CPXV012 peptide on the viruses used in this study and potential inhibitory mechanism(s) are discussed in more detail.

4.1. *Poxviruses*

The CPXV012 peptide was able to inhibit infection with different orthopoxviruses (MVA, VACV, and CPXV) in several cell lines, using various detection methods. Notably, early gene expression, which starts immediately after virus entry, was highly reduced and indicates a block at the entry step. Viral particles of the poxviridae family adopt distinct structures and composition of the outer membrane, such as mature virions (MVs, also named IMVs, intracellular mature virus) [52]. MVs represent the majority of infectious progeny; they are very stable and largely responsible for viral spread. In viral preparations, MVs are the most abundant virion form. Importantly, membranes of MVs have been shown to contain large amounts of PS that is essential for entry [47,53]. We found that CPXV012 peptide binds to PS and thus likely mediates its inhibitory effect on poxvirus infection by interacting with PS in the viral membrane. This was confirmed by the observation that pretreating virions with CPXV012 peptide effectively blocked infection.

4.2. *HSV-1*

Inhibition of HSV-1 by cationic peptides has been previously reported, including the antimicrobial peptide LL-37 that presumably acts by targeting factors within the viral membrane [54–56]. It is believed that HSV-1 derives its final envelope from membranes of the late secretory pathway (e.g., trans-Golgi network) and/or endosomal pathway [57]. The inhibition of HSV-1 by CPXV012 peptide suggests that PS in the HSV-1 envelope is a potent target for antiviral therapy.

4.3. *HBV*

CPXV012 peptide affected HBV infection as measured by a decrease in viral DNA and viral antigen in the supernatant of infected cultures. HBV particles are thought to acquire their envelope from multivesicular bodies [58], which are rich in PS [59]. The role of PS in HBV infection is underlined by studies showing that antibodies specific for PS block HBV infection [60]. Similar to poxviruses, PS may play a role in virus entry through apoptotic mimicry [61,62]. PS may also play a role in the proper folding of HBV envelope proteins [63].

4.4. *HIV*

The presence of PS in the membrane of HIV particles was shown to be an important cofactor for infection and blocking of PS led to a decrease in infection [64,65]. In addition, HIV infection was inhibited by the addition of lipid vesicles composed of PS that competed with the HIV particles for plasma membrane association, while lipid vesicles containing PC had no effect [64]. Therefore, the binding of CPXV012 peptide to PS in the viral envelope could explain the inhibitory effect observed in the EASY-HIT assay.

4.5. RVFV

The inhibition of RVFV infection by CPXV012 peptide suggests that PS is critical for RVFV infection. Although the lipid composition of RVFV is unknown, the viral envelope of the Uukuniemi virus, a related member of the bunyavirus family, is enriched for PS [66]. Like the Uukuniemi virus, RVFV obtains its membrane by budding into the Golgi membrane, suggesting that PS may also be enriched in these virus particles [67,68].

4.6. Viruses Not Affected by CPXV012 Peptide

No effects of CPXV012 peptide were observed on infection with adenovirus, coxsackie B3, measles virus, and VSV. Although the measles virus and VSV are enveloped viruses, the role of PS during infection is limited for these viruses. For the measles virus, different phospholipids were tested for their ability to reassemble with the viral envelope glycoproteins and to restore hemolytic activity. In contrast to phosphatidylethanolamine and PC, PS completely failed to restore hemagglutination activity [69]. For VSV, early studies considered PS expressed on the host cell important for cell entry [70]. In contrast, a more recent study found no correlation between PS surface levels and VSV binding. Moreover, masking of phosphatidylserine with annexin V during infection did not affect VSV binding to cells, or entry of virus particles pseudotyped with VSV-G [64,71]. Thus, the limited role of PS during entry of VSV and measles virus may explain the lack of inhibition by CPXV012 peptide.

Both adenovirus and Coxsackievirus B3 are non-enveloped viruses and the nucleocapsid does not contain any phospholipids (nor glycoproteins) CPXV012 peptide could bind to. CPXV012 peptide binding to the host cell membrane might not be sufficient for the steric impediment of viral ligand–cellular receptor interaction. Supporting this, the adenoviral elongated fiber protein responsible for binding to the host cell is flexible and large in size (9 to 30 nm), as is the coxsackie and adenovirus receptor (CAR) on the host cell [72].

CAR is also important for the infection of free CVB3 particles [73]. In addition, CVB3 may use a PS-dependent entry strategy for the bulk transmission of virus particles [74]. Infected cells release clusters of virus particles wrapped in PS-enriched membranes. These vesicles enhance subsequent virus entry in a process that highly depends on the presence of PS [74]. As CPXV012 peptide had no effect on CVB3 virus infection, PS may not play a significant role in our study. This discrepancy likely results from our virus production protocol that involves repeated freeze-thawing to liberate virus particles. Although necessary to obtain high virus titers, this method also releases CVB3 from PS-enriched vesicles [74].

In conclusion, the CPXV012 peptide-mediated inhibition of virus infection correlates well with the observed interaction between CPXV012 peptide and PS. The broad inhibitory range may thereby be further extended to other viruses for which PS has a crucial function during the viral life cycle. These viruses include clinically and economically important pathogens such as Ebola virus, Lassa virus, dengue virus, and poliovirus [61].

The differences between viral and host membranes make PS an Achilles' heel that can be targeted by the CPXV012 peptide. It will be worth investigating if other cationic peptides, including certain antimicrobial peptides, can also target PS within the viral envelope, and perhaps fulfill a similar function as broad-range antiviral agents. Analyses of the peptide's biophysical and physicochemical properties can shed some light on how to improve the moderate efficiency of CPXV012 and other cationic peptides. Highly efficient antiviral peptides are interesting candidates for prophylactic treatment and antiviral therapy, as they do not require active replication and target viral components that are less likely to develop drug-resistance [75].

Supplementary Materials: The following are available online at http://www.mdpi.com/2073-4409/9/9/1989/s1, Table S1: Inhibitory effect of CPXV012 peptide on enveloped and non-enveloped viruses of different families. Figure S1: CPXV012 peptide does not affect cell viability. Figure S2: CPXV012 peptide (CPX) prevents infection with MVA in different cell lines. Figure S3: CPXV012 peptide does not affect infection with the measles virus.

Author Contributions: Conceptualization, R.D.L. and F.T.; methodology, P.P., S.M.S., K.S., C.B., S.M.d.B. and M.H.; validation, S.T., S.M.S., N.C.B. and U.P.; formal analysis, K.S., R.J.L., S.T. and M.H.; investigation, R.D.L., P.P., F.T., S.M.S., K.S., C.B., S.M.d.B., S.T. and M.H.; resources, E.J.H.J.W., J.W.D., J.A.K., R.J.L. and I.D.; data curation, A.I.C., F.T., R.J.L. and N.C.B.; writing—original draft preparation, R.D.L., P.P. and F.T.; writing—review and editing, R.D.L., P.P., I.D., U.P. and A.I.C.; visualization, R.D.L., J.A.K., P.P. and A.I.C.; project administration, E.J.H.J.W. and I.D.; funding acquisition, P.P. and S.M.S. All authors have read and agreed to the published version of the manuscript.

Funding: P.P. was supported by the European Commission under the Horizon2020 program H2020 MSCA-ITN GA 675278 EDGE. S.M.S. was supported by the seventh framework program of the European Union (Initial Training Network "ManiFold," Grant 317371), and ID was supported by the DFG funding GRK 1949.

Conflicts of Interest: The authors declare no conflict of interest.

References

1. Marston, H.D.; Folkers, G.K.; Morens, D.M.; Fauci, A.S. Emerging viral diseases: Confronting threats with new technologies. *Sci. Transl. Med.* **2014**, *6*, 253ps10. [CrossRef]
2. Zhu, J.D.; Meng, W.; Wang, X.J.; Wang, H.C. Broad-spectrum antiviral agents. *Front. Microbiol.* **2015**, *6*, 517. [CrossRef]
3. Soares, M.M.; King, S.W.; Thorpe, P.E. Targeting inside-out phosphatidylserine as a therapeutic strategy for viral diseases. *Nat. Med.* **2008**, *14*, 1357–1362. [CrossRef]
4. Vigant, F.; Lee, J.; Hollmann, A.; Tanner, L.B.; Akyol Ataman, Z.; Yun, T.; Shui, G.; Aguilar, H.C.; Zhang, D.; Meriwether, D.; et al. A mechanistic paradigm for broad-spectrum antivirals that target virus-cell fusion. *PLoS Pathog.* **2013**, *9*, e1003297. [CrossRef] [PubMed]
5. Wolf, M.C.; Freiberg, A.N.; Zhang, T.; Akyol-Ataman, Z.; Grock, A.; Hong, P.W.; Li, J.; Watson, N.F.; Fang, A.Q.; Aguilar, H.C.; et al. A broad-spectrum antiviral targeting entry of enveloped viruses. *Proc. Natl Acad. Sci. USA* **2010**, *107*, 3157–3162. [CrossRef] [PubMed]
6. Findlay, E.G.; Currie, S.M.; Davidson, D.J. Cationic host defence peptides: Potential as antiviral therapeutics. *BioDrugs* **2013**, *27*, 479–493. [CrossRef] [PubMed]
7. Mulder, K.C.L.; Lima, L.A.; Miranda, V.J.; Dias, S.C.; Franco, O.L. Current scenario of peptide-based drugs: The key roles of cationic antitumor and antiviral peptides. *Front. Microbiol.* **2013**, *4*, 321. [CrossRef]
8. Leventis, P.A.; Grinstein, S. The distribution and function of phosphatidylserine in cellular membranes. *Annu. Rev. Biophys.* **2010**, *39*, 407–427. [CrossRef]
9. Moller-Tank, S.; Maury, W. Phosphatidylserine receptors: Enhancers of enveloped virus entry and infection. *Virology* **2014**, *468–470*, 565–580. [CrossRef]
10. Andrews, N.W.; Almeida, P.E.; Corrotte, M. Damage control: Cellular mechanisms of plasma membrane repair. *Trends Cell Biol.* **2014**, *24*, 734–742. [CrossRef]
11. Cooper, S.T.; McNeil, P.L. Membrane Repair: Mechanisms and Pathophysiology. *Physiol. Rev.* **2015**, *95*, 1205–1240. [CrossRef]
12. Alzhanova, D.; Edwards, D.M.; Hammarlund, E.; Scholz, I.G.; Horst, D.; Wagner, M.J.; Upton, C.; Wiertz, E.J.; Slifka, M.K.; Früh, K. Cowpox Virus Inhibits the Transporter Associated with Antigen Processing to Evade T Cell Recognition. *Cell Host Microbe* **2009**, *6*, 433–445. [CrossRef] [PubMed]
13. Byun, M.; Verweij, M.C.; Pickup, D.J.; Wiertz, E.J.H.J.; Hansen, T.H.; Yokoyama, W.M. Two Mechanistically Distinct Immune Evasion Proteins of Cowpox Virus Combine to Avoid Antiviral CD8 T Cells. *Cell Host Microbe* **2009**, *6*, 422–432. [CrossRef] [PubMed]
14. Lin, J.; Eggensperger, S.; Hank, S.; Wycisk, A.I.; Wieneke, R.; Mayerhofer, P.U.; Tampé, R. A negative feedback modulator of antigen processing evolved from a frameshift in the cowpox virus genome. *PLoS Pathog.* **2014**, *10*, e1004554. [CrossRef] [PubMed]
15. Luteijn, R.D.; Hoelen, H.; Kruse, E.; van Leeuwen, W.F.; Grootens, J.; Horst, D.; Koorengevel, M.; Drijfhout, J.W.; Kremmer, E.; Früh, K.; et al. Cowpox Virus Protein CPXV012 Eludes CTLs by Blocking ATP Binding to TAP. *J. Immunol.* **2014**, *193*, 1578–1589. [CrossRef] [PubMed]

16. Praest, P.; Liaci, A.M.; Förster, F.; Wiertz, E.J.H.J. New insights into the structure of the MHC class I peptide-loading complex and mechanisms of TAP inhibition by viral immune evasion proteins. *Mol. Immunol.* **2019**, *113*, 103–114. [CrossRef]
17. Kremb, S.; Helfer, M.; Heller, W.; Hoffmann, D.; Wolff, H.; Kleinschmidt, A.; Cepok, S.; Hemmer, B.; Durner, J.; Brack-Werner, R. EASY-HIT: HIV full-replication technology for broad discovery of multiple classes of HIV inhibitors. *Antimicrob. Agents Chemother.* **2010**, *54*, 5257–5268. [CrossRef]
18. Gasteiger, G.; Kastenmuller, W.; Ljapoci, R.; Sutter, G.; Drexler, I. Cross-Priming of Cytotoxic T Cells Dictates Antigen Requisites for Modified Vaccinia Virus Ankara Vector Vaccines. *J. Virol.* **2007**, *81*, 11925–11936. [CrossRef]
19. Staib, C.; Drexler, I.; Sutter, G. Construction and isolation of recombinant MVA. *Methods Mol. Biol.* **2004**, *269*, 77–100. [CrossRef]
20. Lucifora, J.; Arzberger, S.; Durantel, D.; Belloni, L.; Strubin, M.; Levrero, M.; Zoulim, F.; Hantz, O.; Protzer, U. Hepatitis B virus X protein is essential to initiate and maintain virus replication after infection. *J. Hepatol.* **2011**, *55*, 996–1003. [CrossRef]
21. Duprex, W.P.; McQuaid, S.; Roscic-Mrkic, B.; Cattaneo, R.; McCallister, C.; Rima, B.K. In vitro and in vivo infection of neural cells by a recombinant measles virus expressing enhanced green fluorescent protein. *J. Virol.* **2000**, *74*, 7972–7979. [CrossRef] [PubMed]
22. Wohlleber, D.; Kashkar, H.; Gartner, K.; Frings, M.K.; Odenthal, M.; Hegenbarth, S.; Borner, C.; Arnold, B.; Hammerling, G.; Nieswandt, B.; et al. TNF-induced target cell killing by CTL activated through cross-presentation. *Cell Rep.* **2012**, *2*, 478–487. [CrossRef] [PubMed]
23. Kortekaas, J.; Oreshkova, N.; Cobos-Jimenez, V.; Vloet, R.P.; Potgieter, C.A.; Moormann, R.J. Creation of a nonspreading Rift Valley fever virus. *J. Virol.* **2011**, *85*, 12622–12630. [CrossRef] [PubMed]
24. Lanke, K.H.; van der Schaar, H.M.; Belov, G.A.; Feng, Q.; Duijsings, D.; Jackson, C.L.; Ehrenfeld, E.; van Kuppeveld, F.J. GBF1, a guanine nucleotide exchange factor for Arf, is crucial for coxsackievirus B3 RNA replication. *J. Virol.* **2009**, *83*, 11940–11949. [CrossRef] [PubMed]
25. Hiemstra, H.S.; Duinkerken, G.; Benckhuijsen, W.E.; Amons, R.; de Vries, R.R.; Roep, B.O.; Drijfhout, J.W. The identification of CD4+ T cell epitopes with dedicated synthetic peptide libraries. *Proc. Natl. Acad. Sci. USA* **1997**, *94*, 10313–10318. [CrossRef] [PubMed]
26. Repetto, G.; del Peso, A.; Zurita, J.L. Neutral red uptake assay for the estimation of cell viability/cytotoxicity. *Nat. Protoc.* **2008**, *3*, 1125–1131. [CrossRef]
27. Livak, K.J.; Schmittgen, T.D. Analysis of relative gene expression data using real-time quantitative PCR and the 2(-Delta Delta C(T)) Method. *Methods* **2001**, *25*, 402–408. [CrossRef]
28. Pfaffl, M.W. A new mathematical model for relative quantification in real-time RT-PCR. *Nucleic Acids Res.* **2001**, *29*, e45. [CrossRef]
29. Rouser, G.; Kritchevsky, G.; Simon, G.; Nelson, G.J. Quantitative analysis of brain and spinach leaf lipids employing silicic acid column chromatography and acetone for elution of glycolipids. *Lipids* **1967**, *2*, 37–40. [CrossRef]
30. Van de Weijer, M.L.; Luteijn, R.D.; Wiertz, E.J.H.J. Viral immune evasion: Lessons in MHC class I antigen presentation. *Semin. Immunol.* **2015**, *27*, 125–137. [CrossRef]
31. Engel, M.F.; Khemtemourian, L.; Kleijer, C.C.; Meeldijk, H.J.; Jacobs, J.; Verkleij, A.J.; de Kruijff, B.; Killian, J.A.; Hoppener, J.W. Membrane damage by human islet amyloid polypeptide through fibril growth at the membrane. *Proc. Natl. Acad. Sci. USA* **2008**, *105*, 6033–6038. [CrossRef] [PubMed]
32. Buchanan, L.E.; Dunkelberger, E.B.; Tran, H.Q.; Cheng, P.N.; Chiu, C.C.; Cao, P.; Raleigh, D.P.; de Pablo, J.J.; Nowick, J.S.; Zanni, M.T. Mechanism of IAPP amyloid fibril formation involves an intermediate with a transient beta-sheet. *Proc. Natl. Acad. Sci. USA* **2013**, *110*, 19285–19290. [CrossRef]
33. Eiriksdottir, E.; Konate, K.; Langel, U.; Divita, G.; Deshayes, S. Secondary structure of cell-penetrating peptides controls membrane interaction and insertion. *Biochim. Biophys. Acta* **2010**, *1798*, 1119–1128. [CrossRef]
34. Shepherd, C.M.; Vogel, H.J.; Tieleman, D.P. Interactions of the designed antimicrobial peptide MB21 and truncated dermaseptin S3 with lipid bilayers: Molecular-dynamics simulations. *Biochem. J.* **2003**, *370*, 233–243. [CrossRef] [PubMed]
35. Hwang, P.M.; Vogel, H.J. Structure-function relationships of antimicrobial peptides. *Biochem. Cell Biol.* **1998**, *76*, 235–246. [CrossRef] [PubMed]

36. Zasloff, M. Antimicrobial peptides of multicellular organisms. *Nature* **2002**, *415*, 389–395. [CrossRef]
37. Baltzer, S.A.; Brown, M.H. Antimicrobial peptides: Promising alternatives to conventional antibiotics. *J. Mol. Microbiol. Biotechnol.* **2011**, *20*, 228–235. [CrossRef]
38. Shai, Y. Mode of action of membrane active antimicrobial peptides. *Pept. Sci.* **2002**, *66*, 236–248. [CrossRef]
39. Currie, S.M.; Gwyer Findlay, E.; McFarlane, A.J.; Fitch, P.M.; Bottcher, B.; Colegrave, N.; Paras, A.; Jozwik, A.; Chiu, C.; Schwarze, J.; et al. Cathelicidins Have Direct Antiviral Activity against Respiratory Syncytial Virus In Vitro and Protective Function In Vivo in Mice and Humans. *J. Immunol.* **2016**, *196*, 2699–2710. [CrossRef]
40. Howell, M.D.; Jones, J.F.; Kisich, K.O.; Streib, J.E.; Gallo, R.L.; Leung, D.Y. Selective killing of vaccinia virus by LL-37: Implications for eczema vaccinatum. *J. Immunol.* **2004**, *172*, 1763–1767. [CrossRef]
41. Tripathi, S.; Tecle, T.; Verma, A.; Crouch, E.; White, M.; Hartshorn, K.L. The human cathelicidin LL-37 inhibits influenza A viruses through a mechanism distinct from that of surfactant protein D or defensins. *J. Gen. Virol.* **2013**, *94*, 40–49. [CrossRef] [PubMed]
42. Marcocci, M.E.; Amatore, D.; Villa, S.; Casciaro, B.; Aimola, P.; Franci, G.; Grieco, P.; Galdiero, M.; Palamara, A.T.; Mangoni, M.L.; et al. The Amphibian Antimicrobial Peptide Temporin B Inhibits In Vitro Herpes Simplex Virus 1 Infection. *Antimicrob. Agents Chemother.* **2018**, *62*, e02367-17. [CrossRef] [PubMed]
43. Holthausen, D.J.; Lee, S.H.; Kumar, V.T.; Bouvier, N.M.; Krammer, F.; Ellebedy, A.H.; Wrammert, J.; Lowen, A.C.; George, S.; Pillai, M.R.; et al. An Amphibian Host Defense Peptide Is Virucidal for Human H1 Hemagglutinin-Bearing Influenza Viruses. *Immunity* **2017**, *46*, 587–595. [CrossRef] [PubMed]
44. Sanchez-Migallon, M.P.; Aranda, F.J.; Gomez-Fernandez, J.C. Role of phosphatidylserine and diacylglycerol in the fusion of chromaffin granules with target membranes. *Arch. Biochem. Biophys.* **1994**, *314*, 205–216. [CrossRef] [PubMed]
45. Teissier, E.; Pecheur, E.I. Lipids as modulators of membrane fusion mediated by viral fusion proteins. *Eur. Biophys. J.* **2007**, *36*, 887–899. [CrossRef]
46. Dowall, S.D.; Graham, V.A.; Corbin-Lickfett, K.; Empig, C.; Schlunegger, K.; Bruce, C.B.; Easterbrook, L.; Hewson, R. Effective binding of a phosphatidylserine-targeting antibody to Ebola virus infected cells and purified virions. *J. Immunol. Res.* **2015**, *2015*, 347903. [CrossRef]
47. Mercer, J.; Helenius, A. Vaccinia virus uses macropinocytosis and apoptotic mimicry to enter host cells. *Science* **2008**, *320*, 531–535. [CrossRef]
48. Moller-Tank, S.; Kondratowicz, A.S.; Davey, R.A.; Rennert, P.D.; Maury, W. Role of the phosphatidylserine receptor TIM-1 in enveloped-virus entry. *J. Virol.* **2013**, *87*, 8327–8341. [CrossRef]
49. Pike, L.J.; Han, X.; Chung, K.N.; Gross, R.W. Lipid rafts are enriched in arachidonic acid and plasmenylethanolamine and their composition is independent of caveolin-1 expression: A quantitative electrospray ionization/mass spectrometric analysis. *Biochemistry* **2002**, *41*, 2075–2088. [CrossRef]
50. Briggs, J.A.; Wilk, T.; Fuller, S.D. Do lipid rafts mediate virus assembly and pseudotyping? *J. Gen. Virol.* **2003**, *84*, 757–768. [CrossRef]
51. Lorizate, M.; Krausslich, H.G. Role of lipids in virus replication. *Cold Spring Harb. Perspect. Biol.* **2011**, *3*, a004820. [CrossRef] [PubMed]
52. Roberts, K.L.; Smith, G.L. Vaccinia virus morphogenesis and dissemination. *Trends Microbiol.* **2008**, *16*, 472–479. [CrossRef] [PubMed]
53. Ichihashi, Y.; Oie, M. The activation of vaccinia virus infectivity by the transfer of phosphatidylserine from the plasma membrane. *Virology* **1983**, *130*, 306–317. [CrossRef]
54. Daher, K.A.; Selsted, M.E.; Lehrer, R.I. Direct inactivation of viruses by human granulocyte defensins. *J. Virol.* **1986**, *60*, 1068–1074. [CrossRef]
55. Gordon, Y.J.; Huang, L.C.; Romanowski, E.G.; Yates, K.A.; Proske, R.J.; McDermott, A.M. Human cathelicidin (LL-37), a multifunctional peptide, is expressed by ocular surface epithelia and has potent antibacterial and antiviral activity. *Curr. Eye Res.* **2005**, *30*, 385–394. [CrossRef]
56. Yasin, B.; Pang, M.; Turner, J.S.; Cho, Y.; Dinh, N.N.; Waring, A.J.; Lehrer, R.I.; Wagar, E.A. Evaluation of the inactivation of infectious Herpes simplex virus by host-defense peptides. *Eur. J. Clin. Microbiol. Infect. Dis.* **2000**, *19*, 187–194. [CrossRef]
57. Crump, C. Virus assembly and egress of HSV. *Adv. Exp. Med. Biol.* **2018**, *1045*, 23–44.
58. Jiang, B.; Himmelsbach, K.; Ren, H.; Boller, K.; Hildt, E. Subviral Hepatitis B Virus Filaments, like Infectious Viral Particles, Are Released via Multivesicular Bodies. *J. Virol.* **2016**, *90*, 3330–3341. [CrossRef]

59. Gyorgy, B.; Szabo, T.G.; Pasztoi, M.; Pal, Z.; Misjak, P.; Aradi, B.; Laszlo, V.; Pallinger, E.; Pap, E.; Kittel, A.; et al. Membrane vesicles, current state-of-the-art: Emerging role of extracellular vesicles. *Cell Mol. Life Sci.* **2011**, *68*, 2667–2688. [CrossRef]
60. De Meyer, S.; Gong, Z.; Depla, E.; Maertens, G.; Yap, S.H. Involvement of phosphatidylserine and non-phospholipid components of the hepatitis B virus envelope in human Annexin V binding and in HBV infection in vitro. *J. Hepatol.* **1999**, *31*, 783–790. [CrossRef]
61. Amara, A.; Mercer, J. Viral apoptotic mimicry. *Nat. Rev. Microbiol.* **2015**, *13*, 461–469. [CrossRef] [PubMed]
62. Vanlandschoot, P.; Leroux-Roels, G. Viral apoptotic mimicry: An immune evasion strategy developed by the hepatitis B virus? *Trends Immunol.* **2003**, *24*, 144–147. [CrossRef]
63. Gomez-Gutierrez, J.; Rodriguez-Crespo, I.; Peterson, D.L.; Gavilanes, F. Reconstitution of hepatitis B surface antigen proteins into phospholipid vesicles. *Biochim. Biophys. Acta* **1994**, *1192*, 45–52. [CrossRef]
64. Callahan, M.K.; Popernack, P.M.; Tsutsui, S.; Truong, L.; Schlegel, R.A.; Henderson, A.J. Phosphatidylserine on HIV envelope is a cofactor for infection of monocytic cells. *J. Immunol.* **2003**, *170*, 4840–4845. [CrossRef]
65. Lorizate, M.; Sachsenheimer, T.; Glass, B.; Habermann, A.; Gerl, M.J.; Krausslich, H.G.; Brugger, B. Comparative lipidomics analysis of HIV-1 particles and their producer cell membrane in different cell lines. *Cell Microbiol.* **2013**, *15*, 292–304. [CrossRef]
66. Renkonen, O.; Kaariainen, L.; Pettersson, R.; Oker-Blom, N. The phospholipid composition of Uukuniemi virus, a non-cubical tick-borne arbovirus. *Virology* **1972**, *50*, 899–901. [CrossRef]
67. Ellis, D.S.; Shirodaria, P.V.; Fleming, E.; Simpson, D.I. Morphology and development of Rift Valley fever virus in Vero cell cultures. *J. Med. Virol.* **1988**, *24*, 161–174. [CrossRef]
68. Kuismanen, E.; Hedman, K.; Saraste, J.; Pettersson, R.F. Uukuniemi virus maturation: Accumulation of virus particles and viral antigens in the Golgi complex. *Mol. Cell Biol.* **1982**, *2*, 1444–1458. [CrossRef]
69. Hall, W.W.; Martin, S.J. Structure and function relationships of the envelope of measles virus. *Med. Microbiol. Immunol.* **1974**, *160*, 143–154. [CrossRef]
70. Schlegel, R.; Tralka, T.S.; Willingham, M.C.; Pastan, I. Inhibition of VSV binding and infectivity by phosphatidylserine: Is phosphatidylserine a VSV-binding site? *Cell* **1983**, *32*, 639–646. [CrossRef]
71. Coil, D.A.; Miller, A.D. Phosphatidylserine is not the cell surface receptor for vesicular stomatitis virus. *J. Virol.* **2004**, *78*, 10920–10926. [CrossRef] [PubMed]
72. Bergelson, J.M.; Coyne, C.B. Picornavirus entry. *Adv. Exp. Med. Biol.* **2013**, *790*, 24–41. [CrossRef]
73. Marjomaki, V.; Turkki, P.; Huttunen, M. Infectious Entry Pathway of Enterovirus B Species. *Viruses* **2015**, *7*, 6387–6399. [CrossRef] [PubMed]
74. Chen, Y.H.; Du, W.; Hagemeijer, M.C.; Takvorian, P.M.; Pau, C.; Cali, A.; Brantner, C.A.; Stempinski, E.S.; Connelly, P.S.; Ma, H.C.; et al. Phosphatidylserine vesicles enable efficient en bloc transmission of enteroviruses. *Cell* **2015**, *160*, 619–630. [CrossRef] [PubMed]
75. Ahmed, A.; Siman-Tov, G.; Hall, G.; Bhalla, N.; Narayanan, A. Human Antimicrobial Peptides as Therapeutics for Viral Infections. *Viruses* **2019**, *11*, 704. [CrossRef] [PubMed]

Review

"Repair Me if You Can": Membrane Damage, Response, and Control from the Viral Perspective

Coralie F. Daussy and Harald Wodrich *

Microbiologie Fondamentale et Pathogénicité, MFP CNRS UMR 5234, University of Bordeaux, 146 rue Leo Saignat, 33076 Bordeaux, France; coralie.daussy@u-bordeaux.fr
* Correspondence: harald.wodrich@u-bordeaux.fr; Tel.: +33-05-57-57-17-63

Received: 30 July 2020; Accepted: 4 September 2020; Published: 7 September 2020

Abstract: Cells are constantly challenged by pathogens (bacteria, virus, and fungi), and protein aggregates or chemicals, which can provoke membrane damage at the plasma membrane or within the endo-lysosomal compartments. Detection of endo-lysosomal rupture depends on a family of sugar-binding lectins, known as galectins, which sense the abnormal exposure of glycans to the cytoplasm upon membrane damage. Galectins in conjunction with other factors orchestrate specific membrane damage responses such as the recruitment of the endosomal sorting complex required for transport (ESCRT) machinery to either repair damaged membranes or the activation of autophagy to remove membrane remnants. If not controlled, membrane damage causes the release of harmful components including protons, reactive oxygen species, or cathepsins that will elicit inflammation. In this review, we provide an overview of current knowledge on membrane damage and cellular responses. In particular, we focus on the endo-lysosomal damage triggered by non-enveloped viruses (such as adenovirus) and discuss viral strategies to control the cellular membrane damage response. Finally, we debate the link between autophagy and inflammation in this context and discuss the possibility that virus induced autophagy upon entry limits inflammation.

Keywords: membrane damage; antiviral autophagy; inflammation; galectin; virus entry; interferon; bacterial invasion; adenovirus; lysophagy; ESCRT machinery

---

1. Introduction

Cellular membranes are selective, permeable barriers, consisting of a phospholipid bilayer. They operate by physically separating two compartments within the cell or separate the contents of the cell from the external environment, and they regulate the exchange of chemicals, ions, and biomolecules across them. Cellular membranes thus define organelles and delineate vesicles in intracellular cargo transport systems, such as the endo-lysosomal or the exocytosis pathway. They are both fluid and dynamic, as well as solid and impermeable if necessary [1]. Therefore, membrane integrity is an essential part of cellular homeostasis, which is constantly challenged by pathogens (bacteria, virus, fungi), protein aggregates, or chemicals. Cells have evolutionary conserved surveillance systems to detect and respond to membrane damage and secure cell survival. As outlined in this review, a hallmark of membrane damage is the transient exposure of complex glycosylated proteins in the cytosolic compartment, which are detected by galectins (Gal). Galectins are small cytosolic carbohydrate binding proteins, which cluster at the site of membrane damage. Galectins also coordinate differential responses such as autophagy, to remove membrane remnants [2,3], and the activation of the endosomal sorting complex required for transport (ESCRT) machinery, favoring membrane repair [4,5]. There is increasing evidence that pathogens, including viruses, take control of the cell response to membrane damage and the associated machinery for their own benefit. Therefore, provoking membrane damage upon infection may not be just an accidental by-product of the pathogen entry process. Instead, eliciting membrane damage could be purposeful and part of a sophisticated entry strategy. This review will discuss this possibility and provide an overview of current knowledge on the relationship between membrane damage, cellular response, and pathogen entry. While briefly summarizing observations made for bacterial pathogens and membrane damage in general, our focus is on the entry of the membrane lytic adenovirus (Ad), and other non-enveloped viruses. We discuss the limited existing knowledge about how viruses manipulate and exploit the cellular membrane damage response. We apologize in advance to the many authors who contribute to the emerging field of membrane damage and whose work we cannot cite due to limited space.

1.1. Virus Inflicted Membrane Damage

Most viruses are taken up by receptor-mediated endocytosis and enter cells through the endo-lysosomal compartment. From here, they have to penetrate into the cytosol to reach their site of replication. For most RNA viruses, replication takes place in the cytosol, while most DNA viruses replicate in the nucleus. The endosome is a dynamic entry compartment, and while it provides some protection from cytosolic innate immune sensors, viruses need to escape before they are either sorted to the lysosome or recycled back to the cell surface [6]. Most enveloped viruses fuse their lipid shell (viral envelope) with the endosomal membrane and release their capsid to the cytosol. This strategy is efficient and topologically simple: it does, *a priori*, not require any kind of membrane rupture [7]. In contrast, non-enveloped viruses, with their hydrophilic capsids, have to cross cell membranes by inflicting membrane damage. This approach risks exposure of the luminal content of the endosome to the cytosol, which can trigger inflammation. To penetrate the cellular membrane, most non-enveloped viruses undergo conformational changes and/or proteolytic processing that allow them to use membrane lytic/modulating factors. Strategies include lipid modification, pore insertion, or large-scale membrane disruption (Figure 1). Still, not much is known about the structural and mechanistic details of the process how non-enveloped viruses inflict membrane damage to cross membranes and even less is known how they engage in the cellular response [8,9].

1.1.1. Adenovirus

Adenoviruses (Ad) are among the best-studied non-enveloped viruses for membrane penetration. Their ~90 nm capsids enter target cells following receptor binding through clathrin-mediated endocytosis. Most Ads rapidly escape the endosomal compartment to avoid lysosomal sorting and use cytosolic motors to reach the nucleus for replication. Adenovirus endosome penetration was shown through co-uptake of non-membrane permeable substrates [10,11]. Observations made with the thermosensitive mutant virus Ad2/5 *ts1* showed that membrane penetration is an essential step in the infection process [12–14]. This *ts1* mutant fails to package the adenoviral protease into the capsid. This failure prevents capsid maturation by cleavage of precursor proteins, resulting in hyperstable and non-infectious particles [15–17]. Adenovirus capsid maturation is required for the release of the internal membrane lytic capsid protein VI upon virus entry. What triggers protein VI release from the capsid is not known, but it may involve disassembly cues during initial cell binding [14,18–21]. Protein VI encodes an amphipathic helix with membrane binding and lytic activity [13,22] and mutations in the amphipathic helix strongly reduced both its membrane lytic activity and viral infectivity [23]. Adenovirus inflicted membrane damage creates openings large enough for cytosolic delivery of 70 kDa dextrans or 25 nm parvovirus particles [10,11]. Electron microscopy images of Ad particles in partially disrupted endosomes show large physical openings in the endosome plugged by the virus [3]. High-resolution fluorescence microscopy images suggest localized protein VI release from the capsid at the membrane damage site [24]. Ceramides have been proposed to increase protein VI membrane affinity, showing the importance of the local lipid composition at the membrane penetration site [19]. Protein VI release also plays a role in virus escape from endosomes by counteracting cellular autophagy, which is normally mounted in response to membrane damage as detailed in Section 2 [3].

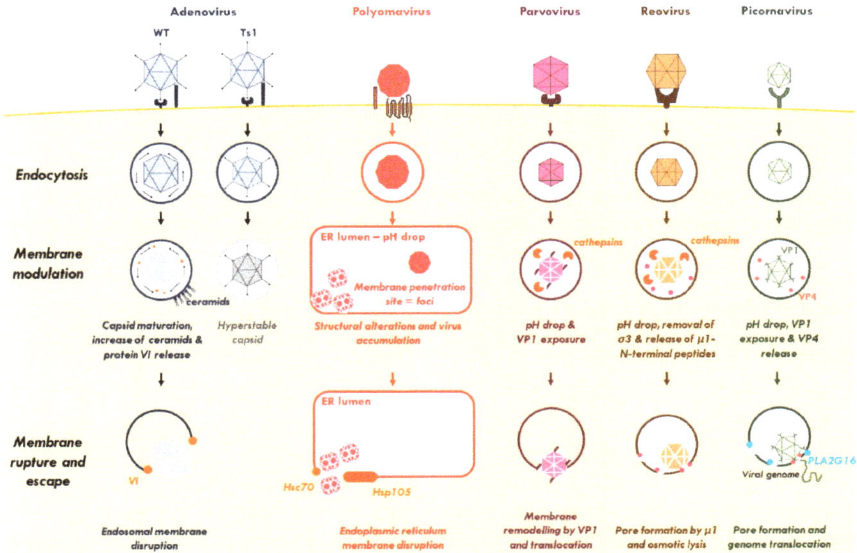

Figure 1. Virus-inflicted membrane damage. After binding to cell-surface receptors, viruses are internalized through endocytosis. Once in the endosome, adenovirus capsid undergoes partial disassembly and releases protein VI. The increase of ceramide concentration enhances the binding of protein VI to the endosomal membrane and its subsequent rupture. Polyomavirus-containing endosomes are targeted to the endoplasmic reticulum (ER) where the virus undergoes conformational changes to penetrate the ER-membrane and escape to the cytosol. Parvovirus and reovirus require a pH drop and the action of endosomal cathepsins to induce conformational rearrangements, disrupt the endosome, and reach the cytosol. After endocytosis and conformational changes, picornaviruses rely on a cellular lipid-modifying enzyme (PLA2G16) to facilitate the translocation of its genome via selective pores across the endosomal membrane. See Section 1 for further details. Abbreviations: ER, endoplasmic reticulum; Hsc70, Heat shock cognate 71 kDa protein; Hsp105, Heat shock protein 105 kDa; PLA2G16, phospholipase A2 group XVI.

1.1.2. Polyomavirus

Polyomaviruses (e.g., BK virus and simian virus 40) are small non-enveloped DNA viruses with a diameter of ~45 nm. Particles enter by endocytosis and are transported to the endoplasmic reticulum (ER), from where they reach the cytosol. To penetrate the membrane, particles undergo conformational changes triggered by pH drop and recruited cellular chaperones [25]. Exposure of the N-terminus of the major capsid protein VP1 uncovers a hidden myristylated domain in the internal capsid protein VP2. This structural alteration makes the virus particle significantly more hydrophobic and primes the capsid for membrane binding and particle translocation [26]. Penetration sites are located at the ER membrane and form distinct, virus-induced foci. Several cellular proteins, including chaperones Heat shock cognate 71 kDa protein (Hsc70) and human heat shock protein 105 kDa (Hsp105), accumulate at these foci likely assisting viral particles to escape into the cytosol [27,28]. The exact nature of the penetration foci and the mechanism of translocation are largely unknown. There are no reports on the size of the resulting membrane damage (if any), but the translocated capsid appears to remain intact, therefore requiring the formation of large openings [29]. However, a more recent report suggests that capsid translocation could be coupled to capsid disassembly for genome release, raising the possibility that the inflicted membrane damage is smaller than initially thought [30].

1.1.3. Parvovirus

With a ~25 nm diameter, parvoviruses (e.g., adeno-associated virus (AAV) and canine parvovirus (CPV)) are rather small, non-enveloped DNA viruses that enter cell by endocytosis. Acidification in the endosome is a crucial step to induce conformational changes required for endosome penetration. The drop in pH allows the deployment of the N-terminus of capsid protein VP1, which has a phospholipase type 2 (PLA2) activity [11,31]. This PLA2 activity is essential for endosome penetration, most likely by transient and localized lipid modification [11,32,33]. For example, CPV entry allows the release of 3 kDa dextrans, but not 10 kDa dextrans from endosomes. The absence of co-release of larger molecules during parvovirus endosomal escape suggests that the resulting membrane damage is limited and does not involve complete endosome lysis [31]. However, the exact mechanism of parvovirus membrane translocation is not known.

1.1.4. Reovirus

Reoviruses (e.g., rotavirus) have non-enveloped capsids with a diameter of ~75 nm and contain RNA genomes. The exact entry of reoviruses is unclear, but it is likely to occur through endocytosis. After uptake, endosomal cathepsins are activated by a drop in pH. The outer capsid is then proteolytically processed and capsid protein $\sigma3$ is removed [34]. This uncovers the membrane lytic capsid protein $\mu1$ and autoproteolytic processing renders it membrane-lytic [35,36]. The fully processed $\mu1$ N-terminal peptides are then myristylated, released from the capsid and insert into the endosomal membrane, where they form size selective pores [37,38]. While these pores are too small (estimated to 4–9 nm) to permit reovirus translocation, it was suggested that endosome lysis could involve osmotic lysis [39,40].

1.1.5. Picornavirus

Picornaviruses (PV, e.g., polioviruses or rhinoviruses) enter their host cells by endocytosis. Unlike other non-enveloped viruses, PV do not translocate their capsid to the cytosol. Instead, they perforate the endosome membrane by creating a pore to translocate their RNA genome to initiate cytosolic replication [41]. Thus, contrary to viruses like Ad, PV inflicted membrane damage has a small diameter [42]. Receptor binding and a drop in pH provide cues of conformational changes in the virus particle. As a result the amphipathic helix of the capsid protein, VP1 is externalized and VP4 is released to allow membrane binding and pore formation [43,44]. The enzymatic activity of the cellular protein phospholipase A2 group XVI (PLA2G16) then facilitates genome translocation into the cytoplasm, but the exact mechanism of pore formation remains unknown [45]. Some PV encode the 2A protein, which are homologues of the cellular PLA2G16, suggesting an essential and conserved role in membrane penetration [46].

1.2. Bacteria Induced Membrane Damage

Pathogen-induced membrane damage has been much more extensively studied in bacterial infections than for viruses. Many intracellular bacteria enter their host cell by phagocytosis, ending up in vacuoles connected with the endo-lysosomal system. To avoid degradation and to proliferate, a number of bacteria escape from their vacuole to the cytosol. Others, such as *Salmonella typhimurium*, first proliferate inside vacuoles and may occasionally escape at a later time point to reach the cytoplasm for hyperproliferation [47]. In all cases, they must destabilize and break the membrane of their vacuole. Numerous studies have shown that this involves bacterial effectors (or their toxins) cooperating with cellular factors. *Shigella flexneri*, for example, is a gram-negative bacteria known to infect epithelial cells. It uses a type 3 secretion system (T3SS) to inject effectors into the cell, resulting in phagocytic uptake. After internalization, *S. flexneri* must escape the vacuole for cytoplasmic proliferation. Initially believed to be mediated via the bacterial T3SS secretion system [48], a subsequent small interfering RNA (siRNA) screening identified the cellular small GTPase Rab11 to be a crucial factor required for vacuole breakdown through the formation of macropinosomes at the invasion site [49,50].

S. typhimurium is another bacterial example for membrane damage and probably the best documented. *S. typhimurium* enters its target cells by endocytosis and establishes itself in a replication vacuole (SCV). A small but significant fraction of the bacteria escapes to reach the cytosol, through their T3SS [51]. TANK binding kinase 1 (TBK1) was suspected to control the integrity of the SCV membrane because in the absence of TBK1, *S. typhimurium* replication was more efficient [52]. More recently, it was shown that TBK1 instead coordinates an autophagic response against bacteria after cytosolic exposure (see Section 2) [53,54]. Some evidence suggests that the accumulation of COPII complexes on the SCV membrane destabilizes the membrane through a mechanism that remains to be deciphered [55].

Listeria monocytogenes encodes Listeriolysin O (LLO), a toxin it uses for vacuole lysis. Listeriolysin O inserts into the vacuole membrane by binding to cholesterol and forms membrane- disrupting pores. The LLO effect is potentiated by cellular factors including gamma-interferon-inducible lysosomal thiol reductase (GILT) and cystic fibrosis transmembrane conductance regulator (CFTR) [56,57]. Once in the cytoplasm, *L. monocytogenes* replicates and controls the polymerization of actin to be able to propagate from cell-to-cell in secondary vesicles. Unlike the first endosomal escape, escaping from secondary vesicles needs destabilization of two membranes. *L monocytogenes* accomplishes this task using LLO as well as the phospholipases PIcA and PIcB [58,59]. The use of a bacterial toxin for membrane rupture is not unique to *L. monocytogenes* and other bacterial toxins, such as perfringolysin *(Clostridium difficile)*, pneumolysin (*Streptococcus pneumonia*), and VacA (*Helicobacter pylori*) may be deployed from bacteria to cause membrane penetration for the same purpose [60–62]. Escape from vacuoles has been demonstrated for several other bacteria including *Mycobacterium tuberculosis*, *Rickettsia prowazekii*, *Burkholderia pseudomallei*, and *Francisella tularensis* [63–73]. The underlying mechanisms are less clear, but generally involve the use of endogenous bacterial genes in conjunction with cellular factors, such as phospholipases, to breach vacuolar membranes. Interestingly for all described examples, these bacterial genes are often associated with pathogenicity linking membrane damage with disease [68].

1.3. Membrane Damage Induced by Other Pathogens

Membrane damage is not restricted to bacteria and viruses. There is increasing evidence for a role of membrane damage during fungal and parasite infection. *Candida albicans* is a polymorphic fungus that can change from a yeast form to a hyphal form [74]. After phagocytosis in the form of yeast, conversion into the hyphal form inside the phagosome involves a change in morphology through stretching, which ultimately leads to the rupture of the phagosomal membrane [75]. This is likely to occur through mechanical pressure (filamentation) rather via a biochemical process [76]. As part of a complex life cycle, the protozoa *Trypanosoma cruzi* invades red blood cells and escapes its entry vacuole to reach the cytoplasm [77]. This vacuolar escape depends in part on the parasite PFP TcTOX protein, which seems to have a similar role as the LLO bacterial toxin of *L. monocytogenes.* Additional factors may be involved in breaching the membrane and the precise mechanism needs further clarification.

1.4. Non-Pathogenic Membrane Damage

Many neurodegenerative diseases and cancer cells are associated with the induction of sterile membrane damage [78]. In neurodegenerative diseases, membrane damage is often a consequence of the formation of large amyloid aggregates, propagating from cell-to-cell. Studies on α-synuclein showed the ability of these amyloid aggregates to induce endosome rupture after endocytosis [79,80]. This capacity is not restricted to α-synuclein, but also extends to Tau and Huntingtin assemblies [80]. The exact reason why cell invasion by amyloid assemblies and their ability to break the membrane of the endosome causes disease is not fully understood, but it is likely to involve dysregulation of autophagy associated with irreparable damage [80]. Lysosomotropic compounds have emerged to artificially trigger lysosome damage and induce cell death, e.g., as therapeutic concept in cancer [81]. Conceptually, the acidic pH inside makes lysosomes susceptible to the accumulation of weak bases that can penetrate them. Once inside, they get protonated and stay trapped within the lysosome where they can induce specific membrane damage [81]. One compound, the L-leucyl-L-leucine methyl ester (LLOMe), is often used as a positive control in endolysosomal rupture experiments in combination with detection of membrane damage markers [82,83]. Endosomal membrane damage can also be caused via a mechanical/physical trigger, especially in the context of intracellular drug delivery, where it is important to achieve efficient endosomal escape. Some lipid formulations used in transfection of siRNAs were shown to be particularly efficient in causing endosomal membrane damage [84,85]. Gold nanoparticles and nanodiamonds were also employed to physically disrupt the endosomal membrane [86,87].

The non-exhaustive summary of possible pathogenic and non-pathogenic membrane insults presented in this section provides insight into the diversity of threats to cellular membrane integrity. This listing also highlights the importance for cellular surveillance of membrane integrity and the ability to respond to membrane damage to secure cell survival, which is discussed in the next section.

2. Membrane Damage Recognition and Cell Response

Extensive damage in endolysosomal vesicles can leak their content and flood the cell with reactive oxygen species (ROS), acid hydrolases (cathepsins), and/or calcium ions provoking injuries to several organelles and ultimately resulting in cell death through necrosis or pyroptosis. A key question, when trying to understand how viruses and other pathogens inflict membrane damage, is how cells sense their membranes have been actively penetrated and how they mount a response. A critical role in the detection of membrane damage is done by cytosolic lectins belonging to the family of Gal. Galectins recognize complex glycans present on lipids or glycosylated proteins. These glycosylated substrates are mainly localized at the cell surface or protruding from the intra-luminal membrane leaflet, including the interior of endosomes and lysosomes. In contrast, they are virtually absent from the cytosol. Hence, membrane damage exposes these glycans to the cytosol and they become easy targets for Gal sensing.

2.1. Galectins and Their Role in Membrane Damage Recognition

Galectins are characterized by the presence of one or two carbohydrate recognition domains (CRD) [88]. They can be monovalent (with a single CRD: Gal-1, 2, 5, 7, 10, 11, 13, 14, 15), bivalent (with two CRDs, Gal -4, 6, 8, 9 and 12) or even chimeric (Gal3) [89,90]. The diversity of Gals allows them to have different affinities for a plethora of substrates, making them ideal pathway mediators. Their cellular functions include regulating cell adhesion, organizing membrane domains, signaling and trafficking, apoptosis and cell cycle regulation. Here, we focus on their role as danger sensors since they quickly accumulate on galactosides exposed to the cytosol, which are recognized as danger signals at sites of membrane breach [91,92]. Although mammals have 15 different forms of Gal, to date only Gal1, 3, 8 and 9 have been described to sense damaged vacuoles [2,92]. They accumulate around membranes ruptured by both pathogen and non-pathogen stimuli, such as protein aggregates [80], liposome transfection [85] or lysosomotropic agents [82]. Upon membrane rupture, glycans and Gals form clusters (puncta) that can be easily detected by immunofluorescence using Gal-specific antibodies or cells expressing Gals tagged with a fluorescent protein [82,92,93]. Most Gals (e.g., Gal1/3) are widely expressed, making them useful targets to assess membrane damage [82,93].

Most observations and functional studies using Gals were made using invasive bacteria models. Galectin3 recognizes the vacuoles from which *S. flexneri* escaped and was the first example of a Gal recruited to pathogen-induced membrane damage [92]. A systematic screen including all Gals identified found Gal3, 8, and 9 as binding to *S. typhimurium* ruptured vacuoles [2]. Galectin 8 was also the first Gal shown to restrict bacterial proliferation and thus to have a direct antimicrobial effect [2]. To date, very few studies have addressed if virus membrane penetration also triggers Gal recruitment. Initial studies with Ad used Gal3 as a marker to show the subcellular localization and the timing of Ad endosome penetration [93,94]. In a subsequent study, it was shown that Ad also recruit Gal8 to the endosome penetration site and that this is essential for the cellular autophagy response [3]. Picornavirus pore formation in the endosomal membrane also recruits Gal8-mediating cellular autophagy and remains so far the only other example showing the recruitment of Gals to virus inflicted membrane damage [45]. Whether cells employ Gals in response to entry of other (non-enveloped) viruses remains to be shown. In any case and as discussed below, Ad and PV found ways to counteract Gal8 mediated antiviral autophagy to efficiently propagate.

2.2. Galectins and Ubiquitin Coordinate Autophagy for Membrane Damage Removal

The easiest way for a cell to deal with damage is to remove it via degradation pathways. An important part in membrane damage removal has been attributed to autophagy [78]. Autophagy is a conserved cytosolic degradation process induced under stress conditions, such as starvation, hypoxia or during membrane damage and pathogen infection [95], which allows the removal and degradation of damaged organelles or proteins from the cytosol into double membrane vesicles called autophagosomes. These autophagosomes subsequently fuse with lysosomes for cargo degradation and recycling [96]. Autophagy is driven by proteins of the "ATG family" (for autophagy-related), a core set of proteins that coordinates formation, elongation and maturation of autophagosomes [96–98]. It is now clearly established that autophagic degradation is highly selective. In mammals, this selectivity is provided by a family of proteins called "autophagic receptors" (including, e.g., the nuclear dot protein 52 kD (NDP52) and p62, reviewed in [99]), which allow targeting of distinct substrates for degradation. Each autophagic degradation process is specified by a "cargo-defined" name such as mitophagy defining the selective degradation of mitochondria, xenophagy for specific degradation of intracellular pathogens or lysophagy naming the degradation of lysosomal membranes upon endolysosomal damage [99].

Membrane damage removal by autophagy can be separated into two steps (Figure 2, lysophagy). In a first step, the cell has to sense and discriminate intact vs. damaged vacuoles. The second step is to assemble and recruit the autophagic machinery to degrade membrane remnants. Key elements of the first step are Gals and ubiquitin. Both are so called "eat-me signals", a name playing with their function to mark membrane remnants for autophagic removal. Certain Gals recognize membrane damage by virtue of complex glycan exposure e.g., during pathogen infection [2,92,93]. To enlist autophagy, Gals have to bind downstream effectors. Thurston et al. reported that Gal8 recruited after SCV rupture forms a complex with the autophagy receptor NDP52 [2]. NDP52 binds FAK family kinase-interacting protein of 200 kDa (FIP200), which serves as seed assembly platform for autophagy initiation [100]. FIP200 assembles a primary kinase complex, the Unc-51 like autophagy activating kinase (ULK1/2) complex, ATG13, and ATG101. Once assembled, this complex activates a downstream Class III PI3-kinase complex with VPS34 and Beclin-1 [100,101]. The Beclin-1 complex modifies membranes and creates a phospholipid patch (also termed phagophore) to support binding of WD-repeat protein interacting with phosphoinositides (WIPI) proteins, creating a membrane anchored landing platform. WIPI proteins recruit the ATG5-ATG12/ATG16L E3 conjugation complex to the phagophore [102]. The ATG5-complex conjugates soluble ATG8 (better known as LC3B) with phosphatidylethanolamine (PE) to be incorporated into the expanding autophagosomal membranes [103]. Additional interactions, such as NDP52-LC3 and FIP200-ATG16L stabilize the developing cargo-autophagosome complex to promote membrane elongation. Gal8 and Gal3 present at the site of membrane damage can also interact with other effectors such as the tripartite motif protein 16 (TRIM16), as shown for ruptured lysosomes using LLOMe or bacteria [104,105]. TRIM16 serves as another assembly hub to initiate autophagy by recruiting the core autophagic machinery (ATG16L1, ULK1, and Beclin-1) directly on damaged membranes [104]. Interestingly, the recruitment of Gal3 to the membrane damage side can also be pro-bacterial by impeding on the autophagic response [106]. Silencing of Gal3 during *L. monocytogenes* infection in murine macrophages increased LC3 recruitment to the vacuoles and reduced bacterial replication. Moreover, treatment of cells with sialidase (which removes sialic acid from glycans) increased Gal3 and decreased Gal8 on the damaged phagosomes. This change in Gal preference directly influenced the autophagic response [106], showing that cytosolic Gals can discriminate the origin of the damaged membrane and orchestrate an adapted cellular response. Deciphering the glycosylation pattern driving Gal specificity could thus be a valuable point for therapeutic intervention. Furthermore, Gal3 also directs other antibacterial functions including the recruitment of interferon (IFN)-inducible guanylate binding proteins (GBPs) to pathogen-containing vacuoles [107].

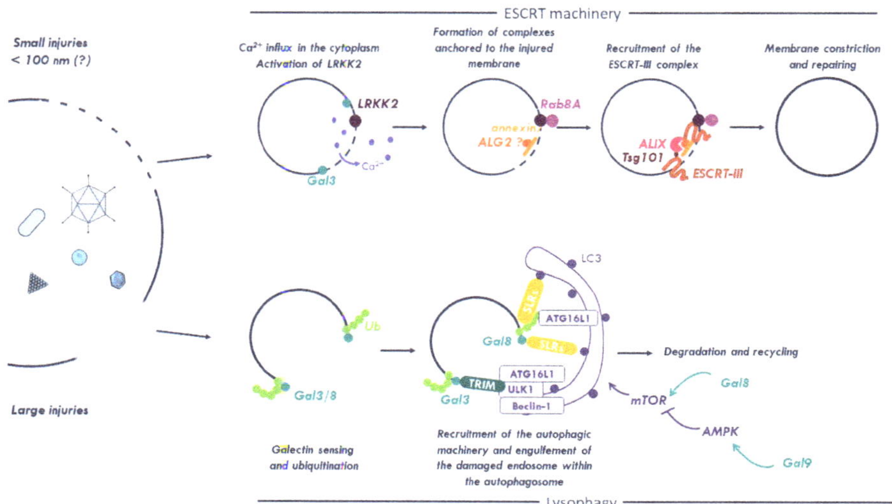

Figure 2. Membrane damage repair and removal. The endosome is constantly challenged by pathogens (bacteria, virus, and fungi), protein aggregates, or chemicals that can disrupt its membrane and provoke injuries of different sizes. Small disruptions (<100 nm) trigger a leakage of Ca^{2+} into the cytoplasm and activate LRKK2. LRKK2 and Ca^{2+} effectors mediate the recruitment of the ESCRT machinery to promote repair of the injured organelle. Galectin-3 (Gal3) is recruited to damage sites and may promote ESCRT assembly. If the injury is too large, the cell will trigger a process of degradation called lysophagy. During lysophagy, damaged vacuoles are sensed and tagged by galectins (Gal) and ubiquitin (Ub). Both signals mediate the recruitment of the autophagic machinery (either directly via autophagic receptors, such as the sequestosome like receptors (SLRs) or indirectly via TRIMs). The membrane remnant is engulfed in a double-membrane vesicle called autophagosome, which fuses with lysosomes for content degradation and recycling. Autophagy is also controlled by metabolic kinase mechanistic target of rapamycin (mTOR) through Gal8. Moreover, Gal9 can also control autophagy induction by directly activating AMP-activated protein kinase (AMPK) in response to endosomal damage to inhibit mTOR. See text for details. Abbreviations: ESCRT, endosomal sorting complexes required for transport; Gal8, galectin8; LRRK2, Leucine-rich repeat kinase 2; ROS, reactive oxygen species; Ub, ubiquitin.

Next to Gals, ubiquitin is the second most important early danger signal of many membrane damage sites following bacterial [2,108–110], viral [3,111], and apathogenic insults [112]. Ubiquitin appears shortly after Gals, and like them, is involved in the recruitment of a variety of autophagic receptors (also known as sequestosome like receptors or SLRs), which contain ubiquitin binding motifs. All SLRs also contain LIR-domains (LC3-interacting region) and bridge the ubiquitylated cargo and LC3 to link the cargo into the growing autophagosomal membrane [99]. Moreover, ubiquitin can directly bind the autophagic machinery through ATG16L1, which may enhance its anchoring to the surface of ubiquitylated endosomes [113]. Two types of ubiquitin chains, K48- and K63 linked, are predominantly found on damaged lysosomes [113]. K63 chains are targeted by SLRs and recruit the autophagy machinery [113,114]. The enzymes driving ubiquitylation of membrane damage, and how the chain type influences the cell response, are not entirely clear, and may involve more than one player. Some of them have been characterized in the context of bacterial infection, where they allow the ubiquitylation of both the bacteria and the membrane remnants to turn them into signaling platforms. The first E3 ubiquitin ligase described to target bacterial membrane damage was the LRR-containing RING E3 ligase (LRSAM) and is required for autophagic clearance of *S. typhimurium* [115]. Another E3 ligase is linear ubiquitin chain assembly complex (LUBAC), which mediates the formation of M1-linked ubiquitin chains and is also involved in autophagic degradation of *S. typhimurium* [116]. The aforementioned TRIM16, recruited via Gal3/8 to endomembrane damage, is also an E3 ubiquitin ligase promoting K63-linkage to ATG proteins and to ubiquitylate lysosome substrates e.g., after LLOMe treatment [104]. FBXO27 (F-Box Protein 27, part of the SCF ubiquitin ligase SKP1/CUL1/F-box protein complex) ubiquitylates several glycoproteins present on the surface of damaged lysosomes (e.g., lysosomal-associated membrane protein LAMP2), which in turn promotes the recruitment of the autophagic machinery [112]. A recent siRNA screen identified a crucial role for Ubiquitin Conjugating Enzyme E2 Q Family Like 1 (UBE2QL1), an E2 ubiquitin conjugating enzyme, in directing lysophagy [117]. Depletion of UBE2QL1 increased steady-state lysosomal membrane damage and prevented efficient ubiquitylation and SLR recruitment, essentially curtailing the autophagy response. It also prevented recruitment of the AAA-ATPase valosin-containing protein, VCP/p97 complex. The VCP/p97 complex is likely to extract membrane proteins to facilitate downstream lysophagy and cooperates with YOD1/OTU1, a K48-specific deubiquitylating enzyme. This observation suggests that selective membrane deubiquitylation, possibly of K48 labeled VCP/p97 substrates, maybe necessary before lysophagy proceeds [78,118]. A similar mechanism appears active in neurodegenerative diseases, where protein aggregates rupture endo-lysosomes and Gal3/8 recruits NDP52 and p62 suggesting that the response might be cell intrinsic [80,119].

Some receptors including NDP52 and p62 are phosphorylated by TBK1, to enhance their ability to bind ubiquitin following recruitment to the membrane damage site [120,121]. TBK1 also phosphorylates WIPI proteins and enhances autophagosome formation in response to bacterial membrane penetration [54,122]. Thus, TBK1 plays an important role in promoting and fine-tuning the antimicrobial autophagy response at the site of membrane damage.

On a larger scale, autophagy is under the control of two different kinases, the metabolic Ser/Thr kinases mechanistic target of rapamycin (mTOR) and AMP-activated protein kinase (AMPK). Both kinases are major regulators of autophagy. They act in an antagonistic way in that mTOR suppresses autophagy by phosphorylating inhibitory sites on autophagy regulators, such as ULK1/2, while phosphorylation by AMPK activates autophagy regulators [123]. Active mTOR can be found on the outer lysosomal membrane and upon lysosome damage translocates to the cytosol and becomes inactivated. Keeping mTOR inactive involves inhibitory phosphorylation by AMPK, connecting both kinases. It was recently shown that Gal8/9 regulate mTOR and AMPK on damaged lysosomes identifying new functions beyond glycan recognition [4]. The study shows that interaction of Gal8 with the lysosome membrane integral neutral amino acid transporter Solute Carrier Family 38 Member 9 (SLC38A9) upon membrane damage reorganizes a signaling complex (the Ragulator-Rag complex) on the lysosomal membrane. The Ragulator-Rag complex normally retains mTOR in an active state bound to the lysosome surface due to nutrient sensing. The reorganization through Gal8 (e.g., following membrane damage) promotes mTOR dissociation from lysosomes. Inactivation of mTOR is then reinforced by the action of another Gal, Gal9, which activates AMPK [4]. Taken together, this work offers the interesting perspective that membrane damage can be controlled by Gals at two different levels. Locally, at the physical site of membrane damage through recruitment of SLRs and assembly of the autophagic machinery. In addition, Gals also act at the cellular level by coordinating the activities of mTOR and AMPK. This hierarchical organization could be used to amplify and disseminate the local response, e.g., dependent on the extent of the damage by modifying major upstream autophagy regulators. Widening the response via mTOR inactivation has several effects on the cell, including the nuclear translocation of transcription factor EB (TFEB), a mTOR target, and master regulator normally sequestered in the cytosol, which drives gene expression programs for neosynthesis of genes involved in the biogenesis of lysosomes [104], in autophagy [124], and in inflammation [125].

2.3. Viral Control of the Autophagy Response to Membrane Damage Removal

Most of the investigations concerning the cellular response to membrane damage use invasive bacteria or models of induced lysosome damage. Although details are still emerging, there seems to be consistency since cells respond in all cases with autophagy directed against the membrane damage site. Viral membrane damage is no exception, although only two viral systems have been investigated. Adenovirus and PV both cause different types of membrane damage upon entry (Figure 3). Depending on the cell system used, Ads penetrate the endosome within 15–20 min and trigger fast selective autophagy in the infected cell. Adenovirus damaged endosomal membranes are detected by Gal3 [93,94,111] and Gal8 [3]. Using co-detection of viral particles and/or the exposed membrane lytic protein VI revealed that the viral association with ruptured membrane is transient, and that the Gal3/8-positive membrane associated with NDP52, p62, ubiquitin and LC3 [3,111]. Gal3 positive membrane remnants were cleared from the cell within 3 h of entry presumably via autophagy [111]. Interestingly wild type Ad was not cleared by autophagy and suppressing autophagy with genetic or pharmacological tools did not affect viral infectivity [3]. Using the mutant Ad-M1, the authors showed that being refractive to autophagic clearance is not simply due to a rapid escape from the endosome, but also involves active control of autophagosome maturation. The mutant Ad-M1 has strongly reduced infectivity compared to wild type Ad. It lacks a PPxY motif encoded in the membrane lytic protein VI, which is required to recruit the ubiquitin ligase neural precursor cell expressed developmentally down-regulated 4-like (NEDD4.2) upon membrane penetration [18]. Unlike wild type Ad, Ad-M1 is unable to escape from endosomes and becomes susceptible to autophagic degradation [3]. Comparing both viruses revealed that wild type viruses do not prevent autophagy initiation. Instead, they use the PPxY motif to prevent autophagosomes from maturing and fusing to lysosomes, a process that was also observed upon NEDD4.2 depletion from cells using viral and non-viral autophagy stimuli [3]. The exact role for NEDD4.2 is not yet clear and may involve regulating autophagy effectors such as ULK1 or protein VI itself [18,126]. However, this example shows how Ad has evolved a short peptide motif that recruits an essential cellular factor to stall the cell response to membrane damage, thereby buying enough time to reach the safety of the cytosol. The PPxY motif is strategically located in the membrane lytic factor making sure it is exposed at the right time and place. Consequently, eliminating membrane damage recognition by depleting either Gal8 or autophagy (e.g., in ATG5 KO cells), fully restores viral infectivity of the PPxY-deficient Ad-M1. In contrast, this does not accelerate viral escape from the endosome, showing that escape mechanism and autophagy suppression are separate processes, and that one favors the other [3].

Unlike Ad, which generates large openings in the endosomal membrane, PV creates pores of limited size within endosomes to translocate their RNA genome across the membrane. A haploid cell-based genetic screen identified the small phospholipase PLA2G16 (Group 16 phospholipase A2) as an essential factor required for genome translocation [45]. Using a counter screen in PLA2G16 deficient cells, the authors found that depletion of several autophagy genes (e.g., ATG5/7/12) and Gal8 restored viral infectivity. Using fluorescent viruses and alternative membrane rupture assays, they could show that Gal8 and PLA2G16 relocated after membrane damage independently of each other. This suggests that PLA2G16, like Gal8, senses virus-induced membrane rupture, but is recruited in a virus- and Gal8-independent manner. PLA2G16 has no effect on the autophagic flux itself, but facilitates the translocation of the genome into the cytoplasm and prevents its clearance by autophagy. When PLA2G16 is silenced, genomes are degraded via autophagic clearance together with the virus, a phenotype that can be reversed if Gal8 or ATG7 is also silenced [45]. This work thus identified PLA2G16 as cellular factor exploited by a virus to control antiviral autophagy directed against PV-induced membrane damage, a strategy analogous to the PPxY motif of Ad. Unlike for Ads, PV protection from autophagic degradation through accelerated cytosolic translocation concerns only the viral genome. It is conceivable that the enzymatic activity of PLA2G16 favors PV genome translocation over autophagic clearance of the membrane damage by local lipid modification, although the exact mode of action is currently unclear. Interesting in both cases, neither Ad nor PV have evolved mechanisms to suppress autophagy initiation suggesting that there is additional benefit in triggering autophagy, possibly linked to counteracting inflammatory signaling (Section 4).

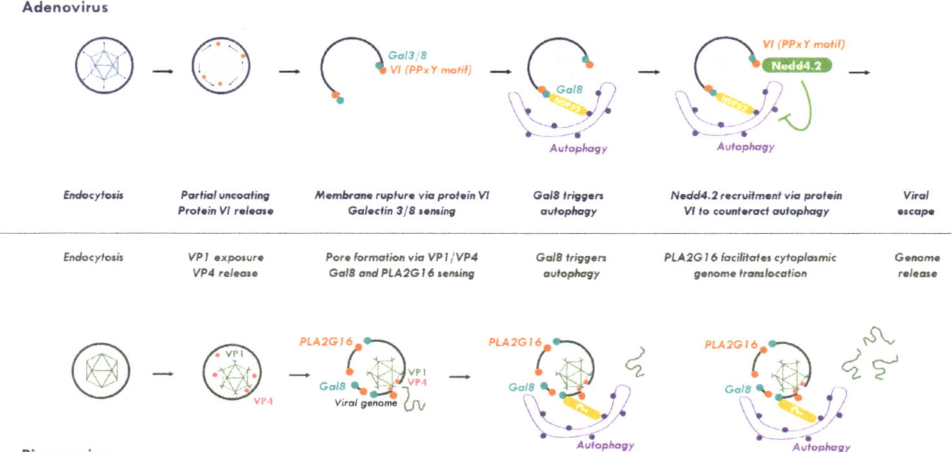

Figure 3. Viral control of the membrane damage response. Top: endocytosed adenoviruses partially uncoat and release the membrane lytic capsid protein VI for endosomal membrane lysis. Membrane damage is sensed by galectins 3 and 8. Galectin 8 recruits autophagic receptors and triggers autophagy. Adenoviruses stall autophagy through a short PPxY peptide motif in protein VI that recruits the ubiquitin ligase Nedd4.2. As a consequence, they avoid degradation and escape into the cytoplasm. Bottom: After endocytosis and acidification of the endosome, picornaviruses undergo conformational changes to expose capsid protein VP1 and release of VP4. Both proteins attach to the endosomal membrane creating membrane-penetrating pores. Membrane damage is then independently sensed by galectin 8 activating autophagy and PLA2G16. PLA2G16 facilitates genome translocation into the cytoplasm preventing autophagic clearance. See Sections 2 and 3 for further details. Abbreviations: Gal, galectin; NDP52, Nuclear dot protein 52; Nedd4.2, neural precursor cell expressed, developmentally down-regulated 4.2; PLA2G16, phospholipase A2 group XVI.

2.4. ESCRT Machinery for Membrane Repair

Cells use autophagy to remove and degrade damaged membranes such as lysosomes (lysophagy) for cell survival [2,104,112,113]. However, the cell response can be more restrained and small membrane injuries were shown to trigger a membrane repair mechanism via the ESCRT system rather than removal [127–129] (Figure 2, ESCRT machinery). The ESCRT complex is divided into five sub-complexes (ESCRT-0, I, II, III, and disassembly proteins) and it was first discovered for its role in regulating endosomal trafficking, but is now known to be involved in numerous other cellular processes such as vesicle formation, vesicle budding and cytokinesis as detailed elsewhere [5,130]. Small membrane lesions (<100 nm), especially at the plasma membrane, trigger calcium ion influx and lead to a rapid recruitment of cytosolic annexin7 and the calcium sensor ALG2 at the membrane damage site. This complex recruits subsequently the ALG2 interactors ALIX and Tsg101, both ESCRT-I proteins that orchestrate the recruitment of the ESCRT-III complex to repair the membrane. ESCRT-III has the ability to form filaments that constrict the membrane and shed the damaged membrane [5,127,128]. A recent report showed that the ESCRT machinery was also quickly recruited to damaged endo-lysosomes to allow their repair following treatment with either LLOMe or silica crystals [131]. Even if the initial recruitment of ALG2 (apoptosis-linked gene-2) is still debated [129,131], endosome repair requires recruitment of the ESCRT-I components ALIX (ALG-2-interacting protein X) and Tsg101 (tumor-susceptibility gene 101), and the subsequent action of the ESCRT-III complex. Repair also involves transient inactivation of lysosomal hydrolases followed by re-acidification [132]. Even with minor damage, Gal3 is still recruited to damaged endo-lysosomes, but follows a slower kinetic than ESCRT recruitment. The ESCRT-III response (monitored by recruitment of CHMP4B/ESCRT-III) occurs on the timescale of minutes preceding lysophagy that mounts within the hour [129,131]. Depletion of either ALIX or Tsg101 or both does not affect Gal3 recruitment, but delays membrane repair in favor of removal/lysophagy [129,131]. A partial explanation for this observation might be that Gal3 initially promotes interactions between ALIX and the downstream ESCRT-III effector CHMP4. At later times, however, Gal3 controls the autophagic response via TRIM16 supported by Gal8/9 that regulate mTOR/AMPK [133]. A very recent study showed that macrophages challenged with either invasive bacteria or LLOMe activates the Parkinson's disease related kinase leucine-rich repeat kinase 2 (LRKK2), which in turn recruits the Rab GTPase Rab8A. Both coordinate the activity of the ESCRT-III complex for membrane repair. In contrast, depletion of LRKK2 and Rab8A change the damage response phenotype from membrane repair to lysophagy [134].

The interplay of Gal3/8/9 with the ESCRT machinery, autophagy and metabolic signaling is also observed during membrane damage upon *M. tuberculosis* and *Coxiella burnetii* infection [129,133]. Interestingly, mycobacterial effectors EsxG/TB9.8 and EsxH/TB10.4 secreted by the ESX-3 T7SS secretion system antagonize the ESCRT response with a kinetic that matches the speed with which the cell responds, showing that bacteria have also developed efficient countermeasures against membrane repair mechanisms [135]. It is still possible that ESCRT and autophagy have overlapping roles in endosome repair because ATG5 was shown to be involved in repairing the endosomal membrane damaged by the type-1 secretion system T1SS after infection with *S. typhimurium* [136]. ESCRT and autophagy also cooperate in the maintenance of the bacterial vacuole in *Dictyostelium discoideum* infected with *Mycobacterium marinum* [137]. Taken together the cell reaction to membrane damage appears conserved, fast, and flexible. This implies an active and hierarchical organization, where the ESCRT-mediated membrane repair is transient and precedes lysophagy [129,133].

2.5. Viral Control of the ESCRT-Response to Membrane Repair

If there is an ESCRT-response against membrane penetration by non-enveloped viruses upon entry has not been investigated. The ESCRT machinery plays a major role in the release process of most enveloped viruses and mediates their non-lytic release from cells (including some non-enveloped viruses). Many viruses use a conserved PPxY peptide motif or the related P(S/T)AP motif encoded in their capsid proteins to recruit ESCRT-I proteins or NEDD4 ubiquitin ligases (for review see [138–140]). These domains are still present in entering virus particles and could easily target the ESCRT machinery during entry. As discussed above, such a conserved PPxY domain able to recruit NEDD4-family ubiquitin ligases is encoded in the Ad membrane lytic protein VI [18], which suppress Gal8-mediated antiviral autophagy upon endosome penetration likely through a NEDD4.2 mediated process [3]. It is not clear if this is functionally related to the ESCRT machinery. A different study showed that upstream of endosomal escape, the binding of Ad to the cell surface results in small (transient) lesions at the plasma membrane in a protein VI-dependent manner that provokes extracellular calcium ion influx and dye penetration into the cell [128]. This local increase in calcium ions triggers exocytosis and local fusion of secretory lysosomes as part of a repair mechanism [128]. Upon membrane fusion, these lysosomes are suggested to release lipid converting acidic sphingomyelinase (ASM) catalyzing sphingomyelin thereby generating high local concentrations of ceramides. Ceramides, when present in the endosomal membrane, increase the affinity of protein VI, helping Ad to penetrate the endosome [19]. Such a mechanism suggests that Ad membrane penetration is a two-step process involving small and large sized membrane damage [19]. As shown recently, sphingomyelins play an important role in initiating the membrane collapse in bacterial vacuoles preceding Gal8 recruitment to exposed glycans [141]. It would be interesting to know if the small lesions observed for Ad in addition to the exocytosis of lysosomes have a similar function and also trigger an ESCRT-response similar to the one described above [127,128]. In this case, capsid-encoded late domains, such as the PPxY motif in protein

VI or perhaps a second motif present in the viral penton protein [142], may exert additional control functions upstream of viral restriction of Gal8-mediated antiviral autophagy in response to endosome damage [3]. More generally, if the ESCRT-response directed against small lesions in endosomes or lysosomes is a cell intrinsic response, several other viruses would have to face and counteract such a response.

2.6. Inflammation to Signal Membrane Damage

Membrane damage, especially in the endo-lysosomal compartment, can be repaired or removed. However, it also enables leakage of soluble contents from the endocytic compartments into the cytosol including protons, ROS, calcium ions, and soluble acid hydrolases such as Cathepsin B. This leakage eventually damages mitochondria further increasing ROS production, which can trigger inflammasome activation [143,144] (Figure 4). Inflammasomes are important cytosolic signaling platform, mediating caspase-1 activation and the secretion of pro-inflammatory cytokines including IL-1β and IL-18 [145]. The best-studied inflammasome is the Nod-like receptor family, pyrin domain-containing 3 (NLRP3)-inflammasome, which can be activated by multiple stimuli, including membrane damage (Figure 4). Its activation can result in cell death by pyroptosis unless cell survival pathways, such as autophagy are activated [146,147]. Not surprisingly, pro-inflammatory signaling and autophagy are an important part of the cellular stress response and are partially overlapping (summarized in [148]).

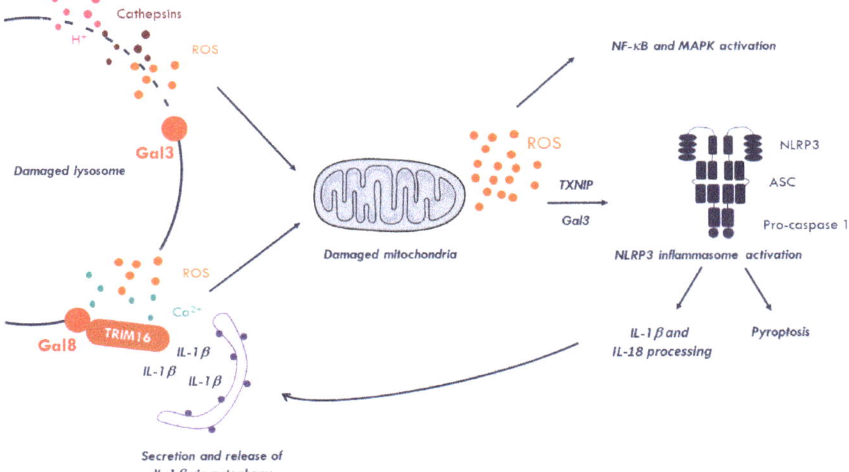

Figure 4. Membrane damage signaling. Endolysosomal rupture triggers the leakage of harmful components (e.g., ROS, H^+, cathepsins or Ca^{2+}), resulting in mitochondrial damage and increased ROS production. Increasing levels of ROS activate nuclear factor-kappa B (NF-κB) and mitogen-activated protein kinase (MAPK) signaling and triggers the release of the thioredoxin-interacting protein (TXNIP) from thioredoxin. Soluble TXNIP mediates Nod-like receptor family, pyrin domain-containing 3 (NLRP3)-inflammasomes activation and association with their adapters (ASC), thereby triggering pro-caspase conversion [149,150]. Activated inflammasomes process pro-inflammatory cytokines (IL-1β and IL-18) and may trigger cell death through pyroptosis. Processed IL-1β in turn can be recruited by Gal8 and TRIM16, that coordinate the autophagic machinery to secrete IL-1β via an unconventional secretory pathway [105,151]. Abbreviations: Gal3, galectin3; Gal8, galectin8; ROS, reactive oxygen species.

Recognizing invading pathogens involves pathogen-associated molecular patterns (PAMPs) intrinsic to the invading microbes and damage-associated molecular patterns (DAMPs). DAMPs are best described as out-of-place detection of cell components due to pathogens or stress. The understanding of pro-inflammatory signaling responding to pathogen-induced membrane damage is limited, partly because it is difficult to discriminate membrane damage signaling from microbial PAMP-activated signals. The latter trigger pathogen recognition receptors including lumenal Toll-like receptors (TLRs), cytosolic nucleic acid sensing receptors such as Cyclic GMP-AMP synthase/Stimulator of interferon genes protein (cGAS/STING), retinoic acid-inducible gene I (RIG-I) like receptors (RLRs) and Nod-like receptors (NLRs). The different receptors exert their pathogen sensing role in association with the endocytic entry compartment [152]. To complicate matters even further, different sensing pathways can converge in overlapping responses such as IFN or pro-inflammatory cytokine expression [153]. As discussed above, cytosolic glycan exposure is considered an important and unique DAMP for membrane damage. Thus, inflammatory responses upon non-microbial membrane damage provides some insight into cause and consequences and can be studied e.g., in the context of neuro-degenerative diseases and neuroinflammation [154]. Some, neurodegenerative diseases are caused by the formation of pathogenic protein aggregates. These include amyloid-β or tau in Alzheimer's disease, α-synuclein in Parkinson disease or aberrant polyglutamine stretches in mutated huntingtin in Huntington's disease (reviewed in [155]). During cell-to-cell transmission, aggregates reach the cytosol by damaging the endocytic compartment, which results in Gal3 accumulation on endosomes but mostly lysosomes [80]. Furthermore, α-synuclein mediated lysosome rupture produces ROS e.g., due to mitochondrial damage upon cathepsin release [79,149]. Unbalanced ROS accumulating in the cytosol can trigger the inflammasome, but also other inflammatory pathways including the nuclear factor-kappa B (NF-κB) pathway ([156,157], reviewed in [158]). Thus, it is important to understand which pathway is activated upon membrane breach. A recent study using several Huntington's disease models showed a role for Gal3 in the regulation of the inflammatory response towards membrane damage caused by the huntingtin mutant. Pharmacological inhibition or depletion of Gal3 in respective microglia cells or a mouse model attenuated pro-inflammatory cytokines including interleukins (IL-1β, IL-6) and TNFα. This reduction correlated with reduced NLRP3 inflammasome levels suggesting a direct regulation via Gal3 [159]. Mechanistically, the authors discuss a role for Gal3 in inhibiting the clearing of damaged lysosomes and a possible, Gal3-mediated crosstalk with the NF-κB pathway [160]. Based on these results, it is possible that Gal3 plays part in maintaining the inflammatory signaling upon membrane damage by keeping the inflammasome activated until overruled by autophagy (see above). Other Gals, such as Gal8/9, also contribute to inflammation upon membrane damage. A recent study showed that Gal8 recruited to damaged lysosomes binds to TRIM16, which sequesters processed IL-1β (Figure 4). TRIM16 coordinates autophagosome assembly and transfers IL-1β to secretory autophagic vesicles to enhance IL-1β secretion [105,145]. If Gal3/Trim16 has a similar role is not known. These examples indicate that Gals may have a dual role by regulating inflammatory responses on top off membrane damage control.

2.7. Inflammation upon Virus Membrane Penetration

Detecting viral nucleic acids such as double stranded RNA or cytosolic DNA is crucial for cells to sense invading viruses. Toll-like receptors are the first line of sensors (10 in humans) and can detect different danger signals including nucleic acids (DNA and RNA) [161]. They localize either to the plasma membrane (TLR1/2/4/5/6/10) or inside the endosomal compartment (TLR3/7/8/9) [162]. Pathogen-associated molecular pattern detection induces a change in TLR conformation, allowing interaction with adapter molecules, such as Myeloid differentiation primary response protein (Myd88) or TIR Domain-Containing Adapter-Inducing Beta Interferon (TRIF), which activate in turn IRF transcription factors and induce IFN-I production. The cytosolic presence of short double-stranded RNA with a 5′ triphosphate end activates RIG-I [163,164], while long double-stranded RNA activates the Melanoma differentiation-associated protein 5 (MDA5) [165]. These two RLRs recruit

MAVS adapter molecules located on the mitochondrial surface were they activate signaling kinases inhibitor of nuclear factor kappa-B kinase subunit alpha (IKKα) and TBK1 [166,167]. Both kinases phosphorylate transcription factors IRF3/7 and NF-κB, which drive expression of inflammatory genes and interferons [167,168]. Reovirus and PV, with their RNA genomes, were shown to activate the RIG-I/MDA5 pathway although probably not during entry [169–173]. In addition, the encephalomyocarditis virus viroporin 2B was shown to activate the NLRP3 inflammasome and to trigger IL-1β release [174,175]. Other PV like human rhinovirus (HRV) or expression of the HRV 2B protein also activated the NLRP3 inflammasome. However, because UV-inactivation of PV strongly decreased inflammasome activation it is unlikely that the PV membrane penetration step plays a dominant role [176]. Moreover, cGAS is a ubiquitous sensor of cytosolic DNA, which recognizes nucleic acids e.g., from entering DNA viruses. In the presence of DNA, it catalyzes the local formation of cGAMP, which binds to the ER associated adapter protein STING. This induces a conformational change in STING, which will then translocate to the Golgi where it serves as phosphorylation platform by recruiting TBK1 and its IRF3 substrate, again resulting in expression of pro-inflammatory cytokines and IFN. The Ad *ts1* mutant was instrumental in identifying inflammatory signaling linked to Ad membrane penetration. As discussed above, Ad *ts1* is endocytosed, but does not rupture the endosome. Wild type virus (but not *ts1*) induces a pro-inflammatory response upon cell entry, characterized by increasing the activity of p38 and Extracellular signal-regulated kinase (ERK) [177]. Likewise, rupture of the endosomal membrane with high doses of wild type Ad in murine macrophages induced the expression of IFN via activation of TBK1/IRF3 in a STING dependent manner [178,179]. Immune-complexed Ad in monocyte-derived DC (MoDC) was shown to release protein VI and to accumulate Gal3 indicative of membrane rupture and to activate the absent in melanoma 2 (AIM2) inflammasome [180]. Using a macrophage model, it was shown that high doses of Ad leads to the rapid production of ROS, release of cathepsins into the cytosol and mitochondrial damage. This resulted in secretion of IL-1β following activation of the inflammasome NLRP3. In their studies neither the Ad-*ts1* mutant nor reovirus was capable of inducing this kind of response [181–183]. The data show that inflammasome activation by Ads requires membrane damage and cathepsin release. However, endosome rupture and cytosolic accumulation also exposes the viral DNA to cGAS/STING and RIG-I detection triggering additional immune responses [184–186]. The immune activation mechanisms induced by Ad membrane rupture are therefore not easy to dissect and it remains to be established if there is signaling hierarchy or cooperativity. Further evidence of membrane modulation in the direct activation of the antiviral response have also been found in enveloped viruses discussed elsewhere [187,188].

3. Crosstalk Between Membrane Damage, Autophagy, and Inflammatory Response

As discussed in the previous sections, membrane damage induces autophagy and inflammation. It is probably fair to assume that autophagy and inflammation elicit some kind of retro-control on each other. Interestingly, several of the inflammatory pathways that are activated upon membrane damage or upon virus entry also result in a net activation of autophagy. Not surprisingly, autophagy was suggested to be involved in regulating and limiting innate and adaptive immunity (reviewed in [189]). An early indication of the direct link between autophagy and inflammation was shown by studying Crohn's disease in ATG16L1 deficient mice [190]. These mice had high levels of IL-1β and IL-18, which are indicative of elevated inflammasome activation. Similarly, IL-1β production was increased in macrophages treated with the autophagy inhibitor 3-MA or macrophages deficient for several ATGs [190,191]. The absence of ATG7 [192] or ATG5 [193,194] also caused an increase in IL-1β production in murine (alveolar) macrophages and induced pyroptosis showing how important autophagy is in limiting inflammation at least in immune cells. Mechanistically autophagy either removes inflammasome activators, inflammasome substrates or removes the activated inflammasome itself. For example, autophagy selectively degrades damaged lysosomes (lysophagy) to remove a source of cytosolic cathepsins, which can cause depolarization of mitochondria (Figure 4). Damaged mitochondria produce large amounts of ROS or oxidized mitochondrial DNA, both strong inflammasome activators. The selective removal of damaged mitochondria (mitophagy), thus, further restricts inflammasome activity [143,194]. Like lysophagy, mitophagy is a process of selective autophagy and uses selective autophagy receptors such as p62 [195] or Fanconi anemia complementation group C (FANCC) [196]. These receptors are recruited to damaged mitochondria by virtue of ubiquitin tags placed by the ubiquitin ligase Parkin [197] described in detail elsewhere [198]. Autophagy also sequesters and removes the immature form of pro-inflammatory IL-1β, an inflammasome substrate [199]. Autophagy also directly degrades activated inflammasomes. For example, the AIM2 inflammasome is ubiquitylated via the TRIM11 E3 ligase upon activation. The SLR, p62 recognizes the ubiquitin signal on AIM2 and mediates selective autophagy degradation [200,201]. A different TRIM, TRIM20, was shown to ubiquitylate pro-caspase 1 as well as the NLRP1/3 receptor [202], suggesting a more universal degradation mechanism of activated immune regulators by TRIMs (reviewed in [203]). In contrast, several inflammasome receptors interact with Beclin-1 and block the induction of autophagy [204]. This observation suggests that NLR receptors may have reciprocal regulatory functions through direct interactions with ATGs controlling autophagy and inflammasome activation [204]. Autophagic processes are not systematically anti-inflammatory. In macrophages and probably other immune cells, Gal8/TRIM16 sequestration of mature IL-1β to autophagic membranes regulates unconventional secretion of IL-1β [205,206]. Consequently, inhibiting autophagy decreased inflammation in response to activators of AIM2 and NLRP3 inflammasomes in some studies [205,207].

Inflammasomes are not the only signaling platforms degraded by autophagy. Virus sensing in the endosome via TLR, cytosolic RIG-I/MDA5 sensing of viral RNAs or cGAS/STING sensing of viral DNAs all assemble specific signaling platforms that activate interferon and/or proinflammatory cytokine expression. Whether some (or all) of these pathogen sensing and signaling platforms are linked to the surveillance of membrane integrity is not clear. It would make sense for cells to aim to understand what penetrates the membrane and establish some mechanistic link between membrane damage detection, sensing, signaling, and removal by integrating autophagic and immunity signaling. This idea is supported by several observations. Autophagy can be induced following the activation of certain TLRs [208,209]. For example, TLR7 and 3 (involved in innate responses to viruses) induce autophagy in vitro after their activation by single-stranded and double-stranded RNA present in viral genomes [210,211]. The precise mechanism is not known, but involves recruitment of Beclin-1 [211]. Autophagy in turn limits TLR signaling by degrading several mediators such as IKK/TBK1 and IRF3/7 (see below, [203]). ATGs (and autophagy) are also involved in preventing unprovoked RIG-I pathway activation by blocking the interaction between RIG-I and mitochondrial antiviral-signaling protein (MAVS). For example, ATG12-ATG5 inhibits RIG-I and MAVS by binding to their caspase activation and recruitment (CARD) domain [212]. A similar function was shown for Beclin-1 blocking RIG-I/MAVS also by binding the CARD domain of MAVS [213]. In contrast, once activated, RIG-I/MAVS trigger the production of IFN and the expression of interferon inducible genes (ISGs). Two ISGs, tetherin and ISG-15, provide a negative feedback loop targeting respectively MAVS and RIG-I for selective autophagy [213–215]. Autophagic restriction of cGAS/STING works in a similar fashion. ATG9 prevents activation of TBK1 by controlling STING [216], while Beclin-1 binds cGAS to prevent the production of cGAMP and thereby IFN production [217]. In addition, at steady state, cGAS is rapidly turned over via K48 ubiquitylation, recognized by p62 and targeted to selective autophagy [218]. This constant turnover is counteracted upon cGAS stimulation by the ISG TRIM14 [218]. STING can also be ubiquitylated presumably by TRIM56 [219] resulting, like for cGAS, in degradation via p62 and selective autophagy [220]. Both pathways, RIG-I/MDA5 and cGAS/STING recruit kinases TBK1 and IKK respectively to phosphorylate transcription factor IRF3/7. Like the different signaling platforms (including the inflammasome), the kinases and transcription factors substrates are also subject to autophagic control. TRIM21 is capable of recruiting the autophagic machinery and initiating de novo autophagy to target IKKβ [221], while TBK1 is degraded via a TRIM27 mediated process [222]. In addition, TRIM21 controls autophagy degradation of active IRF3 dimers [202]. This by far non-exhaustive list of examples shows how intricate and tight innate immunity and the autophagic degradation machinery work together to exacerbate and restrict the inflammatory response.

4. Concluding Remarks: What Is in It for the Virus?

Our understanding of the cellular membrane damage response has much increased over the years. Most of our knowledge stems from membrane damage caused by invasive bacteria or by mechanical disruption through protein aggregates or lysosomal damage. With the study of virus-induced membrane damage, the field grows further and it appears that the cell response is conserved and linked to pro-inflammatory signaling.

Whatever the trigger, cells seem to be able to determine the size and extend of membrane damage. The first membrane damage response "repair me if you can" involves the ESCRT machinery, while larger and sustained damage then triggers autophagy and probably pro-inflammatory signaling. As summarized in this review, Gals, SLRs, ATGs, and TRIMs are four interconnected protein families at the heart of the cellular response to membrane damage, regulating autophagy and controlling the ensuing inflammatory response [148,203]. Galectins detect the membrane damage and guide the ATGs via SLRs to start the autophagic process. So far, only one TRIM, TRIM16, was demonstrated to directly participate in the cell response to membrane damage driven by Gals and autophagy. However, more than half of > 60 TRIMs screened in two independent approaches were found to be positive regulators of autophagy ([223,224]) or virophagy (e.g., TRIM21/23/41) ([223]). In addition, a plethora of articles show that several TRIMs participate in innate immunity, either by regulating the TLRs (e.g., TRIM8/30a/31/32/38/56) and/or the nucleic acid sensors RIG-1/MDA5 and cGAS/STING (e.g., TRIM13/14/25/29/31/32/38/56) or by directly targeting viruses (TRIM 5α/21/22). The respective TRIM targets are almost always subjected to autophagic degradation (reviewed in [203] and references there in). Interestingly, a variety of the TRIMs involved in immune or autophagy regulation were also shown to bind Gal3/8 in a pull down assay (TRIM5α/6/17/20/22/23) [104]. To date, it is not known if Gal binding and autophagy or immune regulation by these TRIMs functionally bridges membrane damage recognition with immune regulation and/or pathogen sensing.

Very few data exist today concerning viral membrane damage especially elicited by non-enveloped viruses during cell entry. One particularity of virus entry is that unlike bacteria, entering viruses are unable to express and use effector molecules against the cell defenses until their genomes are delivered and expressed. Consequently, viral capsid components are the only available means to counteract the cell response. The two examples discussed in this review, Ad and PV, have in common that they stall the membrane damage response by delaying autophagy until they delivered either the capsid (Ad) or their genome (PV) to the cytosol. Moreover, both viruses induce a cellular membrane damage response as shown by using a mutant virus or depletion studies suggesting some advantage connected to the cell response during virus entry. This is an important observation because membrane damage or antiviral autophagy can be deleterious for the virus. For example, following uptake into human macrophages, Ad particles opsonized with antibodies were shown to rupture lysosomes and to trigger inflammasomes, while cytosolic TRIM21 would recognize any opsonized virus that made it to the cytosol and hand it over to cGAS/STING sensing [225]. As discussed above, inflammasome and cGAS/STING are down regulated via autophagy. Likewise, Ad-M1 mutant viruses are unable to stall the membrane damage response upon entry because they lack an essential PPxY peptide motif. As end result, Ad-M1 is degraded via autophagy and presented much more efficiently to MHC class II than wild type Ad in a murine model [3]. Furthermore, during the cell response elicited by bacteria, TBK1 phosphorylates SLRs independent of its IFN promoting role via IRF3 phosphorylation [122]. Without delaying autophagy, PV genomes are degraded with their capsids [45]. In contrast, it is well known that PV exploit autophagy to create a membrane compartment for cytosolic genome replication (reviewed in [226]). Delayed autophagy activation upon genome translocation therefore may facilitate the onset of PV replication. To put this into a more general perspective, it appears possible that non-enveloped viruses actively use the cellular membrane damage response during entry to limit the cell intrinsic immune response (Figure 5). Some viruses may have evolved to escape the endosome using openings that are small enough to be repaired inflicting only local and transient damage without significant immune activation. The opposite scenario with activating autophagy through extended membrane damage and controlling it (like Ad and PV), thus may come as a trade-off. On one hand, it allows endosomal escape and the exploitation of autophagic membranes for replication while autophagy maybe used to accelerate the turnover of molecules involved in immune sensing and pro-inflammatory signaling. Finding the right balance through selecting, stalling, and timing the cell response might be a powerful compensation for not being able to express effector molecules. While this is an exciting scenario that merits further investigation, experimental evidence for a direct link between virus induced membrane damage response and the (autophagic) control of the inflammatory response is currently lacking. The possibility that several additional TRIMs bind Gals offers the perspective that many more (TRIM) substrates could be recruited to membrane damage sites. The purpose of eliciting membrane damage for entering viruses could then be to remove as many cell effector molecules as possible and blunt the cells response to entry.

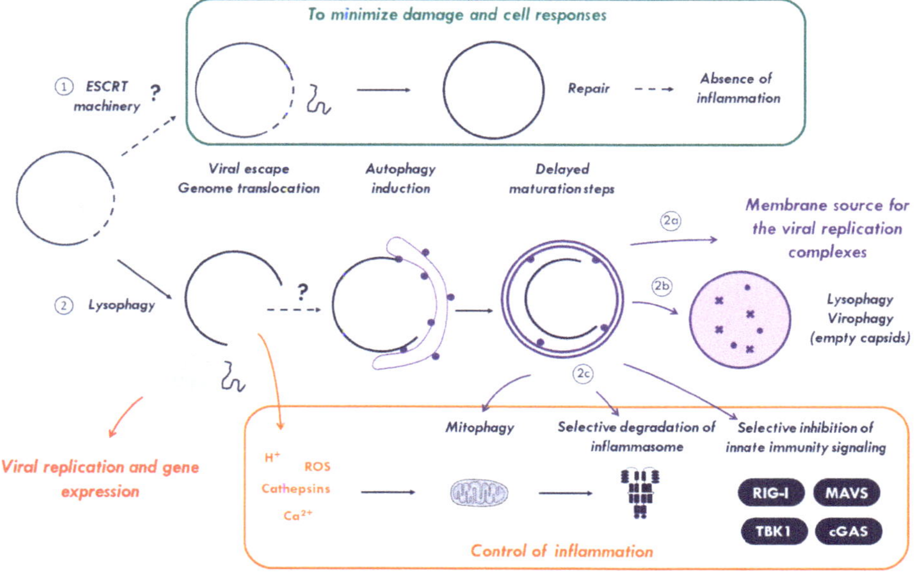

Figure 5. Model for non-enveloped virus membrane penetration. Non-enveloped viruses breach endo-lysosomal membranes for particle escape into the cytosol or cytosolic genome translocation. (1) If the membrane damage is small enough, Ca^{2+}-dependent recruitment of the ESCRT machinery will repair the endosomes limiting time and extend of the damage. This strategy could provide enough time for the virus/genome to reach the cytosol without activating the inflammatory pathways. (2) If this repair fails or the inflicted damage is too large, "eat-me signals" (galectins and ubiquitin) will accumulate at the damage site and initiate autophagy. Viruses use viral factors (e.g., Ad) or cellular factors (e.g., PV) to delay the autophagy response until capsids/viral genomes reach the safety of the cytosol. The ensuing autophagy response than may continue and work in favor of the virus. Activating autophagy via membrane damage could (2a) provide a membranes source for viral replication, and (2b) remove and recycle membrane remnants, damaged lysosomes, and empty viral capsids to avoid excessive cells responses. Additionally, autophagy could (2c) limit excessive inflammation through autophagic degradation of damaged mitochondria and by removing activated inflammasomes and effectors of the innate immunity signaling pathways (see text for details).

Taken together, there remains much to be discovered concerning our understanding how viruses penetrate cell membranes and how cells respond to this insult. Using the few existing examples presented in this review about the cell response to virus inflicted membrane damage, we invite the reader to reflect on the wider role of membrane damage for the virus.

Funding: This research was funded by Agence National de la recherché (ANR), project VirMEDACO grant number ANR-19-CE15-0013-01. H.W. is an INSERM fellow.

Acknowledgments: We acknowledge help from S.C. for manuscript corrections and the members of our team for helpful discussion.

Conflicts of Interest: The authors declare no conflict of interest. The funders had no role in the design of the study; in the collection, analyses, or interpretation of data; in the writing of the manuscript, or in the decision to publish the results.

References

1. Lombard, J. Once upon a time the cell membranes: 175 years of cell boundary research. *Biol. Direct* **2014**, *9*, 1–35. [CrossRef]
2. Thurston, T.L.M.M.; Wandel, M.P.; Von Muhlinen, N.; Foeglein, Á.; Randow, F.; Foeglein, A.; Randow, F.; Foeglein, Á.; Randow, F. Galectin 8 targets damaged vesicles for autophagy to defend cells against bacterial invasion. *Nature* **2012**, *482*, 414–418. [CrossRef] [PubMed]
3. Montespan, C.; Marvin, S.A.; Austin, S.; Burrage, A.M.; Roger, B.; Rayne, F.; Faure, M.; Campell, E.M.; Schneider, C.; Reimer, R.; et al. Multi-layered control of Galectin-8 mediated autophagy during adenovirus cell entry through a conserved PPxY motif in the viral capsid. *PLoS Pathog.* **2017**, *13*. [CrossRef] [PubMed]
4. Jia, J.; Abudu, Y.P.; Claude-Taupin, A.; Gu, Y.; Kumar, S.; Choi, S.W.; Peters, R.; Mudd, M.H.; Allers, L.; Salemi, M.; et al. Galectins Control mTOR in Response to Endomembrane Damage. *Mol. Cell* **2018**, *70*, 120–135.e8. [CrossRef] [PubMed]
5. Vietri, M.; Radulovic, M.; Stenmark, H. The many functions of ESCRTs. *Nat. Rev. Mol. Cell Biol.* **2020**, *21*, 25–42. [CrossRef]
6. Lozach, P.Y.; Huotari, J.; Helenius, A. Late-penetrating viruses. *Curr. Opin. Virol.* **2011**, *1*, 35–43. [CrossRef] [PubMed]
7. Ketter, E.; Randall, G. Virus Impact on Lipids and Membranes. *Annu. Rev. Virol.* **2019**, *6*, 319–340. [CrossRef]
8. Moyer, C.L.; Nemerow, G.R. Viral weapons of membrane destruction: Variable modes of membrane penetration by non-enveloped viruses. *Curr. Opin. Virol.* **2011**, *1*, 44–49. [CrossRef]
9. Stewart, P.L.; Dermody, T.S.; Nemerow, G.R. Structural basis of nonenveloped virus cell entry. *Adv. Protein Chem.* **2003**, *64*, 455–491. [CrossRef]
10. Prchla, E.; Plank, C.; Wagner, E.; Blaas, D.; Fuchs, R. Virus-mediated release of endosomal content in vitro: Different behavior of adenovirus and rhinovirus serotype 2. *J. Cell Biol.* **1995**, *131*, 111–123. [CrossRef]
11. Farr, G.A.; Zhang, L.G.; Tattersall, P. Parvoviral virions deploy a capsid-tethered lipolytic enzyme to breach the endosomal membrane during cell entry. *Proc. Natl. Acad. Sci. USA* **2005**, *102*, 17148–17153. [CrossRef] [PubMed]
12. Greber, U.F.; Webster, P.; Weber, J.; Helenius, A. The role of the adenovirus protease in virus entry into cells. *EMBO J.* **1996**, *15*, 1766–1777. [CrossRef]
13. Wiethoff, C.M.; Wodrich, H.; Gerace, L.; Nemerow, G.R. Adenovirus Protein VI Mediates Membrane Disruption following Capsid Disassembly. *J. Virol.* **2005**, *79*, 1992–2000. [CrossRef]
14. Martinez, R.; Schellenberger, P.; Vasishtan, D.; Aknin, C.; Austin, S.; Dacheux, D.; Rayne, F.; Siebert, A.; Ruzsics, Z.; Gruenewald, K.; et al. The amphipathic helix of adenovirus capsid protein VI contributes to penton release and postentry sorting. *J. Virol.* **2015**, *89*, 2121–2135. [CrossRef] [PubMed]
15. Weber, J. Genetic analysis of adenovirus type 2 III. Temperature sensitivity of processing viral proteins. *J. Virol.* **1976**, *17*, 462–471. [CrossRef] [PubMed]
16. Cotten, M.; Weber, J.M. The adenovirus protease is required for virus entry into host cells. *Virology* **1995**, *213*, 494–502. [CrossRef]
17. Martín, C.S. Latest insights on adenovirus structure and assembly. *Viruses* **2012**, *4*, 847–877. [CrossRef]
18. Wodrich, H.; Henaff, D.; Jammart, B.; Segura-Morales, C.; Seelmeir, S.; Coux, O.; Ruzsics, Z.; Wiethoff, C.M.; Kremer, E.J. A capsid-encoded PPxY-motif facilitates adenovirus entry. *PLoS Pathog.* **2010**, *6*. [CrossRef]

19. Luisoni, S.; Suomalainen, M.; Boucke, K.; Tanner, L.B.; Wenk, M.R.; Guan, X.L.; Grzybek, M.; Coskun, U.; Greber, U.F. Co-option of membrane wounding enables virus penetration into cells. *Cell Host Microbe* **2015**, *18*, 75–85. [CrossRef]
20. Wiethoff, C.M.; Nemerow, G.R. Adenovirus membrane penetration: Tickling the tail of a sleeping dragon. *Virology* **2015**, *479–480*, 591–599. [CrossRef]
21. Hernando-Pérez, M.; Martín-González, N.; Pérez-Illana, M.; Suomalainen, M.; Condezo, G.N.; Ostapchuk, P.; Gallardo, J.; Menéndez, M.; Greber, U.F.; Hearing, P.; et al. Dynamic competition for hexon binding between core protein VII and lytic protein VI promotes adenovirus maturation and entry. *Proc. Natl. Acad. Sci. USA* **2020**, *117*, 13699–13707. [CrossRef] [PubMed]
22. Maier, O.; Galan, D.L.; Wodrich, H.; Wiethoff, C.M. An N-terminal domain of adenovirus protein VI fragments membranes by inducing positive membrane curvature. *Virology* **2010**, *402*, 11–19. [CrossRef] [PubMed]
23. Moyer, C.L.; Wiethoff, C.M.; Maier, O.; Smith, J.G.; Nemerow, G.R. Functional Genetic and Biophysical Analyses of Membrane Disruption by Human Adenovirus. *J. Virol.* **2011**, *85*, 2631–2641. [CrossRef] [PubMed]
24. Pied, N.; Wodrich, H. Imaging the adenovirus infection cycle. *FEBS Lett.* **2019**, *593*, 3419–3448. [CrossRef]
25. Dupzyk, A.; Tsai, B. How polyomaviruses exploit the ERAD machinery to cause infection. *Viruses* **2016**, *8*, 242. [CrossRef]
26. Magnuson, B.; Rainey, E.K.; Benjamin, T.; Baryshev, M.; Mkrtchian, S.; Tsai, B. ERp29 triggers a conformational change in polyomavirus to stimulate membrane binding. *Mol. Cell* **2005**, *20*, 289–300. [CrossRef]
27. Guerrero, C.A.; Bouyssounade, D.; Zárate, S.; Iša, P.; López, T.; Espinosa, R.; Romero, P.; Méndez, E.; López, S.; Arias, C.F. Heat Shock Cognate Protein 70 Is Involved in Rotavirus Cell Entry. *J. Virol.* **2002**, *76*, 4096–4102. [CrossRef]
28. Walczak, C.P.; Ravindran, M.S.; Inoue, T.; Tsai, B. A Cytosolic Chaperone Complexes with Dynamic Membrane J-Proteins and Mobilizes a Nonenveloped Virus out of the Endoplasmic Reticulum. *PLoS Pathog.* **2014**, *10*, e1004007. [CrossRef]
29. Inoue, T.; Tsai, B. A large and intact viral particle penetrates the endoplasmic reticulum membrane to reach the cytosol. *PLoS Pathog.* **2011**, *7*, 1002037. [CrossRef]
30. Spriggs, C.C.; Badieyan, S.; Verhey, K.J.; Cianfrocco, M.A.; Tsai, B. Golgi-associated BICD adaptors couple ER membrane penetration and disassembly of a viral cargo. *J. Cell Biol.* **2020**, *219*. [CrossRef]
31. Suikkanen, S.; Antila, M.; Jaatinen, A.; Vihinen-Ranta, M.; Vuento, M. Release of canine parvovirus from endocytic vesicles. *Virology* **2003**, *316*, 267–280. [CrossRef] [PubMed]
32. Canaan, S.; Zádori, Z.; Ghomashchi, F.; Bollinger, J.; Sadilek, M.; Moreau, M.E.; Tijssen, P.; Gelb, M.H. Interfacial Enzymology of Parvovirus Phospholipases A2. *J. Biol. Chem.* **2004**, *279*, 14502–14508. [CrossRef] [PubMed]
33. Dorsch, S.; Liebisch, G.; Kaufmann, B.; von Landenberg, P.; Hoffmann, J.H.; Drobnik, W.; Modrow, S. The VP1 Unique Region of Parvovirus B19 and Its Constituent Phospholipase A2-Like Activity. *J. Virol.* **2002**, *76*, 2014–2018. [CrossRef]
34. Ebert, D.H.; Deussing, J.; Peters, C.; Dermody, T.S. Cathepsin L and cathepsin B mediate reovirus disassembly in murine fibroblast cells. *J. Biol. Chem.* **2002**, *277*, 24609–24617. [CrossRef] [PubMed]
35. Chandran, K.; Farsetta, D.L.; Nibert, M.L. Strategy for Nonenveloped Virus Entry: A Hydrophobic Conformer of the Reovirus Membrane Penetration Protein μ1 Mediates Membrane Disruption. *J. Virol.* **2002**, *76*, 9920–9933. [CrossRef] [PubMed]
36. Odegard, A.L.; Chandran, K.; Zhang, X.; Parker, J.S.L.; Baker, T.S.; Nibert, M.L. Putative Autocleavage of Outer Capsid Protein μ1, Allowing Release of Myristoylated Peptide μ1N during Particle Uncoating, Is Critical for Cell Entry by Reovirus. *J. Virol.* **2004**, *78*, 8732–8745. [CrossRef] [PubMed]
37. Liemann, S.; Chandran, K.; Baker, T.S.; Nibert, M.L.; Harrison, S.C. Structure of the reovirus membrane-penetration protein, μ1, in a complex with its protector protein, σ3. *Cell* **2002**, *108*, 283–295. [CrossRef]
38. Zhang, L.; Agosto, M.A.; Ivanovic, T.; King, D.S.; Nibert, M.L.; Harrison, S.C. Requirements for the Formation of Membrane Pores by the Reovirus Myristoylated μ1N Peptide. *J. Virol.* **2009**, *83*, 7004–7014. [CrossRef]
39. Agosto, M.A.; Ivanovic, T.; Nibert, M.L. Mammalian reovirus, a nonfusogenic nonenveloped virus, forms size-selective pores in a model membrane. *Proc. Natl. Acad. Sci. USA* **2006**, *103*, 16496–16501. [CrossRef]

40. Ivanovic, T.; Agosto, M.A.; Zhang, L.; Chandran, K.; Harrison, S.C.; Nibert, M.L. Peptides released from reovirus outer capsid form membrane pores that recruit virus particles. *EMBO J.* **2008**, *27*, 1289–1298. [CrossRef]
41. Tuthill, T.J.; Groppelli, E.; Hogle, J.M.; Rowlands, D.J. Picornaviruses. In *Current Topics in Microbiology and Immunology*; Springer: Berlin/Heidelberg, Germany, 2010; Volume 343, pp. 43–89.
42. Schober, D.; Kronenberger, P.; Prchla, E.; Blaas, D.; Fuchs, R. Major and Minor Receptor Group Human Rhinoviruses Penetrate from Endosomes by Different Mechanisms. *J. Virol.* **1998**, *72*, 1354–1364. [CrossRef] [PubMed]
43. Panjwani, A.; Strauss, M.; Gold, S.; Wenham, H.; Jackson, T.; Chou, J.J.; Rowlands, D.J.; Stonehouse, N.J.; Hogle, J.M.; Tuthill, T.J. Capsid Protein VP4 of Human Rhinovirus Induces Membrane Permeability by the Formation of a Size-Selective Multimeric Pore. *PLoS Pathog.* **2014**, *10*, e1004294. [CrossRef] [PubMed]
44. Staring, J.; Raaben, M.; Brummelkamp, T.R. Viral escape from endosomes and host detection at a glance. *J. Cell Sci.* **2018**, *131*. [CrossRef] [PubMed]
45. Staring, J.; Von Castelmur, E.; Blomen, V.A.; Van Den Hengel, L.G.; Brockmann, M.; Baggen, J.; Thibaut, H.J.; Nieuwenhuis, J.; Janssen, H.; Van Kuppeveld, F.J.M.M.; et al. PLA2G16 represents a switch between entry and clearance of Picornaviridae. *Nature* **2017**, *541*, 412–416. [CrossRef] [PubMed]
46. Hughes, P.J.; Stanway, G. The 2A proteins of three diverse picornaviruses are related to each other and to the H-rev107 family of proteins involved in the control of cell proliferation. *J. Gen. Virol.* **2000**, *81*, 201–207. [CrossRef] [PubMed]
47. Knodler, L.A.; Vallance, B.A.; Celli, J.; Winfree, S.; Hansen, B.; Montero, M.; Steele-Mortimer, O. Dissemination of invasive Salmonella via bacterial-induced extrusion of mucosal epithelia. *Proc. Natl. Acad. Sci. USA* **2010**, *107*, 17733–17738. [CrossRef]
48. Mellouk, N.; Enninga, J. Cytosolic access of intracellular bacterial pathogens: The Shigella paradigm. *Front. Cell. Infect. Microbiol.* **2016**, *6*, 1–8. [CrossRef]
49. Mellouk, N.; Weiner, A.; Aulner, N.; Schmitt, C.; Elbaum, M.; Shorte, S.L.; Danckaert, A.; Enninga, J. Shigella subverts the host recycling compartment to rupture its vacuole. *Cell Host Microbe* **2014**, *16*, 517–530. [CrossRef]
50. Weiner, A.; Mellouk, N.; Lopez-Montero, N.; Chang, Y.-Y.Y.; Souque, C.; Schmitt, C.; Enninga, J. Macropinosomes are Key Players in Early Shigella Invasion and Vacuolar Escape in Epithelial Cells. *PLoS Pathog.* **2016**, *12*, e1005602. [CrossRef]
51. Birmingham, C.L.; Smith, A.C.; Bakowski, M.A.; Yoshimori, T.; Brumell, J.H. Autophagy controls Salmonella infection in response to damage to the Salmonella-containing vacuole. *J. Biol. Chem.* **2006**, *281*, 11374–11383. [CrossRef]
52. Radtke, A.L.; Delbridge, L.M.; Balachandran, S.; Barber, G.N.; O'Riordan, M.X.D. TBK1 protects vacuolar integrity during intracellular bacterial infection. *PLoS Pathog.* **2007**, *3*. [CrossRef]
53. Boyle, K.B.; Thurston, T.L.M.; Randow, F. TBK1 directs WIPI2 against Salmonella. *Autophagy* **2016**, *12*, 2508–2509. [CrossRef] [PubMed]
54. Thurston, T.L.; Boyle, K.B.; Allen, M.; Ravenhill, B.J.; Karpiyevich, M.; Bloor, S.; Kaul, A.; Noad, J.; Foeglein, A.; Matthews, S.A.; et al. Recruitment of TBK 1 to cytosol-invading Salmonella induces WIPI 2-dependent antibacterial autophagy. *EMBO J.* **2016**, *35*, 1779–1792. [CrossRef] [PubMed]
55. Santos, J.C.; Duchateau, M.; Fredlund, J.; Weiner, A.; Mallet, A.; Schmitt, C.; Matondo, M.; Hourdel, V.; Chamot-Rooke, J.; Enninga, J. The COPII complex and lysosomal VAMP7 determine intracellular Salmonella localization and growth. *Cell. Microbiol.* **2015**, *17*, 1699–1720. [CrossRef] [PubMed]
56. Singh, R.; Jamieson, A.; Cresswell, P. GILT is a critical host factor for Listeria monocytogenes infection. *Nature* **2008**, *455*, 1244–1247. [CrossRef] [PubMed]
57. Radtke, A.L.; Anderson, K.L.; Davis, M.J.; DiMagno, M.J.; Swanson, J.A.; O'Riordan, M.X. Listeria monocytogenes exploits cystic fibrosis transmembrane conductance regulator (CFTR) to escape the phagosome. *Proc. Natl. Acad. Sci. USA* **2011**, *108*, 1633–1638. [CrossRef] [PubMed]
58. Gedde, M.M.; Higgins, D.E.; Tilney, L.G.; Portnoy, D.A. Role of listeriolysin O in cell-to-cell spread of Listeria monocytogenes. *Infect. Immun.* **2000**, *68*, 999–1003. [CrossRef]
59. Alberti-Segui, C.; Goeden, K.R.; Higgins, D.E. Differential function of Listeria monocytogenes listeriolysin O and phospholipases C in vacuolar dissolution following cell-to-cell spread. *Cell. Microbiol.* **2007**, *9*, 179–195. [CrossRef]

60. Malet, J.K.; Cossart, P.; Ribet, D. Alteration of epithelial cell lysosomal integrity induced by bacterial cholesterol-dependent cytolysins. *Cell. Microbiol.* **2017**, *19*, 1–11. [CrossRef]
61. Palframan, S.L.; Kwok, T.; Gabriel, K. Vacuolating cytotoxin A (VacA), a key toxin for Helicobacter pylori pathogenesis. *Front. Cell. Infect. Microbiol.* **2012**, *2*, 92. [CrossRef]
62. Li, F.Y.; Weng, I.C.; Lin, C.H.; Kao, M.C.; Wu, M.S.; Chen, H.Y.; Liu, F.T. Helicobacter pylori induces intracellular galectin-8 aggregation around damaged lysosomes within gastric epithelial cells in a host O-glycan-dependent manner. *Glycobiology* **2018**, *29*, 151–162. [CrossRef] [PubMed]
63. van der Wel, N.; Hava, D.; Houben, D.; Fluitsma, D.; Van Zon, M.; Pierson, J.; Brenner, M.; Peters, P.J.M. tuberculosis and M. leprae Translocate from the Phagolysosome to the Cytosol in Myeloid Cells. *Cell* **2007**, *129*, 1287–1298. [CrossRef] [PubMed]
64. De Jonge, M.I.; Pehau-Arnaudet, G.; Fretz, M.M.; Romain, F.; Bottai, D.; Brodin, P.; Honoré, N.; Marchal, G.; Jiskoot, W.; England, P.; et al. ESAT-6 from Mycobacterium tuberculosis dissociates from its putative chaperone CFP-10 under acidic conditions and exhibits membrane-lysing activity. *J. Bacteriol.* **2007**, *189*, 6028–6034. [CrossRef] [PubMed]
65. Akimana, C.; Al-Khodor, S.; Kwaik, Y.A. Host factors required for modulation of phagosome biogenesis and proliferation of francisella tularensis within the cytosol. *PLoS ONE* **2010**, *5*. [CrossRef]
66. De Leon, J.; Jiang, G.; Ma, Y.; Rubin, E.; Fortune, S.; Sun, J. Mycobacterium tuberculosis ESAT-6 exhibits a unique membrane-interacting activity that is not found in its ortholog from non-pathogenic Mycobacterium smegmatis. *J. Biol. Chem.* **2012**, *287*, 44184–44191. [CrossRef]
67. Peng, X.; Sun, J. Mechanism of ESAT-6 membrane interaction and its roles in pathogenesis of Mycobacterium tuberculosis. *Toxicon* **2016**, *116*, 29–34. [CrossRef]
68. Pilatz, S.; Breitbach, K.; Hein, N.; Fehlhaber, B.; Schulze, J.; Brenneke, B.; Eberl, L.; Steinmetz, I. Identification of Burkholderia pseudomallei genes required for the intracellular life cycle and in vivo virulence. *Infect. Immun.* **2006**, *74*, 3576–3586. [CrossRef]
69. Whitworth, T.; Popov, V.L.; Yu, X.J.; Walker, D.H.; Bouyer, D.H. Expression of the Rickettsia prowazekii pld or tlyC gene in Salmonella enterica serovar typhimurium mediates phagosomal escape. *Infect. Immun.* **2005**, *73*, 6668–6673. [CrossRef]
70. Renesto, P.; Dehoux, P.; Gouin, E.; Touqui, L.; Cossart, P.; Raoult, D. Identification and Characterization of a Phospholipase D–Superfamily Gene in Rickettsiae. *J. Infect. Dis.* **2003**, *188*, 1276–1283. [CrossRef]
71. Clemens, D.L.; Lee, B.Y.; Horwitz, M.A. Virulent and avirulent strains of Francisella tularensis prevent acidification and maturation of their phagosomes and escape into the cytoplasm in human macrophages. *Infect. Immun.* **2004**, *72*, 3204–3217. [CrossRef]
72. Clemens, D.L.; Ge, P.; Lee, B.Y.; Horwitz, M.A.; Zhou, Z.H. Atomic structure of T6SS reveals interlaced array essential to function. *Cell* **2015**, *160*, 940–951. [CrossRef] [PubMed]
73. Chong, A.; Wehrly, T.D.; Nair, V.; Fischer, E.R.; Barker, J.R.; Klose, K.E.; Celli, J. The early phagosomal stage of Francisella tularensis determines optimal phagosomal escape and Francisella pathogenicity island protein expression. *Infect. Immun.* **2008**, *76*, 5488–5499. [CrossRef] [PubMed]
74. Berman, J. Morphogenesis and cell cycle progression in Candida albicans. *Curr. Opin. Microbiol.* **2006**, *9*, 595–601. [CrossRef] [PubMed]
75. Westman, J.; Moran, G.; Mogavero, S.; Hube, B.; Grinsteina, S. crossm Candida albicans Hyphal Expansion Causes Phagosomal. *MBio* **2018**, *9*, 1–14. [CrossRef]
76. Vylkova, S.; Carman, A.J.; Danhof, H.A.; Collette, J.R.; Zhou, H.; Lorenz, M.C. The fungal pathogen candida albicans autoinduces hyphal morphogenesis by raising extracellular pH. *MBio* **2011**, *2*. [CrossRef]
77. De Souza, W.; De Carvalho, T.M.U.; Barrias, E.S. Review on Trypanosoma cruzi: Host cell interaction. *Int. J. Cell Biol.* **2010**, *2010*. [CrossRef]
78. Papadopoulos, C.; Kirchner, P.; Bug, M.; Grum, D.; Koerver, L.; Schulze, N.; Poehler, R.; Dressler, A.; Fengler, S.; Arhzaouy, K.; et al. VCP/p97 cooperates with YOD 1, UBXD 1 and PLAA to drive clearance of ruptured lysosomes by autophagy. *EMBO J.* **2017**, *36*, 135–150. [CrossRef]
79. Freeman, D.; Cedillos, R.; Choyke, S.; Lukic, Z.; McGuire, K.; Marvin, S.; Burrage, A.M.; Sudholt, S.; Rana, A.; O'Connor, C.; et al. Alpha-Synuclein Induces Lysosomal Rupture and Cathepsin Dependent Reactive Oxygen Species Following Endocytosis. *PLoS ONE* **2013**, *8*. [CrossRef]

80. Flavin, W.P.; Bousset, L.; Green, Z.C.; Chu, Y.; Skarpathiotis, S.; Chaney, M.J.; Kordower, J.H.; Melki, R.; Campbell, E.M. Endocytic vesicle rupture is a conserved mechanism of cellular invasion by amyloid proteins. *Acta Neuropathol.* **2017**, *134*, 629–653. [CrossRef]
81. Villamil Giraldo, A.M.; Appelqvist, H.; Ederth, T.; Öllinger, K. Lysosomotropic agents: Impact on lysosomal membrane permeabilization and cell death. *Biochem. Soc. Trans.* **2014**, *42*, 1460–1464. [CrossRef]
82. Aits, S.; Kricker, J.; Liu, B.; Ellegaard, A.-M.M.; Hämälistö, S.; Tvingsholm, S.; Corcelle-Termeau, E.; Høgh, S.; Farkas, T.; Jonassen, A.H.; et al. Sensitive detection of lysosomal membrane permeabilization by lysosomal galectin puncta assay. *Autophagy* **2015**, *11*, 1408–1424. [CrossRef] [PubMed]
83. Uchimoto, T.; Nohara, H.; Kamehara, R.; Iwamura, M.; Watanabe, N.; Kobayashi, Y. Mechanism of apoptosis induced by a lysosomotropic agent, L-leucyl-L- leucine methyl ester. *Apoptosis* **1999**, *4*, 357–362. [CrossRef] [PubMed]
84. Du Rietz, H.; Hedlund, H.; Wilhelmson, S.; Nordenfelt, P.; Wittrup, A. Imaging small molecule-induced endosomal escape of siRNA. *Nat. Commun.* **2020**, *11*, 1809. [CrossRef] [PubMed]
85. Wittrup, A.; Ai, A.; Liu, X.; Hamar, P.; Trifonova, R.; Charisse, K.; Manoharan, M.; Kirchhausen, T.; Lieberman, J. Visualizing lipid-formulated siRNA release from endosomes and target gene knockdown. *Nat. Biotechnol.* **2015**, *33*, 870–876. [CrossRef] [PubMed]
86. Fraire, J.C.; Houthaeve, G.; Liu, J.; Raes, L.; Vermeulen, L.; Stremersch, S.; Brans, T.; García-Díaz Barriga, G.; De Keulenaer, S.; Van Nieuwerburgh, F.; et al. Vapor nanobubble is the more reliable photothermal mechanism for inducing endosomal escape of siRNA without disturbing cell homeostasis. *J. Control. Release* **2020**, *319*, 262–275. [CrossRef] [PubMed]
87. Chu, Z.; Miu, K.; Lung, P.; Zhang, S.; Zhao, S.; Chang, H.C.; Lin, G.; Li, Q. Rapid endosomal escape of prickly nanodiamonds: Implications for gene delivery. *Sci. Rep.* **2015**, *5*, 11661. [CrossRef] [PubMed]
88. Barondes, S.H.; Castronovo, V.; Cooper, D.N.W.; Cummings, R.D.; Drickamer, K.; Felzi, T.; Gitt, M.A.; Hirabayashi, J.; Hughes, C.; Kasai, K.I.; et al. Galectins: A family of animal β-galactoside-binding lectins. *Cell* **1994**, *76*, 597–598. [CrossRef]
89. Hirabayashi, J.; Kasai, K.I. The family of metazoan metal-independent β-galactoside-binding lectins: Structure, function and molecular evolution. *Glycobiology* **1993**, *3*, 297–304. [CrossRef]
90. Wang, W.H.; Lin, C.Y.; Chang, M.R.; Urbina, A.N.; Assavalapsakul, W.; Thitithanyanont, A.; Chen, Y.H.; Liu, F.T.; Wang, S.F. The role of galectins in virus infection—A systemic literature review. *J. Microbiol. Immunol. Infect.* **2019**. [CrossRef]
91. Baum, L.G.; Garner, O.B.; Schaefer, K.; Lee, B. Microbe-host interactions are positively and negatively regulated by galectin-glycan interactions. *Front. Immunol.* **2014**, *5*, 284. [CrossRef]
92. Paz, I.; Sachse, M.; Dupont, N.; Mounier, J.; Cederfur, C.; Enninga, J.; Leffler, H.; Poirier, F.; Prevost, M.C.; Lafont, F.; et al. Galectin-3, a marker for vacuole lysis by invasive pathogens. *Cell. Microbiol.* **2010**, *12*, 530–544. [CrossRef] [PubMed]
93. Maier, O.; Marvin, S.A.; Wodrich, H.; Campbell, E.M.; Wiethoff, C.M. Spatiotemporal Dynamics of Adenovirus Membrane Rupture and Endosomal Escape. *J. Virol.* **2012**, *86*, 10821–10828. [CrossRef] [PubMed]
94. Martinez, R.; Burrage, A.M.; Wiethoff, C.M.; Wodrich, H. High temporal resolution imaging reveals endosomal membrane penetration and escape of adenoviruses in real time. *Methods Mol. Biol.* **2013**, *1064*, 211–226. [CrossRef] [PubMed]
95. Klionsky, D.J.; Codogno, P. The mechanism and physiological function of macroautophagy. *J. Innate Immun.* **2013**, *5*, 427–433. [CrossRef] [PubMed]
96. Bento, C.F.; Renna, M.; Ghislat, G.; Puri, C.; Ashkenazi, A.; Vicinanza, M.; Menzies, F.M.; Rubinsztein, D.C. Mammalian Autophagy: How Does It Work? *Annu. Rev. Biochem.* **2016**, *85*, 685–713. [CrossRef]
97. Suzuki, K.; Noda, T.; Ohsumi, Y. Interrelationships among Atg proteins during autophagy inSaccharomyces cerevisiae. *Yeast* **2004**, *21*, 1057–1065. [CrossRef]
98. Klionsky, D.J.; Abdelmohsen, K.; Abe, A.; Abedin, M.J.; Abeliovich, H.; Acevedo Arozena, A.; Adachi, H.; Adams, C.M.; Adams, P.D.; Adeli, K.; et al. Guidelines for the use and interpretation of assays for monitoring autophagy (3rd edition). *Autophagy* **2016**, *12*, 1–222. [CrossRef]
99. Johansen, T.; Lamark, T. Selective autophagy mediated by autophagic adapter proteins. *Autophagy* **2011**, *7*, 279–296. [CrossRef]

100. Ravenhill, B.J.; Boyle, K.B.; von Muhlinen, N.; Ellison, C.J.; Masson, G.R.; Otten, E.G.; Foeglein, A.; Williams, R.; Randow, F. The Cargo Receptor NDP52 Initiates Selective Autophagy by Recruiting the ULK Complex to Cytosol-Invading Bacteria. *Mol. Cell* **2019**, *74*, 320–329.e6. [CrossRef]
101. Vargas, J.N.S.; Wang, C.; Bunker, E.; Hao, L.; Maric, D.; Schiavo, G.; Randow, F.; Youle, R.J. Spatiotemporal Control of ULK1 Activation by NDP52 and TBK1 during Selective Autophagy. *Mol. Cell* **2019**, *74*, 347–362.e6. [CrossRef]
102. Roberts, R.; Ktistakis, N.T. Omegasomes: PI3P platforms that manufacture autophagosomes. *Essays Biochem.* **2013**, *55*, 17–27. [CrossRef] [PubMed]
103. Tanida, I.; Ueno, T.; Kominami, E. LC3 conjugation system in mammalian autophagy. *Int. J. Biochem. Cell Biol.* **2004**, *36*, 2503–2518. [CrossRef]
104. Chauhan, S.; Kumar, S.; Jain, A.; Ponpuak, M.; Mudd, M.H.; Kimura, T.; Choi, S.W.; Peters, R.; Mandell, M.; Bruun, J.A.; et al. TRIMs and Galectins Globally Cooperate and TRIM16 and Galectin-3 Co-direct Autophagy in Endomembrane Damage Homeostasis. *Dev. Cell* **2016**, *39*, 13–27. [CrossRef] [PubMed]
105. Kimura, T.; Jia, J.; Kumar, S.; Choi, S.W.; Gu, Y.; Mudd, M.; Dupont, N.; Jiang, S.; Peters, R.; Farzam, F.; et al. Dedicated SNARE s and specialized TRIM cargo receptors mediate secretory autophagy. *EMBO J.* **2017**, *36*, 42–60. [CrossRef]
106. Weng, I.C.; Chen, H.L.; Lo, T.H.; Lin, W.H.; Chen, H.Y.; Hsu, D.K.; Liu, F.T. Cytosolic galectin-3 and -8 regulate antibacterial autophagy through differential recognition of host glycans on damaged phagosomes. *Glycobiology* **2018**, *28*, 392–405. [CrossRef] [PubMed]
107. Feeley, E.M.; Pilla-Moffett, D.M.; Zwack, E.E.; Piro, A.S.; Finethy, R.; Kolb, J.P.; Martinez, J.; Brodsky, I.E.; Coers, J. Galectin-3 directs antimicrobial guanylate binding proteins to vacuoles furnished with bacterial secretion systems. *Proc. Natl. Acad. Sci. USA* **2017**, *114*, E1698–E1706. [CrossRef] [PubMed]
108. Herhaus, L.; Dikic, I. Regulation of Salmonella-host cell interactions via the ubiquitin system. *Int. J. Med. Microbiol.* **2018**, *308*, 176–184. [CrossRef]
109. Dupont, N.; Lacas-Gervais, S.; Bertout, J.; Paz, I.; Freche, B.; Van Nhieu, G.T.; van der Goot, F.G.; Sansonetti, P.J.; Lafont, F. Shigella Phagocytic Vacuolar Membrane Remnants Participate in the Cellular Response to Pathogen Invasion and Are Regulated by Autophagy. *Cell Host Microbe* **2009**, *6*, 137–149. [CrossRef]
110. Mostowy, S.; Sancho-Shimizu, V.; Hamon, M.A.; Simeone, R.; Brosch, R.; Johansen, T.; Cossart, P. p62 and NDP52 proteins target intracytosolic Shigella and Listeria to different autophagy pathways. *J. Biol. Chem.* **2011**, *286*, 26987–26995. [CrossRef]
111. Luisoni, S.; Bauer, M.; Prasad, V.; Boucke, K.; Papadopoulos, C.; Meyer, H.; Hemmi, S.; Suomalainen, M.; Greber, U. Endosomophagy clears disrupted early endosomes but not virus particles during virus entry into cells. *Matters* **2016**, 1–9. [CrossRef]
112. Yoshida, Y.; Yasuda, S.; Fujita, T.; Hamasaki, M.; Murakami, A.; Kawawaki, J.; Iwai, K.; Saeki, Y.; Yoshimori, T.; Matsuda, N.; et al. Ubiquitination of exposed glycoproteins by SCFFBXO27 directs damaged lysosomes for autophagy. *Proc. Natl. Acad. Sci. USA* **2017**, *114*, 8574–8579. [CrossRef]
113. Fujita, N.; Morita, E.; Itoh, T.; Tanaka, A.; Nakaoka, M.; Osada, Y.; Umemoto, T.; Saitoh, T.; Nakatogawa, H.; Kobayashi, S.; et al. Recruitment of the autophagic machinery to endosomes during infection is mediated by ubiquitin. *J. Cell Biol.* **2013**, *203*, 115–128. [CrossRef] [PubMed]
114. Tan, J.M.M.; Wong, E.S.P.; Kirkpatrick, D.S.; Pletnikova, O.; Ko, H.S.; Tay, S.P.; Ho, M.W.L.; Troncoso, J.; Gygi, S.P.; Lee, M.K.; et al. Lysine 63-linked ubiquitination promotes the formation and autophagic clearance of protein inclusions associated with neurodegenerative diseases. *Hum. Mol. Genet.* **2008**, *17*, 431–439. [CrossRef] [PubMed]
115. Huett, A.; Heath, R.J.; Begun, J.; Sassi, S.O.; Baxt, L.A.; Vyas, J.M.; Goldberg, M.B.; Xavier, R.J. The LRR and RING domain protein LRSAM1 is an E3 ligase crucial for ubiquitin-dependent autophagy of intracellular salmonella typhimurium. *Cell Host Microbe* **2012**, *12*, 778–790. [CrossRef] [PubMed]
116. Noad, J.; Von Der Malsburg, A.; Pathe, C.; Michel, M.A.; Komander, D.; Randow, F. LUBAC-synthesized linear ubiquitin chains restrict cytosol-invading bacteria by activating autophagy and NF-κB. *Nat. Microbiol.* **2017**, *2*, 1–10. [CrossRef] [PubMed]
117. Koerver, L.; Papadopoulos, C.; Liu, B.; Kravic, B.; Rota, G.; Brecht, L.; Veenendaal, T.; Polajnar, M.; Bluemke, A.; Ehrmann, M.; et al. The ubiquitin-conjugating enzyme UBE 2 QL 1 coordinates lysophagy in response to endolysosomal damage. *EMBO Rep.* **2019**, *20*. [CrossRef]

118. Seczynska, M.; Dikic, I. Removing the waste bags: How p97 drives autophagy of lysosomes. *EMBO J.* **2017**, *36*, 129–131. [CrossRef]
119. Falcon, B.; Noad, J.; McMahon, H.; Randow, F.; Goedert, M. Galectin-8-mediated selective autophagy protects against seeded tau aggregation. *J. Biol. Chem.* **2018**, *293*, 2438–2451. [CrossRef]
120. Richter, B.; Sliter, D.A.; Herhaus, L.; Stolz, A.; Wang, C.; Beli, P.; Zaffagnini, G.; Wild, P.; Martens, S.; Wagner, S.A.; et al. Phosphorylation of OPTN by TBK1 enhances its binding to Ub chains and promotes selective autophagy of damaged mitochondria. *Proc. Natl. Acad. Sci. USA* **2016**, *113*, 4039–4044. [CrossRef]
121. Thurston, T.L.M.; Ryzhakov, G.; Bloor, S.; von Muhlinen, N.; Randow, F. The TBK1 adaptor and autophagy receptor NDP52 restricts the proliferation of ubiquitin-coated bacteria. *Nat. Immunol.* **2009**, *10*, 1215–1221. [CrossRef]
122. Pilli, M.; Arko-Mensah, J.; Ponpuak, M.; Roberts, E.; Master, S.; Mandell, M.A.; Dupont, N.; Ornatowski, W.; Jiang, S.; Bradfute, S.B.; et al. TBK-1 Promotes Autophagy-Mediated Antimicrobial Defense by Controlling Autophagosome Maturation. *Immunity* **2012**, *37*, 223–234. [CrossRef] [PubMed]
123. Kim, J.; Kundu, M.; Viollet, B.; Guan, K.L. AMPK and mTOR regulate autophagy through direct phosphorylation of Ulk1. *Nat. Cell Biol.* **2011**, *13*, 132–141. [CrossRef] [PubMed]
124. Settembre, C.; Fraldi, A.; Medina, D.L.; Ballabio, A.; Children, T. Signals from the lysosome. *Nat. Rev. Mol. Cell Biol.* **2013**, *14*, 283–296. [CrossRef] [PubMed]
125. Brady, O.A.; Martina, J.A.; Puertollano, R. Emerging roles for TFEB in the immune response and inflammation. *Autophagy* **2018**, *14*, 181–189. [CrossRef]
126. Nazio, F.; Carinci, M.; Valacca, C.; Bielli, P.; Strappazzon, F.; Antonioli, M.; Ciccosanti, F.; Rodolfo, C.; Campello, S.; Fimia, G.M.; et al. Fine-tuning of ULK1 mRNA and protein levels is required for autophagy oscillation. *J. Cell Biol.* **2016**, *215*, 841–856. [CrossRef]
127. Jimenez, A.J.; Maiuri, P.; Lafaurie-Janvore, J.; Divoux, S.; Piel, M.; Perez, F. ESCRT machinery is required for plasma membrane repair. *Science* **2014**, *343*. [CrossRef]
128. Scheffer, L.L.; Sreetama, S.C.; Sharma, N.; Medikayala, S.; Brown, K.J.; Defour, A.; Jaiswal, J.K. Mechanism of Ca^{2+}-triggered ESCRT assembly and regulation of cell membrane repair. *Nat. Commun.* **2014**, *5*, 5646. [CrossRef]
129. Radulovic, M.; Schink, K.O.; Wenzel, E.M.; Nähse, V.; Bongiovanni, A.; Lafont, F.; Stenmark, H. ESCRT-mediated lysosome repair precedes lysophagy and promotes cell survival. *EMBO J.* **2018**, *37*, 1–15. [CrossRef]
130. Hurley, J.H. ESCRTs are everywhere. *Embo J.* **2015**, *34*, 2398–2407. [CrossRef]
131. Skowyra, M.L.; Schlesinger, P.H.; Naismith, T.V.; Hanson, P.I. Triggered recruitment of ESCRT machinery promotes endolysosomal repair. *Science* **2018**, *360*. [CrossRef]
132. Repnik, U.; Distefano, M.B.; Speth, M.T.; Wui Ng, M.Y.; Progida, C.; Hoflack, B.; Gruenberg, J.; Griffiths, G. L-leucyl-L-leucine methyl ester does not release cysteine cathepsins to the cytosol but inactivates them in transiently permeabilized lysosomes. *J. Cell Sci.* **2017**, *130*, 3124–3140. [CrossRef] [PubMed]
133. Jia, J.; Abudu, Y.P.; Claude-Taupin, A.; Gu, Y.; Kumar, S.; Choi, S.W.; Peters, R.; Mudd, M.H.; Allers, L.; Salemi, M.; et al. *Galectins Control MTOR and AMPK in Response to Lysosomal Damage to Induce Autophagy*; Taylor and Francis Inc.: Abingdon, UK, 2019; Volume 15, pp. 169–171.
134. Herbst, S.; Campbell, P.; Harvey, J.; Bernard, E.M.; Papayannopoulos, V.; Wood, N.W.; Morris, H.R.; Gutierrez, M.G. LRRK2 activation controls the repair of damaged endomembranes in macrophages. *EMBO J.* **2020**, *44*, 1–14. [CrossRef]
135. Mittal, E.; Skowyra, M.L.; Uwase, G.; Tinaztepe, E.; Mehra, A.; Köster, S.; Hanson, P.I.; Philips, J.A. Mycobacterium tuberculosis type VII secretion system effectors differentially impact the ESCRT endomembrane damage response. *MBio* **2018**, *9*. [CrossRef] [PubMed]
136. Kreibich, S.; Emmenlauer, M.; Fredlund, J.; Dehio, C. Autophagy Proteins Promote Repair of Endosomal Membranes Damaged by the Salmonella Type Three Secretion System 1 Graphical Abstract Highlights d RNAi screen implicates autophagy in Salmonella-containing vacuole (SCV) maintenance d Autophagy promotes repair of TTSS-1-damaged SCV membranes and TTSS-2 expression d In parallel, autophagy eliminates cytosolic Salmonella d Autophagy could play a general role in maintaining endosomal membrane integrity. *Cell Host Microbe* **2015**, *18*, 527–537. [CrossRef] [PubMed]

137. López-Jiménez, A.T.; Cardenal-Muñoz, E.; Leuba, F.; Gerstenmaier, L.; Barisch, C.; Hagedorn, M.; King, J.S.; Soldati, T. The ESCRT and autophagy machineries cooperate to repair ESX-1-dependent damage at the Mycobacterium-containing vacuole but have opposite impact on containing the infection. *PLoS Pathog.* **2018**, *14*, e1007501. [CrossRef] [PubMed]
138. Hurley, J.H.; Emr, S.D. The ESCRT complexes: Structure and mechanism of a membrane-trafficking network. *Annu. Rev. Biophys. Biomol. Struct.* **2006**, *35*, 277–298. [CrossRef]
139. Robinson, M.; Schor, S.; Barouch-Bentov, R.; Einav, S. Viral journeys on the intracellular highways. *Cell. Mol. Life Sci.* **2018**, *75*, 3693–3714. [CrossRef]
140. Votteler, J.; Sundquist, W.I. Virus budding and the ESCRT pathway. *Cell Host Microbe* **2013**, *14*, 232–241. [CrossRef]
141. Ellison, C.J.; Kukulski, W.; Boyle, K.B.; Munro, S.; Randow, F. Transbilayer Movement of Sphingomyelin Precedes Catastrophic Breakage of Enterobacteria-Containing Vacuoles. *Curr. Biol.* **2020**, *30*, 2974–2983.e6. [CrossRef]
142. Gout, E.; Gutkowska, M.; Takayama, S.; Reed, J.C.; Chroboczek, J. Co-chaperone BAG3 and adenovirus penton base protein partnership. *J. Cell. Biochem.* **2010**, *111*, 699–708. [CrossRef]
143. Zhao, M.; Antunes, F.; Eaton, J.W.; Brunk, U.T. Lysosomal enzymes promote mitochondrial oxidant production, cytochrome c release and apoptosis. *Eur. J. Biochem.* **2003**, *270*, 3778–3786. [CrossRef] [PubMed]
144. Deus, C.M.; Yambire, K.F.; Oliveira, P.J.; Raimundo, N. Mitochondria–Lysosome Crosstalk: From Physiology to Neurodegeneration. *Trends Mol. Med.* **2020**, *26*, 71–88. [CrossRef] [PubMed]
145. Christgen, S.; Place, D.E.; Kanneganti, T.D. Toward targeting inflammasomes: Insights into their regulation and activation. *Cell Res.* **2020**, *30*, 315–327. [CrossRef]
146. Boya, P.; Kroemer, G. Lysosomal membrane permeabilization in cell death. *Oncogene* **2008**, *27*, 6434–6451. [CrossRef]
147. Aits, S.; Jäättelä, M. Lysosomal cell death at a glance. *J. Cell Sci.* **2013**, *126*, 1905–1912. [CrossRef] [PubMed]
148. Deretic, V.; Saitoh, T.; Akira, S. Autophagy in infection, inflammation and immunity. *Nat. Rev. Immunol.* **2013**, *13*, 722–737. [CrossRef] [PubMed]
149. Zhou, R.; Tardivel, A.; Thorens, B.; Choi, I.; Tschopp, J. Thioredoxin-interacting protein links oxidative stress to inflammasome activation. *Nat. Immunol.* **2010**, *11*, 136–140. [CrossRef] [PubMed]
150. Zhou, R.; Yazdi, A.S.; Menu, P.; Tschopp, J. A role for mitochondria in NLRP3 inflammasome activation. *Nature* **2011**, *469*, 221–226. [CrossRef]
151. Kimura, T.; Jia, J.; Claude-Taupin, A.; Kumar, S.; Won Choi, S.; Gu, Y.; Mudd, M.; Dupont, N.; Jiang, S.; Peters, R.; et al. Cellular and molecular mechanism for secretory autophagy. *Autophagy* **2017**, *13*, 1084–1085. [CrossRef]
152. Brubaker, S.W.; Bonham, K.S.; Zanoni, I.; Kagan, J.C. Innate Immune Pattern Recognition: A Cell Biological Perspective. *Annu. Rev. Immunol.* **2015**, *33*, 257–290. [CrossRef]
153. Kawai, T.; Akira, S. Toll-like Receptors and Their Crosstalk with Other Innate Receptors in Infection and Immunity. *Immunity* **2011**, *34*, 637–650. [CrossRef] [PubMed]
154. Gabandé-Rodríguez, E.; Pérez-Cañamás, A.; Soto-Huelin, B.; Mitroi, D.N.; Sánchez-Redondo, S.; Martínez-Sáez, E.; Venero, C.; Peinado, H.; Ledesma, M.D. Lipid-induced lysosomal damage after demyelination corrupts microglia protective function in lysosomal storage disorders. *EMBO J.* **2019**, *38*, 1–22. [CrossRef] [PubMed]
155. Davis, A.A.; Leyns, C.E.G.; Holtzman, D.M. Intercellular Spread of Protein Aggregates in Neurodegenerative Disease. *Annu. Rev. Cell Dev. Biol.* **2018**, *34*, 545–568. [CrossRef] [PubMed]
156. Schoonbroodt, S.; Ferreira, V.; Best-Belpomme, M.; Boelaert, J.R.; Legrand-Poels, S.; Korner, M.; Piette, J. Crucial Role of the Amino-Terminal Tyrosine Residue 42 and the Carboxyl-Terminal PEST Domain of IκBα in NF-κB Activation by an Oxidative Stress. *J. Immunol.* **2000**, *164*, 4292–4300. [CrossRef] [PubMed]
157. Takada, Y.; Mukhopadhyay, A.; Kundu, G.C.; Mahabeleshwar, G.H.; Singh, S.; Aggarwal, B.B. Hydrogen peroxide activates NF-κB through tyrosine phosphorylation of IκBα and serine phosphorylation of p65. Evidence for the involvement of IκBα kinase and Syk protein-tyrosine kinase. *J. Biol. Chem.* **2003**, *278*, 24233–24241. [CrossRef]
158. Anderson, F.L.; Coffey, M.M.; Berwin, B.L.; Havrda, M.C. Inflammasomes: An Emerging Mechanism Translating Environmental Toxicant Exposure Into Neuroinflammation in Parkinson's Disease. *Toxicol. Sci.* **2018**, *166*, 3–15. [CrossRef]

159. Siew, J.J.; Chen, H.M.; Chen, H.Y.; Chen, H.L.; Chen, C.M.; Soong, B.W.; Wu, Y.R.; Chang, C.P.; Chan, Y.C.; Lin, C.H.; et al. Galectin-3 is required for the microglia-mediated brain inflammation in a model of Huntington's disease. *Nat. Commun.* **2019**, *10*, 1–18. [CrossRef]
160. Streetly, M.J.; Maharaj, L.; Joel, S.; Schey, S.A.; Gribben, J.G.; Cotter, F.E. GCS-100, a novel galectin-3 antagonist, modulates MCL-1, NOXA, and cell cycle to induce myeloma cell death. *Blood* **2010**, *115*, 3939–3948. [CrossRef]
161. Kawasaki, T.; Kawai, T. Toll-like receptor signaling pathways. *Front. Immunol.* **2014**, *5*, 1–8. [CrossRef]
162. O'Neill, L.A.J.; Golenbock, D.; Bowie, A.G. The history of Toll-like receptors-redefining innate immunity. *Nat. Rev. Immunol.* **2013**, *13*, 453–460. [CrossRef]
163. Pichlmair, A.; Schulz, O.; Tan, C.P.; Näslund, T.I.; Liljeström, P.; Weber, F.; Reis E Sousa, C. RIG-I-mediated antiviral responses to single-stranded RNA bearing 5′-phosphates. *Science* **2006**, *314*, 997–1001. [CrossRef] [PubMed]
164. Takahasi, K.; Yoneyama, M.; Nishihori, T.; Hirai, R.; Kumeta, H.; Narita, R.; Gale, M.; Inagaki, F.; Fujita, T. Nonself RNA-Sensing Mechanism of RIG-I Helicase and Activation of Antiviral Immune Responses. *Mol. Cell* **2008**, *29*, 428–440. [CrossRef]
165. Kato, H.; Takeuchi, O.; Sato, S.; Yoneyama, M.; Yamamoto, M.; Matsui, K.; Uematsu, S.; Jung, A.; Kawai, T.; Ishii, K.J.; et al. Differential roles of MDA5 and RIG-I helicases in the recognition of RNA viruses. *Nature* **2006**, *441*, 101–105. [CrossRef] [PubMed]
166. Zeng, W.; Sun, L.; Jiang, X.; Chen, X.; Hou, F.; Adhikari, A.; Xu, M.; Chen, Z.J. Reconstitution of the RIG-I pathway reveals a signaling role of unanchored polyubiquitin chains in innate immunity. *Cell* **2010**, *141*, 315–330. [CrossRef] [PubMed]
167. Loo, Y.M.; Gale, M. Immune Signaling by RIG-I-like Receptors. *Immunity* **2011**, *34*, 680–692. [CrossRef]
168. Pichlmair, A.; Reis e Sousa, C. Innate Recognition of Viruses. *Immunity* **2007**, *27*, 370–383. [CrossRef]
169. Goubau, D.; Schlee, M.; Deddouche, S.; Pruijssers, A.J.; Zillinger, T.; Goldeck, M.; Schuberth, C.; Van Der Veen, A.G.; Fujimura, T.; Rehwinkel, J.; et al. Antiviral immunity via RIG-I-mediated recognition of RNA bearing 59-diphosphates. *Nature* **2014**, *514*, 372–375. [CrossRef]
170. Gitlin, L.; Barchet, W.; Gilfillan, S.; Cella, M.; Beutler, B.; Flavell, R.A.; Diamond, M.S.; Colonna, M. Essential role of mda-5 in type I IFN responses to polyriboinosinic: Polyribocytidylic acid and encephalomyocarditis picornavirus. *Proc. Natl. Acad. Sci. USA* **2006**, *103*, 8459–8464. [CrossRef]
171. Feng, Q.; Hato, S.V.; Langereis, M.A.; Zoll, J.; Virgen-Slane, R.; Peisley, A.; Hur, S.; Semler, B.L.; van Rij, R.P.; van Kuppeveld, F.J.M. MDA5 Detects the Double-Stranded RNA Replicative Form in Picornavirus-Infected Cells. *Cell Rep.* **2012**, *2*, 1187–1196. [CrossRef]
172. Wang, J.P.; Cerny, A.; Asher, D.R.; Kurt-Jones, E.A.; Bronson, R.T.; Finberg, R.W. MDA5 and MAVS Mediate Type I Interferon Responses to Coxsackie B Virus. *J. Virol.* **2010**, *84*, 254–260. [CrossRef]
173. Slater, L.; Bartlett, N.W.; Haas, J.J.; Zhu, J.; Message, S.D.; Walton, R.P.; Sykes, A.; Dahdaleh, S.; Clarke, D.L.; Belvisi, M.G.; et al. Co-ordinated role of TLR3, RIG-I and MDA5 in the innate response to rhinovirus in bronchial epithelium. *PLoS Pathog.* **2010**, *6*. [CrossRef] [PubMed]
174. Ito, M.; Yanagi, Y.; Ichinohe, T. Encephalomyocarditis Virus Viroporin 2B Activates NLRP3 Inflammasome. *PLoS Pathog.* **2012**, *8*. [CrossRef]
175. Rajan, J.V.; Rodriguez, D.; Miao, E.A.; Aderem, A. The NLRP3 Inflammasome Detects Encephalomyocarditis Virus and Vesicular Stomatitis Virus Infection. *J. Virol.* **2011**, *85*, 4167–4172. [CrossRef]
176. da Costa, L.S.; Outlioua, A.; Anginot, A.; Akarid, K.; Arnoult, D. RNA viruses promote activation of the NLRP3 inflammasome through cytopathogenic effect-induced potassium efflux. *Cell Death Dis.* **2019**, *10*. [CrossRef] [PubMed]
177. Smith, J.S.; Xu, Z.; Tian, J.; Palmer, D.J.; Ng, P.; Byrnes, A.P. The role of endosomal escape and mitogen-activated protein kinases in adenoviral activation of the innate immune response. *PLoS ONE* **2011**, *6*, e26755. [CrossRef] [PubMed]
178. Nociari, M.; Ocheretina, O.; Murphy, M.; Falck-Pedersen, E. Adenovirus Induction of IRF3 Occurs through a Binary Trigger Targeting Jun N-Terminal Kinase and TBK1 Kinase Cascades and Type I Interferon Autocrine Signaling. *J. Virol.* **2009**, *83*, 4081–4091. [CrossRef]
179. Stein, S.C.; Falck-Pedersen, E. Sensing Adenovirus Infection: Activation of Interferon Regulatory Factor 3 in RAW 264.7 Cells. *J. Virol.* **2012**, *86*, 4527–4537. [CrossRef]

180. Eichholz, K.; Bru, T.; Tran, T.T.P.; Fernandes, P.; Welles, H.; Mennechet, F.J.D.; Manel, N.; Alves, P.; Perreau, M.; Kremer, E.J. Immune-Complexed Adenovirus Induce AIM2-Mediated Pyroptosis in Human Dendritic Cells. *PLoS Pathog.* **2016**, *12*, e1005871. [CrossRef]
181. Barlan, A.U.; Griffin, T.M.; Mcguire, K.A.; Wiethoff, C.M. Adenovirus Membrane Penetration Activates the NLRP3 Inflammasome. *J. Virol.* **2011**, *85*, 146–155. [CrossRef]
182. McGuire, K.A.; Barlan, A.U.; Griffin, T.M.; Wiethoff, C.M. Adenovirus Type 5 Rupture of Lysosomes Leads to Cathepsin B-Dependent Mitochondrial Stress and Production of Reactive Oxygen Species. *J. Virol.* **2011**, *85*, 10806–10813. [CrossRef]
183. Tibbles, L.A.; Spurrell, J.C.L.; Bowen, G.P.; Liu, Q.; Lam, M.; Zaiss, A.K.; Robbins, S.M.; Hollenberg, M.D.; Wickham, T.J.; Muruve, D.A. Activation of p38 and ERK Signaling during Adenovirus Vector Cell Entry Lead to Expression of the C-X-C Chemokine IP-10. *J. Virol.* **2002**, *76*, 1559–1568. [CrossRef] [PubMed]
184. Anghelina, D.; Lam, E.; Falck-Pedersen, E. Diminished Innate Antiviral Response to Adenovirus Vectors in cGAS/STING-Deficient Mice Minimally Impacts Adaptive Immunity. *J. Virol.* **2016**, *90*, 5915–5927. [CrossRef] [PubMed]
185. Maler, M.D.; Nielsen, P.J.; Stichling, N.; Cohen, I.; Ruzsics, Z.; Wood, C.; Engelhard, P.; Suomalainen, M.; Gyory, I.; Huber, M.; et al. Key role of the scavenger receptor MARCO in mediating adenovirus infection and subsequent innate responses of macrophages. *MBio* **2017**, *8*. [CrossRef] [PubMed]
186. Watkinson, R.E.; McEwan, W.A.; Tam, J.C.H.; Vaysburd, M.; James, L.C. TRIM21 Promotes cGAS and RIG-I Sensing of Viral Genomes during Infection by Antibody-Opsonized Virus. *PLoS Pathog.* **2015**, *11*, e1005253. [CrossRef]
187. Hare, D.; Mossman, K.L. Novel paradigms of innate immune sensing of viral infections. *Cytokine* **2013**, *63*, 219–224. [CrossRef]
188. Collins, S.E.; Mossman, K.L. Danger, diversity and priming in innate antiviral immunity. *Cytokine Growth Factor Rev.* **2014**, *25*, 525–531. [CrossRef]
189. Deretic, V. Autophagy as an innate immunity paradigm: Expanding the scope and repertoire of pattern recognition receptors. *Curr. Opin. Immunol.* **2012**, *24*, 21–31. [CrossRef]
190. Saitoh, T.; Fujita, N.; Jang, M.H.; Uematsu, S.; Yang, B.G.; Satoh, T.; Omori, H.; Noda, T.; Yamamoto, N.; Komatsu, M.; et al. Loss of the autophagy protein Atg16L1 enhances endotoxin-induced IL-1β production. *Nature* **2008**, *456*, 264–268. [CrossRef]
191. Santeford, A.; Wiley, L.A.; Park, S.; Bamba, S.; Nakamura, R.; Gdoura, A.; Ferguson, T.A.; Rao, P.K.; Guan, J.L.; Saitoh, T.; et al. Impaired autophagy in macrophages promotes inflammatory eye disease. *Autophagy* **2016**, *12*, 1876–1885. [CrossRef]
192. Pu, Q.; Gan, C.; Li, R.; Li, Y.; Tan, S.; Li, X.; Wei, Y.; Lan, L.; Deng, X.; Liang, H.; et al. Atg7 Deficiency Intensifies Inflammasome Activation and Pyroptosis in Pseudomonas Sepsis. *J. Immunol.* **2017**, *198*, 3205–3213. [CrossRef]
193. Bechelli, J.; Vergara, L.; Smalley, C.; Buzhdygan, T.P.; Bender, S.; Zhang, W.; Liu, Y.; Popov, V.L.; Wang, J.; Garg, N.; et al. Atg5 supports Rickettsia australis infection in macrophages in vitro and in vivo. *Infect. Immun.* **2019**, *87*. [CrossRef] [PubMed]
194. Nakahira, K.; Haspel, J.A.; Rathinam, V.A.K.; Lee, S.J.; Dolinay, T.; Lam, H.C.; Englert, J.A.; Rabinovitch, M.; Cernadas, M.; Kim, H.P.; et al. Autophagy proteins regulate innate immune responses by inhibiting the release of mitochondrial DNA mediated by the NALP3 inflammasome. *Nat. Immunol.* **2011**, *12*, 222–230. [CrossRef] [PubMed]
195. Zhong, Z.; Umemura, A.; Sanchez-Lopez, E.; Liang, S.; Shalapour, S.; Wong, J.; He, F.; Boassa, D.; Perkins, G.; Ali, S.R.; et al. NF-κB Restricts Inflammasome Activation via Elimination of Damaged Mitochondria. *Cell* **2016**, *164*, 896–910. [CrossRef] [PubMed]
196. Sumpter, R.; Sirasanagandla, S.; Fernández, Á.F.; Wei, Y.; Dong, X.; Franco, L.; Zou, Z.; Marchal, C.; Lee, M.Y.; Clapp, D.W.; et al. Fanconi Anemia Proteins Function in Mitophagy and Immunity. *Cell* **2016**, *165*, 867–881. [CrossRef]
197. Nguyen, T.N.; Padman, B.S.; Lazarou, M. Deciphering the Molecular Signals of PINK1/Parkin Mitophagy. *Trends Cell Biol.* **2016**, *26*, 733–744. [CrossRef]
198. Chen, G.; Kroemer, G.; Kepp, O. Mitophagy: An Emerging Role in Aging and Age-Associated Diseases. *Front. Cell Dev. Biol.* **2020**, *8*, 1–15. [CrossRef]

199. Harris, J.; Hartman, M.; Roche, C.; Zeng, S.G.; O'Shea, A.; Sharp, F.A.; Lambe, E.M.; Creagh, E.M.; Golenbock, D.T.; Tschopp, J.; et al. Autophagy controls IL-1β secretion by targeting Pro-IL-1β for degradation. *J. Biol. Chem.* **2011**, *286*, 9587–9597. [CrossRef]
200. Shi, C.-S.S.; Shenderov, K.; Huang, N.-N.N.; Kabat, J.; Abu-Asab, M.; Fitzgerald, K.A.; Sher, A.; Kehrl, J.H. Activation of autophagy by inflammatory signals limits IL-1β production by targeting ubiquitinated inflammasomes for destruction. *Nat. Immunol.* **2012**, *13*, 255–263. [CrossRef]
201. Liu, T.; Tang, Q.; Liu, K.; Xie, W.; Liu, X.; Wang, H.; Wang, R.F.; Cui, J. TRIM11 Suppresses AIM2 Inflammasome by Degrading AIM2 via p62-Dependent Selective Autophagy. *Cell Rep.* **2016**, *16*, 1988–2002. [CrossRef]
202. Kimura, T.; Jain, A.; Choi, S.W.; Mandell, M.A.; Schroder, K.; Johansen, T.; Deretic, V. TRIM-mediated precision autophagy targets cytoplasmic regulators of innate immunity. *J. Cell Biol.* **2015**, *210*, 973–989. [CrossRef]
203. Van Gent, M.; Sparrer, K.M.J.; Gack, M.U. TRIM proteins and their roles in antiviral host defenses. *Annu. Rev. Virol.* **2018**, *5*, 385–405. [CrossRef] [PubMed]
204. Jounai, N.; Kobiyama, K.; Shiina, M.; Ogata, K.; Ishii, K.J.; Takeshita, F. NLRP4 Negatively Regulates Autophagic Processes through an Association with Beclin1. *J. Immunol.* **2011**, *186*, 1646–1655. [CrossRef] [PubMed]
205. Dupont, N.; Jiang, S.; Pilli, M.; Ornatowski, W.; Bhattacharya, D.; Deretic, V. Autophagy-based unconventional secretory pathway for extracellular delivery of IL-1β. *EMBO J.* **2011**, *30*, 4701–4711. [CrossRef] [PubMed]
206. Zhang, M.; Kenny, S.J.; Ge, L.; Xu, K.; Schekman, R. Translocation of interleukin-1β into a vesicle intermediate in autophagy-mediated secretion. *Elife* **2015**, *4*. [CrossRef] [PubMed]
207. Wang, L.J.; Huang, H.Y.; Huang, M.P.; Liou, W.; Chang, Y.T.; Wu, C.C.; Ojcius, D.M.; Chang, Y.S. The microtubule-associated protein EB1 links AIM2 inflammasomes with autophagy-dependent secretion. *J. Biol. Chem.* **2014**, *289*, 29322–29333. [CrossRef]
208. Deretic, V. Multiple regulatory and effector roles of autophagy in immunity. *Curr. Opin. Immunol.* **2009**, *21*, 53–62. [CrossRef]
209. Xu, Y.; Liu, X.-D.; Gong, X.; Eissa, N.T. Signaling pathway of autophagy associated with innate immunity. *Autophagy* **2008**, *4*, 110–112. [CrossRef]
210. Delgado, M.A.; Elmaoued, R.A.; Davis, A.S.; Kyei, G.; Deretic, V. Toll-like receptors control autophagy. *EMBO J.* **2008**, *27*, 1110–1121. [CrossRef]
211. Shi, C.-S.; Kehrl, J.H. MyD88 and Trif target Beclin 1 to trigger autophagy in macrophages. *J. Biol. Chem.* **2008**, *283*, 33175–33182. [CrossRef]
212. Jounai, N.; Takeshita, F.; Kobiyama, K.; Sawano, A.; Miyawaki, A.; Xin, K.Q.; Ishii, K.J.; Kawai, T.; Akira, S.; Suzuki, K.; et al. The Atg5-Atg12 conjugate associates with innate antiviral immune responses. *Proc. Natl. Acad. Sci. USA* **2007**, *104*, 14050–14055. [CrossRef]
213. Jin, S.; Tian, S.; Chen, Y.; Zhang, C.; Xie, W.; Xia, X.; Cui, J.; Wang, R. USP 19 modulates autophagy and antiviral immune responses by deubiquitinating Beclin-1. *EMBO J.* **2016**, *35*, 866–880. [CrossRef]
214. Jin, S.; Tian, S.; Luo, M.; Xie, W.; Liu, T.; Duan, T.; Wu, Y.; Cui, J. Tetherin Suppresses Type I Interferon Signaling by Targeting MAVS for NDP52-Mediated Selective Autophagic Degradation in Human Cells. *Mol. Cell* **2017**, *68*, 308–322.e4. [CrossRef] [PubMed]
215. Du, Y.; Duan, T.; Feng, Y.; Liu, Q.; Lin, M.; Cui, J.; Wang, R. LRRC25 inhibits type I IFN signaling by targeting ISG15-associated RIG-I for autophagic degradation. *EMBO J.* **2018**, *37*, 351–366. [CrossRef] [PubMed]
216. Saitoh, T.; Fujita, N.; Hayashi, T.; Takahara, K.; Satoh, T.; Lee, H.; Matsunaga, K.; Kageyama, S.; Omori, H.; Noda, T.; et al. Atg9a controls dsDNA-driven dynamic translocation of STING and the innate immune response. *Proc. Natl. Acad. Sci. USA* **2009**, *106*, 20842–20846. [CrossRef] [PubMed]
217. Liang, Q.; Seo, G.J.; Choi, Y.J.; Kwak, M.J.; Ge, J.; Rodgers, M.A.; Shi, M.; Leslie, B.J.; Hopfner, K.P.; Ha, T.; et al. Crosstalk between the cGAS DNA sensor and beclin-1 autophagy protein shapes innate antimicrobial immune responses. *Cell Host Microbe* **2014**, *15*, 228–238. [CrossRef] [PubMed]
218. Chen, M.; Meng, Q.; Qin, Y.; Liang, P.; Tan, P.; He, L.; Zhou, Y.; Chen, Y.; Huang, J.; Wang, R.F.; et al. TRIM14 Inhibits cGAS Degradation Mediated by Selective Autophagy Receptor p62 to Promote Innate Immune Responses. *Mol. Cell* **2016**, *64*, 105–119. [CrossRef]
219. Tsuchida, T.; Zou, J.; Saitoh, T.; Kumar, H.; Abe, T.; Matsuura, Y.; Kawai, T.; Akira, S. The ubiquitin ligase TRIM56 regulates innate immune responses to intracellular double-stranded DNA. *Immunity* **2010**, *33*, 765–776. [CrossRef]

220. Prabakaran, T.; Bodda, C.; Krapp, C.; Zhang, B.; Christensen, M.H.; Sun, C.; Reinert, L.; Cai, Y.; Jensen, S.B.; Skouboe, M.K.; et al. Attenuation of c GAS-STING signaling is mediated by a p62/ SQSTM 1-dependent autophagy pathway activated by TBK1. *EMBO J.* **2018**, *37*. [CrossRef]
221. Niida, M.; Tanaka, M.; Kamitani, T. Downregulation of active IKKβ by Ro52-mediated autophagy. *Mol. Immunol.* **2010**, *47*, 2378–2387. [CrossRef]
222. Zheng, Q.; Hou, J.; Zhou, Y.; Yang, Y.; Xie, B.; Cao, X. Siglec1 suppresses antiviral innate immune response by inducing TBK1 degradation via the ubiquitin ligase TRIM27. *Cell Res.* **2015**, *25*, 1121–1136. [CrossRef]
223. Sparrer, K.M.J.; Gableske, S.; Zurenski, M.A.; Parker, Z.M.; Full, F.; Baumgart, G.J.; Kato, J.; Pacheco-Rodriguez, G.; Liang, C.; Pornillos, O.; et al. TRIM23 mediates virus-induced autophagy via activation of TBK1. *Nat. Microbiol.* **2017**, *2*, 1543–1557. [CrossRef] [PubMed]
224. Mandell, M.A.; Jain, A.; Arko-Mensah, J.; Chauhan, S.; Kimura, T.; Dinkins, C.; Silvestri, G.; Münch, J.; Kirchhoff, F.; Simonsen, A.; et al. TRIM proteins regulate autophagy and can target autophagic substrates by direct recognition. *Dev. Cell* **2014**, *30*, 394–409. [CrossRef] [PubMed]
225. Labzin, L.I.; Bottermann, M.; Rodriguez-Silvestre, P.; Foss, S.; Andersen, J.T.; Vaysburd, M.; Clift, D.; James, L.C. Antibody and DNA sensing pathways converge to activate the inflammasome during primary human macrophage infection. *EMBO J.* **2019**, *38*. [CrossRef] [PubMed]
226. Klein, K.A.; Jackson, W.T. Picornavirus subversion of the autophagy pathway. *Viruses* **2011**, *3*, 1549–1561. [CrossRef]

Article

Ebola Virus Nucleocapsid-Like Structures Utilize Arp2/3 Signaling for Intracellular Long-Distance Transport

Katharina Grikscheit [1,2], Olga Dolnik [1], Yuki Takamatsu [1,3], Ana Raquel Pereira [4] and Stephan Becker [1,2,*]

1 Institute of Virology, Philipps University Marburg, Hans-Meerwein-Str. 2, 35043 Marburg, Germany; katharina.grikscheit@posteo.de (K.G.); dolnik@staff.uni-marburg.de (O.D.); yukiti@niid.go.jp (Y.T.)
2 German Center for Infection Research (DZIF), Partner Site: Giessen-Marburg-Langen, Hans-Meerwein-Str. 2, 35043 Marburg, Germany
3 Department of Virology I, National Institute of Infectious Diseases, Tokyo 208-0011, Japan
4 Oxford Nano Imager, Linacre House, Banbury Rd, Oxford OX2 8TA, UK; pereira.arr@gmail.com
* Correspondence: becker@staff.uni-marburg.de

Received: 18 May 2020; Accepted: 16 July 2020; Published: 19 July 2020

Abstract: The intracellular transport of nucleocapsids of the highly pathogenic Marburg, as well as Ebola virus (MARV, EBOV), represents a critical step during the viral life cycle. Intriguingly, a population of these nucleocapsids is distributed over long distances in a directed and polar fashion. Recently, it has been demonstrated that the intracellular transport of filoviral nucleocapsids depends on actin polymerization. While it was shown that EBOV requires Arp2/3-dependent actin dynamics, the details of how the virus exploits host actin signaling during intracellular transport are largely unknown. Here, we apply a minimalistic transfection system to follow the nucleocapsid-like structures (NCLS) in living cells, which can be used to robustly quantify NCLS transport in live cell imaging experiments. Furthermore, in cells co-expressing LifeAct, a marker for actin dynamics, NCLS transport is accompanied by pulsative actin tails appearing on the rear end of NCLS. These actin tails can also be preserved in fixed cells, and can be visualized via high resolution imaging using STORM in transfected, as well as EBOV infected, cells. The application of inhibitory drugs and siRNA depletion against actin regulators indicated that EBOV NCLS utilize the canonical Arp2/3-Wave1-Rac1 pathway for long-distance transport in cells. These findings highlight the relevance of the regulation of actin polymerization during directed EBOV nucleocapsid transport in human cells.

Keywords: Ebola virus; actin cytoskeleton; nucleocapsid transport; Arp2/3 complex

1. Introduction

The dynamic actin cytoskeleton is commonly utilized by pathogens for entry, exit and their intracellular assembly [1]. While for some intracellular bacteria and DNA viruses, the mechanism of hijacking the actin cytoskeleton is described in detail, the remodeling of actin through highly pathogenic RNA viruses such as filoviruses remains poorly understood [1,2]. Recently, it has been demonstrated that the intracytoplasmic transport of the highly pathogenic Marburg virus (MARV) and the Ebola virus (EBOV) depends on actin polymerization, but the detailed mechanisms and cellular interaction partners remain largely unknown [3–5].

MARV and EBOV belong to the family of *Filoviridae*, which are filamentous, enveloped viruses containing a single strand, negative-sense RNA genome, and which cause severe fevers in humans with very high fatality rates [6]. Strikingly, filoviral RNA encodes for only seven structural proteins, which are multifunctional, and diversely hijack, disable or reorganize cellular pathways [7]. For example,

EBOV VP24 and VP35 are able to block the cell's interferon system, resulting in the viruses' escape from the innate immune response [8].

A hallmark of the filoviral life cycle is the formation of perinuclear inclusion bodies, where transcription and replication occur and de novo viral nucleocapsids are formed [9]. These nucleocapsids are mainly composed of the nucleoprotein (NP) that encapsidates the viral RNA in highly organized helical structures (approximately 1000 nm long and 50 nm in diameter) [10]. Furthermore, nucleocapsids contain the NP-binding proteins VP24 and VP35, as well as polymerase L and the transcription initiation factor VP30 [10–12].

Following their assembly in the inclusion bodies, individual nucleocapsids have to be transported over long distances through the cytoplasm to reach the cell periphery [3,5,9]. It has been shown that they accumulate in filopodia structures. Finally, nucleocapsids acquire their envelope at the plasma membrane, containing the viral trans-membrane glycoprotein (GP) and the viral matrix protein VP40, triggering the budding of the filamentous infectious virus particles. [8]. The directed transport is required for the rapid distancing away from the inclusion bodies, and to transfer functional nucleocapsids to the plasma membrane. It has been shown that individual single mutations within the nucleocapsid proteins critically affect long-distance transport, thereby leading to a significant delay and reduction of the release of filamentous nucleocapsids [13–15].

While other viruses utilize the microtubule network for intracellular distribution, [16] the transport of nucleocapsids of MARV as well as EBOV is entirely blocked by the pharmacological inhibition of actin polymerization [3,4]. However, the role of actin nucleators and the regulation of their upstream effectors in this process remains elusive. One major obstacle in studying filoviruses is the requirement of high containment laboratories, restricting cell biological experiments that could accelerate our understanding of the critical steps in the virus life cycle. To overcome this problem, we recently established a novel transfection-based system that only requires three viral proteins to produce nucleocapsid-like structures (NCLS), which highly resemble EBOV nucleocapsids in their structure as well as their intracellular dynamics [5]. Through the co-expression of GFP-tagged VP30, which faithfully labels NCLS, it is possible to monitor intracellular transport mechanisms with real time imaging under normal laboratory conditions (biosafety level 1) conditions (Figure 1A,B and Figure 2A).

The host cell contains a highly dynamic actin cytoskeleton that is required for many essential cellular processes, such as cytokinesis, contractility and motility [17,18]. Filamentous actin (called F-actin) is assembled from monomeric globular G-actin, and this polymerization, as well as depolymerization, is highly regulated through a plethora of different cellular factors [19–21]. With this high power of regulation, the cell is able to assemble and coordinate diverse structures, such as the strong cortical actin network stabilizing the cell cortex, short and highly dynamic filaments that are involved in vesicular trafficking, and filament networks that form membrane protrusions such as the filopodia and lamellipodia required for cell motility or cell–cell contact formation [22–25]. The Arp2/3 (actin-related protein) complex efficiently nucleates the actin filaments typically attached to mother filaments to form the highly branched networks that (amongst other things) enable lamellipodia to rapidly adapt during cell migration [26]. This active protein complex has to be tightly regulated, and requires activation for polymerization [27]. So-called nucleation promoting factors (NFPs), such as WAVE1 and WASP proteins, directly induce Arp2/3 complex activity [27,28], and are themselves activated downstream of RhoGTPase signaling in a highly spatial–temporal manner [29].

The Arp2/3 complex is highly conserved, and different pathogens, including *Listeria monocytogenes* and vaccinia virus, utilize its activity for viral intracellular transport steps [30]. For example, the membrane-integrated *Listeria* protein ActA mimics the NPF WASP, thereby recruiting and activating the Arp2/3 complex [31–33]. The Arp2/3 complex in turn induces local actin polymerization, resulting in so-called actin comet tails that efficiently propel the bacterium through the cytoplasm, pushing it into neighboring cells. Furthermore, actin comet tails have been previously observed at EBOV nucleocapsids [3]; however, the mechanism by which viral nucleocapsids use actin dynamics for their transport has not been described in detail.

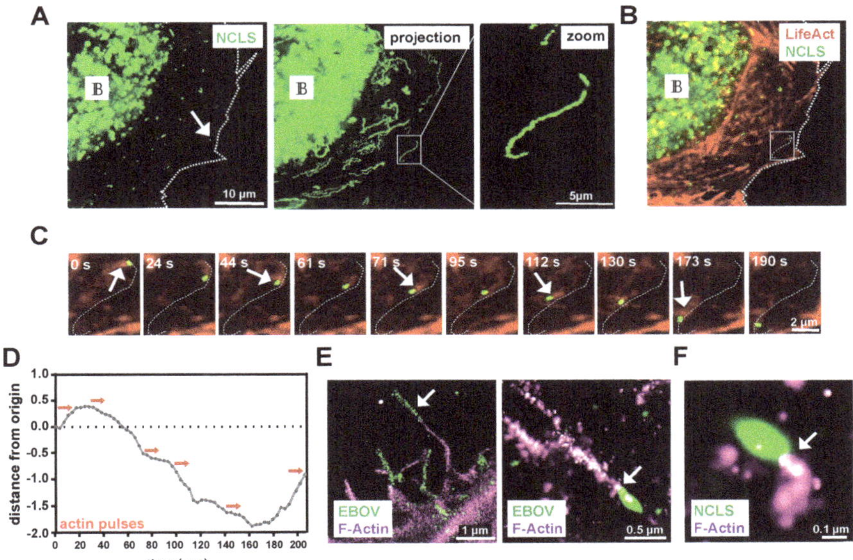

Figure 1. Ebola virus (EBOV) nucleocapsid-like structures induce polar actin tails during transport. (**A**) Exemplary still image from a movie of a Huh7 cell expressing NP, VP24, VP35 and VP30-GFP (left panel). During imaging, NCLS are tracked for 2 min producing long-distance trajectories as depicted in the maximum intensity projection of the movie (right panel). The white arrow highlights an individual track. IB labels inclusion bodies. (**B**) Co-imaging of actin using LifeAct-CLIP (stained with Alexa-657 dye) reveals a dense and dynamic actin network. White inset is magnified in (**C**) showing the time course of an individual NCLS (white arrow, Figure 1**B**), revealing a pulsative actin tail located on one site of the subviral particle during movement (see also Movie S1). (**D**) Graph showing the movement of this NCLS over time. The red arrows indicate actin pulses observed. (**E**,**F**) STORM images showing NCLS or EBOV nucleocapsids immunolabelled with NP (green) and stained with Phalloidin (magenta). (**D**) The left panel shows a filamentous virus particle likely prior to budding (white arrow). Individual intracellular nucleocapsids reveal a preserved actin tail (white arrow). (**E**) Huh7 cells were transfected with NP, VP24 and VP35, and fixed after 24 h. The samples were stained with anti-NP and Phalloidin, and prepared for STORM microscopy. The zoomed image shows an individual NCLS with a preserved actin tail (white arrow).

In this study, we employ our recently established live cell imaging approach to delineate the cellular pathways by which EBOV exploits host actin signaling, and extend the previously applied manual quantification approach to a semi-automatic high throughput method. Using small inhibitory compounds and siRNA-mediated knockdown, we demonstrate that Arp2/3 complex activity downstream of Rac1 is critically involved in the directed long-distance transport of EBOV nucleocapsid structures. Furthermore, through co-visualization of NCLS transport with the actin marker LifeAct, we detected pulsative actin tails accompanying the movement of NCLS through the cytoplasm, which requires Arp2/3 activity.

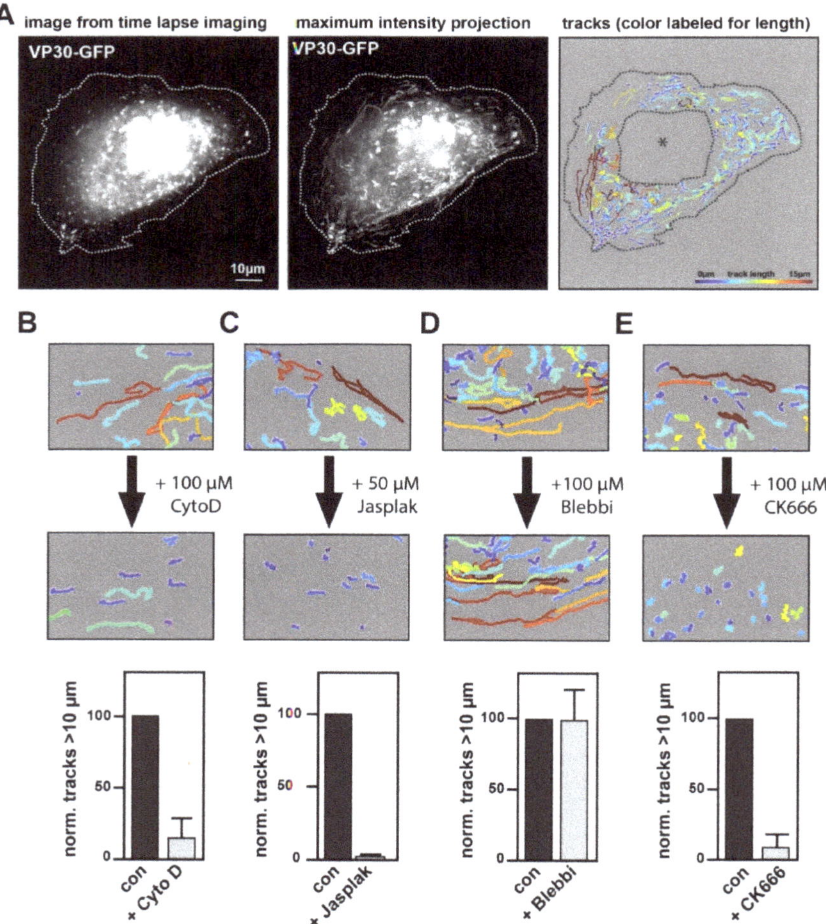

Figure 2. Analysis of NCLS transport using actin-modulating drugs. (**A**) Exemplary still image from a movie of a Huh7 cell expressing NP, VP24, VP35 and VP30-GFP (left panel). For live cell imaging, NCLS are tracked through the cell over 2 min with images captured every 900 ms. While the middle panel shows the maximum-intensity projection of this movie, the right panel depicts the same movie quantified using the spot algorithm (Imaris). The tracks are color-coded for length. To exclude artefacts, we semi-manually deleted areas with strong accumulation of VP30-GFP, such as the inclusion bodies (asterisk). (**B–E**) Huh7 cells transfected with NCLS (VP30-GFP) were recorded for 2 min. This was followed by a short incubation with cytoskeletal-modulating drugs after which the cells were re-recorded. Per experiment three cells were recorded and all tracks > 20 s were quantified. The pictures show a magnified area after quantification with Imaris. Cells were incubated with (**B**) 100 µM Cytochalasin D, (**C**) 50 µM Jasplakinolide, (**D**) 100 µM para-nitro-Blebbistatin or (**E**) 100 µM CK666 following the first imaging. The graphs depict the normalized number of tracks >10 µm, error bars show the SD of $n = 3$ experiments.

2. Materials and Methods

2.1. Cells and Viruses

Huh7 (human hepatoma) cells were cultured in DMEM (Life Technologies, Carlsbad, CA, USA) supplemented with 10% (*v*/*v*) fetal calf serum (FCS) (PAN Biotech), 5 mM L-glutamine, 50 U/mL penicillin and 50 µg/mL streptomycin (Life Technologies) and grown at 37 °C with 5% CO_2. For live cell imaging experiments, cells were kept in phenol-free Leibovitz's medium (Life Technologies) with PS/Q, non-essential amino acid solution and 20% (*v*/*v*) FCS.

The virus used in this study was based on EBOV Zaire (Strain Mayinga; GenBank accession no AF27200 (National Center for Biotechnology Information, Bethesda, MD USA)). The experiments with infectious EBOV virus were performed in the BSL-4 facility at the University of Marburg.

2.2. Transfections, Plasmids, siRNA and Inhibitors

The plasmids coding for EBOV proteins (pCAGGS-NP, -VP35, -VP24) and pCAGGS-VP30-GFP were described previously [34]. Transient transfections of Huh7 cells were carried out using TransIT-LT1 (Mirus, Madison, WI, USA) according to the manufacturers' instructions with 3 µL reagent per 1 µg plasmid DNA [5]. Transfection of siRNAs was performed using DharmaFECT (Horizon, Waterbeach, UK) using 25 nM siRNA and 5 µL transfection reagent in Opti-MEM (Thermo Fisher, Waltham, MA, USA). Huh7 cells were transfected with siRNA in a 6-well µ-slide, seeded to a 4-well µ-slide (IBIDI) and then transfected with the plasmids encoding for the viral proteins (200 ng/µL VP30-GFP, 30 ng/µL VP24, 200 ng/µL VP35 and 200 ng/µL NP). Then, NCLS movement was monitored 24 h later through detection of VP30-GFP. The FlexiTube siRNA used were purchased from Qiagen and diluted to 10 µM. The siRNAs used in the study were HS-ACTR3_5 (AAAGTGGGTGATCAAGCTCAA), HS_RAC1_6 (ATGCATTTCCTGGAGAATATA), HS_WASF1_3 (CAAGAACGTGTGGACCGTTTA) and HS-CDC42_16 (5′-CATCAGATTTTGAAAATATTTAA 3′). For treatment with inhibitors, individual transfected cells (NP, VP24, VP35 and VP30-GFP) were monitored, then the inhibitor was added to the cells in the appropriate dilution and after a short incubation the very same cell was imaged. Cytochalasin D, Jasplakinolide and NSC 23766 were obtained from Sigma Aldrich, and CK666 was purchased from EMD Millipore.

2.3. Antibodies and Reagents for Microscopy

The following primary antibodies were used in this study: polyclonal anti-chicken EBOV NP [35], polyclonal rabbit anti-Wave1 (Sigma Aldrich, St. Louis, MO, USA), mouse anti-Arp3 (Sigma, A5979), mouse anti-Rac1 (Cytoskeleton) and mouse anti-Tubulin (Sigma). The corresponding secondary antibodies were anti-mouse-HRP and anti-rabbit-HRP (DAKO, Jena, Germany), and anti-chicken-Alexa555 (Invitrogen, Carlsbad, CA, USA). Western Blot analyses were performed as described previously [5]. SNAP/CLIP technology was used to visualize actin by using LifeAct-CLIP. LifeAct-CLIP was cloned using standard cloning procedures and then transfected with the NCLS system. For co-visualization of actin and NCLS transport, transfected cells cells were incubated with the dye CLIP-647 (1:500 diluted in media, NEB) for 30 min prior to the experiment, then washed and replaced with Leibovitz medium for live cell imaging.

2.4. Confocal Microscopy, Live Cell Imaging and STORM Microscopy

Huh7 cells were fixed using 4% PFA/DMEM, permeabilized with 0.3%Triton-X100/PBS and blocked with blocking buffer containing 3% glycerol, 2% BSA, 0.2% Tween and 0.05% NaN_3. Primary antibodies as well as secondary antibodies were diluted in blocking buffer and both were incubated for 1 h at RT. The actin cytoskeleton was visualized with Phalloidin conjugated with Alexa647 (Cytoskeleton). For STORM analysis of transfected cells, cells were fixed with 4% PFA/PBS, permeabilized with 0.3% Triton-X100 and then stained with Phalloidin-Alexa647 (1:50) in PBS for 48 h at 4 °C, before being immunolabelled for NP using anti-chicken NP (1:1000) in blocking buffer followed by an incubation with anti-chicken-Alexa555 (Invitrogen, 1:250). Every instance of Phalloidin or antibody staining was followed by a post-fixation step using 4% PFA/PBS. Cells infected with EBOV were fixed with 4% PFA/PBS and permeabilized, then stained with Phalloidin for 1 h in the BSL4 lab, followed by 48 h incubation with 4% PFA/DMEM. Cells were then blocked and immunolabeled as described above and re-stained with Phalloidin for 48 h. For STORM analysis, the coverslip was mounted in switching buffer (1 M MEA-HCl in Glucose buffer containing catalase, TCEP and glucose oxidase).

Dual color live-cell imaging was performed on a SP8 confocal laser scanning microscope (Leica) equipped with a 64 × 1.4 oil objective, and movies for single color live cell imaging were recorded using a Nikon ECLIPSE TE2000-E microscope with a 64x-oil objective. The movies for quantification were acquired every 900 ms for 120 s. All live cell imaging was performed at 37 °C. STORM images were acquired using the Oxford Nanoimaging system (ONI). The light was collected by the 100× objective and imaged onto the EM-CCD camera at 30 ms per frame. Images were processed using the NimOS localization software (Oxford Nanoimaging).

2.5. Particle Tracking, Quantification and Statistics

Image processing was performed with Imaris (Bitplane, Oxford Instruments, Abingdon, UK), FIJI (NIH) and Photoshop CS6 (Adobe, San Jose, CA, USA). For figure design, we used Adobe Illustrator. NCLS tracking was performed using Imaris (Bitplane). To this end, VP30-GFP signal was monitored every 900 ms over 2 min and then moving NCLS were analyzed using the "spots feature", a presetting that has been optimized for cell and particle tracking within the software. Within the "spots" feature, we determined an algorithm that automatically classifies objects of over 0.4-μm diameter appearing continuously for at least 20 s as NCLS. The positions of the structures were then calculated in each image of the 2-min movie. From these data, the software is able to calculate trajectories using an autoregressive motion algorithm. After the computational calculation and visualization of all tracks, we semi-manually filtered out particle aggregations, as well as GFP signals derived from inclusion bodies, by deleting spots within certain areas within the cell (see Figure 2A), resulting in approximately 400–800 tracks per cell. We filtered the data (e.g., for tracks >10 μm or straightness > 0.7) which were then used for quantification. This algorithm was then applied to all files of a set of experiments. Each experiment was performed in at least three independent repetitions. Statistical analyses were performed with Prism x (GraphPad, San Diego, CA, USA).

3. Results

3.1. Nucleocapsid Movement Is Accompanied by Pulsative Actin Tails

Here, we employ a novel live cell imaging system to monitor the long-distance movement of EBOV NCLS, in which only three viral proteins are expressed, namely, NP, VP24 and VP35 (Figure 1A, left panel) [5]. Using VP30-GFP, we monitor the intracellular transport of NCLS for two minutes, and the corresponding maximum-intensity projection reveals that a robust number of NCLS (6.9% ± 2.7%) are transported in a highly directed fashion over long distances (>10μm–20μm), now referred to as long-distance trajectories (Figure 1A, right panel). However, due to defective assembly or saturation of the cellular transport machinery, we also detect many VP30-GFP-positive NCLS that do not move, or travel only short distances (Figure S1A).

Actin tail formation at EBOV nucleocapsids was detected in an earlier study [3]. Therefore, we aimed here to determine whether or not the established transfection system can recapitulate the detection of actin tails at NCLS as well. In cells that co-express the actin marker LifeAct to visualize the actin cytoskeleton (Figure 1B), we observed that NCLS induces actin tails during their movement through the cell (movie S1), thereby further highlighting the relevance of this transfection system to the study of the transport of EBOV nucleocapsids. Here, we further assessed the movement of an individual NCLS-track over time, starting from its origin and following it through the cell (inset of Figure 1B is magnified in Figure 1C, white arrows). In Figure 1D, the red arrows indicate actin pulses during movement, which appear to correlate with the altered movement of NCLS, such as changes in direction (Figure 1C, Movie S1). While with the live cell imaging setting we used approximately 80% of the tracks that moved in a directed fashion revealed detectable actin tails at their rear end (Figure 3E, control siRNA), we cannot exclude the possibility that smaller NCLS induce actin polymerization below the detection limit, due to the high intracellular background level of LifeAct.

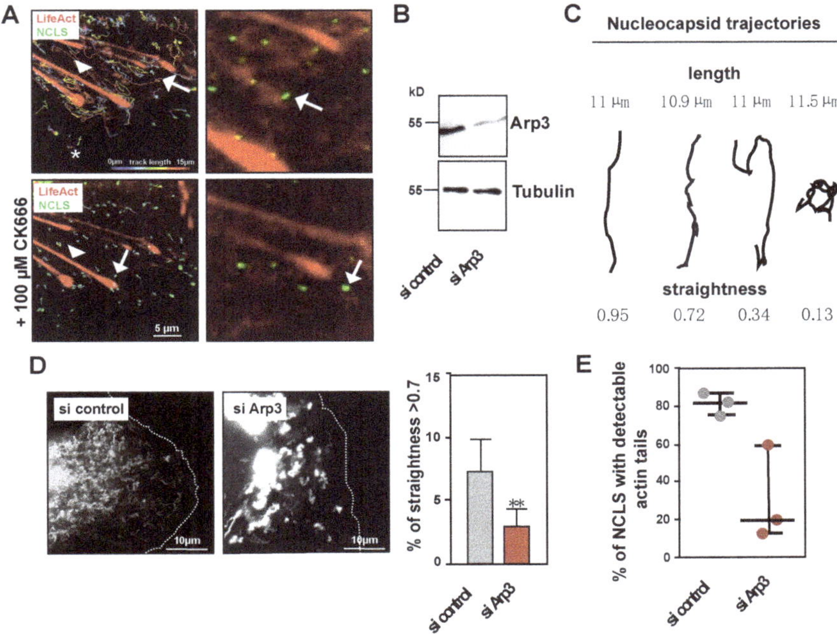

Figure 3. NCLS induction of actin tails depends on Arp2/3 activity. (**A**) Huh7 cells transfected with NP, VP24 and VP35 and LifeAct-CLIP were monitored over time. The left panels show an overlay image of the tracks as calculated by Imaris (color coded for length) and LifeAct-CLIP (still image from movie). Note, CK666 incubation does not interfere with filopodia (asterisk) or stress fibers (arrow head) (see Movie S2). The zoomed image (right panels) show still images of the movie, revealing that CK666 treatment abolishes actin tail formation (white arrows). (**B**) Western Blot showing the depletion of Arp3 after siRNA treatment. (**C**) Cartoon showing different movements of trajectories and their corresponding straightness value. (**D**) Left panel shows maximum intensity projections of time lapse imaging of Huh7 cells transfected either with control or Arp3 siRNA. The right panel shows the graphical analysis of live cell imaging comparing all tracks > 10 µm in cells transfected either with control or Arp3 siRNA. Note, the relative amount of tracks with a straightness > 0.7, representing directed long-distance transport, is reduced in cells transfected with Arp3 siRNA (taken from $n = 3$ experiments, * $p = 0.0027$, Mann–Whitney Test). (**E**) Graph shows the percentage of trajectories with pulsative actin tails in live cell imaging ($n = 3$, error bars indicate mean with standard deviation).

While live cell imaging was applied to capture rapid and transient phenomena, the image resolution remained low. To precisely determine the localization of actin tails at filoviral nucleocapsids, we decided to use a high resolution, single molecule detection technique (STORM). To this end, we either transfected cells with NP, VP35 and VP24, or infected cells with EBOV and fixed them after 24 h. This was followed by the staining of the actin cytoskeleton and immunolabelling for the viral NP protein. Intracellular NCLS or EBOV nucleocapsids appear as elongated NP-positive structures (Figure 1E,F). However, compared to NCLS in transfected cells, filamentous nucleocapsids in cells infected with EBOV were commonly detected along the plasma membrane and within filopodia, likely representing virus particles prior to or during budding, thereby supporting previous observations from live-cell imaging and electron microscopy (Figure 1D, left panel) [3]. Co-staining with Phalloidin revealed actin tail structures in the vicinity of the nucleocapsids in infected cells and transfected cells (Figure 1D,E), which were less frequent in infected cells. This finding is likely a consequence of the very high spatial-temporal dynamics of actin tail formation, which do not allow a quantifiable detection in fixed cells. Further, in the minimalistic transfection system, NCLS remain in the cytosol and are not released, thus the detection of NCLS with actin tails may occur more frequently. Additionally, it cannot be excluded that in infected cells, the long fixation procedure required for the handling of BSL4 samples might interfere with the fixation of these transient actin structures.

3.2. Characterization of NCLS Transport Dynamics

To measure and characterize the long-distance movement of NCLS in a semi-automated manner, we analyzed our time-lapse movies with a tracking tool within the software Imaris. To this end, we set up an algorithm to follow NCLS, and filtered for trajectories > 10 μm that show mean velocities comparable to those measured in previous reports (here, 167 ± 58 nm/s compared to 187 ± 67 nm/s as in Takamatsu et al., 2018 [5]) (Figure 2A).

These studies demonstrated that the long-distance movement of EBOV nucleocapsids, as well as NCLS, is sensitive to the inhibition of actin polymerization [3,5]. Here, we recapitulated these experiments by monitoring NCLS movement followed by an incubation with an actin-modulating drug, and then re-recorded the very same cell. Using this system, we ensured that the imaged cells were capable of showing long-distance trajectories, and would consequently enable us to analyze alterations in NCLS movement. As observed previously, the application of the inhibitor of actin polymerization, Cytochalasin D, results in a rapid and strong decrease in any long-distance trajectories of NCLS, quantified as a highly reduced number of tracks over 10 μm (Figure 2B) [3,5]. Expanding on previous results, we further incubated the cells with the actin polymerization and stabilization agent Jasplakinolide, resulting in a similar phenotype, thereby indicating that indeed not only actin polymerization, but also depolymerization, is required for the long-distance transport of NCLS (Figure 2C). Treatment with Blebbistatin (here we used para-nitro-Blebbistatin to avoid phototoxicity in live cell experiments) did not interfere with long-distance movement, suggesting that myosin II activity is not required for the long-distance transport of NCLS (Figure 2D). Taken together, these findings reinforce the importance of actin regulation and dynamics in the long-distance transport of NCLS.

3.3. Arp2/3 Complex Activity Is Required for Actin Tail Formation and Directed Long-Distance Transport

Previous experiments showed that inhibitors of Arp2/3 affected the long-distance movement of EBOV nucleocapsids [3]. To confirm and further elaborate on these results, we used the minimalistic transfection system to analyze how treatment with an Arp2/3 inhibitor (CK666) affects NCLS movement [36]. Movie S2 shows that blocking Arp2/3 activity rapidly abolishes any directed long-distance movement of NCLS, and that minimal movement likely derives from Brownian motion (Figure 2E, Movie S2). When dissecting different pools of tracks, we observed that treatment with CK666 strongly decreases tracks >10 μm, subsequently resulting in increases in tracks < 5 μm, which also show altered mean speeds (Figure S1A,B). In cells treated with CK666, we observed a slight reduction in the overall NCLS (−19 ± 7%), which might have resulted from the decreased NCLS transport from the inclusion bodies into the cytoplasm (Figure S1C). Furthermore, Movie S2 reveals that the abolishment of NCLS transport, using CK666, coincides with the loss of actin tail formation (Figure 3A, tracks are color-coded for length and actin is visualized with stills from the confocal movie). Time-lapse imaging further demonstrates that Arp2/3-independent structures, such as filopodia or stress fibers, are not compromised during the observed time frame, thereby highlighting a specific role for Arp2/3-dependent actin polymerization in the directed long-distance transport of NCLS (Movie S2).

Next, we continued with the depletion of one protein within the Arp2/3 complex, namely Arp3, using siRNA transfection to further evaluate its role in actin tail-dependent NCLS transport (Figure 3B). The siRNA depletion of Arp3 likely results in the destabilization or reduction of functional Arp2/3 complexes, and does not entirely block all long-distance movement of NCLS, as observed under treatment with CK666. Thus, we assessed whether the remaining tracks showed altered directionality in cells depleted for Arp3, through the calculation of trajectory straightness (ratio of distance to length) (Figure 3C). We determined that tracks with a straightness > 0.7 (Figure 3C), and indeed the depletion of Arp3, interferes with the total number of straight tracks, further highlighting a role for Arp2/3 activity in the long-distance transport of NCLS (Figure 3D,E). Next, we compared the formation of actin tails in control and Arp3-depleted cells. While approximately 80% of the NCLS that travel over long distances showed pulsative actin tails, cells depleted for Arp3 showed a decreased level of detectable actin polymerization at their rear end (Figure 3E,F). Taken together, these results indicate that Arp2/3 activity is required for the efficient directed long-distance transport of NCLS, and that the Arp2/3 complex is directly involved in the formation of actin tails in Huh7 cells.

3.4. Identification of Rac1/WAVE1/Arp2/3 Signaling Network that Is Involved in Long-Distance Transport of NCLS

The Arp2/3 complex resides, in inactive status, within the cell, and requires activation by NFPs, such as N-WASP or Wave proteins, that bind and activate the Arp2/3 complex via their conserved VCA-domain (Figure 4A) [29]. As it was recently demonstrated that the inhibition of *N*-WASP using Wiskostatin did not interfere with the long-distance transport of EBOV nucleocapsids [3], we focused on Wave1, an alternative activator of the Arp2/3 complex. Interestingly, siRNA directed against WASF1 (Wave1) faithfully recapitulated the phenotype of Arp3 siRNA treatment, leading to the reduced straightness of the NCLS (Figure 4B). This indicates that activation of the Arp2/3 complex downstream of Wave1 is involved in the directionality of long-distance trajectories. Wave1 activity itself can be controlled by the RhoGTPase Rac1 (Figure 4A–C) [37,38]. Again, the treatment of Huh7 cells with siRNA against Rac1 resulted in a decreased level of directed long-distance trajectories (Figure 4D,F,G). In contrast, the depletion of Cdc42, another member of the RhoGTPase family, did not affect the straightness of trajectories (Figure S1D), thereby further reinforcing the inference that Rac1 activity is involved in NCLS transport (Figure 4F,H). For further confirmation, we applied the Rac1 inhibitor NSC 23766, which also interfered with long-distance transport, confirming the relevance of Rac1 activity in the regulation of the long-distance transport of EBOV NCLS (Figure 4H).

Taken together, these findings indicate that EBOV nucleocapsids exploit the canonical Rac1/WAVE1/Arp2/3 signaling pathway for long-distance transport of NCLS inside the cell. This signaling pathway classically induces lamellipodial protrusions, in which Rac1 acts coordinately with other upstream signals to activate actin regulators. Future studies shall reveal whether other actin-regulatory mechanisms are also involved in the viral life cycle, and how alterations in long-distance transport affect virus propagation.

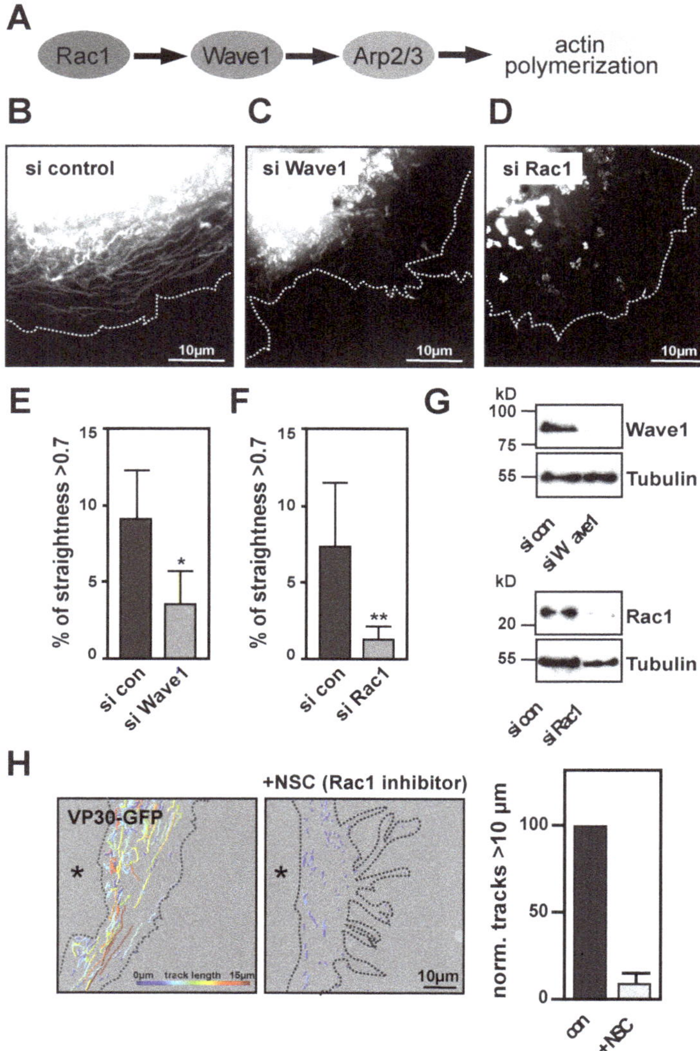

Figure 4. Identification of a Rac1/Wave1/Arp2/3 signaling. (**A**) Cartoon depicting canonical Arp2/3 signaling downstream of Rac1 [39]. (**B–D**) Images show maximum intensity projection of time-lapse images of cells recorded for 2 min; images were captured every 900 ms. Cells were transfected with control siRNA (**B**), siRNA against (**C**) Wave1 (*WASF1*) or (**D**) Rac1. (**E**,**F**) Analysis of live cell imaging comparing relative straightness of tracks > 10 µm and a straightness > 0.7 after either Wave1 siRNA (**E**) or Rac1 siRNA (**F**) transfection. Cells were transfected with siRNA against (**B**) Wave1 (*WASF1*) or (**C**) Rac1. Note that siRNA treatment against Wave1 and Rac1 results in a decrease in straight long-distance trajectories (taken from three independent experiments, at least 5 cells per experiments, ns = non-significant, ** $p > 0.001$, * $p > 0.01$). (**G**) Western Blots showing effective siRNA depletion of Wave1 and Rac1. (**H**) Incubation of Huh7 cells transfected with NCLS (VP30-GFP) was first recorded for 2 min, then they were incubated with 100 nM Rac1 inhibitor (NSC23766) for 1 h and then reimaged. The overview images reveal that NCS23766 also inhibits long-distance transport. Tracks are color-coded for mean speed (6 cells in $n = 3$, $p < 0.0079$, *t*-test).

4. Discussion

In this study, we have identified the Rac1/Wave1/Arp2/3 pathway as being involved in the actin-dependent transport of EBOV NCLS in human cell culture. Arp2/3 activity is also essential to inducing propulsive and polar-localized actin tails at the rear ends of NCLS, which can be robustly visualized in live cell imaging. Furthermore, the association of actin tails with EBOV NCLS and nucleocapsids can also be observed via super resolution in fixed samples.

Actin comet tails were initially characterized in *Listeria monocytogenes* in the late 1980s [40]. Since then, diverse types of intracellular pathogens have been identified as inducing actin polymerization at their surface. For instance, vaccinia virus protein A36 is able to recruit NCK and GRB2, which in turn recruit N-WASP to stimulate the Arp2/3 complex, or virus protein p78/83 mimics N-WASP and directly activates Arp2/3-induced actin polymerization [30,41]. Here, we showed that EBOV NCLS induce actin tails at one end, and actin tail formation is sensitive to the inhibition of Arp2/3 using CK666, indicating that the hijacking of the highly abundant and conserved Arp2/3 complex for induction of actin tails is a common mechanism in diverse types of pathogens without a common origin [30].

We further show that directed long-distance transport is regulated via Arp2/3 activity, yet the siRNA depletion of Arp3 in cells forming NCLS does not entirely block intracellular transport. Furthermore, the effects of siRNA treatment on track straightness appear more pronounced after Rac1 knockdown, when compared to the depletion of Arp3 or Wave1. These findings also suggest that other actin regulators downstream of Rac1 either compensate Arp2/3 activity, or synergistically regulate actin polymerization. One candidate could be the scaffolding protein IQGAP1 that interconnects multiple pathways of actin dynamics and interacts with Rac1. In cells infected with MARV, IQGAP1 was recruited to inclusion and to the read end of nucleocapsids, and the down-regulation of IQGAP1 resulted in the impaired release of MARV, suggesting a role for other major actin regulators in this process [14].

In addition, we also gained unprecedented evidence that Wave1 upstream of Arp2/3 is involved in regulating the long-distance transport of NCLS in Huh7 cells. Given that nucleocapsid transport in EBOV-infected cells does not depend on *N*-WASP [3], we concluded that other NFPs could be involved in the upstream regulation of Arp2/3. Further supporting this notion, N-WASP is typically involved in endocytotic events, where it regulates actin polymerization in a manner reminiscent of actin tails that propel pathogens through cells [42]. Consequently, enveloped viruses, such as vaccinia and EBOV, might profit from mimicking or recruiting N-WASP during their life cycle.

In recent years, additional NPFs, like WASH, WHAMM and JMY, have been described to promote actin tail formations at intracellular membranes like endosomes and autophagosomes [43–45]. In contrast, filoviral nucleocapsids travel in the cytosol without a membrane, and are highly structured protein complexes that probably utilize endocytosis-independent pathways for their transport, thereby likely avoiding membranous structures prior to their arrival at the plasma membrane. Wave1 and Rac1 signaling is considered primarily relevant to actin polymerization in lamellipodial cell protrusions [43], thus it is not surprising that long-distance NCLS transport is best observed in areas with a high activity of this pathway in Huh7 cells. Future studies using high resolution microscopy or electron microscopic approaches shall reveal whether and how actin regulators are actively recruited to EBOV nucleocapsids. It could be that at different stages of the nucleocapsid transport, from inclusion bodies to the budding sites, different actin tail-inducing machineries are exploited by the virus.

As described for other viruses, the long-distance transport of EBOV NCLS is accompanied by the induction of polar actin polymerization, likely resulting in the directionality of movement also observed in other actin tail-inducing pathogens and in in vitro reconstructions [41,46,47]. How this polar induction of actin polymerization is initiated is not understood in detail. One hypothesis derives from studies in *Listeria*, where it was shown that the surface protein ActA accumulates locally, likely during cell wall growth, which subsequently results in polar interactions with actin regulators [48]. In contrast, filoviral nucleocapsids are not enveloped when they leave inclusion bodies, and only encounter the viral proteins GP and VP40, which are transported independently, when they reach the plasma membrane to form infectious particles [49–52]. Thus, it remains unclear how this polar actin polymerization is induced in filoviral nucleocapsids, and which viral protein might be relevant for the actin polymerization. One explanation derives from a study investigating the structure of MARV capsids using cryo-electron microscopy, revealing that nucleocapsid assembly itself results in polar structures [10]. Here, it was demonstrated that MARV nucleocapsids are highly oriented towards the plasma membrane, with the pointed end of the nucleocapsid directed towards the plasma membrane prior to budding [11,53]. Thus, these findings support the notion that polar nucleocapsid assembly itself might result in specific protein conformations, thereby exposing the binding sites for cellular proteins that induce actin polymerization.

Taken together, our minimalistic transfection-based system reveals new opportunities to study the cellular transport of filoviral nucleocapsids, and to reliably quantify the dynamics of subviral structures. Future studies have to further characterize how filoviral proteins modulate highly spatial–temporal RhoGTPase signaling, and identify the direct interaction partners and structural changes that are required for long-distance transport during the EBOV life cycle.

Supplementary Materials: The following are available online at http://www.mdpi.com/2073-4409/9/7/1728/s1, Figure S1: Analysis of tracks in cells treated with the Arp2/3 inhibitor CK666 in subgroups, Movie S1: NCLS induce actin tails during long-distance movement. Huh7 cells expressing EBOV proteins NP, VP35, VP24 and VP30-GFP (NCLS) and LifeAct-CLIP were monitored by time-lapse microscopy. Here, a single NCLS structure is followed showing a pulsative actin tail, Movie S2: NCLS movement and actin tails depend on Arp2/3 activity. A Huh7 cell expressing EBOV proteins NP, VP35, VP24 and VP30-GFP (NCLS) and LifeAct-CLIP was monitored by time lapse confocal microscopy. Note that NCLS capsids transported over long distances show actin tails (white arrow). The same cells were then incubated with the Arp2/3 inhibitor CK666 and reimaged. All long–distance movement is abolished while actin structures like filopodia and stress fibers remain unaffected.

Author Contributions: Conceptualization, K.G. and S.B.; methodology, K.G., Y.T., O.D., A.R.P.; formal analysis, K.G.; investigation, K.G., O.D., A.R.P.; writing—original draft preparation, K.G., S.B.; writing—review and editing, K.G., S.B., O.D.; visualization, K.G.; supervision, S.B.; funding acquisition, S.B. All authors have read and agreed to the published version of the manuscript.

Funding: This research was funded by German Centre for Infectious Research (DZIF; number FKZ8009801908) and by the Deutsche Forschungsgemeinschaft (DFG) through the Sonderforschungsbereich 1021 project B03.

Acknowledgments: We thank Astrid Herwig and Martina Weik for their technical support. We also wish to thank Katrin Roth (Marburg Imaging Facility) and Andreas Rausch (THM Giessen) for their support and helpful discussion.

Conflicts of Interest: The authors declare no conflict of interest.

References

1. Taylor, M.P.; Koyuncu, O.O.; Enquist, L.W. Subversion of the actin cytoskeleton during viral infection. *Nat. Rev. Microbiol.* **2011**, *9*, 427–439. [CrossRef]
2. Newsome, T.P.; Marzook, N.B. Viruses that ride on the coat-tails of actin nucleation. *Semin. Cell. Dev. Biol.* **2015**, *46*, 155–163. [CrossRef]
3. Schudt, G.; Dolnik, O.; Kolesnikova, L.; Biedenkopf, N.; Herwig, A.; Becker, S. Transport of Ebolavirus Nucleocapsids Is Dependent on Actin Polymerization: Live-Cell Imaging Analysis of Ebolavirus-Infected Cells. *J. Infect. Dis.* **2015**, *212*, 160–166. [CrossRef] [PubMed]
4. Schudt, G.; Kolesnikova, L.; Dolnik, O.; Sodeik, B.; Becker, S. Live-cell imaging of Marburg virus-infected cells uncovers actin-dependent transport of nucleocapsids over long distances. *Proc. Natl. Acad. Sci. USA* **2013**, *113*, 14402–14407. [CrossRef] [PubMed]
5. Takamatsu, Y.; Kolesnikova, L.; Becker, S. Ebola virus proteins NP, VP35, and VP24 are essential and sufficient to mediate nucleocapsid transport. *Proc. Natl. Acad. Sci. USA* **2018**, *115*, 1075–1080. [CrossRef]
6. Baseler, L.; Chertow, D.S.; Johnson, K.M.; Feldmann, H.; Morens, D.M. The Pathogenesis of Ebola Virus Disease. *Annu. Rev. Pathol.* **2017**, *12*, 387–418. [CrossRef]
7. Cantoni, D.; Rossman, J.S. Ebolaviruses: New roles for old proteins. *PLoS Negl. Trop. Dis.* **2018**, *12*, e0006349. [CrossRef]

8. Messaoudi, I.; Amarasinghe, G.K.; Basler, C.F. Filovirus pathogenesis and immune evasion: Insights from Ebola virus and Marburg virus. *Nat. Rev. Microbiol.* **2015**, *13*, 663–676. [CrossRef]
9. Hoenen, T.; Shabman, R.S.; Groseth, A.; Herwig, A.; Weber, M.; Schudt, G.; Dolnik, O.; Basler, C.F.; Becker, S.; Feldmann, H. Inclusion bodies are a site of ebolavirus replication. *J. Virol.* **2012**, *86*, 11779–11788. [CrossRef]
10. Wan, W.; Kolesnikova, L.; Clarke, M.; Koehler, A.; Noda, T.; Becker, S.; Briggs, J.A.G. Structure and assembly of the Ebola virus nucleocapsid. *Nature* **2017**, *551*, 394–397. [CrossRef] [PubMed]
11. Bharat, T.A.; Noda, T.; Riches, J.D.; Kraehling, V.; Kolesnikova, L.; Becker, S.; Kawaoka, Y.; Briggs, J.A. Structural dissection of Ebola virus and its assembly determinants using cryo-electron tomography. *Proc. Natl. Acad. Sci. USA* **2012**, *109*, 4275–4280. [CrossRef]
12. Biedenkopf, N.; Hartlieb, B.; Hoenen, T.; Becker, S. Phosphorylation of Ebola virus VP30 influences the composition of the viral nucleocapsid complex: Impact on viral transcription and replication. *J. Biol. Chem.* **2013**, *288*, 11165–11174. [CrossRef] [PubMed]
13. Takamatsu, Y.; Kolesnikova, L.; Schauflinger, M.; Noda, T.; Becker, S. The Integrity of the YxxL Motif of Ebola Virus VP24 Is Important for the Transport of Nucleocapsid–Like Structures and for the Regulation of Viral RNA Synthesis. *J. Virol.* **2020**, *94*, e02170–e02219. [CrossRef] [PubMed]
14. Dolnik, O.; Kolesnikova, L.; Welsch, S.; Strecker, T.; Schudt, G.; Becker, S. Interaction with Tsg101 is necessary for the efficient transport and release of nucleocapsids in marburg virus-infected cells. *PLoS Pathog.* **2014**, *10*, e1004463. [CrossRef]
15. Dolnik, O.; Kolesnikova, L.; Stevermann, L.; Becker, S. Tsg101 is recruited by a late domain of the nucleocapsid protein to support budding of Marburg virus–like particles. *J. Virol.* **2010**, *84*, 7847–7856. [CrossRef]
16. Naghavi, M.H.; Walsh, D. Microtubule Regulation and Function during Virus Infection. *J. Virol.* **2017**, *91*, e00538–e00617. [CrossRef]
17. Pollard, T.D. Cell Motility and Cytokinesis: From Mysteries to Molecular Mechanisms in Five Decades. *Annu. Rev. Cell. Dev. Biol.* **2019**, *35*, 1–28. [CrossRef]
18. Murrell, M.; Oakes, P.W.; Lenz, M.; Gardel, M.L. Forcing cells into shape: The mechanics of actomyosin contractility. *Nat. Rev. Mol. Cell. Biol.* **2015**, *16*, 486–498. [CrossRef] [PubMed]
19. KCampellone, G.; Welch, M.D. A nucleator arms race: Cellular control of actin assembly. *Nat. Rev. Mol. Cell. Biol.* **2010**, *11*, 237–251. [CrossRef]
20. Pollard, T.D. Actin and Actin-Binding Proteins. *Cold Spring Harb. Perspect Biol.* **2016**, *18*, a018226. [CrossRef]
21. Lappalainen, P. Actin-binding proteins: The long road to understanding the dynamic landscape of cellular actin networks. *Mol. Biol. Cell.* **2016**, *27*, 2519–2522. [CrossRef] [PubMed]
22. Grikscheit, K.; Grosse, R. Formins at the Junction. *Trends Biochem. Sci.* **2016**, *41*, 148–159. [CrossRef] [PubMed]
23. Zech, T.; Calaminus, S.D.; Machesky, L.M. Actin on trafficking: Could actin guide directed receptor transport? *Cell. Adh. Migr.* **2012**, *6*, 476–481. [CrossRef] [PubMed]
24. Schaks, M.; Giannone, G.; Rottner, K. Actin dynamics in cell migration. *Essays Biochem.* **2019**, *63*, 483–495.
25. Schuh, M. An actin-dependent mechanism for long-range vesicle transport. *Nat. Cell. Biol.* **2011**, *13*, 1431–1436. [CrossRef]
26. Goley, E.D.; Welch, M.D. The ARP2/3 complex: An actin nucleator comes of age. *Nat. Rev. Mol. Cell. Biol.* **2006**, *7*, 713–726. [CrossRef]
27. Rodal, A.A.; Sokolova, O.; Robins, D.B.; Daugherty, K.M.; Hippenmeyer, S.; Riezman, H.; Grigorieff, N.; Goode, B.L. Conformational changes in the Arp2/3 complex leading to actin nucleation. *Nat. Struct. Mol. Biol.* **2005**, *12*, 26–31. [CrossRef]
28. Rodnick-Smith, M.; Luan, Q.; Liu, S.L.; Nolen, B.J. Role and structural mechanism of WASP-triggered conformational changes in branched actin filament nucleation by Arp2/3 complex. *Proc. Natl. Acad. Sci. USA* **2016**, *113*, E3834–E3843. [CrossRef]
29. Rotty, J.D.; Wu, C.; Bear, J.E. New insights into the regulation and cellular functions of the ARP2/3 complex. *Nat. Rev. Mol. Cell. Biol.* **2013**, *14*, 7–12. [CrossRef]
30. Welch, M.D.; Way, M. Arp2/3-mediated actin-based motility: A tail of pathogen abuse. *Cell. Host Microbe* **2013**, *14*, 242–255. [CrossRef]

31. Welch, M.D.; Rosenblatt, J.; Skoble, J.; Portnoy, D.A.; Mitchison, T.J. Interaction of human Arp2/3 complex and the Listeria monocytogenes ActA protein in actin filament nucleation. *Science* **1998**, *281*, 105–108. [CrossRef]
32. Jeng, R.L.; Goley, E.D.; D'Alessio, J.A.; Chaga, O.Y.; Svitkina, T.M.; Borisy, G.G.; Heinzen, R.A.; Welch, M.D. A Rickettsia WASP-like protein activates the Arp2/3 complex and mediates actin-based motility. *Cell. Microbiol.* **2004**, *6*, 761–769. [CrossRef]
33. Boujemaa–Paterski, R.; Gouin. E.; Hansen, G.; Samarin, S.; le Clainche, C.; Didry, D.; Dehoux, P.; Cossart, P.; Kocks, C.; Carlier, M.F.; et al. Listeria protein ActA mimics WASp family proteins: It activates filament barbed end branching by Arp2/3 complex. *Biochemistry* **2001**, *40*, 11390–11404. [CrossRef]
34. Hoenen, T.; Groseth, A.; Kolesnikova, L.; Theriault, S.; Ebihara, H.; Hartlieb, B.; Bamberg, S.; Feldmann, H.; Ströher, U.; Becker, S. Infection of naive target cells with virus-like particles: Implications for the function of ebola virus VP24. *J. Virol.* **2006**, *80*, 7260–7264. [CrossRef]
35. Biedenkopf, N.; Lier, C.; Becker, S. Dynamic Phosphorylation of VP30 Is Essential for Ebola Virus Life Cycle. *J. Virol.* **2016**, *90*, 4914–4925. [CrossRef]
36. Hetrick, B.; Han, M.S.; Helgeson, L.A.; Nolen, B.J. Small molecules CK-666 and CK-869 inhibit actin-related protein 2/3 complex by blocking an activating conformational change. *Chem. Biol.* **2013**, *20*, 701–712. [CrossRef]
37. Chen, B.; Chou, H.T.; Brautigam, C.A.; Xing, W.; Yang, S.; Henry, L.; Doolittle, L.K.; Walz, T.; Rosen, M.K. Rac1 GTPase activates the WAVE regulatory complex through two distinct binding sites. *Elife* **2017**, *6*, e29795. [CrossRef]
38. Eden, S.; Rohatgi, R.; Podtelejnikov, A.V.; Mann, M.; Kirschner, M.W. Mechanism of regulation of WAVE1-induced actin nucleation by Rac1 and Nck. *Nature* **2002**, *418*, 790–793. [CrossRef]
39. Pollitt, A.Y.; Insall, R.H. WASP and SCAR/WAVE proteins: The drivers of actin assembly. *J. Cell. Sci.* **2009**, *122*, 2575–2578. [CrossRef]
40. Tilney, L.G.; Portnoy, D.A. Actin filaments and the growth, movement, and spread of the intracellular bacterial parasite, Listeria monocytogenes. *J. Cell. Biol.* **1989**, *109*, 1597–1608. [CrossRef]
41. Mueller, J.; Pfanzelter, J.; Winkler, C.; Narita, A.; le Clainche, C.; Nemethova, M.; Carlier, M.F.; Maeda, Y.; Welch, M.D.; Ohkawa, T.C.; et al. Electron tomography and simulation of baculovirus actin comet tails support a tethered filament model of pathogen propulsion. *PLoS Biol.* **2014**, *12*, e1001765. [CrossRef]
42. Rottner, K.; Hänisch, J.; Campellone, K.G. WASH, WHAMM and JMY: Regulation of Arp2/3 complex and beyond. *Trends Cell. Biol.* **2010**, *20*, 650–661. [CrossRef] [PubMed]
43. Derivery, E.; Helfer, E.; Henriot, V.; Gautreau, A. Actin polymerization controls the organization of WASH domains at the surface of endosomes. *PLoS ONE* **2012**, *7*, e39774. [CrossRef] [PubMed]
44. Kast, D.J.; Dominguez, R. WHAMM links actin assembly via the Arp2/3 complex to autophagy. *Autophagy* **2015**, *11*, 1702–1704. [CrossRef]
45. Hu, X.; Mullins, R.D. LC3 and STRAP regulate actin filament assembly by JMY during autophagosome formation. *J. Cell Biol.* **2019**, *218*, 251–266. [CrossRef]
46. Shenoy, V.B.; Tambe, D.T.; Prasad, A.; Theriot, J.A. A kinematic description of the trajectories of Listeria monocytogenes propelled by actin comet tails. *Proc. Natl. Acad. Sci. USA* **2007**, *104*, 8229–8234. [CrossRef]
47. Dayel, M.J.; Akin, O.; Landeryou, M.; Risca, V.; Mogilner, A.; Mullins, R.D. In silico reconstitution of actin-based symmetry breaking and motility. *PLoS Biol.* **2009**, *7*, e1000201. [CrossRef]
48. Lacayo, C.I.; Soneral, P.A.; Zhu, J.; Tsuchida, M.A.; Footer, M.J.; Soo, F.S.; Lu, Y.; Xia, Y.; Mogilner, A.; Theriot, J.A. Choosing orientation: Influence of cargo geometry and ActA polarization on actin comet tails. *Mol. Biol. Cell.* **2012**, *23*, 614–629. [CrossRef]
49. Becker, S.; Klenk, H.D.; Mühlberger, E. Intracellular transport and processing of the Marburg virus surface protein in vertebrate and insect cells. *Virology* **1996**, *225*, 145–155. [CrossRef]
50. Mittler, E.; Schudt, G.; Halwe, S.; Rohde, C.; Becker, S. A Fluorescently Labeled Marburg Virus Glycoprotein as a New Tool to Study Viral Transport and Assembly. *J. Infect. Dis.* **2018**, *218*, S318–S326. [CrossRef]
51. Yamayoshi, S.; Noda, T.; Ebihara, H.; Goto, H.; Morikawa, Y.; Lukashevich, I.S.; Neumann, G.; Feldmann, H.; Kawaoka, Y. Ebola virus matrix protein VP40 uses the COPII transport system for its intracellular transport. *Cell. Host Microbe* **2008**, *3*, 168–177. [CrossRef]

52. Johnson, K.A.; Taghon, G.J.; Scott, J.L.; Stahelin, R.V. The Ebola Virus matrix protein, VP40, requires phosphatidylinositol 4,5-bisphosphate (PI(4,5)P2) for extensive oligomerization at the plasma membrane and viral egress. *Sci. Rep.* **2016**, *6*, 19125. [CrossRef]
53. Welsch, S.; Kolesnikova, L.; Krähling, V.; Riches, J.D.; Becker, S.; Briggs, J.A. Electron tomography reveals the steps in filovirus budding. *PLoS Pathog* **2010**, *6*, e1000875. [CrossRef]

Article

The Cellular Protein CAD is Recruited into Ebola Virus Inclusion Bodies by the Nucleoprotein NP to Facilitate Genome Replication and Transcription

Janine Brandt, Lisa Wendt, Bianca S. Bodmer, Thomas C. Mettenleiter and Thomas Hoenen *

Institute of Molecular Virology and Cell Biology, Friedrich-Loeffler-Institut, 17493 Greifswald-Insel Riems, Germany; janine.brandt@fli.de (J.B.); lisa.wendt@fli.de (L.W.); bianca.bodmer@fli.de (B.S.B.); ThomasC.Mettenleiter@fli.de (T.C.M.)

* Correspondence: thomas.hoenen@fli.de

Received: 30 March 2020; Accepted: 29 April 2020; Published: 1 May 2020

Abstract: Ebola virus (EBOV) is a zoonotic pathogen causing severe hemorrhagic fevers in humans and non-human primates with high case fatality rates. In recent years, the number and extent of outbreaks has increased, highlighting the importance of better understanding the molecular aspects of EBOV infection and host cell interactions to control this virus more efficiently. Many viruses, including EBOV, have been shown to recruit host proteins for different viral processes. Based on a genome-wide siRNA screen, we recently identified the cellular host factor carbamoyl-phosphate synthetase 2, aspartate transcarbamylase, and dihydroorotase (CAD) as being involved in EBOV RNA synthesis. However, mechanistic details of how this host factor plays a role in the EBOV life cycle remain elusive. In this study, we analyzed the functional and molecular interactions between EBOV and CAD. To this end, we used siRNA knockdowns in combination with various reverse genetics-based life cycle modelling systems and additionally performed co-immunoprecipitation and co-immunofluorescence assays to investigate the influence of CAD on individual aspects of the EBOV life cycle and to characterize the interactions of CAD with viral proteins. Following this approach, we could demonstrate that CAD directly interacts with the EBOV nucleoprotein NP, and that NP is sufficient to recruit CAD into inclusion bodies dependent on the glutaminase (GLN) domain of CAD. Further, siRNA knockdown experiments indicated that CAD is important for both viral genome replication and transcription, while substrate rescue experiments showed that the function of CAD in pyrimidine synthesis is indeed required for those processes. Together, this suggests that NP recruits CAD into inclusion bodies via its GLN domain in order to provide pyrimidines for EBOV genome replication and transcription. These results define a novel mechanism by which EBOV hijacks host cell pathways in order to facilitate genome replication and transcription and provide a further basis for the development of host-directed broad-spectrum antivirals.

Keywords: Ebola virus; filovirus; inclusion bodies; CAD; pyrimidine synthesis

1. Introduction

Ebola virus (EBOV) is a zoonotic pathogen belonging to the genus *Ebolavirus* within the order *Filoviridae*, and is the causative agent of severe hemorrhagic fevers in humans and non-human primates with high case fatality rates [1,2]. Increasing numbers of EBOV outbreaks in Africa highlight the importance of understanding the molecular mechanisms of the EBOV life cycle and virus-host cell interactions better in order to develop new countermeasures against this virus. EBOV possesses a non-segmented single-stranded RNA genome of negative polarity that forms a helical nucleocapsid in the center of virions together with the ribonucleoprotein (RNP) complex proteins. During assembly of the nucleocapsid, the RNA genome is tightly coated with the viral nucleoprotein (NP), which protects it from degradation and recognition by the cellular immune response [3]. During EBOV infection, NP-associated RNA genomes serve as templates for mRNA transcription and genome replication [4]. For viral replication, NP interacts with the polymerase cofactor VP35, which acts as a linker between NP and the RNA-dependent RNA polymerase L [5]. NP, VP35, and L are sufficient to facilitate EBOV genome replication, while for viral transcription the transcriptional activator VP30 is additionally required [6,7]. EBOV replication and transcription takes place in cytoplasmic inclusion bodies, which represent a characteristic feature of EBOV infections in cells [8,9]. Their formation can be driven by the expression of NP alone [5,10,11]. Due to the limited number of viral genes, successful genome replication and transcription is highly dependent on host cell factors, which play an important role during the EBOV life cycle. For instance, the host factor STAU1 has been shown to interact with multiple EBOV RNP components, and to redistribute into NP-induced or virus-induced inclusion bodies, suggesting that STAU1 plays a crucial role during viral RNA synthesis by facilitating the interaction between the viral genome and RNP proteins [12]. EBOV has also been shown to recruit SMYD3 into inclusion bodies, which modulates NP-VP30 interaction and enhances mRNA transcription [13]. Similarly, RBBP6 was found to influence EBOV replication by disrupting the interaction between NP and VP30 [14]. Importin-α7 was described as being required for the efficient formation of inclusion bodies [15]. Furthermore, several cellular kinases and phosphatases are known to localize in inclusion bodies to support EBOV replication and transcription [16–18]. Finally, we previously showed that EBOV NP recruits the nuclear RNA export factor 1 (NXF1) into inclusion bodies to facilitate viral mRNA export from these structures into the cytoplasm [19]. Despite this recent progress in our understanding of the interplay between host factors and EBOV, there remains a considerable need to identify and, more importantly, characterize further host factors required for EBOV replication to identify novel targets for antiviral drug development.

We previously performed a genome-wide siRNA screen using a minigenome system to identify potential host-directed targets [20]. In this system, a minigenome, i.e., a truncated version of the EBOV genome lacking all viral open reading frames (ORF) and consisting of a reporter gene (e.g., a luciferase or green fluorescent protein) flanked by the viral non-coding terminal leader and trailer regions, is expressed from a plasmid in mammalian cells together with the plasmids encoding the viral RNP proteins [6]. For initial transcription of the minigenome RNAs from the minigenome-encoding plasmids most existing EBOV minigenome systems use a T7 RNA polymerase (T7) promoter, and therefore require expression of T7 polymerase, which is usually provided via a T7-expressing plasmid that is cotransfected with the plasmids encoding the RNP proteins [6,21]. However, recently, an EBOV minigenome system using the cellular RNA polymerase II (Pol-II) for initial minigenome RNA transcription has also been established and shown to be more efficient at least in some cell types [22]. After initial transcription and encapsidation by RNP proteins, minigenome RNAs are recognized as authentic templates by the viral polymerase due to their leader and trailer regions, and are replicated and transcribed into mRNAs, which results in expression of the reporter protein. Thus, minigenome assays allow us to study viral genome replication and transcription, as well as viral protein expression, outside of maximum containment laboratories, simplifying the identification of host factors involved in these processes. By using this system, we recently identified the trifunctional protein carbamoyl-phosphate synthetase 2, aspartate transcarbamylase, and dihydroorotase (CAD) as being important for the EBOV life cycle [20].

CAD is an important component of the pyrimidine pathway that catalyzes the first three steps during the de novo biosynthesis of pyrimidine nucleotides using its four distinct enzymatic domains [23–25]. The first domain, glutaminase (GLN), initiates the pathway by catalyzing the hydrolysis of glutamine. This is followed by the synthesis of carbamoyl phosphate facilitated by the carbamoyl-phosphate synthetase (CPS). Carbamoyl phosphate is in turn the substrate for the aspartate transcarbamylase (ATC), which catalyzes the reaction of aspartate with carbamoyl phosphate to carbamoyl aspartate [26,27]. Finally, carbamoyl aspartate is converted to dihydroorotate by dihydroorotase (DHO) [28]. In response to cell growth and proliferation, CAD activity is upregulated by phosphorylation through MAP kinases at position Thr-456, while in resting cells Thr-456 is dephosphorylated [29]. Furthermore, CAD is known to primarily localize in the cytoplasm of resting cells, but in response to cell growth and Thr-456 phosphorylation a small fraction is translocated into nuclear compartments, suggesting a cellular function of CAD in the nucleus [30,31]. However, little is known about the role of CAD during virus infection, and particularly the role of CAD in the EBOV life cycle still needed to be further analyzed. Therefore, we wanted to characterize the interaction of CAD with EBOV on both a biochemical and functional level. Based on our results, we suggest that CAD is important for both genome replication and transcription due to its function in pyrimidine synthesis and that it is recruited into NP-induced and virus-induced inclusion bodies to facilitate the de novo biosynthesis of pyrimidine nucleotides.

2. Materials and Methods

2.1. Cell Lines

Human embryonic kidney cells (HEK 293T, Collection of Cell Lines in Veterinary Medicine CCLV-RIE 1018), African green monkey kidney cells (Vero E6, kindly provided by Stephan Becker, Philipps University Marburg), and human hepatocellular carcinoma cells (Huh7, kindly provided by Stephan Becker, Philipps University Marburg) were cultured in Dulbecco's modified Eagle's medium (DMEM; Thermo Fisher Scientific, Darmstadt, Germany) supplemented with 10% fetal calf serum (FCS), 100 U/mL penicillin, 100 μg/mL streptomycin (PS; Thermo Fisher Scientific), and 1× GlutaMAX (Thermo Fisher Scientific). All cells were incubated at 37°C and 5% CO_2.

2.2. Plasmids and Cloning

Minigenome assay components, including expression plasmids coding for the EBOV RNP proteins, T7 polymerase, firefly luciferase, and a classical T7-driven monocistronic minigenome (pT7-1cis-EBOV-vRNA-nLuc) have been previously described [20,32]. A NanoLuc luciferase-expressing T7-driven replication-deficient minigenome was cloned from a classical minigenome expressing NanoLuc luciferase as a reporter by deletion of 55 nucleotides (nt) in the antigenomic replication promoter as previously described [32]. Based on this, a Pol-II-driven replication-deficient minigenome was generated by PCR to amplify a linear version of the replication-deficient minigenome flanked by hammerhead and hepatitis delta virus ribozymes using primers #4571 (5′-AGC TTA CGT GAC TAC TTC CTT CGG ATG CCC AGG TCG GAC CGC G-3′) and #4572 (5′-GAC CGG TAG AAA ACT GAT GAG TCC GTG AGG ACG AAA CGG AGT CTA GAC TCC GTC TTT TCC AGG AAT CCT TTT TGC AAC GTT TAT TCT G-3′). The linearized construct was subsequently inserted into pCAGGS. The CAD gene was cloned from 293T cells into pCAGGS, and deletion mutants and domains of CAD were then generated using PCR-based approaches. All constructs were first cloned into pCAGGS, followed by subcloning into a pCAGGS plasmid encoding an N-terminal FLAG/HA-tag (DYKDDDDKLDGGYPYDVPDYA) immediately upstream of a BsmBI cloning site, allowing a seamless insertion of the open reading frame of interest. The expression plasmid for N-terminally myc-tagged VP35 was constructed by cloning a myc-tag (EQKLISEEDL) immediately before the VP35 ORF. Detailed cloning strategies are available on request.

2.3. Antibodies

The anti-FLAG (clone M2) antibody used for immunofluorescence analyses (IFA), co-immuno precipitation (coIP), and Western blot analyses was purchased from Sigma-Aldrich (Munich, Germany) [F1804], and the anti-c-myc antibody used for IFA analysis was obtained from Thermo Fisher Scientific [A-21281]. Primary antibodies against NP (rabbit anti-EBOV NP polyclonal antibody), GAPDH (mouse anti-GAPDH clone 0411), and CAD (rabbit anti-CAD clone EP710Y) were ordered from IBT Bioservices (San Jose, USA; anti-NP [0301-012]), Santa Cruz (Heidelberg, Germany; anti-GAPDH [sc47724]), or Abcam (Cambridge, UK; anti-CAD [ab40800]). Secondary antibodies used for IFA analysis against mouse (Alexa Fluor 488 anti-mouse [A-11029]), rabbit (Alexa Fluor 568 anti-rabbit [A-11036]), and chicken IgY (Alexa Fluor 647 anti-chicken [A-21449]) were obtained from Thermo Fisher Scientific. For Western blotting, secondary antibodies against mouse (IRDye 680RD anti-mouse [926-68070]) and rabbit IgG (IRDye 800CW anti-rabbit [926-68071]) were purchased from Li-COR (Bad Homburg, Germany), while anti-mouse IgG (Kappa light chain) Alexa Fluor 680 [115-625-174] used for coIP analyses was ordered from Dianova (Hamburg, Germany).

2.4. Viruses

Zaire ebolavirus rec/COD/1976/Mayinga-rgEBOV (GenBank accession number KF827427.1), which is identical in sequence to the EBOV Mayinga isolate with the exception of four silent mutations as genetic markers [33], was used for all infection experiments. rgEBOV was propagated in VeroE6 cells and virus titers were determined by 50% tissue culture infectious dose (TCID50) assay. All work with the infectious virus was performed under BSL-4 conditions at the Friedrich-Loeffler-Institut (Federal Research Institute of Animal Health, Greifswald Insel-Riems, Germany) following approved standard operating procedures.

2.5. Chemical Compounds

100mM uridine or cytidine (both Sigma-Aldrich) stock solutions were prepared in dimethyl sulfoxide (DMSO) and further diluted in cell culture medium. Diluted pyrimidines or DMSO corresponding to 1% of the supernatant volume in 12-well plates was added to the cells at the time of transfection and after medium changes. All concentrations indicated in the figures are final concentrations.

2.6. siRNA Knockdown with EBOV Minigenomes and Pyrimidine Complementation

For siRNA knockdown of endogenous CAD, 293T cells were reverse transfected (i.e., transfected in suspension and subsequently seeded into plates) with 12 pmol pre-designed silencer select siRNAs (CAD-siRNA#1: s2320 [5′-GAG GGU CUC UUC UUA AGU A-3′]; CAD-siRNA#2: 117891 [5′-GCU AGC UGA GAA AAA CUU U-3′]; Negative Control siRNA #2; all Thermo Fisher Scientific) or a self-designed EBOV-anti-L siRNA [5′-UUU AUA UAC AGC UUC GUA CUU-3′] ordered from Eurofins Genomics (Ebersberg, Germany). Transfection was performed in 12-well plates using Lipofectamine RNAiMax (Thermo Fisher Scientific) following the manufacturer's instructions. 48 h post-siRNA transfection, the cells were transfected using Transit-LT1 (Mirus Bio LLC, Madison, USA) with all minigenome assay components, i.e., pCAGGS-based expression plasmids for NP (62.5 ng), VP35 (62.5 ng), VP30 (37.5 ng), L (500 ng), codon-optimized T7-polymerase (125 ng), firefly luciferase (as a control, 125 ng), and the T7-driven monocistronic minigenome (pT7-EBOV-1cis-vRNA-nLuc; 125 ng). For analyses of vRNA and mRNA levels the control firefly luciferase was replaced with GFP (200 ng), and for the replication-deficient minigenome assay a Pol-II-driven replication-deficient minigenome (pCAGGS-EBOV-1cis-vRNA-nLuc-RdM) was used. Transfections were performed using Transit LT1 as previously described [32]. All samples were harvested 48 h post-transfection for either determination of reporter activity or RNA isolation (see below). For measuring the luciferase activity, cells were lysed for 10 min in 1x Lysis Juice (PJK, Kleinblittersdorf, Germany) at room temperature and lysates were cleared of cell debris by centrifugation for 3 min at 10,000× g. Then, 40 μL of the cleared lysates were added to either 40 μL of Beetle Juice (PJK) or NanoGlo Luciferase Assay Reagent (Promega, Madison, USA) in opaque 96-well plates and luminescence was measured using a Glomax Multi (Promega) microplate reader. NanoLuc luciferase activities were normalized to firefly luciferase activities.

2.7. RNA Isolation and RT-qPCR

RNA isolation from minigenome cell lysates was performed following the manufacturer's instructions using a NucleoSpin RNA kit (Machery-Nagel, Düren, Germany). After RNA purification, all samples were treated with DNase (TURBO DNA-free kit; Thermo Fisher Scientific) following the manufacturer's instructions to avoid plasmid contamination. For cDNA generation, RNA samples were incubated with an oligo(dT)-primer for mRNA quantification or with a strand-specific primer (5′-AGT GTG AGC TTC TAA AGC AAC C-3′) for vRNA quantification using the RevertAid Reverse Transcriptase (Thermo Fisher Scientific) following the manufacturer's instructions. The subsequent qPCR was performed using a PowerUp SYBR Green Master Mix (Thermo Fisher Scientific) with 1 µL of cDNA and primers targeting either the reporter gene (5′-TTC AGA ATC TCG GG GTG TCC-3′, 5′-CGT AAC CCC GTC GAT TAC CA-3′), or GFP as a control (5′-CTT GTA CAG CTC GTC CAT GC-3′, 5′-CGA CAA CCA CTA CCT GAG CAC-3′). Values for vRNA and mRNA levels were normalized to control GFP mRNA levels.

2.8. Immunofluorescence Analysis

Huh7 cells, which are more suitable for IFA than 293T cells, were seeded on coverslips in 12-well plates and transfected 24 h later with 500 ng pCAGGS-EBOV-NP and 500 ng pCAGGS-FLAG-HA-CAD (or CAD mutants) and, in selected experiments, additionally with 500 ng pCAGGS-myc-VP35 as indicated. For a mock control, cells were transfected with pCAGGS. Transfection was performed using polyethylenimine (Sigma-Aldrich) following the manufacturer's instructions. 48 h post-transfection, cells were fixed using 4% paraformaldehyde (Roth, Karlsruhe, Germany) in DMEM for 20 min and then treated with 1 M glycine (in phosphate-buffered saline^{++} (PBS with 0.9M Ca^{2+} and 0.5M Mg^{2+})) for 10 min. Then, cells were permeabilized with 0.1% Triton X-100 in PBS for another 10 min and incubated with 10% fetal calf serum (FCS) in PBS for 45 min. Primary antibodies (rabbit anti-EBOV-NP 1:500; mouse anti-FLAG 1:2500; chicken anti-myc 1:1200) were diluted in PBS with 10% FCS and cells were incubated for 1 h at room temperature with the prepared antibody solutions. Secondary antibodies (Alexa Fluor 488 anti-mouse 1:1200; Alexa Fluor 568 anti-rabbit 1:500; Alexa Fluor 647 anti-chicken 1:1200) were prepared as described for the primary antibodies. After 45 min of staining, cells were washed with PBS and water before mounting with ProLong Diamond Antifade mountant with 4′,6-diamidino-2-phenylindole (DAPI) (Thermo Fisher Scientific). Slides were analyzed by confocal laser scanning microscopy using a Leica SP5.

2.9. Infection of Transfected Huh7 Cells

To investigate the localization of CAD during EBOV infection, Huh7 cells were seeded in 8-well chambered slides (ibidi, Martinsried, Germany) and transfected as described above (immunofluorescence analysis) with 500 ng pCAGGS-FLAG/HA-CAD. At 48 h post-transfection, the transfected cells were infected with EBOV at an MOI of 1, and the samples were fixed 16 h post-infection in 10% formalin twice overnight prior to removal from the BSL4 facility and immunofluorescence analysis.

2.10. Co-Immunoprecipitation of Viral Proteins

CoIPs were performed as previously described [19]. Briefly, 293T cells were seeded in 6-well plates and transfected with expression plasmids encoding FLAG/HA-tagged CAD and EBOV-NP using Transit LT-1 (Mirus Bio LLC) following the manufacturer's instructions. The medium was changed after 24 h and the cells were harvested 48 h post-transfection. For coIP, cells were lysed in 1 mL coIP lysis buffer (1% NP-40; 50 mM Tris pH 7.4; 167 mM NaCl in water) with protease inhibitors (cOmplete; Roche, Mannheim, Germany). To investigate a possible RNA dependency of the interaction between CAD and NP, 100 µg/mL RNase A (Machery-Nagel) were added to the samples. Subsequently, the samples were incubated rotating at 15 RPM for 2 h at 4 °C. Then, 150 µL of the cleared lysates were taken as an input control (representing a sixth of the complete pre-immune lysate and 20% of the sample used for immunoprecipitation) and subjected to acetone precipitation. The remaining 750 µL of cell lysate were mixed with the prepared bead-antibody solution (Dynabeads Protein G, Thermo Fisher Scientific; 1 µL anti-FLAG M2 antibody per 10 µL beads). Immunoprecipitation was performed for 10 min, as recommended by the manufacturer, at room temperature and rotation at 15 RPM. Then, samples were transferred to new tubes and boiled for 10 min at 99 °C. Input and coIP samples were analyzed by SDS-PAGE and Western blotting.

2.11. Western Blotting

For validation of CAD knockdown efficiency and analyses of coIP input and lysates, samples were subjected to SDS-PAGE and Western blotting as previously described [34]. FLAG-tagged CAD was detected using a monoclonal anti-FLAG antibody (1:2000), while NP, wild type CAD, and GAPDH were detected using anti-NP (1:1000), anti-CAD (1:250), and anti-GAPDH (1:1000) antibodies. As secondary antibodies, 680RD-coupled goat-anti-mouse, goat-anti-mouse-Alexa Fluor 680, and 800CW-coupled goat-anti-rabbit antibodies (1:14000) were used. Fluorescent signals were detected and quantified using an Odyssey CLx infrared imaging system (Li-Cor Biosciences). For knockdown quantification, CAD signals were normalized to GAPDH signals.

2.12. Statistical Analyses

One-way ANOVA with a Dunnett's multiple comparisons test was performed using the GraphPad Prism 8.1.0 software.

3. Results

3.1. CAD Knockdown Affects Both EBOV Genome Replication and Transcription

Using a genome-wide siRNA screen, we previously identified CAD to be important for EBOV RNA synthesis and/or viral protein expression [20]. However, since only the effect of CAD knockdown on the sum of these processes had been tested, we now analyzed the role of CAD on individual aspects of the EBOV life cycle. As a first step, we assessed the efficiency of endogenous CAD knockdown using two different siRNAs via quantitative Western blotting, which revealed a 60% to 80% reduction in endogenous CAD expression levels for the two siRNAs (Figure 1A,B).

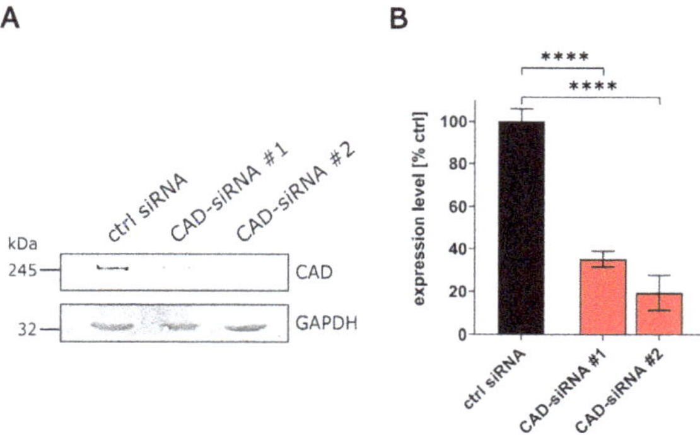

Figure 1. Quantification of CAD knockdown. (**A**) Analysis of CAD knockdown. 293T cells were transfected with siRNAs targeting CAD (CAD-siRNA), or a negative control (ctrl siRNA). The cells were harvested 48 h post-transfection and the lysates were subjected to SDS-PAGE and Western blotting. (**B**) Quantification of CAD knockdown. The Western blot signals for CAD knockdown (as shown in Figure 1A) were measured and normalized to the GAPDH signals. The negative control (ctrl siRNA) was set to 100% and the efficiency of CAD knockdown was calculated (**** $p \leq 0.0001$).

Next, we performed a classical minigenome assay (Figure 2A) in connection with an siRNA knockdown of CAD. As previously shown, knockdown of CAD led to a 40 to 53-fold reduction in reporter activity, verifying an influence of CAD on EBOV viral RNA synthesis and protein expression (Figure 2B) [20]. In order to identify whether CAD knockdown affects transcription and/or protein expression independent of replication, we next used a replication-deficient minigenome system [32]. In contrast to a replication-competent minigenome, the replication-deficient minigenome lacks 55 nt in the antigenomic replication promoter leading to a block of minigenome vRNA replication, while minigenome transcription still takes place [32]. However, when using this system, which is based on T7-driven initial transcription of minigenomes, we observed a very low dynamic range between our controls, which made it difficult to evaluate a possible influence of CAD knockdown (Figure S1). Therefore, in order to increase the dynamic range of this system, we generated a Pol-II-driven replication-deficient minigenome that resulted in a ~10-fold higher dynamic range (Figure S1). Using this system, CAD knockdown resulted in a clear reduction in reporter activity, indicating that CAD is important for EBOV transcription and/or protein expression independent of viral genome replication (Figure 2C).

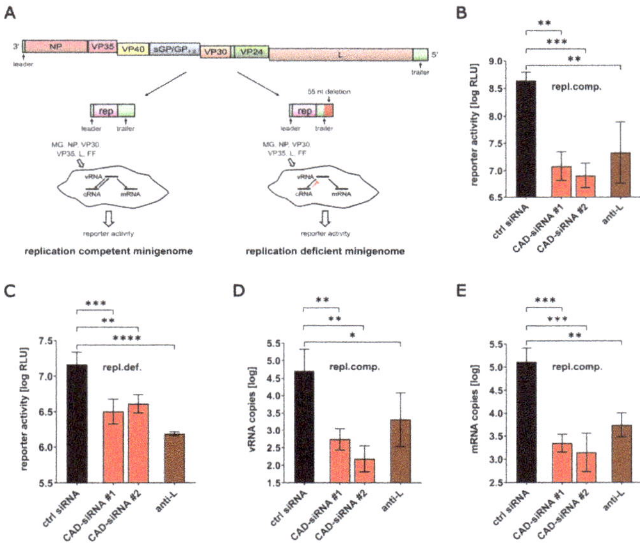

Figure 2. Influence of CAD knockdown on the Ebola virus life cycle. (**A**) Replication-competent and -deficient minigenome systems. The full-length genome structure of EBOV, as well as replication-competent and -deficient minigenomes derived from this full-length genome, are shown. Abbrevations: MG: minigenome, rep: reporter; FF: Firefly luciferase. Figure modified from [35] under CC BY 4.0 license. (**B**) Influence of CAD knockdown on EBOV RNA synthesis. 293T cells were transfected with siRNAs targeting either CAD (CAD-siRNA), EBOV-L (anti-L), or a negative control (ctrl siRNA). 48 h post-transfection, cells were transfected with all the components required for a replication-competent minigenome assay (repl.comp.). Another 48 h later, cells were harvested and the reporter activity was measured. (**C**) Analysis of CAD knockdown on EBOV transcription and gene expression. 293T cells were transfected with siRNAs targeting either CAD (CAD-siRNA), EBOV-L (anti-L), or a negative control (ctrl siRNA). 48 h post-transfection, cells were transfected with all the components required for a replication-deficient minigenome assay (repl.def.). Another 48 h later, cells were harvested and the reporter activity was measured. (**D**) Impact of CAD knockdown on EBOV replication. Cells were treated as described in 2B. After cell harvesting, RNA was extracted from the cell lysates and RT-qPCR for vRNA was performed. (**E**) Influence of CAD knockdown on EBOV mRNA levels. Cells were treated as described in 2B. After cell harvesting, RNA was extracted from cell lysates and RT-qPCR for mRNA was performed. The means and standard deviations of 3 independent experiments are shown for each panel. Asterisks indicate p-values from a one-way ANOVA (* $p \leq 0.05$; ** $p \leq 0.01$; *** $p \leq 0.001$; **** $p \leq 0.0001$; ns: $p > 0.05$).

To further dissect the influences of CAD on viral genome replication, mRNA transcription, and later steps of viral protein expression, we performed classical minigenome assays in the context of an siRNA knockdown of CAD and measured vRNA and mRNA levels in cell lysates using RT-qPCR. For this, we used either an oligo-dT primer for reverse transcription of mRNAs, or a strand-specific primer for reverse transcription of vRNA, followed by qPCR against the reporter gene. CAD siRNA-treated cells showed a strong reduction in both vRNA and mRNA levels in comparison to the control cells, demonstrating that CAD is important for both EBOV transcription and viral genome replication (Figure 2D,E).

3.2. The Effect of CAD Knockdown Can Be Compensated for by Exogenous Pyrimidines

As CAD is an important component for pyrimidine synthesis [23], we wanted to investigate the effect of providing exogenous pyrimidines on EBOV transcription and replication during siRNA knockdown of CAD. To this end, we performed an siRNA-mediated knockdown of CAD with EBOV minigenomes and treated the cells with 1 mM of either uridine or cytidine. Complementation of uridine resulted in reporter activities similar to the positive controls, indicating that the effect of CAD knockdown on EBOV genome replication and transcription is due to a lack of pyrimidines (Figure 3). When providing cytidine, a similar rescue effect was seen, albeit less pronounced, possibly because cytidine is not metabolized into uridine, whereas exogenous uridine can be metabolized into cytidine during natural pyrimidine synthesis.

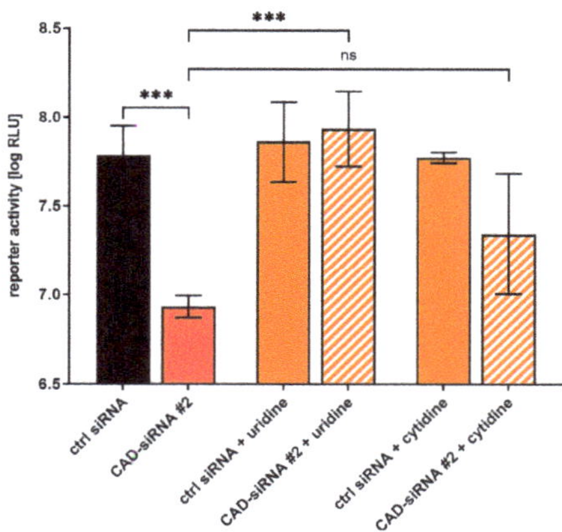

Figure 3. Supplementation of pyrimidines compensates for the effect of CAD knockdown. 293T cells were transfected with siRNAs targeting CAD (CAD-siRNA) or a negative control (ctrl siRNA). 48 h post-transfection, the cells were transfected with all the components required for a replication-competent minigenome assay and treated with 1 mM pyrimidines, either uridine or cytidine. Another 48 h later, the cells were harvested and the reporter activity was measured. The means and standard deviations of 3 independent experiments are shown. Asterisks indicate p-values from a one-way ANOVA (*** $0.0001 < p \leq 0.001$; ns: $p > 0.05$).

3.3. CAD Colocalizes with NP-Induced Inclusion Bodies

Similar to other negative-sense RNA viruses, EBOV and in particular its nucleoprotein NP is known to induce the formation of cytoplasmic inclusion bodies, which are sites of viral genome replication and transcription [8,9]. Since we had shown that CAD is important for EBOV replication and transcription, we wanted to investigate whether the presence of inclusion bodies has an influence on the intracellular distribution of CAD, and in particular whether recruitment of CAD into NP-induced inclusion bodies can be detected. As previously reported, expression of only NP resulted in the formation of inclusion bodies, predominantly in the perinuclear region [5,10,11], while sole expression of CAD led to an even distribution throughout the cytoplasm, with small amounts of CAD present in the nucleus [30] (Figure 4A). During coexpression of NP and CAD we observed relocalization of CAD into NP-induced inclusion bodies (with clear accumulation in inclusion bodies in 70% of the cells, clear exclusion in 0%, and an unclear phenotype in 30%). When we additionally coexpressed VP35, which is involved in nucleocapsid formation during EBOV infection, together with NP [36], we observed a similar relocalization (Figure 4B). To confirm these results, we also performed experiments with infectious EBOV and stained the samples for NP as an inclusion body marker and CAD (Figure 5). Colocalization of CAD and inclusion bodies was still detectable, albeit not as apparent as under conditions of recombinant overexpression of NP and VP35. Taken together, these results suggest that CAD is recruited into viral inclusion bodies to provide sufficient amounts of pyrimidines for EBOV genome replication and transcription.

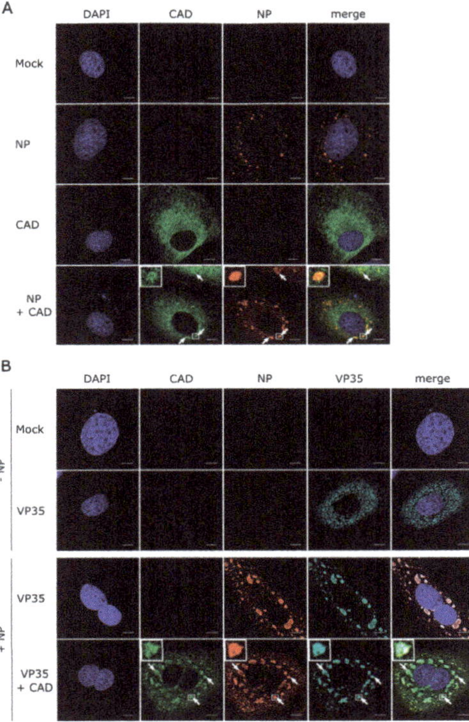

Figure 4. Recruitment of CAD into NP-induced inclusion bodies. (**A**) Colocalization between CAD and NP-induced inclusion bodies. Huh7 cells were transfected with plasmids encoding FLAG/HA-CAD and EBOV-NP as indicated. 48 h post-transfection, the cells were fixed with 4% paraformaldehyde and permeabilized with 0.1% Triton X-100. FLAG-tagged CAD (shown in green) was detected using an anti-FLAG antibody and NP (shown in red) was stained with anti-EBOV NP antibodies. (**B**) Recruitment of CAD into inclusion bodies occurs in the presence of VP35. Huh7 cells were transfected with plasmids encoding FLAG/HA-CAD, EBOV-NP, and myc-EBOV-VP35 as indicated. 48 h post-transfection, the cells were fixed with 4% PFA and permeabilized with 0.1% Triton X-100. FLAG-tagged CAD (shown in green) was detected using an anti-FLAG antibody, NP (shown in red) was stained with anti-EBOV NP antibodies, and myc-tagged VP35 (shown in turquoise) with an anti-myc antibody. The nuclei were stained with DAPI (shown in blue), and the cells were visualized by confocal laser scanning microscopy. The scale bars indicate 10 µm. The arrows point out colocalization, and the insets show magnifications of the indicated areas. Merge shows an overlay of all three channels.

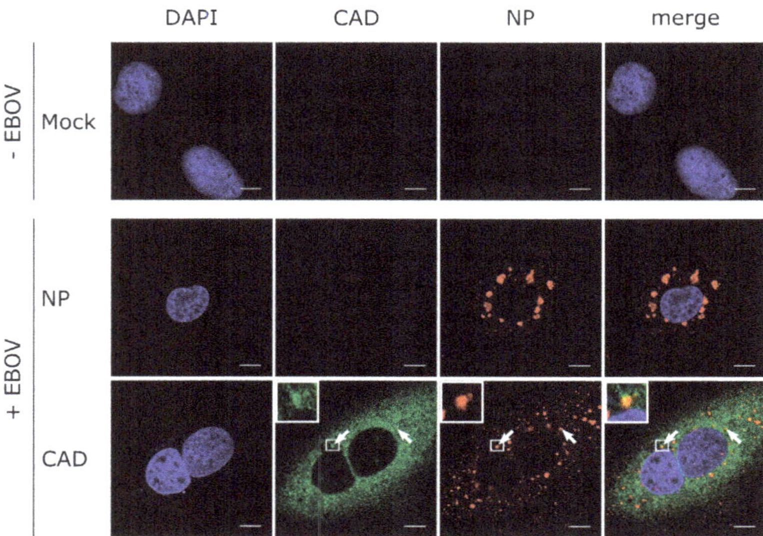

Figure 5. CAD localizes in EBOV inclusion bodies. Huh7 cells were transfected with a plasmid encoding FLAG/HA-CAD. 48 h post-transfection, the cells were infected with rgEBOV at an MOI of 1. After incubation for 16 h, the cells were fixed with 10% formalin and permeabilized with Triton X-100. CAD (shown in green) was detected with an anti-FLAG antibody and NP (shown in red) with an anti-NP antibody. The nuclei were stained with DAPI (shown in blue), and the cells were visualized by confocal laser scanning microscopy. Scale bars indicate 10 μm. The arrows point out colocalization, and the insets show magnifications of the indicated areas. Merge shows an overlay of all three channels.

3.4. The GLN Domain of CAD Is Required for its Accumulation in Inclusion Bodies

To assess the contribution of individual domains of CAD in its recruitment into NP-induced inclusion bodies, we focused on the GLN and the CPS domains. When we expressed deletion mutants lacking these domains, they showed a similar intracellular distribution compared to wild-type CAD when expressed alone in cells. During coexpression of NP and CAD-ΔCPS, we observed recruitment of this mutant into NP-driven inclusion bodies (with clear accumulation in inclusion bodies in 50% of the cells, clear exclusion in 0%, and an unclear phenotype in 50%), indicating that the CPS domain of CAD is not required for its accumulation in inclusion bodies (Figure 6). In stark contrast, when NP was expressed together with CAD-ΔGLN, colocalization with inclusion bodies was abolished (with clear accumulation in inclusion bodies in 0% of the cells, clear exclusion in 68%, and unclear phenotype in 32%), suggesting that the GLN domain is required for recruitment and accumulation in NP-induced inclusion bodies.

3.5. CAD Interacts with NP in an RNA-Independent Manner

As NP recruits CAD into EBOV inclusion bodies, we next assessed whether CAD interacts with NP. To this end, we performed coIP assays using FLAG-CAD expressed in the presence of NP by precipitating CAD with an anti-FLAG antibody and then detecting NP by Western blotting. We could readily co-precipitate NP with CAD, indicating that CAD is able to interact with NP (Figure 6). Because NP is an RNA-binding protein [37], we also tested whether this interaction between CAD and NP is RNA-dependent by treating the samples prior to coIP with RNase A. Under these conditions, we were still able to co-precipitate NP with CAD, demonstrating that the interaction between CAD and NP is not dependent on the presence of RNA (Figure 7).

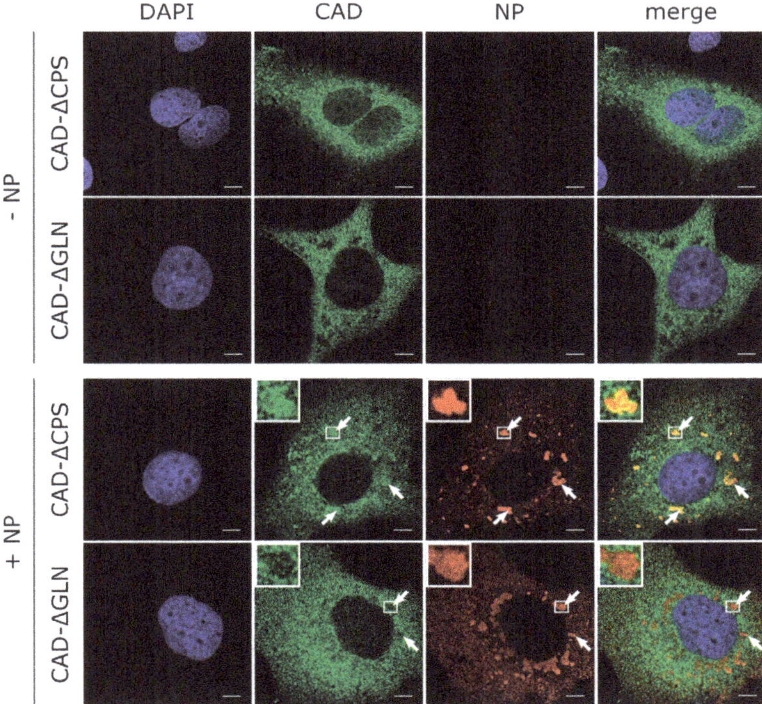

Figure 6. Recruitment of CAD deletion mutants into inclusion bodies. Huh7 cells overexpressing FLAG/HA-CAD-ΔGLN, FLAG/HA-CAD-ΔCPS and EBOV-NP, as indicated, were fixed with 4% PFA and permeabilized with 0.1% Triton X-100 48 h post-transfection. FLAG-tagged CAD (shown in green) was detected using an anti-FLAG antibody and NP (shown in red) was stained with EBOV anti-NP antibodies. The nuclei were stained with DAPI (shown in blue), and the cells were visualized by confocal laser scanning microscopy. Scale bars indicate 10 μm. The arrows point out inclusion bodies, and the insets show magnifications of the indicated areas. Merge shows an overlay of all three channels.

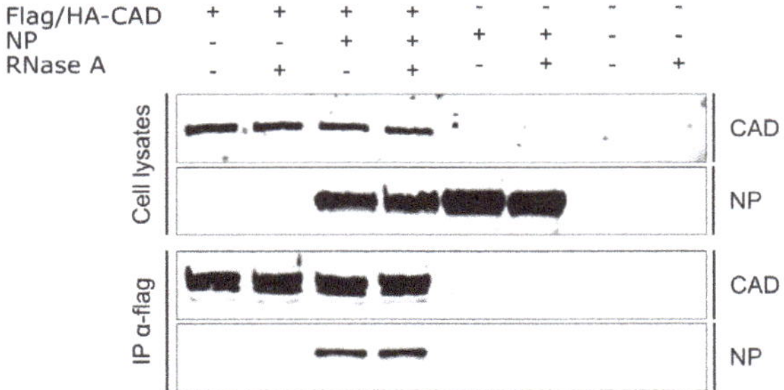

Figure 7. Interaction of CAD with NP. 293T cells were transfected with plasmids encoding FLAG/HA-CAD and EBOV-NP. 48 h post-transfection, the cells were lysed and treated with RNase A (100 μg/mL) or remained untreated. FLAG/HA-CAD was precipitated using anti-FLAG antibodies, and input and precipitates were analyzed via SDS-PAGE and Western blotting using anti-FLAG and anti-NP antibodies. In the CAD IP sample, several bands for CAD are visible, possibly due to posttranslational modifications that are not visible in the lysates because of the overall lower CAD signals in those samples.

4. Discussion

In this work, we identified CAD, an essential component of the de novo pyrimidine synthesis pathway, to be important for both EBOV genome replication and transcription, and demonstrated that the function of CAD in pyrimidine synthesis is responsible for this effect. Knockdown of CAD was also shown to affect replication and transcription of other viruses, e.g., hepatitis C viruses [38]. Furthermore, inhibitors of CAD, e.g., the antinucleoside N-phosphonacetyl-L-aspartate (PALA), which transiently inhibits the aspartate transcarbamylase activity of CAD, were effective in vitro against various viruses, including vaccinia virus and arenaviruses [39,40]. The fact that these compounds exhibit antiviral activity against a broad range of viruses qualifies CAD as a promising indirect antiviral target. However, whether PALA shows antiviral efficiency against EBOV remains to be investigated. Further, whether targeting viral RNA synthesis by inhibition of CAD will be synergistic with other inhibitors of EBOV RNA synthesis, such as remdesivir [41], will have to be addressed in future studies.

Our results are consistent with the fact that several pyrimidine synthesis inhibitors are effective against EBOV in vitro, underlining the importance of the pyrimidine pathway for these viruses [20,42]. Examples are the FDA-approved drug leflunomide and its active metabolite teriflunomide, as well as SW835, a racemic version of GSK983, which has been described to exhibit a broad-spectrum antiviral activity [20,42,43]. These compounds all impair de novo pyrimidine biosynthesis through inhibition of dihydroorotate dehydrogenase (DHODH), an enzyme downstream of CAD in the pyrimidine pathway. Interestingly, treatment with these inhibitors seems to have similar inhibitory effects on EBOV minigenome assays compared to the effect we observe for CAD knockdown, although CAD activity is not directly affected [20,42]. Provision of pyrimidines or upstream metabolites, e.g., orotic acid, reversed antiviral activity of all pyrimidine pathway inhibitors in EBOV minigenome assays, which is consistent with our observation that supplementation with pyrimidines restores reporter activity after CAD knockdown. Interestingly, inhibition of DHODH by using SW835 not only showed pyrimidine depletion, but also stimulated ISG (interferon-stimulated gene) expression, which contributes to the innate immune response [42]. However, currently, the mechanism behind this stimulation of the innate immune response by DHODH inhibitors remains incompletely understood and needs to be further analyzed. Further supporting the importance of CAD for the EBOV lifecycle is the fact that the de novo pyrimidine synthesis activity of CAD is a prerequisite for cell division, which has been suggested to be necessary for productive infection of cells with EBOV [44].

We were further able to show that CAD is recruited to EBOV inclusion bodies, which represent the site of EBOV replication and transcription [8,9]. Since we observed CAD recruitment into NP-induced inclusion bodies during expression of NP alone and detected an interaction of CAD with NP using CoIP studies, we suggest that this recruitment is mediated via an interaction of CAD with NP. So far, knowledge regarding direct interactions between CAD and the proteins of other viruses is limited, but Angeletti et al., showed that CAD recruits the preterminal protein (pTP) of adenoviruses to the site of adenovirus replication in the nuclear matrix via direct interaction. This interaction is believed to be required for anchorage of the adenovirus replication complex at the nuclear matrix in close proximity of the cellular factors required to segregate replicated and genomic viral DNA [45,46].

In the context of its cellular function, CAD has been shown to localize primarily in the cytoplasm, although small amounts can also be detected in the nucleus of dividing cells. Redistribution of CAD into nuclear compartments during cell growth and proliferation is believed to be in response to phosphorylation by MAP kinases at position Thr-456, which results in upregulation of the enzymatic activity of CAD [30]. Since NP is known to recruit a number of factors, including kinases and phosphatases, into inclusion bodies [16–18], it is possible that recruited CAD is activated in inclusion bodies in order to provide pyrimidines for EBOV replication and transcription. However, CAD lacking the CPS domain, which contains Thr-456, was still recruited into NP-induced inclusion bodies, excluding selective recruitment of Thr-456-phosphorylated and thus activated CAD into inclusion bodies.

Overall, we have shown that CAD is recruited into NP-induced and virus-induced inclusion bodies to provide sufficient amounts of pyrimidines for EBOV genome replication and transcription. Furthermore, we demonstrated that the GLN domain of CAD is required for recruitment into inclusion bodies. These findings increase our understanding of EBOV and its host cell interactions, and provide a basis for future identification of molecular targets for the development of novel therapies against this virus.

Supplementary Materials: The following are available online at http://www.mdpi.com/2073-4409/9/5/1126/s1: Figure S1. Comparison of T7 and Pol-II-driven replication-deficient minigenomes.

Author Contributions: Conceptualization, J.B., L.W., T.H.; investigation, J.B., L.W., B.S.B., T.H.; supervision, T.C.M., T.H.; visualization, J.B., T.H.; funding acquisition, T.H.; writing—original draft preparation, J.B., T.H. All authors have read and agreed to the published version of the manuscript.

Funding: Funding was provided by the Deutsche Forschungsgemeinschaft (DFG), grant number 389002253, (J.B., L.W.), and by the Friedrich-Loeffler-Institut, intramural funding (T.H.), and funding as part of the VISION consortium (B.S.B.).

Acknowledgments: The authors would like to thank Logan Banadyga (Public Health Agency of Canada) for his help in establishing the coIP procedure and Stephan Becker (Philipps University Marburg) for providing cell lines. We further thank Luca Zaeck (Friedrich-Loeffler-Institut) for technical assistance with the confocal microscope, as well as Allison Groseth and Sandra Diederich (Friedrich-Loeffler-Institut) for technical assistance with the BSL4 work.

Conflicts of Interest: The authors declare no conflict of interest. The funders had no role in the design of the study; in the collection, analyses, or interpretation of the data; in the writing of the manuscript, or in the decision to publish the results.

References

1. Feldmann, H.; Geisbert, T.W. Ebola haemorrhagic fever. *Lancet* **2011**, *377*, 849–862. [CrossRef]
2. Burk, R.; Bollinger, L.; Johnson, J.C.; Wada, J.; Radoshitzky, S.R.; Palacios, G.; Bavari, S.; Jahrling, P.B.; Kuhn, J.H. Neglected filoviruses. *FEMS Microbiol. Rev.* **2016**, *40*, 494–519. [CrossRef]
3. Bharat, T.A.; Noda, T.; Riches, J.D.; Kraehling, V.; Kolesnikova, L.; Becker, S.; Kawaoka, Y.; Briggs, J.A. Structural dissection of Ebola virus and its assembly determinants using cryo-electron tomography. *Proc. Natl. Acad. Sci. USA* **2012**, *109*, 4275–4280. [CrossRef]
4. Ruigrok, R.W.; Crepin, T.; Kolakofsky, D. Nucleoproteins and nucleocapsids of negative-strand RNA viruses. *Curr. Opin. Microbiol.* **2011**, *14*, 504–510. [CrossRef]
5. Becker, S.; Rinne, C.; Hofsass, U.; Klenk, H.D.; Muhlberger, E. Interactions of Marburg virus nucleocapsid proteins. *Virology* **1998**, *249*, 406–417. [CrossRef]
6. Muhlberger, E.; Weik, M.; Volchkov, V.E.; Klenk, H.D.; Becker, S. Comparison of the transcription and replication strategies of marburg virus and Ebola virus by using artificial replication systems. *J. Virol.* **1999**, *73*, 2333–2342. [CrossRef]
7. Weik, M.; Modrof, J.; Klenk, H.D.; Becker, S.; Muhlberger, E. Ebola virus VP30-mediated transcription is regulated by RNA secondary structure formation. *J. Virol.* **2002**, *76*, 8532–8539. [CrossRef]
8. Hoenen, T.; Shabman, R.S.; Groseth, A.; Herwig, A.; Weber, M.; Schudt, G.; Dolnik, O.; Basler, C.F.; Becker, S.; Feldmann, H. Inclusion bodies are a site of ebolavirus replication. *J. Virol.* **2012**, *86*, 11779–11788. [CrossRef]
9. Lier, C.; Becker, S.; Biedenkopf, N. Dynamic phosphorylation of Ebola virus VP30 in NP-induced inclusion bodies. *Virology* **2017**, *512*, 39–47. [CrossRef]
10. Boehmann, Y.; Enterlein, S.; Randolf, A.; Muhlberger, E. A reconstituted replication and transcription system for Ebola virus Reston and comparison with Ebola virus Zaire. *Virology* **2005**, *332*, 406–417. [CrossRef]
11. Groseth, A.; Charton, J.E.; Sauerborn, M.; Feldmann, F.; Jones, S.M.; Hoenen, T.; Feldmann, H. The Ebola virus ribonucleoprotein complex: A novel VP30-L interaction identified. *Virus Res.* **2009**, *140*, 8–14. [CrossRef] [PubMed]
12. Fang, J.; Pietzsch, C.; Ramanathan, P.; Santos, R.I.; Ilinykh, P.A.; Garcia-Blanco, M.A.; Bukreyev, A.; Bradrick, S.S. Staufen1 Interacts with Multiple Components of the Ebola Virus Ribonucleoprotein and Enhances Viral RNA Synthesis. *mBio* **2018**, *9*. [CrossRef] [PubMed]

13. Chen, J.; He, Z.; Yuan, Y.; Huang, F.; Luo, B.; Zhang, J.; Pan, T.; Zhang, H.; Zhang, J. Host factor SMYD3 is recruited by Ebola virus nucleoprotein to facilitate viral mRNA transcription. *Emerg. Microbes Infect.* **2019**, *8*, 1347–1360. [CrossRef] [PubMed]
14. Batra, J.; Hultquist, J.F.; Liu, D.; Shtanko, O.; Von Dollen, J.; Satkamp, L.; Jang, G.M.; Luthra, P.; Schwarz, T.M.; Small, G.I.; et al. Protein Interaction Mapping Identifies RBBP6 as a Negative Regulator of Ebola Virus Replication. *Cell* **2018**, *175*, 1917–1930. [CrossRef]
15. Gabriel, G.; Feldmann, F.; Reimer, R.; Thiele, S.; Fischer, M.; Hartmann, E.; Bader, M.; Ebihara, H.; Hoenen, T.; Feldmann, H. Importin-alpha7 Is Involved in the Formation of Ebola Virus Inclusion Bodies but Is Not Essential for Pathogenicity in Mice. *J. Infect. Dis.* **2015**, *212* Suppl 2, S316–S321. [CrossRef]
16. Morwitzer, M.J.; Tritsch, S.R.; Cazares, L.H.; Ward, M.D.; Nuss, J.E.; Bavari, S.; Reid, S.P. Identification of RUVBL1 and RUVBL2 as Novel Cellular Interactors of the Ebola Virus Nucleoprotein. *Viruses* **2019**, *11*, 372. [CrossRef]
17. Kruse, T.; Biedenkopf, N.; Hertz, E.P.T.; Dietzel, E.; Stalmann, G.; Lopez-Mendez, B.; Davey, N.E.; Nilsson, J.; Becker, S. The Ebola Virus Nucleoprotein Recruits the Host PP2A-B56 Phosphatase to Activate Transcriptional Support Activity of VP30. *Mol. Cell* **2018**, *69*, 136–145. [CrossRef]
18. Takamatsu, Y.; Krahling, V.; Kolesnikova, L.; Halwe, S.; Lier, C.; Baumeister, S.; Noda, T.; Biedenkopf, N.; Becker, S. Serine-Arginine Protein Kinase 1 Regulates Ebola Virus Transcription. *mBio* **2020**, *11*. [CrossRef]
19. Wendt, L.; Brandt, J.; Bodmer, B.S.; Reiche, S.; Schmidt, M.L.; Traeger, S.; Hoenen, T. The Ebola Virus Nucleoprotein Recruits the Nuclear RNA Export Factor NXF1 into Inclusion Bodies to Facilitate Viral Protein Expression. *Cells* **2020**, *9*, 187. [CrossRef]
20. Martin, S.; Chiramel, A.I.; Schmidt, M.L.; Chen, Y.C.; Whitt, N.; Watt, A.; Dunham, E.C.; Shifflett, K.; Traeger, S.; Leske, A.; et al. A genome-wide siRNA screen identifies a druggable host pathway essential for the Ebola virus life cycle. *Genome Med.* **2018**, *10*, 58. [CrossRef]
21. Uebelhoer, L.S.; Albarino, C.G.; McMullan, L.K.; Chakrabarti, A.K.; Vincent, J.P.; Nichol, S.T.; Towner, J.S. High-throughput, luciferase-based reverse genetics systems for identifying inhibitors of Marburg and Ebola viruses. *Antiviral Res.* **2014**, *106*, 86–94. [CrossRef] [PubMed]
22. Nelson, E.V.; Pacheco, J.R.; Hume, A.J.; Cressey, T.N.; Deflube, L.R.; Ruedas, J.B.; Connor, J.H.; Ebihara, H.; Muhlberger, E. An RNA polymerase II-driven Ebola virus minigenome system as an advanced tool for antiviral drug screening. *Antiviral Res.* **2017**, *146*, 21–27. [CrossRef] [PubMed]
23. Coleman, P.F.; Suttle, D.P.; Stark, G.R. Purification from hamster cells of the multifunctional protein that initiates de novo synthesis of pyrimidine nucleotides. *J. Biol. Chem.* **1977**, *252*, 6379–6385. [PubMed]
24. Jones, M.E. Pyrimidine nucleotide biosynthesis in animals: Genes, enzymes, and regulation of UMP biosynthesis. *Annu. Rev. Biochem.* **1980**, *49*, 253–279. [CrossRef]
25. Lee, L.; Kelly, R.E.; Pastra-Landis, S.C.; Evans, D.R. Oligomeric structure of the multifunctional protein CAD that initiates pyrimidine biosynthesis in mammalian cells. *Proc. Natl. Acad. Sci. USA* **1985**, *82*, 6802–6806. [CrossRef]
26. Christopherson, R.I.; Jones, M.E. The overall synthesis of L-5,6-dihydroorotate by multienzymatic protein pyr1-3 from hamster cells. Kinetic studies, substrate channeling, and the effects of inhibitors. *J. Biol. Chem.* **1980**, *255*, 11381–11395.
27. Irvine, H.S.; Shaw, S.M.; Paton, A.; Carrey, E.A. A reciprocal allosteric mechanism for efficient transfer of labile intermediates between active sites in CAD, the mammalian pyrimidine-biosynthetic multienzyme polypeptide. *Eur. J. Biochem.* **1997**, *247*, 1063–1073. [CrossRef]
28. Evans, D.R.; Guy, H.I. Mammalian pyrimidine biosynthesis: Fresh insights into an ancient pathway. *J. Biol. Chem.* **2004**, *279*, 33035–33038. [CrossRef]
29. Sigoillot, F.D.; Berkowski, J.A.; Sigoillot, S.M.; Kotsis, D.H.; Guy, H.I. Cell cycle-dependent regulation of pyrimidine biosynthesis. *J. Biol. Chem.* **2003**, *278*, 3403–3409. [CrossRef]
30. Sigoillot, F.D.; Kotsis, D.H.; Serre, V.; Sigoillot, S.M.; Evans, D.R.; Guy, H.I. Nuclear localization and mitogen-activated protein kinase phosphorylation of the multifunctional protein CAD. *J. Biol. Chem.* **2005**, *280*, 25611–25620. [CrossRef]
31. Chaparian, M.G.; Evans, D.R. Intracellular location of the multidomain protein CAD in mammalian cells. *FASEB J.* **1988**, *2*, 2982–2989. [CrossRef] [PubMed]
32. Hoenen, T.; Jung, S.; Herwig, A.; Groseth, A.; Becker, S. Both matrix proteins of Ebola virus contribute to the regulation of viral genome replication and transcription. *Virology* **2010**, *403*, 56–66. [CrossRef] [PubMed]

33. Shabman, R.S.; Hoenen, T.; Groseth, A.; Jabado, O.; Binning, J.M.; Amarasinghe, G.K.; Feldmann, H.; Basler, C.F. An upstream open reading frame modulates ebola virus polymerase translation and virus replication. *PLoS Pathog.* **2013**, *9*, e1003147. [CrossRef] [PubMed]
34. Kamper, L.; Zierke, L.; Schmidt, M.L.; Muller, A.; Wendt, L.; Brandt, J.; Hartmann, E.; Braun, S.; Holzerland, J.; Groseth, A.; et al. Assessment of the function and intergenus-compatibility of Ebola and Lloviu virus proteins. *J. Gen. Virol.* **2019**, *100*, 760–772. [CrossRef] [PubMed]
35. Schmidt, M.L.; Hoenen, T. Characterization of the catalytic center of the Ebola virus L polymerase. *PLoS Negl. Trop. Dis.* **2017**, *11*, e0005996. [CrossRef] [PubMed]
36. Huang, Y.; Xu, L.; Sun, Y.; Nabel, G.J. The assembly of Ebola virus nucleocapsid requires virion-associated proteins 35 and 24 and posttranslational modification of nucleoprotein. *Mol. Cell.* **2002**, *10*, 307–316. [CrossRef]
37. Noda, T.; Hagiwara, K.; Sagara, H.; Kawaoka, Y. Characterization of the Ebola virus nucleoprotein-RNA complex. *J. Gen. Virol.* **2010**, *91*, 1478–1483. [CrossRef]
38. Borawski, J.; Troke, P.; Puyang, X.; Gibaja, V.; Zhao, S.; Mickanin, C.; Leighton-Davies, J.; Wilson, C.J.; Myer, V.; Cornellataracido, I.; et al. Class III phosphatidylinositol 4-kinase alpha and beta are novel host factor regulators of hepatitis C virus replication. *J. Virol.* **2009**, *83*, 10058–10074. [CrossRef]
39. Ortiz-Riano, E.; Ngo, N.; Devito, S.; Eggink, D.; Munger, J.; Shaw, M.L.; de la Torre, J.C.; Martinez-Sobrido, L. Inhibition of arenavirus by A3, a pyrimidine biosynthesis inhibitor. *J. Virol.* **2014**, *88*, 878–889. [CrossRef]
40. Katsafanas, G.C.; Grem, J.L.; Blough, H.A.; Moss, B. Inhibition of vaccinia virus replication by N-(phosphonoacetyl)-L-aspartate: Differential effects on viral gene expression result from a reduced pyrimidine nucleotide pool. *Virology* **1997**, *236*, 177–187. [CrossRef]
41. Warren, T.K.; Jordan, R.; Lo, M.K.; Ray, A.S.; Mackman, R.L.; Soloveva, V.; Siegel, D.; Perron, M.; Bannister, R.; Hui, H.C.; et al. Therapeutic efficacy of the small molecule GS-5734 against Ebola virus in rhesus monkeys. *Nature* **2016**, *531*, 381–385. [CrossRef] [PubMed]
42. Luthra, P.; Naidoo, J.; Pietzsch, C.A.; De, S.; Khadka, S.; Anantpadma, M.; Williams, C.G.; Edwards, M.R.; Davey, R.A.; Bukreyev, A.; et al. Inhibiting pyrimidine biosynthesis impairs Ebola virus replication through depletion of nucleoside pools and activation of innate immune responses. *Antiviral Res.* **2018**, *158*, 288–302. [CrossRef] [PubMed]
43. Deans, R.M.; Morgens, D.W.; Okesli, A.; Pillay, S.; Horlbeck, M.A.; Kampmann, M.; Gilbert, L.A.; Li, A.; Mateo, R.; Smith, M.; et al. Parallel shRNA and CRISPR-Cas9 screens enable antiviral drug target identification. *Nat. Chem. Biol.* **2016**, *12*, 361–366. [CrossRef] [PubMed]
44. Kota, K.P.; Benko, J.G.; Mudhasani, R.; Retterer, C.; Tran, J.P.; Bavari, S.; Panchal, R.G. High content image based analysis identifies cell cycle inhibitors as regulators of Ebola virus infection. *Viruses* **2012**, *4*, 1865–1877. [CrossRef] [PubMed]
45. Angeletti, P.C.; Engler, J.A. Adenovirus preterminal protein binds to the CAD enzyme at active sites of viral DNA replication on the nuclear matrix. *J. Virol.* **1998**, *72*, 2896–2904. [CrossRef] [PubMed]
46. Fredman, J.N.; Engler, J.A. Adenovirus precursor to terminal protein interacts with the nuclear matrix in vivo and in vitro. *J. Virol.* **1993**, *67*, 3384–3395. [CrossRef]

Article

The Ebola Virus Nucleoprotein Recruits the Nuclear RNA Export Factor NXF1 into Inclusion Bodies to Facilitate Viral Protein Expression

Lisa Wendt [1], Janine Brandt [1], Bianca S. Bodmer [1], Sven Reiche [2], Marie Luisa Schmidt [1,†], Shelby Traeger [3,‡] and Thomas Hoenen [1,*]

1 Institute of Molecular Virology and Cell Biology, Friedrich-Loeffler-Institut, 17493 Greifswald, Germany; lisa.wendt@fli.de (L.W.); janine.brandt@fli.de (J.B.); bianca.bodmer@fli.de (B.S.B.); marie.schmidt@charite.de (M.L.S.)

2 Department of Experimental Animal Facilities and Biorisk Management, Friedrich-Loeffler-Institut, 17493 Greifswald, Germany; sven.reiche@fli.de

3 Laboratory of Virology, Division of Intramural Research, National Institute for Allergy and Infectious Diseases, National Institutes of Health, Hamilton, MT 59840, USA; shelbytraeger8@gmail.com

* Correspondence: thomas.hoenen@fli.de

† Current affiliation: Institute of Virology, Charité-Universitätsmedizin Berlin, Corporate Member of Freie Universität Berlin, Humboldt-Universität zu Berlin, and Berlin Institute of Health, 10117 Berlin, Germany.

‡ Current affiliation: Bellingham Technical College, Bellingham, WA 98225, USA.

Received: 20 December 2019; Accepted: 9 January 2020; Published: 11 January 2020

Abstract: Ebola virus (EBOV) causes severe outbreaks of viral hemorrhagic fever in humans. While virus-host interactions are promising targets for antivirals, there is only limited knowledge regarding the interactions of EBOV with cellular host factors. Recently, we performed a genome-wide siRNA screen that identified the nuclear RNA export factor 1 (NXF1) as an important host factor for the EBOV life cycle. NXF1 is a major component of the nuclear mRNA export pathway that is usurped by many viruses whose life cycles include nuclear stages. However, the role of NXF1 in the life cycle of EBOV, a virus replicating in cytoplasmic inclusion bodies, remains unknown. In order to better understand the role of NXF1 in the EBOV life cycle, we performed a combination of co-immunoprecipitation and double immunofluorescence assays to characterize the interactions of NXF1 with viral proteins and RNAs. Additionally, using siRNA-mediated knockdown of NXF1 together with functional assays, we analyzed the role of NXF1 in individual aspects of the virus life cycle. With this approach we identified the EBOV nucleoprotein (NP) as a viral interaction partner of NXF1. Further studies revealed that NP interacts with the RNA-binding domain of NXF1 and competes with RNA for this interaction. Co-localization studies showed that RNA binding-deficient, but not wildtype NXF1, accumulates in NP-derived inclusion bodies, and knockdown experiments demonstrated that NXF1 is necessary for viral protein expression, but not for viral RNA synthesis. Finally, our results showed that NXF1 interacts with viral mRNAs, but not with viral genomic RNAs. Based on these results we suggest a model whereby NXF1 is recruited into inclusion bodies to promote the export of viral mRNA:NXF1 complexes from these sites. This would represent a novel function for NXF1 in the life cycle of cytoplasmically replicating viruses, and may provide a basis for new therapeutic approaches against EBOV, and possibly other emerging viruses.

Keywords: Ebola virus; filovirus; inclusion bodies; NXF1; liquid organelles; mRNA export

---

1. Introduction

Ebola virus (EBOV) belongs to the genus *Ebolavirus* within the family *Filoviridae* and causes a severe hemorrhagic fever, called Ebola virus disease, in humans with high case fatality rates of about

40–60% [1,2]. Ongoing and past outbreaks of Ebola virus disease in Africa highlight the importance of a better understanding of the EBOV life cycle in order to develop new therapeutic approaches. During the viral life cycle the EBOV nucleoprotein (NP) encapsidates the negative-stranded RNA genome and is essential for viral replication and transcription [3]. NP interacts with the transcriptional activator viral protein 30 (VP30), which bridges NP and the RNA-dependent RNA polymerase L [4–6]. Furthermore, NP interacts with the polymerase cofactor VP35 [5,6]. This interaction regulates the oligomerization and RNA-binding of NP, and also bridges NP to L [5–9]. NP, VP35, VP30, and L, together with the RNA genome, form the ribonucleoprotein complex (RNP) and are sufficient to mediate viral replication and transcription [3], which takes place in cytoplasmic inclusion bodies [10]. The formation of these inclusion bodies is driven by expression of NP, which is localized in these structures not only during infection, but also after sole expression of this protein [5,6]. However, only limited knowledge exists regarding host factors that interact with the viral proteins and RNAs found in these structures. One such host factor that has been identified is importin-α7, which seems to be involved in inclusion body formation [11]. Marburg virus, a close relative of EBOV, was shown to recruit components of the endosomal sorting complex required for transport (ESCRT) to inclusion bodies to facilitate the trafficking of nucleocapsids to the plasma membrane for viral assembly and budding [12,13]. Kinases and phosphatases such as PP2A-B56 are also known to be recruited to inclusion bodies, and are important in regulating the activity of VP30 in viral RNA synthesis, which is dependent on its phosphorylation status [14,15]. Similarly, RBBP6 appears to regulate the balance of replication and transcription by binding to VP30, and also Staufen1 was described to influence viral RNA synthesis [16,17]. Finally, EBOV VP35 appears to sequester cellular stress granule proteins within inclusion bodies in order to prevent stress granule formation [18].

To obtain a comprehensive picture of the pro- and anti-viral factors that are important for EBOV RNA synthesis (i.e., genome replication and transcription) and/or protein expression, we recently performed a genome-wide siRNA screen [19]. As primary readout we used a minigenome assay (reviewed in [20]). In this assay RNA minigenomes, i.e., miniature versions of the EBOV genome with all viral genes removed and replaced with a reporter gene, are expressed in mammalian cells together with the viral RNP proteins. Since the minigenomes still contain the regulatory terminal leader and trailer regions of the EBOV genome that carry the replication and transcription promoters, the RNP proteins recognize these minigenomes as authentic templates. This results in their replication and transcription, and ultimately reporter protein expression reflecting these aspects of the viral life cycle. In our siRNA screen we showed that the nuclear RNA export factor 1 (NXF1) is necessary for the EBOV life cycle, and also for the life cycle of the highly pathogenic New World arenavirus Junín virus. These data suggest a mechanism of action that may be conserved among several cytoplasmically replicating negative-stranded RNA viruses (NSVs) that share commonalties in their replication cycles, such as replication in cytoplasmic inclusion bodies [10,21–24]. Importantly, we also could show that NXF1 is important for the life cycle of EBOV in context of infectious virus [19]. However, the precise function of NXF1 in the EBOV life cycle remained unclear.

NXF1 is a crucial component of the nuclear mRNA export pathway, where it exports cellular mRNAs from the nucleus by interacting with nucleoporins [25,26]. For this interaction NXF1 forms a stable heterodimer with p15 (also called NXT1) via its NTF2-like domain (NTF2) [27,28]. The NTF2-like domain is one of the five functional domains found in NXF1, with the other ones being the RNA-binding domain (RBD), the pseudo RNA recognition motif (RRM), a region of leucine-rich repeats (LRR), and the ubiquitin-associated domain (UBA) [27,29,30]. NTF2 and UBA promote interaction with nucleoporins, while the three amino-terminal domains (RBD, RRM, and LRR) are important for mRNA binding [26,27,31]. Recently, NXF1 was shown to be loaded onto cellular mRNAs co-transcriptionally, a process that depends on several other export adapters that promote the interaction between NXF1 and mRNA, in order to enable mRNA export from the nucleus [29,31,32]. However, the NXF1-driven nuclear mRNA export pathway is also utilized by many viruses that replicate in the nucleus [33–36]. For instance, herpesviruses and influenza viruses usurp this host pathway to export their viral RNAs, and particularly their mRNAs, from the nucleus [33,34,37–39]. Furthermore, NXF1 is known to export retroviral RNAs containing constitutive transport elements and Hepatitis B virus pregenomic RNA [35,36]. Thus far, the only non-nuclear function of NXF1 has been described for gammaretroviruses, which use NXF1 for loading of gag RNA onto polysomes, rather than for mRNA export [40]. However, since EBOV replicates in cytoplasmic inclusion bodies, the function of NXF1 during an EBOV infection must differ from those previously described [10]. Therefore, we characterized the interaction of NXF1 with EBOV on a biochemical and functional level. Based on our results we propose a model in which NXF1 is recruited to inclusion bodies where it interacts with NP and subsequently with viral mRNAs, which it then exports from inclusion bodies.

2. Materials and Methods

2.1. Cells

HEK 293T (human embryonic kidney) cells (Collection of Cell Lines in Veterinary Medicine CCLV-RIE 1018) and Huh7 (human hepatocarcinoma) cells (kindly provided by Stephan Becker, Philipps University Marburg, Marburg, Germany) were maintained in Dulbecco's modified Eagle's medium (DMEM; ThermoFisher Scientific, Waltham, MA, USA) supplemented with 10% fetal bovine serum (FBS), 100 U/mL penicillin and 100 µg/mL streptomycin (PS; ThermoFisher Scientific) and 1× GlutaMAX (ThermoFisher Scientific). All cells were grown at 37 °C with 5% CO_2.

2.2. Plasmids and Antibodies

Expression plasmids for the RNP proteins, the T7 polymerase and the T7-driven monocistronic minigenomes (pT7-1cis-vRNA-rLuc, pT7-1cis-vRNA-nLuc) have been previously described [19,41]. Plasmids for the expression of N-terminally flag/HA-tagged constructs were generated by insertion of the respective genes (NXF1, p15, NP) into a pCAGGS-flag/HA vector. All mutant versions and individual domains of NXF1 were first inserted into pCAGGS and subsequently subcloned into the pCAGGS-flag/HA plasmid. EBOV VP40 and VP24 were also cloned into pCAGGS. The PolII-driven replication-deficient minigenome expressing NanoLuc luciferase as its reporter was generated by flanking the minigenome in vRNA-orientation with hammerhead and Hepatitis Delta Virus ribozymes and insertion of the minigenome into pCAGGS. Deletion of 55 nt in the antigenomic replication promotor to generate a replication-deficient version of the PolII-driven minigenome was performed as previously described [41]. Detailed cloning strategies are available on request.

The anti-flag (mouse anti-flag mAb, clone M2, cat. no. F1804) antibody used for co-immunoprecipitation assays (coIPs) and Western blot analyses was purchased from Sigma-Aldrich. Primary antibodies against NP (rabbit anti-EBOV NP pAb, cat. no. 0301-012), VP35 (rabbit anti-ZEBOV VP35 affinity purified polyclonal antibody, cat. no. 0301-040) and VP40 (mouse anti-EBOV VP40 mAb, clone 3G5, cat. no. 0201-016) were obtained from IBT Bioservices, and the antibody against NXF1 (mouse anti-NXF1 mAb, clone 53H8, cat. no. ab50609) was obtained from Abcam. For VP30, a mouse monoclonal antibody (clone 5F11 A1) was produced by the FLI biobank as previously described [42] and following approved protocols using affinity-purified flag/HA-tagged VP30 expressed in mammalian cells as the immunogen. Secondary antibodies against mouse (Peroxidase AffiniPure Goat Anti-Mouse IgG, light chain specific, cat. no. 115-035-174, used for Western blots with exception of that shown in Figure 4) and rabbit (Peroxidase AffiniPure Goat Anti-Rabbit IgG H+L, cat. no. 111-035-003) were obtained from Jackson Immunoresearch or Abcam (anti-mouse IgG for IP, HRP, cat. no. ab131368, used for the Western blot shown in Figure 4).

2.3. Co-Immunoprecipitation of Viral Proteins

For coIPs, 293T cells were seeded into 6-well plates and transfected at 30% confluency with expression plasmids encoding the flag/HA-tagged host proteins (NXF1, p15, mutated versions or individual domains of NXF1) and for their putative interaction partners (p15, NXF1 or EBOV proteins) using Transit LT-1 (Mirus Bio LLC, Madison, WI, USA) following the manufacturer's instructions. Supernatant was exchanged after 24 h and coIP was performed 48 h post transfection (p.t.). To this end, cells were washed in PBS and pelleted before being lysed in 1 mL coIP lysis buffer (1% NP-40; 50 mM Tris; 150 mM NaCl; pH 7.4) with protease inhibitor (cOmplete; Roche, Basel, Switzerland) and incubated with rotation at 15 rpm for 2 h at 4 °C. Initial experiments investigating an interaction between viral proteins and NXF1 (Figure 1A) where performed in the absence of RNase treatment. However, since RNase treatment increases the interaction between NXF1 and NP (see Figure 2), in all further experiments 100 μg/mL RNase A (Machery-Nagel, Düren, Germany) was added to the samples prior to incubation. Lysates were separated from cell debris by centrifugation (11,000× *g*, 10 min, 4 °C) and 150 μL was used for the input control (representing a sixth of the whole lysate and 20% of the sample subjected to IP) and subjected to acetone precipitation. To this end 750 μL acetone pre-chilled to −20 °C was added to the samples, and they were incubated for at least 30 min at −20 °C. Proteins were then pelleted by centrifugation (20,000× *g*, 15 min, 4 °C), resuspended in 60 μL 1× SDS sample buffer and boiled for 10 min at 99 °C. The remaining 750 μL of lysate was added to the prepared bead-antibody solution (Dynabeads Protein G, ThermoFisher Scientific; 1 μL anti-flag M2 antibody per 10 μL beads), resuspended, and the IP was performed for 10 min (as recommended by the manufacturer to minimize non-specific binding) at room temperature with rotation at 15 rpm. Beads were washed three times with PBS containing 0.02% Tween-20 and transferred to new tubes. The beads were resuspended in 60 μL 1× SDS sample buffer and boiled for 10 min at 99 °C. Input and coIP samples were analyzed via SDS-PAGE and semi-dry Western blot, using an ethanol-containing blotting buffer (5.8 g Tris(hydroxymethyl)aminomethan; 2.9 g Glycin; 200 mL Ethanol; ad 1 mL H_2O) together with Nitrocellulose membranes and a blotting current of 0.8 mA/cm^2 for 1 to 2 h, and 7% skim milk powder in PBS for blocking.

2.4. Immunofluorescence Analysis

Huh7 cells were seeded onto coverslips in 12-well plates and transfected the next day with 500 ng pCAGGS-EBOV-NP, 500 ng pCAGGS-p15 and 500 ng pCAGGS-flag/HA-NXF1 (or pCAGGS-flag/HA-NXF1-10RA or pCAGGS-flag/HA-NXF1-ΔRBD) using Polyethylenimine (Sigma-Aldrich) following the manufacturer's instructions, or remained untransfected (mock). Forty-eight hours p.t., cells were fixed using 4% PFA in DMEM, treated with 0.1 M glycine and permeabilized with 0.1% Triton X-100 (in PBS++ (PBS with 1 mM Ca^{2+} and 1 mM Mg^{2+}) with 10% fetal calf serum). Primary antibodies (rabbit anti-EBOV-NP 1:500; mouse anti-flag 1:2000) were diluted in PBS with 10% fetal calf serum, and cells were incubated for 1 h at 4 °C with the antibody solutions. The same procedure was used for the secondary antibodies (Alexa Fluor-568 anti-rabbit 1:1500, Alexa Fluor-488 anti-mouse 1:1200; both ThermoFisher Scientific). After staining, cells were washed with PBS and water before mounting with ProLong Diamond Antifade mountant with DAPI (ThermoFisher Scientific). Slides were analyzed by confocal laser scanning microscopy using a Leica SP5.

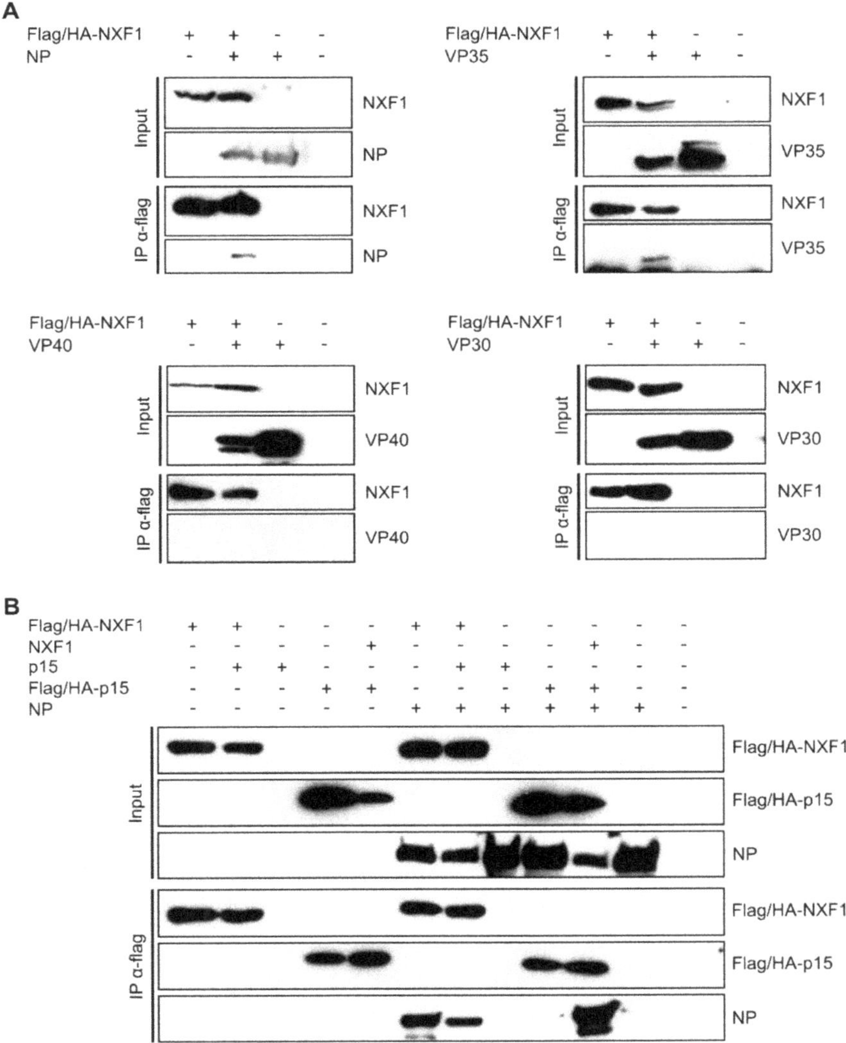

Figure 1. Interaction of NXF1 with nucleoprotein (NP) and VP35. (**A**) Immunoprecipitation of NXF1 and coIP of viral proteins. 293T cells were transfected with the plasmids encoding for flag/HA-NXF1 and one of the viral proteins NP, VP35, VP40 or VP30. Forty-eight hours p.t. flag/HA-NXF1 was precipitated using anti-flag antibodies, and input and precipitates were analyzed via SDS-PAGE and subsequent Western blot. NXF1 was detected with anti-flag antibodies, and anti-NP, anti-VP35, anti-VP40 or anti-VP30 antibodies were used for the detection of the respective viral protein. In case of the VP35 blot, the band at the lower edge of the blot that is partially cut off is caused by the light chain of the IP antibody. (**B**) Influence of p15 on the interaction between NXF1 and NP. Flag/HA-tagged NXF1 or p15 was precipitated from transfected 293T cells using anti-flag antibodies. Input and precipitate samples were subjected to SDS-PAGE and analyzed by Western blot using anti-flag antibodies for the detection of p15 and NXF1, whereas an anti-NP antibody was used for the detection of NP.

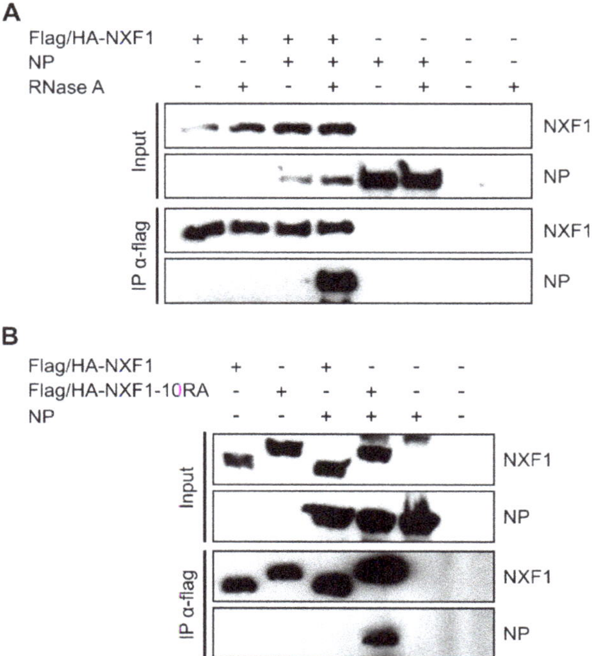

Figure 2. RNA dependence of the interaction between NXF1 and NP. (**A**) Absence of single-stranded RNA enhances interaction between NP and NXF1. 293T cell lysates containing overexpressed flag/HA-NXF1 and NP were treated with RNase A (100 µg/mL) or remained untreated before flag/HA-NXF1 was precipitated using anti-flag antibodies. Input and precipitates were analyzed via SDS-PAGE and Western blot using anti-flag and anti-NP antibodies. (**B**) RNA binding-deficient NXF1-10RA displays a stronger interaction with NP than wildtype NXF1. 293T cells were transfected with plasmids encoding for either flag/HA-NXF1 or flag/HA-NXF1-10RA together with NP, and 48 h p.t. cells were lysed and subjected to coIP analyses. Pulldowns were performed using anti-flag antibodies, and precipitates as well as input samples were analyzed via SDS-PAGE and Western blot using anti-flag and anti-NP antibodies.

2.5. siRNA Knockdown and Minigenome Assay

293T cells were reverse transfected with 12 pmol pre-designed silencer select siRNAs (NXF1 #1: s20532; NXF1 #2: s20533; Negative Control siRNA No. 2; all ThermoFisher Scientific) or an anti-L siRNA (5′-UUU AUA UAC AGC UUC GUA CUU-3′) using Lipofectamine RNAiMax (ThermoFisher Scientific) following the manufacturer's instructions in 12-well plates. Forty-eight hours p.t., cells were transfected with all plasmids for the respective minigenome assay using Transit LT-1 as described before [41]. For the classical T7-driven replication-competent minigenome system (including the analysis of mRNA levels), the cells were transfected with the plasmids encoding the T7-polymerase, the T7-driven monocistronic minigenome (1cis-vRNA-nLuc), the RNP proteins L, NP, VP30 and VP35 and firefly luciferase or green fluorescent protein (GFP) as a control. For the replication-deficient minigenome system, cells were transfected with the plasmids encoding the PolII-driven replication-deficient minigenome (pCAGGS-1cis-vRNA-nLuc-RdM), the RNP proteins L, NP, VP30 and VP35 and firefly luciferase as a control. A further 48 h later cells were harvested for either determination of reporter activity or RNA isolation (see below). For measuring the reporter activity, cells were lysed for 10 min in 1× Lysis Juice (PJK, Kleinblittersdorf, Germany) and lysates were cleared by centrifugation (3 min at 10,000× *g*, room temperature). Forty µL of cleared lysate was added to either 40 µL of Beetle Juice

(PJK) or Nano-Glo Luciferase Assay Reagent (Promega, Madison, WI, USA) in opaque 96-well plates. Luminescence was measured using a Glomax Multi (Promega) microplate reader. Minigenome reporter activity was normalized to control firefly luciferase activity, which acts as a measure for plasmid-driven gene expression.

2.6. Infection Experiments

siRNA-transfected 293T cells in 12-well format were infected with 5000 TCID50 rgEBOV-luc2 [43] 24–48 h p.t. Twenty-four hours post infection (p.i.) cells were lysed in 300 µL GloLysis buffer (Promega) for 10 min and 50 µL samples were mixed with either 50 µL BrightGlo (Promega) or 50 µL CellTiterGlo (Promega) in white opaque 96-well plates and measured on a Tecan F200Pro. Firefly luciferase activity reflecting viral gene expression was normalized to luciferase activity in the CellTiterGlo assay (which quenches firefly luciferase activity and instead measures intracellular ATP levels as proxy for cell viability using an UltraGlo luciferase). For mRNA quantification, identically transfected and infected cells were harvested in 700 µL RLT buffer (Qiagen, Hilden, Germany), and after addition of 700 µL 70% EtOH and removal from the BSL4-laboratory RNA was extracted using the RNEasy kit (Qiagen) following the manufacturer's instructions. mRNA quantification was performed as described below.

For immunofluorescence analysis, Huh7 cells in ibidi 8-well chamber slides were infected with rgEBOV-wt [44] at an MOI of 1. Sixteen hours p.i., cells were washed with PBS and fixed using 10% formalin. The subsequent procedure for immunofluorescence staining was performed as described above. All experiments involving infectious EBOV were done in the BSL4 laboratory of the Friedrich-Loeffler-Institut following approved SOPs.

2.7. RNA Co-Immunoprecipitation

For mRNA-coIPs 293T cells were transfected using Transit LT-1 (Mirus Bio LLC) following the manufacturer's instructions with the plasmids encoding for flag/HA-NXF1 or its mutants (1500 ng; amounts per well), GFP (250 ng) as a transfection control, as well as all T7 monocistronic minigenome components (T7—250 ng, L—1000 ng, NP—125 ng, VP30—65 ng, VP35—125 ng, monocistronic EBOV minigenome encoding Renilla luciferase—250 ng), while as a control either L (-L) or flag/HA-NXF1 (-NXF1) was omitted. For vRNA-coIPs 293T cells were transfected with the plasmids encoding for the T7 polymerase (250 ng) and the T7-driven monocistronic minigenome encoding Renilla luciferase (250 ng) together with either flag/HA-NXF1 (1500 ng) or a combination of flag/HA-NP (1500 ng), VP35 (1250 ng) and VP24 (1250 ng). As controls, either the minigenome (-1cis) or the flag/HA-tagged protein (1cis w/o flag-protein) was omitted. The coIP procedure was then carried out 48 h p.t. as described above for the protein coIPs, but the beads were resuspended in 140 µL PBS instead of 1× SDS sample buffer and not boiled before they were subjected to RNA isolation and RT-qPCR.

2.8. RNA Isolation and RT-qPCR

RNA isolation from minigenome-transfected cell lysates was performed using the NucleoSpin RNA kit (Machery-Nagel), and RNA isolation from coIP precipitates was performed using the QIAmp Viral RNA Mini kit (Qiagen), both according to the manufacturers' instructions. The beads were separated from the precipitate-buffer solution with the use of a magnetic rack prior to loading of the spin columns to avoid clogging of the columns. All RNA samples were treated with DNase (TURBO DNA-free kit; ThermoFisher Scientific) following the manufacturer's instructions to avoid plasmid contamination. The resulting RNA samples were then used for the generation of cDNA with either oligo(dT)-primers for total mRNA or strand-specific primers for the reporter gene in vRNA-orientation (Renilla: luc(−) 5′-CCA AAC AAG CAC CCC AAT CA-3′) using RevertAid Reverse Transcriptase (ThermoFisher Scientific) following the manufacturer's instructions. The subsequent qPCR was performed using the PowerUp SYBR Green Master Mix (ThermoFisher Scientific) with 1 µL of cDNA and primers targeting either the reporter genes (NanoLuc: nluc(−) 5′-TTC AGA ATC TCG GG GTG TCC-3′, nluc(+): 5′-CGT AAC CCC GTC GAT TAC CA-3′; Renilla: luc(−), luc(+) 5′-TGT

GCC ACA TAT TGA GCC AG-3′; firefly: luc2(−) 5′-CAG TCG TCG TGC TGG AAC AC-3′, luc2(+) 5′-GTC CAA CTT GCC GGT CAG TC-3′), or as a control for the minigenome experiments GFP (GFP(+) 5′-CTT GTA CAG CTC GTC CAT GC-3′, GFP(−) 5′-CGA CAA CCA CTA CCT GAG CAC-3′) and for the infection experiments GAPDH (GAPDH(−) 5′-CCA GGT GGT CTC CTC TGA CTT CAA-3′, GAPDH(+) 5′-ATA CCA GGA AAT GAG CTT GAC A-3′). Values for luciferase reporter mRNA levels were normalized to GFP/GAPDH mRNA levels and values for the mRNA coIP samples were normalized to their respective -L controls.

2.9. Statistical Analyses

One-way ANOVA with Dunett's multiple comparisons test (Figure 7A–E) or Sidak's multiple comparisons test (Figure 7F–G) were performed using the GraphPad Prism 8 software.

3. Results & Discussion

3.1. NXF1 Interacts with NP and VP35

To assess whether NXF1 interacts with EBOV proteins, we performed co-immunoprecipitation (coIP) assays using overexpressed flag/HA-NXF1 in the presence of either NP, VP35, VP30, or the EBOV matrix protein VP40, to find a possible interaction partner (Figure 1A). Using this approach, we could co-precipitate NP as well as VP35 with NXF1 (Figure 1A, second lanes), indicating that NXF1 interacts with both viral proteins. In contrast, VP40 and VP30 could not be co-precipitated with NXF1, indicating that most likely there is no interaction between NXF1 and these viral factors. Since NXF1 is known to form stable heterodimers with p15 in order to promote the interaction of NXF1 with nucleoporins and thereby fulfill its cellular function in nuclear export, we also performed coIPs of NP with NXF1 in presence and absence of overexpressed p15 [25,28]. The amount of NP co-precipitated with NXF1 was not increased in presence of p15 (Figure 1B, sixth and seventh lane). Further, p15 alone failed to co-precipitate NP, while co-precipitation could be recovered by NXF1 overexpression (Figure 1B, ninth and 10th lane). These data show that p15 is dispensable for the interaction between NXF1 and NP, and that the formation of the NXF1:p15 heterodimer is not necessary to mediate an interaction with NP. This is in contrast to other viruses interacting with NXF1 such as Hepatitis B virus, where HBc was shown to interact with both NXF1 and p15 [35].

3.2. NP and Cellular RNAs Compete for Interaction with NXF1

Since NP, VP35, and NXF1 are all known to be RNA-binding proteins, we next analyzed the dependence of their interaction on RNA by treating samples with RNase A to digest single-stranded RNAs before they were used for coIPs [25,45,46]. In case of VP35 the interaction with NXF1 was clearly dependent on the presence of RNA (Figure S1), raising the possibility that the observed interaction is merely due to unspecific binding of the same RNA, so that we did not further investigate this interaction, but rather focused on the interaction between NP and NXF1. Here, to our surprise, the experiments revealed that the interaction between NP and NXF1 was stronger in the presence of RNase A (Figure 2A, third and fourth lane), indicating a negative impact of single-stranded RNA on their interaction. To confirm this result, we performed coIPs with a mutant version of NXF1 that has been described to have lost its RNA-binding capability due to an exchange of ten arginines to alanines within its RNA-binding domain [29,31]. Using this mutant (NXF1-10RA) we observed that much more NP was co-precipitated than when using wildtype NXF1 (Figure 2B, third and fourth lane), similar to the results seen after RNase A treatment. Indeed, it was not possible to find conditions where both the very strong band for NP co-precipitated by NXF1-10RA and the much weaker band for NP co-precipitated by wildtype NXF1 could be visualized together with reasonable band intensities. This suggests that the observed differences in NXF1:NP interaction in the presence/absence of RNA are linked to the RNA-binding activity of NXF1. The more prominent interaction between NXF1 and NP in the absence of RNA-binding by NXF1 further suggests that there is competition between NP and single-stranded RNA for the binding of NXF1. Finally, the strong interaction between NXF1-10RA and NP also shows that the amino acids within NXF1 that are important for its RNA-binding capability are dispensable for the interaction with NP.

3.3. Interaction with NP Is Mediated by the RNA-Binding Domain of NXF1

To identify the domain of NXF1 that facilitates the interaction with NP, we generated different mutants of NXF1, each lacking one of its five described domains (Figure 3A) [29]. Using these mutants in coIP studies, we could co-precipitate NP with four of the five deletion mutants (ΔRRM, ΔLRR, ΔNTF2, ΔUBA), demonstrating that none of these domains is essential for the interaction (Figure 3B, fourth lanes). In contrast, NXF1-ΔRBD failed to co-precipitate NP, indicating that it is the RBD that is required for the interaction between these proteins.

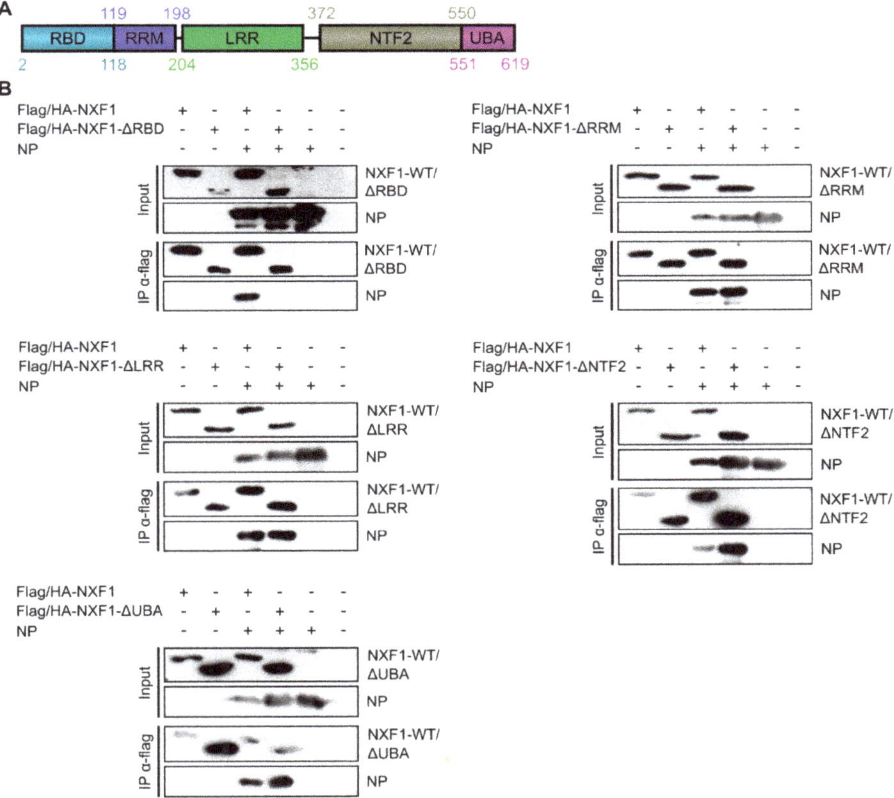

Figure 3. Identification of NXF1 domains necessary for the interaction with NP. (**A**) Domain organization of NXF1. Numbers indicate amino acid positions. (**B**) The RNA-binding domain of NXF1 is essential for its interaction with NP. 293T cells were transfected with plasmids encoding flag/HA-NXF1 or flag/HA-tagged NXF1 mutants together with NP. Forty-eight hours p.t., immunoprecipitation was performed using anti-flag antibodies. Subsequent SDS-PAGE and Western blot analyses with input and precipitates were performed using anti-flag and anti-NP antibodies.

To further analyze whether the RBD is also sufficient to mediate an interaction with NP, we also performed coIPs of NP with only the RBD of NXF1, as well as with NXF1-RRM and NXF1-RBD-RRM as controls. In these experiments, NP could be co-precipitated with the RBD and the RBD-RRM, but not with the RRM alone (Figure 4, sixth, eighth and seventh lane). This demonstrates that the RBD is not only required, but also sufficient for the interaction between NP and NXF1, while the RRM is neither required or sufficient for this interaction, nor does it interfere with it.

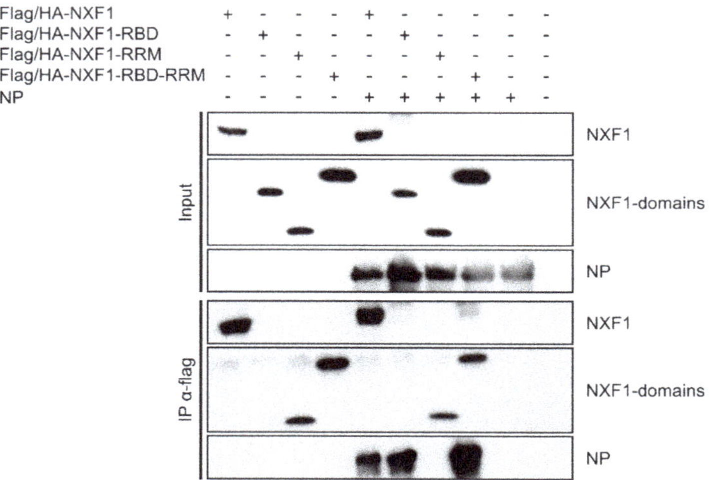

Figure 4. Role of the RNA-binding domain (RBD) and RNA recognition motif (RRM) in the interaction between NXF1 and NP. Immunoprecipitations of lysates from 293T cells transfected with flag/HA tagged NXF1, NXF1-RBD, NXF1-RRM or NXF1-RBD-RRM together with NP using anti-flag antibodies were performed 48 h p.t. Input and precipitate samples were subjected to SDS-PAGE and Western blot, and NXF1 and NP were detected using anti-flag and anti-NP antibodies, respectively.

3.4. NXF1 Is Recruited into NP-Derived Inclusion Bodies

Genome replication and transcription of EBOV is known to take place in inclusion bodies within the cytoplasm [10,14], and overexpression of NP alone leads to formation of such inclusion bodies [6]. In contrast, NXF1 is mainly localized in and around the nucleus [25,26]. Therefore, we wanted to analyze the influence of NP overexpression (and thus inclusion body formation) on the intracellular localization of NXF1. As previously shown [5,6], NP formed the typical inclusion bodies in the cytoplasm when expressed alone (Figure 5), and thus was used as a marker for these structures, while overexpression of NXF1 alone showed a localization in the nucleus and in the perinuclear region. This is in line with previous publications, which have described both a nuclear localization, and particularly a nuclear rim staining mediated by the NTF2 and UBA domains of NXF1 [27], as well as a cytoplasmic localization of NXF1, e.g., inside stress granules [47]. Co-expression of NP and wildtype NXF1 led to no dramatic changes in the intracellular distribution of NXF1, although small amounts of NXF1 could be detected in the NP-derived inclusion bodies. In contrast, the intracellular localization of NXF1-10RA changed upon co-expression of NXF1-10RA, and it clearly accumulated in NP-derived inclusion bodies. Similarly, NXF1-ΔRBD also accumulated in inclusion bodies, indicating that the accumulation of RNA binding-deficient NXF1 in inclusion bodies is not due to the identified direct interaction of NP with the RBD of NXF1, but that NXF1 is recruited via another mechanism. Importantly, while in the coIP experiments NXF1-10RA showed a stronger interaction with NP than wildtype NXF1, this was not the case for NXF1-ΔRBD, which is no longer able to interact with NP. Therefore, the fact that both of these NXF1 mutants accumulate in inclusion bodies suggests that the loss of their ability to bind RNA, but not changes to their ability to bind NP, dictates this phenotype. Together, these data suggest that NXF1 can enter inclusion bodies, but that only RNA binding-deficient NXF1 is retained and, therefore, accumulates in these structures, whereas RNA binding-competent NXF1 rapidly leaves them again.

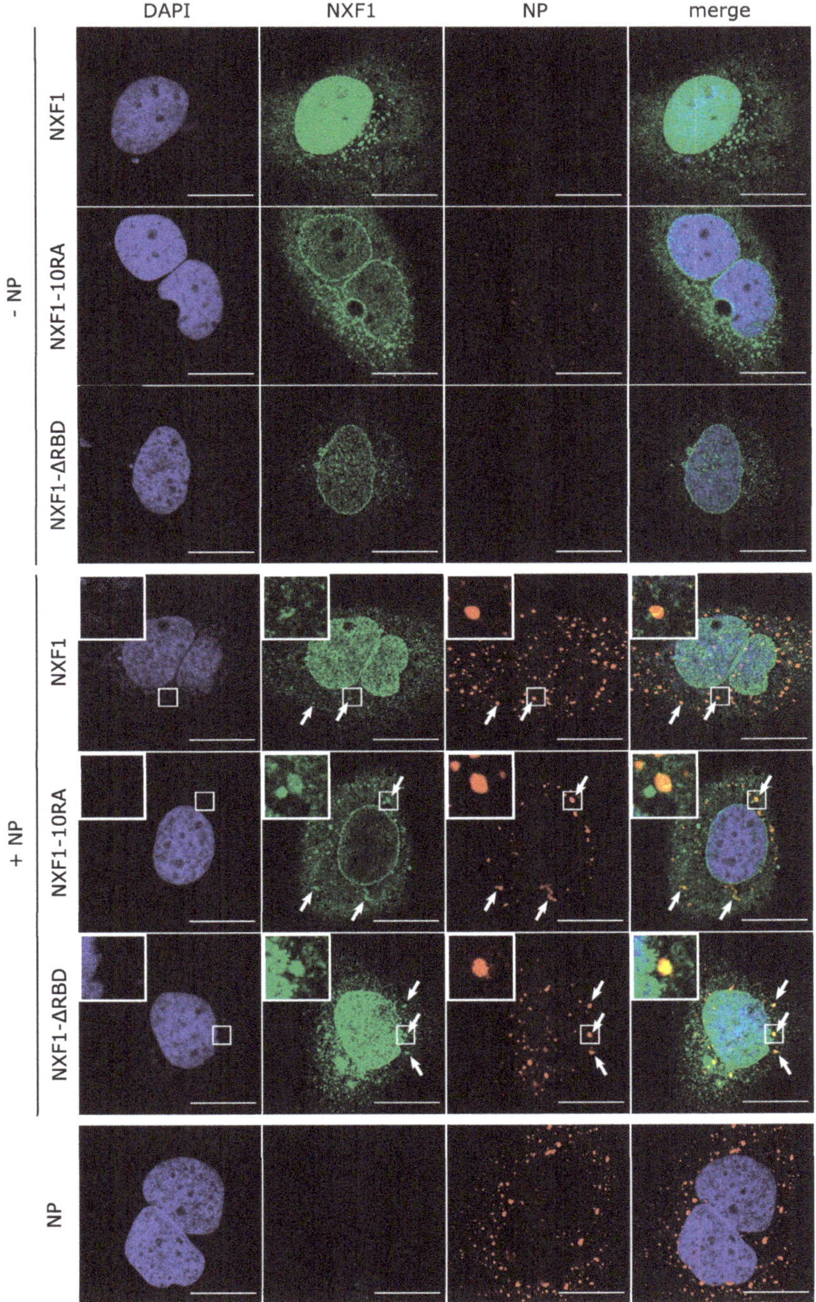

Figure 5. Recruitment of NXF1 into NP-derived inclusion bodies. Huh7 cells were transfected with plasmids encoding flag/HA-NXF1, flag/HA-NXF1-10RA or flag/HA-NXF1-ΔRBD together with NP as indicated. Forty-eight hours p.t., cells were fixed with 4% PFA and permeabilized with Triton X-100. NXF1 (shown in green) was detected with anti-flag antibodies and NP (shown in red) with anti-NP antibodies. Nuclei were stained with DAPI (shown in blue), and cells were visualized by confocal laser scanning microscopy. Scale bars indicate 20 µm11Arrows point out colocalization, and insets show magnifications of indicated areas.

To further substantiate these results, we analyzed the localization of endogenous NXF1 in context of infectious EBOV (Figure 6). As shown above with overexpressed NXF1, endogenous NXF1 was mostly localized in the nucleus. In contrast, NP was only detectable in the cytoplasmic inclusion bodies. Staining of both NXF1 and NP in the context of an EBOV infection showed a subtle but clearly visible relocalization of NXF1 into inclusion bodies, supporting the results obtained using overexpression of NP and NXF1.

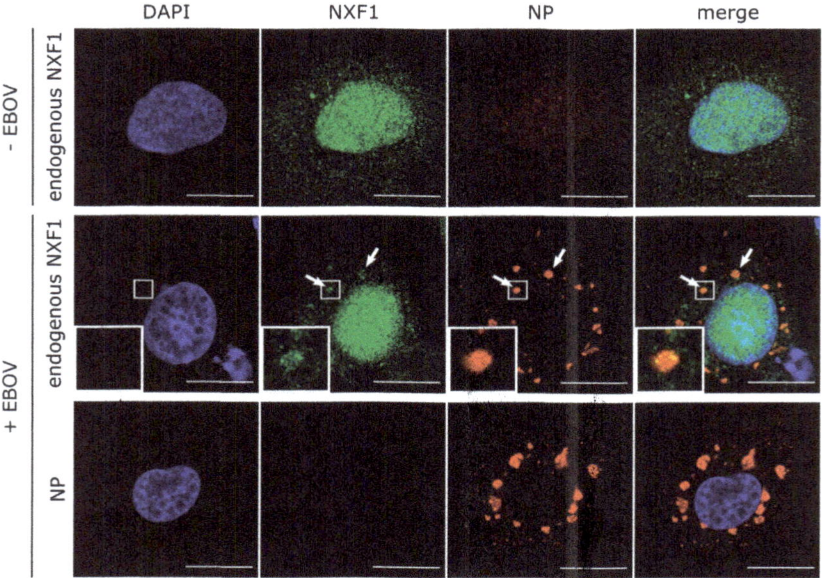

Figure 6. Recruitment of endogenous NXF1 into EBOV inclusion bodies. Huh7 cells were infected with rgEBOV. Sixteen hours p.i., cells were fixed with 10% formalin and permeabilized with Triton X-100. NXF1 (shown in green) was detected with anti-NXF1 antibodies and NP (shown in red) with anti-NP antibodies. Nuclei were stained with DAPI (shown in blue), and cells were visualized by confocal laser scanning microscopy. Scale bars indicate 20 µm. Arrows point out colocalization, and insets show magnifications of indicated areas.

3.5. NXF1 Influences Protein Expression but Not Viral mRNA Transcription

As we had previously identified NXF1 in a genome-wide siRNA screen as being necessary for viral RNA synthesis and/or viral protein expression [19], we wanted to further dissect its impact on viral RNA synthesis and protein expression by assessing which of the different steps of these processes are influenced by NXF1. Therefore, we first performed classical EBOV minigenome assays in the context of an siRNA-mediated knockdown of endogenous NXF1. As shown before, the knockdown of NXF1 led to a reduction of reporter activity, confirming a role of NXF1 in viral RNA synthesis and/or gene expression (Figure 7A). In contrast to this, overexpression of NXF1 did not influence reporter activity (Figure S2), suggesting that the endogenous levels of NXF1 are sufficient to fulfill its function in the virus life cycle. To distinguish between a function of NXF1 in viral genome replication or in transcription/protein expression, we next used a replication-deficient minigenome, which lacks 55 nt in the trailer region, thus abolishing viral genome replication while still allowing transcription of the minigenome as previously described [41]. Using this minigenome system, NXF1 knockdown resulted in an 11.2 to 33.5-fold reduction in reporter activity, demonstrating that NXF1 is important for either viral transcription or protein expression, independent of viral genome replication (Figure 7B). In order to discriminate between an effect on viral mRNA transcription and later steps of viral protein expression such as mRNA transport from inclusion bodies or mRNA translation we next analyzed the influence of NXF1 knockdown on minigenome-derived viral mRNA levels. To this end, we again performed minigenome assays in NXF1 siRNA-treated cells, and analyzed the levels of viral mRNA in the cell lysates via RT-qPCR (Figure 7C). As the mRNA levels in the NXF1 knockdown cells were comparable to the control cells, NXF1 seems not to be important for viral transcription, but rather for either mRNA transport out of inclusion bodies towards ribosomes or efficient mRNA translation.

To confirm these findings in context of infectious EBOV, we repeated these experiments under BSL4 conditions using a recombinant EBOV that expresses firefly luciferase from an additional transcriptional unit, and thus allows precise quantification of viral gene expression [43]. In case of NXF1 knockdown we observed a significant reduction in luciferase activity (Figure 7D), and thus in viral gene expression compared to the negative control. In contrast, when we determined the amount of luciferase mRNAs (Figure 7E), there was no reduction but rather an increase in the mRNA levels upon NXF1 knockdown compared to the negative control, although this increase was not statistically significant. In contrast, knockdown of the viral polymerase L resulted in reductions in both reporter activity and mRNA levels by 62%. This indicates that also in context of a viral infection NXF1 is important for a late step in viral protein expression, i.e., either mRNA transport or efficient mRNA translation, but not for mRNA synthesis.

To further determine whether the influence of NXF1 on reporter protein expression can be explained by binding of viral mRNAs by NXF1, coIPs of viral mRNA in the context of a minigenome assay were performed (Figure 7F). To this end, NXF1 was precipitated and mRNA was isolated from the precipitate. When viral mRNA was produced (i.e., in the sample where the viral polymerase L was present; NXF1 +L in Figure 7F), we detected viral mRNA that could be co-precipitated together with NXF1. This was in contrast to the background signals observed in the control sample without polymerase (NXF1 -L) or in the negative control samples (-NXF1, mock IP, -RT). In contrast, the mutant lacking the RNA-binding domain (NXF1-ΔRBD) failed to co-precipitate the viral mRNA. However, coIP with NXF1-ΔRRM led to similar amounts of viral mRNA in the precipitate compared to wildtype NXF1. These results indicate that NXF1 interacts with the minigenome-derived viral mRNA, and that this interaction is mediated by the RBD of NXF1.

Finally, to check whether the interaction of NXF1 with the viral mRNA is specific for mRNA, or whether NXF1 binds viral RNA (vRNA) as well, we performed coIPs of minigenome vRNA together with NXF1 (Figure 7G). For this experiment the vRNA was generated by ectopically expressed T7 bacteriophage RNA polymerase in the absence of any viral proteins. As a positive control, vRNA was co-expressed with flag/HA-NP, VP35 and VP24, since these three proteins have been described as being minimally required for the formation of vRNA-containing nucleocapsids, and as such are necessary for co-precipitation of vRNA together with NP [7,48]. RNA was isolated from the precipitates and subsequent RT-qPCR showed that vRNA could indeed be co-precipitated with NP under these conditions (compare RNP +1cis vs. RNP -1cis in Figure 7G). In contrast, NXF1 failed to efficiently co-precipitate minigenome vRNA, suggesting a specific interaction with viral mRNAs.

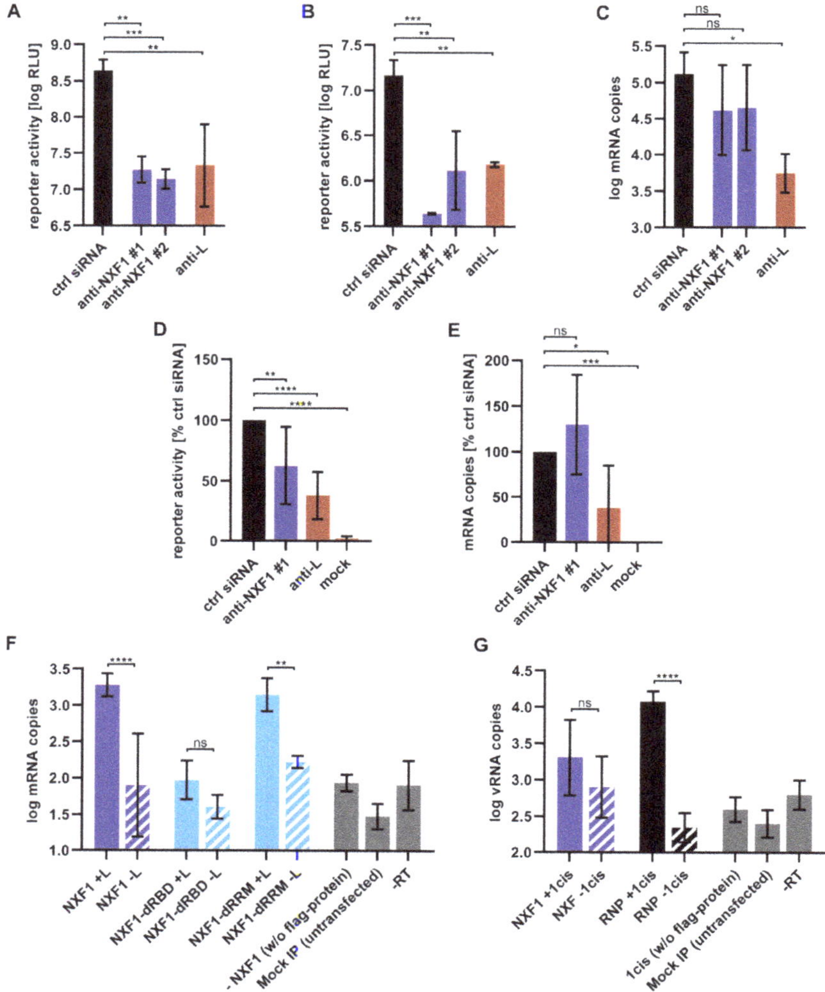

Figure 7. Influence of NXF1 on viral RNA synthesis and protein expression and interaction of NXF1 with viral RNAs. (**A**) Analysis of viral RNA synthesis and protein expression in NXF1 knockdown cells. 293T cells were transfected with siRNAs targeting either NXF1 (anti-NXF1) or L (anti-L) or with a negative control (ctrl) siRNA. Forty-eight hours p.t., cells were transfected with the plasmids required for a replication-competent minigenome assay. Forty-eight hours later, reporter activity was measured. (**B**) Analysis of viral transcription and protein expression in NXF1 knockdown cells in the absence of genome replication. 293T cells were transfected with siRNAs and 48 h p.t. with the plasmids required for a replication-deficient minigenome assay. After a further 48 h, reporter activity was analyzed. (**C**) Analysis of viral mRNA levels in NXF1 knockdown cells. Forty-eight hours after siRNA transfection of 293T cells, the cells were transfected with all components for the replication-competent minigenome assay. Forty-eight hours later RNA was isolated and RT-qPCR for viral mRNA was performed. (**D**) Analysis of viral gene expression after infection of NXF1 knockdown cells infected with recombinant EBOV expressing firefly luciferase. Reporter activity was determined one day p.i. and is

shown relative to the reporter activity of cells treated with negative control siRNA. (**E**) Influence of NXF1 knockdown on viral mRNA synthesis. RNA from cells from panel **D** was isolated and luciferase mRNA was quantified using RT-qPCR. (**F**) Interaction of NXF1 with viral mRNA. 293T cells were transfected with plasmids encoding flag/HA-NXF1 or flag/HA-NXF1-mutants as indicated, in addition to the replication-competent minigenome, T7-polymerase, and the RNP proteins. As a control L was omitted (-L). Further negative controls (shown in grey) were as indicated. Forty-eight hours p.t., pulldowns were performed using anti-flag antibodies and RNA was isolated from the precipitates. mRNA levels in the precipitates were analyzed by subsequent RT-qPCR. (**G**) Interaction of NXF1 with viral genomic RNA. IP with anti-flag antibodies was performed with cell lysates from 293T cells transfected with plasmids encoding the replication-competent minigenome, T7-polymerase, and either flag/HA-NXF1, or flag/HA-NP, VP35, and VP24 (RNP). In controls the minigenome was omitted (-1cis). Further negative controls (shown in grey) were as indicated. RNA was isolated from the precipitates and analyzed via RT-qPCR. Means and standard deviations for two independent experiments (six biological replicates) are shown in panels **D** and **E**, and for three independent experiments in all other panels. Asterisks indicate p values from one-way ANOVA (*: $p \leq 0.05$; **: $p \leq 0.01$; ***: $p \leq 0.001$; ****: $p \leq 0.0001$; ns: $p > 0.05$).

3.6. A Model for the Function of NXF1 in the EBOV Life Cycle

Our results show that NXF1 is recruited to inclusion bodies, which we and others have shown to be sites of EBOV replication and transcription [10,14]. Furthermore, it is able to interact with NP, with the RBD of NXF1 being both necessary and sufficient for this interaction. In the context of its endogenous function, NXF1 has been shown to interact with many cellular proteins, including export adaptor proteins such as Aly and THOC5, to promote efficient mRNA binding (reviewed in [49]). Binding of these adaptor proteins, which are both components of the transcription and export complex (TREX), leads to a structural rearrangement of NXF1 [29,31]. This conformational change leads to exposure of the RBD resulting in an enhanced RNA-binding activity and subsequently to a handover of mRNA from Aly to NXF1. Since the interaction of NP with NXF1 resembles that of Aly and NXF1, as it involves the RBD, but not the amino acids that are important for the interaction with RNA [31], we propose a model in which a binding of NP to NXF1 results in a handover of viral mRNA from NP to NXF1 (Figure 8). Indeed, we were able to show that NXF1 binds mRNAs that were produced in inclusion bodies. We further propose that NXF1 then rapidly exports the bound RNA from inclusion bodies. This is in line with our observations that only RNA binding-deficient NXF1 accumulates in these structures, while wildtype NXF1 can only be observed in inclusion bodies in minute amounts, as well as the fact that the interaction of NP and NXF1 is weakened when NXF1 binds to RNA.

This model could explain how EBOV is able to overcome a number of fundamental challenges it faces during its life cycle, and which it shares with other cytoplasmically replicating NSVs. First, due to the RNA nature of its genome EBOV has to perform both genome replication and transcription, which exhibit fundamental differences: while both processes involve the generation of RNAs from a viral RNA template, during genome replication an accurate and complete copy of the RNA template has to be produced that is then encapsidated by NP to form a new RNP. This is in contrast to the subgenomic RNA fragments produced during transcription, which are processed by the addition of a cap structure and of a poly(A)-tail, and are not encapsidated by NP. While the production of subgenomic RNAs as well as their capping and poly(A)-tailing are all carried out by the viral polymerase L (reviewed in [50]), it is not clear how viral mRNAs evade encapsidation by NP, which is present in abundant quantities in inclusion bodies. However, co-transcriptional binding of viral mRNAs by NXF1 could shield the mRNA from such encapsidation by NP and/or displace already bound NP.

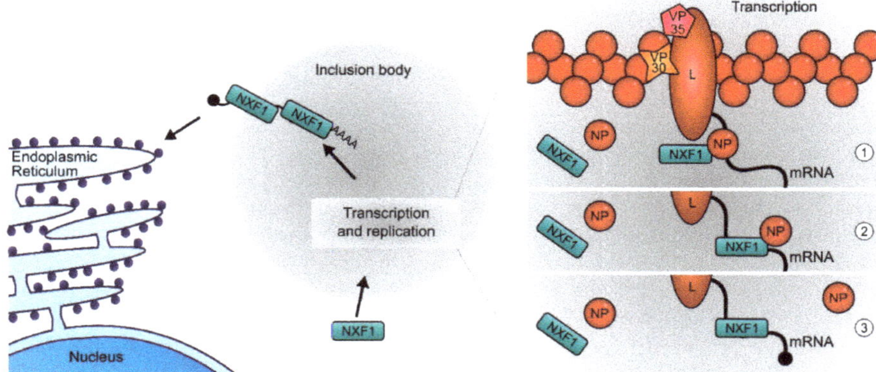

Figure 8. Model for the role of NXF1 and its interaction with NP and the viral mRNA during the EBOV life cycle. NXF1 enters inclusion bodies, which are the site of viral genome replication and transcription. Once in the inclusion bodies, NP recruits NXF1 to nascent mRNA during viral transcription (**1**). NXF1 then takes over the viral mRNA from NP (**2**), and thereby avoids encapsidation of the mRNA (**3**). After the completion of transcription, NXF1 together with the viral mRNA is exported from the inclusion bodies and transports it to ribosomes for translation.

A second challenge during the EBOV life cycle is the fact that viral RNA synthesis occurs in inclusion bodies. These are increasingly appreciated for their importance in the genome replication of cytoplasmically replicating NSVs [12,21–23,51], and have for some of these viruses been described to show properties of liquid organelles [24,52]. Similar properties have been observed for EBOV inclusion bodies [10], suggesting that inclusion bodies of NSVs might generally represent liquid organelles. Such liquid organelles consist in most cases of RNA and RNA-binding proteins, and are formed by liquid-liquid phase separation, with multivalent low affinity interactions and intrinsically disordered protein regions being important drivers of their formation (reviewed in [53,54]). RNAs are important constituents of these liquid organelles and sometimes even the seed for their formation; however, in the case of mRNAs they have to be exported in order to reach ribosomes for translation. It is not clear how naked viral RNAs such as mRNAs would be able to escape this liquid organelle environment. To this end, their decoration with NXF1 might help viral mRNA to be exported from inclusion bodies by phase separation between the RNA:NXF1 complexes and other inclusion body constituents.

Finally, in contrast to most cellular mRNAs, the mRNAs of cytoplasmically replicating NSVs do not undergo splicing. Besides the removal of introns, splicing also leads to the association of mRNAs with a multitude of cellular proteins including NXF1, and has been shown to greatly improve translation from both cellular and viral mRNAs, independent of any effects on nuclear mRNA export [55–57]. Interestingly, gammaretroviruses usurp the NXF1 pathway to increase the translation efficiency of their unspliced mRNAs [40], and thus it is possible that the recruitment of NXF1 to EBOV mRNAs might serve a similar purpose.

Nevertheless, at this time point we cannot fully exclude that NXF1 fulfills another, not yet described function in the late stages of viral protein expression, and future studies will be required to prove or disprove the model proposed by us. Further, it will be interesting to dissect mechanistic details of the proposed export of viral mRNAs from inclusion bodies, as well as the precise localization of NXF1 in inclusion bodies, and specifically whether there is a compartmentalization of different aspects of the virus life cycle inside inclusion bodies that correlates with the localization of NXF1 and other host factors. Finally, it is intriguing that when we first identified NXF1 as a factor involved in viral RNA synthesis and/or protein expression, we could show that it is also important for these processes for Junín virus, a viral hemorrhagic fever-causing arenavirus [19] that, like EBOV, performs its genome replication in inclusion bodies [21]. It will be interesting to see whether NXF1 fulfills the same role in the arenavirus life cycle as in the filovirus life cycle, and given that the fundamental nature of the challenges described above are shared between many if not all cytoplasmically replicating negative-sense RNA viruses, whether other such viruses also are dependent on NXF1.

Supplementary Materials: The following are available online at http://www.mdpi.com/2073-4409/9/1/187/s1, Figure S1: RNA dependence of the interaction between NXF1 and VP35; Figure S2: Influence of NXF1 overexpression on viral RNA synthesis and protein expression.

Author Contributions: Conceptualization: L.W., J.B. and T.H.; Investigation: L.W., J.B., B.S.B. and T.H.; Resources: S.R., M.L.S. and S.T.; Supervision: T.H.; Visualization: L.W. and T.H.; Funding acquisition: T.H.; Writing—original draft preparation: L.W. and T.H. All authors have read and agreed to the published version of the manuscript.

Funding: Funding was provided by the Deutsche Forschungsgemeinschaft (DFG), grant number 389002253, (L.W. and J.B.); by the Friedrich-Loeffler-Institut, intramural funding (S.R., T.H. and M.L.S.) and funding as part of the VISION consortium (B.S.B.); and in part by the Intramural Research Program of the NIH, NIAID (S.T.).

Acknowledgments: The authors would like to thank Allison Groseth (Friedrich-Loeffler-Institut) for helpful discussions, critical reading of this manuscript, and technical assistance. We further thank Sandra Diederich and Luca Zaeck (Friedrich-Loeffler-Institut) for technical assistance, Logan Banadyga (Public Health Agency of Canada) for his help in establishing the coIP procedure, and Stephan Becker (Philipps University Marburg) for providing the Huh7 cells.

Conflicts of Interest: The authors declare no conflict of interest. The funders had no role in the design of the study; in the collection, analyses, or interpretation of data; in the writing of the manuscript, or in the decision to publish the results.

References

1. Burk, R.; Bollinger, L.; Johnson, J.C.; Wada, J.; Radoshitzky, S.R.; Palacios, G.; Bavari, S.; Jahrling, P.B.; Kuhn, J.H. Neglected filoviruses. *FEMS Microbiol. Rev.* **2016**, *40*, 494–519. [CrossRef] [PubMed]
2. Feldmann, H.; Sanchez, A.; Geisbert, T.W. Filoviridae: Marburg and Ebola Viruses. In *Fields Virology*, 6th ed.; Knipe, D.M., Howley, P.M., Eds.; Lippincott Williams & Wilkins: Philadelphia, PA, USA, 2013; pp. 923–956.
3. Mühlberger, E.; Weik, M.; Volchkov, V.E.; Klenk, H.D.; Becker, S. Comparison of the transcription and replication strategies of marburg virus and Ebola virus by using artificial replication systems. *J. Virol.* **1999**, *73*, 2333–2342. [CrossRef] [PubMed]
4. Modrof, J.; Mühlberger, E.; Klenk, H.D.; Becker, S. Phosphorylation of VP30 impairs ebola virus transcription. *J. Biol. Chem.* **2002**, *277*, 33099–33104. [CrossRef] [PubMed]
5. Groseth, A.; Charton, J.E.; Sauerborn, M.; Feldmann, F.; Jones, S.M.; Hoenen, T.; Feldmann, H. The Ebola virus ribonucleoprotein complex: A novel VP30-L interaction identified. *Virus Res.* **2009**, *140*, 8–14. [CrossRef] [PubMed]
6. Becker, S.; Rinne, C.; Hofsass, U.; Klenk, H.D.; Muhlberger, E. Interactions of Marburg virus nucleocapsid proteins. *Virology* **1998**, *249*, 406–417. [CrossRef]
7. Huang, Y.; Xu, L.; Sun, Y.; Nabel, G.J. The assembly of Ebola virus nucleocapsid requires virion-associated proteins 35 and 24 and posttranslational modification of nucleoprotein. *Mol. Cell* **2002**, *10*, 307–316. [CrossRef]
8. Kirchdoerfer, R.N.; Abelson, D.M.; Li, S.; Wood, M.R.; Saphire, E.O. Assembly of the Ebola Virus Nucleoprotein from a Chaperoned VP35 Complex. *Cell Rep.* **2015**, *12*, 140–149. [CrossRef]
9. Leung, D.W.; Borek, D.; Luthra, P.; Binning, J.M.; Anantpadma, M.; Liu, G.; Harvey, I.B.; Su, Z.; Endlich-Frazier, A.; Pan, J.; et al. An Intrinsically Disordered Peptide from Ebola Virus VP35 Controls Viral RNA Synthesis by Modulating Nucleoprotein-RNA Interactions. *Cell Rep.* **2015**, *11*, 376–389. [CrossRef]
10. Hoenen, T.; Shabman, R.S.; Groseth, A.; Herwig, A.; Weber, M.; Schudt, G.; Dolnik, O.; Basler, C.F.; Becker, S.; Feldmann, H. Inclusion bodies are a site of ebolavirus replication. *J. Virol.* **2012**, *86*, 11779–11788. [CrossRef]
11. Gabriel, G.; Feldmann, F.; Reimer, R.; Thiele, S.; Fischer, M.; Hartmann, E.; Bader, M.; Ebihara, H.; Hoenen, T.; Feldmann, H. Importin-alpha7 Is Involved in the Formation of Ebola Virus Inclusion Bodies but Is Not Essential for Pathogenicity in Mice. *J. Infect. Dis.* **2015**, *212* (Suppl. 2), S316–S321. [CrossRef]
12. Dolnik, O.; Stevermann, L.; Kolesnikova, L.; Becker, S. Marburg virus inclusions: A virus-induced microcompartment and interface to multivesicular bodies and the late endosomal compartment. *Eur. J. Cell Biol.* **2015**, *94*, 323–331. [CrossRef] [PubMed]

13. Jasenosky, L.D.; Neumann, G.; Lukashevich, I.; Kawaoka, Y. Ebola virus VP40-induced particle formation and association with the lipid bilayer. *J. Virol.* **2001**, *75*, 5205–5214. [CrossRef] [PubMed]
14. Lier, C.; Becker, S.; Biedenkopf, N. Dynamic phosphorylation of Ebola virus VP30 in NP-induced inclusion bodies. *Virology* **2017**, *512*, 39–47. [CrossRef] [PubMed]
15. Kruse, T.; Biedenkopf, N.; Hertz, E.P.T.; Dietzel, E.; Stalmann, G.; Lopez-Mendez, B.; Davey, N.E.; Nilsson, J.; Becker, S. The Ebola Virus Nucleoprotein Recruits the Host PP2A-B56 Phosphatase to Activate Transcriptional Support Activity of VP30. *Mol. Cell* **2018**, *69*, 136–145.e136. [CrossRef] [PubMed]
16. Batra, J.; Hultquist, J.F.; Liu, D.; Shtanko, O.; Von Dollen, J.; Satkamp, L.; Jang, G.M.; Luthra, P.; Schwarz, T.M.; Small, G.I.; et al. Protein Interaction Mapping Identifies RBBP6 as a Negative Regulator of Ebola Virus Replication. *Cell* **2018**, *175*, 1917–1930.e1913. [CrossRef]
17. Fang, J.; Pietzsch, C.; Ramanathan, P.; Santos, R.I.; Ilinykh, P.A.; Garcia-Blanco, M.A.; Bukreyev, A.; Bradrick, S.S. Staufen1 Interacts with Multiple Components of the Ebola Virus Ribonucleoprotein and Enhances Viral RNA Synthesis. *MBio* **2018**, *9*. [CrossRef]
18. Nelson, E.V.; Schmidt, K.M.; Deflube, L.R.; Doganay, S.; Banadyga, L.; Olejnik, J.; Hume, A.J.; Ryabchikova, E.; Ebihara, H.; Kedersha, N.; et al. Ebola Virus Does Not Induce Stress Granule Formation during Infection and Sequesters Stress Granule Proteins within Viral Inclusions. *J. Virol.* **2016**, *90*, 7268–7284. [CrossRef]
19. Martin, S.; Chiramel, A.I.; Schmidt, M.L.; Chen, Y.C.; Whitt, N.; Watt, A.; Dunham, E.C.; Shifflett, K.; Traeger, S.; Leske, A.; et al. A genome-wide siRNA screen identifies a druggable host pathway essential for the Ebola virus life cycle. *Genome Med.* **2018**, *10*, 58. [CrossRef]
20. Wendt, L.; Bostedt, L.; Hoenen, T.; Groseth, A. High-throughput screening for negative-stranded hemorrhagic fever viruses using reverse genetics. *Antiviral. Res.* **2019**, *170*, 104569. [CrossRef]
21. Baird, N.L.; York, J.; Nunberg, J.H. Arenavirus infection induces discrete cytosolic structures for RNA replication. *J. Virol.* **2012**, *86*, 11301–11310. [CrossRef]
22. Lahaye, X.; Vidy, A.; Pomier, C.; Obiang, L.; Harper, F.; Gaudin, Y.; Blondel, D. Functional characterization of Negri bodies (NBs) in rabies virus-infected cells: Evidence that NBs are sites of viral transcription and replication. *J. Virol.* **2009**, *83*, 7948–7958. [CrossRef] [PubMed]
23. Rincheval, V.; Lelek, M.; Gault, E.; Bouillier, C.; Sitterlin, D.; Blouquit-Laye, S.; Galloux, M.; Zimmer, C.; Eleouet, J.F.; Rameix-Welti, M.A. Functional organization of cytoplasmic inclusion bodies in cells infected by respiratory syncytial virus. *Nat. Commun.* **2017**, *8*, 563. [CrossRef] [PubMed]
24. Zhou, Y.; Su, J.M.; Samuel, C.E.; Ma, D. Measles Virus Forms Inclusion Bodies with Properties of Liquid Organelles. *J. Virol.* **2019**. [CrossRef]
25. Katahira, J.; Strasser, K.; Podtelejnikov, A.; Mann, M.; Jung, J.U.; Hurt, E. The Mex67p-mediated nuclear mRNA export pathway is conserved from yeast to human. *EMBO J.* **1999**, *18*, 2593–2609. [CrossRef] [PubMed]
26. Bachi, A.; Braun, I.C.; Rodrigues, J.P.; Pante, N.; Ribbeck, K.; von Kobbe, C.; Kutay, U.; Wilm, M.; Gorlich, D.; Carmo-Fonseca, M.; et al. The C-terminal domain of TAP interacts with the nuclear pore complex and promotes export of specific CTE-bearing RNA substrates. *RNA* **2000**, *6*, 136–158. [CrossRef]
27. Fribourg, S.; Braun, I.C.; Izaurralde, E.; Conti, E. Structural basis for the recognition of a nucleoporin FG repeat by the NTF2-like domain of the TAP/p15 mRNA nuclear export factor. *Mol. Cell* **2001**, *8*, 645–656. [CrossRef]
28. Guzik, B.W.; Levesque, L.; Prasad, S.; Bor, Y.C.; Black, B.E.; Paschal, B.M.; Rekosh, D.; Hammarskjold, M.L. NXT1 (p15) is a crucial cellular cofactor in TAP-dependent export of intron-containing RNA in mammalian cells. *Mol. Cell Biol.* **2001**, *21*, 2545–2554. [CrossRef]
29. Viphakone, N.; Hautbergue, G.M.; Walsh, M.; Chang, C.T.; Holland, A.; Folco, E.G.; Reed, R.; Wilson, S.A. TREX exposes the RNA-binding domain of Nxf1 to enable mRNA export. *Nat. Commun.* **2012**, *3*, 1006. [CrossRef]
30. Herold, A.; Suyama, M.; Rodrigues, J.P.; Braun, I.C.; Kutay, U.; Carmo-Fonseca, M.; Bork, P.; Izaurralde, E. TAP (NXF1) belongs to a multigene family of putative RNA export factors with a conserved modular architecture. *Mol. Cell Biol.* **2000**, *20*, 8996–9008. [CrossRef]
31. Hautbergue, G.M.; Hung, M.L.; Golovanov, A.P.; Lian, L.Y.; Wilson, S.A. Mutually exclusive interactions drive handover of mRNA from export adaptors to TAP. *Proc. Natl. Acad. Sci. USA* **2008**, *105*, 5154–5159. [CrossRef]

32. Viphakone, N.; Sudbery, I.; Griffith, L.; Heath, C.G.; Sims, D.; Wilson, S.A. Co-transcriptional Loading of RNA Export Factors Shapes the Human Transcriptome. *Mol. Cell* **2019**. [CrossRef] [PubMed]
33. Larsen, S.; Bui, S.; Perez, V.; Mohammad, A.; Medina-Ramirez, H.; Newcomb, L.L. Influenza polymerase encoding mRNAs utilize atypical mRNA nuclear export. *Virol. J.* **2014**, *11*, 154. [CrossRef] [PubMed]
34. Tunnicliffe, R.B.; Hautbergue, G.M.; Kalra, P.; Jackson, B.R.; Whitehouse, A.; Wilson, S.A.; Golovanov, A.P. Structural basis for the recognition of cellular mRNA export factor REF by herpes viral proteins HSV-1 ICP27 and HVS ORF57. *PLoS Pathog.* **2011**, *7*, e1001244. [CrossRef] [PubMed]
35. Yang, C.C.; Huang, E.Y.; Li, H.C.; Su, P.Y.; Shih, C. Nuclear export of human hepatitis B virus core protein and pregenomic RNA depends on the cellular NXF1-p15 machinery. *PLoS ONE* **2014**, *9*, e106683. [CrossRef]
36. Grüter, P.; Tabernero, C.; von Kobbe, C.; Schmitt, C.; Saavedra, C.; Bachi, A.; Wilm, M.; Felber, B.K.; Izaurralde, E. TAP, the human homolog of Mex67p, mediates CTE-dependent RNA export from the nucleus. *Mol. Cell* **1998**, *1*, 649–659. [CrossRef]
37. Chen, I.H.B.; Sciabica, K.S.; Sandri-Goldin, R.M. ICP27 Interacts with the RNA Export Factor Aly/REF To Direct Herpes Simplex Virus Type 1 Intronless mRNAs to the TAP Export Pathway. *J. Virol.* **2002**, *76*, 12877–12889. [CrossRef]
38. Satterly, N.; Tsai, P.L.; van Deursen, J.; Nussenzveig, D.R.; Wang, Y.; Faria, P.A.; Levay, A.; Levy, D.E.; Fontoura, B.M. Influenza virus targets the mRNA export machinery and the nuclear pore complex. *Proc. Natl. Acad. Sci. USA* **2007**, *104*, 1853–1858. [CrossRef]
39. Tunnicliffe, R.B.; Hautbergue, G.M.; Wilson, S.A.; Kalra, P.; Golovanov, A.P. Competitive and cooperative interactions mediate RNA transfer from herpesvirus saimiri ORF57 to the mammalian export adaptor ALYREF. *PLoS Pathog.* **2014**, *10*, e1003907. [CrossRef]
40. Bartels, H.; Luban, J. Gammaretroviral pol sequences act in cis to direct polysome loading and NXF1/NXT-dependent protein production by gag-encoded RNA. *Retrovirology* **2014**, *11*, 73. [CrossRef]
41. Hoenen, T.; Jung, S.; Herwig, A.; Groseth, A.; Becker, S. Both matrix proteins of Ebola virus contribute to the regulation of viral genome replication and transcription. *Virology* **2010**, *403*, 56–66. [CrossRef]
42. Bussmann, B.M.; Reiche, S.; Jacob, L.H.; Braun, J.M.; Jassoy, C. Antigenic and cellular localisation analysis of the severe acute respiratory syndrome coronavirus nucleocapsid protein using monoclonal antibodies. *Virus. Res.* **2006**, *122*, 119–126. [CrossRef]
43. Hoenen, T.; Groseth, A.; Callison, J.; Takada, A.; Feldmann, H. A novel Ebola virus expressing luciferase allows for rapid and quantitative testing of antivirals. *Antivir. Res.* **2013**, *99*, 207–213. [CrossRef]
44. Shabman, R.S.; Hoenen, T.; Groseth, A.; Jabado, O.; Binning, J.M.; Amarasinghe, G.K.; Feldmann, H.; Basler, C.F. An upstream open reading frame modulates ebola virus polymerase translation and virus replication. *PLoS Pathog.* **2013**, *9*, e1003147. [CrossRef] [PubMed]
45. Noda, T.; Hagiwara, K.; Sagara, H.; Kawaoka, Y. Characterization of the Ebola virus nucleoprotein-RNA complex. *J. Gen. Virol.* **2010**, *91*, 1478–1483. [CrossRef] [PubMed]
46. Basler, C.F.; Wang, X.; Muhlberger, E.; Volchkov, V.; Paragas, J.; Klenk, H.D.; Garcia-Sastre, A.; Palese, P. The Ebola virus VP35 protein functions as a type I IFN antagonist. *Proc. Natl. Acad. Sci. USA* **2000**, *97*, 12289–12294. [CrossRef] [PubMed]
47. Hochberg-Laufer, H.; Schwed-Gross, A.; Neugebauer, K.M.; Shav-Tal, Y. Uncoupling of nucleo-cytoplasmic RNA export and localization during stress. *Nucleic Acids Res.* **2019**, *47*, 4778–4797. [CrossRef] [PubMed]
48. Banadyga, L.; Hoenen, T.; Ambroggio, X.; Dunham, E.; Groseth, A.; Ebihara, H. Ebola virus VP24 interacts with NP to facilitate nucleocapsid assembly and genome packaging. *Sci. Rep.* **2017**, *7*, 7698. [CrossRef] [PubMed]
49. Heath, C.G.; Viphakone, N.; Wilson, S.A. The role of TREX in gene expression and disease. *Biochem. J.* **2016**, *473*, 2911–2935. [CrossRef]
50. Hume, A.J.; Muhlberger, E. Distinct Genome Replication and Transcription Strategies within the Growing Filovirus Family. *J. Mol. Biol.* **2019**. [CrossRef]
51. Zhang, S.; Chen, L.; Zhang, G.; Yan, Q.; Yang, X.; Ding, B.; Tang, Q.; Sun, S.; Hu, Z.; Chen, M. An amino acid of human parainfluenza virus type 3 nucleoprotein is critical for template function and cytoplasmic inclusion body formation. *J. Virol.* **2013**, *87*, 12457–12470. [CrossRef]
52. Nikolic, J.; Le Bars, R.; Lama, Z.; Scrima, N.; Lagaudriere-Gesbert, C.; Gaudin, Y.; Blondel, D. Negri bodies are viral factories with properties of liquid organelles. *Nat. Commun.* **2017**, *8*, 58. [CrossRef] [PubMed]

53. Weber, S.C.; Brangwynne, C.P. Getting RNA and protein in phase. *Cell* **2012**, *149*, 1188–1191. [CrossRef] [PubMed]
54. Gomes, E.; Shorter, J. The molecular language of membraneless organelles. *J. Biol. Chem.* **2019**, *294*, 7115–7127. [CrossRef] [PubMed]
55. Sadek, J.; Read, G.S. The Splicing History of an mRNA Affects Its Level of Translation and Sensitivity to Cleavage by the Virion Host Shutoff Endonuclease during Herpes Simplex Virus Infections. *J. Virol.* **2016**, *90*, 10844–10856. [CrossRef] [PubMed]
56. Nott, A.; Meislin, S.H.; Moore, M.J. A quantitative analysis of intron effects on mammalian gene expression. *RNA* **2003**, *9*, 607–617. [CrossRef]
57. Nott, A.; Le Hir, H.; Moore, M.J. Splicing enhances translation in mammalian cells: An additional function of the exon junction complex. *Genes Dev.* **2004**, *18*, 210–222. [CrossRef]

Review

Activation of the PI3K-AKT Pathway by Old World Alphaviruses

Eline Van Huizen and Gerald M. McInerney *

Department of Microbiology, Tumor and Cell Biology, Karolinska Institutet, 17177 Stockholm, Sweden; vanhuizeneline@gmail.com

* Correspondence: gerald.mcinerney@ki.se

Received: 16 March 2020; Accepted: 9 April 2020; Published: 15 April 2020

Abstract: Alphaviruses can infect a broad range of vertebrate hosts, including birds, horses, primates, and humans, in which infection can lead to rash, fever, encephalitis, and arthralgia or arthritis. They are most often transmitted by mosquitoes in which they establish persistent, asymptomatic infections. Currently, there are no vaccines or antiviral therapies for any alphavirus. Several Old World alphaviruses, including Semliki Forest virus, Ross River virus and chikungunya virus, activate or hyperactivate the phosphatidylinositol-3-kinase (PI3K)-AKT pathway in vertebrate cells, potentially influencing many cellular processes, including survival, proliferation, metabolism and autophagy. Inhibition of PI3K or AKT inhibits replication of several alphaviruses either in vitro or in vivo, indicating the importance for viral replication. In this review, we discuss what is known about the mechanism(s) of activation of the pathway during infection and describe those effects of PI3K-AKT activation which could be of advantage to the alphaviruses. Such knowledge may be useful for the identification and development of therapies.

Keywords: metabolism; apoptosis; autophagy

---

1. Introduction

1.1. PI3K-AKT Pathway

Viruses activate metabolic pathways in order to meet their needs for production of appropriate macromolecules. Once such pathway is the multifunctional phosphatidylinositol 3-kinases (PI3K)-AKT pathway. Due to its central importance in metabolism but also other cellular functions, this pathway is a common target for viruses [1–4]. In this review, we describe the activation of the PI3K-AKT pathway by alphaviruses and the consequent cellular effects.

PI3Ks are a large family of kinases that is divided in three classes: class 1 (1A and 1B), class II, and class III. The PI3K-AKT pathway involves class 1A PI3Ks. These PI3Ks are heterodimers with a catalytic subunit (p110) and a regulatory subunit (p85). Class 1A PI3Ks are activated by receptor tyrosine kinases (RTK) or G protein-coupled receptors (GPCR) after binding of growth factors, either directly or indirectly via activation of the small GTPase RAS [5] (Figure 1). Once the regulatory action of p85 has been relieved, active class 1A PI3Ks phosphorylate phosphatidylinositol-4,5-biphosphate (PIP2, or PtdIns(4,5)P_2) to phosphatidylinositol-3,4,5-triphosphate (PIP3, or PtdIns(3,4,5)P_3) at the plasma membrane (PM). PIP3-enriched membranes are a docking site for the PI3K-dependent kinase-1 (PDK1) and mTORC2 (or PDK2). Furthermore, the serine/threonine kinase AKT (or protein kinase B, PKB) is relocated to membrane sites with PIP3, where it can be activated via phosphorylation by PDK1 and mTORC2 [6,7]. A number of downstream targets of AKT have been identified, several of which are multifunctional nodes, integrating AKT signalling with signalling through other pathways [7]. One of these is the serine/threonine kinase mammalian target of rapamycin (mTOR) in the multi-protein complex mTORC1. AKT signalling leads to the activation of mTORC1, which promotes cell growth by inducing lipid biogenesis through activation of the transcription factors SREBP1 and PPARγ and by promoting protein synthesis by activating the S6 kinase (S6K) and by inactivating the translational inhibitor 4E-BP1. mTORC1 also inhibits autophagy by blocking ULK1 [6]. The serine/threonine kinase glycogen synthase kinase 3 (GSK3) is another multifunctional target of AKT. Through phosphorylation, active GSK3 inhibits most of its substrates. Upon phosphorylation by AKT, GSK3 itself is inhibited and thereby the downstream targets are positively regulated. These targets include the prosurvival BCL-2 family member MCL-1 and the transcription factor c-Myc, which is required for expression of many genes involved in proliferation. Other GSK3-targets, such as glycogen synthase, are involved in (regulation of) cellular metabolism [7]. The third multifunctional target of AKT are the forkhead box O (FoxO) transcription factors. Phosphorylation of FoxO transcription factors by AKT leads to acute translocation out of the nucleus. Thus, AKT signalling suppresses the expression of FoxO targets. These include targets involved in the induction of apoptosis, cell-cycle arrest, catabolic metabolism and growth inhibition. Thus, PI3K-AKT signalling via these multifunctional targets promotes cell survival, growth and proliferation and steers cellular metabolism towards anabolism.

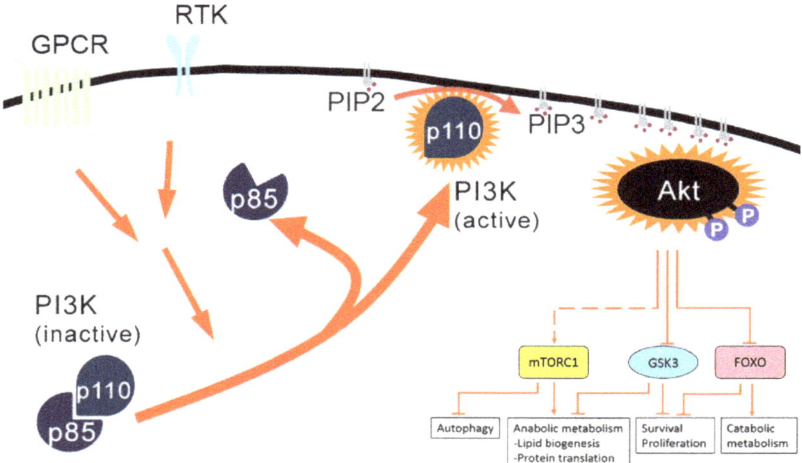

Figure 1. Activation of PI3K-AKT pathway by external signals via G-protein coupled receptors (GPCR) or receptor tyrosine kinases (RTK) leads to dissociation of the p85 regulatory subunit from the active p110 PI3K subunit. The active subunit catalyses the conversion of PIP2 to PIP3 at the plasma membrane, leading to the recruitment and activation of the AKT kinase. Via multiple downstream effector pathways, cellular states of growth, proliferation, heightened metabolic activity and survival are promoted. For more details, see text.

The PI3K-AKT pathway is regulated in many ways. The RAF-MEK-ERK pathway also promotes cell survival and growth and the two pathways have co-regulated proteins and negatively regulate each other. For example, MEK promotes membrane localisation of the phosphatase and tensin homologue (PTEN), where it dephosphorylates PIP3 and inhibits AKT activation [6]. Also, post-translational modifications (including (de)phosphorylation and acetylation) of AKT play important roles in regulation. An important immediate negative feedback loop is provided by mTORC1. Through a variety of downstream targets of mTORC1, AKT signalling is inhibited [7].

1.2. Alphaviruses

Alphaviruses, belonging to the family *Togaviridae*, are enveloped, icosahedral, positive-sense single-stranded RNA viruses. They are arthropod-borne, typically establishing persistent, asymptomatic infections in their vectors, which are usually mosquitoes, and acute infections in a broad range of vertebrate hosts, including humans, primates, horses and birds. They are globally distributed and can be distinguished into Old World and New World alphaviruses by geographical distribution and by clinical symptoms. Much of what is known of the biology of the Old World alphaviruses stems from decades of research on model viruses Semliki Forest (SFV) and Sindbis viruses (SINV), but the group also contains important human pathogens such as chikungunya virus (CHIKV) and Ross River virus (RRV). Currently, no vaccines or antiviral therapies have been approved for any alphavirus.

The viral genome is a single strand of positive-sense RNA with a 5′ 7-methyl-GpppA cap and a 3′ poly(A) tail. It contains two open reading frames (ORFs). The first ORF encodes the non-structural polyprotein (nsP) P1234 and is translated immediately after infection and the second, encoding the structural proteins, is translated later and to very high levels. All nsPs fulfil a role in viral RNA synthesis [8]. Those roles are quite well understood although the function of nsP3 remained obscure for a long time but is starting to be more understood [9,10]. nsP3 has three domains: the macrodomain, the alphavirus unique or zinc-binding domain (AUD/ZBD) and the hypervariable region (HVD). The macrodomain and AUD/ZBD are conserved among alphaviruses, while the HVD shows high variability in sequence and length even between closely related alphaviruses. The nsP3 HVD is a hub for interactions with multiple cellular proteins and can modulate multiple cellular pathways during infection [11].

1.3. Effect of Alphaviruses on PI3K-AKT Pathway During Infection

Several alphaviruses influence the PI3K-AKT(-mTOR) pathway during infection. For four of the Old World alphaviruses (SFV, RRV, CHIKV and SINV) PI3K-AKT activation has been studied and will be discussed in the following sections. Briefly, activation appears to manifest itself as either extremely strong or weak, depending on the virus. The complexity of PI3K-AKT pathway activation, cross-talk with other pathways and the numerous effectors, make it very difficult to appreciate what outcomes are important for replication of the different members of the alphavirus genus.

1.4. SFV and RRV Hyperactivate the PI3K-AKT-mTOR Pathway and Downstream Effectors

Both SFV and RRV have been shown to very strongly activate the PI3K-AKT-mTOR pathway, via the presence of YXXM motifs in the C-terminal HVD of nsP3 (Y369-E370-P371-M372 for SFV, and Y356-E357-T358-M359 for RRV) [12] (Figure 2). YXXM motifs are known activators of class 1A PI3Ks [13,14] and, after localisation to the PM in nascent viral RNA replication complexes, the YXXM motifs of SFV and RRV nsP3 bind to one or both of the SH2 domains of p85, thereby releasing the enzymatic p110 subunit, leading to activation of AKT [12]. Strikingly, the hyperactivation of the pathway even occurs when cells are starved of amino acids and growth factors prior to infection. The presence of PM-associated nsP3 is sufficient to induce AKT phosphorylation [15], suggesting interaction with a PM-localised cellular factor important for this activation. By comparison with other YXXM-containing proteins, we propose that the tyrosine residue is phosphorylated by a PM-localised cellular tyrosine kinase, but the identity of such a kinase(s) is not known.

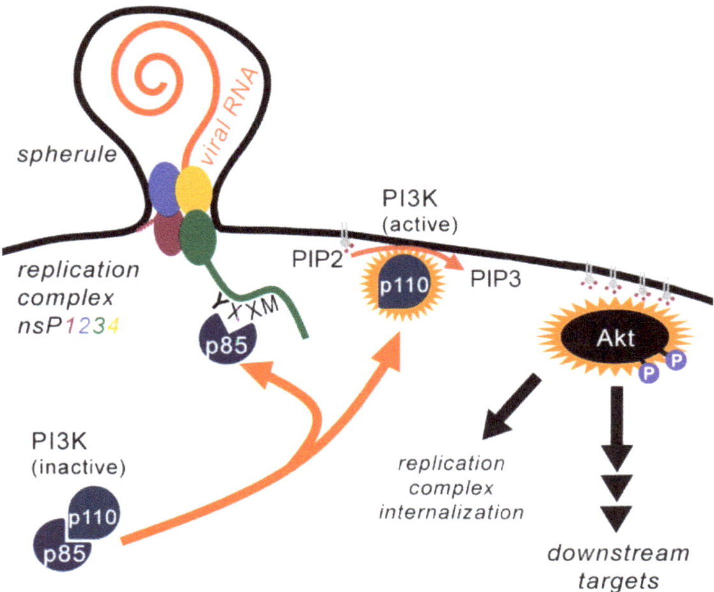

Figure 2. In Semliki Forest virus (SFV) and Ross River virus (RRV)-infected cells, viral replication complexes form at the PM. Via recruitment of p85 to a YXXM motif in nsP3, p110 is freed from regulation and can catalyse the conversion of PIP2 to PIP3 and recruitment and activation of AKT to the PM. Via multiple downstream effector pathways, the internalization of viral replication complexes as well as a heightened cellular metabolic activity and likely other yet undescribed states are promoted. For more details, see text.

All alphaviruses encode nsP3, but the HVD greatly varies in length and sequence between alphaviruses. We analysed the nsP3 sequence of all alphaviruses and found that YXXM motifs are only present in SFV, RRV and the equine alphaviruses Getah virus and Sagiyama virus [12]. There are no studies showing whether the latter viruses also hyperactivate AKT in their respective host cells. Interestingly, some viruses from other families also activate AKT via YXX(X)M motifs, including influenza virus [16] and herpes simplex virus type 1 [17].

SFV infection leads to the induction of strong phosphorylation of AKT and two downstream targets of mTORC1, S6 and 4E-BP1 [15]. Furthermore, it induces phosphorylation of the AKT substrates phosphofructokinase 2 (PFK2), Rab GTPase-activating protein AS160 (or TBC1D4) and ATP citrate lyase (ACL) [12]. These latter substrates all play a role in cell metabolism. Treatment of infected cells with wortmannin, a specific PI3K inhibitor, reduces but does not eliminate phosphorylation of AKT, PFK2, AS160 and ACL [12,15]. This suggests that SFV-induced AKT hyperactivation is mediated by PI3K-dependent and independent mechanisms. S6 phosphorylation is not reduced upon wortmannin treatment, probably due to the fact that mTORC1 is also stimulated via other pathways [15].

For both SFV and RRV, wortmannin treatment inhibits viral replication at late stages only, suggesting that PI3K-AKT activation and consequent changes in cellular metabolism are more important for later stages of viral replication. For RRV however, there is no difference in release of new virions when cells are infected with an RRV mutant that cannot hyperactivate AKT. However, in a murine model of RRV infection, wildtype RRV causes a more severe disease than the non-hyperactivating mutant [12]. Probably, AKT activation is not essential for in vitro replication, but does contribute to infection in vivo. Effects of the pathway on SFV pathogenesis in vivo have not been studied.

1.5. CHIKV Activates PI3K-AKT Moderately

CHIKV infection of a number of different cell types and species also leads to activation of PI3K-AKT [18–21], and PI3K expression was found to be upregulated in total white blood cells isolated from CHIKV infected patients with both mild and severe disease [22]. As the PI3K inhibitor wortmannin completely blocks CHIKV-induced activation of AKT [15], the activation is likely mediated entirely via PI3K. However, when directly compared to SFV, the PI3K-AKT activation by CHIKV is significantly slower and less extensive [15]. In line with that, CHIKV nsP3 does not have a YXXM motif and likely does not hyperactivate by binding to SH2 domains of p85. The mechanism of this moderate activation of PI3K-AKT is not known. Src family kinases (SFK), which have been shown to be important for CHIKV-replication and with which CHIKV might interact [19], directly interact with AKT and are required for its activation [23,24]. It has been shown that microRNAs that target proteins in the PI3K-AKT pathway are modulated by CHIKV [21] and may play some role in the activation. Nevertheless, despite less extensive activation than SFV or RRV, the pathway appears to be important for replication and pathogenesis of CHIKV. Inhibition of PI3K has been shown to inhibit CHIKV replication, translation of CHIKV proteins and the amount of new infectious viruses [19,25].

In contrast, another study showed that inhibition of PI3K did not affect CHIKV replication, whereas direct inhibition of AKT or inhibition of protein kinase A (PKA), a kinase which regulates AKT-activation independent of PI3K, did inhibit CHIKV replication [18]. This might suggest that CHIKV could activate AKT independently of PI3K, through PKA. Another AKT inhibitor, GSK 690693, did not influence CHIKV replication [19]. CHIKV infection transiently inhibits activation of mTORC1 by increasing reactive oxygen species (ROS) levels [25,26]. Inhibition of mTORC1 with either rapamycin, TORISEL or RAD001 is reported to not affect CHIKV replication [18], and might even favour it [19,25]. Treatment of CHIKV infected human fibroblasts with dasitinib strongly reduced viral titres by blocking translation of subgenomic RNA [19]. This might be through PI3K-AKT activation, but also through MAPK activation, as MAPK signalling pathways were shown to be required for optimal replication of CHIKV [27]. However, inhibition of both mTORC1 and mTORC2 by Torin 1 decreases CHIKV replication, suggesting that CHIKV needs mTORC2, though it does not depend on mTORC1 [19]. mTORC2 activates AKT, but also other kinases, such as protein kinase C isotypes and serum and glucocorticoid-induced kinase 1 (GSK1) [28]. SFK and mTORC1/mTORC2 seem to be important for more alphaviruses, as Mayaro virus (MAYV), o'nyong'nyong virus (ONNV) and RRV titres are also decreased upon inhibition with dasitinib or Torin 1 [19]. The contradicting results of different inhibitors on CHIKV replication might be due to the relatively weak activation of the PI3K-AKT pathway, which may be more sensitive to variation.

1.6. SINV Differentially Activates PI3K-AKT in Different Species

Lacking a YXXM motif, SINV also does not hyperactivate the PI3K-AKT pathway [12], but rather appears to induce low or transient activation in cell lines from different species [12,29,30]. LY294002, which inhibits class I PI3Ks [31], inhibits SINV RNA replication, but not virus titres in human cells [30]. Specific mTORC1 inhibition does not decrease SINV replication [30], and might even increase it [25]. Taken together, these reports do not suggest a strong requirement for the PI3K-AKT-mTOR activation for efficient SINV infection and replication.

1.7. Benefits of Activation of PI3K Pathway for Alphaviruses

As described above, activation of the PI3K-AKT pathway can have many effects on diverse cellular processes, including cell survival, growth and proliferation and metabolism, and many more functions specific for certain cell types [7]. In the subsequent sections, we discuss what is known about the effects of alphavirus-induced PI3K-AKT activation on vertebrate cells.

2. Metabolic Change

Alphaviruses Influence Cellular Metabolism

PI3K-AKT activation by SFV leads to a change in cellular metabolism, probably through increased phosphorylation of AKT targets. It increases glycolysis and causes an increased glucose flux through the tricarboxylic acid cycle (TCA). Consequently, more citrate is exported to the cytoplasm and more fatty acids are de novo synthesised. Furthermore, there is increased activation of the pentose phosphate pathway (PPP), which results in increased synthesis of nucleotides, such as UMP [12]. Treating cells for 16h with the glucose analogue 2-deoxyglucose (2-DG), which inhibits glycolysis and prevents entrance of glucose into the PPP, followed by SFV infection, strongly reduces production of new virions [12,32]. This effect is still seen when 2-DG is administered at 3 hpi, which indicates that glycolysis is not required for early events in replication [32]. Inhibition of the PPP specifically through the glucose-6-phosphate dehydrogenase inhibitor (DHEA), reduces production of new virions almost 10-fold. Furthermore, treatment of infected cells with wortmannin or the AKT-inhibitor MK-2206 dramatically inhibits glycolysis and reduces the release of new virions. This indicates the importance of the metabolic shift for SFV replication [12].

RRV infection increases glycolysis moderately in a manner that appears to be independent of the hyperactivation of PI3K-AKT [12]. In contrast, wildtype RRV increases fatty acid levels significantly, whereas RRV-YF (which does not hyperactivate PI3K-AKT) does not. It is not shown whether inhibition of fatty acid synthesis hampers RRV replication. RRV-YF is attenuated in an in vivo mouse model, but this attenuation is not directly attributed to effects on metabolism [12].

Conversely, although SINV does not induce sustained AKT phosphorylation in human cells, it does activate glycolysis at 8 hpi [12]. SINV also increases glycolytic flux in mouse neuroblastoma cells (Neuro 2a), perhaps to compensate for virus-induced mitochondrial dysfunction [33]. Inhibition of glycolysis or PPP in SINV-infected cells strongly decreases the release of progeny virus in both human and African green monkey kidney (Vero) cells [12,32]. Similar to SFV, glycolysis is not required for virus entry, but is important later during the viral life cycle [32]. It does not seem that SINV infection increases fatty acid synthesis. Thus, SINV induces a slightly different metabolic profile, but needs glycolysis and PPP for replication in cells of various species. SINV does not induce sustained AKT-activation in human cells, and probably has evolved other mechanisms to stimulate those metabolic pathways [12].

Metabolic pathways have not been well studied in the context of CHIKV infection. There are some indications that CHIKV activates glycolysis. PKM2, an isoenzyme of the glycolytic enzyme pyruvate kinase, is overexpressed in CHIKV-infected mouse muscles [34]. Furthermore, in human hepatic cells, CHIKV induces upregulation of PDHA1, a subunit of the pyruvate dehydrogenase complex involved in transforming pyruvate to acetyl-CoA in the TCA. However, other glycolytic and TCA-enzymes, such as alpha-enolase and isocitrate dehydrogenase, are downregulated in CHIKV-infected cells [35]. Regarding the other alphaviruses, only MAYV is known to increase glycolytic flux strongly, through activation of the enzyme 6-phosphofructo 1-kinase (PFK1) [36]. However, it is not shown whether MAYV activates AKT.

To conclude, all abovementioned alphaviruses influence cellular metabolism. However, this is not always entirely AKT-dependent, and all alphaviruses modulate cellular metabolism slightly differently. Processes such as increased glycolysis are essential for some viruses, but dispensable for others, at least in vitro. Further study will be necessary to gain a more complete understanding of the importance of metabolic alterations to alphavirus infections.

Many viruses influence cellular metabolism in order to create optimal circumstances for replication. Most viruses stimulate glycolysis and the PPP and fatty acid synthesis. Through these mechanisms, viruses can increase the availability of energy and building blocks for replication and progeny production [37]. During acute infection, viruses need much energy and building blocks for viral replication and production of viruses. Thus, it is very plausible that alphaviruses as well as other viruses causing acute infection, activate PI3K-AKT to influence cellular mechanism and ensure this.

However, PI3K-AKT-dependent increased glucose uptake and glycolysis was also shown to promote an interferon-induced antiviral response [38]. Increased glycolysis is beneficial for the virus but comes at a cost.

3. Interaction with Autophagy

3.1. Various Viruses Influence Autophagy

Macroautophagy, the process by which intracellular components, such as long-lived proteins and organelles are degraded and recycled is also regulated at least in part by the PI3K-AKT pathway [39]. It is a constitutive process but can be further induced through activation of the kinase Ulk1 by AMP activated protein kinase (AMPK). Activation of the PI3K-AKT-mTORC1 pathway inhibits autophagy through inhibition of the interaction between Ulk1 and AMPK by mTORC1 [40].

Autophagy in the context of virus infection has been well studied, although it is not always clear if the effects of the pathway are pro- or antiviral [41,42]. The benefits of autophagy seem to be different for different viruses, and dependent on the time after infection. Some viruses, such as ZIKV and rotavirus require the process early in infection [43–46], whereas this is detrimental for others. Conversely, some viruses, such as influenza A virus induce autophagy late after infection to increase replication [47]. Finally, there are viruses in which autophagy does not appear to influence replication. Autophagy has been studied in the context of CHIKV, SFV and SINV infection, though most of these studies did not investigate the direct role of PI3K-AKT activation. These alphaviruses appear to have different effects on autophagy, though there also might be variation due to different cell types used and different experimental conditions.

3.2. CHIKV Induce Autophagy, Whereas SFV Blocks It

CHIKV infection increases the amount of autophagosomes in various human cell lines: human embryonic kidney (HEK) 293 cells [48], HeLa cells [49,50], human glioblastoma cells (U-87) [51], and murine embryonic fibroblasts (MEF) [26] from early in infection and continuously increasing during the course of infection. CHIKV induces oxidative stress, leading to increased intracellular ROS and NO levels. This inhibits mTORC1, which results in increased autophagy [26]. CHIKV-induced autophagy has the effect to reduce or delay CHIKV-induced cell death [26]. The correlation between increased autophagy and efficient CHIKV replication might be species dependent [26,51,52]. An explanation for the difference between species came with the discovery that in human cells, CHIKV nsP2 interacts with the autophagy receptor NDP52, but not in murine cells [50]. The situation appears more complicated in vivo, where one study reported that induction of autophagy attenuates disease, when compared between wildtype and autophagy-deficient mice [26], while another showed that inhibition of autophagy with Tat-beclin 1 peptide reduces viral titres and improves clinical outcomes in CHIKV-infected mice [52].

SFV also increases the amount of autophagosomes in the infected cell [53]. However, in contrast to CHIKV, SFV infection does not appear to strongly affect the production of autophagosomes but rather blocks their degradation in a manner dependent on expression of the viral glycoproteins. Since SFV replication was unaffected by ATG5 depletion, it is concluded that the pathway plays no role in replication in vitro [53].

In conclusion, many acute viruses influence AKT and mTORC1 activation to influence autophagy. For most alphaviruses it is not clear how PI3K-AKT activation relates to autophagy. SFV strongly activates PI3K-AKT-mTORC1, but specifically blocks autophagosome degradation, which seems contradictory as active mTORC1 blocks both autophagosome formation and degradation. CHIKV activates PI3K-AKT moderately, but inhibits mTORC1 via ROS and NO, resulting in increased autophagy.

4. Promotion of Cell Survival

Unknown Whether Alphaviruses Activate PI3K-AKT to Promote Cell Survival

Many AKT substrates play a role in promoting cell survival and proliferation, but there are currently no studies addressing the effects of AKT on cell death in alphavirus infection. The New World alphavirus Venezuelan equine encephalitis virus stimulates early growth factor response genes in infected astrocytes, but this was shown not to be dependent on MAPK or PI3K signaling [54]. Rubella virus (RV) which also belongs to the *Togaviridae* family, induces caspase-dependent apoptosis, leading to a cytopathic effect, but delays and reduces this process through PI3K-AKT activation [55]. However, in our preliminary experiments, we did not detect any significant delay in cell death in wildtype SFV infection compared to mutants, which do not hyperactivate the pathway (unpublished).

5. The Strange Case of the Trafficking of Replication Complexes

Some Alphaviruses Stimulate Internalisation of Replication Complexes

Early in infection, alphavirus replication complexes (RC) assemble in membrane invaginations/spherules at the PM in which viral replication takes place [56,57]. A striking effect of PI3K-AKT hyperactivation by SFV and RRV is the trafficking of RC from the PM. First, RC localise in small and scattered cytoplasmic vesicles and then in large acidic perinuclear vacuoles (called cytopathic vacuoles of type I (CPV-I)). In SFV infection, RCs are relocalised from the PM to CPV-I at 8 hpi [56]. When PI3K is inhibited or when the YXXM motif in nsP3 is mutated, trafficking of RC does not take place [12,15,56]. It is not known which downstream targets of AKT mediate RC trafficking. Remarkably, viral RNA synthesis is not hindered when PI3K is inhibited, indicating that RC at the PM alone can sustain RNA replication [15]. Therefore, it is difficult to determine whether internalisation of RCs has benefits for the virus.

In other alphaviruses, the effect is less clear. CHIKV activates PI3K-AKT, although to a lesser extent than SFV, and CHIKV RCs mostly remain at the PM [15]. RC of SINV, which does not activate PI3K-AKT in human cells, but does activate it in murine cells, partially relocalises into the cytoplasm in baby hamster kidney cells (BHK-21), though the majority of RCs remain PM membrane associated [57]. Altogether, it seems that only strong AKT activation is required to induce trafficking of alphavirus RCs and replication happens to a similar extent in RC at the PM as in CPV-I.

6. Remarks in Conclusion

Alphaviruses are pathogens of growing importance, causing increasing numbers of cases throughout the world but for which there are no vaccines or antiviral therapies. We have reviewed the literature showing that several alphaviruses activate the PI3K-AKT pathway in vertebrate cells. The extent of activation differs (Table 1), as SFV and RRV induce hyperactivation and CHIKV causes moderate activation. SINV does not induce sustained AKT activation in humans but activates AKT in mice. The mechanism of hyperactivaton by SFV and RRV is known [12], and future work will reveal the mechanism of the moderate activation by CHIKV, SINV and mutants of SFV and RRV which do not hyperactivate. Inhibition of PI3K-AKT most clearly inhibits replication of SFV and RRV, while for CHIKV there are contradicting results and SINV does not seem to rely on PI3K-AKT for replication.

Table 1. Mechanisms and Effects of PI3K-AKT activation.

	Mechanism of PI3K-AKT Activation	Effects of PI3K-AKT Activation on			
		Metabolism	**Autophagy**	**Apoptosis**	**Trafficking RC**
SFV	Strong activation via YXXM motif in nsP3	Increases glycolysis and fatty acid synthesis	Blocks degradation of autophagosomes	Small, not significant delay	RCs traffic from PM to CPV-I
RRV	Strong activation via YXXM motif in nsP3	Increases fatty acid synthesis	Unknown	Small, not significant delay	RCs traffic from PM to CPV-I
CHIKV	Moderate activation by unknown mechanism	Unknown	Increases production of autophagosomes	Unknown	RC mostly remain at PM
SINV	Weak or transient activation by unknown mechanism	Unknown	Unknown	Unknown	RC mostly remain at PM

Thus, PI3K-AKT activation has several (potential) benefits for alphaviruses, of which changing the cellular metabolism seems to be the most important. However, PI3K-AKT-mTORC1-S6K mediates an interferon-induced antiviral response [58]. Furthermore, AKT contributes to the activation of NF-κB, which leads to production of inflammatory cytokines [2]. This is not studied in alphavirus-infected cells, but it is very important to determine to which extent PI3K-AKT is needed for antiviral immunity and when PI3K-AKT activation is beneficial for the virus, but detrimental for the host. In development of antiviral therapies, proteins from the PI3K-AKT pathway might be good targets. However, too strong inhibition of PI3K-AKT will probably damage the host, as cells will become much more prone to apoptosis and innate immune pathways may be induced. Therefore, targeting proteins that play a role in the processes that AKT stimulates and are beneficial for alphaviruses, such as anabolic metabolism, is also an important subject for further research.

Author Contributions: Both authors researched and wrote the paper. G.M.M. communicated with the journal. All authors have read and agreed to the published version of the manuscript.

Funding: This research and the APC was funded by the Swedish Research Council (621-2014-4718) and the Swedish Cancer Foundation (CAN 2015–751).

Acknowledgments: The authors thank Siwen Long and Laura Perez Vidakovics for critical reading of the manuscript. The work in the GM lab is supported by Karolinska Institutet, The Swedish Research Council and the Swedish Cancer Foundation.

Conflicts of Interest: The authors declare no conflict of interest.

References

1. Diehl, N.; Schaal, H. Make yourself at home: Viral hijacking of the PI3K/Akt signaling pathway. *Viruses* **2013**, *5*, 3192–3212. [CrossRef] [PubMed]
2. Dunn, E.F.; Connor, J.H. HijAkt: The PI3K/Akt pathway in virus replication and pathogenesis. *Prog. Mol. Biol. Transl. Sci.* **2012**, *106*, 223–250. [PubMed]
3. Buchkovich, N.J.; Yu, Y.; Zampieri, C.A.; Alwine, J.C. The TORrid affairs of viruses: Effects of mammalian DNA viruses on the PI3K-Akt-mTOR signalling pathway. *Nat. Rev. Microbiol.* **2008**, *6*, 266–275. [CrossRef] [PubMed]
4. Cooray, S. The pivotal role of phosphatidylinositol 3-kinase-Akt signal transduction in virus survival. *J. Gen. Virol.* **2004**, *85*, 1065–1076. [CrossRef]
5. Vanhaesebroeck, B.; Guillermet-Guibert, J.; Graupera, M.; Bilanges, B. The emerging mechanisms of isoform-specific PI3K signalling. *Nat. Rev. Mol. Cell Biol.* **2010**, *11*, 329–341. [CrossRef]
6. Ersahin, T.; Tuncbag, N.; Cetin-Atalay, R. The PI3K/AKT/mTOR interactive pathway. *Mol. Biosyst.* **2015**, *11*, 1946–1954. [CrossRef]
7. Manning, B.D.; Toker, A. AKT/PKB Signaling: Navigating the Network. *Cell* **2017**, *169*, 381–405. [CrossRef]

8. Rupp, J.C.; Sokoloski, K.J.; Gebhart, N.N.; Hardy, R.W. Alphavirus RNA synthesis and non-structural protein functions. *J. Gen. Virol.* **2015**, *96*, 2483–2500. [CrossRef]
9. Gotte, B.; Panas, M.D.; Hellstrom, K.; Liu, L.; Samreen, B.; Larsson, O.; Ahola, T.; McInerney, G.M. Separate domains of G3BP promote efficient clustering of alphavirus replication complexes and recruitment of the translation initiation machinery. *PLoS Pathog.* **2019**, *15*, e1007842. [CrossRef] [PubMed]
10. Lark, T.; Keck, F.; Narayanan, A. Interactions of Alphavirus nsP3 Protein with Host Proteins. *Front. Microbiol.* **2017**, *8*, 2652. [CrossRef] [PubMed]
11. Gotte, B.; Liu, L.; McInerney, G.M. The Enigmatic Alphavirus Non-Structural Protein 3 (nsP3) Revealing Its Secrets at Last. *Viruses* **2018**, *10*, 105. [CrossRef] [PubMed]
12. Mazzon, M.; Castro, C.; Thaa, B.; Liu, L.; Mutso, M.; Liu, X.; Mahalingam, S.; Griffin, J.L.; Marsh, M.; McInerney, G.M. Alphavirus-induced hyperactivation of PI3K/AKT directs pro-viral metabolic changes. *PLoS Pathog.* **2018**, *14*, e1006835. [CrossRef]
13. Wu, H.; Windmiller, D.A.; Wang, L.; Backer, J.M. YXXM motifs in the PDGF-beta receptor serve dual roles as phosphoinositide 3-kinase binding motifs and tyrosine-based endocytic sorting signals. *J. Biol. Chem.* **2003**, *278*, 40425–40428. [CrossRef] [PubMed]
14. Songyang, Z.; Shoelson, S.E.; Chaudhuri, M.; Gish, G.; Pawson, T.; Haser, W.G.; King, F.; Roberts, T.; Ratnofsky, S.; Lechleider, R.J.; et al. SH2 domains recognize specific phosphopeptide sequences. *Cell* **1993**, *72*, 767–778. [PubMed]
15. Thaa, B.; Biasiotto, R.; Eng, K.; Neuvonen, M.; Gotte, B.; Rheinemann, L.; Mutso, M.; Utt, A.; Varghese, F.; Balistreri, G.; et al. Differential Phosphatidylinositol-3-Kinase-Akt-mTOR Activation by Semliki Forest and Chikungunya Viruses Is Dependent on nsP3 and Connected to Replication Complex Internalization. *J. Virol.* **2015**, *89*, 11420–11437. [CrossRef] [PubMed]
16. Shin, Y.K.; Liu, Q.; Tikoo, S.K.; Babiuk, L.A.; Zhou, Y. Influenza A virus NS1 protein activates the phosphatidylinositol 3-kinase (PI3K)/Akt pathway by direct interaction with the p85 subunit of PI3K. *J. Gen. Virol.* **2007**, *88*, 13–18. [CrossRef] [PubMed]
17. Strunk, U.; Saffran, H.A.; Wu, F.W.; Smiley, J.R. Role of herpes simplex virus VP11/12 tyrosine-based motifs in binding and activation of the Src family kinase Lck and recruitment of p85, Grb2, Shc. *J. Virol.* **2013**, *87*, 11276–11286. [CrossRef]
18. Sharma, A.; Bhomia, M.; Yeh, T.J.; Singh, J.; Maheshwari, R.K. Miltefosine inhibits Chikungunya virus replication in human primary dermal fibroblasts. *F1000Res* **2018**, *7*, 9. [CrossRef]
19. Broeckel, R.; Sarkar, S.; May, N.A.; Totonchy, J.; Kreklywich, C.N.; Smith, P.; Graves, L.; DeFilippis, V.R.; Heise, M.T.; Morrison, T.E.; et al. Src Family Kinase Inhibitors Block Translation of Alphavirus Subgenomic mRNAs. *Antimicrob. Agents Chemother.* **2019**, *63*, e02325-18. [CrossRef]
20. Das, I.; Basantray, I.; Mamidi, P.; Nayak, T.K.; Pratheek, B.M.; Chattopadhyay, S. Heat shock protein 90 positively regulates Chikungunya virus replication by stabilizing viral non-structural protein nsP2 during infection. *PLoS ONE* **2014**, *9*, e100531. [CrossRef]
21. Sharma, A.; Balakathiresan, N.S.; Maheshwari, R.K. Chikungunya Virus Infection Alters Expression of MicroRNAs Involved in Cellular Proliferation, Immune Response and Apoptosis. *Intervirology* **2015**, *58*, 332–341. [CrossRef] [PubMed]
22. Wikan, N.; Khongwichit, S.; Phuklia, W.; Ubol, S.; Thonsakulprasert, T.; Thannagith, M.; Tanramluk, D.; Paemanee, A.; Kittisenachai, S.; Roytrakul, S.; et al. Comprehensive proteomic analysis of white blood cells from chikungunya fever patients of different severities. *J. Transl. Med.* **2014**, *12*, 96. [CrossRef] [PubMed]
23. Jiang, T.; Qiu, Y. Interaction between Src and a C-terminal proline-rich motif of Akt is required for Akt activation. *J. Biol. Chem.* **2003**, *278*, 15789–15793. [CrossRef] [PubMed]
24. Yori, J.L.; Lozada, K.L.; Seachrist, D.D.; Mosley, J.D.; Abdul-Karim, F.W.; Booth, C.N.; Flask, C.A.; Keri, R.A. Combined SFK/mTOR inhibition prevents rapamycin-induced feedback activation of AKT and elicits efficient tumor regression. *Cancer Res.* **2014**, *74*, 4762–4771. [CrossRef] [PubMed]
25. Joubert, P.E.; Stapleford, K.; Guivel-Benhassine, F.; Vignuzzi, M.; Schwartz, O.; Albert, M.L. Inhibition of mTORC1 Enhances the Translation of Chikungunya Proteins via the Activation of the MnK/eIF4E Pathway. *PLoS Pathog.* **2015**, *11*, e1005091. [CrossRef]
26. Joubert, P.E.; Werneke, S.W.; De la Calle, C.; Guivel-Benhassine, F.; Giodini, A.; Peduto, L.; Levine, B.; Schwartz, O.; Lenschow, D.J.; Albert, M.L. Chikungunya virus-induced autophagy delays caspase-dependent cell death. *J. Exp. Med.* **2012**, *209*, 1029–1047. [CrossRef]

27. Varghese, F.S.; Thaa, B.; Amrun, S.N.; Simarmata, D.; Rausalu, K.; Nyman, T.A.; Merits, A.; McInerney, G.M.; Ng, L.F.; Ahola, T. The antiviral alkaloid berberine reduces chikungunya virus-induced mitogen-activated protein kinase (MAPK) signaling. *J. Virol.* **2015**, *90*, 9743–9757. [CrossRef]
28. Oh, W.J.; Jacinto, E. mTOR complex 2 signaling and functions. *Cell Cycl.* **2011**, *10*, 2305–2316. [CrossRef]
29. Scherbik, S.V.; Brinton, M.A. Virus-induced Ca2+ influx extends survival of west nile virus-infected cells. *J. Virol.* **2010**, *84*, 8721–8731. [CrossRef]
30. Mohankumar, V.; Dhanushkodi, N.R.; Raju, R. Sindbis virus replication, is insensitive to rapamycin and torin1, suppresses Akt/mTOR pathway late during infection in HEK cells. *Biochem. Biophys. Res. Commun.* **2011**, *406*, 262–267. [CrossRef]
31. Gharbi, S.I.; Zvelebil, M.J.; Shuttleworth, S.J.; Hancox, T.; Saghir, N.; Timms, J.F.; Waterfield, M.D. Exploring the specificity of the PI3K family inhibitor LY294002. *Biochem. J.* **2007**, *404*, 15–21. [CrossRef] [PubMed]
32. Findlay, J.S.; Ulaeto, D. Semliki Forest virus and Sindbis virus, but not vaccinia virus, require glycolysis for optimal replication. *J. Gen. Virol.* **2015**, *96*, 2693–2696. [CrossRef] [PubMed]
33. Silva da Costa, L.; Da Silva, A.P.P.; Da Poian, A.T.; El-Bacha, T. Mitochondrial bioenergetic alterations in mouse neuroblastoma cells infected with Sindbis virus: Implications to viral replication and neuronal death. *PLoS ONE* **2012**, *7*, e33871. [CrossRef] [PubMed]
34. Dhanwani, R.; Khan, M.; Lomash, V.; Rao, P.V.; Ly, H.; Parida, M. Characterization of chikungunya virus induced host response in a mouse model of viral myositis. *PLoS ONE* **2014**, *9*, e92813. [CrossRef] [PubMed]
35. Thio, C.L.; Yusof, R.; Abdul-Rahman, P.S.; Karsani, S.A. Differential proteome analysis of chikungunya virus infection on host cells. *PLoS ONE* **2013**, *8*, e61444. [CrossRef] [PubMed]
36. El-Bacha, T.; Menezes, M.M.; Silva, M.C.A.e.; Sola-Penna, M.; Da Poian, A.T. Mayaro virus infection alters glucose metabolism in cultured cells through activation of the enzyme 6-phosphofructo 1-kinase. *Mol. Cell Biochem.* **2004**, *266*, 191–198. [CrossRef] [PubMed]
37. Sanchez, E.L.; Lagunoff, M. Viral activation of cellular metabolism. *Virology* **2015**, *479–480*, 609–618.
38. Burke, J.D.; Platanias, L.C.; Fish, E.N. Beta interferon regulation of glucose metabolism is PI3K/Akt dependent and important for antiviral activity against coxsackievirus B3. *J. Virol.* **2014**, *88*, 3485–3495. [CrossRef]
39. Chun, Y.; Kim, J. Autophagy: An Essential Degradation Program for Cellular Homeostasis and Life. *Cells* **2018**, *7*, E278. [CrossRef]
40. Kim, J.; Kundu, M.; Viollet, B.; Guan, K.L. AMPK and mTOR regulate autophagy through direct phosphorylation of Ulk1. *Nat. Cell Biol.* **2011**, *13*, 132–141. [CrossRef]
41. Echavarria-Consuegra, L.; Smit, J.M.; Reggiori, F. Role of autophagy during the replication and pathogenesis of common mosquito-borne flavi- and alphaviruses. *Open Biol.* **2019**, *9*, 190009. [CrossRef]
42. Choi, Y.; Bowman, J.W.; Jung, J.U. Autophagy during viral infection—A double-edged sword. *Nat. Rev. Microbiol.* **2018**, *16*, 341–354. [CrossRef] [PubMed]
43. Liang, Q.; Luo, Z.; Zeng, J.; Chen, W.; Foo, S.S.; Lee, S.A.; Ge, J.; Wang, S.; Goldman, S.A.; Zlokovic, B.V.; et al. Zika Virus NS4A and NS4B Proteins Deregulate Akt-mTOR Signaling in Human Fetal Neural Stem Cells to Inhibit Neurogenesis and Induce Autophagy. *Cell Stem Cell* **2016**, *19*, 663–671. [CrossRef] [PubMed]
44. Cao, B.; Parnell, L.A.; Diamond, M.S.; Mysorekar, I.U. Inhibition of autophagy limits vertical transmission of Zika virus in pregnant mice. *J. Exp. Med.* **2017**, *214*, 2303–2313. [CrossRef] [PubMed]
45. Zhou, Y.; Geng, P.; Liu, Y.; Wu, J.; Qiao, H.; Xie, Y.; Yin, N.; Chen, L.; Lin, X.; Liu, Y.; et al. Rotavirus-encoded virus-like small RNA triggers autophagy by targeting IGF1R via the PI3K/Akt/mTOR pathway. *Biochim. Biophys. Acta Mol. Basis Dis.* **2018**, *1864*, 60–68. [CrossRef]
46. Yin, Y.; Dang, W.; Zhou, X.; Xu, L.; Wang, W.; Cao, W.; Chen, S.; Su, J.; Cai, X.; Xiao, S.; et al. PI3K-Akt-mTOR axis sustains rotavirus infection via the 4E-BP1 mediated autophagy pathway and represents an antiviral target. *Virulence* **2018**, *9*, 83–98. [CrossRef]
47. Wang, R.; Zhu, Y.; Zhao, J.; Ren, C.; Li, P.; Chen, H.; Jin, M.; Zhou, H. Autophagy Promotes Replication of Influenza A Virus In Vitro. *J. Virol.* **2016**, *2019*, 93. [CrossRef]
48. Krejbich-Trotot, P.; Gay, B.; Li-Pat-Yuen, G.; Hoarau, J.J.; Jaffar-Bandjee, M.C.; Briant, L.; Gasque, P.; Denizot, M. Chikungunya triggers an autophagic process which promotes viral replication. *Virol. J.* **2011**, *8*, 432. [CrossRef]
49. Khongwichit, S.; Wikan, N.; Abere, B.; Thepparit, C.; Kuadkitkan, A.; Ubol, S.; Smith, D.R. Cell-type specific variation in the induction of ER stress and downstream events in chikungunya virus infection. *Microb. Pathog.* **2016**, *101*, 104–118. [CrossRef]

50. Judith, D.; Mostowy, S.; Bourai, M.; Gangneux, N.; Lelek, M.; Lucas-Hourani, M.; Cayet, N.; Jacob, Y.; Prevost, M.C.; Pierre, P.; et al. Species-specific impact of the autophagy machinery on Chikungunya virus infection. *EMBO Rep.* **2013**, *14*, 534–544. [CrossRef]
51. Abraham, R.; Mudaliar, P.; Padmanabhan, A.; Sreekumar, E. Induction of cytopathogenicity in human glioblastoma cells by chikungunya virus. *PLoS ONE* **2013**, *8*, e75854. [CrossRef]
52. Shoji-Kawata, S.; Sumpter, R.; Leveno, M.; Campbell, G.R.; Zou, Z.; Kinch, L.; Wilkins, A.D.; Sun, Q.; Pallauf, K.; MacDuff, D.; et al. Identification of a candidate therapeutic autophagy-inducing peptide. *Nature* **2013**, *494*, 201–206. [CrossRef] [PubMed]
53. Eng, K.E.; Panas, M.D.; Murphy, D.; Hedestam, G.B.K.; McInerney, G.M. Accumulation of autophagosomes in Semliki Forest virus infected cells is dependent on the expression of the viral glycoproteins. *J. Virol.* **2012**, *86*, 5674–5685. [CrossRef] [PubMed]
54. Dahal, B.; Lin, S.C.; Carey, B.D.; Jacobs, J.L.; Dinman, J.D.; Van Hoek, M.L.; Adams, A.A.; Kehn-Hall, K. EGR1 upregulation following Venezuelan equine encephalitis viaarus infection is regulated by ERK and PERK pathways contributing to cell death. *Virology* **2020**, *539*, 121–128. [CrossRef] [PubMed]
55. Cooray, S.; Jin, L.; Best, J.M. The involvement of survival signaling pathways in rubella-virus induced apoptosis. *Virol. J.* **2005**, *2*, 1. [CrossRef]
56. Spuul, P.; Balistreri, G.; Kaariainen, L.; Ahola, T. Phosphatidylinositol 3-kinase-, actin-, microtubule-dependent transport of Semliki Forest Virus replication complexes from the plasma membrane to modified lysosomes. *J. Virol.* **2010**, *84*, 7543–7557. [CrossRef]
57. Frolova, E.I.; Gorchakov, R.; Pereboeva, L.; Atasheva, S.; Frolov, I. Functional Sindbis virus replicative complexes are formed at the plasma membrane. *J. Virol.* **2010**, *84*, 11679–11695. [CrossRef]
58. Kaur, S.; Sassano, A.; Joseph, A.M.; Majchrzak-Kita, B.; Eklund, E.A.; Verma, A.; Brachmann, S.M.; Fish, E.N.; Platanias, L.C. Dual regulatory roles of phosphatidylinositol 3-kinase in IFN signaling. *J. Immunol.* **2008**, *181*, 7316–7323. [CrossRef]

Article

The RNA Replication Site of Tula Orthohantavirus Resides within a Remodelled Golgi Network

Katherine A. Davies [1], Benjamin Chadwick [1], Roger Hewson [2], Juan Fontana [1], Jamel Mankouri [1] and John N. Barr [1,*]

1 School of Molecular and Cellular Biology, University of Leeds, Leeds LS2 9JT, UK; katherine.davies@phe.gov.uk (K.A.D.); bmbmch@leeds.ac.uk (B.C.); J.Fontana@leeds.ac.uk (J.F.); j.mankouri@leeds.ac.uk (J.M.)
2 National Infection Service, Public Health England, Porton Down, Salisbury SP4 0JG, UK; Roger.Hewson@phe.gov.uk
* Correspondence: j.n.barr@leeds.ac.uk; Tel.: +44-113-3438069

Received: 30 May 2020; Accepted: 23 June 2020; Published: 27 June 2020

Abstract: The family *Hantaviridae* within the *Bunyavirales* order comprises tri-segmented negative sense RNA viruses, many of which are rodent-borne emerging pathogens associated with fatal human disease. In contrast, hantavirus infection of corresponding rodent hosts results in inapparent or latent infections, which can be recapitulated in cultured cells that become persistently infected. In this study, we used Tula virus (TULV) to investigate the location of hantavirus replication during early, peak and persistent phases of infection, over a 30-day time course. Using immunofluorescent (IF) microscopy, we showed that the TULV nucleocapsid protein (NP) is distributed within both punctate and filamentous structures, with the latter increasing in size as the infection progresses. Transmission electron microscopy of TULV-infected cell sections revealed these filamentous structures comprised aligned clusters of filament bundles. The filamentous NP-associated structures increasingly co-localized with the Golgi and with the stress granule marker TIA-1 over the infection time course, suggesting a redistribution of these cellular organelles. The analysis of the intracellular distribution of TULV RNAs using fluorescent in-situ hybridization revealed that both genomic and mRNAs co-localized with Golgi-associated filamentous compartments that were positive for TIA. These results show that TULV induces a dramatic reorganization of the intracellular environment, including the establishment of TULV RNA synthesis factories in re-modelled Golgi compartments.

Keywords: hantavirus; Tula virus; replication; factory; RNA synthesis; Golgi; stress granules

1. Introduction

The genus *Orthohantavirus*, within the *Hantaviridae* family of the *Bunyavirales* order, comprises over 36 species of enveloped segmented negative stranded RNA viruses [1] that are found throughout the globe, with many capable of causing devastating human disease [2]. Orthohantaviruses segregate into either New World (NW) and Old World (OW) clades based on their country of isolation, with viruses of the OW clade being widespread throughout Eurasia, whereas NW viruses are found within the Americas. Orthohantaviruses are mostly associated with a specific rodent host, in which they cause persistent infections [3]. In contrast, many OW and NW orthohantaviruses are associated with pathologically distinct disease outcomes in humans. Typically, viruses of the OW clade, such as the Hantaan virus (HTNV) and the Seoul virus (SEOV), are associated with a haemorrhagic fever with renal syndrome (HFRS), whereas NW viruses including Andes virus (ANDV) and Sin Nombre virus (SNV) are the causative agents of hantavirus cardiopulmonary syndrome (HCPS). Both syndromes are characterised by extensive vascular leakage, with human mortality rates ranging from 0.1–10% for HFRS to around 40% for HCPS [4]. In rodent hosts, few cytopathic effects are apparent and the animals develop asymptomatic or subclinical persistent infections [5]. This outcome is mirrored in cell culture systems that become persistently infected [6], reflecting the ability of hantaviruses to evade pathogen surveillance and innate immune defences.

The orthohantavirus genome comprises three virion-associated negative sense RNA segments (vRNAs) named small (S), medium (M) and large (L) based on their relative sizes [7]. The S segment encodes the nucleocapsid protein (NP), which coats the vRNAs forming helical ribonucleoprotein (RNP) complexes that act as templates for RNA synthesis. The M segment encodes a glycoprotein precursor (GPC) that is processed by cellular proteases into Gn and Gc proteins that form envelope spikes with primary roles in virus attachment and entry. The L segment encodes the large (L) protein that is the catalytic component of the RNA-dependent RNA polymerase (RdRp), responsible for two distinct RNA synthesis activities; mRNA transcription to yield a single 5′ capped mRNA from each vRNA template, and RNA replication to amplify the three vRNA segments via complementary (cRNA) intermediates. Hantavirus transcription involves the acquisition of short 5′ capped oligoribonucleotides from host cell mRNAs by an RdRp-associated endonuclease, which are then used by the RdRp to prime mRNA synthesis using a 'prime-and-realign' mechanism [8]. The source of these primers for hantaviruses is thought to be processing bodies (PBs), that are non-enveloped cytoplasmic RNP granules where non-sense mRNAs are stored awaiting degradation, and with which the SNV NP has been shown to closely associate [9]. NP is thought to bind the mRNA cap and protect it from degradation by the PB resident de-capping enzymes including DCP1a and DCP2, host factors that are required for other members of the order *Bunyavirales* [10].

The replication of hantaviruses, as with all bunyaviruses, is thought to occur within cytoplasmic viral factories [11]. Specifically, the replication factory of the prototypical bunyavirus, Bunyamwera virus (BUNV) has been visualized and described in detail. In mammalian cells, BUNV establishes large virus factories around the Golgi stack, recruiting organelles, cytoskeletal proteins and other cellular factors that together form elongated membranous structures known as viral tubes [12]. These tubes are lined with multiple membrane-bound polymerase molecules that are the sites of viral RNA synthesis, subsequently leading to the generation of RNPs for virion assembly. RNPs are thought to travel to adjacent assembly sites in the Golgi with the aid of actomyosin motors [12], where they interact with the viral glycoproteins and bud into the Golgi lumen. Virions subsequently mature as they traffic within Golgi-derived vesicles to the plasma membrane for release. BUNV replication and assembly in insect cells, in which the virus establishes a persistent infection, similarly occurs in the Golgi [13]. However, it was found that in mosquito cells, viruses do not accumulate intracellularly and exit the cell immediately after assembly, with NP and L proteins accumulated in non-membrane-bound cytoplasmic aggregates [13].

In contrast to BUNV, the precise intracellular site of hantavirus replication has not been defined. As NP is the major protein component of RNPs, its location is expected to coincide with sites of RNA synthesis and previous work has shown that NP from the NW hantaviruses, ANDV and Black Creek Canal virus (BCCV), accumulates within membranous structures surrounding the Golgi [14,15]. Consistent with this, NP from the OW hantavirus Tula virus (TULV) localizes in perinuclear regions [16] and HTNV NP resides within both perinuclear regions and the endoplasmic reticulum (ER)–Golgi intermediate compartment [17,18]. Further clues as to the location of sites of hantaviral RNA synthesis were revealed by the finding that the GFP-tagged L protein of TULV co-localized with the Golgi [19] and that a recombinant L protein from ANDV accumulated in perinuclear regions, co-localizing with both its cognate NP and also PB marker DCP1a [20]. Overall, with some notable NW hantavirus exceptions [21,22], hantavirus assembly is thought to occur by budding into the Golgi [7], suggesting the above description of BUNV replication and assembly sites may also broadly apply to those of hantaviruses.

In this study, we used confocal immunofluorescence microscopy alongside fluorescent single molecule in-situ hybridization (FISH) and electron microscopy (EM) to identify and characterise the cellular sites where NP as well as S segment RNAs were localised throughout an extended infection time course of the model OW hantavirus, TULV. We revealed that at peak and persistent stages of infection, NP is present as large filamentous and tubular structures, co-localizing with TIA-1, a marker for stress granules (SGs), and the Golgi, and that the FISH analysis strongly suggested that these TIA-1-containing and Golgi-derived structures are the sites of TULV RNA synthesis.

2. Methods

2.1. Virus and Cells

Tula virus (TULV) strain Moravia/5302v/95 was obtained from The National Collection of Pathogenic Viruses, PHE, UK. Stocks were shown to be mycoplasma free using MycoAlert (Lonza group AG, Basel, Switzerland). Vero E6 cells are a clone of Vero 76 cells isolated from *Cercopithecus aethiops* (African green monkey) kidney cells, and were obtained from the European Collection of Authenticated Cell Cultures (ECACC 85020206). These cells were selected as previous findings show that TULV-infection allows the establishment of a persistently infected state [23]. Virus titres were calculated by indirect staining using the anti-NP antibody in conjunction with an Alexa Fluor 488 secondary antibody (Thermo Fisher Scientific, Waltham, MA USA) as previously described [23].

2.2. Antibody Production and Validation

Antisera specific for TULV NP were generated in house as previously described [23], using bacterially expressed SEOV NP core domain as an antigen, purified as previously described, and used to immunize a sheep (AltaBiosciences, Redditch, UK). Antibodies were validated by Western blot analysis.

2.3. TULV Infections of Vero E6 Cells

Briefly, Vero E6 cells were plated out at 1×10^6 cells per 25 cm^2 flask or at 1×10^5 cells/well of a 12-well plate onto glass coverslips, sterilised by submerging into 100% ethanol and washing with sterile PBS. Cells were maintained in a humidified incubator at 37 °C with 5% CO_2. TULV infections were carried out at a multiplicity of infection (MOI) of 0.1–0.5 in serum-free DMEM. Virus was adsorbed for 90 min before the addition of DMEM supplemented with 2% foetal bovine serum, 100 U/mL penicillin and 100 µg/mL streptomycin. For long-term infections, TULV infections were carried out in 25 cm^2 flask and seeded onto coverslips 24 h prior to fixation. Fixation was carried out using 4% paraformaldehyde at 36 h post infection, 7 days post infection and 30 days post infection to represent early, peak and persistent infections.

2.4. Indirect Immunofluorescence

Mock-infected (media alone) or TULV-infected monolayers on coverslips were permeabilized in ice-cold methanol before washing in 1X PBS and blocked in 5% BSA in 1×PBS for 30 min. Primary staining was carried out using the following antibodies and reagents: in house anti-TULV NP [23], Golgi (Santa Cruz Biotechnology Inc, Dallas, TX AE6, USA), ER (Thermo Fisher Scientific Concanavalin A), β-actin (New England Biolabs (NEB), Ipswich, MA, USA D6A8), β-tubulin (NEB 9F3), vimentin (NEB D21H3), Rabs 5, 7 and 11 (NEB 2143, D95F2, D4F5), LAMP1 (NEB D2D11), clathrin heavy chain (NEB 4796), TIA-1 (Santa Cruz G-3) and DCP1a (Santa Cruz 56-Y). Incubations were carried out for 2 h at room temperature, followed by thorough washing in 1X PBS. Secondary antibody staining was carried out using donkey anti-goat Alexa Fluor 488 (Thermo Fisher Scientific A-11055), donkey anti-rabbit Alexa Fluor 555 (Thermo Fisher Scientific A-31572), donkey anti-mouse Alexa Fluor 555 (Thermo Fisher Scientific A-31570), donkey anti-rabbit Alexa Fluor 647 (Thermo Fisher Scientific A-31573), donkey anti-mouse Alexa Fluor 647 (Thermo Fisher Scientific A-31571) at a 1:500 dilution for 1 h. Nuclear staining was carried out by incubating with 300 nM DAPI for 3–5 min at room temperature. Cells to be visualised by confocal microscopy were mounted onto slides using Vectashield (Vector laboratories, Burlingame, CA, USA) hard set anti-fade mounting media and cured for 15 min before storing at 4 °C overnight. Cells to be further analysed by RNA-FISH were fixed in 4% formaldehyde for 20 min and stored in 1×PBS supplemented with 0.1 % diethylprocarbonate (DEPC).

2.5. TEM Preparation of TULV-Infected Vero E6 Cells

Vero E6 cells were fixed by adding 5% glutaraldehyde (*v/v* in 0.1 M phosphate buffer (PB) Agar Scientific, Stansted, UK) to an equal amount of cell media in the tissue culture flask, resulting in a final glutaraldehyde concentration of 2.5% (*v/v*). Cells were left for 2.5 h to fix at room temperature. After fixation, the cells were removed from the tissue culture flask by scraping and washed in 0.1 M PB for 30 min, twice. The cells were then post-fixed in 0.1 M PB with 1% osmium tetroxide (*w/v*) (Agar Scientific) for 1 h followed by four washes in 0.1 M PB. Subsequently, the post-fixed cells were dehydrated using an ascending ethanol series (40%, 60%, 80% and 2×100%) for 20 min, and then twice in propylene oxide (Agar Scientific) for 20 min. Dehydrated cells were incubated with a 1:1 mixture of propylene oxide and araldite epoxy resin (araldite CY212, dodecenyl succinic anhydride, and tri-dimethylaminomethyl phenol (Agar Scientific)) overnight, after which the mixture was replaced by pure araldite and incubated for 8 h. To ensure adequate embedding, new resin was added four times, after which the cells were polymerized at 60 °C for 24 to 48 h.

Embedded cells were then sectioned in the ultra-thin range (80 to 100 nm) using a EM UC7 Ultramicrotome (Leica, Wetzlar, Germany). Ultra-thin sections were then placed upon 200 mesh carbon-coated copper-formvar grids and stained with lead citrate and uranyl acetate (UA). Sectioned cells were stained in 8% saturated (*w/v*) UA (Agar Scientific) for 1 h. UA-stained grids were then washed in dH_2O for 5 min (×5). Afterwards, the sections were stained with lead citrate (Agar Scientific) for 10 min and washed in 0.02 M NaOH for 1 min (×3), and in dH_2O for 1 min (×5). Sections were then imaged at 120 kV using a FEI Tecnai G2-Spirit electron microscope (Thermo Fisher Scientific) with a Gatan US4000/SP 4k × 4k CCD camera (Gatan, Pleasanton, CA, USA).

2.6. Fluorescent In-Situ Hybridisation (FISH) Probes

Fluorescent in-situ hybridisation DNA probes were designed against the sense and anti-sense TULV S segment ORF sequence using the Stellaris Probe Designer version 4.2 (LGC Biosearch Technologies, Novato, CA, USA). The Stellaris Probe Designer provided 48 oligonucleotide probe sequences that were 20 nucleotides in length with a minimum spacing of 2 nucleotides. Target TULV S segment ORF sequence was obtained from sequencing viral RNA isolated from TULV-infected cell culture medium (Genewiz Inc, South Plainfield, NJ, USA). Probes were assessed for cross-hybridization using NCBI Nucleotide Blast against *C. aethiops* (taxid:9534) mRNAs and against the TULV M and L segment sense and anti-sense sequences (strain: Tula/Moravia/5302v/95). Any probes with more than 15 consecutive complimentary nucleotides were discarded. TULV S segment sense and anti-sense probes were labelled with Quasar670, and all probe sequences are shown Table S1. Strand sets were kept separate throughout this study.

2.7. Fluorescent In-Situ Hybridisation (FISH)

Probe hybridisation was carried out according to manufacturers' instructions. Briefly, each set of strand-specific FISH probes were reconstituted to 12.5 mM in RNase free 1× TE buffer (10 mM Tris-HCl, 1 mM EDTA pH 8.0 (G Biosciences) diluted in UltraPure DNase/RNase-Free Distilled water (Thermo Fisher Scientific) and stored separately at −20 °C. For each coverslip, 1 µL probe stock was used per 100 µL hybridisation buffer (Stellaris RNA FISH hybridisation buffer diluted with deionised formamide (Severn Biotech Ltd, Kidderminster, UK) in a 9:1 ratio).

Cells were incubated in Stellaris RNA FISH wash buffer A diluted with deionized formamide and UltraPure DNase/RNase-Free Distilled water in a 2:7:1 ratio for 5 min at room temperature. Additionally, 100 µL hybridisation buffer with probes was dispensed into a droplet onto parafilm in a humidified chamber. Coverslips were incubated cell-side down onto probe mix for 4 h at 37 °C, protected from light. Cells were incubated in 1 mL Stellaris RNA FISH wash buffer A for 30 min at

37 °C, protected from light and then incubated in 1 mL Stellaris RNA FISH wash buffer A with 300 nM DAPI for 5 min at room temperature. The buffer was removed and the cells were incubated in 1 mL Stellaris RNA FISH wash buffer B for 5 min before mounting onto glass microscope slides using Vectashield hard-set antifade mounting media (Vector laboratories). Slides were cured for 45 min at room temperature, protected from light, and imaged within 24 h.

2.8. Quantitative RT-PCR

TULV RNA was isolated from infected cells or supernatant using the QIAamp viral RNA mini kit (Qiagen, Hilden, Germany), and the quantitative reverse transcriptase real-time PCR was carried out using the 1-step RT-qPCR kit (Promega, Madison, WI, USA), both as previously described [23]. For each qRT-PCR experiment, in vitro transcribed TULV S segment RNAs were diluted to generate 10^8, 10^6, 10^4, 10^2 genome copies/5 μL and used to build a standard curve from which the sample relative genome copies per mL could be calculated, as previously described [23].

2.9. Inhibition of Intermediate Filaments

Vero E6 were infected with TULV at a MOI of 0.5 in a 6-well plate. Infected cells were harvested for immunofluorescence and q-RT PCR analysis at 30 days post infection. Prior to harvesting, the cells were treated with the following cytoskeletal inhibitors; 17 μM nocodazole (Sigma-Aldrich, St. Louis, MO, USA) for 60 min and 400 nM okadaic acid (OA) (Sigma-Aldrich) for 30, 60 or 90 min. Infected Vero E6 cells treated with OA were also recovered by removing the inhibitor and incubating overnight with fresh media before harvesting. Mock treatments were carried out using an equal volume of solvent (dimethyl sulfoxide (DMSO)/ethanol).

2.10. Laser Scanning Confocal Microscopy

Laser scanning confocal imaging was carried out using the Zeiss LSM700 confocal microscope using the 40×/1.3 Oil DIC Plan Apochromat, M27 objective lens with the diode 405 nm, 488 nm, 55 nm and 639 nm lasers or the Zeiss LSM880 + Airyscan upright confocal microscope using the Plan-Apochromat 40×/1.4 Oil DIC objective lens with the diode 405 nm, argon 488 nm, DPSS 561 nm and HeNe 633 nm lasers. Image analysis was carried out in Fiji Is Just ImageJ (FIJI).

3. Results

3.1. TULV NP Forms Both Punctate and Tubular Structures during Infection

We first analysed the intracellular distribution of TULV NP during an extended time course of up to 30 days, with time points of 36 h post infection (hpi), 7 days post infection (dpi) and 30 dpi chosen to represent the early, peak and persistent stages of TULV replication, respectively, as previously reported [23]. At the 36 hpi time point, NP mostly localised within discrete perinuclear puncta, with a minor component forming larger filamentous and tube-like structures (Figure 1A, top row). At this early time point, NP compartments had an average area of 0.54 μm^2 (14 cells and 1248 compartments analysed) with the largest structure within a single focal plane having an area of 11.8 μm^2. After 7 dpi, the NP compartments had an average area of 0.96 μm^2 (31 cells and 2959 compartments analysed) with the largest structure having an area of 154 μm^2 (Figure 1A, middle row). At 30 dpi, when fewer cells were present within the culture, but when surviving cells were still producing infectious TULV [23], most of the NP was present as tubes/filaments, reaching a maximum area of 513 μm^2 (15 cells and 8615 compartments analysed), with only low levels of punctate NP visible (Figure 1A, bottom row). Across the time course, the length of the filamentous/tubular structures increased, and their diameters remained relatively constant (Figure 1A).

To reveal further information of the three-dimensional morphology of the large NP-stained structures, successive Z axis sections of persistently infected cells (0.25 μm per stack) were reconstructed to a generate a three-dimensional image. These revealed that the long tubular and filamentous structures were connected and appeared to wrap around the nucleus (Figure 1B and Movie S1).

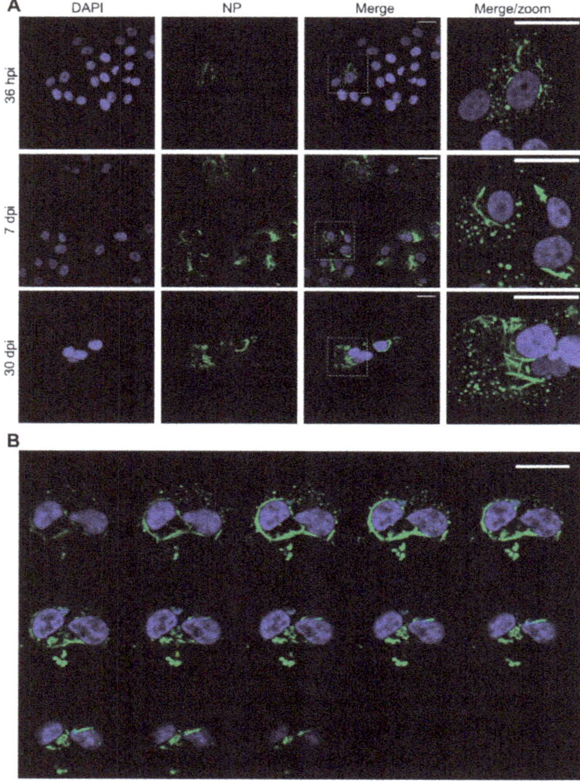

Figure 1. Confocal images showing the characteristic morphology of Tula virus (TULV) nucleocapsid protein (NP) -stained structures at early, peak and persistent stages of TULV infection in Vero E6 cells. (**A**) Vero E6 cells were infected with TULV at a multiplicity of infection of 0.1, and the images were taken at 36 h post infection (hpi), 7 days post infection (dpi) and 30 dpi using 40× magnification. Nuclei were stained with DAPI (blue) and TULV NP, detected using NP antisera (green), revealing the characteristic NP punctate and filamentous/tubular structures in the perinuclear region. (**B**) Successive Z sections of TULV-infected cells at 30 dpi show NP filamentous/tubular structures extending around the nucleus. The scale bars represent 30 μm.

3.2. Imaging of TULV-Induced Filamentous Cytoplasmic Structures Using Transmission Electron Microscopy

To further characterise these TULV NP-associated filamentous tubular structures, we then used transmission electron microscopy (TEM) to visualize the sections of TULV-infected Vero E6 cells at 36 hpi, 7 dpi and 30 dpi time points (Figure 2). During the early 36 hpi time point, no clear differences were detected between the TULV- and mock-infected cells (data not shown), most likely due to the relatively small size of TULV NP-associated structures, previously viewed (Figure 1). However, after 7 days, long filamentous and tubular structures were observed in 6% of the infected cells (eight out of 134 cells; Figure 2B), that were absent in non-infected cells, with no filaments detected in any of 61 non-infected cells that were imaged (Figure 2A). The relatively small percentage of infected cells containing filaments is explained because the EM images represent thin sections (~100 nm thick), suggesting that the filaments are concentrated in specific locations within the infected cell. Such filaments comprised clusters of small bundles, with each comprising multiple slender filamentous structures. Filament bundles were around ~100 to 1000 nm in length and spread over a large area within the cell, forming clusters that reached up to 4 µm in length. While filaments within the bundles were aligned in similar directions, the separate clusters often protruded in different orientations.

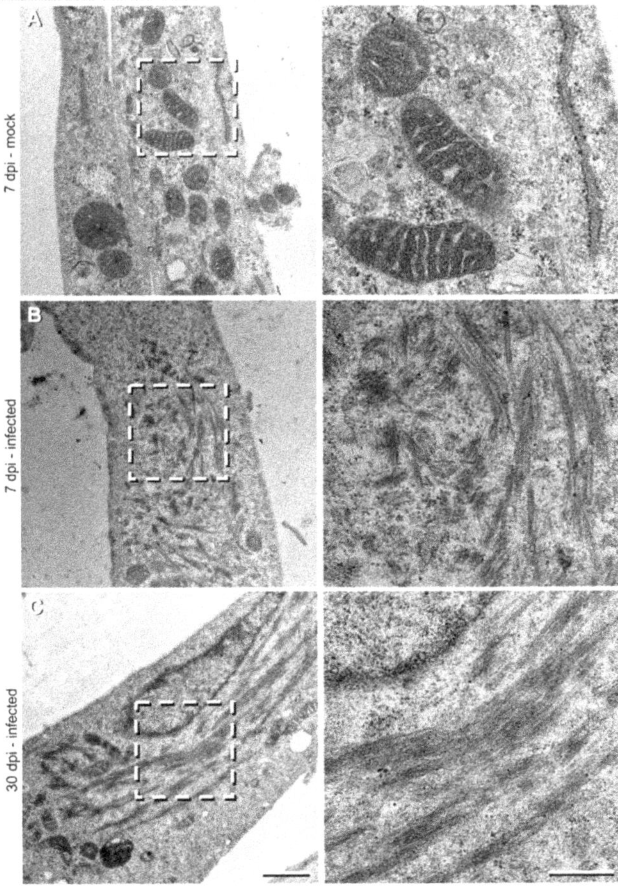

Figure 2. Transmission electron microscopy (TEM) images showing the filamentous structures within the TULV-infected Vero E6 cells. Vero E6 cells, (**A**) mock-infected, or infected with TULV at a MOI of 0.1 at either (**B**) 7 days post infection (dpi) or (**C**) 30 dpi. Higher magnification images from the marked inset regions are shown in the right column. Scale bars represent 500 nm and 1 µm on low and high magnification micrographs, respectively.

At 30 dpi, the filamentous clusters were present, but had approximately doubled in their frequency (to 10%; out of ~60 imaged cells) and had dramatically increased in size to ~5 µm in total length (Figure 2C). Individual bundles of tubular filaments were still evident, but these were less frequently observed. Instead, the bundles were longer (500–3000 nm) and shared a similar orientation. The dimensions and overall morphology of these TULV-induced structures at both the 7 dpi and 30 dpi time points coincided with those filamentous tubular structures stained for TULV NP and visualized by light microscopy, shown above (Figure 1).

3.3. Co-Localization of TULV NP with Cytoskeletal Components

Previous studies have shown that HTNV NP associates with actin [15], tubulin and vimentin [18], the latter forming cages surrounding HTNV NP in the vicinity of the ER-Golgi intermediate compartment (ERGIC). Consequently, we then determined the distribution of the punctate and filamentous forms of TULV NP in relation to these cytoskeletal components.

At 36 hpi, no association of either punctate or filamentous NP with actin, tubulin, or vimentin were observed (Figure 3A). However, at 7 dpi, vimentin was redistributed and co-localized with NP in the larger filamentous/tubular structures (Figure 3B, line scan 1) although vimentin/NP co-localization was less apparent in punctate compartments (Figure 3B, line scan 2). At 30 dpi, vimentin appeared to surround the NP structures (Figure 3C). Neither actin nor tubulin co-localized with NP at any time point.

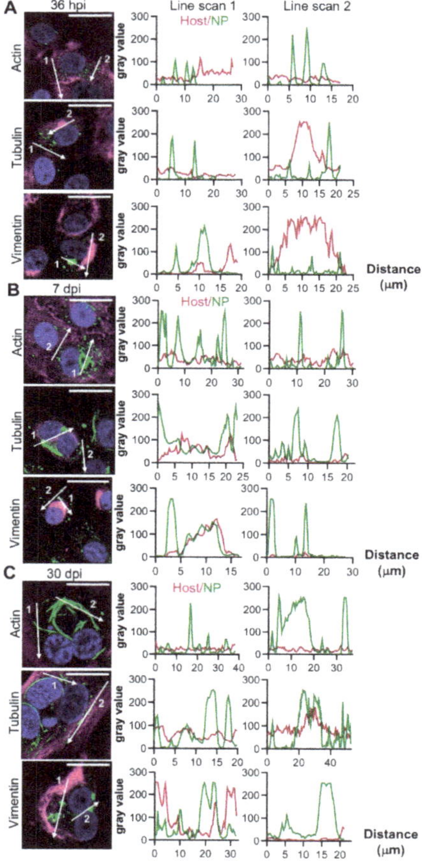

Figure 3. Co-localisation between the TULV NP and cytoskeletal filaments at the early, peak and persistent stages of TULV infection in Vero E6 cells. The spatial distribution of the cytoskeletal markers actin, tubulin and vimentin (magenta) was observed alongside TULV NP (green) in Vero E6 cells infected with TULV at a MOI of 0.1 by confocal microscopy at 40× magnification at (**A**) 36 h post infection (hpi), (**B**) 7 days post infection (dpi), and (**C**) 30 dpi. The scale bar represents 30 μm. Fluorescent line scans were taken using Fiji software with the designated number labels corresponding to adjacent line scan plots. Nuclei were stained with DAPI (blue), the TULV NP was detected using NP antisera and the cytoskeleton was detected using primary antibodies against β-actin, β-tubulin and vimentin. Images shown in panels A, B and C show the representative analysis of 87, 109 and 58 cells, respectively.

To further characterise the association between vimentin and NP, the infected cells were treated with okadaic acid (OA), an inhibitor of protein phosphatase 1 and 2A [24] that promotes vimentin disassembly [25]. OA treatment led to the disappearance of the typical NP structures, which rapidly reformed upon OA removal (Figure S1). In addition, OA treatment of TULV-infected cultures at 30 dpi led to reduced TULV genome copy numbers, suggesting that the disruption of the vimentin network was detrimental to TULV RNA replication (Figure S2). In contrast, nocodazole (NOC) caused no changes to the tubulin networks (Figure S1) and did not influence TULV replication (Figure S2).

3.4. Stress Granules (SGs) Are Recruited to Filamentous Structures during TULV Infection

PBs and stress granules (SGs) are non-membrane bound, liquid phase compartments that have been shown to stimulate the RNA synthesis of some negative sense RNA viruses and closely associate with their corresponding replication factories [26–29]. To examine the possible association of TULV NP and these compartments, the infected cells were stained with the PB marker DCP1a, a PB-resident mRNA de-capping enzyme, or the SG marker TIA-1, an RNA binding protein that scaffolds SG formation.

At the early, peak and persistent stages of the time course visualized, DCP1a-stained puncta were abundant (Figure 4A), but were spatially-distinct from the large filamentous/tubular NP-stained structures (Figure 4B,C) and with fewer than 4% co-localizing with NP (Figure 4D).

In contrast, the staining pattern of TIA-1 was different; at 36 hpi, only low numbers of TIA-1 puncta were detectable and their co-localization with NP was relatively low (Figure 4A). However, at 7 dpi and 30 dpi, the abundance of TIA-1 increased in dense cytoplasmic structures that co-localized with both filamentous/tubular NP compartments, as well as NP puncta (Figure 4B–C). The line scans associated with these later time points clearly revealed an overlap of the TIA-1 signal with both the filamentous/tubular and punctate NP signals, with their peak intensities coinciding (Figure 4B,C), corroborated by almost 30% of the TIA-1-stained compartments co-localizing with NP puncta (Figure 4E). Taken together, these results suggest that SG formation is simulated during the later stages of TULV infection, and that SGs are recruited to NP positive filamentous/tubular compartments.

3.5. TULV NP Associates with the Golgi during Infection

For many hantaviruses of both OW and NW clades, their corresponding NPs accumulate at sites within and adjacent to the Golgi, suggesting that this compartment may play an important role in hantavirus RNA synthesis. To better understand the intracellular distribution of TULV NP, the proximity of known ER and Golgi markers were assessed relative to NP in TULV-infected cells.

At 36 hpi, the Golgi appeared with the classical morphology of elongated and stacked vesicular compartments that were distinct from NP (Figure 5A). In contrast, at both 7 dpi and 30 dpi, there was extensive overlap with NP and the Golgi, in both large filamentous/tubular NP structures, and also punctate NP, with closely coinciding peaks on the associated line scans (Figure 5B,C). No such co-localisation between NP and the ER marker concanavalin A was observed at any of the specified time points (Figure 5A–C). In addition, NP showed minimal co-staining with clathrin, Rab5, Rab7, Rab11 or LAMP 1 markers, implying that the punctate NP structures were not endosomal in nature (Figure S3).

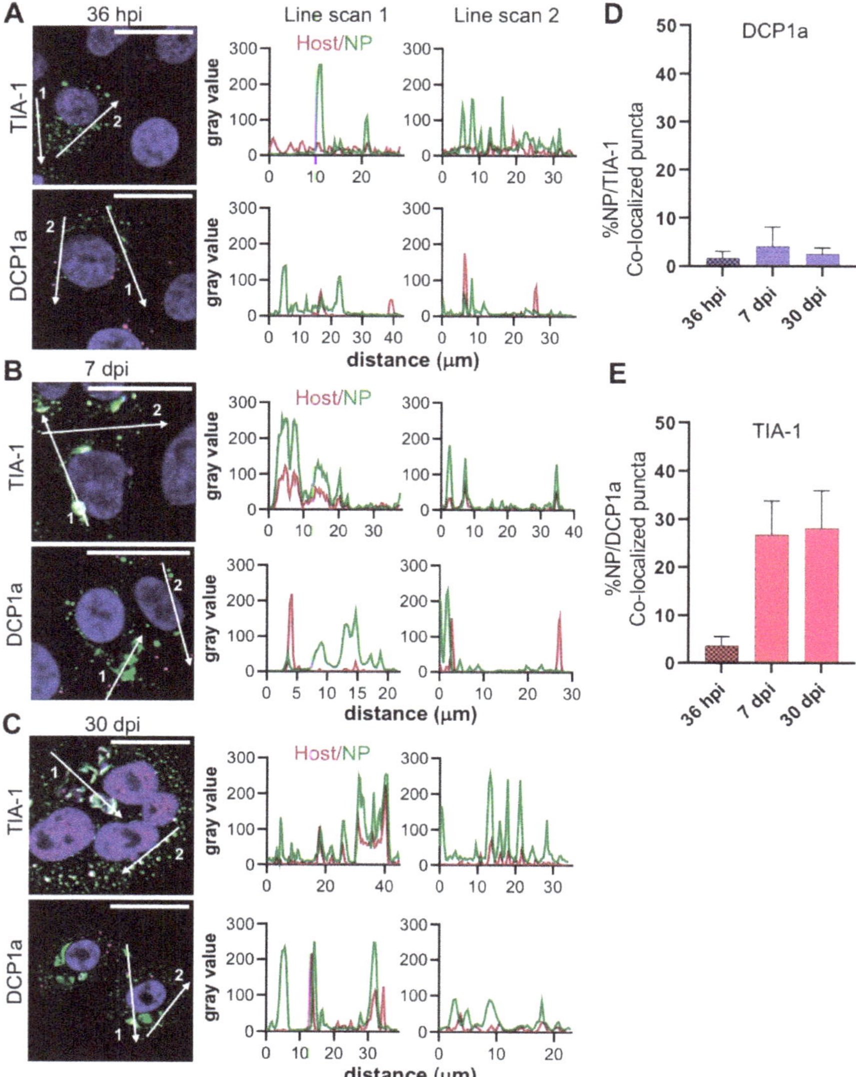

Figure 4. Co-localisation between TULV NP and either the stress granules (SGs) or the processing bodies (PBs) at the early, peak and persistent stages of TULV infection in Vero E6 cells. The spatial distribution of the SG marker TIA-1 and the PB marker DCP1a (magenta) was observed alongside TULV NP (green) in Vero E6 cells infected with TULV at a MOI of 0.1 by confocal microscopy at 40× magnification at the following time points; (**A**) 36 h post infection (hpi), (**B**) 7 days post infection (dpi) and (**C**) 30 dpi. The scale bars represent 30 μm. Fluorescent line scans were taken using Fiji software with the designated number labels corresponding to the adjacent line scan plots. Nuclei were stained with DAPI (blue), TULV NP was detected using NP antisera, and the SGs and PBs were detected using specific antisera against TIA-1 and DCP1a, respectively. Percentage co-localization of D) TIA-1 and E) DCP1a puncta co-localizing with NP at all three time points were quantified and represented as separate histograms. Images shown in panels A, B and C show a representative analysis of 7, 25 and 23 cells, respectively.

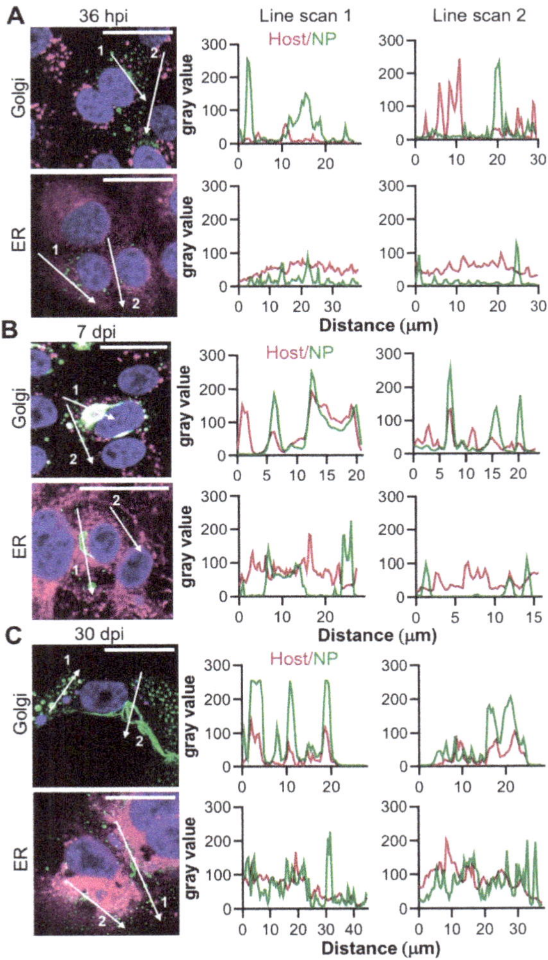

Figure 5. Co-localisation between the TULV NP and either the endoplasmic reticulum (ER) or the Golgi compartments at the early, peak and persistent stages of TULV infection in Vero E6 cells. The spatial distribution of the endoplasmic reticulum and the Golgi network (both magenta) was observed alongside TULV NP (green) in Vero E6 cells infected with TULV at an MOI of 0.1 by confocal microscopy at 40 X magnification at (**A**) 36 h post infection (hpi), (**B**) 7 days post infection (dpi) and (**C**) 30 dpi. The scale bars represent 30 μm. Fluorescent line scans were taken using Fiji software with the designated number labels corresponding to the adjacent line scan plots. Nuclei were stained with DAPI (blue), TULV NP was detected using NP antisera, with the ER detected using concanavalin-A and the Golgi was detected by specific antisera. Images shown in panels A, B and C show a representative analysis of the 19, 27 and 13 cells, respectively.

3.6. Products of TULV RNA Synthesis co-Localize with NP

The results presented thus far show that NP exhibits a close co-localization with the Golgi and SGs, within large filamentous and tubular structures. As NP is the major protein component of TULV RNPs, which act as templates for TULV mRNA transcription and RNA replication, we next investigated whether these NP-associated compartments were also stained for the products of TULV RNA synthesis.

Two independent FISH probe sets were designed comprising multiple non-overlapping fluorescently labelled short oligonucleotides with near complete coverage of the TULV S segment RNA target. One set was designed to detect negative sense TULV S segment RNAs, comprising the vRNA genome alone, whereas a second set was designed to detect positive sense S segment RNAs. The positive sense probe set would unavoidably hybridize to both S segment mRNAs and cRNAs due to their nested genetic organisation, however, as most positive sense RNAs generated during bunyavirus infection are mRNA transcripts [30], the majority of the detected positive sense signal could be attributed to mRNA transcripts, with only a minor contribution from cRNA replication products.

At the early, peak and persistent time points, the cells were fixed and stained for TULV NP along with the individual probe sets. Only the cells staining positive for NP were also stained with the FISH-probe sets, showing that the probes were specific for viral RNA targets. Interestingly, the ability of the probe sets to detect negative sense RNAs showed that the association of NP with RNA within assembled RNPs did not prevent probe binding. In addition, the TULV RNA signal was not detected outside regions positive for NP, strongly suggesting that, as expected, NP is required for all TULV RNA synthesis activities.

At 36 hpi, 7 dpi and 30 dpi time points, the signal corresponding to both the positive and negative sense RNAs co-localized with NP, with the extremely close co-localization of NP and RNA within the filamentous/tubular objects (Figure 6A–C). RNA and NP also co-localized within some NP puncta of various sizes, although with a less consistent level of abundance. Line scan analysis of the filamentous/tubular and punctate NP structures supported these observations, with consistently high levels of the NP and RNA signal within the filamentous/tubular structures, but often lower levels of RNA in NP puncta.

At all of the time points assessed, there were no consistent differences between the distribution of positive and negative sense RNAs in relation to the NP structures, suggesting that the synthesis of mRNA/cRNAs and vRNAs is not segregated.

3.7. Concurrent Detection of TULV and Cellular Components in Persistently Infected Cells

To confirm the above findings, and unequivocally show that the staining of viral and host components were located within the same compartments, 30 dpi persistently infected cells were co-stained for either NP/RNA/TIA-1 or NP/RNA/Golgi, concurrently (Figure 7).

Merging the signals from all three channels representing NP/RNA/TIA-1 confirmed that the NP, TIA-1 and both polarities of TULV RNA were simultaneously present within both filamentous and punctate structures. As shown above (Figure 6), all RNA staining coincided with that of NP, however, some of these regions were devoid of TIA-1. Similarly, merging the signals for NP/RNA/Golgi revealed high levels of co-localization for all three markers within the filamentous/tubular compartments (Figure 7B). Along with the results shown in Figure 5, these findings strongly suggest that TULV infection leads to the redistribution of the Golgi to coincide with TULV RNAs, and the formation of structures that we suggest most likely represent the viral replication factories, the major sites of TULV RNA synthesis.

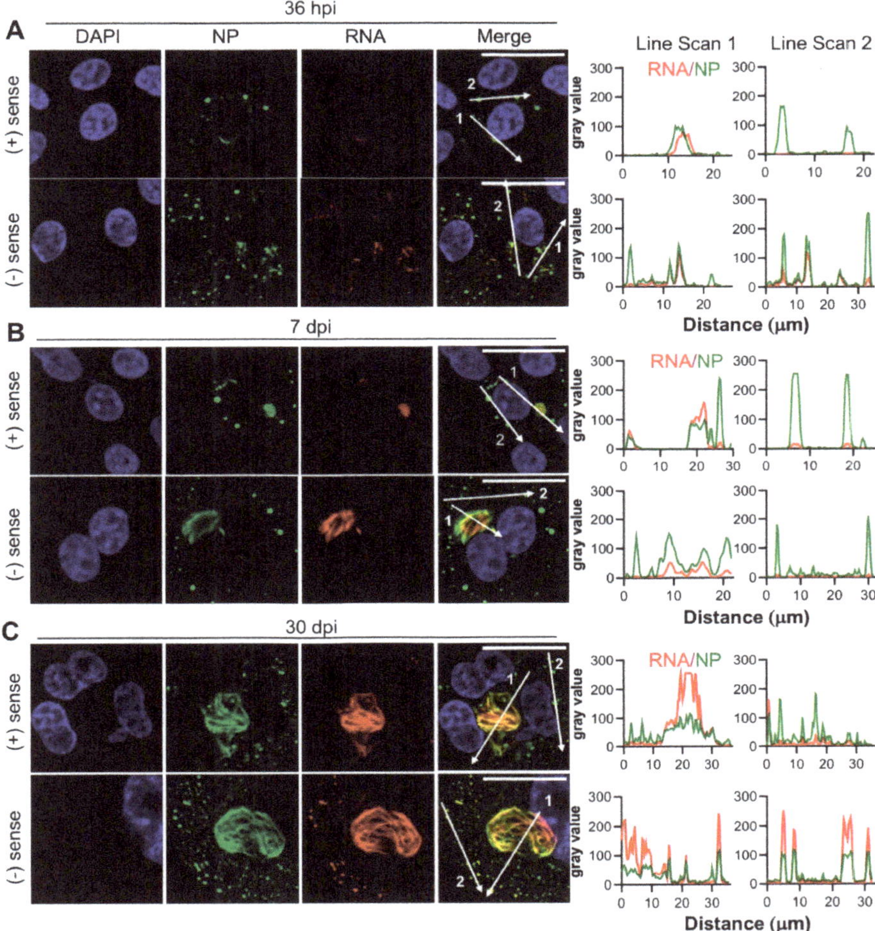

Figure 6. Co-localisation between the TULV NP and the TULV small (S) segment RNAs at the early, peak and persistent stages of TULV infection in Vero E6 cells. The spatial distribution of the TULV NP (green) with positive and negative sense S segment RNA (red) detected with strand-specific probe sets was observed using confocal microscopy at (**A**) 36 h post infection (hpi), (**B**) 7 days post infection (dpi) and (**C**) 30 dpi. The size bars represent 30 μm. Fluorescent line scans were taken using Fiji software with the designated number labels corresponding to the adjacent line scan plots. Nuclei were stained with DAPI (blue), TULV NP was detected using NP antisera and S segment RNA was detected using sense or anti-sense RNA FISH probes. Images presented in panels A, B and C show a representative analysis of 14, 25 and 8 cells, respectively.

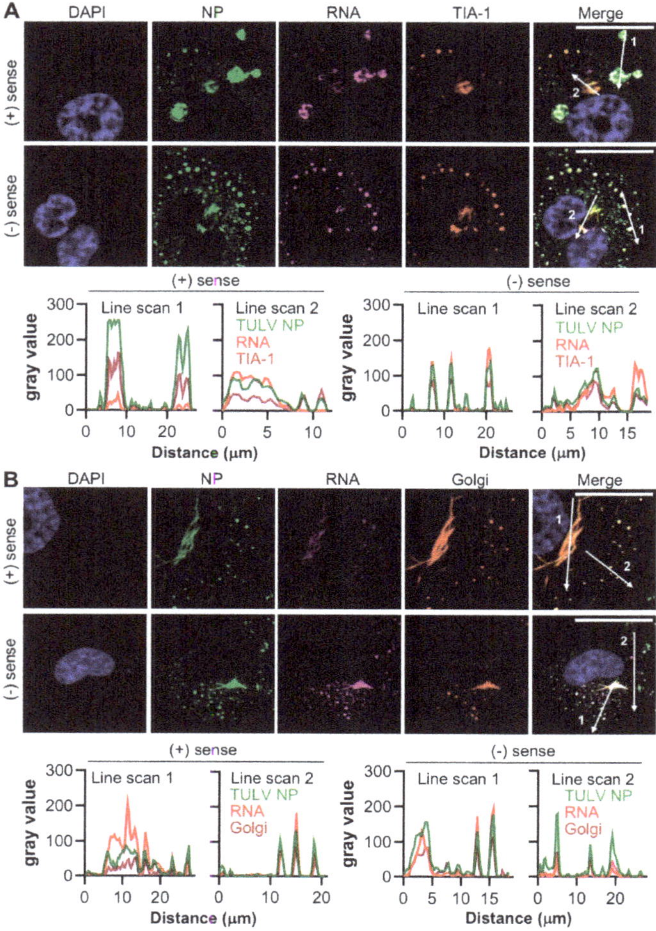

Figure 7. Co-localisation between the TULV NP, the TULV S segment positive and the negative sense RNAs and host cell components TIA-1 and the Golgi marker in Vero E6 cells at 30 days post infection. The spatial distribution of the TULV NP (green), the S segment RNA (magenta) and the host cell markers (**A**) TIA-1, or (**B**) Golgi (both red) were observed during the persistent infections using confocal microscopy at 40× magnification. Fluorescent line scans were taken using Fiji software with designated number labels corresponding to the adjacent line scan plots. The nuclei were stained with DAPI (blue), TULV NP was detected using NP antisera, with TIA-1 and Golgi detected using specific antisera. Scale bar represents 30 μm. TULV NP was detected using NP antisera, and S segment RNAs were detected using separate sets of positive sense or negative-sense RNA FISH probes. Images shown in panels A and B show a representative analysis of 8 and 20 cells, respectively.

4. Discussion

Many hantaviruses establish persistent infections within their rodent hosts by manipulating the host cellular environment to allow the abundant synthesis of viral components for an extended time period, while also evading or minimizing the cellular innate immune response [6,7]. To better understand the intracellular events during long-term hantavirus infections, we used immunofluorescent microscopy to investigate the intracellular distribution of TULV NP and RNA, alongside multiple host cell components during an extended time course of up to 30 days, during which time the persistent infection by TULV has been previously demonstrated [23]. This was combined with EM imaging to understand the morphological changes that TULV induces in infected cells. Our findings revealed that as the infection progressed, the NP formed increasingly large ultrastructural assemblies, mostly with an overall filamentous and tubular appearance, but also as discrete puncta (Figure 1). At all the time points examined, these NP assemblies closely co-localized with TULV RNA, regardless of its negative or positive sense polarity (Figure 6). This close association between all hantavirus RNA and NP is consistent with the known structure of the hantavirus RNP segments, in which hantaviral vRNAs and cRNAs are always found in close association with NP, forming extended helical assemblies [31]. Interestingly, our observations also suggest that TULV mRNAs, which will likely comprise the majority of positive sense TULV-specific RNA in the infected cell [30], also remain in close proximity with NP. This may reflect the organisation of the replication factories, with newly transcribed S, M and L segment mRNAs remaining in the vicinity of their respective NP enwrapped vRNA templates. Alternatively, it may reflect the previous findings that the hantavirus NP binds the 5′ mRNA cap at a separate RNA binding site [32], and this interaction is proposed to enable mRNA circularization, so as to promote eIF4F independent translation [33]. The lack of distinction between the sites of positive sense and negative sense RNA staining at the resolution used here also suggests that TULV RNA replication and mRNA transcription processes are not physically separated into distinct compartments.

Co-staining experiments using markers for intracellular compartments revealed that NP and TULV RNA co-localized within structures that were closely associated with the Golgi (Figure 5). At the early stages of infection (36 hpi), NP, RNA and Golgi components localized within both punctate and irregular filamentous structures within and surrounding compartments that exhibited a canonical Golgi morphology. At the later time points of 7 dpi and 30 dpi, NP and RNA were detected with increased abundance within a single filamentous/tubular structure of large dimensions, which also closely co-localized with the Golgi (Figure 5B–C). The morphology of this filamentous/tubular NP/RNA/Golgi stained structure was atypical of the Golgi compartment, with a clear loss of the typical fused compact ribbon structure associated with Golgi cisternae, suggesting that the Golgi was re-modelled, most likely by TULV-related activities. As both TULV NP and RNA components localised to these compartments (Figure 7), it is difficult to escape the conclusion that this remodelled Golgi is the site of TULV RNA synthesis and accumulation, and most likely represents TULV replication factories.

The visualization of long filamentous structures at later 7 dpi and 30 dpi time points in TULV-infected cells only, by TEM, was in agreement with our findings using light microscopy (Figure 2). These TULV-infected cell sections revealed the increased detail of the filamentous/tubular NP/RNA/Golgi stained structures, and confirmed their atypical morphology. Their apparent absence in sections of 36 hpi infected cells likely reflects their low abundance and the correspondingly high likelihood that they would not appear in the selected sections. Few studies have employed EM to study hantavirus-infected cells, with most analyses focusing on the morphology of cell-associated hantavirus particles [34–36]. However, the filamentous structures we observed here are reminiscent of filamentous and granular structures induced by the Sin Nombre virus [21], which were also in close proximity to the Golgi. However, in this study we showed that these filamentous structures also harbour abundant viral RNA.

If the filamentous/tubular NP-stained structures are the sites of RNA synthesis and accumulation, then the function of the punctate NP structures that are seen throughout the infection time course remains unclear. As they mostly stain weakly for RNA, we propose that they are unlikely to be sites of on-going abundant RNA synthesis. One possibility is that they may represent Golgi-derived particles, released by Golgi fragmentation, although an alternative possibility that we cannot rule out, is that the smallest of these structures may represent exocytic vesicles containing assembled viruses in transit towards the plasma membrane for release.

In contrast to the SG marker TIA-1, the levels of co-localization between the NP and the PB marker DCP1a were considerably lower (Figure 4), and this was unexpected given the previous reports of a close association between DCP1a and the NP of the NW SNV, where NP is proposed to bind and protect capped RNAs for later delivery to the transcribing RdRp [9]. This may represent fundamental differences between the host–pathogen interactions of NW and OW hantavirus species, and may suggest that OW hantaviruses such as TULV do not acquire or process host cell cap moieties in the same way. Given the high degree of association we report between TULV NP and TIA-1, an alternative cap resource may be SGs.

The observation that the SG marker TIA-1 showed a punctate distribution at the early stages of virus infection, but was filamentous/tubular distribution at later times, suggests that it was re-distributed into the Golgi-derived filamentous/tubular compartments (Figure 4A–D). To the best of our knowledge, SGs and the Golgi have not been previously reported to extensively co-localize for a native function, and thus we propose that this redistribution is in response to TULV infection. Whether this represents a pro-viral or anti-viral response is currently unknown. Considerable evidence suggests that SGs represent innate immune signalling platforms [37] and one possibility is that their sequestration into TULV replication factories may prevent or delay a host cell anti-viral response to infection. Such a mechanism of immune evasion may contribute to the establishment of the characteristic state of hantavirus persistence.

Supplementary Materials: The following are available online at http://www.mdpi.com/2073-4409/9/7/1569/s1, Table S1. DNA oligonucleotide probes designed to hybridize with TULV S segment RNAs in FISH experiments. Sequences shown are written 5′ to 3′ and represent the actual probe sequences, Movie S1. TULV-infected Vero E6 cells at 30 dpi were stained using TULV NP antisera and DAPI, and successive Z plane images taken, which were compiled into a movie, with rotation about a single axis, Figure S1. The effect of cytoskeletal disassembly and reorganisation on the localisation of TULV NP in Vero E6 cells persistently infected with TULV. Vero E6 cells were infected with TULV at an MOI of 0.5, and at 30 dpi cytoskeleton filaments were disassembled by treatment with A) 400 nM okadaic acid (OA) for periods of 60 and 90 min. Distribution of vimentin and TULV NP were examined using LSCM at 40 X magnification, stained with specific antisera NP antisera (green) and vimentin (magenta) with nuclei stained with DAPI (blue). Vimentin filaments were allowed to reassemble by the removal of OA, and the distribution of NP and vimentin examined as above. The scale bar represents 30 µm. B) The effect of microtubule depolymerisation on TULV NP localisation was also examined in Vero E6 cells at 30 dpi. Microtubules were depolymerised by treating Vero E6 cells with 17 µM nocodazole (NOC) for 60 min, with the distribution of tubulin and TULV NP examined using LSCM at 40 X magnification, stained with specific antisera NP antisera (green) and tubulin (magenta) with nuclei stained with DAPI (blue). The scale bar represents 30 µm, Figure S2. Okadaic acid treatment of TULV-infected Vero E6 cells decreases viral genome copies. TULV-infected Vero E6 cells at 30 dpi were treated with 400 nM okadaic acid (OA) for 30, 60 or 90 min or 17 µM nocodazole (NOC) for 60 min. After harvesting, the TULV genome copies in cell lysates were quantified by qRT-PCR., Figure S3. Co-localisation of cellular trafficking proteins with TULV NP in TULV-infected Vero E6 cells during early and persistent time points. The spatial distribution of Rab5, Rab7, Rab11, LAMP1 and clathrin (magenta) was observed alongside TULV NP (green) in TULV-infected Vero E6 cells at A) 36 hpi and B) 30 dpi by LSCM at 40 X magnification using specific antisera against TULV NP, Rab5, Rab7, Rab11, LAMP1 and clathrin. Nuclei are stained with DAPI (blue). The scale bar represents 30 µm. Fluorescent line scans taken using Fiji software.

Author Contributions: Conceptualization: K.A.D., B.C., J.N.B., J.M., J.F. and R.H.; methodology: K.A.D. and B.C.; validation: K.A.D. and B.C.; formal analysis: K.A.D. and BC.; investigation: K.A.D. and B.C.; resources: J.N.B., J.M., J.F. and R.H.; data curation: K.A.D. and B.C.; writing—original draft preparation: J.N.B.; writing—review and editing; J.N.B., K.A.D., B.C., J.M., J.F. and R.H.; supervision: J.N.B., J.F.; project administration: J.N.B.; funding acquisition, J.N.B., R.H. and J.F. All authors have read and agreed to the published version of the manuscript.

Funding: KAD was funded by a faculty PhD studentship from The University of Leeds. BC is supported by a studentship from the MRC Discovery Medicine North (DiMeN) Doctoral Training Partnership (MR/N013840/1). This work was supported by the Academic Fellow scheme at the University of Leeds (to J. Fontana and J.M.). Electron Microscope and ultra-microtome were funded by Wellcome Trust (090932/Z/09/Z and 208395/Z/17/Z).

Acknowledgments: We would like to thank Martin Fuller for support with resin embedding and ultra-thin sectioning.

Conflicts of Interest: The authors declare no conflict of interest.

References

1. Abudurexiti, A.; Adkins, S.; Alioto, D.; Alkhovsky, S.V.; Avsic-Zupanc, T.; Ballinger, M.J.; Bente, D.A.; Beer, M.; Bergeron, E.; Blair, C.D.; et al. Taxonomy of the order Bunyavirales: Update 2019. *Arch. Virol.* **2019**, *164*, 1949–1965. [CrossRef]

2. Spiropoulou, C.F.; Bente, D.A. Orthohantavirus, Orthonairovirus, Orthobunyavirus and Phlebovirus. In *Fields Virology; Emerging Viruses*, 7th ed.; Knipe, D., Howley, P., Eds.; Lippincott, Williams & Wilkins: Philadelphia, PA, USA, 2020; Volume 1, p. 750.
3. Schountz, T.; Prescott, J. Hantavirus Immunology of Rodent Reservoirs: Current Status and Future Directions. *Viruses* **2014**, *6*, 1317–1335. [CrossRef]
4. Jonsson, C.B.; Figueiredo, L.T.M.; Vapalahti, O. A Global Perspective on Hantavirus Ecology, Epidemiology, and Disease. *Clin. Microbiol. Rev.* **2010**, *23*, 412–441. [CrossRef]
5. Meyer, B.J.; Schmaljohn, C.S. Persistent hantavirus infections: Characteristics and mechanisms. *Trends Microbiol.* **2000**, *8*, 61–67. [CrossRef]
6. Meyer, B.J.; Schmaljohn, C.S. Accumulation of Terminally Deleted RNAs May Play a Role in Seoul Virus Persistence. *J. Virol.* **2000**, *74*, 1321–1331. [CrossRef] [PubMed]
7. Barr, J.N.; Weber, F.; Schmaljohn, C.S. Bunyavirlales; The viruses and thier replication. In *Fields Virology; Emerging Viruses*, 7th ed.; Knipe, D., Howley, P., Eds.; Lippincott, Williams & Wilkins: Philadelphia, PA, USA, 2020; Volume 1, p. 706.
8. Garcin, D.; Lezzi, M.; Dobbs, M.; Elliott, R.M.; Schmaljohn, C.; Kang, C.Y.; Kolakofsky, D. The 5′ ends of Hantaan virus (Bunyaviridae) RNAs suggest a prime-and-realign mechanism for the initiation of RNA synthesis. *J. Virol.* **1995**, *69*, 5754–5762. [CrossRef]
9. Mir, M.A.; Duran, W.A.; Hjelle, B.L.; Ye, C.; Panganiban, A.T. Storage of cellular 5′ mRNA caps in P bodies for viral cap-snatching. *Proc. Natl. Acad. Sci. USA* **2008**, *105*, 19294–19299. [CrossRef]
10. Hopkins, K.C.; McLane, L.M.; Maqbool, T.; Panda, D.; Gordesky-Gold, B.; Cherry, S. A genome-wide RNAi screen reveals that mRNA decapping restricts bunyaviral replication by limiting the pools of Dcp2-accessible targets for cap-snatching. *Genes Dev.* **2013**, *27*, 1511–1525. [CrossRef]
11. Novoa, R.R.; Calderita, G.; Arranz, R.; Fontana, J.; Granzow, H.; Ortiz, C.R. Virus factories: Associations of cell organelles for viral replication and morphogenesis. *Boil. Cell* **2012**, *97*, 147–172. [CrossRef]
12. Fontana, J.; Lopez-Montero, N.; Elliott, R.M.; Fernández, J.J.; Ortiz, C.R. The unique architecture of Bunyamwera virus factories around the Golgi complex. *Cell. Microbiol.* **2008**, *10*, 2012–2028. [CrossRef] [PubMed]
13. Lopez-Montero, N.; Ortiz, C.R. Self-protection and survival of arbovirus-infected mosquito cells. *Cell. Microbiol.* **2010**, *13*, 300–315. [CrossRef] [PubMed]
14. Rowe, R.K.; Suszko, J.W.; Pekosz, A. Roles for the recycling endosome, Rab8, and Rab11 in hantavirus release from epithelial cells. *Virology* **2008**, *382*, 239–249. [CrossRef] [PubMed]
15. Ravkov, E.V.; Compans, R.W. Hantavirus Nucleocapsid Protein Is Expressed as a Membrane-Associated Protein in the Perinuclear Region. *J. Virol.* **2001**, *75*, 1808–1815. [CrossRef] [PubMed]
16. Kaukinen, P.; Vaheri, A.; Plyusnin, A. Non-covalent interaction between nucleocapsid protein of Tula hantavirus and small ubiquitin-related modifier-1, SUMO-1. *Virus Res.* **2003**, *92*, 37–45. [CrossRef]
17. Maeda, A.; Lee, B.H.; Yoshimatsu, K.; Saijo, M.; Kurane, I.; Arikawa, J.; Morikawa, S. The Intracellular Association of the Nucleocapsid Protein (NP) of Hantaan Virus (HTNV) with Small Ubiquitin-like Modifier-1 (SUMO-1) Conjugating Enzyme 9 (Ubc9). *Virology* **2003**, *305*, 288–297. [CrossRef] [PubMed]
18. Ramanathan, H.N.; Chung, D.-H.; Plane, S.J.; Sztul, E.; Chu, Y.-K.; Guttieri, M.C.; McDowell, M.; Ali, G.; Jonsson, C.B. Dynein-Dependent Transport of the Hantaan Virus Nucleocapsid Protein to the Endoplasmic Reticulum-Golgi Intermediate Compartment. *J. Virol.* **2007**, *81*, 8634–8647. [CrossRef] [PubMed]
19. Kukkonen, S.K.J.; Vaheri, A.; Plyusnin, A. Tula hantavirus L protein is a 250 kDa perinuclear membrane-associated protein. *J. Gen. Virol.* **2004**, *85*, 1181–1189. [CrossRef]
20. Heinemann, P.; Schmidt-Chanasit, J.; Günther, S. The N Terminus of Andes Virus L Protein Suppresses mRNA and Protein Expression in Mammalian Cells. *J. Virol.* **2013**, *87*, 6975–6985. [CrossRef]
21. Goldsmith, C.S.; Elliott, L.H.; Peters, C.J.; Zaki, S.R. Ultrastructural characteristics of Sin Nombre virus, causative agent of hantavirus pulmonary syndrome. *Arch. Virol.* **1995**, *140*, 2107–2122. [CrossRef]
22. Ravkov, E.V.; Nichol, S.T.; Compans, R.W. Polarized entry and release in epithelial cells of Black Creek Canal virus, a New World hantavirus. *J. Virol.* **1997**, *71*, 1147–1154. [CrossRef]
23. Davies, K.; Afrough, B.; Mankouri, J.; Hewson, R.; Edwards, T.A.; Barr, J.N. Tula orthohantavirus nucleocapsid protein is cleaved in infected cells and may sequester activated caspase-3 during persistent infection to suppress apoptosis. *J. Gen. Virol.* **2019**, *100*, 1208–1221. [CrossRef] [PubMed]

24. Yatsunami, J.; Fujiki, H.; Suganuma, M.; Yoshizawa, S.; Eriksson, J.E.; Olson, M.O.; Goldman, R.D. Vimentin is hyperphosphorylated in primary human fibroblasts treated with okadaic acid. *Biochem. Biophys. Res. Commun.* **1991**, *177*, 1165–1170. [CrossRef]
25. Lee, W.-C.; Yu, J.-S.; Yang, S.-D.; Lai, Y.-K. Reversible hyperphosphorylation and reorganization of vimentin intermediate filaments by okadaic acid in 9L rat brain tumor cells. *J. Cell. Biochem.* **1992**, *49*, 378–393. [CrossRef] [PubMed]
26. Dinh, P.X.; Beura, L.K.; Das, P.B.; Panda, D.; Das, A.; Pattnaik, A.K. Induction of Stress Granule-Like Structures in Vesicular Stomatitis Virus-Infected Cells. *J. Virol.* **2012**, *87*, 372–383. [CrossRef]
27. Lindquist, M.E.; Lifland, A.W.; Utley, T.J.; Santangelo, P.J.; Crowe, J.E. Respiratory Syncytial Virus Induces Host RNA Stress Granules To Facilitate Viral Replication. *J. Virol.* **2010**, *84*, 12274–12284. [CrossRef]
28. Nelson, E.V.; Schmidt, K.M.; Deflubé, L.R.; Doğanay, S.; Banadyga, L.; Olejnik, J.; Hume, A.; Ryabchikova, E.; Ebihara, H.; Kedersha, N.; et al. Ebola Virus Does Not Induce Stress Granule Formation during Infection and Sequesters Stress Granule Proteins within Viral Inclusions. *J. Virol.* **2016**, *90*, 7268–7284. [CrossRef]
29. Nikolic, J.; Civas, A.; Lama, Z.; Lagaudriere-Gesbert, C.; Blondel, D. Rabies Virus Infection Induces the Formation of Stress Granules Closely Connected to the Viral Factories. *PLOS Pathog.* **2016**, *12*, e1005942. [CrossRef]
30. Barr, J.N.; Elliott, R.M.; Dunn, E.F.; Wertz, G.W. Segment-specific terminal sequences of Bunyamwera bunyavirus regulate genome replication. *Virology* **2003**, *311*, 326–338. [CrossRef]
31. Arragain, B.; Reguera, J.; Desfosses, A.; Gutsche, I.; Schoehn, G.; Malet, H. High resolution cryo-EM structure of the helical RNA-bound Hantaan virus nucleocapsid reveals its assembly mechanisms. *eLife* **2019**, *8*. [CrossRef]
32. Mir, M.A.; Sheema, S.; Haseeb, A.; Haque, A. Hantavirus Nucleocapsid Protein Has Distinct m7G Cap- and RNA-binding Sites. *J. Boil. Chem.* **2010**, *285*, 11357–11368. [CrossRef]
33. A Mir, M.; Panganiban, A.T. A protein that replaces the entire cellular eIF4F complex. *EMBO J.* **2008**, *27*, 3129–3139. [CrossRef] [PubMed]
34. Gu, S.H.; Kumar, M.; Sikorska, B.; Hejduk, J.; Markowski, J.; Markowski, M.; Liberski, P.P.; Yanagihara, R. Isolation and partial characterization of a highly divergent lineage of hantavirus from the European mole (Talpa europaea). *Sci. Rep.* **2016**, *6*, 21119. [CrossRef] [PubMed]
35. Elliott, L.H.; Goldsmith, C.S.; Childs, J.; Humphrey, C.D.; Nichol, S.T.; Peters, C.J.; Zaki, S.R.; Ksiazek, T.G.; Rollin, P.E.; Krebs, J.W.; et al. Isolation of the Causative Agent of Hantavirus Pulmonary Syndrome. *Am. J. Trop. Med. Hyg.* **1994**, *51*, 102–108. [CrossRef] [PubMed]
36. Xu, F.; Yang, Z.; Wang, L.; Lee, Y.-L.; Yang, C.-C.; Xiao, S.-Y.; Xiao, H.; Wen, L. Morphological Characterization of Hantavirus HV114 by Electron Microscopy. *Intervirology* **2007**, *50*, 166–172. [CrossRef]
37. McCormick, C.; Khaperskyy, D.A. Translation inhibition and stress granules in the antiviral immune response. *Nat. Rev. Immunol.* **2017**, *17*, 647–660. [CrossRef]

Article

The Combination of IFN β and TNF Induces an Antiviral and Immunoregulatory Program via Non-Canonical Pathways Involving STAT2 and IRF9

Mélissa K. Mariani [1], Pouria Dasmeh [2,3], Audray Fortin [1], Elise Caron [1], Mario Kalamujic [1], Alexander N. Harrison [1,4], Diana I. Hotea [1,2], Dacquin M. Kasumba [1,2], Sandra L. Cervantes-Ortiz [1,5], Espérance Mukawera [1], Adrian W. R. Serohijos [2,3] and Nathalie Grandvaux [1,2,*]

1 CRCHUM—Centre Hospitalier de l'Université de Montréal, Montréal, QC H2X 0A9, Canada
2 Department of Biochemistry and Molecular Medicine, Faculty of Medicine, Université de Montréal, Montréal, QC H3T 1J4, Canada
3 Centre Robert Cedergren en Bioinformatique et Génomique, Université de Montréal, Montréal, QC H3T 1J4, Canada
4 Department of Microbiology and Immunology, McGill University, Montréal, QC H3A 2B4, Canada
5 Department of Microbiology, Infectiology and Immunology, Faculty of Medicine, Université de Montréal, Montréal, QC H3T 1J4, Canada
* Correspondence: nathalie.grandvaux@umontreal.ca; Tel.: +1-514-890-8000 (ext. 35292)

Received: 17 July 2019; Accepted: 14 August 2019; Published: 17 August 2019

Abstract: Interferon (IFN) β and Tumor Necrosis Factor (TNF) are key players in immunity against viruses. Compelling evidence has shown that the antiviral and inflammatory transcriptional response induced by IFNβ is reprogrammed by crosstalk with TNF. IFNβ mainly induces interferon-stimulated genes by the Janus kinase (JAK)/signal transducer and activator of transcription (STAT) pathway involving the canonical ISGF3 transcriptional complex, composed of STAT1, STAT2, and IRF9. The signaling pathways engaged downstream of the combination of IFNβ and TNF remain elusive, but previous observations suggested the existence of a response independent of STAT1. Here, using genome-wide transcriptional analysis by RNASeq, we observed a broad antiviral and immunoregulatory response initiated in the absence of STAT1 upon IFNβ and TNF costimulation. Additional stratification of this transcriptional response revealed that STAT2 and IRF9 mediate the expression of a wide spectrum of genes. While a subset of genes was regulated by the concerted action of STAT2 and IRF9, other gene sets were independently regulated by STAT2 or IRF9. Collectively, our data supports a model in which STAT2 and IRF9 act through non-canonical parallel pathways to regulate distinct pool of antiviral and immunoregulatory genes in conditions with elevated levels of both IFNβ and TNF.

Keywords: interferon; tumor necrosis factor; STAT; interferon regulatory factor; antiviral; autoimmunity; inflammation

1. Introduction

Interferon (IFN) β plays a critical role in the first line of defense against viruses, through its ability to induce a broad antiviral transcriptional response in virtually all cell types [1]. IFNβ also possesses key immunoregulatory functions that determine the outcome of the adaptive immune response against pathogens [1,2]. Over the years, in vitro and in vivo studies aimed at characterizing the mechanisms and the functional outcomes of IFNβ signaling were mostly performed in relation to single cytokine stimulation. This unlikely reflects physiological settings, as a plethora of cytokines are secreted in a specific situation. As a consequence, a cell rather simultaneously responds to a cocktail of cytokines to foster the appropriate transcriptional program. Response to IFNβ is no exception and is very context-dependent, particularly regarding the potential crosstalk with other cytokines. Elevated levels of IFNβ and Tumor Necrosis Factor (TNF) are found during the host response to viruses. Aberrant increased levels of both cytokines is also associated with a number of autoinflammatory and autoimmune diseases such as Systemic Lupus Erythematosus (SLE), psoriasis, and Sjögren's syndrome [3]. While the cross-regulation of IFNβ and TNF is well documented [4–6], the functional crosstalk between these two cytokines remains poorly known.

IFNβ typically acts through binding to the IFNAR receptor (IFNAR1 and IFNAR2) leading to the Janus kinase (JAK)/signal transducer and activator of transcription (STAT) pathway involving JAK1- and Tyk2-mediated phosphorylation of STAT1 and STAT2, and to a lesser extent other STAT members in a cell-specific manner [7,8]. Phosphorylated STAT1 and STAT2, together with IFN Regulatory Factor (IRF) 9, form the IFN-stimulated gene factor 3 (ISGF3) complex that binds to the consensus IFN-stimulated response element (ISRE) sequences in the promoter of hundreds of IFN stimulated genes (ISGs) [9]. Formation of the ISGF3 complex is considered a hallmark of the engagement of the type I IFN response and, consequently, the requirement of STAT1 in a specific setting has become a marker of the engagement of type I IFN signaling [7,10]. However, in recent years, this paradigm has started to be challenged with accumulating evidence demonstrating the existence of non-canonical JAK-STAT signaling that mediates type I IFN responses [8,11].

Synergism between IFNβ and TNF was shown to enhance the antiviral response to Vesicular Stomatitis Virus (VSV), Myxoma virus, and paramyxovirus infections [12–14]. Gene expression analyses showed that IFNβ and TNF synergistically regulate hundreds of genes induced by individual cytokines alone, but also drive a specific delayed transcriptional program composed of genes that are either not responsive to IFNβ or TNF separately or are only responsive to either one of the cytokine [4,13]. The signaling mechanisms engaged downstream of the costimulation with IFNβ and TNF remained elusive, but it is implicitly assumed that the fate of the gene expression response requires that both IFNβ- and TNF-induced signaling pathways exhibit significant crosstalk. Analysis of the enrichment of specific transcription factors binding sites in the promoters of a panel of genes synergistically induced by IFNβ and TNF failed to give a clue about the specificity of the transcriptional regulation of these genes [13]. We previously showed that the *DUOX2* gene belongs to the category of delayed genes that are remarkably induced to high levels in response to the combination of IFNβ and TNF in lung epithelial cells [8]. We found that *DUOX2* expression required STAT2 and IRF9 but not STAT1, suggesting that STAT2 and IRF9 activities might segregate in an alternative STAT1-independent pathway that could be involved in gene regulation downstream of IFNβ and TNF [14].

In the present study, we aimed to fully characterize the transcriptional profile of the delayed response to IFNβ and TNF that occurs independently of STAT1 and evaluate the role of STAT2 and IRF9 in the regulation of this response. We found that the costimulation by IFNβ and TNF induces a broad set of antiviral and immunoregulatory genes in the absence of STAT1. We also report the differential regulation of distinct subsets of IFNβ− and TNF-induced genes by STAT2 and IRF9. While IFNβ and TNF act in part through the concerted action of STAT2 and IRF9, specific sets of genes were only regulated by either STAT2 or IRF9. Altogether, our findings uncovered non-canonical STAT2 and/or IRF9-dependent pathways that coexist to regulate distinct pools of antiviral and immunoregulatory genes in a context of IFNβ and TNF crosstalk.

2. Materials and Methods

2.1. Cell Culture and Stimulation

A549 cells (American Type Culture Collection, ATCC) were grown in F-12 nutrient mixture (Ham) medium supplemented with 10% heat-inactivated fetal bovine serum (HI-FBS) and 1% L-glutamine. The 2ftGH fibrosarcoma cell line and the derived STAT1-deficient U3A cell line, a generous gift from Dr. G. Stark, Cleveland, USA [15], were grown in DMEM medium supplemented with 10% HI-FBS or

HI-Fetal Clone III (HI-FCl) and 1% L-glutamine. U3A cells stably expressing STAT1 were generated by transfection of the STAT1 alpha flag pRc/CMV plasmid (Addgene plasmid #8691; a generous gift from Dr. J. Darnell, Rockfeller University, USA [16,17]) and selection with 800 μg/ml Geneticin (G418). Monoclonal populations of U3A stably expressing STAT1 cells were isolated. A pool of two clones, referred to as U3A-STAT1, was used in the experiments to mitigate the clonal effects. U3A-STAT1 cells were maintained in culture in DMEM supplemented with 10% HI-FCl, 1% Glu, and 200 μg/mL G418. All cell lines were cultured without antibiotics except for the selection of stable cells. All media and supplements were from Gibco (Life Technologies, Grand Island, NY, USA), with the exception of HI-FCl, which was from HyClone (Logan, UT, USA). Mycoplasma contamination was excluded by regular analysis using the MycoAlert Mycoplasma Detection Kit (Lonza, Basel, Switzerland). Cells were stimulated with IFNβ (1000 U/mL, PBL Assay Science, Piscataway, NJ, USA), TNF (10 ng/mL, R&D Systems, Minneapolis, MN, USA) or IFNβ (1000 U/mL) +TNF (10 ng/mL) for the indicated times.

2.2. siRNA Transfection

The sequences of non-targeting control (Ctrl) and STAT2- and IRF9-directed RNAi oligonucleotides (Dharmacon, Lafayette, CO, USA) have previously been described in [14]. U3A cells at 30% confluency were transfected using the Oligofectamine transfection reagent (Life Technologies-Thermofisher, Carlsbad, CA, USA). RNAi transfection was pursued for 48 h before stimulation.

2.3. Immunoblot Analysis

Cells were lysed on ice using Nonidet P-40 lysis buffer as fully detailed in [18]. Whole-cell extracts (WCE) were quantified using the Bradford protein assay (Bio-Rad, Hercules, CA, USA), resolved by SDS-PAGE and transferred to nitrocellulose membrane before analysis by immunoblot. Membranes were incubated with the following primary antibodies: anti-actin Cat #MAB1501 from Millipore (Burlington, MA, USA), anti-IRF9 Cat #610285 from BD Transduction Laboratories (San Jose, CA, USA), and anti-STAT1-P-Tyr701 Cat #9171, anti-STAT2-P-Tyr690 Cat #4441, anti-STAT1 Cat #9172, anti-STAT2 Cat #4594, all from Cell Signaling (Danvers, MA, USA), before further incubation with horseradish peroxidase (HRP)-conjugated secondary antibodies (KPL, Gaithersburg, MD, USA or Jackson Immunoresearch Laboratories, West Grove, PA, USA). Antibodies were diluted in PBS containing 0.5% Tween and either 5% nonfat dry milk or BSA. Immunoreactive bands were visualized by enhanced chemiluminescence (Western Lightning Chemiluminescence Reagent Plus, Perkin-Elmer Life Sciences Waltham, MA, USA) using a LAS4000mini CCD camera apparatus (GE Healthcare, Mississauga, ON, Canada).

2.4. RNA Isolation and qRT-PCR Analyses

Total RNA was prepared using the RNAqueous-96 Isolation Kit (Invitrogen-Thermo Fisher, Carlsbad, CA, USA) following the manufacturer's instructions. Total RNA (1 μg) was subjected to reverse transcription using the QuantiTect Reverse Transcription Kit (Qiagen, Toronto, ON, Canada). Quantitative PCR were performed using either Fast start SYBR Green Kit (Roche, Indianapolis, IN, USA) for *MX1*, *IDO*, *APOBEC3G*, *CXCL10*, *NOD2*, *PKR*, *IRF1*, *IFIT1* and *IL8* or TaqMan Gene Expression Assays (Life Technologies-Thermo Fisher) for *DUOX2*, *IFI27*, *SERPINB2*, *IL33*, *CCL20*, *ISG20*. Sequences of oligonucleotides and probes used in PCR reactions are described in Supplemental Table S4. Data collection was performed on a Rotor-Gene 3000 Real Time Thermal Cycler (Corbett Research, Mortlake, Australia). Gene inductions were normalized over S9 levels, measured using Fast start SYBR Green Kit or TaqMan probe as necessary. Fold induction of genes was determined using the $\Delta\Delta Ct$ method [19]. All qRT-PCR data are presented as the mean ± standard error of the mean (SEM).

2.5. RNA-Sequencing (RNASeq)

Total RNA prepared as described above was quantified using a NanoDrop Spectrophotometer ND-1000 (NanoDrop Technologies, Inc., Wilmington, DE, USA) and its integrity was assessed using a 2100 Bioanalyzer (Agilent Technologies, Santa Clara, CA, USA). Libraries were generated from 250 ng of total RNA using the NEBNext poly(A) magnetic isolation module and the KAPA stranded RNA-Seq library preparation kit (Kapa Biosystems, Wilmington, MA, USA), as per the manufacturer's recommendations. TruSeq adapters and PCR primers were purchased from IDT. Libraries were quantified using the Quant-iT™ PicoGreen® dsDNA Assay Kit (Molecular Probes, Eugene, OR, USA) and the Kapa Illumina GA with Revised Primers-SYBR Fast Universal kit (Kapa Biosystems). Average size fragment was determined using a LabChip GX (Perkin-Elmer Life Sciences, Waltham, MA, USA) instrument. Massively parallel sequencing was carried out on an Illumina HiSeq 2500 sequencer (Illumina Inc., San Diego, CA, USA). Read counts were obtained using HTSeq. Reads were trimmed from the 3′ end to have a Phred score of at least 30. Illumina sequencing adapters were removed from the reads and all reads were required to have a length of at least 32bp. Trimming and clipping was performed using Trimmomatic [20]. The filtered reads were aligned to the Homo-sapiens assembly GRCh37 reference genome. Each read set was aligned using STAR [21] and merged using Picard (http://broadinstitute.github.io/picard/). For all samples, the sequencing resulted in more than 29 million clean reads (ranging from 29 to 44 million reads) after removing low quality reads and adaptors. The reads were mapped to the total of 63,679 gene biotypes including 22,810 protein-coding genes. The non-specific filter for 1 count-per million reads (CPM) in at least three samples was applied to the reads and 14,254 genes passed this criterion.

2.6. Bioinformatics Analysis

Differential transcripts analysis. A reference-based transcript assembly was performed, which allows the detection of known and novel transcripts isoforms, using Cufflinks [22], merged using Cuffmerge (cufflinks/AllSamples/merged.gtf) and used as a reference to estimate transcript abundance and perform differential analysis using Cuffdiff and Cuffnorm tool to generate a normalized data set that includes all the samples. The fragments per kilobase million (FPKM) values calculated by Cufflinks were used as input. The transcript quantification engine of Cufflinks, Cuffdiff, was used to calculate transcript expression levels in more than one condition and test them for significant differences. To identify a transcript as being differentially expressed, Cuffdiff tests the observed log-fold-change in its expression against the null hypothesis of no change (i.e., the true log-fold-change is zero). Because of measurement errors, technical variability, and cross-replicate biological variability might result in an observed log-fold-change that is non-zero, Cuffdiff assesses significance using a model of variability in the log-fold-change under the null hypothesis. This model is described in detail in [23]. The differential gene expression analysis was performed using DESeq [24] and edgeR [25] within the R Bioconductor packages. Genes were considered differentially expressed between two group if they met the following requirement: fold change (FC) > ±1.5, $p < 0.05$, false discovery rate (FDR) < 0.05.

Enrichment of gene ontology (GO). GO enrichment analysis amongst differentially expressed genes (DEGs) was performed using Goseq [26] against the background of full human genome (hg19). GO-terms with adjusted p value < 0.05 were considered significantly enriched.

Clustering of DEGs. We categorized the DEGs according to their response upon silencing of siSTAT2 and siIRF9; categories are listed as A to I (Figure 2E). Then to determine relationship between these categories, we calculated the distance of centers of different categories. For each gene, we transformed siSTAT2 and siIRF9 FC to deviation from the mean FC of the category the respective gene belongs to using the equation: $FC_{new} = FC_{old} - \varepsilon\ (FC_{category})$. The parameter ε was estimated to give the perfect match between predefined categories (A to I) and clustering based on Euclidean distance. Results were plotted as a heatmap.

Modular transcription analysis. The *tmod* package in R [27] was used for modular transcription analysis. In brief, each transcriptional module is a set of genes that shows coherent expression across many biological samples [28,29]. Modular transcription analysis then calculates significant enrichment of a set of foreground genes, here DEGs, in pre-defined transcriptional module compared to a reference set. For transcriptional modules, we used a combined list of 606 distinct functional modules encompassing 12,712 genes, defined by Chaussabel et al. [30] and Li et al. [31], as the reference set in *tmod* package (Supplemental Table S5). The hypergeometric test devised in *tmodHGtest* was used to calculate enrichments and p-values employing Benjamini-Hochberg correction [32] for multiple sampling. All the statistical analyses and graphical presentations were performed in R [33].

2.7. Virus Titration by Plaque Assay

Quantification of VSV infectious virions was achieved through methylcellulose plaque forming unit assays. U3A and U3A-STAT1 cells were either left untreated or stimulated with IFNβ or IFNβ + TNF for 30 h. Cells were then infected with Vesicular Stomatitis Virus (VSV)-GFP (kindly provided by Dr. J. Bell, University of Ottawa, Canada) at a multiplicity of infection (MOI) of 5 for 1 h in serum-free medium (SFM). Cells were then washed twice with SFM and further cultured in DMEM medium containing 2% HI-FCl. The supernatants were harvested at 12 h post-infection and serial dilutions were used to infect confluent Vero cells (ATCC) for 1 h in SFM. The medium was then replaced with 1% methylcellulose in DMEM containing 10% HI-FCl. Two days post-infection, GFP-positive plaques were detected using a Typhoon Trio apparatus and quantified using the ImagequantTL software (GE Healthcare, Mississauga, ON, Canada).

2.8. Luciferase Gene Reporter Assay

U3A or U3A-STAT1 cells at 90% confluency were cotransfected with 100 ng of one of the following CXCL10 promoter containing firefly luciferase reporter plasmids (generously donated by Dr. David Proud, Calgary, [34]), CXCL10prom-972pb-pGL4 (full length −875/+97 promoter), CXCL10prom-376pb-pGL4 (truncated −279/+97 promoter), CXCL10prom972pb-ΔISRE(3)-pGL4 (full length promoter with ISRE(3) site mutated), together with 50 ng of pRL-null renilla-luciferase expressing plasmid (internal control). Transfection was performed using Lipofectamine 2000 (Life Technologies-Thermo Fisher) using a 1:2 DNA to lipofectamine ratio. At 8 h post-transfection, cells were stimulated for 16 h with either IFNβ or IFNβ + TNF. Firefly and renilla luciferase activities were quantified using the Dual-luciferase reporter assay system (Promega Corporation, Madison, WI, USA). Luciferase activities were calculated as the luciferase/renilla ratio and were expressed as fold over the non-stimulated condition.

2.9. Statistical Analyses

Statistical analyses of qRT-PCR and luciferase assay results were performed using the Prism v7 or v8 software (GraphPad Software, San Diego, CA, USA) using the tests indicated in the figure legends. Statistical significance was evaluated using the following p values: $p < 0.05$ (*), $p < 0.01$ (**), $p < 0.001$ (***) or $p < 0.0001$ (****). Differences with a p-value < 0.05 were considered significant. Statistical analysis of the RNASeq data is described in the Bioinformatics analysis section above.

2.10. Data and Software Availability

The entire set of RNAseq data has been submitted to the Gene Expression Omnibus (GEO) database (http://www.ncbi.nlm.nih.gov/geo) under accession number GEO: GSE111195.

3. Results

3.1. Distinct Induction Profiles of Antiviral and Immunoregulatory Genes in Response to IFNβ, TNF and IFNβ + TNF

First, to validate previous observations, we sought to determine the induction profile by the combination of IFNβ and TNF of a selected panel of immunoregulatory and antiviral genes that had previously been shown to be regulated by IFNβ or TNF alone. A549 cells, in which we previously documented the synergistic action of IFNβ and TNF on the *DUOX2* gene, were stimulated either with IFNβ, TNF, or IFNβ + TNF for various times between 3–24 h and the relative mRNA expression levels were quantified by qRT-PCR. Analysis of the expression of the selected genes revealed distinct profiles of response to IFNβ, TNF, or IFNβ + TNF (Figure 1). *IDO, DUOX2, CXCL10, APOBEC3G, ISG20,* and *IL33* exhibited synergistic induction in response to IFNβ + TNF compared to IFNβ or TNF alone. Expression in response to IFNβ + TNF increased over time, with maximum expression levels observed between 16 and 24 h. While *NOD2* and *IRF1* induction following stimulation with IFNβ + TNF was also significantly higher than upon IFNβ or TNF single cytokine stimulation, they exhibited a steady-state induction profile starting as early as 3 h. *MX1* and *PKR,* two typical IFNβ-inducible ISGs, were found induced by IFNβ + TNF similarly to IFNβ alone. *CCL20* responded to IFNβ + TNF with a kinetic and amplitude similar to TNF, but was not responsive to IFNβ alone. *IL8* expression was not induced by IFNβ, but was increased by TNF starting at 3 h and remained steady until 24 h. In contrast to other genes, *IL8* induction in response to IFNβ + TNF was significantly decreased compared to TNF alone. Overall, these results confirm previous reports that a subset of antiviral and immunoregulatory genes is greatly increased in response to IFNβ + TNF compared to either IFNβ or TNF alone.

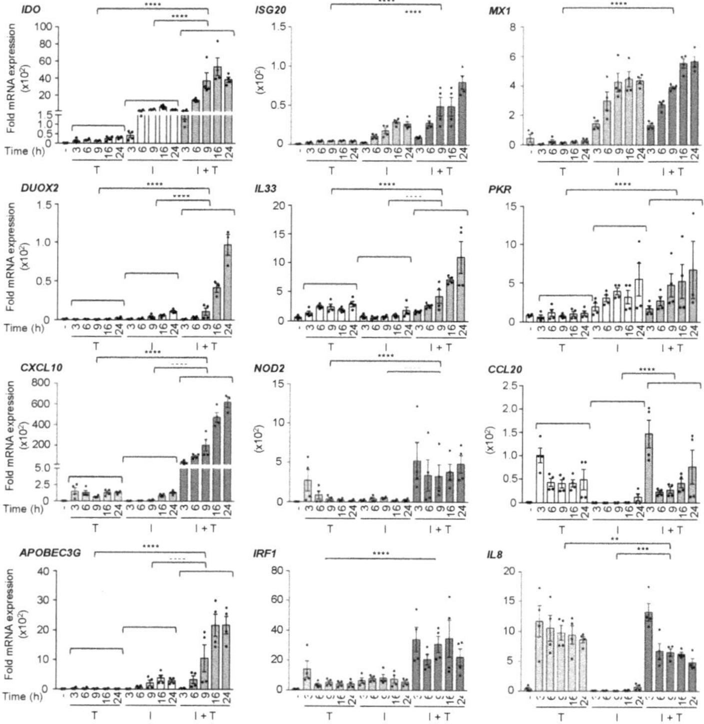

Figure 1. Expression of a panel of antiviral and immunoregulatory genes in response to IFNβ, TNF, and IFNβ + TNF. A549 cells were stimulated with either TNF (T), IFNβ (I), or costimulated with IFNβ + TNF (I + T) for the indicated times. Quantification of mRNA was performed by qRT-PCR and expressed as fold expression after normalization to the S9 mRNA levels using the ΔΔCt method. Mean +/− SEM, $n \geq 3$. Statistical comparison of TNF vs. IFNβ + TNF and IFNβ vs. IFNβ + TNF was conducted using two-way ANOVA with Tukey's post-test. $p < 0.01$ (**), $p < 0.001$ (***) or $p < 0.0001$ (****).

3.2. *Workflow for Genome-Wide Characterization of the Delayed Transcriptional Program Induced by IFNβ+TNF in the Absence of STAT1*

In a previous study, we provided evidence supporting the existence of a STAT1-independent, but STAT2- and IRF9-dependent, pathway engaged downstream of IFNβ + TNF [14]. Here, using the STAT1-deficient human U3A cell line [15], we aimed to fully characterize the STAT1-independent transcriptional program induced by IFNβ + TNF. The human U3A cell line was derived by mutagenesis from the 2ftGH cells [15], in which a synergistic response to IFNβ + TNF can be observed on a subset of genes (Figure 2A). Two hallmarks of STAT2 and IRF9 activation, i.e., STAT2 Tyr690 phosphorylation and induction of IRF9, were observed in the U3A cells following stimulation with IFNβ + TNF, although to reduced levels compared to the parental 2ftGH cells expressing endogenous STAT1 (Figure 2B). This observation implies that the activation of STAT2 and IRF9 depends to a large extent on the STAT1-dependent canonical pathway, but that a significant response occurs in the absence of STAT1. Therefore, the human U3A cell model is suitable for specifically studying STAT1-independent, but STAT2- and IRF9-dependent, gene expression in response to IFNβ + TNF.

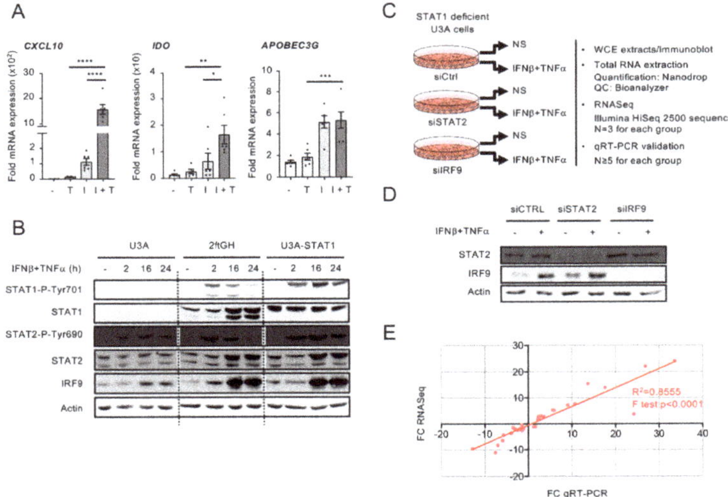

Figure 2. Experimental design used to study the STAT1-independent delayed transcriptional program induced by the combination of IFNβ and TNF. (**A**) 2ftGH cells were stimulated with either TNF (T), IFNβ (I), or costimulated with IFNβ + TNF (I + T) for 24 h. Quantification of mRNA was performed by qRT-PCR and expressed as fold expression after normalization to the S9 mRNA levels using the ΔΔCt method. Mean +/− SEM, $n \geq 5$. Statistical comparison was conducted using one-way ANOVA with Tukey's post-test. $p < 0.05$ (*), $p < 0.01$ (**), $p < 0.001$ (***), or $p < 0.0001$ (****). (**B**) U3A (STAT1-deficient), 2ftGH (parental STAT1-positive) cells and U3A-STAT1 cells (U3A cells stably reconstituted with STAT1) were left untreated or stimulated with IFNβ + TNF for the indicated times. WCE (whole cell extracts) were analyzed by SDS-PAGE followed by immunoblot using anti STAT1-P-Tyr701, total STAT1, STAT2-P-Tyr690, total STAT2, IRF9, or actin antibodies. (**C–E**) U3A cells were transfected with siCTRL, siSTAT2, or siIRF9 before being left untreated (NS) or stimulated with IFNβ + TNF for 24 h. (**C**) The schematic describes the workflow of sample preparation and analysis. (**D**) WCE were analyzed by SDS-PAGE followed by immunoblot using anti STAT2, IRF9, and actin antibodies. (**E**) Graph showing the correlation between fold-changes (FC) measured by RNASeq and qRT-PCR for 13 randomly selected genes. Data from siCTRL NS vs. siCTRL IFNβ +TNF, siSTAT2 IFNβ +TNF vs. siCTRL IFNβ + TNF, siIRF9 IFNβ + TNF vs. siCTRL IFNβ + TNF conditions were used.

To profile the genome wide transcriptional program induced by the combination of IFNβ and TNF in the absence of STAT1 and define the role of STAT2 and IRF9, the U3A cells were transfected with Control (Ctrl)−, STAT2- or IRF9-RNAi and further left untreated or stimulated with IFNβ (1000 U/mL) + TNF (10 ng/mL) for 24 h (Figure 2C). Efficient silencing was confirmed by immunoblot (Figure 2D). Total RNA was isolated and analyzed by RNA sequencing (n = 3 for each group detailed in Figure 2B) on an Illumina HiSeq2500 platform. To validate the expression profile of genes in RNASeq results, the fold changes (FC) of 13 genes randomly selected were analyzed by qRT-PCR in each experimental groups, i.e., siCTRL non-stimulated (NS) vs. siCTRL IFNβ + TNF, siCTRL IFNβ + TNF vs. siSTAT2 IFNβ + TNF and siCTRL IFNβ + TNF vs. siIRF9 IFNβ + TNF. A positive linear relationship between RNASeq and qRT-PCR FC was observed (Figure 2E).

3.3. A Broad Antiviral and Immunoregulatory Transcriptional Signature is Induced by IFNβ + TNF in the Absence of STAT1

To identify differentially expressed genes (DEGs) upon IFNβ + TNF stimulation in the absence of STAT1, comparison between non-stimulated (NS) and IFNβ + TNF-stimulated control cells was performed (Figure 3A). In total, 612 transcripts, including protein-coding transcripts, pseudogenes and long non-coding RNA (lncRNA), were significantly different (FC > 1.5, $p < 0.05$, FDR < 0.05) in IFNβ + TNF vs. NS. Among these, 482 DEGs were upregulated and 130 were downregulated (Figure 3B; See Supplemental Table S1 for a complete list of DEGs).

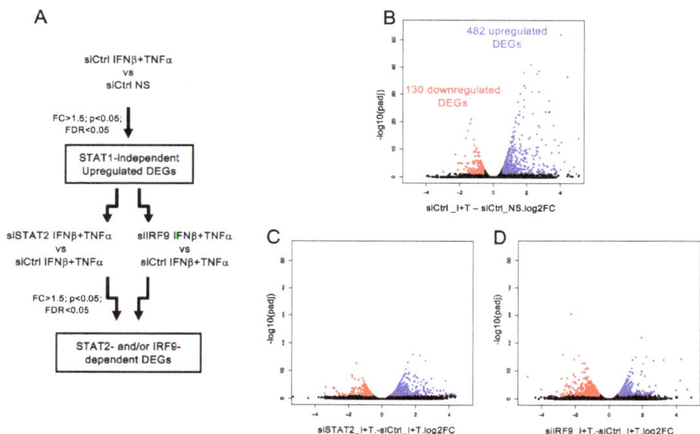

Figure 3. Analysis of STAT1-independent IFNβ + TNF-induced DEGs. (**A**) Diagram describing the bioinformatics analysis strategy used to determine STAT1-independent differentially expressed genes (DEGs) and their regulation by STAT2 and IRF9. (**B**) Volcano plots of the fold-change (FC) vs. adjusted p-value of IFNβ + TNF (I + T) vs. non-stimulated (NS) siCtrl conditions. (**C**) Volcano plots of the fold-change vs. adjusted p-value of siSTAT2 IFNβ + TNF vs. siCTRL IFNβ + TNF (I + T) conditions. (**D**) Volcano plots of the fold-change vs. adjusted p-value of siIRF9 IFNβ + TNF vs. siCTRL IFNβ + TNF conditions.

To identify the Biological Processes (BP) and Molecular Functions (MF) induced by IFNβ + TNF independently of STAT1, we further analyzed the upregulated DEGs. The top 40 upregulated DEGs are shown in Figure 4A. We subjected the complete list of upregulated DEGs (Supplemental Table S1) through Gene Ontology (GO) enrichment analysis. The top enriched GO BP ($p < 10^{-10}$) and MF, are depicted in Figure 4B (See Supplemental Table S2 for a complete list of enriched GO). The majority of the top enriched BPs were related to cytokine production and function (response to cytokine, cytokine-mediated signaling pathway, cytokine production, and regulation of cytokine production), immunoregulation (immune response, immune system process, innate immune response, and regulation of immune system process) and host defense response (defense response, response to other organism, 2′-5′-oligoadenylate synthetase activity and dsRNA binding). Fourteen MF categories were found enriched in IFNβ + TNF. The top enriched MFs were related to cytokine and chemokine functions (cytokine activity, cytokine receptor binding, chemokine activity, and Interleukin 1-receptor binding). Other enriched MF included peptidase related functions (endopeptidase inhibitor activity, peptidase regulator activity, and serine-type endopeptidase activity).

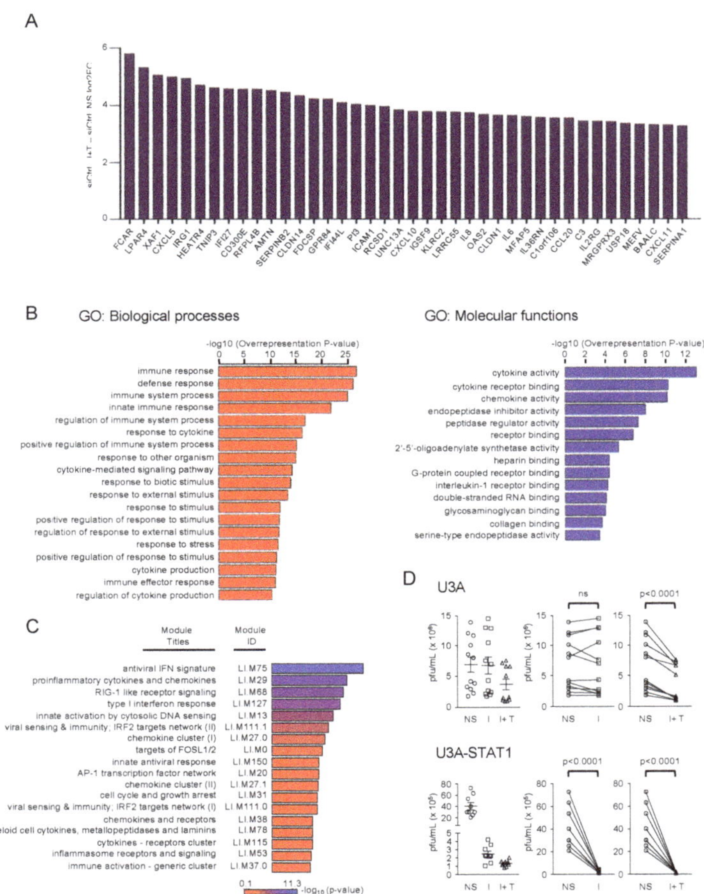

Figure 4. Functional characterization of STAT1-independent IFNβ + TNF-induced DEGs. (**A**) Top forty IFNβ + TNF- upregulated DEGs. (**B**) Gene ontology (GO) enrichment analysis of the differentially upregulated genes in IFNβ + TNF vs. non-stimulated siCtrl conditions based on the Biological Processes and Molecular Function categories. Top enriched terms are shown and the full list is available in Supplemental Table S2. (**C**) Modular transcription analysis of upregulated DEGs. Eighteen enriched modules are shown. The full list of enriched modules is available in Supplemental Table S3. (**D**) U3A and STAT1-rescued U3A-STAT1 cells were stimulated with IFNβ (I) or IFNβ + TNF (I + T) for 30 h before infection with VSV at a MOI of 5 for 12 h. The release of infectious viral particles was quantified by plaque forming unit (pfu) assay. The left graphs show dot-plots of all stimulations. Statistical comparisons were performed on the "before and after" plots (displayed on the right of dot-plot graphs) using ratio paired *t*-tests.

To gain deeper insight into the relevance of the STAT1-independent IFNβ + TNF-induced transcripts towards immunological and host defense responses, we conducted a modular transcription analysis of upregulated DEGs against 606 immune-related functional modules. These modules were previously defined from co-clustered gene sets built via an unbiased data-driven approach as detailed in the material and methods section [30,31]. STAT1-independent IFNβ + TNF-induced DEGs showed significant enrichment in 37 modules (See Supplemental Table S3 for a complete list of enriched modules). Six of these modules were associated with virus sensing/Interferon antiviral response, including LI.M75 (antiviral IFN signature), LI.M68 (RIG-I-like receptor signaling), LI.M127 (type I interferon response), LI.M111.0 and LI.M111.1 (IRF2 target network), and LI.M150 (innate antiviral response) (Figure 4C). Additionally, six modules were associated with immunoregulatory functions, including LI.M29 (proinflammatory cytokines and chemokines), LI.M27.0 and LI.M27.1 (chemokine cluster I and II), LI.M38 (chemokines and receptors), LI.M115 (cytokines receptors cluster), and LI.M37.0 (immune activation - generic cluster) (Figure 4C). Module analysis also showed enriched AP-1 transcription factor-related network modules, LI.M20 and LI.M0, and cell cycle and growth arrest LI.M31 module.

GO enrichment and modular transcription analyses were suggestive of an antiviral response being induced by IFNβ + TNF. To demonstrate the physiological relevance of STAT1-independent gene expression, we evaluated the capacity of IFNβ + TNF to restrict virus replication in the absence of STAT1. Previous study has shown that infection by Vesicular Stomatitis Virus (VSV) is sensitive to IFNβ, but not to TNF, in 2ftGH-derived cells [35]. Thus, U3A cells were stimulated with IFNβ alone or IFNβ + TNF and further infected with (VSV). While VSV replicated similarly in untreated U3A cells and in cells treated with IFNβ, treatment with IFNβ + TNF significantly limited VSV replication (Figure 4D). As a control of the efficiency of IFNβ alone treatment, we observed a significant antiviral response when STAT1 expression was stably restored to endogenous levels in U3A cells (U3A-STAT1) (Figures 2B and 4D). Collectively, these data demonstrate that the combination of IFNβ + TNF induces an efficient antiviral and immunoregulatory transcriptional response in the absence of STAT1.

3.4. Clustering of STAT1-Independent IFNβ + TNF Induced DEGs According to their Regulation by STAT2 and IRF9

Having shown that IFNβ + TNF induce a broad antiviral and immunoregulatory transcriptional response independently of STAT1, we next sought to gain insight into the role of STAT2 and IRF9. To do so, we compared transcripts levels in siSTAT2_IFNβ + TNF vs. siCTRL_IFNβ + TNF and siIRF9_IFNβ + TNF vs. siCTRL_IFNβ + TNF conditions (Figure 3A and Supplemental Table S1). Volcano plots revealed that a fraction of IFNβ + TNF-induced DEGs were significantly (FC > 1.5, $p < 0.05$, FDR < 0.05) downregulated or upregulated upon silencing of STAT2 (Figure 3C) or IRF9 (Figure 3D). Nine distinct theoretical categories of DEGs can be defined based on their potential individual behavior across siSTAT2 and siIRF9 groups (Categories A–I, Figure 5A): a gene can either be downregulated upon STAT2 and IRF9 silencing, indicative of a positive regulation by STAT2 and IRF9 (Category A); conversely, a gene negatively regulated by STAT2 and IRF9 would exhibit upregulation upon STAT2 and IRF9 silencing (Category B); genes that do not exhibit significant differential expression in siSTAT2 and siIRF9 groups would be classified as STAT2 and IRF9 independent (Category C); IRF9-independent genes could exhibit positive (Category D) or negative (Category E) regulation by STAT2; conversely, STAT2-independent genes might be positively (Category F) or negatively (Category G) regulated by IRF9; lastly, STAT2 and IRF9 could have opposite effects on a specific gene regulation (Category H and I). Based on a priori clustering of RNASeq data we found that in the absence of STAT1 IFNβ + TNF-induced DEGs clustered into only seven of the nine possible categories (Figure 5B). The top 15 upregulated DEGs of each category are shown in Figure 5C. The complete list of genes in each category is available in Supplemental Table S1. Amongst the 482 DEGs, 163 genes exhibited either inhibition or upregulation following silencing of STAT2 and/or IRF9 (Categories B–G). A large majority of upregulated DEGs, i.e., 319 out of the 482 DEGs, were not significantly affected by either STAT2 or

IRF9 silencing, and were therefore classified as STAT2/IRF9-independent (Figure 5B). No genes were found in Category H and only one gene was found in Category I.

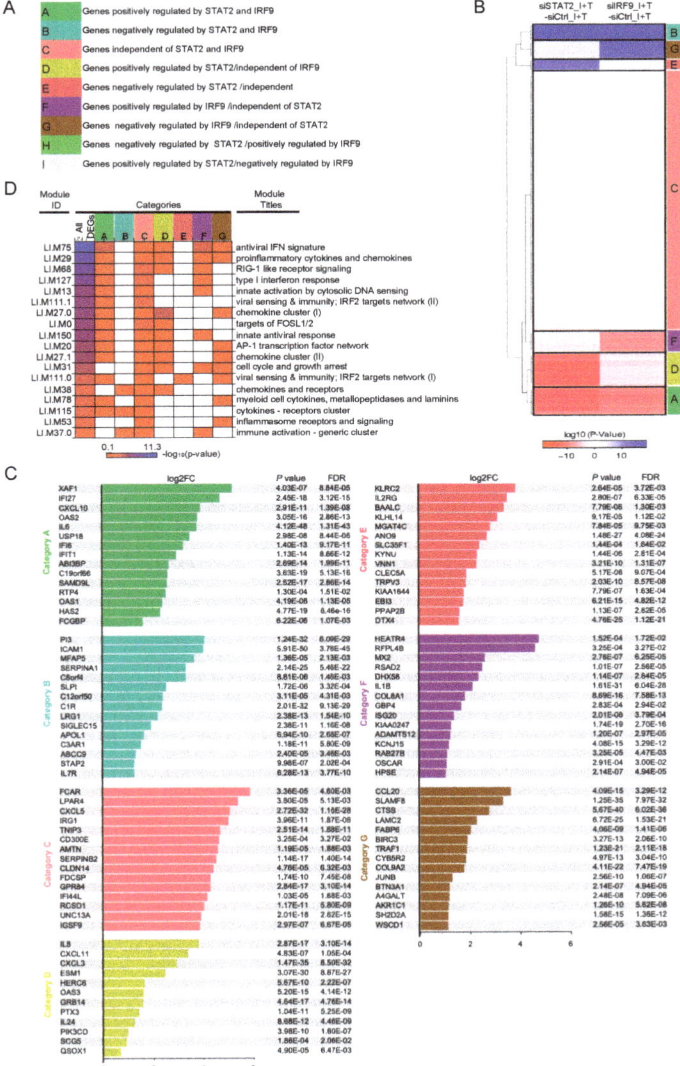

Figure 5. Clustering of IFNβ + TNF-induced DEGs according to their regulation by STAT2 and IRF9. (**A**) Theoretical categories in which IFNβ + TNF-induced DEGs can be segregated based on their potential individual regulation by STAT2 and IRF9. (**B**) Hierarchical clustering of the categories of DEG responses according to their regulation by STAT2 and IRF9. Euclidean distance metric is used for the construction of distance matrix and the categories are used as a priori input into clustering algorithm as detailed in Materials and Methods. (**C**) Top 15 induced DEGs (Log2FC, siCTRL NS vs. siCTRL IFNβ + TNF) in each category identified in (**B**) are displayed. Note that category D contains only twelve genes, so all genes are shown. The full list of genes is available in Supplemental Table S1. (**D**) Diagram showing enriched transcription modules in each gene category.

To functionally interpret these clusters, we applied the modular transcription analysis to each of the categories to assess for the specific enrichment of the functional modules found associated with IFNβ + TNF-upregulated DEGs (Figure 5D). First, most modules, except LI.M31 (cell cycle and growth arrest), LI.M38 (chemokines and receptors), LI.M37.0 (immune activation - generic cluster), and LI.M53 (inflammasome receptors and signaling), were enriched in the category of DEGs positively regulated by STAT2 and IRF9 (Category A). Conversely, the cluster negatively regulated by STAT2 and IRF9 (Category B) exclusively contains enriched LI.M38 (chemokines and receptors), LI.M37.0 (immune activation - generic cluster) and LI.M115 (cytokines receptors cluster). The cluster of IRF9-independent genes that are negatively regulated by STAT2 (Category E) only exhibited enrichment in the virus sensing/IRF2 target network LI.M111.0 module, while the IRF9-independent/STAT2-positively regulated cluster (Category D) encompasses antiviral and immunoregulatory functions. The STAT2-independent but IRF9 positively regulated transcripts (Category F) were mainly enriched in modules associated with the IFN antiviral response, including LI.M75 (antiviral IFN signature), LI.M68 (RIG-I-like receptor signaling), LI.M127 (type I interferon response), and LI.M150 (innate antiviral response). Finally, the STAT2-independent but IRF9 negatively regulated cluster (Category G) was mostly enriched in modules associated with immunoregulatory functions, including LI.M29 (proinflammatory cytokines and chemokines), LI.M27.0 and LI.M27.1 (chemokine cluster I and II), LI.M78 (myeloid cell cytokines), but also with cell cycle and growth arrest (LI.M31) and inflammasome receptors and signaling (LI.M53). Of note, all modules were found enriched in the cluster of genes induced independently of STAT2 and IRF9 (Category C), pointing to a broad function of this pathway(s) in the regulation of the antiviral and immunoregulatory program elicited by IFNβ + TNF. Altogether, these observations reveal that STAT2 and IRF9 are involved in the regulation of a subset of the genes induced in response to the co-stimulation by IFNβ and TNF in the absence of STAT1. Importantly, our results also reveal that STAT2 and IRF9 act in part in a concerted manner, but also independently in distinct non-canonical pathways, to regulate specific subsets of the IFNβ + TNF-induced DEGs.

3.5. Differential Regulation of CXCL10 in Response to IFNβ and IFNβ+TNF

The canonical ISGF3 complex mediates ISGs transcriptional regulation through binding to ISRE consensus sequences [7]. Identification of DEGs upregulated by IFNβ+TNF in a STAT1-independent, but STAT2 and IRF9-dependent, manner raised the question of the ISRE site usage compared to the ISGF3 pathway. To address this question, we studied the regulation of the *CXCL10* gene promoter that was found induced by STAT2 and IRF9 in the absence of STAT1 in response to IFNβ and TNF (Supplemental Table S1 and Figure 5C), but is also inducible by IFNβ alone in the presence of STAT1 (Figures 1 and 2A). The *CXCL10* promoter contains three ISRE sites. We used full-length wild-type (972bp containing the three ISRE sites), truncated (376bp containing only the ISRE (3) site) or mutated (972bp containing a mutated ISRE (3) site) *CXCL10* promoter luciferase (*CXCL10*prom-Luc) reporter constructs (Figure 6A) to determine the ISRE consensus site(s) requirement. U3A and STAT1-rescued U3A-STAT1 cells were transfected with the *CXCL10*prom-Luc constructs and further stimulated with IFNβ or IFNβ+TNF to monitor the canonical ISGF3-dependent and the STAT1-independent responses. In STAT1-deficient U3A cells, IFNβ was unable to activate the *CXCL10*prom reflecting the dependence on the ISFG3 pathway. In contrast, induction of *CXCL10*prom was significantly induced when cells were stimulated with IFNβ +TNF. This induction in the absence of STAT1 was not altered by the deletion of ISRE (1) and (2) sites, but was significantly impaired when the ISRE (3) site was mutated (Figure 6B). STAT1 expression rescue led to a higher induction of *CXCL10*prom by IFNβ + TNF. Additionally, IFNβ-mediated induction of *CXCL10*prom was restored, albeit to a much lower extent than IFNβ +TNF. The activation of the *CXCL10*prom in the presence of STAT1 involved both the distal ISRE (1) and/or ISRE (2) sites and the proximal ISRE (3) site (Figure 6B). Altogether, this shows that the STAT1-independent, but STAT2/IRF9-dependent, pathway mediates gene expression through a restricted ISRE site usage compared to the ISGF3-dependent regulation.

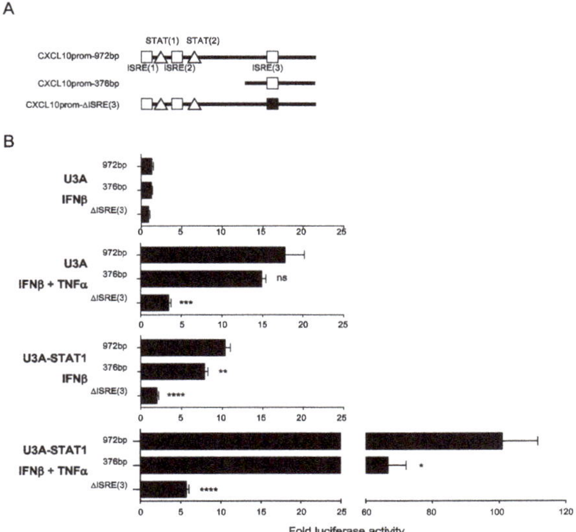

Figure 6. Analysis of the *CXCL10* promoter regulation in response to IFNβ + TNF vs. IFNβ. (**A**) Schematic representation of the *CXCL10* promoter (CXCL10prom) luciferase constructs used in this study indicating the main transcription factors consensus binding sites. (**B**) U3A and U3A-STAT1 cells were transfected with the indicated CXCL10prom-luciferase reporter constructs and either left untreated or stimulated with IFNβ or IFNβ + TNF. Relative luciferase activities were measured at 16 h post-stimulation and expressed as fold over the corresponding unstimulated condition. Mean +/− SEM, n = 6. Statistical analyses were performed using an unpaired t-test comparing each promoter to the CXCL10prom-972bp construct. $p < 0.01$ (**), $p < 0.001$ (***), and $p < 0.0001$ (****).

4. Discussion

Previous studies have described that IFNβ and TNF synergize to elicit a specific delayed transcriptional program that differs from the one induced by either cytokine alone [13,36]. The mechanisms underlying the transcriptional induction of genes specifically regulated by IFNβ and TNF remain poorly defined. The present study was specifically designed to document the functional relevance of a previously observed delayed gene expression induced by IFNβ in the presence of TNF in the absence of STAT1 [14] and to document the role of STAT2 and IRF9 in this response.

The observation that STAT2 and IRF9 activation in response to IFNβ + TNF is reduced in STAT1-deficient U3A cells compared to the wild-type 2ftGH parental cells and that IFNβ + TNF-mediated induction of the STAT2- and IRF9-dependent *CXCL10* promoter exhibits partial dependence on STAT1 support a model in which a canonical ISGF3 pathway is engaged downstream of the costimulation. Importantly, it also implied the existence of a STAT2- and/or IRF9-dependent transcriptional response occurring in the absence of STAT1. The human STAT1-deficient U3A cell model offered a unique opportunity to specifically pinpoint this STAT1-independent response. In this model, genome wide RNA sequencing highlighted that the transcriptional program induced by IFNβ + TNF in the absence of STAT1 encompasses a wide range of immunoregulatory and antiviral functions. The functional relevance of this response was confirmed by the observation that the treatment with IFNβ + TNF induced an antiviral state capable of restricting VSV replication in the absence of STAT1. This points to a significant role of the STAT1-independent pathway in the establishment of the antiviral state induced by the synergistic action of IFNβ and TNF that enhances the restriction of VSV (Figure 4D and [12]), Myxoma virus [13], and paramyxoviruses [14]. Although previous reports have shown that type I IFNs, mostly IFNα, alone can trigger STAT1-independent responses [8], we neither observed establishment of an IFNβ-induced antiviral state against VSV, nor activation of the *CXCL10* promoter in the absence of STAT1 in our model (Figures 4D and 6).

We previously reported that IFNβ + TNF induces the *DUOX2* gene via a STAT2- and IRF9-dependent pathway in the absence of STAT1 [14]. To what extent this pathway contributes to the STAT1-independent transcriptional response elicited by IFNβ + TNF remained to be addressed. Here, we demonstrate that IFNβ + TNFα-induced DEGs segregate into seven categories that reflect distinct contributions of STAT2 and/or IRF9, thereby highlighting an unexpected heterogeneity of the STAT1-independent pathways engaged downstream of IFNβ + TNF. Importantly, only one anecdotic gene was found in categories implying inverse regulation by STAT2 and IRF9 (categories H and I) pointing to convergent functions of STAT2 and IRF9 when both are engaged in gene regulation. We can rule out that these distinct regulation mechanisms reflect specific induction profiles by IFNβ + TNF as *CXCL10, IL33, CCL20,* and *ISG20* all exhibit synergistic induction by IFNβ + TNF, but are differentially regulated by STAT2 and/or IRF9; while *CXCL10* is dependent on STAT2 and IRF9, *IL33* is independent on STAT2 and IRF9, and *CCL20* and *ISG20* are STAT2-independent but IRF9-dependent (Figure 5; Supplemental Table S1). Consistent with our previous observation [14], we found several STAT1-independent genes positively regulated by STAT2 and IRF9 (Category A). DEGs in this category encompass most of the functions induced in response to IFNβ + TNF, with the notable exception of cell cycle and growth arrest and inflammasome and receptor signaling functions. Genes negatively regulated by STAT2 and IRF9 were also identified (Category B). Formation of an alternative STAT2/IRF9-containing complex mediating gene expression in the absence of STAT1 [37–41] has been reported, but with limited DNA-binding affinity for the typical ISRE sequence [37]. The existence of a STAT2/IRF9 complex is also supported by our recent observation of a high affinity of IRF9 for STAT2 with an equilibrium dissociation constant (Kd) of 10 nM [42]. A recent report of experiments, performed in murine bone marrow-derived macrophages proposes a model in which murine STAT2/IRF9 complex drives basal expression of ISGs, while IFNβ-inducible expression of ISGs depends on a switch to the ISGF3 complex [43]. This differs from our results as silencing of either STAT2 or IRF9 did not alter basal gene expression (Supplemental Figure S1). Further analysis of the *CXCL10* promoter demonstrates a restricted usage of ISRE sites by the STAT2/IRF9 pathway compared to the ISGF3 pathway. Further large-scale studies will be needed to identify the parameters allowing binding of ISGF3, but not STAT2/IRF9, to specific ISRE sequences upon IFNβ + TNF.

The observation that some IFNβ + TNF-induced genes were solely dependent on STAT2 (either positively or negatively) but not on IRF9, (Categories D and E) is a rare genome wide demonstration of gene regulation by STAT2 independently of STAT1 and IRF9. Previous reports have identified ISGF3-independent, STAT2-dependent genes but the association with IRF9 was not formally excluded [44–47]. STAT2 was shown to associate with STAT3 and STAT6, but it is not clear whether IRF9 is also part of these alternative complexes [44,46]. Transcriptional module analyses demonstrated that the functional distribution of genes negatively regulated by STAT2 is very limited compared to other categories; only a virus-sensing module was enriched in this category. In contrary, IRF9-independent genes positively regulated by STAT2 mediate broader antiviral and immunoregulatory functions.

ISGF3-independent functions of IRF9 have been proposed based on the study of IRF9 deficiencies [11,48]. However, IRF9 target genes in these contexts have been barely documented. Intriguingly, Li et al. [49] studied IFNα-induced genes and their dependency on the ISGF3 subunits. While they confirmed previous studies showing that IFNα can trigger a delayed and sustained ISG response via an ISGF3-independent pathway, it is very striking that they did not find STAT1- and STAT2-independent but IRF9-dependent genes. All identified IRF9-dependent genes were either STAT2- or STAT1-dependent. This result greatly differs with our study. Here, we found several IFNβ + TNF-induced DEGs independent of STAT1 and STAT2, but positively or negatively regulated by IRF9 (Categories F and G). Typically, IRF9 is considered a positive regulator of gene transcription. However, our findings are consistent with recent reports documenting the role of IRF9 in the negative regulation of the TRIF/NF-κB transcriptional response [50] or the expression of SIRT1 in acute myeloid leukemia cells [51]. The molecular mechanisms underlying gene regulation by IRF9 without association with either STAT1 or STAT2 remain to be elucidated. To the best of our knowledge, no alternative IRF9-containing complex has yet been described.

Our analysis showed that a large number of genes were induced by IFNβ + TNF independently of STAT2 and IRF9 (Category C). All transcriptional modules were enriched in this category pointing to a major role of this pathway in the establishment of a host defense and immunoregulatory response. The STAT2 and IRF9 independent genes does not solely reflects induction by TNF alone. For instance, *APOBEC3G* that is amongst the STAT2- and IRF9- independent genes is not induced by TNF alone (Figure 2A). While NF-κB, a downstream effector of the TNF receptor, is an obvious candidate for the regulation of these DEGs, this might fall short in explaining the synergistic action of IFNβ + TNF as we did not observe enhanced NF-κB activation compared to TNF alone [14]. Alternatively, the potential role of AP-1 is supported by the finding that the AP-1 transcription network module is enriched amongst IFNβ + TNF-induced DEGs. However, this module is not restricted to genes regulated independently of STAT2 and IRF9. It is also worth noting that two modules of IRF2-target genes were enriched, although again not specifically in the STAT2- and IRF9-independent category. A similar crosstalk was reported between IFNα and TNF in macrophages resulting in increased colocalized recruitment of IRF1 and p65 to the promoter of a subset of genes [52]. However, while IRF1 was found synergistically induced by IFNβ + TNF at early stages (Figure 1), we did not observe significant induction of IRF1 in the absence of STAT1 by RNASeq (Supplemental Table S1) or qRT-PCR (data not shown), suggesting that IRF1 is unlikely to be involved in our system. Further studies will be required to uncover these STAT2- and IRF9-independent pathways.

This study provides novel insight into the molecular pathways leading to delayed antiviral and immunoregulatory gene expression in conditions where elevated levels of both IFNβ and TNF are present. Altogether our results demonstrate that in addition to the engagement of an ISGF3-dependent canonical response, a broad transcriptional program is elicited independently of STAT1, and support a model in which STAT2 and IRF9 contribute to the regulation of this response through non-canonical parallel pathways involving their concerted or independent action (Figure 7).

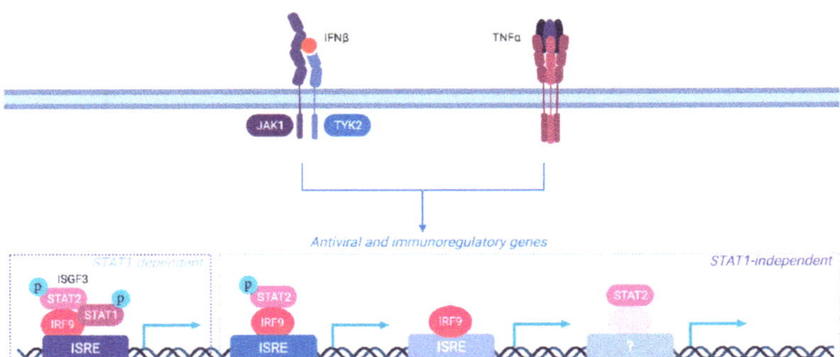

Figure 7. Role of distinct STAT2 and/or IRF9-dependent pathways in the regulation of distinct subset of antiviral and immunoregulatory genes in response to IFNβ and TNF. Our data supports a model in which multiple pathways participate to the synergistic action of IFNβ + TNF. While the STAT1-dependent pathway, likely ISFG3, is engaged downstream of IFNβ and TNF, STAT1-independent pathways are also involved in the control of the delayed gene expression. STAT2 and IRF9 act not only in a concerted fashion, likely as a complex, but also independently. IRF9 is known to act as the DNA-binding subunit of the ISGF3 complex and therefore likely mediates binding of STAT2/IRF9 complexes and of alternative complexes devoid of STAT2. The mechanisms of STAT2-dependent regulation of gene expression remains to be characterized.

Consistent with accumulating evidence [8], these distinct STAT2 and IRF9 actions most likely result from the formation of specific complexes that coexist with ISGF3 upon IFNβ and TNF stimulation. Studies are underway to biochemically solve the complexity of the dynamic and specific mechanisms of activation of the alternative STAT2 and/or IRF9-containing complexes in a wild-type cell context to further characterize the transcriptional response induced by IFNβ and TNF.

Supplementary Materials: The following are available online at http://www.mdpi.com/2073-4409/8/8/919/s1, Table S1: List of I+T_ DEG; Table S2: List of GO_I+T _DEG; Table S3: Module analysis_DEG; Table S4: MatetMeth_qPCR; Table S5: list_all_modules; Figure S1: qRT-PCR analysis of STAT2 and IRF9 regulated DEGs.

Author Contributions: Conceptualization, M.K.M. and N.G.; formal analysis, M.K.M., N.G., P.D., and A.W.R.S.; investigation, M.K.M., A.F., E.C., M.K., A.N.H., D.M.K., D.I.H., S.L.C.-O., and E.M.; writing—original draft preparation, N.G.; writing—review and editing, P.D., D.M.K. and D.I.H.

Funding: This research was funded by grants from NATURAL SCIENCES AND ENGINEERING RESEARCH COUNCIL OF CANADA (NSERC RGPIN/355306-2012 and RGPIN/2018-04279) and Research Chair in signaling in virus infection and oncogenesis from the Université de Montréal to N.G. A.W.R.S. acknowledges funds from Université de Montréal, NSERC and Merck Foundation. S.L.C.-O. was the recipient of a MITACS Globalinks studentship. N.G. was recipient of a Tier II Canada Research Chair.

Acknowledgments: We are very thankful to D. Proud (University of Calgary, Canada), G. Stark (Cleveland clinic, USA), and J. Bell (University of Ottawa, Canada) for sharing reagents used in this study. RNASeq analyses were performed at McGill University and Génome Québec Innovation Centre. Figure 6 was generated using biorender.com.

Conflicts of Interest: The authors declare no conflict of interest. The funders had no role in the design of the study; in the collection, analyses, or interpretation of data; in the writing of the manuscript, or in the decision to publish the results.

References

1. McNab, F.; Mayer-Barber, K.; Sher, A.; Wack, A.; O'Garra, A. Type I interferons in infectious disease. *Nat. Rev. Immunol.* **2015**, *15*, 87–103. [CrossRef] [PubMed]
2. Tomasello, E.; Pollet, E.; Vu Manh, T.P.; Uze, G.; Dalod, M. Harnessing Mechanistic Knowledge on Beneficial Versus Deleterious IFN-I Effects to Design Innovative Immunotherapies Targeting Cytokine Activity to Specific Cell Types. *Front. Immunol.* **2014**, *5*, 526. [CrossRef] [PubMed]
3. Lee-Kirsch, M.A. The Type I Interferonopathies. *Annu. Rev. Med.* **2017**, *68*, 297–315. [CrossRef] [PubMed]
4. Yarilina, A.; Park-Min, K.H.; Antoniv, T.; Hu, X.; Ivashkiv, L.B. TNF activates an IRF1-dependent autocrine loop leading to sustained expression of chemokines and STAT1-dependent type I interferon-response genes. *Nat. Immunol.* **2008**, *9*, 378–387. [CrossRef] [PubMed]
5. Palucka, A.K.; Blanck, J.P.; Bennett, L.; Pascual, V.; Banchereau, J. Cross-regulation of TNF and IFN-alpha in autoimmune diseases. *Proc. Natl. Acad. Sci. USA* **2005**, *102*, 3372–3377. [CrossRef] [PubMed]
6. Tliba, O.; Tliba, S.; Da Huang, C.; Hoffman, R.K.; DeLong, P.; Panettieri, R.A., Jr.; Amrani, Y. Tumor necrosis factor alpha modulates airway smooth muscle function via the autocrine action of interferon beta. *J. Biol. Chem.* **2003**, *278*, 50615–50623. [CrossRef] [PubMed]
7. Au-Yeung, N.; Mandhana, R.; Horvath, C.M. Transcriptional regulation by STAT1 and STAT2 in the interferon JAK-STAT pathway. *JAKSTAT* **2013**, *2*, e23931. [CrossRef]
8. Fink, K.; Grandvaux, N. STAT2 and IRF9: Beyond ISGF3. *JAKSTAT* **2013**, *2*, e27521. [CrossRef]
9. Schneider, W.M.; Chevillotte, M.D.; Rice, C.M. Interferon-stimulated genes: A complex web of host defenses. *Annu. Rev. Immunol.* **2014**, *32*, 513–545. [CrossRef]
10. Levy, D.E.; Marie, I.J.; Durbin, J.E. Induction and function of type I and III interferon in response to viral infection. *Curr. Opin. Virol.* **2011**, *1*, 476–486. [CrossRef]
11. Majoros, A.; Platanitis, E.; Kernbauer-Holzl, E.; Rosebrock, F.; Muller, M.; Decker, T. Canonical and Non-Canonical Aspects of JAK-STAT Signaling: Lessons from Interferons for Cytokine Responses. *Front. Immunol.* **2017**, *8*, 29. [CrossRef] [PubMed]
12. Mestan, J.; Brockhaus, M.; Kirchner, H.; Jacobsen, H. Antiviral activity of tumour necrosis factor. Synergism with interferons and induction of oligo-2′, 5′-adenylate synthetase. *J. Gen. Virol.* **1988**, *69*, 3113–3120. [CrossRef] [PubMed]

13. Bartee, E.; Mohamed, M.R.; Lopez, M.C.; Baker, H.V.; McFadden, G. The addition of tumor necrosis factor plus beta interferon induces a novel synergistic antiviral state against poxviruses in primary human fibroblasts. *J. Virol.* **2009**, *83*, 498–511. [CrossRef] [PubMed]
14. Fink, K.; Martin, L.; Mukawera, E.; Chartier, S.; De Deken, X.; Brochiero, E.; Miot, F.; Grandvaux, N. IFNbeta/TNFalpha synergism induces a non-canonical STAT2/IRF9-dependent pathway triggering a novel DUOX2 NADPH oxidase-mediated airway antiviral response. *Cell Res.* **2013**, *23*, 673–690. [CrossRef] [PubMed]
15. McKendry, R.; John, J.; Flavell, D.; Muller, M.; Kerr, I.M.; Stark, G.R. High-frequency mutagenesis of human cells and characterization of a mutant unresponsive to both alpha and gamma interferons. *Proc. Natl. Acad. Sci. USA* **1991**, *88*, 11455–11459. [CrossRef] [PubMed]
16. Horvath, C.M.; Wen, Z.; Darnell, J.E., Jr. A STAT protein domain that determines DNA sequence recognition suggests a novel DNA-binding domain. *Genes Dev.* **1995**, *9*, 984–994. [CrossRef]
17. Schindler, C.; Fu, X.Y.; Improta, T.; Aebersold, R.; Darnell, J.E., Jr. Proteins of transcription factor ISGF-3: One gene encodes the 91-and 84-kDa ISGF-3 proteins that are activated by interferon alpha. *Proc. Natl. Acad. Sci. USA* **1992**, *89*, 7836–7839. [CrossRef]
18. Robitaille, A.C.; Mariani, M.K.; Fortin, A.; Grandvaux, N. A High Resolution Method to Monitor Phosphorylation-dependent Activation of IRF3. *J. Vis. Exp.* **2016**, *107*, e53723. [CrossRef]
19. Dussault, A.A.; Pouliot, M. Rapid and simple comparison of messenger RNA levels using real-time PCR. *Biol. Proced. Online* **2006**, *8*, 1–10. [CrossRef]
20. Bolger, A.M.; Lohse, M.; Usadel, B. Trimmomatic: A flexible trimmer for Illumina sequence data. *Bioinformatics* **2014**, *30*, 2114–2120. [CrossRef]
21. Dobin, A.; Davis, C.A.; Schlesinger, F.; Drenkow, J.; Zaleski, C.; Jha, S.; Batut, P.; Chaisson, M.; Gingeras, T.R. STAR: Ultrafast universal RNA-seq aligner. *Bioinformatics* **2013**, *29*, 15–21. [CrossRef]
22. Roberts, A.; Pimentel, H.; Trapnell, C.; Pachter, L. Identification of novel transcripts in annotated genomes using RNA-Seq. *Bioinformatics* **2011**, *27*, 2325–2329. [CrossRef]
23. Trapnell, C.; Hendrickson, D.G.; Sauvageau, M.; Goff, L.; Rinn, J.L.; Pachter, L. Differential analysis of gene regulation at transcript resolution with RNA-seq. *Nat. Biotechnol.* **2013**, *31*, 46–53. [CrossRef]
24. Anders, S.; Huber, W. Differential expression analysis for sequence count data. *Genome Biol.* **2010**, *11*, R106. [CrossRef]
25. Robinson, M.D.; McCarthy, D.J.; Smyth, G.K. edgeR: A Bioconductor package for differential expression analysis of digital gene expression data. *Bioinformatics* **2010**, *26*, 139–140. [CrossRef]
26. Young, M.D.; Wakefield, M.J.; Smyth, G.K.; Oshlack, A. Gene ontology analysis for RNA-seq: Accounting for selection bias. *Genome Biol.* **2010**, *11*, R14. [CrossRef]
27. Weiner, J.; Domaszewska, T. tmod: An R package for general and multivariate enrichment analysis. *PeerJ Prepr.* **2016**, *4*, e2420v1. [CrossRef]
28. Bar-Joseph, Z.; Gerber, G.K.; Lee, T.I.; Rinaldi, N.J.; Yoo, J.Y.; Robert, F.; Gordon, D.B.; Fraenkel, E.; Jaakkola, T.S.; Young, R.A.; et al. Computational discovery of gene modules and regulatory networks. *Nat. Biotechnol.* **2003**, *21*, 1337–1342. [CrossRef]
29. Chaussabel, D.; Baldwin, N. Democratizing systems immunology with modular transcriptional repertoire analyses. *Nat. Rev. Immunol.* **2014**, *14*, 271–280. [CrossRef]
30. Chaussabel, D.; Quinn, C.; Shen, J.; Patel, P.; Glaser, C.; Baldwin, N.; Stichweh, D.; Blankenship, D.; Li, L.; Munagala, I.; et al. A modular analysis framework for blood genomics studies: Application to systemic lupus erythematosus. *Immunity* **2008**, *29*, 150–164. [CrossRef]
31. Li, S.; Rouphael, N.; Duraisingham, S.; Romero-Steiner, S.; Presnell, S.; Davis, C.; Schmidt, D.S.; Johnson, S.E.; Milton, A.; Rajam, G.; et al. Molecular signatures of antibody responses derived from a systems biology study of five human vaccines. *Nat. Immunol.* **2014**, *15*, 195–204. [CrossRef]
32. Benjamini, Y.; Hochberg, Y. Controlling the False Discovery Rate: A Practical and Powerful Approach to Multiple Testing. *J. R. Stat. Soc. Ser. B (Methodol.)* **1995**, *57*, 289–300. [CrossRef]
33. Ross, I.; Gentleman, R. R: A language for data analysis and graphics. *J. Comput. Graph. Stat.* **1996**, *5*, 299–314.
34. Zaheer, R.S.; Koetzler, R.; Holden, N.S.; Wiehler, S.; Proud, D. Selective transcriptional down-regulation of human rhinovirus-induced production of CXCL10 from airway epithelial cells via the MEK1 pathway. *J. Immunol.* **2009**, *182*, 4854–4864. [CrossRef]

35. Davis, A.M.; Hagan, K.A.; Matthews, L.A.; Bajwa, G.; Gill, M.A.; Gale, M., Jr.; Farrar, J.D. Blockade of virus infection by human CD4+ T cells via a cytokine relay network. *J. Immunol.* **2008**, *180*, 6923–6932. [CrossRef]
36. Bartee, E.; McFadden, G. Human cancer cells have specifically lost the ability to induce the synergistic state caused by tumor necrosis factor plus interferon-beta. *Cytokine* **2009**, *47*, 199–205. [CrossRef]
37. Bluyssen, H.A.; Levy, D.E. Stat2 is a transcriptional activator that requires sequence-specific contacts provided by stat1 and p48 for stable interaction with DNA. *J. Biol. Chem.* **1997**, *272*, 4600–4605. [CrossRef]
38. Martinez-Moczygemba, M.; Gutch, M.J.; French, D.L.; Reich, N.C. Distinct STAT structure promotes interaction of STAT2 with the p48 subunit of the interferon-alpha-stimulated transcription factor ISGF3. *J. Biol. Chem.* **1997**, *272*, 20070–20076. [CrossRef]
39. Gupta, S.; Jiang, M.; Pernis, A.B. IFN-alpha activates Stat6 and leads to the formation of Stat2:Stat6 complexes in B cells. *J. Immunol.* **1999**, *163*, 3834–3841.
40. Lou, Y.J.; Pan, X.R.; Jia, P.M.; Li, D.; Xiao, S.; Zhang, Z.L.; Chen, S.J.; Chen, Z.; Tong, J.H. IRF-9/STAT2 [corrected] functional interaction drives retinoic acid-induced gene G expression independently of STAT1. *Cancer Res.* **2009**, *69*, 3673–3680. [CrossRef]
41. Abdul-Sater, A.A.; Majoros, A.; Plumlee, C.R.; Perry, S.; Gu, A.D.; Lee, C.; Shresta, S.; Decker, T.; Schindler, C. Different STAT Transcription Complexes Drive Early and Delayed Responses to Type I IFNs. *J. Immunol.* **2015**, *195*, 210–216. [CrossRef] [PubMed]
42. Rengachari, S.; Groiss, S.; Devos, J.; Caron, E.; Grandvaux, N.; Panne, D. Structural basis of STAT2 recognition by IRF9 reveals molecular insights into ISGF3 function. *Proc. Natl. Acad. Sci. USA* **2018**, *115*, E601–E609. [CrossRef] [PubMed]
43. Platanitis, E.; Demiroz, D.; Schneller, A.; Fischer, K.; Capelle, C.; Hartl, M.; Gossenreiter, T.; Muller, M.; Novatchkova, M.; Decker, T. A molecular switch from STAT2-IRF9 to ISGF3 underlies interferon-induced gene transcription. *Nat. Commun.* **2019**, *10*, 2921. [CrossRef] [PubMed]
44. Ghislain, J.J.; Fish, E.N. Application of genomic DNA affinity chromatography identifies multiple interferon-alpha-regulated Stat2 complexes. *J. Biol. Chem.* **1996**, *271*, 12408–12413. [CrossRef] [PubMed]
45. Brierley, M.M.; Marchington, K.L.; Jurisica, I.; Fish, E.N. Identification of GAS-dependent interferon-sensitive target genes whose transcription is STAT2-dependent but ISGF3-independent. *FEBS J.* **2006**, *273*, 1569–1581. [CrossRef]
46. Wan, L.; Lin, C.W.; Lin, Y.J.; Sheu, J.J.; Chen, B.H.; Liao, C.C.; Tsai, Y.; Lin, W.Y.; Lai, C.H.; Tsai, F.J. Type I IFN induced IL1-Ra expression in hepatocytes is mediated by activating STAT6 through the formation of STAT2: STAT6 heterodimer. *J. Cell Mol. Med.* **2008**, *12*, 876–888. [CrossRef]
47. Perry, S.T.; Buck, M.D.; Lada, S.M.; Schindler, C.; Shresta, S. STAT2 mediates innate immunity to Dengue virus in the absence of STAT1 via the type I interferon receptor. *PLoS Pathog.* **2011**, *7*, e1001297. [CrossRef]
48. Suprunenko, T.; Hofer, M.J. The emerging role of interferon regulatory factor 9 in the antiviral host response and beyond. *Cytokine Growth Factor Rev.* **2016**, *29*, 35–43. [CrossRef]
49. Li, W.; Hofer, M.J.; Songkhunawej, P.; Jung, S.R.; Hancock, D.; Denyer, G.; Campbell, I.L. Type I interferon-regulated gene expression and signaling in murine mixed glial cells lacking signal transducers and activators of transcription 1 or 2 or interferon regulatory factor 9. *J. Biol. Chem.* **2017**, *292*, 5845–5859. [CrossRef]
50. Zhao, X.; Chu, Q.; Cui, J.; Huo, R.; Xu, T. IRF9 as a negative regulator involved in TRIF-mediated NF-kappaB pathway in a teleost fish, Miichthys miiuy. *Mol. Immunol.* **2017**, *85*, 123–129. [CrossRef]
51. Tian, W.L.; Guo, R.; Wang, F.; Jiang, Z.X.; Tang, P.; Huang, Y.M.; Sun, L. IRF9 inhibits human acute myeloid leukemia through the SIRT1-p53 signaling pathway. *FEBS Lett.* **2017**, *591*, 2951. [CrossRef]
52. Park, S.H.; Kang, K.; Giannopoulou, E.; Qiao, Y.; Kang, K.; Kim, G.; Park-Min, K.H.; Ivashkiv, L.B. Type I interferons and the cytokine TNF cooperatively reprogram the macrophage epigenome to promote inflammatory activation. *Nat. Immunol.* **2017**, *18*, 1104–1116. [CrossRef]

Article

Fluorescent TAP as a Platform for Virus-Induced Degradation of the Antigenic Peptide Transporter

Magda Wąchalska [1], Małgorzata Graul [1], Patrique Praest [2], Rutger D. Luteijn [2], Aleksandra W. Babnis [1], Emmanuel J. H. J. Wiertz [2], Krystyna Bieńkowska-Szewczyk [1] and Andrea D. Lipińska [1,*]

1 Laboratory of Virus Molecular Biology, Intercollegiate Faculty of Biotechnology, University of Gdańsk, Abrahama 58, 80–307 Gdańsk, Poland; magda.wachalska@phdstud.ug.edu.pl (M.W.); malgorzata.graul@biotech.ug.edu.pl (M.G.); aleksandra.babnis@gmail.com (A.W.B.); krystyna.bienkowska-szewczyk@biotech.ug.edu.pl (K.B.-S.)

2 Department of Medical Microbiology, University Medical Center Utrecht, Heidelberglaan 100, 3584CX Utrecht, The Netherlands; P.Praest-2@umcutrecht.nl (P.P.); rdluteijn@berkeley.edu (R.D.L.); e.wiertz@umcutrecht.nl (E.J.H.J.W.)

* Correspondence: andrea.lipinska@biotech.ug.edu.pl; Tel.: +48-585236383

Received: 20 November 2019; Accepted: 5 December 2019; Published: 7 December 2019

Abstract: Transporter associated with antigen processing (TAP), a key player in the major histocompatibility complex class I-restricted antigen presentation, makes an attractive target for viruses that aim to escape the immune system. Mechanisms of TAP inhibition vary among virus species. Bovine herpesvirus 1 (BoHV-1) is unique in its ability to target TAP for proteasomal degradation following conformational arrest by the UL49.5 gene product. The exact mechanism of TAP removal still requires elucidation. For this purpose, a TAP-GFP (green fluorescent protein) fusion protein is instrumental, yet GFP-tagging may affect UL49.5-induced degradation. Therefore, we constructed a series of TAP-GFP variants using various linkers to obtain an optimal cellular fluorescent TAP platform. Mel JuSo (MJS) cells with CRISPR/Cas9 TAP1 or TAP2 knockouts were reconstituted with TAP-GFP constructs. Our results point towards a critical role of GFP localization on fluorescent properties of the fusion proteins and, in concert with the type of a linker, on the susceptibility to virally-induced inhibition and degradation. The fluorescent TAP platform was also used to re-evaluate TAP stability in the presence of other known viral TAP inhibitors, among which only UL49.5 was able to reduce TAP levels. Finally, we provide evidence that BoHV-1 UL49.5-induced TAP removal is p97-dependent, which indicates its degradation via endoplasmic reticulum-associated degradation (ERAD).

Keywords: TAP-GFP; fluorescent TAP platform; antigen presentation; MHC I; immune evasion; BoHV-1 UL49.5

1. Introduction

The co-existence of a host and a virus depends on a subtle balance between the pathogen replication and the host immune response. Virus-derived peptides, originating mainly from the proteasomal degradation, are presented by the major histocompatibility complex class I (MHC I) molecules, leading to the recognition of an infected cell by cytotoxic CD8^{+} T lymphocytes (CTLs) (reviewed in [1]). The transporter associated with antigen processing (TAP) plays a pivotal role in MHC I-restricted antigen presentation, which makes it an attractive target for viruses that aim to escape the immune system.

TAP is a heterodimer belonging to the ATP-binding cassette (ABC) family transporters. It consists of two subunits, TAP1 (ABCB2) and TAP2 (ABCB3) [2]. The core of each subunit is formed by an N-terminally-located transmembrane domain (TMD), composed of six transmembrane helices

(TMs), responsible for peptide recognition and binding [3], and a highly conserved C-terminal nucleotide-binding domain (NDB), which can bind and hydrolyze ATP [4]. Acquiring both substrates, ATP and the peptide, occurs independently [3]. This induces conformational rearrangements within TAP, resulting in a switch from an inward-open to an outward-facing conformation and release of the peptide into the lumen of endoplasmic reticulum (ER). Afterward, ATP hydrolysis triggers the release of phosphate and restores the resting state of TAP [5]. The presence of core-flanking TMD0 domains (four TMs in TAP1 and three TMs in TAP2) at the N termini of TAP1/TAP2 is not necessary for peptide transport; however, it is crucial for assembly of the peptide-loading complex (PLC) and subsequent exposure of antigenic peptides to CTLs [6].

During co-evolution with their hosts, several herpesviruses and a single known (to date) poxvirus have specialized in TAP inhibition via diverse mechanisms (reviewed in [7]). Herpes simplex virus 1 and 2 (HSV-1 and HSV-2) encode the ICP47 protein, which competes for the peptide-binding site and, through its characteristic structure, tethers the TAP-ICP47 complex in an inward-facing conformation [8–10]. In contrast, the US6 protein of human cytomegalovirus (HCMV) [11–13] and the cowpox virus (CPXV) strain Brighton Red-encoded CPXV012 protein can inhibit ATP binding to NDBs while leaving peptide binding unaffected [14–16]. Mechanisms of TAP inhibition by herpesvirus UL49.5 protein family encoded by members of the *Varicellovirus* genus are still not fully understood and seem to differ in detail between virus species. Most of the UL49.5 orthologs inhibit conformational rearrangements within TAP [17]. Bovine herpesvirus 1 (BoHV-1) UL49.5 seems to be unique in its ability to target bovine, human, and murine TAP for proteasomal degradation following the conformational arrest [7,18,19]. Varicella-zoster virus (VZV)-encoded UL49.5 can bind TAP, yet it exhibits no inhibitory properties [20]. TAP degradation activity was also described for the murine gammaherpesvirus-68 MK3 protein [21] and the rodent herpesvirus Peru pK3 ortholog [22], which both carry a cytoplasmic RING (really interesting new gene) finger domain and can act towards the murine transporter. The recently described poxvirus molluscum contagiosum virus MC80 protein can destabilize human TAP; however, in contrast to BoHV-1 UL49.5, the transporter is not the primary target of the inhibitor [23].

Recently, fluorescent tags and gene fusion technology have become indispensable in a wide range of biochemical and cell biology applications, nevertheless in some circumstances designing a functional fluorescent fusion protein remains challenging. Numerous studies have shown that a choice of a linker may have a significant impact on proper folding, yield, and functionality of the fusion protein and its interaction with other proteins. Flexible linkers are usually applied to provide a certain degree of movement, while rigid linkers are preferable to separate two bioactive domains spatially [24].

To investigate the mechanism of TAP inhibition or removal, a TAP-GFP (green fluorescent protein) fusion protein was instrumental, yet GFP-tagging was observed to abolish the susceptibility of TAP to degradation induced by the BoHV-1-encoded UL49.5 [18]. Here, we report the construction of a series of full-length TAP1 and TAP2 variants carrying either N- or C-terminal GFP with different types of linkers and evaluate the impact of the TAP-GFP fusion design on their fluorescence and functionality, as well as susceptibility to virus-induced inhibition and degradation. Such a fluorescent TAP platform may constitute a platform to explain the molecular mechanism of UL49.5 activity and potentially contribute to better characterization of the transporter itself.

2. Materials and Methods

2.1. Cells and Viruses

Madin-Darby bovine kidney (MDBK) cells (ATCC, Manassas, VA, USA, CCL-22), human melanoma Mel JuSo (MJS) cells, MJS TAP1 CRISPR/Cas9 knock-out (TAP1 KO), MJS TAP2 CRISPR/Cas9 knock-out (TAP2 KO) [25], and U937 (ATCC, CRL-1593) were cultured in RPMI 1640 (Corning, Corning, NY, USA) supplemented with 10% fetal bovine serum (FBS, Thermo Scientific (Thermo Scientific, Waltham, MA, USA)) and Antibiotic Antimycotic Solution (Thermo Scientific). Lenti-X HEK293T and GP2-293 cells (both from Takara/Clontech, Kusatsu, Japan) used for lentivirus and retrovirus production, respectively, were cultured in Iscove's modified Dulbecco's medium (IMDM, Lonza, Basel, Switzerland) supplemented as above. HEK293T (ATCC, CRL-3216) cells were cultured in Dulbecco's modified Eagle's medium (DMEM, high glucose, Lonza) supplemented as above. BoHV-1 field strain Lam (Institute for Animal Health and Science, Lelystad, The Netherlands) was propagated and titrated on MDBK cells.

2.2. DNA Constructs

All TAP constructs were cloned in lentiviral vectors downstream of an EF1α promoter.

For unmodified (wild-type, wt) TAP1 or TAP2 reconstitution, dual promoter lentiviral vectors described in [25] (pPuroR-GFP-TAP1 and pZeoR-mAmetrine-TAP2) were used. mAmetrine and marker GFP genes were removed from these vectors. Fragments of TAP1 and TAP2 sequences were ordered as synthetic genes designed for cloning in pEGFP-N3 or pEGFP-C1 (Takara/Clontech). For TAP1-N-GFP (TAP1 with the N-terminal GFP, random linker), TAP1-C-GFP (TAP1 with the C-terminal GFP, random linker), TAP2-N-GFP (TAP2 with the N-terminal GFP, random linker), and TAP2-C-GFP (TAP2 with the C-terminal GFP, random linker), fusion genes were re-cloned in the original lentiviral vectors. The amino acid sequences of random linkers resulting from the cloning procedure are depicted in Figure 1A. Fragments coding for TAP1 with helical linker sequences were ordered as synthetic genes designed for cloning in pEGFP-N3 or pEGFP-C1. TAP1-HN-GFP (TAP1 with the N-terminal GFP, helical linker) or TAP1-HC-GFP (TAP1 with the C-terminal GFP, helical linker) were re-cloned in the lentiviral vector pCDH-EF1α-MCS-(PGK-Puro) (System Biosciences, Palo Alto, CA, USA).

Genes coding for viral TAP inhibitors were cloned in retroviral vectors downstream of a retroviral promoter. The BoHV-1 UL49.5 gene was cloned from pLZRS-BoHV-1 UL49.5-IRES-GFP [18] in BamHI-EcoRI sites of pLZRS-IRES-ΔNGFR [26]. The VZV UL49.5 gene was amplified from the pLZRS-VZV UL49.5-IRES-GFP vector [20] using KOD Hot Start DNA polymerase (Merck, Darmstadt, Germany) and the following primers: forward 5′-CGGGATCCCACCATGGGATCAATTACC-3′ and reverse 5′-CCGGAATTCTTACCACGTGCTGCGTAATAC-3′. The PCR product was verified by DNA sequencing and introduced into BamHI and EcoRI sites of the pBABEpuro vector [27]. Synthetic genes encoding: myc-tagged HSV-1 ICP47 (Gene ID: 2703441), myc-tagged HCMV US6 (Gene ID: 3077555), or the FLAG-N-CPXV012 (Gene ID: 1485887) variantlacking six N-terminal amino acid residues [28] were introduced into BamHI and EcoRI restriction sites of pBABEpuro.

2.3. Retroviral and Lentiviral Transduction

For the production of recombinant lentiviruses, third-generation packaging vectors based on the pRSV-Rev and pCgpV plasmids (Cell Biolabs, San Diego, CA, USA), the obtained lentiviral expression vectors, and pCMV-VSV-G (Cell Biolabs) for pseudotyping were co-transfected into Lenti-X HEK293T cells using CalPhos mammalian transfection kit (Takara/Clontech). For recombinant retroviruses, a transfer plasmid (pBABEpuro-based or pLZRS-IRES-ΔNGFR-based) and pCMV-VSV-G were co-transfected into GP2-293 packaging cells as above. Twenty-four hours after transfection the medium was refreshed; for lentiviruses it was supplemented with 1 mM sodium butyrate (Sigma-Aldrich, Saint Louis, MO, USA). Virus-containing supernatants were collected after 48 h, concentrated with PEGit (System Biosciences), and used for transduction in the presence of 0.01 mg mL^{-1} polybrene (Sigma-Aldrich). MJS cells with TAP1 or TAP2 knock-outs were stably reconstituted with the wt or fluorescent TAP1 or TAP2 constructs using lentivirus vectors and cell-sorting for GFP- and MHC I-positive cells. The cells were subsequently transduced with a retrovirus coding for BoHV-1 UL49.5 and sorted for nerve growth receptor (NGFR)-positive cells or with a retrovirus coding for HSV-1 ICP47, HCMV US6, VZV UL49.5, or CPXV012, and selected with puromycin (2 μg mL^{-1}) (Sigma-Aldrich).

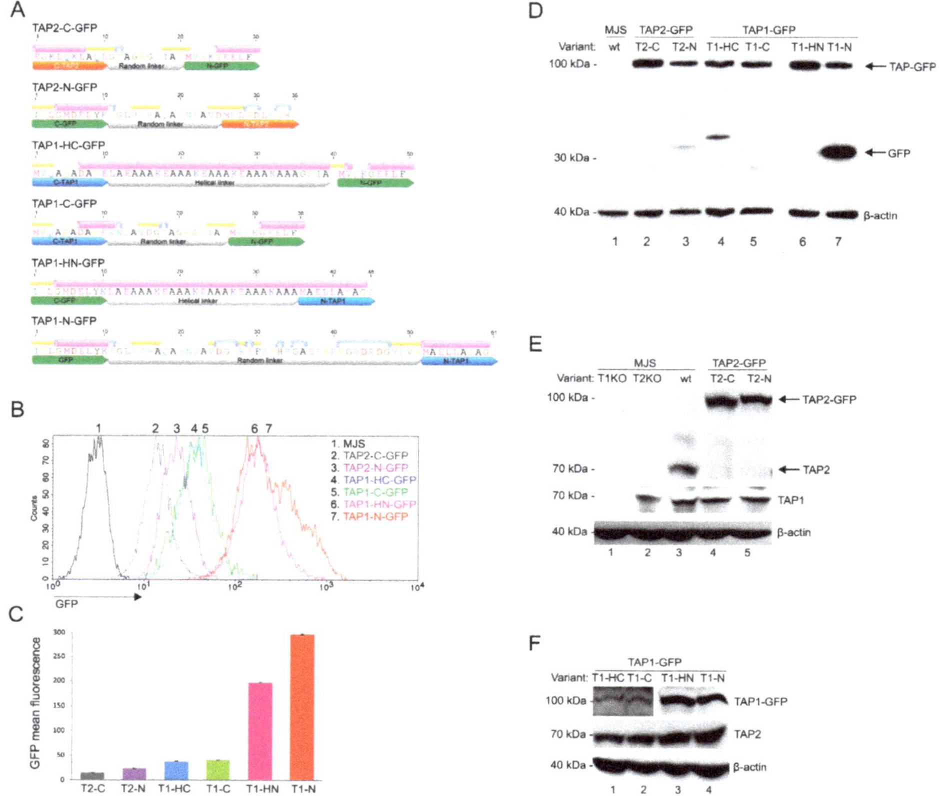

Figure 1. Construction and characterization of fluorescent transporter associated with antigen processing (TAP)-green fluorescent protein (GFP) variants. Mel JuSo (MJS) cells with CRISPR/Cas9 TAP1 or TAP2 knockouts (T1KO, T2KO) were stably reconstituted with fluorescent TAP1 or TAP2 constructs using lentivirus vectors and cell sorting. (**A**) Schematic representation of TAP-GFP constructs. Secondary structures of linkers flanked by ten amino acid residues of fused proteins were determined by the Geneious software; α-helices are depicted in pink, coiled regions in gray, β-strands in yellow, and turns in blue. (**B**) Representative histograms of TAP-GFP fluorescence intensity. (**C**) Comparative TAP-GFP analysis by flow cytometry. The mean fluorescence intensity of three independent measurements is represented as bars with standard deviations. The statistical significance was assessed by *t*-test; $p \leq 0.001$. (**D–F**) Expression of TAP-GFP variants in stable cell lines was determined by SDS-PAGE and immunoblotting using: (**D**) anti-GFP monoclonal antibody (Mab) (**E**) anti-TAP2 MAb (**F**) anti-TAP1 MAb. β-actin was used as a loading control. Abbreviations: T2-C: TAP2-C-GFP (TAP2 with the C-terminal GFP, random linker); T2-N: TAP2-N-GFP (TAP2 with the N-terminal GFP, random linker); T1-C: TAP1-C-GFP (TAP1 with the C-terminal GFP, random linker); T1-HC: TAP1-HC-GFP (TAP1 with the C-terminal GFP, helical linker); T1-N: TAP1-N-GFP (TAP1 with the N-terminal GFP, random linker); T1-HN: TAP1-HN-GFP (TAP1 with the N-terminal GFP, helical linker).

2.4. Plasmid Transfection

HEK293T cells were transfected with plasmids encoding fluorescent TAP variants using JetPRIME (Polyplus-transfection, Illkirch, France) according to the manufacturer's protocol and analyzed after 24 h by flow cytometry.

2.5. Generation of A TAP1/TAP2 Double Knock-Out U937 Cells for Reconstitution with Fluorescent TAP Variants

U937 TAP1/TAP2 KO cells were generated with a strategy described for MJS TAP1/TAP2 KO in [25] Briefly, U937 cells were transfected with pSico-CRISPR-PuroR containing the TAP2-targeting crRNA sequence 5′-GGAAGAAGAAGGCGGCAACG-3′. The cells were selected with puromycin (4 µg mL^{-1}), and cloned by limiting dilution. Individual clones were analyzed by flow cytometry to identify clones with low cell surface MHC I expression, followed by immunoblotting and DNA sequencing of the genomic target site. A clone lacking TAP2 was subsequently transfected with a pSicoR-CRISPR-PuroR vector containing the TAP1-targeting crRNA sequence 5′-GGGGTCCTCAGGGCAACGGT-3′. After selection with puromycin and cell cloning, the clones were analyzed for TAP1 expression by immunoblotting and DNA sequencing of the genomic target site. Genomic DNA sequence analysis revealed a 16-bp deletion around the TAP2 gRNA target site and multiple short deletions altering the whole TAP1 gene sequence downstream of the target site. A monoclonal cell line lacking TAP1 and TAP2 was used for reconstitution with a combination of unmodified and fluorescent TAP-encoding sequences delivered by lentivirus vectors. Reconstituted U937 cells were sorted for GFP and high MHC I expression. The cells were subsequently transduced with the BoHV-1 UL49.5-encoding retrovirus and sorted for NGFR.

2.6. Antibodies

Antibodies used for immunoblotting: mouse anti-TAP1 monoclonal antibody MAb 143.5 (kindly provided by R. Tampé, Institute of Biochemistry, The Johann Wolfgang Goethe University, Frankfurt, Germany); mouse anti-TAP2 MAb 435.3 (a kind gift from P. van Endert, INSERM U25, Institute Necker, Paris, France); rabbit anti-TAP1 (Enzo Life Sciences, Farmingdale, NY, USA); rat anti-GFP 3H9 (Chromotek, Planegg, Germany); mouse anti-myc tag 9B11 (Cell Signaling, Danvers, MA, USA); rabbit anti-β-actin (Novus Biologicals, Centennial, CO, USA); rabbit anti-β-catenin (Santa Cruz Biotechnology, Dallas, TX, USA); rabbit antibodies (H11) against a synthetic peptide derived from the N-terminal domain of BoHV-1 UL49.5 [26] and mouse anti-OctA (FLAG) G-8 (Santa Cruz Biotechnology); and mouse anti-HC10 [19] and rabbit anti-ERp57 H-220 (Santa Cruz Biotechnology). Probes used for immunofluorescence: Alexa 633-conjugated concanavalin A (ConA) (Thermo Scientific). Antibodies used for flow cytometry: mouse anti-MHC I W6/32 (Novus Biologicals); mouse anti-NGFR (Sigma-Aldrich); and Alexa 633-conjugated goat anti-mouse IgG (Thermo Scientific).

2.7. Flow Cytometry

Cell surface expression of MHC I was determined by indirect immunofluorescence using primary anti-MHC I antibodies (1:1000) and secondary antibodies (1:1000), all in phosphate buffered saline (PBS) buffer containing 1% bovine serum albumin and 0.02% sodium azide. For cell sorting, anti-NGFR antibodies (1:1000) and secondary antibodies were used. Cells were analyzed using a FACS Calibur flow cytometer (Becton Dickinson, Franklin Lakes, NJ, USA) and CellQuest software (version 5.2.1, Becton Dickinson)); for cell sorting, the sorting option of FACS Calibur was applied.

2.8. Immunoblotting and Immunoprecipitation

For immunoblotting, the cells were lysed in Cell Lytic M buffer (Sigma-Aldrich); for immunoprecipitation, the cells were lysed in a buffer containing 1% digitonin (Merck), 50 mM Tris-HCl, pH 7.5, 5 mM $MgCl_2$, and 150 mM NaCl. The buffers were supplemented with the cOmplete mini protease inhibitor cocktail (Roche, Basel, Switzerland). Cell lysates were analyzed by SDS-PAGE and immunoblotting as previously described [26] or incubated with GFP-Trap (Chromotek) according to the manufacturer's protocol to isolate protein complexes.

2.9. Peptide Transport Assay

The peptide transport assay was performed as described before [26]. Briefly, the cells were permeabilized with 2 IU mL^{-1} of Streptolysin O (Sigma-Aldrich) at 37 °C for 15 min. The cells (2×10^6 cells/sample) were subsequently incubated with 600 pmol of the fluorescein-conjugated synthetic peptide CVNKTERAY (JPT Peptide Technologies, Berlin, Germany) in the presence or absence of ATP (10 mM final concentration) at 37 °C for 10 min. Peptide translocation was terminated by adding 1 mL of ice-cold lysis buffer containing 1% Triton X-100. After 20 min of lysis, cell debris was removed by centrifugation, and supernatants were collected and incubated with 100 µL of concanavalin A (ConA)-Sepharose (Sigma-Aldrich) at 4 °C for 1 h to isolate the glycosylated peptides. After extensive washing of the beads, the peptides were eluted with elution buffer (500 mM mannopyranoside, 10 mM EDTA, 50 mM Tris-HCl, pH 8.0) by rigorous shaking for 1 h at 25 °C. Eluted peptides were separated from ConA by centrifugation at 12,000× *g*. The fluorescence intensity was measured using a fluorescence plate reader (EnVision, PerkinElmer, Waltham, MA, USA) with excitation and emission wavelengths of 485 nm and 530 nm, respectively.

2.10. Immunofluorescence

MJS cells were grown on microcover glass, fixed with 4% paraformaldehyde in PBS, permeabilized with 0.2% Triton X-100 in PBS, and stained with Alexa 633-ConA (1:1000), prepared in PBS containing 1% bovine serum albumin (Sigma-Aldrich). GFP booster (1:100, Chromotek) was used for MJS-TAP2-C-GFP to enhance the green fluorescence. The blue signal was electronically converted into the red during the analysis of images using Leica TCS SP8X confocal laser scanning microscope (Leica, Wetzlar, Germany).

3. Results

3.1. Construction and Characterization of Fluorescent TAP-GFP Variants

In order to develop a universal fluorescent platform for virus-induced TAP degradation, we constructed six versions of TAP-GFP fusion: four with two types of linkers, a random linker or a helical one, placed at the N- or C-terminus of TAP1, and two with a random linker at the N- or C-terminus of TAP2 (Figure 1A). A number of studies regarding fusion protein linker design have suggested that the most important features of a proper linker are its hydrophilicity and flexibility [24]. The random linkers used to join TAP and GFP have resulted from the cloning procedure into one of the pEGFP plasmid series. The analysis of their amino acid sequence revealed they were unstructured; thus, they should be more flexible. In some cases, flexible linkers may result in undesired interactions or interference between the fusion partners. In such cases, rigid linkers are preferable to separate two independently active domains spatially. An example of a rigid linker is the α helix-forming peptide AEAAAKEAAAKEAAAKA, stabilized by salt bridges between glutamate and lysine residues [29]. The distance between two separated domains can be regulated by changing the number of EAAAK motif repetitions. By using fluorescence resonance energy transfer (FRET) measurement, the helical linker with four repetitions of this motif has been demonstrated as the most efficient in separating two fluorescent proteins [30]. Therefore, this linker was selected for our studies to generate N- and C-terminal fusion of TAP1 and GFP. To introduce fluorescent TAP-GFP into human melanoma MJS cells, we first generated, by using CRISPR/Cas9-based technology, TAP1 or TAP2 knock-outs (KO). This enabled stable reconstitution of the cells with a fluorescent TAP1 or TAP2 using lentiviral vectors. The cells were subsequently sorted based on GFP to high purity (>98%). MJS cells have been shown to be permissive for BoHV-1 infection and are widely used in the research on modulation of antigen presentation by viruses [31–33].

First, we analyzed the GFP fluorescence intensity of our constructs (Figure 1B,C). Flow cytometry analysis revealed that N-terminal fusions of GFP to TAP1 exhibited the highest fluorescence, followed by TAP1 C-terminal fusion constructs. The type of linker seemed to have no crucial impact on the fluorescence intensity, although, for TAP1-N-GFP, a population of brighter green fluorescent cells could be observed when compared to TAP1-HN-GFP. TAP2 constructs exposed the lowest fluorescence, with a similar tendency of N-terminal fusion outperforming the C-terminal one. In addition, for TAP1-N-GFP, we could observe a more heterogeneous distribution of GFP fluorescence than for the other variants.

During the transient expression of fluorescent constructs in plasmid-transfected HEK293T cells, we could observe a similar range of GFP fluorescence, which indicates that differences in GFP fluorescence depend on the properties of individual fusion proteins rather than result from random incorporation of a lentivirus into a host genome (Figure S1). TAP2-C-GFP was performing noticeably better in transfected HEK293T than in the stable MJS cell line, and this time TAP1-HN-GFP slightly outperformed TAP1-N-GFP.

Next, we characterized the expression of TAP-GFP constructs by SDS-PAGE and immunoblotting. TAP1 and TAP2 have a similar apparent molecular mass of approximately 75 kDa. The fusion to GFP should yield a single protein at 100 kDa. All constructs were detected in cell lysates with anti-GFP antibodies at the predicted molecular weight of 100 kDa; only for TAP1-N-GFP did we observe an additional 30-kDa band corresponding to, most probably, cleaved free GFP or cleaved TAP-GFP fragment (Figure 1D). Fluorescent TAP could also be detected with anti-TAP1 or anti-TAP2 antibodies (Figure 1E,F). The reconstitution of TAP1KO cells with a fluorescent variant of TAP1 resulted in increased stability of endogenous TAP2 as it could be detected in higher amounts than in TAP1KO cells (compare lane 1 in Figure 1E with lanes 1–4 in Figure 1F). This is in agreement with a previous study reporting that unlike TAP1, TAP2 is unstable when expressed without the other half of the transporter [34]. The higher sensitivity of TAP2 can most probably explain why no band corresponding to TAP2 could be identified in MJS TAP1KO cells (Figure 1E, lane 1), while TAP1 could be easily detected in MJS TAP2KO cells (Figure 1E, lane 2).

3.2. TAP-GFP Localizes in the ER and Forms a Functional Transporter

TAP localizes in the cells predominantly in the ER [35]. To assess if the fluorescent TAP constructs acquired their proper localization, we stained the cells with Alexa 633-conjugated concanavalin A (ConA), an ER-cis Golgi marker, and analyzed fluorescence distribution by confocal laser scanning microscopy (Figure 2). In all cell lines, TAP-GFP (green) localized predominantly in the ER (red), as it co-localized with the ER marker (yellow). No granular localization patterns that would indicate TAP aggregates could be visualized.

Based on the currently available evidence, a lack of at least one fully functional TAP subunit results in a suppressed peptide translocation and production of empty or suboptimally loaded unstable MHC I molecules, mostly retained in the ER [36]. GFP fusion could potentially affect TAP structure and activity. Therefore, the functionality of fluorescent TAP constructs was tested by measuring cell surface expression of MHC I by flow cytometry (Figure 3). As expected, MJS cells with TAP1KO or TAP2KO exhibited strongly reduced levels of MHC I. To have a proper control for testing the effect of fluorescent tagging on TAP performance, especially during overexpression of a TAP subunit, we generated control cell lines reconstituted with unmodified TAP subunits (MJS TAP1KO + TAP1 and MJS TAP2KO + TAP2). Overexpression of individual TAP subunits might result in the formation of their homodimers, affecting the interpretation of our further experiments [19,37]. Therefore, as an additional control, we transduced TAPKO cells with vectors enabling overexpression of the existing half-transporter (MJS TAP1KO + TAP2 and MJS TAP2KO + TAP1). As expected, MHC I surface levels were reduced in those cells to a similar extent as in TAP1KO or TAP2KO cells. On the other hand, reconstitution of the missing TAP subunit with its fluorescent variant rescued surface MHC I to levels slightly higher than on parental ("wild-type") MJS cells, but similar to the ones observed on the cells reconstituted with the non-fluorescent TAP.

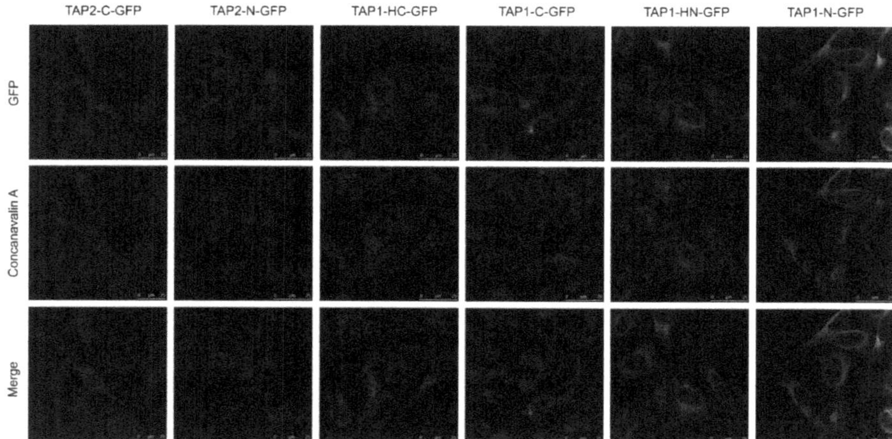

Figure 2. TAP-GFP variants demonstrate endoplasmic reticulum (ER) localization. MJS cells reconstituted with TAP-GFP variants were fixed, permeabilized, and stained with Alexa 633-conjugated concanavalin A (ER marker). Colocalization with GFP was analyzed using fluorescent confocal microscopy. For MJS TAP2-C-GFP, the GFP booster was used to enhance very weak green fluorescence.

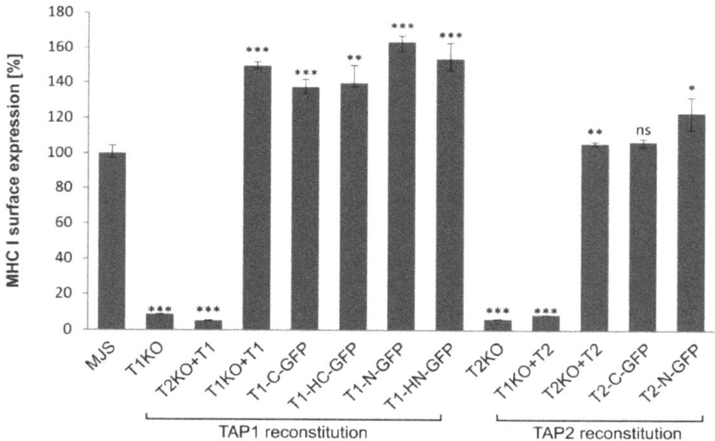

Figure 3. TAP-GFP forms a functional transporter. MJS cells with CRISPR/Cas9 TAP1 or TAP2 knockouts (T1KO or T2KO) were stably reconstituted with wild-type or fluorescent TAP1 or TAP2 variants. Surface expression of major histocompatibility complex class I (MHC I) was assessed by flow cytometry using specific antibodies (W6/32). MHC I expression on MJS cells with TAP reconstitution is presented as the percentage of MHC I mean fluorescence intensity on MJS cells (set as 100%). The analysis was performed in triplicates. The statistical significance of differences between MHC I on MJS cell with TAP reconstitution and MJS wild-type (wt) cells was estimated by *t*-test; *** $p \leq 0.001$, ** $p \leq 0.01$, * $p \leq 0.05$, ns: not significant.

3.3. The Sensitivity of TAP-GFP Variant to UL49.5-Mediated MHC I Downregulation and TAP Degradation

To examine the sensitivity of fluorescent TAP to BoHV-1-encoded UL49.5-mediated inhibition, UL49.5 was introduced to stable cell lines with TAP-GFP variants using a retrovirus vector. The cells were subsequently sorted for high purity based on the expression of truncated NGFR as a marker, and analyzed by flow cytometry for surface expression of MHC I (Figure 4A). According to the obtained data, all fluorescent transporters were prone to UL49.5-induced inhibition, which was illustrated as the downregulation of MHC I, to a similar extent for all the tested variants.

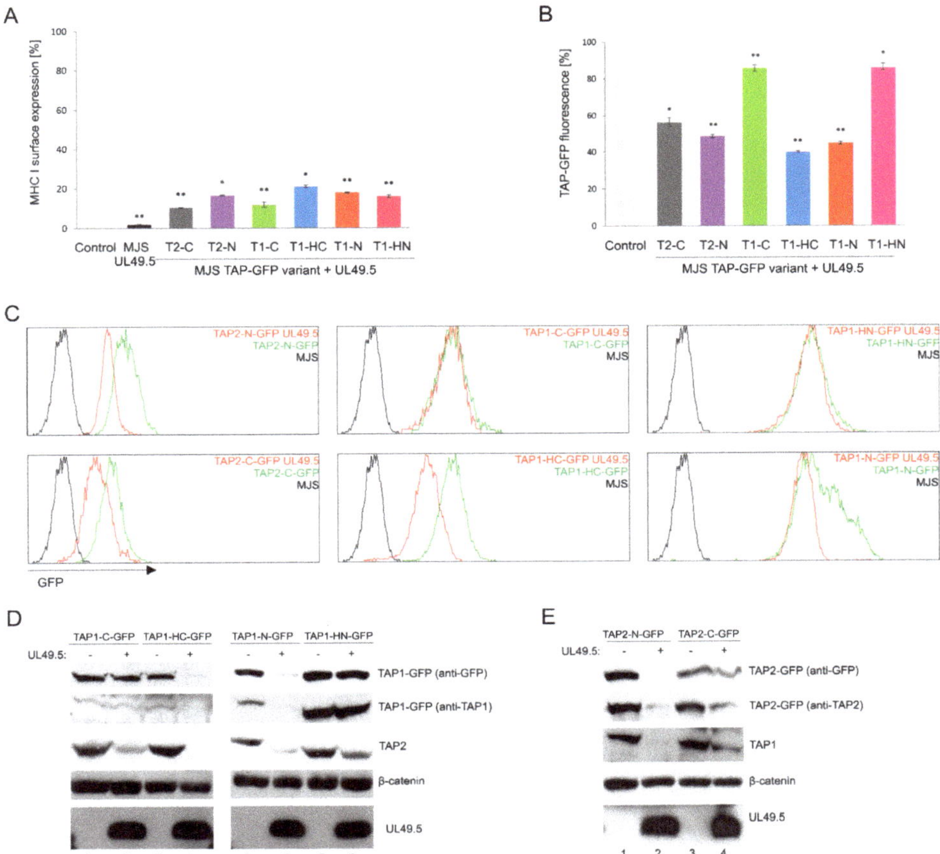

Figure 4. TAP-GFP variants differ in their sensitivity to UL49.5-mediated inhibition and degradation. MJS cells with fluorescent TAP1 or TAP2 variants were transduced with a retrovirus encoding bovine herpesvirus 1 (BoHV-1) UL49.5. (**A**) Surface expression of MHC I was assessed by flow cytometry using specific antibodies (W6/32). MHC I expression is presented as the percentage of mean fluorescence intensity; fluorescence of parental cells without UL49.5 was set as 100%. The analysis was performed in triplicates. The statistical significance of differences between MJS TAP-GFP and MJS TAP-GFP UL49.5 cell lines was estimated by *t*-test; ** $p \leq 0.001$ * $p \leq 0.005$. (**B**) GFP mean fluorescence intensity is presented as the percentage of GFP fluorescence of parental cells (set as 100%). The analysis was performed in triplicates. The statistical significance of differences between MJS TAP-GFP and MJS TAP-GFP UL49.5 cell lines was estimated by *t*-test; ** $p \leq 0.001$ * $p \leq 0.005$. (**C**) The effect of UL49.5 on GFP level in MJS TAP-GFP cells was assessed by flow cytometry. (**D**,**E**) Degradation of TAP-GFP variants in the presence of BoHV-1 UL49.5 in stable cell lines was determined by SDS-PAGE and immunoblotting using: anti-GFP, anti-TAP1, anti-TAP2, or anti-UL49.5 antibodies. β-catenin was used as a loading control.

Next, we analyzed the susceptibility of TAP-GFP variants to UL49.5-triggered degradation (Figure 4B,C). Flow cytometry analysis revealed a reduction of GFP mean fluorescence intensity by approximately 50% in the cells co-expressing UL49.5 with TAP1-N-GFP, TAP1-HC-GFP, and both TAP2 variants, in comparison to the control cells expressing the fluorescent transporter without UL49.5. There were no significant changes in GFP fluorescence observed in the cells expressing TAP1-C-GFP or TAP1-HN-GFP with UL49.5.

Susceptibility of TAP1-HC-GFP and TAP2-N-GFP constructs to UL49.5-dependent TAP inhibition and degradation was also confirmed in a reconstituted U937 cell line (Figure S2), where downregulation of surface MHC I, as well as a decrease of mean GFP fluorescence to the similar extent as in MJS cells, could be denoted.

To verify if changes in GFP fluorescence correspond with the decreased TAP protein level, immunoblotting analysis of cell lysates was performed (Figure 4D,E). A similar level of BoHV-1 UL49.5 protein could be observed in all cell lines. Decreased amounts of TAP1-N-GFP, TAP1-HC-GFP, and both TAP2-GFP fusion proteins could be detected in the presence of UL49.5, with the use of both anti-GFP and specific anti-TAP antibodies. Of note, in the case of non-degradable fluorescent TAP1 constructs (TAP1-HN-GFP and TAP1-C-GFP), the level of endogenous TAP2 was decreased (compare TAP2 in Figure 4D in lanes 1–2 and 5–6), what may suggest partial degradation of the untagged TAP subunit only. Degradation in TAP1-HC-GFP and TAP2-N-GFP cell lines was observed for both, exogenous fluorescent and endogenous TAP subunits. TAP1-N-GFP and TAP2-C-GFP characterized with only partial degradation.

3.4. Interaction of TAP-GFP with UL49.5 and the Peptide-Loading Complex

To assess the interaction of a non-degradable TAP with BoHV-1-encoded UL49.5 and selected components of PLC, we chose the TAP1-HN-GFP construct as demonstrating high expression. Among proteins co-immunoprecipitating with the fluorescent transporter (using GFP-Trap), we could identify UL49.5 and endogenous TAP2, as well as the known components of PLC: MHC I heavy chain and ERp57 (Figure 5). These results highlight the fact that UL49.5 is still capable of interacting with the non-degradable TAP1-HN-GFP. Unspecific binding of UL49.5 was excluded by immunoprecipitation from MJS wt cell lysate (with unmodified TAP) expressing UL49.5.

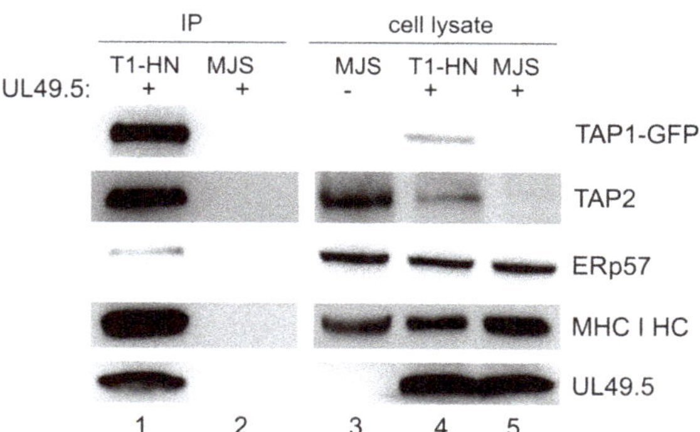

Figure 5. TAP1-GFP interacts with UL49.5 and the peptide-loading complex. TAP1-HN-GFP (T1-HN) was immunoprecipitated by GFP-Trap from lysates of MJS cells expressing TAP1-HN-GFP and UL49.5 or wt MJS with UL49.5 only. Co-precipitating proteins were analyzed by SDS-PAGE and immunoblotting using antibodies against GFP, TAP2, ERp57, MHC I HC, and UL49.5. Right panel: cell lysates were loaded on SDS-PAGE directly and analyzed by immunoblotting.

3.5. Application of the TAP2-N-GFP Variant as a Platform to Study BoHV-1 UL49.5 Activity in Virus-Infected Cells

In the more detailed studies on our fluorescent TAP platform, we focused on a single TAP-GFP variant, selecting TAP2-N-GFP. TAP1-N-GFP was excluded based on the presence of a free form of GFP. The constructs resistant to UL49.5-triggered degradation were eliminated as well, as was TAP2-C-GFP due to its very week, nearly undetectable basic green fluorescence and only partial degradation in the presence of UL49.5.

First, we confirmed that UL49.5-mediated MHC I downregulation relies on the inhibition of antigenic peptide translocation. TAP transport assay was performed in TAP2-N-GFP MJS cells and TAP2-N-GFP cells expressing UL49.5 or (as controls) in MJS wt, MJS TAP2KO cells, and MJS TAP2KO cells reconstituted with a non-fluorescent TAP2. The assay was based on cytoplasm-to-ER translocation of fluorescein-conjugated substrate peptides, in the presence of ATP or EDTA, as a passive diffusion control (Figure 6). Reconstitution with fluorescent TAP restored peptide transport when compared to the parental TAP2KO cells to a level of almost 50% higher than for MJS wt cells. This might result from high expression of exogenous TAP2 as a very similar transport activity was denoted for the non-fluorescent TAP2 reconstitution. The presence of UL49.5 inhibited peptide translocation to the level of TAP2KO cells.

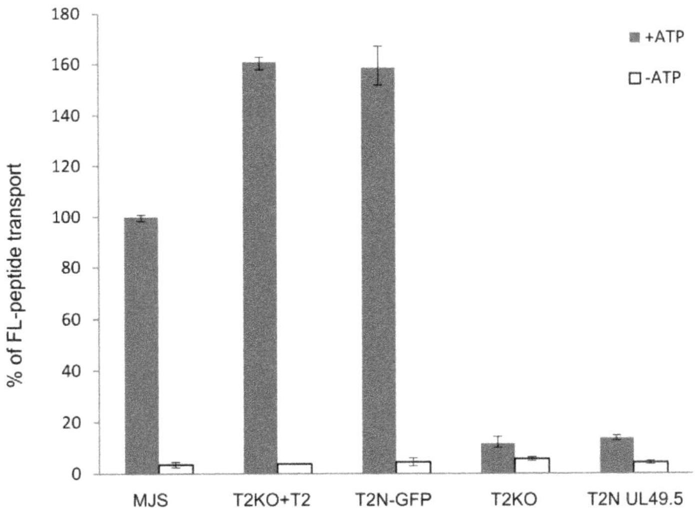

Figure 6. UL49.5-induced inhibition of peptide transport. MJS cells with CRISPR/Cas9 TAP2 knockout (T2KO) were stably reconstituted with wild-type TAP2 (T2KO+T2) or fluorescent TAP2 (T2N-GFP), subsequently transduced with a retrovirus encoding BoHV-1 UL49.5, and sorted (T2N UL49.5). Transport activity of TAP was analyzed using fluorescein-labeled peptide CVNKTERAY in the presence of ATP (gray bars) or EDTA (white bars). Peptide transport is expressed as a percentage of translocation, relative to the translocation observed in control MJS cells (set at 100%). The experiment was performed in triplicates. The statistical significance of differences between MJS controls and reconstituted or KO variants was estimated by *t*-test; for all the samples $p \leq 0.005$.

Finally, to test whether we can apply our fluorescent TAP platform to quantify the TAP level during virus infection, we infected MJS TAP2-N-GFP cells with BoHV-1 and analyzed TAP-derived GFP fluorescence by flow cytometry (Figure 7A,B). TAP2-N-GFP and endogenous TAP1 levels were assessed in infected cell lysates by immunoblotting (Figure 7C,D). The viral infection resulted in a 30% decrease of mean GFP fluorescence intensity, while approximately 70% and 50% reduction of TAP2-N-GFP and TAP1 protein levels, respectively, could be demonstrated by immunoblotting.

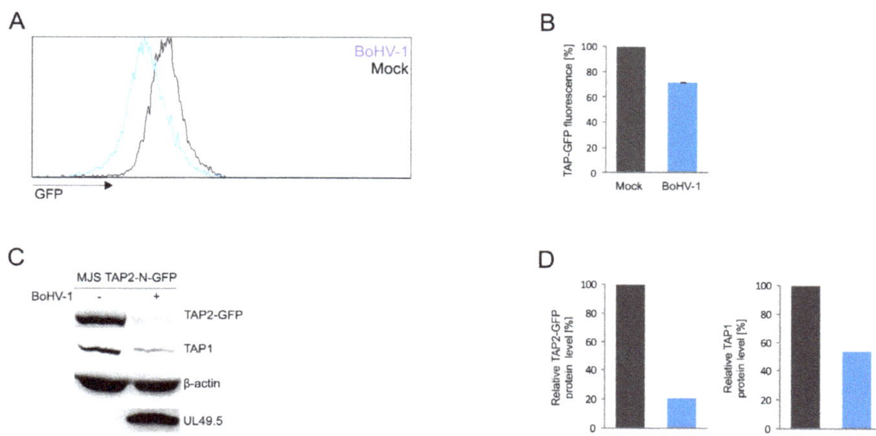

Figure 7. BoHV-1 infection results in TAP2-GFP degradation. MJS TAP2-N-GFP cells were infected with BoHV-1 at a multiplicity of infection (moi) = 10. Twenty-four hours post-infection, cells were collected and analyzed. (**A**) TAP2-GFP fluorescence was assessed by flow cytometry; histograms from a representative analysis are shown, and (**B**) depicted as the percentage of fluorescence in mock-infected MJS TAP2-N-GFP cells (set as 100%). The analysis was performed in triplicates. The statistical significance was assessed by *t*-test; $p \leq 0.0005$ (**C**) TAP2-GFP degradation was determined by SDS-PAGE and immunoblotting using anti-GFP, anti-TAP1, or anti-UL49.5 antibodies; β-actin was used as a loading control. (**D**) The relative amount of TAP2-GFP detected by immunoblotting was normalized to β-actin.

3.6. Only UL49.5 among Different Viral TAP Inhibitors Can Induce Human TAP Degradation

The effect of viral TAP inhibitors on TAP stability was reported, usually, in separate studies. TAP levels were analyzed in those reports by immunoblotting. To compare the sensitivity of fluorescent TAP to different viral inhibitors, we selected several representatives with distinct modes of action. Those were: competition for peptide (HSV-1 ICP47) or ATP (HCMV US6 and CPXV012) binding, as well as conformational arrest and degradation (BoHV-1 UL49.5) or a TAP-binding protein with no activity towards the transporter (VZV-encodedUL49.5). MJS TAP2-N-GFP cells were transduced with a retrovirus vector encoding a viral TAP inhibitor and subsequently selected to high purity. The presence of inhibitors was determined by immunoblotting (Figure 8A). Downregulation of MHC I surface expression could be observed during flow cytometry analysis, as expected, for all but the VZV-encoded protein (Figure 8B). Mean GFP fluorescence intensity measurement illustrated that BoHV-1 UL49.5 was unique in causing TAP-GFP degradation. ICP47 even seemed to slightly stabilize TAP-GFP (Figure 8C).

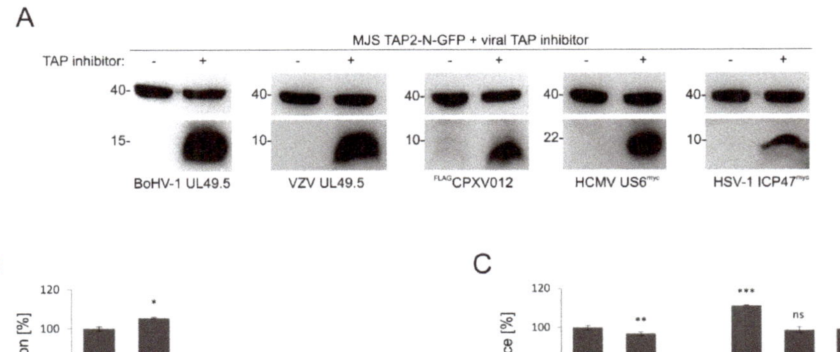

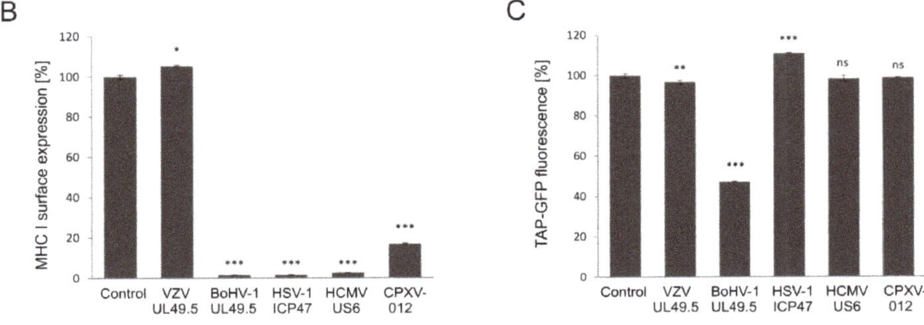

Figure 8. BoHV-1 UL49.5-mediated TAP degradation is unique among viral inhibitors of human TAP. MJS TAP2-N-GFP cells were transduced with a retrovirus encoding BoHV-1 UL49.5, varicella-zoster virus (VZV) UL49.5, human cytomegalovirus (HCMV) US6, herpes simplex virus 1 (HSV-1) ICP47, and cowpox virus CPXV012 and selected with puromycin. (**A**) The presence of viral TAP inhibitors was confirmed by SDS-PAGE and immunoblotting using anti-β, anti-BHV-1 UL49.5, anti-VZV UL49.5, anti-c-myc for HSV-1 ICP47, and HCMV US6 or anti-FLAG antibodies for CPXV012; β-actin (upper panels) was used as a loading control. Size markers are in kDa. (**B**) Surface expression of MHC I was assessed by flow cytometry using specific antibodies (W6/32). MHC I expression is presented as the percentage of MHC I on MJS TAP2-N-GFP cells (set as 100%). The analysis was performed in triplicates. The statistical significance of differences between MJS TAP2-N-GFP cells and cells with viral inhibitor was assessed by *t*-test; *** $p \leq 0.001$ * $p \leq 0.05$. (**C**) The mean fluorescence intensity of GFP was analyzed by flow cytometry and presented as the percentage of GFP fluorescence of MJS TAP2-N-GFP cells (set as 100%). The analysis was performed in triplicates. The statistical significance of differences between MJS TAP2-N-GFP cells and cells with a viral inhibitor was assessed by *t*-test; *** $p \leq 0.001$ ** $p \leq 0.01$ ns: not significant.

3.7. UL49.5-Induced TAP-GFP Degradation Is p97-Dependent

The fluorescent TAP platform can be potentially applied to search for cellular proteins involved in the activity of BoHV-1 UL49.5. According to previous studies, UL49.5-induced TAP degradation is proteasome-dependent [18]. Since TAP is an ER-resident protein, it is presumed to be removed via one of the endoplasmic reticulum-associated degradation (ERAD) pathways. To verify this hypothesis, we used NMS-873 (NMS), an allosteric inhibitor of p97/VCP (valosin-containing protein) a major AAA ATPase belonging to ERAD [38]. MJS TAP2-N-GFP cells with or without UL49.5 were treated with a 2 μM concentration of NMS for 24 h and analyzed by flow cytometry. Inhibition of p97 drastically rescued the mean fluorescence intensity of TAP2 in the presence of UL49.5 to 170% of mean fluorescence intensity in cells treated with DMSO, while in TAP2-N-GFP cells without UL49.5 we could observe only minimal increase to 115% (Figure 9A,B). Compatible results were obtained during immunoblotting analysis of cell lysates (Figure 9C, compare TAP2-GFP in lanes 1–2 and 3–4). Interestingly, inhibition of p97 seemed to stabilize also the expression of UL49.5 (Figure 9C, lanes 1–2). However, apparently, this did not result in a more pronounced TAP-GFP degradation when p97 was blocked.

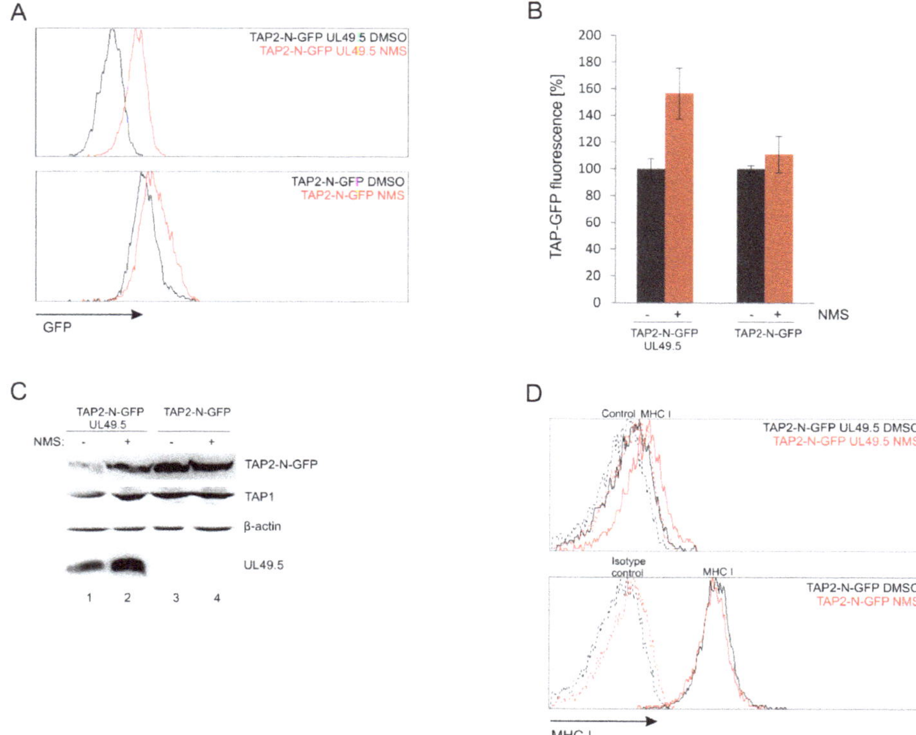

Figure 9. UL49.5 induced TAP-GFP degradation is p97-dependent. MJS TAP2-N-GFP UL49.5 cells were treated with p97 inhibitor NMS-873 (NMS) at 2 μM concentration for 24 hours. (**A**) Flow cytometry analysis of GFP fluorescence in NMS-treated and control (DMSO-treated) cells. (**B**) Relative GFP fluorescence in NMS-treated cells calculated as a percentage of GFP fluorescence in the control cells. The analysis was performed in triplicates. (**C**) The level of TAP-GFP in the presence of p97 inhibitor was determined by SDS-PAGE and immunoblotting using anti-GFP, anti-TAP1 or anti-UL49.5 antibodies; β-actin was used as a loading control. (**D**) Surface expression of MHC I assessed by flow cytometry using specific antibodies (W6/32).

Next, by using flow cytometry, we determined how inhibition of p97 influences MHC I cell surface expression in the presence (and also absence) of UL49.5. NMS-873-treated cells had only slightly improved surface MHC I levels (Figure 9D). This effect was reminiscent of MHC I levels in the presence of some UL49.5 point mutants, which lost the ability to induce TAP degradation, like, for instance, mutants in the C-terminal RGRG motif [32].

4. Discussion

Quantitative studies on protein stability and degradation may require proper tools and platforms that grant full functionality of the protein of interest and, at the same time, assess the protein levels accurately. Here, we tested the application potential of various full-length TAP to GFP fusion constructs for the studies on TAP stability in the presence of four inhibiting proteins encoded by viruses, to obtain the most suitable fluorescent TAP platform. All the tested TAP-GFP variants were functional. Nevertheless, our results point toward a critical role of GFP localization on fluorescence intensity of the tagged transporter, which in concert with the type a linker used to separate TAP and GFP may regulate its susceptibility to virally induced degradation. By using this platform, we also provide evidence that BoHV-1 UL49.5-induced TAP degradation is p97-dependent.

In a study on HSV-1 ICP47, a truncated fluorescent TAP complex (the so-called 6+6 transmembrane TAP core C-terminally fused to mVenus or mCerulean) was used to determine the effect of the viral protein on TAP thermostability [39]. However, for the studies on UL49.5, which was our primary protein of interest, the full-length TAP should constitute a better platform, as N-terminal TMD0s are required for maximum efficiency of UL49.5 binding and inhibition [19]. Full-length fluorescent TAP has been successfully used in several basic studies, some of which were to elucidate the association of H2L^d molecules with the TAP complex [40], follow lateral mobility of TAP in living cells [41], or illustrate its cellular distribution [42,43]. In those reports, the addition of a relatively large GFP tag to a much larger multiple membrane-spanning partner protein was tolerated to grant proper localization and functionality of the transporter. However, when exploited in a study on varicellovirus immune evasion, GFP-tagged TAP (C-terminal fusion using a random linker) failed to be degraded by BoHV-1 UL49.5, contrary to non-fluorescent wt TAP [18]. Since BoHV-1-encoded UL45.9 has been, so far, the only known viral inhibitor which can cause human TAP degradation apart from its inhibition, further investigation into this mechanism seemed very intriguing, and for this purpose, construction of fluorescent TAP was instrumental.

Designing an optimal fluorescent TAP construct was hampered by the lack of complete structural information about TAP-UL49.5 interaction, and thus, it required an experimental evaluation of different TAP-GFP variants. The latest structural study on BoHV-1 UL49.5 revealed its 3D structure, while subsequent molecular docking experiments proposed three different possible orientations of TAP-UL49.5 complex in which UL49.5 was suggested to interact simultaneously with both TAP subunits [44]. However, these models were predicted based on the structure of ICP47-arrested TAP conformation [10], and therefore the actual UL45.9-TAP binding model needs to be further confirmed.

Fluorescence analysis of constructed TAP-GFP variants in stable MJS cell lines provides evidence that the tag location, rather than the type of a linker used to separate TAP and GFP, has a pivotal impact on fluorescence intensity. N-terminal fusions generally granted stronger fluorescence (Figure 1B,C and Figure S1). It is worth mentioning here that both ends of TAP1 are present in the cytoplasm, while TAP2 incorporates its C terminus in the cytoplasm, and the N terminus localizes to the ER lumen [45,46]. Together with the fact that fluorescence of both TAP1 fusions was more intense than of TAP2, our results lead to speculations that it is the structure of both TMD0 and C-terminal NBDs that determines the fluorescent potential of the tagged constructs. For some constructs, especially for TAP1-N-GFP, we could observe additional protein products reacting with GFP-specific antibodies (Figure 1D), which might correspond to cleaved GFP and could also, most probably, explain higher and heterogeneous GFP signal of this construct observed by flow cytometry. The reason for the presence of free GFP in the case of TAP1-N-GFP is not fully understood. The length of this linker exceeds the size of other tested linkers, so it has a higher chance of affecting the stability of the protein. Another explanation might be the presence of a sequence recognized by cellular proteases, but the ExPASy PeptideCutter software analysis (https://web.expasy.org/peptide_cutter) did not reveal any significant candidates.

Fluorescent tagging did not affect the subcellular localization and function of the transporter, even upon overexpression of only one TAP subunit (Figures 2 and 3). Both TAP1 and TAP2 lack an N-terminal signal sequence for ER targeting [47], and the exact ER-targeting or ER-retention signals have not been identified to date. This encouraged us to design N-terminal GFP fusions with no additional signaling sequences preceding the tag. The localization of our constructs resembles patterns previously described for other recombinant fluorescent TAP proteins [38–41]. Our results stay in line with the studies on truncated TAP1/TAP2 [43] or functional dissection of transmembrane regions of TAP [6], which have indicated that the transmembrane segments themselves determine ER-localization. It is interesting that even genetically separated TMD0 and the core domains of TAP1 and TAP2 were previously found in the ER (TMD0 additionally localizing to the ER-Golgi intermediate compartment (ERGIC)), when co-expressed [6].

Replacing endogenous TAP1 or TAP2 with TAP-GFP or the untagged subunit restored MHC I on a cell surface equally well and to a level higher than on MJS wt cells (especially in the case of TAP1 constructs). One possible explanation could be the stronger stabilization of endogenous TAP2 by overexpression of TAP1. This effect might be especially noticeable in MJS cells since many melanoma-derived lines have lower endogenous expression of TAP, normally limiting MHC I surface levels [48]. Transduction of MJS TAP1KO or TAP2KO cells with the endogenously present subunit of the transporter did not increase MHC I level, which stays in line with the current view that although TAP1 and TAP2 can form homodimers under certain conditions, they are not functional in antigen presentation [19,35].

One of the most important results of this work provides evidence that all TAP-GFP variants were susceptible to UL49.5-induced inhibition to a similar extent, as assessed by surface MHC I downregulation (Figure 4A). However, only some of them were prone to degradation (both TAP2 fusions, TAP1-N-GFP and TAP1-HC-GFP, Figure 4B,D,E). TAP1-C-GFP remained resistant to UL49.5 what stays in agreement with the previous report [18]. As an interpretation of these data, we can suggest that the helical linker, in contrast to the random one, located at the C terminus of TAP1, effectively separates TAP from GFP to enable undisturbed TAP-UL49.5 interaction, resulting eventually in TAP degradation. Alternatively, it may also permit better access to ERAD components. In the case of the fluorescent TAP2 subunit, the location of GFP, despite the presence of random linkers, did not affect degradation, which could arise from structural differences between the TAP subunits. The TAP2 construct with GFP located in the ER lumen (TAP2-N-GFP) manifested more prominent degradation than the one with GFP in the cytoplasm. An additional observation from this experiment demonstrates that even in the case of a non-degradable fluorescent TAP variant, the second untagged endogenous TAP subunit seems to be sensitive to UL49.5-induced degradation (Figure 4D,E). This, in our opinion, supports the idea of reduced access to ERAD components in TAP1-C-GFP, whereas the access of the second untagged destabilized subunit remains, in this case, undisturbed. It is still unsolved whether UL49.5 can bind single TAP subunits, and the current mechanism points out at the heterodimer as the primary target [19]. As in MJS cells with non-degradable TAP variants, we could observe very efficient MHC I reduction, and the PLC composition in those cells seemed to be intact (Figure 5), at least with regard to the interaction of TAP with ERp57 and MHC I; our data confirm the previous report by [18], demonstrating that abolished degradation does not exclude inhibition. It even seems that TAP degradation might be only an auxiliary event, a "finish-off" effect, in the mechanism of UL49.5 action.

For further studies, we selected and validated TAP2-N-GFP as the most promising variant. TAP transport assay performed on this cell line confirmed that changes in MHC I surface levels reflect TAP transport efficiency (Figure 6). TAP transport in reconstituted cell lines, either with wt or a fluorescent version of the TAP subunit, was higher than in wt MJS. Then we demonstrated that results obtained in a stable cell line model system reflect a situation that occurs upon BoHV-1 infection, which was illustrated as loss of GFP fluorescence observed by flow cytometry and reduction of protein level shown by immunoblotting (Figure 7).

A former pulse-chase experiment with the use of proteasome inhibitor postulated co-degradation of TAP with UL49.5 [18]. In line with this working model, our data show that inhibition of p97 increases levels of both TAP and UL49.5, and demonstrate for the first time that UL49.5-induced TAP degradation requires functional p97. Most of known ER-resident substrates of this ATPase, which retrotranslocates proteins back to the cytoplasm, are ubiquitinated and targeted for proteasomal degradation [38], indicating that UL49.5 mediated TAP degradation occurs via ERAD.

Finally, the TAP2-N-GFP construct was verified as a platform for different viral TAP inhibitors, representing distinct mechanisms of transport inhibition and, most probably, binding another TAP conformation [28,39,49]. BoHV-1, HSV-1, and HCMV-encoded proteins were capable of drastic reduction of surface MHC I; CPXV012 contributed to a slightly weaker but still significant downregulation of MHC I, whereas VZV UL49.5, as expected, did not cause any changes. In terms of degradation, only BoHV-1 UL49.5 was able to decrease TAP-GFP levels, while ICP47 seemed even to stabilize TAP, which is in accordance with its reported effect on TAP thermostability. We believe that the fluorescent TAP platform provides more quantitative data in this respect when compared to previous immunoblotting analyses, which generally are more technically error-prone.

5. Conclusions

In this study, we were able to validate the application potential of fluorescent TAP as a platform for viral immune evasion studies. Our results indicate TAP-GFP variants susceptible to BoHV-1 UL49.5-induced degradation, demonstrate that this degradation is p97-dependent, and emphasize the importance of linker design in fusion protein construction. The fluorescent TAP platform can be now applied in further research on BoHV-1 UL49.5, for instance in the genome-wide search for cellular proteins responsible for UL49.5-induced degradation, where the fluorescent signal can be measured and indicate even small changes in TAP levels. TAP-GFP could be also exploited to identify the active motifs or amino acid residues of UL49.5 affecting TAP stability. The same platform with viral inhibitors can be applied, in a similar way as in the study by [28], to identify TAP conformation recognized by UL49.5.

Supplementary Materials: The following are available online at http://www.mdpi.com/2073-4409/8/12/1590/s1. Figure S1: Comparison of GFP fluorescence of TAP-GFP variants in HEK293T cells; Figure S2: Susceptibility of TAP-GFP expressed in reconstituted U937 cells to UL49.5-mediated inhibition and degradation.

Author Contributions: Conceptualization, M.W. and A.D.L.; methodology, M.W. and A.D.L.; investigation, M.W., M.G., P.P., A.W.B., R.D.L., and A.D.L.; resources, A.D.L. and E.J.H.J.W.; writing—original draft preparation, A.D.L. and M.W.; writing—review and editing, A.D.L., P.P., K.B.-S., and E.J.H.J.W.; visualization, M.W. and A.D.L.; supervision, A.D.L., K.B.-S., and E.J.H.J.W.; project administration, A.D.L.; funding acquisition, P.P. and A.D.L.

Funding: This study was funded by Polish National Science Center, grant number UMO-2014/14/E/NZ6/00164 to A.D.L. P.P. was funded by the European Commission under the Horizon2020 program H2020 MSCA-ITN GA 675278 EDGE.

Acknowledgments: We would like to thank Michał Rychłowski, Laboratory of Virus Molecular Biology, University of Gdańsk, Poland, for help with fluorescent confocal microscopy and Robert-Jan Lebbink, Department of Medical Microbiology, University Medical Center Utrecht, The Netherlands, for help in preparation of this manuscript.

Conflicts of Interest: The authors declare no conflict of interest.

References

1. Jensen, P.E. Recent advances in antigen processing and presentation. *Nat. Immunol.* **2007**, *8*, 1041–1048. [CrossRef] [PubMed]
2. Dean, M.; Annilo, T. Evolution of the ATP-binding cassette (ABC) transporter superfamily in vertebrates. *Annu. Rev. Genom. Hum. Genet.* **2005**, *6*, 123–142. [CrossRef] [PubMed]
3. van Endert, P.M.; Tampé, R.; Meyer, T.H.; Tisch, R.; Bach, J.-F.; McDevitt, H.O. A sequential model for peptide binding and transport by the transporters associated with antigen processing. *Immunity* **1994**, *1*, 491–500. [CrossRef]
4. Neefjes, J.J.; Momburg, F.; Hämmerling, G.J. Selective and ATP-dependent translocation of peptides by the MHC-encoded transporter. *Science* **1993**, *261*, 769–771. [CrossRef] [PubMed]
5. Geng, J.; Sivaramakrishnan, S.; Raghavan, M. Analyses of conformational states of the transporter associated with antigen processing (TAP) protein in a native cellular membrane environment. *J. Biol. Chem.* **2013**, *288*, 37039–37047. [CrossRef]
6. Koch, J.; Guntrum, R.; Heintke, S.; Kyritsis, C.; Tampé, R. Functional dissection of the transmembrane domains of the transporter associated with antigen processing (TAP). *J. Biol. Chem.* **2004**, *279*, 10142–10147. [CrossRef]
7. Verweij, M.C.; Horst, D.; Griffin, B.D.; Luteijn, R.D.; Davison, A.J.; Ressing, M.E.; Wiertz, E.J.H.J. Viral Inhibition of the transporter associated with antigen processing (TAP): A striking example of functional convergent evolution. *PLoS Pathog.* **2015**, *11*, e1004743. [CrossRef]
8. Früh, K.; Ahn, K.; Djaballah, H.; Sempé, P.; Peter, M.; van Endert, P.M.; Tampé, R.; Peterson, P.A.; Yang, Y. A viral inhibitor of peptide transporters for antigen presentation. *Nature* **1995**, *375*, 415–418. [CrossRef]

9. Tomazin, R.; Hill, A.B.; Jugovic, P.; York, I.; van Endert, P.; Ploegh, H.L.; Andrews, D.W.; Johnson, D.C. Stable binding of the herpes simplex virus ICP47 protein to the peptide binding site of TAP. *EMBO J.* **1996**, *15*, 3256–3266. [CrossRef]
10. Oldham, M.L.; Grigorieff, N.; Chen, J. Structure of the transporter associated with antigen processing trapped by herpes simplex virus. *eLife* **2016**, *5*, e21829. [CrossRef]
11. Ahn, K.; Gruhler, A.; Galocha, B.; Jones, T.R.; Wiertz, E.J.H.J.; Ploegh, H.L.; Peterson, P.A.; Yang, Y.; Früh, K. The ER-luminal domain of the HCMV glycoprotein US6 inhibits peptide translocation by TAP. *Immunity* **1997**, *6*, 613–621. [CrossRef]
12. Kyritsis, C.; Gorbulev, S.; Hutschenreiter, S.; Pawlitschko, K.; Abele, R.; Tampé, R. Molecular mechanism and structural aspects of transporter associated with antigen processing inhibition by the cytomegalovirus protein US6. *J. Biol. Chem.* **2001**, *276*, 48031–48039. [CrossRef] [PubMed]
13. Halenius, A.; Momburg, F.; Reinhard, H.; Bauer, D.; Lobigs, M.; Hengel, H. Physical and functional interactions of the cytomegalovirus US6 glycoprotein with the transporter associated with antigen processing. *J. Biol. Chem.* **2006**, *281*, 5383–5390. [CrossRef]
14. Alzhanova, D.; Edwards, D.M.; Hammarlund, E.; Scholz, I.G.; Horst, D.; Wagner, M.J.; Upton, C.; Wiertz, E.J.; Slifka, M.K.; Früh, K. Cowpox virus inhibits the transporter associated with antigen processing to evade T cell recognition. *Cell Host Microbe* **2009**, *6*, 433–445. [CrossRef]
15. Byun, M.; Verweij, M.C.; Pickup, D.J.; Wiertz, E.J.H.J.; Hansen, T.H.; Yokoyama, W.M. Two mechanistically distinct immune evasion proteins of cowpox virus combine to avoid antiviral CD8 T Cells. *Cell Host Microbe* **2009**, *6*, 422–432. [CrossRef]
16. Luteijn, R.D.; Hoelen, H.; Kruse, E.; van Leeuwen, W.F.; Grootens, J.; Horst, D.; Koorengevel, M.; Drijfhout, J.W.; Kremmer, E.; Früh, K.; et al. Cowpox virus protein CPXV012 eludes CTLs by blocking ATP binding to TAP. *J. Immunol.* **2014**, *193*, 1578–1589. [CrossRef]
17. Verweij, M.C.; Lipinska, A.D.; Koppers-Lalic, D.; van Leeuwen, W.F.; Cohen, J.I.; Kinchington, P.R.; Messaoudi, I.; Bienkowska-Szewczyk, K.; Ressing, M.E.; Rijsewijk, F.A.M.; et al. The capacity of UL49.5 proteins to inhibit TAP is widely distributed among members of the genus varicellovirus. *J. Virol.* **2011**, *85*, 2351–2363. [CrossRef]
18. Koppers-Lalic, D.; Reits, E.A.J.; Ressing, M.E.; Lipinska, A.D.; Abele, R.; Koch, J.; Rezende, M.M.; Admiraal, P.; van Leeuwen, D.; Bienkowska-Szewczyk, K.; et al. Varicelloviruses avoid T cell recognition by UL49.5-mediated inactivation of the transporter associated with antigen processing. *Proc. Natl. Acad. Sci. USA* **2005**, *102*, 5144–5149. [CrossRef]
19. Verweij, M.C.; Koppers-Lalic, D.; Loch, S.; Klauschies, F.; de la Salle, H.; Quinten, E.; Lehner, P.J.; Mulder, A.; Knittler, M.R.; Tampé, R.; et al. The varicellovirus UL49.5 protein blocks the transporter associated with antigen processing (TAP) by inhibiting essential conformational transitions in the 6+6 transmembrane TAP core complex. *J. Immunol.* **2008**, *181*, 4894–4907. [CrossRef]
20. Koppers-Lalic, D.; Verweij, M.C.; Lipińska, A.D.; Wang, Y.; Quinten, E.; Reits, E.A.; Koch, J.; Loch, S.; Rezende, M.M.; Daus, F.; et al. Varicellovirus UL49.5 proteins differentially affect the function of the transporter associated with antigen processing, TAP. *PLoS Pathog.* **2008**, *4*, e1000080. [CrossRef]
21. Boname, J.; May, J.; Stevenson, P. The murine gamma-herpesvirus-68 MK3 protein causes TAP degradation independent of MHC class I heavy chain degradation. *Eur. J. Immunol.* **2005**, *35*, 171–179. [CrossRef] [PubMed]
22. Herr, R.A.; Wang, X.; Loh, J.; Virgin, H.W.; Hansen, T.H. Newly discovered viral E3 ligase pK3 induces endoplasmic reticulum-associated degradation of class I major histocompatibility proteins and their membrane-bound chaperones. *J. Biol. Chem.* **2012**, *287*, 14467–14479. [CrossRef] [PubMed]
23. Harvey, I.B.; Wang, X.; Fremont, D.H. Molluscum contagiosum virus MC80 sabotages MHC-I antigen presentation by targeting tapasin for ER-associated degradation. *PLoS Pathog.* **2019**, *15*, e1007711. [CrossRef] [PubMed]
24. Chen, X.; Zaro, J.L.; Shen, W.-C. Fusion protein linkers: Property, design and functionality. *Adv. Drug Deliver. Rev.* **2013**, *65*, 1357–1369. [CrossRef] [PubMed]
25. Praest, P.; Luteijn, R.D.; Brak-Boer, I.G.J.; Lanfermeijer, J.; Hoelen, H.; Ijgosse, L.; Costa, A.I.; Gorham, R.D.; Lebbink, R.J.; Wiertz, E.J.H.J. The influence of TAP1 and TAP2 gene polymorphisms on TAP function and its inhibition by viral immune evasion proteins. *Mol. Immunol.* **2018**, *101*, 55–64. [CrossRef]

26. Lipińska, A.D.; Koppers-Lalic, D.; Rychlowski, M.; Admiraal, P.; Rijsewijk, F.A.M.; Bienkowska-Szewczyk, K.; Wiertz, E.J.H.J. Bovine herpesvirus 1 UL49.5 protein inhibits the transporter associated with antigen processing despite complex formation with glycoprotein M. *J. Virol.* **2006**, *80*, 5822–5832. [CrossRef]
27. Morgenstern, J.P.; Land, H. Advanced mammalian gene transfer: High titre retroviral vectors with multiple drug selection markers and a complementary helper-free packaging cell line. *Nucleic Acid Res.* **1990**, *18*, 3587–3596. [CrossRef]
28. Lin, J.; Eggensperger, S.; Hank, S.; Wycisk, A.I.; Wieneke, R.; Mayerhofer, P.U.; Tampé, R. A negative feedback modulator of antigen processing evolved from a frameshift in the cowpox virus genome. *PLoS Pathog.* **2014**, *10*, e1004554. [CrossRef]
29. Marqusee, S.; Baldwin, R.L. Helix stabilization by Glu-Lys+ salt bridges in short peptides of de novo design. *Proc. Natl. Acad. Sci. USA* **1987**, *84*, 8898–8902. [CrossRef]
30. Arai, R.; Ueda, H.; Kitayama, A.; Kamiya, N.; Nagamune, T. Design of the linkers which effectively separate domains of a bifunctional fusion protein. *Protein Eng. Des. Sel.* **2001**, *14*, 529–532. [CrossRef]
31. Koppers-Lalic, D.; Rychlowski, M.; van Leeuwen, D.; Rijsewijk, F.A.M.; Ressing, M.E.; Neefjes, J.J.; Bienkowska-Szewczyk, K.; Wiertz, E.J.H.J. Bovine herpesvirus 1 interferes with TAP-dependent peptide transport and intracellular trafficking of MHC class I molecules in human cells. *Arch. Virol.* **2003**, *148*, 2023–2037. [CrossRef] [PubMed]
32. Verweij, M.C.; Lipińska, A.D.; Koppers-Lalic, D.; Quinten, E.; Funke, J.; van Leeuwen, H.C.; Bieńkowska-Szewczyk, K.; Koch, J.; Ressing, M.E.; Wiertz, E.J.H.J. Structural and functional analysis of the TAP-inhibiting UL49.5 proteins of varicelloviruses. *Mol. Immunol.* **2011**, *48*, 2038–2051. [CrossRef] [PubMed]
33. Praest, P.; de Buhr, H.; Wiertz, E.J.H.J. A flow cytometry-based approach to unravel viral interference with the MHC class I antigen processing and presentation pathway. In *Antigen Processing*; van Endert, P., Ed.; Springer: New York, NY, USA, 2019; Volume 1988, pp. 187–198.
34. Keusekotten, K.; Leonhardt, R.M.; Ehses, S.; Knittler, M.R. Biogenesis of functional antigenic peptide transporter TAP requires assembly of pre-existing TAP1 with newly synthesized TAP2. *J. Biol. Chem.* **2006**, *281*, 17545–17551. [CrossRef] [PubMed]
35. Kleijmeer, M.J.; Kelly, A.; Geuze, H.J.; Slot, J.W.; Townsend, A.; Trowsdale, J. Location of MHC-encoded transporters in the endoplasmic reticulum and cis-Golgi. *Nature* **1992**, *357*, 342–344. [CrossRef] [PubMed]
36. Van Kaer, L.; Ashton-Rickardt, P.G.; Ploegh, H.L.; Tonegawa, S. TAP1 mutant mice are deficient in antigen presentation, surface class I molecules, and CD4−8+ T cells. *Cell* **1992**, *71*, 1205–1214. [CrossRef]
37. Antoniou, A.N.; Ford, S.; Pilley, E.S.; Blake, N.; Powis, S.J. Interactions formed by individually expressed TAP1 and TAP2 polypeptide subunits. *Immunology* **2002**, *106*, 182–189. [CrossRef]
38. Xia, D.; Tang, W.K.; Ye, Y. Structure and function of the AAA+ ATPase p97/Cdc48p. *Gene* **2016**, *583*, 64–77. [CrossRef]
39. Herbring, V.; Bäucker, A.; Trowitzsch, S.; Tampé, R. A dual inhibition mechanism of herpesviral ICP47 arresting a conformationally thermostable TAP complex. *Sci. Rep.* **2016**, *6*, 36907. [CrossRef]
40. Marguet, D.; Spiliotis, E.T.; Pentcheva, T.; Lebowitz, M.; Schneck, J.; Edidin, M. Lateral diffusion of GFP-tagged H2Ld molecules and of GFP-TAP1 reports on the assembly and retention of these molecules in the endoplasmic reticulum. *Immunity* **1999**, *11*, 231–240. [CrossRef]
41. Reits, E.A.J.; Vos, J.C.; Grommé, M.; Neefjes, J. The major substrates for TAP in vivo are derived from newly synthesized proteins. *Nature* **2000**, *404*, 774–778. [CrossRef]
42. Kobayashi, A.; Maeda, T.; Maeda, M. Membrane localization of transporter associated with antigen processing (TAP)-like (ABCB9) visualized in vivo with a fluorescence protein-fusion technique. *Biol. Pharm. Bull.* **2004**, *27*, 1916–1922. [CrossRef] [PubMed]
43. Ghanem, E.; Fritzsche, S.; Al-Balushi, M.; Hashem, J.; Ghuneim, L.; Thomer, L.; Kalbacher, H.; van Endert, P.; Wiertz, E.; Tampe, R.; et al. The transporter associated with antigen processing (TAP) is active in a post-ER compartment. *J. Cell Sci.* **2010**, *123*, 4271–4279. [CrossRef] [PubMed]
44. Karska, N.; Graul, M.; Sikorska, E.; Zhukov, I.; Ślusarz, M.J.; Kasprzykowski, F.; Lipińska, A.D.; Rodziewicz-Motowidło, S. Structure determination of UL49.5 transmembrane protein from bovine herpesvirus 1 by NMR spectroscopy and molecular dynamics. *BBA Biomembr.* **2019**, *1861*, 926–938. [CrossRef] [PubMed]

45. Vos, J.C.; Spee, P.; Momburg, F.; Neefjes, J. Membrane topology and dimerization of the two subunits of the transporter associated with antigen processing reveal a three-domain structure. *J. Immunol.* **1999**, *163*, 6679–6685.
46. Schrodt, S.; Koch, J.; Tampé, R. Membrane topology of the transporter associated with antigen processing (TAP1) within an assembled functional peptide-loading complex. *J. Biol. Chem.* **2006**, *281*, 6455–6462. [CrossRef]
47. Lankat-Buttgereit, B.; Tampé, R. The transporter associated with antigen processing (TAP): A peptide transport and loading complex essential for cellular immune response. In *ABC Proteins: From Bacteria to Man*; Holland, B., Cole, S.P.C., Kuchler, K., Higgins, C.F., Eds.; Academic Press: Cambridge, MA, USA, 2003; pp. 533–550.
48. Kageshita, T.; Hirai, S.; Ono, T.; Hicklin, D.J.; Ferrone, S. Down-regulation of HLA class I antigen-processing molecules in malignant melanoma. *Am. J. Pathol.* **1999**, *154*, 745–754. [CrossRef]
49. Matschulla, T.; Berry, R.; Gerke, C.; Döring, M.; Busch, J.; Paijo, J.; Kalinke, U.; Momburg, F.; Hengel, H.; Halenius, A. A highly conserved sequence of the viral TAP inhibitor ICP47 is required for freezing of the peptide transport cycle. *Sci. Rep.* **2017**, *7*, 2933. [CrossRef]

Article

A New ERAP2/Iso3 Isoform Expression Is Triggered by Different Microbial Stimuli in Human Cells. Could It Play a Role in the Modulation of SARS-CoV-2 Infection?

Irma Saulle [1,2], Claudia Vanetti [1,2], Sara Goglia [1], Chiara Vicentini [1], Enrico Tombetti [1], Micaela Garziano [1], Mario Clerici [2,3] and Mara Biasin [1,*]

1 Department of Biomedical and Clinical Sciences-L. Sacco, University of Milan, 20157 Milan, Italy; irma.saulle@unimi.it (I.S.); claudia.vanetti@unimi.it (C.V.); sarag9623@gmail.com (S.G.); chiara.vicentini@studenti.unimi.it (C.V.); enrico.tombetti@unimi.it (E.T.); micaela.garziano@unimi.it (M.G.)
2 Department of Pathophysiology and Transplantation, University of Milan, 20122 Milan, Italy; mario.clerici@unimi.it
3 Don C. Gnocchi Foundation ONLUS, IRCCS, 20148 Milan, Italy
* Correspondence: mara.biasin@unimi.it; Tel.: +39-0250319679

Received: 23 July 2020; Accepted: 20 August 2020; Published: 24 August 2020

Abstract: Following influenza infection, rs2248374-G ERAP2 expressing cells may transcribe an alternative spliced isoform: ERAP2/Iso3. This variant, unlike ERAP2-wt, is unable to trim peptides to be loaded on MHC class I molecules, but it can still dimerize with both ERAP2-wt and ERAP1-wt, thus contributing to profiling an alternative cellular immune-peptidome. In order to verify if the expression of ERAP2/Iso3 may be induced by other pathogens, PBMCs and MDMs isolated from 20 healthy subjects were stimulated with flu, LPS, CMV, HIV-AT-2, SARS-CoV-2 antigens to analyze its mRNA and protein expression. In parallel, Calu3 cell lines and PBMCs were in vitro infected with growing doses of SARS-CoV-2 (0.5, 5, 1000 MOI) and HIV-1_{BAL} (0.1, 1, and 10 ng p24 HIV-1_{Bal}/1 $\times 10^6$ PBMCs) viruses, respectively. Results showed that: (1) ERAP2/Iso3 mRNA expression can be prompted by many pathogens and it is coupled with the modulation of several determinants (cytokines, interferon-stimulated genes, activation/inhibition markers, antigen-presentation elements) orchestrating the anti-microbial immune response (Quantigene); (2) ERAP2/Iso3 mRNA is translated into a protein (western blot); (3) ERAP2/Iso3 mRNA expression is sensitive to SARS-CoV-2 and HIV-1 concentration. Considering the key role played by ERAPs in antigen processing and presentation, it is conceivable that these enzymes may be potential targets and modulators of the pathogenicity of infectious diseases and further analyses are needed to define the role played by the different isoforms.

Keywords: ERAP2; ERAP2/Iso3; microbial infections; alternative splicing; SARS-CoV-2; host cell response

1. Introduction

ERAP1 and ERAP2 (endoplasmic reticulum aminopeptidases 1 and 2) are two IFNγ- and TNFα-inducible, ubiquitously-expressed human enzymes, which belong to the M1 family of zinc aminopeptidases [1]. In the endoplasmic reticulum (ER), ERAPs shape the antigenic repertoire by trimming the N-terminus of precursor peptides previously generated in the cytoplasm by the proteasome. In this way, ERAPs generate optimal-length peptides for loading onto MHC class I groove to be presented to CD8+ T lymphocytes [2,3]. Despite maintaining marked differences in their enzymatic specificity these two enzymes can act together in a concerted way, through the formation of homo- or heterodimers, thus allowing the generation of a variegated and more immunogenic antigenic repertoire [4]. In particular, ERAP1–ERAP2 heterodimer generation has been demonstrated to improve the shaping of peptides suitable for MHC class I molecule binding [5].

ERAPs are encoded by two genes, sharing ~49% sequence homology and situated on chromosome 5q15 in opposite directions, which are highly polymorphic [6]. Since their leading role in the antigen processing pathway, several studies have investigated any potential link between ERAP polymorphic variants and alterations in their functioning, which could result in MHC-I-associated disorder onset [2,7,8] as well as into variations in susceptibility/progression to microbial infections [9].

As for ERAP2, the most relevant single nucleotide polymorphism (SNP) is the non-coding rs2248374 (A/G) which identifies two haplotypes, hereafter referred to as HapA (A allele for rs2248374) and HapB (G allele for rs2248374). In HapB, the G allele for this SNP primes the transcription of a spliced ERAP2 variant (ERAP2/Iso2), presenting an extended exon 10 (56 extra nucleotides) and two in-frame TAG stop codons, which in turn lead to its nonsense-mediated decay (NMD) [10]. Conversely, HapA is translated into a 965-amino-acid protein and is associated with Crohn's disease [11], HLA-A29-associated birdshot uveitis [12], ankylosing spondylitis [13,14] and juvenile idiopathic arthritis [15], as well as natural resistance to HIV infection [9,16,17]. Since these variants are maintained by a balanced selection to a frequency of approximately 50% (HapB: 53% and HapA: 47%), nearly 25% of the population fails to express the ERAP2 protein [18]. This observation raises a logic question: in which peculiar setting does balancing selection operate to conserve the apparently loss-of-function HapB and the disease-causing HapA in the human population? Quite recently, Ye and co-workers provided an exhaustive explanation to this apparent paradox [19]. Indeed, for the first time, they documented the transcription of two novel short isoforms (ERAP2/Iso3, ERAP2/Iso4) from flu-infected monocyte-derived dendritic cells isolated from homozygous HapB-carrying subjects [19]. The two short isoforms are transcribed from HapB and differ from the full-length one—ERAP2/Iso1, transcribed from HapA—since their transcription begins in correspondence of exon 9 and undergoes alternative splicing of an extended exon 10. Besides, they diverge from each other by alternative splicing at a secondary splice site at exon 15 (Figure 1). Of note, while ERAP2/Iso4 is predicted to harbor a premature termination codon that could lead to NMD, ERAP2/Iso3, is expected to be translated into a protein [19]. Such protein misses the catalytic domain [19], but could still critically contribute to profile the cellular immune-peptidome as it preserves the capacity to dimerize with ERAP1 and possibly ERAP2-wild type (wt) [20].

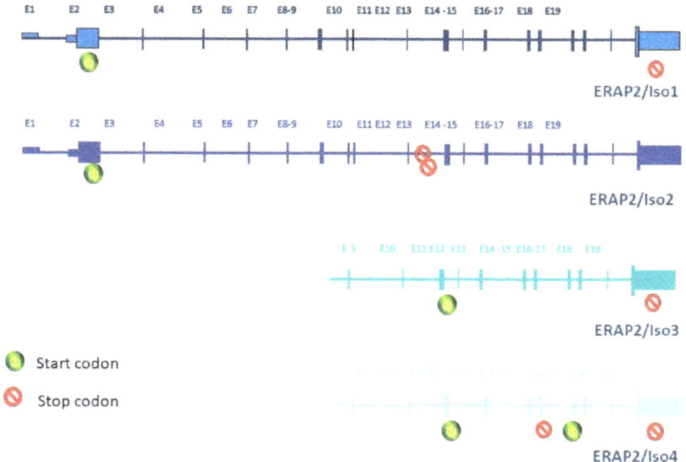

Figure 1. Genetics of ERAP2 isoforms regulation. Structures of transcripts derived from each ERAP2 isoform are represented. Start and stop codons for each isoform are reported.

Based on these premises, the aim of our study was to investigate if the transcription of ERAP2/Iso3 is either flu-specific or if it can be induced by other kind of stimuli, such as other viruses, bacteria, or inflammatory triggers. Given the foremost role played by ERAPs in the field of both acquired and innate immunity, the characterization of the different isoforms produced in a particular pathological setting, such as the one caused by microbial infections, could lead to the identification of new molecular targets to be exploited in the setting up of innovative therapeutically or vaccinal approaches.

2. Methods

2.1. Study Population

Twenty healthy controls (HC) were enrolled in the study in order to investigate the induction of ERAP isoform transcription in response to common recall antigens such as influenza-antigens (flu), Lipopolisaccaride (LPS), Citomegalovirus (CMV), in addition to Aldrithiol-2 (AT-2)-inactivated R5-tropic human immunodeficiency virus-$1_{Ba\text{-}L}$ (HIV-AT-2) and acute respiratory syndrome coronavirus 2 (SARS-CoV-2) inactivated virus (i-SARS-CoV-2). The Ethical Committee of the Fondazione IRCCS Ca' Granda Ospedale Maggiore Policlinico approved the study (Prot. N°0028257). All the donors signed an informed consent form, in agreement with the Declaration of Helsinki of 1975, revised in 2013.

2.2. Viruses

The laboratory-adapted HIV-1 strain used in the experiments was the R5 tropic HIV-1_{BaL} (courtesy of Drs. S. Gartner, M. Popovic, and R. Gallo; NIH AIDS Research and Reference Reagent Program) provided through the EU program EVA Centre for AIDS Reagents (NIBSC, Potter Bars, UK). The virus was inactivated with AT-2, which is able to change the zinc finger nucleocapsid proteins of HIV-1, therefore deactivating the viral infectivity as previously described [21].

SARS-CoV-2 (Virus Human 2019-nCoV strain 2019-nCoV/Italy-INMI1, Rome, Italy) was expanded on Calu-3 cells (ATCC® HTB-55™) and $TCID_{50}$ was calculated as previously reported [22]. SARS-CoV-2 inactivation (i-SARS-CoV-2) was obtained by incubation at 65 °C for 30 min [23].

2.3. ERAP2 Genotyping Analyses

Whole blood was collected by all the subjects enrolled in the study by venipuncture in Vacutainer tubes containing EDTA (BD Vacutainer, San Diego, CA, USA). Total DNA was extracted by DNA purification Maxwell® RSC Instrument (Promega, Fitchburg, WI, USA) and quantified using the Nanodrop 2000 Instrument (Thermo Scientific, Waltham, MA, USA). Two-hundred ng of DNA were used to perform an SNP genotyping assay for ERAP2 rs2549782 (G/T) (TaqMan SNP Genotyping Assay; Applied Biosystems, Foster City, CA, USA), which is in linkage disequilibrium with the non-coding rs2248374 (A/G). Analyses were performed on Peripheral blood mononuclear cells (PBMCs) from all the subjects recruited in the study as well as on lung adenocarcinoma cells (Calu3; ATCC® HTB-55™). Allelic discrimination real-time PCR method was used to analyze the results.

2.4. Isolation of PBMCs and Monocyte-Derived Macrophages (MDMs) Differentiation

PBMCs, obtained from density gradient centrifugation on Ficoll (Cedarlane Laboratories Limited, Hornby, ON, Canada), were counted by automated cell counter ADAM-MC (NanoEnTek Inc., Seoul, Korea), which distinguishes viable from non-viable cells.

Flow cytometer analyses was used to quantify the percentage of CD14+ monocytes in PBMCs isolated from 3 HeteroAB HC. MDMs were obtained as previously described [24]. Briefly, 5×10^5 adherent monocytes were incubated for 5 days in RPMI with 20% of fetal bovine serum (FBS) (Euroclone, Milan, Italy) and 100 ng/mL macrophage-colony stimulating factor (M-CSF) (R&D Systems, Minneapolis, MN, USA). Optical microscope (ZOE™ Fluorescent Cell Imager, Bio-Rad, Hercules, CA, USA) observation allows to verify MDM differentiation.

2.5. Cell Cultures for Microbial Antigen Stimulation

PBMCs were resuspended at the concentration of 1×10^6 PBMCs/mL in RPMI 1640 medium (Euroclone, Milan, Italy) containing 10% fetal bovine serum (FBS), 1% levo-glutammin LG and 2% penstreptomicin. Subsequently, PBMCs and MDMs from HC were stimulated with antigens from different pathogens: 32 µg/mL of CMV grade 2 Antigen (Microbix Biosystem, Mississauga, ON, Canada), 1 µg/mL of LPS, 1 ng/mL of HIV-AT-2 equivalents and 5 multiplicity of infection (MOI) of i-SARS-CoV-2 inactivated virus. Two live UV-inactivated influenza viruses (flu) were used as well: an influenza A virus (A/RX73 and A/Puerto Rico/8/34 strains; 1:800) and the 1998–1999 formula of flu vaccine (1:5000; Wyeth Laboratories Inc., Marietta, PA, USA). Cells were stimulated even with non-microbial stimuli: 100U of IFNα and 1 µg/mL of IL-1β (Sigma, Saint Louis, MO, USA). Unstimulated PBMCs were cultured as control as well. Cells were harvested 10 (PBMCs) and 36 (MDMs) h post-treatment for RNA and protein analyses, respectively.

2.6. In Vitro Infection of PBMCs and Calu3 Cells with SARS-CoV-2

2.5×10^5 Calu3 cells (ATCC® HTB-55™) were cultured in DMEM medium (Euroclone, Milan, Italy) supplemented with 2% FBS in a 24-well plate. DMEM containing 100 U/mL penicillin and 100 µg/mL streptomycin was used as inoculum in the mock-infected cells. Cell cultures were incubated with serial dilutions of virus supernatant in duplicate, (1000 MOI, 5 MOI, 0.5 MOI) for three h at 37 °C and 5% CO_2. Cells were washed two times with lukewarm PBS and refilled with the proper growth medium (10%FBS). Optical microscope observation (ZOE™ Fluorescent Cell Imager, Bio-Rad, Hercules, CA, USA) was performed daily to investigate the cytopathic effect. The infected cells were harvested for SARS-CoV-2 detection in the supernatant and mRNA collection at 48 h. Each culture condition was run in triplicate. All the procedures were performed in agreement with the GLP guidelines adopted in our laboratory.

Maxwell® RSC Viral Total Nucleic Acid Purification Kit (Promega, Fitchburg, WI, USA) was used to extract RNA from Calu3 cell culture supernatants by the Maxwell® RSC Instrument (Promega, Fitchburg, WI, USA). Viral RNA was quantified, by single-step RT PCR -time PCR (GoTaq® 1-Step RT-qPCR) (Promega, Fitchburg, WI, USA) on a CFX96 (Bio-Rad, Hercules, CA, USA) by using TaqMan probes which target two portions of SARS-CoV-2 nucleocapsid (N) gene (N1 and N2). Specifically, we used the 2019-nCoV CDC qPCR Probe Assay emergency kit (IDT, Coralville, IA, USA), which allows also to amplify the human RNase P gene. Viral copy number quantification was performed by generating a standard curve from the quantified 2019-nCoV_N positive Plasmid Control (IDT, Coralville, IA, USA).

2.7. In Vitro HIV-Infection Assay

3×10^6 PBMCs from all the subjects included in the study were in vitro HIV-infected as previously described [25] with 10, 1, and 0.1 ng p24 HIV-1_{Bal}/1×10^6 PBMCs. After 5 days, supernatants were collected for p24 antigen ELISA (Cell Biolabs, San Diego, CA, USA), whereas PBMCs collected at 2 days post-infection were used for RNA extraction and gene expression analyses.

2.8. Gene Expression Analysis

RNA extracted from 1×10^6 PBMCs, and Calu3 cell lines were retrotranscribed as previously described [16]. cDNA quantification for ERAPs was performed on antigen-stimulated and HIV-infected PBMCs as well as on SARS-CoV-2 infected Calu3 cells through a real-time PCR (CFX96 connect, Bio-Rad, Hercules, CA, USA) and an SYBR Green PCR mix (Bio-Rad, Hercules, CA, USA); all the reactions were carried out in duplicate. Results are shown as the media of the relative expression units to the glyceraldehyde-3-phosphate dehydrogenase (GAPDH) and β-actin housekeeping genes calculated by the $2^{-\Delta\Delta Ct}$ equation. The following thermal protocol was used: initial denaturation (95 °C, 15 min) followed by 40 cycles of 15 s at 95 °C (denaturation), 20 s at 60 °C (annealing) and 20 s at 72 °C (extension).

Furthermore, a melting curve analysis was assessed for amplicon characterization. Ct values of 35 or higher were let off the analyses.

2.9. Quantigene Plex Gene Expression Assay

Gene expression of 8×10^5 PBMCs was performed by quantiGene Plex assay (Thermo Scientific, Waltham, MA, USA) which provides a fast and high-throughput solution for multiplexed gene expression quantitation, allowing the simultaneous measurement of 70 custom selected genes of interest in a single well of a 96-well plate. The QuantiGene Plex assay is hybridization-based and incorporates branched DNA (bDNA) technology, which uses signal amplification for direct measurement of RNA transcripts. The assay does not require RNA purification.

2.10. Western Blot Analyses

Cultured MDMs were removed by non-enzymatic cell dissociation solution (Sigma, Saint Louis, Missouri, USA), counted by the automated cell counter ADAM-MC (Digital Bio) and used for protein extraction by RIPA buffer (Sigma, Saint Louis, MO, USA). Extracted proteins were stored at −80 °C for further analyses. For WB analyses, samples from 3 HeteroAB subjects were sub-pooled (50 µg per pool). Equal amounts of proteins were separated by 4–20% SDS-polyacrylamide gel electrophoresis (Criterion TGX Stain-free precast gels and Criterion Cell system; Bio-Rad) and transferred onto nitrocellulose membrane using a Bio-Rad Trans-Blot Turbo System. Membranes were probed using a 1:1000 dilution of primary antibody goat anti-ERAP1 (AF2334; R&D Systems, Minneapolis, MN, USA) goat anti-ERAP2 polyclonal antibody (AF3830; R&D Systems, Minneapolis, MN, USA), rabbit anti-GAPDH polyclonal (VPA00187); Bio-Rad, Hercules, CA, USA] and a 1:10,000 dilution of secondary antibody conjugated with alkaline phosphatase anti-goat IgG (A4187; Sigma; goat anti-rabbit (STAR208P); Bio-Rad]. Membranes were incubated with the appropriate antibody and, after being excited using the Clarity Western ECL substrate, bands were visualized with a ChemiDoc MP imaging system (Bio-Rad) and quantified for densitometry with the Bio-Rad Image Lab software.

2.11. Statistical Analyses

Data are shown as mean and standard deviation. Analysis and figures were performed by GRAPHPAD PRISM version 5 (Graphpad software, La Jolla, CA, USA) and SPSS Statistics, version 25 (IBM software, Armonk, NY, USA). Gene expressions of ERAP2/iso1, ERAP2/iso1, and ERAP1 in PBMCs, monocyte-derived macrophages, and Calu-3 cells upon stimulation were compared to that of untreated cells by Wilcoxon test and Mann–Whitney test, as appropriate. P-values were corrected for false discovery rate (FDR) by using Microsoft R software, p-values ≤ 0.005 were considered to be significant.

3. Results

3.1. ERAP2 Allelic Variants Analyses

Analysis of ERAP2 SNP prevalence was aligned with European population distribution reported in the U.S. National Library of Medicine Database [https://www.ncbi.nlm.nih.gov/snp/rs2549782?fbclid=IwAR1ZdwC747PDWvtAzt6hZBV5j7oFZiPkLjY-JdSee1Plzvym7fhJVQc1Aks (data not shown). Among the 20 genotyped HC: 6 were HomoA, 8 heterozygous, 6 HomoB. Subsequent analyses were performed only on PBMCs and MDMs isolated from HomoB and heteroAB donors to exclude confounding results. Indeed, HomoA individuals express negligible levels of ERAP2/Iso2 and Iso3.

Calu3 cell lines were heterozygous for rs2248374 ERAP2 genotype.

3.2. mRNA Expression of ERAPs in PBMCs from Subjects Carrying Different ERAP2 Genotypes Following Microbial Stimulation

To verify whether the expression of ERAP2/Iso3 isoform is exclusively flu-specific or it may be triggered by other stimuli, we analyzed its expression on PBMCs isolated from HomoB and HeteroAB HC following HIV-AT-2, i-SARS-CoV-2, CMV, LPS, flu, IFNα, and IL-1β. As expected, the expression of the newly identified ERAP2/Iso3 was significantly augmented in PBMCs from all the subjects included in the study in response to flu ($p < 0.002$). However, even following CMV ($p < 0.004$), LPS ($p < 0.002$), HIV-AT-2 ($p < 0.003$), i-SARS-CoV-2 ($p < 0.002$) and IFNα ($p < 0.003$) stimulation, we observed a significant increase of its expression (Figure 2A). Conversely, IL-1β addition to cell culture did not result in ERAP2/Iso3 induction (Figure 2A). ERAP2/Iso1 expression was observed only in PBMCs from HeteroAB subjects and was induced following all the microbial-stimuli employed plus IFNα but not in response to IL-1β (Figure 2B). However, statistical significance was observed exclusively following HIV-AT-2 ($p < 0.025$), i-SARS-CoV-2 ($p < 0.03$) and CMV ($p < 0.02$). Likewise, ERAP1 expression was induced by all the stimuli employed excepting IL-1β and reached statistical significance following flu ($p < 0.002$), HIV-AT-2 ($p < 0.03$) CMV ($p < 0.002$), i-SARS-CoV-2 ($p < 0.008$) LPS ($p < 0.05$). We also observed a reduction following IL-1β ($p < 0{,}001$) stimulations (Figure 2C).

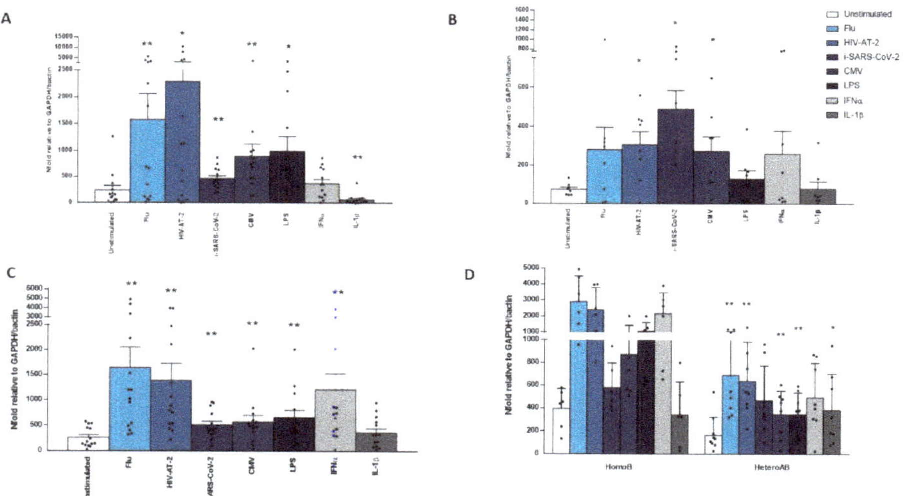

Figure 2. ERAP1, ERAP2/Iso1 and ERAP2/Iso3 mrna expression is increased following microbial stimulation. pbmcs isolated from 8 heteroab and 6 homob individuals were in vitro stimulated with microbial antigens (flu, CMV) inactivated viruses (i-SARS-CoV-2, HIV-AT-2) bacterial by-products (LPS) or inflammatory stimuli (IFNα, IL-1β) for ten h. mRNA expression for ERAP2/Iso3 (**A**), ERAP2/Iso1 (**B**) and ERAP1 (**C**) were assessed by RT-Real-Time PCR. Results are shown as the media of the relative expression units to the glyceraldehyde-3-phosphate dehydrogenase (GAPDH) and β-actin reference genes calculated by the $2^{-\Delta\Delta Ct}$ equation. (**D**) The microbial-dependent genetic control of ERAP2/Iso3 expression is underlined by the observation that its abundances are nearly doubled in HapB homozygotes compared to heterozygotes. Results are expressed as mean ± ES. * = $p < 0.05$; ** = $p < 0.01$.

Notably, the microbial-dependent genetic control of ERAP2 isoform usage is sustained by further evidence. Indeed, there was a significant correlation between whole microbial transcript quantity and ERAP2/Iso3 transcript abundances in heterozygotes compared to HapB homozygotes (Figure 2D).

3.3. Gene Expression of Immune Selected Effectors in PBMCs Following Microbial Stimulation

To verify if the increased expression of ERAPs in response to microbial stimulation could be extended to other factors involved in the orchestration of the immune response, mRNA expression of 70 selected effectors was investigated by the innovative Quantigene Plex Gene expression technology. The genes whose mRNA expression was upregulated are implicated in almost all phases of immune response, including: chemokines, cytokines and cytokine receptors, pathogen recognition receptor, inflammasome, cholesterol metabolism, interferon stimulated genes, adhesion molecules, activation/inhibition markers, antigen presentation factors (Figure 3). Notably, the gene expression pattern was partially shared by all the stimuli employed and partially pathogen-specific as summarized in Figure 3. In particular, following i-SARS-CoV-2-stimulation a significantly higher transcription rate was observed for: CCL2 ($p < 0.05$); CCL5 ($p < 0.035$), HMGCS1 ($p < 0.002$), PYCARD ($p < 0.002$), CASP1 ($p < 0.001$), CD44 ($p < 0.016$), CD274 ($p > 0.008$), IL-8 ($p < 0.003$); IL-1β ($p < 0.006$), ABCA1 ($p < 0.02$), IL-6R ($p < 0.004$), CCL3 ($p < 0.01$), IFNγ ($p < 0.05$); TAP1 ($p < 0.05$).

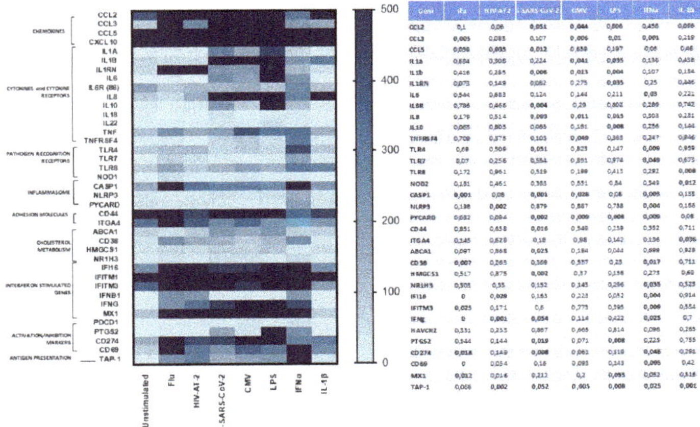

Gene	flu	HIV-AT2	i-SARS-CoV-2	CMV	LPS	IFNa	IL-1b
CCL2	0,1	0,06	0,051	0,044	0,006	0,456	0,098
CCL3	0,005	0,085	0,107	0,006	0,01	0,001	0,219
CCL5	0,058	0,035	0,012	0,658	0,197	0,06	0,46
IL1a	0,634	0,306	0,224	0,041	0,035	0,136	0,458
IL1b	0,416	0,285	0,006	0,013	0,004	0,107	0,184
IL1RN	0,073	0,149	0,062	0,275	0,035	0,25	0,446
IL6	0,544	0,683	0,124	0,144	0,211	0,03	0,221
IL6R	0,786	0,466	0,004	0,29	0,802	0,289	0,742
IL8	0,179	0,514	0,003	0,011	0,015	0,303	0,281
IL10	0,065	0,805	0,065	0,181	0,008	0,256	0,144
TNFRSF4	0,709	0,375	0,103	0,049	0,365	0,247	0,846
TLR4	0,69	0,509	0,051	0,823	0,147	0,009	0,959
TLR7	0,07	0,256	0,554	0,891	0,974	0,049	0,675
TLR8	0,172	0,961	0,519	0,189	0,415	0,292	0,006
NOD2	0,181	0,461	0,383	0,551	0,84	0,549	0,012
CASP1	0,001	0,08	0,001	0,028	0,08	0,009	0,155
NLRP3	0,138	0,002	0,879	0,887	0,738	0,004	0,158
PYCARD	0,032	0,094	0,002	0,009	0,006	0,009	0,08
CD44	0,851	0,658	0,016	0,540	0,259	0,352	0,711
ITGA4	0,345	0,628	0,18	0,98	0,142	0,136	0,036
ABCA1	0,097	0,868	0,025	0,184	0,044	0,699	0,928
CD36	0,007	0,265	0,369	0,557	0,25	0,017	0,711
HMGCS1	0,517	0,875	0,002	0,37	0,138	0,273	0,69
NR1H3	0,305	0,35	0,152	0,145	0,296	0,035	0,523
IFI16	0	0,029	0,163	0,225	0,052	0,004	0,914
IFITM3	0,025	0,171	0,6	0,775	0,395	0,006	0,554
IFNg	0	0,001	0,054	0,114	0,422	0,025	0,7
HAVCR2	0,331	0,255	0,867	0,665	0,814	0,096	0,265
PTGS2	0,544	0,144	0,019	0,071	0,008	0,225	0,755
CD274	0,018	0,149	0,008	0,061	0,116	0,048	0,291
CD69	0	0,054	0,16	0,093	0,143	0,005	0,42
MX1	0,012	0,016	0,212	0,2	0,093	0,052	0,516
TAP-1	0,066	0,002	0,052	0,005	0,008	0,025	0,001

Figure 3. mRNA expression of genes involved in the anti-microbial immune response was modulated in response to different pathogens. Quantigene Plex Gene expression technology was applied to quantify gene expression on PBMCs isolated from 8 HeteroAB and 6 HomoB individuals and stimulated with microbial antigens (flu, CMV) inactivated viruses (i-SARS-CoV-2, HIV-AT-2) bacterial by-products (LPS) or inflammatory stimuli (IFNα, IL-1β). Gene expression (mean values) is shown as a color scale from white to blue (Heatmap). Only statistically significant p values from T-test comparison between unstimulated and stimulated PBMCs are shown in table.

Unlike ERAP2 expression, no differences were observed in mRNA expression levels of all the analyzed genes in PBMCs from HomoB and HeteroAB subjects in response to all the stimuli engaged (data not shown).

3.4. mRNA Expression of ERAPs in In Vitro SARS-CoV-2 Infected Calu3 Cell Lines and HIV-Infected PBMCs

To verify whether ERAP2/Iso3 expression varies in response to growing viral concentrations we adopted two in vitro model of infection. Thus, Calu3 cell lines were infected with different SARS-CoV-2 viral input and after 48 hours' viral replication as well as ERAP mRNA expression were assessed. As expected, SARS-CoV-2 replication increased according to the rising viral input as assessed analyzing both N1 (MOI 0.5 vs. 5: $p < 0.01$; MOI 0.5 vs. 1000: $p < 0.001$) and N2 (MOI 0.5 vs. 5: $p < 0.01$; MOI 0.5 vs. 1000: $p < 0.001$) (Figure 4A). Images of cellular cytopathic effect on SARS-CoV-2 infected cells at 48 h showed that despite robust SARS-CoV-2 replication in Calu3 cells, substantial cell death was detected only in cells infected with 1000 MOI (Figure 4B). Notably, this increase was coupled with a progressive rise of ERAP2/Iso3 expression compared to the uninfected condition (MOI 0.5: $p < 0.04$; MOI 5: $p < 0.01$; MOI 1000: $p < 0.01$). Likewise, ERAP2/Iso1 and ERAP1 expression was induced in a viral-dose dependent manner, although statistical significance was observed only for ERAP2/Iso1 (MOI 0.5: $p < 0.05$) (Figure 4C).

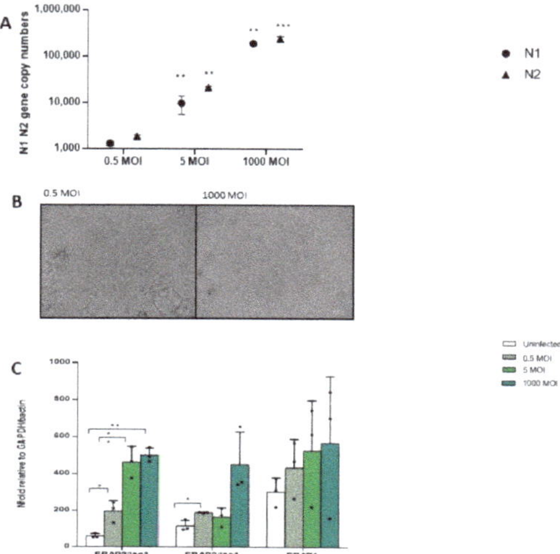

Figure 4. In vitro SARS-CoV-2 infection assay on Calu3 cells. (**A**) SARS-CoV-2 replication was assessed on Calu3 cells infected at 0.5, 5, and 1000 MOI 48 h post-infection. Viral copy number quantification was performed by generating a standard curve from the quantified 2019-nCoV_N positive plasmid control for Nucleocapsid (N) region 1 and 2, showing a significant increase in response to the increased viral input. (**B**) SARS-CoV-2-induced cytopathic effects were assessed in Calu3 infected cells. Representative images of SARS-CoV-2 infected-cells at 0.5 and 1000 MOI are reported. At 48 h post-infection, typical cytopathic effects, including cell rounding, detachment, degeneration, and syncytium formation were seen only in cells infected at 1000 MOI. Cells were imaged by optical microscope observation (ZOE™ Fluorescent Cell Imager, Bio-Rad, Hercules, CA, USA). (**C**) ERAP2/Iso3, Iso1, and ERAP1 mRNA expression by in vitro SARS-CoV-2 infected Calu3 cells increased according to the rising viral input. Mean values ± ES are reported. * = $p < 0.05$; ** = $p < 0.01$; *** = $p < 0.001$. MOI = multiplicity of infection.

In vitro HIV-1 infection of PBMCs isolated from HapB HC produced similar results. Indeed, as the viral input raised, viral replication quantified through p24 concentration analyses at 5 days post infection increased (0.1 vs. 1 ng p24 HIV-1_{Bal}/1 × 10^6 PBMCs: $p < 0.02$; 0.1 vs. 10 ng p24 HIV-1_{Bal}/1 × 10^6 PBMCs: $p < 0.001$) (Figure 5A). Likewise, ERAP2/Iso3 gene expression increased compared to the uninfected condition (0.1 ng p24 HIV-1_{Bal}/1 × 10^6 PBMCs: $p < 0.003$; 0.1 ng p24 HIV-1_{Bal}/1 × 10^6 PBMCs: $p < 0.04$; 10 ng p24 HIV-1_{Bal}/1 × 10^6 PBMCs: $p < 0.03$) (Figure 5B). ERAP2/Iso1 (0.1 ng p24 HIV-1_{Bal}/1 × 10^6 PBMCs: $p < 0.05$) and ERAP1 (0.1 ng p24 HIV-1_{Bal}/1 × 10^6 PBMCs: $p < 0.05$) mRNA levels showed a similar trend (Figure 5B).

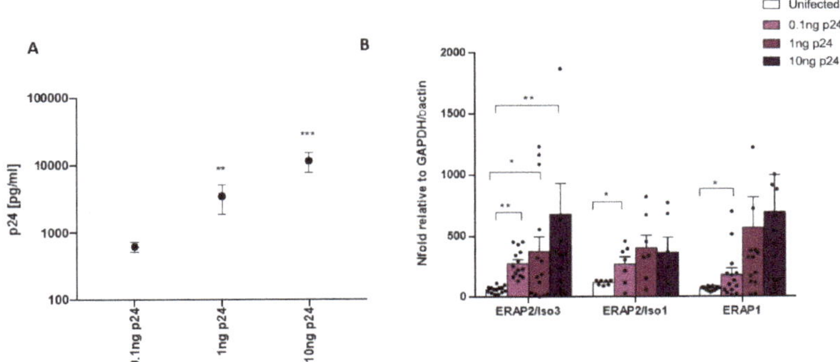

Figure 5. In vitro HIV-1 infection assay on PBMCs. (**A**) HIV-1 replication was assessed by p24 quantification on PBMCs isolated from 8 HeteroAB and 6 HomoB individuals infected with 0.1, 1, and 10 ng p24 HIV-1Bal/1 $\times 10^6$ PBMCs 5 days post infection. Results showed a significant increase in viral replication according to the increased viral input. (**B**) ERAP2/Iso3, Iso1, and ERAP1 mRNA expression by in vitro HIV-1-infected PBMCs increased according to the rising viral input. Mean values ± ES are reported. * = $p < 0.05$; ** = $p < 0.01$; *** = $p < 0.001$.

3.5. ERAP2/Iso3 Protein Production by MDMs from HeteroAB Subjects Following Microbial Specific Stimulation

To verify if the short ERAP2/Iso3 isoform would function as an RNA or is translated into a protein product we performed a western blot assay on MDMs from 3 HeteroAB subjects triggered with different microbial antigens. Remarkably, the antibody which recognizes the full-length ERAP2 (Iso1) was able to detect also one short protein isoform (~50 kDa) in CMV, flu, HIV-AT-2, i-SARS-CoV-2, LPS, IFNα stimulated cells from HeteroAB subjects, suggesting the translation of the short microbial-specific ERAP2/Iso3 (Figure 6). Conversely, following IL-1β stimulation, only ERAP2/Iso1 isoform was detected. The production of ERAP1 protein was observed in all the stimulated conditions except IL-1β (Figure 6). As the proteins extracted from the 3 HeteroAB subjects were sub-pooled for WB analyses, statistical evaluation of the results was not possible.

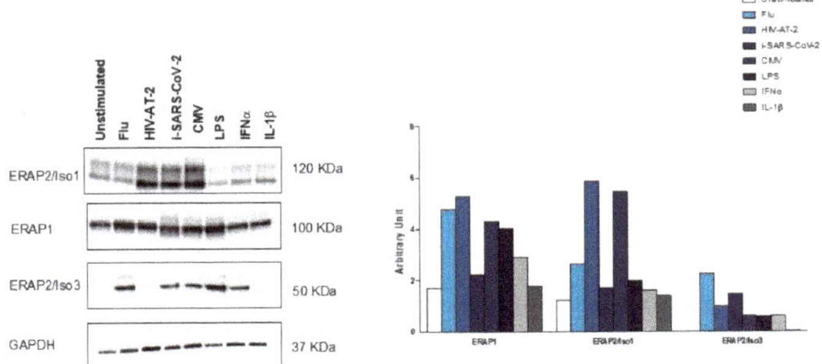

Figure 6. ERAP1, ERAP2/Iso1, and ERAP2/Iso3 production in pathogen stimulated monocyte-derived macrophages (MDMs). MDMs differentiated from 3 HeteroAB participants stimulated for 36 h with microbial antigens (flu, CMV) inactivated viruses (i-SARS-CoV-2, HIV-AT-2) bacterial by-products (LPS) or inflammatory stimuli (IFNα, IL-1β) were tested for protein using primary antibodies specific to a ERAP1 (goat polyclonal), ERAP2 and β-actin. Proteins extracted from the 3 HeteroAB subjects were sub-pooled for WB analyses. Histograms representing ERAP1, ERAP2/Iso1, and ERAP2/Iso3 densitometric quantification. Quantification was performed by Quantity One 4.6.6 software (Bio-Rad) and normalization was permed on GAPDH.

4. Discussion

Given the documented role of ERAP2 in antigen presentation [20] and viral infections [21], we examined the genetic control of ERAP2 transcripts in the human antimicrobial response. In particular, based on the results recently reported by Ye and co-workers [19], we investigated if the expression of the recently characterized ERAP2/Iso3 is flu-specific or if it can be triggered by other microbial stimuli. Our results suggest that: (1) ERAP2/Iso3 mRNA expression is not restricted to flu-infection but it can be prompted by other pathogens including HIV, SARS-CoV-2, CMV, Bacteria (LPS); (2) ERAP2/Iso3 mRNA is translated into a protein following microbial induction; (3) ERAP2/Iso3 mRNA expression is sensitive to viral concentration.

Remarkably, Ye et al. did not detect ERAP2/Iso3 expression in IFNB1 stimulated cells leading to the conclusion that the transcription of novel ERAP2 isoforms is likely initiated by viral sensing pathways upstream of type 1 interferon signaling [19]. Conversely, in our cell culture condition, IFNα-stimulation was able to induce ERAP2/Iso3 expression in a consistent way, suggesting the participation of type 1 interferon cascade to the induction of ERAP2/Iso3 transcription, as already documented for the wild type forms of ERAP1 and 2 [26]. The different cell kinds and IFN subtypes adopted in the two experimental settings could at least partially justify the discrepancies reported in the two studies, but further analyses are needed to clarify the pathways and molecules induced by microbial exposure, directly responsible for ERAP2/Iso3 synthesis. For example, in our study IL-1β stimulus was not able to trigger the expression of any ERAP variants, further strengthening the assumption of a specific response of ERAP gene transcription controlled by pathogen exposure. Furthermore, as IFNα is also strongly associated to various type of I IFN conditions such as Sjogren's disease [27], systemic lupus erythematosus [28] and Scleroderma [29] it would be valuable to verify if ERAP2/Iso3 expression varies in patients affected by these pathologies and/or following administration of IFNα therapy.

The discovery of these new ERAP2 isoforms is of great importance and gives a plausible explanation to the maintenance of one of the major gene expression quantitative trait loci (eQTL) and alternate isoform usage in most tissues and cell types [19,30]. Indeed, until last year, the preservation of HapB at intermediate frequency in human population was almost unexplained, as its transcript was believed to be addressed to NMD [9,10]. The identification of this previously uncharacterized short isoform, ERAP2/Iso3, transcribed from HapB, results in the partial rescue of ERAP2 expression, suggesting its involvement in the anti-microbial response.

The mechanism of action and functional impact of this new genetic variant to host defense; however, is still indefinite. The lack of the aminopeptidase domain proves that it does not directly participate in the shaping of antigenic peptides to be presented to CD8+ T cells. However, in 2005 Saveanu et al. conducted an immunoblot analysis detecting the presence of ERAP1-ERAP2 heterodimers and possibly homodimers of each enzyme [4]. This crystallographic dimer is described as mediated by domains I and II of the enzyme that is missing from ERAP2/Iso3 [31]. However, this observation has been reiterated by Ye et al. who hypothesized that ERAP2 isoforms could have a dominant-negative effect on either ERAP1 and possibly ERAP2 wt, through the formation of hetero- or homodimeric complexes [15]. Despite still speculative the fact that ERAP2/Iso3 may exert such an effect cannot be ruled out: this, in turn, could lead to an altered peptide processing, which could confer an advantage/disadvantage against infections by presenting a more/less immunogenic antigen repertoire. In relation to this aspect, previous studies have demonstrated that the functional skills of ERAP monomers, homo, or heterodimers may significantly differ in terms of both substrate specificity and trimming efficiency [32]. In particular, ERAP1/2 dimerization creates complexes with superior peptide-trimming efficacy and a higher affinity towards ERAP1 preferential substrates. This is allowed by the adoption of a modified physical conformation by ERAP1, caused by its interaction with ERAP2, which mainly works as an enhancer of ERAP1 role upon dimer assembling [33]. Further studies are needed to verify if also ERAP2/Iso3 physical interaction with the wt ERAP variants prompts an allosteric effect able to modify basic enzymatic parameters and to improve their substrate-binding affinity. However, the observation that ERAP2/Iso3 mRNA is translated into a protein allows to speculate on the generation of a new ERAP member which further contributes to enrich the non-redundant, yet a complete and potent system of aminopeptidases, warranting an efficient trimming of various kinds of precursors.

The results obtained in this study definitely establish a link between invading pathogens and ERAP2/Iso3 expression which further strengthens the significance of the results reported by Ye and collaborators. Supporting this hypothesis, we observed that ERAP2/Iso3 expression progressively increased in response to growing doses of viral input in both SARS-CoV-2 and HIV-1 in vitro infection assay, as if the production of this genetic variant, as well as one of the other elements within the ERAP family, were directly dependent on the viral dose of exposure. Additionally, the observation that the increased expression of ERAP2/Iso3 in response to pathogen exposure is accompanied by the modulation of many other determinants (chemokines, cytokines, pathogen recognition receptor, inflammasome, interferon-stimulated genes, adhesion molecules, activation/inhibition markers, antigen presentation elements) orchestrating the anti-microbial immune response, further supports its direct intervention in this defensive pathway. Notably, as its expression is increased by a wide range of microbial stimuli including viral antigens, inactivated virus as well as bacterial by-products, it is possible to assume that ERAP2/Iso3 expression is not pathogen-specific, but it's secondary to the activation of an antimicrobial cascade commonly shared by different pathogens. Meanwhile, we cannot exclude that ERAPs responses observed following microbial stimulations and in vitro viral infections result from an erroneous transcription and translation due to pathogen-induced cellular stress. Indeed, as ERAP1 and ERAP2 expression has been demonstrated to be prompted by IFNγ stimulation, ERAPs production could be secondary to the innate immune response of the cells to infections, rather than to an immune response or an immune evasion mechanism. Further analyses will be necessary to verify this hypothesis.

The involvement of ERAPs in modulating viral infections is widely recognized as recently reviewed in [34]. Several studies, indeed, have demonstrated the intervention of ERAP genetic variants in the life cycle of HCV, flu, CMV, HPV, HIV, and other pathogens at different levels. In particular, studies performed by our research group have established an association between ERAP2/Iso1 and HIV-infection in terms of both susceptibility [17,35] and progression [36]. However, to our knowledge ERAP expression and/or genetic variants have been correlated to the recent coronavirus disease 2019 (COVID-19), provoked by SARS-CoV-2, only by two recent studies [34,37,38]. In the first one by Stamatakis et al. ERAP2 trimming ability has been investigated in SARS-CoV-2 infection together with ERAP1 and IRAP, and it has been proved as the most stable of the enzymes generating optimal length antigenic peptides for HLA binding [38]. In the second one by Lu et al. by examining 193 deaths from 1,412 confirmed infections in a group of 5,871 UK Biobank participants tested for the virus, rs150892504 variant in ERAP2 gene came up as potentially being implicated in risk from SARS-CoV-2 infection. Although rs150892504 variant is not in linkage disequilibrium with rs2248374, this finding suggests the involvement of ERAP2 in the modulation of SARS-CoV-2 infection. Such an assumption is further supported by other intriguing observations. This virus enters the cells by spike protein binding to ACE2 (angiotensin-converting enzyme 2), which is responsible for the conversion of angiotensin I to angiotensin I-9 and of angiotensin II to angiotensin I-7, an effective vasodilator, thus working as a negative regulator of the renin-angiotensin system (RAS) Our results, for the first time demonstrate that SARS-CoV-2 exposure triggers the expression of ERAP1, ERAP2/Iso1 and also the recently detected ERAP2/Iso3 in a dose-dependent mode, suggesting its participation in the control of the anti-SARS-CoV-2 response. This observation is far more important, considering that besides their involvement in the antigen presentation pathway, ERAPs display several key anti-SARS-CoV-2 functions. Indeed, they intervene in the RAS, where ERAP1 efficiently cleaves angiotensin II to angiotensin III and IV, and ERAP2 cuts angiotensin III to angiotensin IV, thus influencing both ACE2 virus receptor bio-availability and blood pressure levels [39]. Furthermore, ERAPs modulate the proteolytic cleavage of IL-6 receptor (IL-6Rα) [40] a function which can improve the clinical conditions in COVID-19 patients, as recently documented following its pharmacological inhibition by Tocilizumab [41]. Last but not least, Ranjit and colleagues recently demonstrated that sex-specific differences in ERAP1 modulation influence blood pressure and RAS responses [42], and a male bias in mortality has emerged in the COVID-19 pandemic since the very beginning. As ERAPs genetic variants have been demonstrated to orchestrate and condition the result of infection of coronavirus in other animal species [43,44] detailed studies investigating SARS-CoV-2-host interplay is absolutely mandatory.

Another query which needs to be addressed in the near future concerns the cellular localization of ERAP2/Iso3 isoform. In particular, it would be interesting to verify if, as with the wt ERAP variants [17], even ERAP2/Iso3 may be secreted into the extracellular milieu following inflammatory stimulation, with which substrates it may interact, which functions may eventually exert in this environment and if its administration can interfere with viral replication.

Considering the key role played by ERAPs in antigen processing and presentation, it is plausible that these aminopeptidases may be potential targets and controllers of the pathogenicity of infectious diseases, shaping the susceptibility and response to microbial infections. The recent acquisition of ERAP intervention even in the modulation of innate immunity further reinforces this assumption.

Given the growing number of viral epidemics, the identification of molecular mechanisms driven by factors such as ERAPs that can interfere, control or modulate viral replication is unequivocally needed as they could be widely exploited for the inception of future, still unknown viral infections.

Author Contributions: Each author has approved the submitted version and agrees to be personally accountable for the author's own contributions and for ensuring that questions related to the accuracy or integrity of any part of the work. Conceptualization, M.B. and I.S.; Methodology, C.V. (Claudia Vanetti), S.G. and C.V. (Chiara Vicentini); Formal Analysis, I.S., M.G.; Statistical Analyses, E.T.; Investigation, S.G. and C.V. (Chiara Vicentini); Data Curation, C.V. (Claudia Vanetti), M.G.; Writing—Original Draft Preparation, I.S.; Writing—Review & Editing, M.B.; Supervision, M.B. and M.C.; Funding Acquisition, M.C. All authors have read and agreed to the published version of the manuscript.

Funding: This research was partially funded by Falk Renewables, (REC18GZUCC N. 27767).

Conflicts of Interest: The authors declare no conflict of interest.

References

1. Neefjes, J.; Jongsma, M.L.M.; Paul, P.; Bakke, O. Towards a systems understanding of MHC class I and MHC class II antigen presentation. *Nat. Rev. Immunol.* **2011**, *11*, 823–836. [CrossRef] [PubMed]
2. Cifaldi, L.; Romania, P.; Lorenzi, S.; Locatelli, F.; Fruci, D. Role of Endoplasmic Reticulum Aminopeptidases in Health and Disease: From Infection to Cancer. *Int. J. Mol. Sci.* **2012**, *13*, 8338–8352. [CrossRef] [PubMed]
3. Compagnone, M.; Fruci, D. Peptide Trimming for MHC Class I Presentation by Endoplasmic Reticulum Aminopeptidases. *Methods Mol. Biol. Clifton NJ* **2019**, *1988*, 45–57. [CrossRef]
4. Saveanu, L.; Carroll, O.; Lindo, V.; Del Val, M.; Lopez, D.; Lepelletier, Y.; Greer, F.; Schomburg, L.; Fruci, D.; Niedermann, G.; et al. Concerted peptide trimming by human ERAP1 and ERAP2 aminopeptidase complexes in the endoplasmic reticulum. *Nat. Immunol.* **2005**, *6*, 689–697. [CrossRef]
5. Evnouchidou, I.; van Endert, P. Peptide trimming by endoplasmic reticulum aminopeptidases: Role of MHC class I binding and ERAP dimerization. *Hum. Immunol.* **2019**, *80*, 290–295. [CrossRef]
6. Yao, Y.; Liu, N.; Zhou, Z.; Shi, L. Influence of ERAP1 and ERAP2 gene polymorphisms on disease susceptibility in different populations. *Hum. Immunol.* **2019**, *80*, 325–334. [CrossRef]
7. Stamogiannos, A.; Koumantou, D.; Papakyriakou, A.; Stratikos, E. Effects of polymorphic variation on the mechanism of Endoplasmic Reticulum Aminopeptidase 1. *Mol. Immunol.* **2015**, *67*, 426–435. [CrossRef]
8. López de Castro, J.A. How ERAP1 and ERAP2 Shape the Peptidomes of Disease-Associated MHC-I Proteins. *Front. Immunol.* **2018**, *9*. [CrossRef]
9. Cagliani, R.; Riva, S.; Biasin, M.; Fumagalli, M.; Pozzoli, U.; Lo Caputo, S.; Mazzotta, F.; Piacentini, L.; Bresolin, N.; Clerici, M.; et al. Genetic diversity at endoplasmic reticulum aminopeptidases is maintained by balancing selection and is associated with natural resistance to HIV-1 infection. *Hum. Mol. Genet.* **2010**, *19*, 4705–4714. [CrossRef]
10. Andrés, A.M.; Dennis, M.Y.; Kretzschmar, W.W.; Cannons, J.L.; Lee-Lin, S.-Q.; Hurle, B.; Schwartzberg, P.L.; Williamson, S.H.; Bustamante, C.D.; Nielsen, R.; et al. Balancing Selection Maintains a Form of ERAP2 that Undergoes Nonsense-Mediated Decay and Affects Antigen Presentation. *PLoS Genet.* **2010**, *6*. [CrossRef]

11. Jostins, L.; Ripke, S.; Weersma, R.K.; Duerr, R.H.; McGovern, D.P.; Hui, K.Y.; Lee, J.C.; Philip Schumm, L.; Sharma, Y.; Anderson, C.A.; et al. Host–microbe interactions have shaped the genetic architecture of inflammatory bowel disease. *Nature* **2012**, *491*, 119–124. [CrossRef] [PubMed]
12. Kuiper, J.J.W.; Van Setten, J.; Ripke, S.; Van 'T Slot, R.; Mulder, F.; Missotten, T.; Baarsma, G.S.; Francioli, L.C.; Pulit, S.L.; De Kovel, C.G.F.; et al. A genome-wide association study identifies a functional ERAP2 haplotype associated with birdshot chorioretinopathy. *Hum. Mol. Genet.* **2014**, *23*, 6081–6087. [CrossRef] [PubMed]
13. Wiśniewski, A.; Kasprzyk, S.; Majorczyk, E.; Nowak, I.; Wilczyńska, K.; Chlebicki, A.; Zoń-Giebel, A.; Kuśnierczyk, P. ERAP1-ERAP2 haplotypes are associated with ankylosing spondylitis in Polish patients. *Hum. Immunol.* **2019**, *80*, 339–343. [CrossRef] [PubMed]
14. Robinson, P.C.; Costello, M.E.; Leo, P.; Bradbury, L.A.; Hollis, K.; Cortes, A.; Lee, S.; Joo, K.B.; Shim, S.-C.; Weisman, M.; et al. ERAP2 is associated with ankylosing spondylitis in HLA-B27-positive and HLA-B27-negative patients. *Ann. Rheum. Dis.* **2015**, *74*, 1627–1629. [CrossRef] [PubMed]
15. Chiaroni-Clarke, R.C.; Munro, J.E.; Chavez, R.A.; Pezic, A.; Allen, R.C.; Akikusa, J.D.; Piper, S.E.; Saffery, R.; Ponsonby, A.-L.; Ellis, J.A. Independent confirmation of juvenile idiopathic arthritis genetic risk loci previously identified by immunochip array analysis. *Pediatr. Rheumatol.* **2014**, *12*, 53. [CrossRef]
16. Biasin, M.; Sironi, M.; Saulle, I.; de Luca, M.; la Rosa, F.; Cagliani, R.; Forni, D.; Agliardi, C.; lo Caputo, S.; Mazzotta, F.; et al. Endoplasmic reticulum aminopeptidase 2 haplotypes play a role in modulating susceptibility to HIV infection. *AIDS Lond. Engl.* **2013**, *27*, 1697–1706. [CrossRef]
17. Saulle, I.; Ibba, S.V.; Torretta, E.; Vittori, C.; Fenizia, C.; Piancone, F.; Minisci, D.; Lori, E.M.; Trabattoni, D.; Gelfi, C.; et al. Endoplasmic Reticulum Associated Aminopeptidase 2 (ERAP2) Is Released in the Secretome of Activated MDMs and Reduces in vitro HIV-1 Infection. *Front. Immunol.* **2019**, *10*. [CrossRef]
18. Evnouchidou, I.; Birtley, J.; Seregin, S.; Papakyriakou, A.; Zervoudi, E.; Samiotaki, M.; Panayotou, G.; Giastas, P.; Petrakis, O.; Georgiadis, D.; et al. A Common Single Nucleotide Polymorphism in Endoplasmic Reticulum Aminopeptidase 2 Induces a Specificity Switch That Leads to Altered Antigen Processing. *J. Immunol.* **2012**. [CrossRef]
19. Ye, C.J.; Chen, J.; Villani, A.-C.; Gate, R.E.; Subramaniam, M.; Bhangale, T.; Lee, M.N.; Raj, T.; Raychowdhury, R.; Li, W.; et al. Genetic analysis of isoform usage in the human anti-viral response reveals influenza-specific regulation of ERAP2 transcripts under balancing selection. *Genome Res.* **2018**, *28*, 1812–1825. [CrossRef]
20. Saveanu, L.; Carroll, O.; Hassainya, Y.; Endert, P.V. Complexity, contradictions, and conundrums: Studying post-proteasomal proteolysis in HLA class I antigen presentation. *Immunol. Rev.* **2005**, *207*, 42–59. [CrossRef]
21. Rossio, J.L.; Esser, M.T.; Suryanarayana, K.; Schneider, D.K.; Bess, J.W.; Vasquez, G.M.; Wiltrout, T.A.; Chertova, E.; Grimes, M.K.; Sattentau, Q.; et al. Inactivation of Human Immunodeficiency Virus Type 1 Infectivity with Preservation of Conformational and Functional Integrity of Virion Surface Proteins. *J. Virol.* **1998**, *72*, 7992–8001. [CrossRef] [PubMed]
22. Hui, K.P.Y.; Cheung, M.-C.; Perera, R.A.P.M.; Ng, K.-C.; Bui, C.H.T.; Ho, J.C.W.; Ng, M.M.T.; Kuok, D.I.T.; Shih, K.C.; Tsao, S.-W.; et al. Tropism, replication competence, and innate immune responses of the coronavirus SARS-CoV-2 in human respiratory tract and conjunctiva: An analysis in ex-vivo and in-vitro cultures. *Lancet Respir. Med.* **2020**, *8*, 687–695. [CrossRef]
23. Batéjat, C.; Grassin, Q.; Manuguerra, J.-C.; Leclercq, I. Heat inactivation of the Severe Acute Respiratory Syndrome Coronavirus 2. *bioRxiv* **2020**. [CrossRef]
24. Merlini, E.; Tincati, C.; Biasin, M.; Saulle, I.; Cazzaniga, F.A.; d'Arminio Monforte, A.; Cappione, A.J.I.; Snyder-Cappione, J.; Clerici, M.; Marchetti, G.C. Stimulation of PBMC and Monocyte-Derived Macrophages via Toll-Like Receptor Activates Innate Immune Pathways in HIV-Infected Patients on Virally Suppressive Combination Antiretroviral Therapy. *Front. Immunol.* **2016**, *7*. [CrossRef] [PubMed]
25. Saulle, I.; Ibba, S.V.; Vittori, C.; Fenizia, C.; Mercurio, V.; Vichi, F.; Caputo, S.L.; Trabattoni, D.; Clerici, M.; Biasin, M. Sterol metabolism modulates susceptibility to HIV-1 Infection. *AIDS Lond. Engl.* **2020**. [CrossRef]
26. Wu, T.G.; Rose, W.A.; Albrecht, T.B.; Knutson, E.P.; König, R.; Perdigão, J.R.; Nguyen, A.P.A.; Fleischmann, W.R. IFN-alpha-induced murine B16 melanoma cancer vaccine cells: Induction and accumulation of cell-associated IL-15. *J. Interf. Cytokine Res. Off. J. Int. Soc. Interf. Cytokine Res.* **2007**, *27*, 13–22. [CrossRef]
27. Nordmark, G.; Ronnblom, M.-L.E.; Primary, L. Sjogren's Syndrome and the Type I Interferon System. Available online: https://www.eurekaselect.com/101237/article (accessed on 15 August 2020).

28. Kirou, K.A.; Gkrouzman, E. Anti-interferon alpha treatment in SLE. *Clin. Immunol.* **2013**, *148*, 303–312. [CrossRef]
29. Raschi, E.; Chighizola, C.B.; Cesana, L.; Privitera, D.; Ingegnoli, F.; Mastaglio, C.; Meroni, P.L.; Borghi, M.O. Immune complexes containing scleroderma-specific autoantibodies induce a profibrotic and proinflammatory phenotype in skin fibroblasts. *Arthritis Res. Ther.* **2018**, *20*, 187. [CrossRef]
30. Lappalainen, T.; Sammeth, M.; Friedländer, M.R.; Ac't Hoen, P.; Monlong, J.; Rivas, M.A.; Gonzàlez-Porta, M.; Kurbatova, N.; Griebel, T.; Ferreira, P.G.; et al. Transcriptome and genome sequencing uncovers functional variation in humans. *Nature* **2013**, *501*, 506–511. [CrossRef]
31. The Crystal Structure of Human Endoplasmic Reticulum Aminopeptidase 2 Reveals the Atomic Basis for Distinct Roles in Antigen Processing | Biochemistry. Available online: https://pubs.acs.org/doi/10.1021/bi201230p (accessed on 17 August 2020).
32. de Castro, J.A.L.; Stratikos, E. Intracellular antigen processing by ERAP2: Molecular mechanism and roles in health and disease. *Hum. Immunol.* **2019**, *80*, 310–317. [CrossRef]
33. Evnouchidou, I.; Weimershaus, M.; Saveanu, L.; Endert, P. van ERAP1–ERAP2 Dimerization Increases Peptide-Trimming Efficiency. *J. Immunol.* **2014**. [CrossRef] [PubMed]
34. Saulle, I.; Vicentini, C.; Clerici, M.; Biasin, M. An Overview on ERAP Roles in Infectious Diseases. *Cells* **2020**, *9*, 720. [CrossRef] [PubMed]
35. Forni, D.; Cagliani, R.; Tresoldi, C.; Pozzoli, U.; Gioia, L.D.; Filippi, G.; Riva, S.; Menozzi, G.; Colleoni, M.; Biasin, M.; et al. An Evolutionary Analysis of Antigen Processing and Presentation across Different Timescales Reveals Pervasive Selection. *PLoS Genet.* **2014**, *10*, e1004189. [CrossRef] [PubMed]
36. Lori, E.M.; Cozzi-Lepri, A.; Tavelli, A.; Mercurio, V.; Ibba, S.V.; Lo Caputo, S.; Castelli, F.; Castagna, A.; Gori, A.; Marchetti, G.; et al. Evaluation of the effect of protective genetic variants on cART success in HIV-1-infected patients. *J. Biol. Regul. Homeost. Agents* **2020**, *34*. [CrossRef]
37. Genetic Risk Factors for Death with SARS-CoV-2 from the UK Biobank | medRxiv. Available online: https://www.medrxiv.org/content/10.1101/2020.07.01.20144592v1 (accessed on 15 August 2020).
38. Stamatakis, G.; Samiotaki, M.; Mpakali, A.; Panayotou, G.; Stratikos, E. Generation of SARS-CoV-2 S1 spike glycoprotein putative antigenic epitopes in vitro by intracellular aminopeptidases. *bioRxiv* **2020**. [CrossRef]
39. Hisatsune, C.; Ebisui, E.; Usui, M.; Ogawa, N.; Suzuki, A.; Mataga, N.; Takahashi-Iwanaga, H.; Mikoshiba, K. ERp44 Exerts Redox-Dependent Control of Blood Pressure at the ER. *Mol. Cell* **2015**, *58*, 1015–1027. [CrossRef]
40. Cui, X.; Rouhani, F.N.; Hawari, F.; Levine, S.J. An Aminopeptidase, ARTS-1, Is Required for Interleukin-6 Receptor Shedding. *J. Biol. Chem.* **2003**, *278*, 28677–28685. [CrossRef]
41. Michot, J.-M.; Albiges, L.; Chaput, N.; Saada, V.; Pommeret, F.; Griscelli, F.; Balleyguier, C.; Besse, B.; Marabelle, A.; Netzer, F.; et al. Tocilizumab, an anti-IL-6 receptor antibody, to treat COVID-19-related respiratory failure: A case report. *Ann. Oncol. Off. J. Eur. Soc. Med. Oncol.* **2020**, *31*, 961–964. [CrossRef]
42. Ranjit, S.; Wong, J.Y.; Tan, J.W.; Sin Tay, C.; Lee, J.M.; Yin Han Wong, K.; Pojoga, L.H.; Brooks, D.L.; Garza, A.E.; Maris, S.A.; et al. Sex-specific differences in endoplasmic reticulum aminopeptidase 1 modulation influence blood pressure and renin-angiotensin system responses. *JCI Insight* **2019**, *4*. [CrossRef]
43. Golovko, L.; Lyons, L.A.; Liu, H.; Sørensen, A.; Wehnert, S.; Pedersen, N.C. Genetic susceptibility to feline infectious peritonitis in Birman cats. *Virus Res.* **2013**, *175*, 58–63. [CrossRef]
44. Cong, F.; Liu, X.; Han, Z.; Shao, Y.; Kong, X.; Liu, S. Transcriptome analysis of chicken kidney tissues following coronavirus avian infectious bronchitis virus infection. *BMC Genom.* **2013**, *14*, 743. [CrossRef] [PubMed]

Review

Infection of Mammals and Mosquitoes by Alphaviruses: Involvement of Cell Death

Lucie Cappuccio [1,2] and Carine Maisse [1,*]

1 IVPC UMR754 INRA, Univ Lyon, Université Claude Bernard Lyon 1, EPHE, 69007 Lyon, France; lucie.cappuccio@univ-lyon1.fr
2 Interspecies Transmission of Arboviruses and Therapeutics Research Unit, Institut Pasteur of Shanghai, Chinese Academy of Sciences, Shanghai 200031, China
* Correspondence: carine.paradisi@univ-lyon1.fr

Received: 5 October 2020; Accepted: 2 December 2020; Published: 5 December 2020

Abstract: Alphaviruses, such as the chikungunya virus, are emerging and re-emerging viruses that pose a global public health threat. They are transmitted by blood-feeding arthropods, mainly mosquitoes, to humans and animals. Although alphaviruses cause debilitating diseases in mammalian hosts, it appears that they have no pathological effect on the mosquito vector. Alphavirus/host interactions are increasingly studied at cellular and molecular levels. While it seems clear that apoptosis plays a key role in some human pathologies, the role of cell death in determining the outcome of infections in mosquitoes remains to be fully understood. Here, we review the current knowledge on alphavirus-induced regulated cell death in hosts and vectors and the possible role they play in determining tolerance or resistance of mosquitoes.

Keywords: alphaviruses; apoptosis; cell death; mosquito; tolerance

---

1. Alphaviruses

Viruses belonging to the *Alphavirus* genus can be found in an ecological but not taxonomic group called arboviruses (an acronym for "arthropod-borne viruses" [1]). These viruses are transmitted by a hematophagous arthropod to a vertebrate host during a blood meal; in the case of alphaviruses, the predominant vectors are mosquitoes [2].

Alphaviruses are small, enveloped viruses of approximately 70 nm of diameter. The positive-sense, single-strand RNA contains two open reading frames (ORFs) that encode four non-structural proteins (nsp1–4) and five structural proteins (capsid, E3, E2, 6K, and E1) [3]. Alphaviruses include approximately 30 members, and infection results in clinical symptoms range from mild to severe [3]. Historically, alphaviruses were divided into New World and Old World alphaviruses, following their global distribution, evolution, pathogenicity, tissue, and cellular tropism or interactions with respective hosts. Old World alphaviruses (chikungunya virus (CHIKV); Sindbis virus (SINV); Semliki Forest virus (SFV); Ross River virus (RRV), etc.) were mainly found in Asia, Africa, and Europe, while New World alphaviruses (Eastern, Western, Venezuelan, and Equine Encephalitis Viruses (EEVs); Mayaro virus (MAYV)) were found in North and South America. However, with the global spreading of these viruses and their vectors, this division between New and Old World has become obsolete. Alphaviruses are now divided into three categories: aquatic viruses, arthritogenic viruses, and encephalitic viruses [4,5]. Infections by arthritogenic viruses in humans are characterized by rashes, fever, joint and muscle pain, and encephalitis for some of them (e.g., SINV, SFV). In some cases, incapacitating arthralgia and myalgia can last for months or years after infection (e.g., CHIKV, RRV, MAYV). Encephalitic virus infections are characterized by debilitating febrile disease and encephalomyelitis, leading to death in some cases (e.g., EEEV, VEEV) [3,6].

In mammals, skin cells are the first cells targeted by an arbovirus, such as an alphavirus, when inoculated by an infected mosquito. They are not clearly defined for each alphavirus but may be dermal fibroblasts [7], dermal dendritic cells, enterocytes or keratinocytes [8,9]—they constitute the first line of defense. Viruses will then reach other organs, such as joints, muscles [10] or the brain, where they will trigger pathology through induced cell death in the acute phase or long-lasting inflammation during the chronic phase [6].

In mosquito vectors, the arboviral infection is persistent and lasts the insect's whole life. In comparison with their effect in the vertebrate host, alphaviruses do not seem to cause significant pathology in the mosquito vector. Even if some fitness costs have been described for some arboviruses [11–13], many other studies have concluded that arboviral infection is mainly silent and that mosquito vectors are tolerant to arboviruses [14,15]. From the oral acquisition of a viremic bloodmeal to the transmission to a new uninfected vertebrate host, alphaviruses replicate in arthropod cells and must cope with antiviral pathways. In mosquito, the first limit threshold to cross is the gut epithelium (i.e., midgut barrier), where the virus replicates to join the hemocoel (blood-containing body cavity) thus allowing viral spread to the whole body. To be transmitted again through blood feeding, the virus must penetrate the basal lamina of the salivary glands (salivary gland barrier) to join the acinar cells and replicate inside [16].

Interestingly, among the approximately 112 mosquito genera, the *Aedes* and *Culex* genera (such as *Aedes Albopictus*, *Aedes Aegypti*, *Culex Quinquefasciatus* or *Culex Pipiens*) seem to be the main vectors able to transmit viruses to humans [17]. Indeed, some mosquitoes may bite preferentially animals or may not be viral-transmission competent. As will be described below, part of the competence is linked to the different tissue barriers that can be crossed or not during the viral dissemination in the mosquito; this depends on innate immune response and cell death regulation in the infected cells [15,17,18].

2. Cell Death in Mammals

Cell death pathways can be divided into two opposite processes: accidental cell death (ACD) and regulated cell death (RCD). If ACD is a consequence of a severe and rapid injury (osmotic forces, pH variations, lytic viral replication), RCD is based on dedicated molecular machinery, implying that it can be modulated by pharmacological, genetic or infectious interventions [19].

Regulated cell death occurs under two different circumstances. The first one is programmed cell death (PCD) [20], which occurs during embryonic development or in the event of tissue homeostasis. The second one regroups different RCD pathways that occur following an external or internal, prolonged, and intense stress event. This contributes to tissue homeostasis and protection by eliminating useless or potentially dangerous cells (i.e., malignant or infected cells).

Dying cells present different and well described morphological features that have been used so far to classify cell death processes into three main types [19]: apoptosis, autophagy-dependent cell death, and necrosis. Apoptosis is characterized by chromatin condensation, nuclear fragmentation, cytoplasmic shrinkage, membrane blebbing, and the formation of "apoptotic bodies", which are subsequently destroyed by professional or non-surrounding phagocytes. Autophagy-dependent cell death is essentially defined by its distinctive features of extensive cytoplasm vacuolization, leading to phagocytosis and degradation by lysosomes. Necrosis is mainly characterized by swelling, plasma membrane disruption, and cytoplasmic content efflux in the extracellular environment, without evident phagocytosis or lysosomal degradation by the neighboring cells.

Intuitively, necrosis is associated to ACD, but it is nowadays clear that some RCD can also lead to non-apoptotic cell death as recently described in necroptosis [21], pyroptosis [22], and ferroptosis [23].

We will focus here on the cell death pathways that have been shown to be involved in antiviral response so far, without considering subtypes such as attachment dependence (i.e., anoïkis) and entotic cell death, parthanos, etc. For complete reviews see References [19,24].

2.1. Intrinsic Apoptosis

"Intrinsic apoptosis is a form of RCD initiated by perturbations of the intracellular or extracellular microenvironment, demarcated by mitochondrial outer membrane permeabilization (MOMP) and precipitated by executioner caspases, mainly caspase 3 (CASP3)" (Nomenclature Committee on Cell Death (NCCD) [19]). It can be induced by numerous dysregulations including DNA damage, endoplasmic reticulum (ER) stress, reactive oxygen species (ROS) overload or infection. The main characteristic is that cells still present plasma membrane integrity and metabolic activity, leading, in vivo, to the removal of apoptotic bodies by surrounding phagocytic cells that recognize phosphatidylserine (PS) at the cell surface. In vitro, unless the cultured cells present phagocytic capacities, apoptosis usually ends by a "secondary necrosis", exposing damaged plasma membrane [25].

The decisive step of intrinsic apoptosis is the irreversible and extensive MOMP, leading to the release of numerous pro-apoptotic factors contained in the intermembrane space [26]. Mitochondrial outer membrane permeabilization is controlled by a family of 20 pro- or anti-apoptotic proteins: the B cells lymphoma 2 (Bcl-2) family proteins, which share one to four Bcl-2 homology domains (BH1 to BH4) [27]. All of them are finely regulated at the transcriptional and/or post-translational level (degradation, phosphorylation, localization, oligomerization, etc.) in order to integrate the extracellular or intracellular signals, which will potentially lead to apoptosis.

In mammals, only three of them (Bax, Bak, and Bok) have been described as able to form pores in the mitochondrial outer membrane (MOM) and other cellular membranes after oligomerization. These proteins are thus considered as "effectors" that can be activated, transcriptionally or post-translationally, after a cellular stress to induce MOMP [28,29]. Moreover, a pool of BH3-only proteins, described as "activators" promotes MOMP induction by interacting with Bax and Bak, allowing the conformational changes necessary for pore formation. These proteins can be post-translationally modified (e.g., Bid, cleaved in the pro-apoptotic truncated form "t-Bid") [30,31] or transcriptionally activated (e.g., PUMA, Noxa and Bim). In particular, the transcription factor p53 represents one of the links between DNA damage or oxidative stress and intrinsic apoptosis. Indeed, after a stress signal, post-translational modifications induce p53 stabilization and translocation in the nucleus. p53 will induce pro-apoptotic Bcl-2 family proteins transcription (i.e., Bax, Bak, PUMA, and Noxa) [32,33]. In the absence of stress conditions, other members of the Bcl-2 family (Bcl-2, Bcl-Xl, Bcl-W, Mcl-1, and Bfl-1) constantly block MOMP [34]. They contain all four BH domains and are inserted in the MOM or ER membrane, interacting with and inhibiting the effectors members (Bax, Bak, and Bok) or the BH3-only activators (PUMA, Noxa, Bim, and tBid) [35,36]. In addition, these anti-apoptotic proteins have been shown to regulate Ca^{2+} homeostasis in the ER [37,38] and cellular redox equilibrium [39,40]. Finally, it has been shown that some BH3-only proteins (Bad, Bmf, and Hrk) carry out their pro-apoptotic effect without interacting with the effector proteins but by inhibiting the pro-survival ones [41].

Interestingly, it is now clear that mitochondria and ER are physically connected, forming platforms called mitochondria associated (ER) membranes (MAMs) [42]. Mitochondria associated membranes regulate numerous cellular processes such as calcium (Ca^{2+}) homeostasis, autophagy, lipid metabolism, apoptosis, and the rapid exchange of biological molecules [43]. They are involved in inflammasome formation and activation and participate in the antiviral response through the mitochondrial antiviral protein (MAV)/RNA sensors (retinoic acid-inducible gene I (RIG-I) or melanoma differentiation-associated protein 5 (MDA5)) complex activation [44].

Mitochondrial outer membrane permeabilization induces the release of intermembrane space elements, among which cytochrome C (CYC), Omi, and DIABLO (also called SMAC) [45–47]. Furthermore, following MOMP, the mitochondrial transmembrane potential ($\Delta\Psi m$) is usually lost, mostly due to CYC release in the cytosol and the consequent stop of the respiratory chain [48].

In the cytosol, the association of CYC with Apaf1 and pro-caspase 9 (CASP9) forms a complex called apoptosome that will activate CASP9 in an ATP-dependent process [45]. In turn, the activated CASP9 will then activate the executioner caspases (i.e., mainly CASP3 and -7) that are involved in the final cellular destruction: poly (ADP-ribose) polymerase (PARP) cleavage, DNA fragmentation,

PS exposure, apoptotic bodies formation [49–52]. Omi and DIABLO enhance cell death by inhibiting the inhibitor of apoptosis protein (IAP) family, which includes XIAP, c-IAP1, and c-IAP2. XIAP is constitutively bound to executioner CASP3 and -7 and, thus, blocks their activity [53,54]. c-IAP1 and c-IAP2, for their part, are two E3 ubiquitin ligases. They upregulate the CASP8 inhibitor c-Flip, induce caspases' degradation through ubiquitination, and promote NF-κB pro-survival pathway through receptor interacting serine/threonine kinase 1 (RIPK1) ubiquitination. These functions have mostly been maintained throughout evolution, from insects to mammals [55–57] (Figure 1).

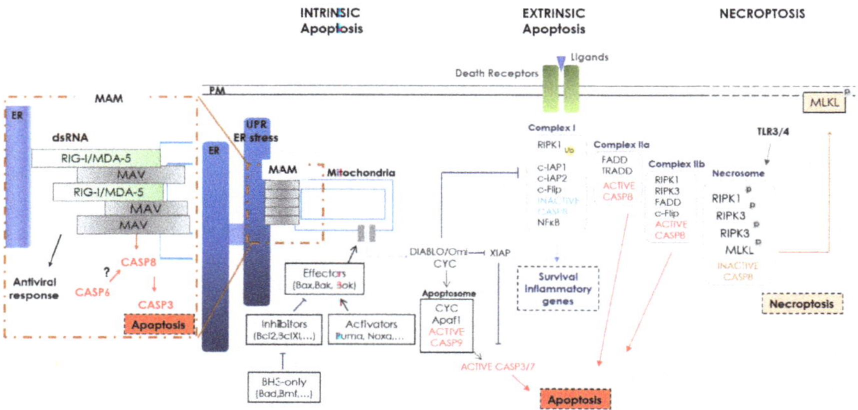

Figure 1. Apoptosis and necroptosis in mammals. PM: plasma membrane, ER: endoplasmic reticulum, MAM: mitochondria-associated membranes, CYC: cytochrome c, IAP: inhibitors of apoptosis proteins, RIPK1/3: receptor interacting serine/threonine kinase 1/3, Ub: uiquitin, p: phosphorylation, MLKL: mixed-lineage kinase domain-Like, FADD: Fas-associated protein with death domain, TRADD: tumor necrosis factor receptor super family (TNFR1)-associated death domain protein, TLR3/4: Toll-like receptor.

Finally, executioner caspases can positively or negatively regulate the emission of multiple damage-associated molecular patterns (DAMPs) by dying cells, including immunostimulatory [58] as well as immunosuppressive [59] factors.

2.2. Extrinsic Apoptosis

"Extrinsic apoptosis is a type of RCD initiated by perturbations of the extracellular microenvironment that are detected by plasma membrane receptors, propagated by CASP8 (with the optional involvement of MOMP), and precipitated by executioner caspases, mainly CASP3" (NCCD [19]).

Extrinsic apoptosis is mainly carried out by the activation of two main receptors types: death receptors and dependence receptors.

Dependence receptors are a functional family of around 20 receptors, characterized by the induction of a positive signal when bound by their ligand (survival, proliferation, differentiation, etc.) while they activate RCD in the absence of the ligand. Among them can be found netrin-1 receptors (deleted in colorectal carcinoma (DCC) [60], uncoordinated 5 homologs (UNC5Hs, UNC5H1,2,3,4 also called UNC5A,B,C,D) [61], the neogenin receptor [62], the low affinity neurotrophin receptor, p75 neurotrophin receptor (p75NTR) [63], and receptors with tyrosine kinase activity (e.g., rearranged during transfection (RET) [64], tropomyosin receptor kinase A and C (TrkA and TrkC) [65], and c-kit (CD117) [66]). Their physiological role is mainly cell guidance and they are mostly involved in tumor progression when dysregulated [67]. Even if Netrin-1 plays a role in inflammation regulation [68–71], dependence receptors have not been involved, so far, in antiviral response [67]. As such, they will not be described in this review.

Death receptors include Fas (CD95, APO-1), the tumor necrosis factor receptor super family TNFRSF1A (TNFR1), 10a (TNF-related apoptosis-inducing ligand receptor TRAILR1, DR4), and 10b (TRAILR2, DR5) [72,73]. The general mechanism is that ligand binding induces the receptors' oligomerization and the subsequent recruitment through their death domains (DDs), of adapter proteins in the intra-cellular side, to form a "death-inducing signaling complex" (DISC).

Fas ligand or TRAIL binding drives the oligomerization of their receptors, the recruitment of Fas-associated protein with death domain (FADD) through the DD and the subsequent formation of the DISC through interaction with CASP8 via death effector domain (DED) and different isoforms of cFlip [74,75].

Bound TNFR1 interacts with TNFR1-dssociated death domain protein (TRADD), through its DD, which enables the formation of "Complex I". The subsequent formations of "Complex II" ("IIa" or "IIb") operate as molecular platforms to regulate the activation and functions of CASP8 (or CASP10, in some cases) [76,77]. CASP8 activation leads to RCD following two different pathways. In Type I cells, CASP8 directly activates CASP3 and -7 thus inducing the execution of the apoptotic pathway. In Type II cells, where CASP3 and -7 are sequestrated by XIAP, extrinsic apoptosis occurs through the cleavage by CASP8 of Bid, a BH3-only protein, and the release of its truncated form, t-Bid. t-Bid acts as an activator on Bax and Bak to provoke MOMP and the subsequent CASP9-dependent RCD, described above in intrinsic apoptosis [78].

CASP8 activation is the key process of extrinsic apoptosis; its regulation is complex and also involved in inflammation and antiviral response [77]. cFlip is one of the key components that promotes or inhibits CASP8 oligomerization and ensuing activation by autoproteolytic cleavage. As cFlip is transcriptionally regulated by NF-κB, it can also participate in a pro-survival pathway induced by TNFR1 in some conditions [79]. It is indeed increasingly clear that the activation of death receptors by their ligands does not necessarily lead to RCD but can also activate pro-survival signals. Specifically, the TNFR1-induced pathway depends on the RIPK1 ubiquitination level, which directly influences the formation of pro-survival versus pro-death complexes.

Briefly, in Complex I, RIPK1 polyubiquitination by cIAP1 and cIAP2 leads to NF-κB activation, pro-survival, and inflammatory genes transcription, where a high level of cFlip is correlated to survival. Subsequently, deubiquitinated RIPK1 is released from Complex I and forms Complex IIa in the cytosol with FADD, TRADD, cFlip, and CASP8. If cFLIP concentration is low, this complex leads to the degradation of RIPK1 and RIPK3, allowing CASP8 dimerization and activation and the subsequent apoptotic cell death through CASP3. In a context of high cFlip concentration, CASP8/cFlip heterodimers are formed and apoptosis is blocked [56,57,78]. Moreover, in the absence of cIAP (after MOMP and IAP inhibitors release for instance), phosphorylated RIPK1 leads to non-canonical NF-κB activation and subsequent association with RIPK3, FADD, and cFlip to activate CASP8 (Complex IIb) [73].

Finally, another possible pathway is induced when CASP8 is inhibited by chemical caspase inhibitors or virally encoded proteins. In this case, deubiquitinated RIPK1 and RIPK3 bind in microfilaments, "amyloid-like" complexes called necrosomes (most likely trimers or tetramers) [80,81]. The auto- and transphosphorylation of RIPK1 and RIPK3 and the recruitment of mixed lineage kinase domain-Like (MLKL) to the plasma membrane, triggering membrane permeabilization, initiate what is called necroptosis [82]. It is of interest to note that MLKL oligomers also lead to PS exposure, a feature usually considered as a hallmark of apoptosis [80].

If CASP8 is inhibited, RIPK3 phosphorylation can be triggered by some activated PRR (pathogen recognition receptors), such as TL3 and TL4 [83], nucleic acid sensors, such as RIG-I and MDA5, and some adhesion receptors [84,85]. In addition, IFNα and β receptor subunit 1 (IFNAR1) and IFNγ receptor 1 (IFNGR1) are also able to trigger necroptosis through TRIF and ISGF3 activation [86] (Figure 1).

2.3. Inflammasome Activation and Pyroptosis

"Pyroptosis is a form of inflammatory RCD that critically depends on the formation of plasma membrane pores by members of the gasdermin (GSDM) protein family, often as a consequence of inflammatory caspase (CASP1, 4 or 5) activation" (NCCD [19]).

Pyroptotic cells present PS exposure, chromatin condensation, TUNEL staining but no DNA laddering, and a slight MOMP. Final GSDM-dependent membrane permeabilization allows the release of pro-inflammatory cytokines (IL1β and IL18, both NF-κB-target genes), maturated by interleukin-1β-converting enzyme (ICE/CASP1)-dependent cleavage. Other factors are also released, thus participating in the defense against pathogens through inflammation and the induction of an adaptive response [19,87,88]. In fact, pyroptosis seems to be mainly involved in the innate immunity against intracellular pathogens [88]. Pyroptosis was first thought to be restricted to monocyte/macrophage lineage, but it has been observed in other cells [89].

The inflammasome is activated by different DAMPs or pathogen-associated molecular pattern (PAMPs). It is a multiprotein complex that, like the apoptosome (intrinsic apoptosis) or the DISC (extrinsic apoptosis), acts as a caspase-activating platform. It is formed by a receptor (NOD-like receptor (NLR) family or non-NLR (AIM2)), an adapter protein (ASC, apoptosis-associated speck-like protein containing a CARD) and CASP1 that cleaves pro-IL18, pro-IL1β and GSDM. However, it is now clear that pyroptosis can also be activated by other caspases such as CASP3 [90] (Figure 2).

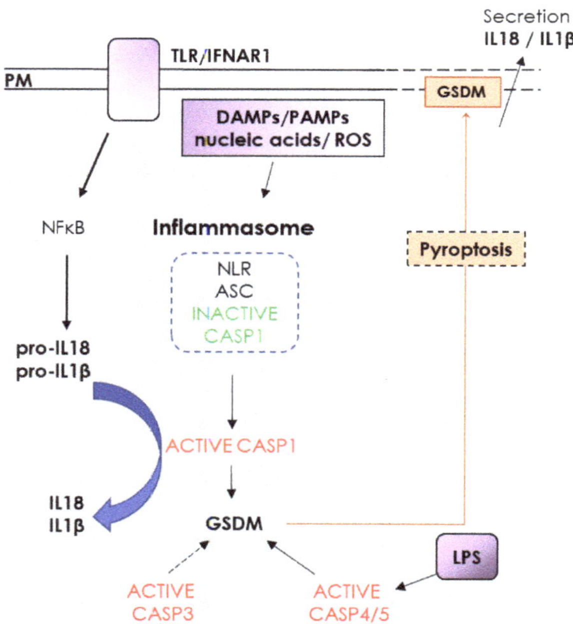

Figure 2. Inflammasome activation and pyroptosis. PM: plasma membrane, TLR: Toll-like receptor, IFNAR1: Interferon associated-receptor 1, DAMP: damage-associated molecular patterns, PAMPs: pathogen-associated molecular patterns, ROS: reactive oxygen species, NLR: NOD-like receptor, IL: interleukin, LPS: lipopolysaccharide.

The recent description of necroptosis and pyroptosis processes has amplified the complexity of RCD understanding. Contrary to general consensus so far, it is now clear that intrinsic as well as extrinsic regulated cell death can be immunogenic, thus participating in the establishment of the adaptive immune response [91,92]. This has been recently underlined by the description of the concomitant activation of apoptosis, necroptosis, and pyroptosis in a context of bacterial or viral infection in macrophages, leading to inflammatory cell death. The phenomenon has been named PANoptosis and would involve molecules of the three RCD pathways (i.e., CASP8, RIPK3, and CASP1) in a single complex called PANoptosome [93].

2.4. Autophagy-Dependent Cell Death

Autophagy-dependent cell death is a type of RCD that depends on components of the macroautophagy machinery [19]. Macroautophagy is a particular form of autophagy where double-membrane vesicles (autophagosomes) sequester a large part of organelles and cytoplasm, leading to their lysis and, in some cases, to cell death. Morphologically, dying cells present an accumulation of autophagosomes and autolysosomes in the cytoplasm, a feature extremely different from apoptotic or necrotic RCD. However, it seems increasingly clear that autophagy usually inhibits rather than induces cell death and has to be considered as a way for the cell to maintain homeostasis after stress signals (hypoxia, ROS, starvation, PRR activation, etc.) [94]. Even if in some cases inhibition of specific autophagy proteins can delay RCD [95], pharmacological or genetic inhibition of macroautophagy components usually accelerates the death of cells rather than protects them. Autophagy may degrade damaged mitochondria or pro-apoptotic complexes thus preventing cell death.

It is of interest to note that autophagy leads to the degradation by lysosomes of components of endogenous or exogenous origin, which are accessible in the cytoplasm. This has to be distinguished from vesicular trafficking, which starts in the plasma membrane and also leads to lysosomal degradation (i.e., phagocytosis or receptor-mediated endocytosis). Macroautophagy and vesicular trafficking pathways interact at numerous regulation points, especially in the late phases of the pathways. Autophagosomes or late endosomes fusion with lysosomes actually require the same machinery. Thus, numerous proteins involved in physiological or lytic autophagy are also essential for viral penetration, from receptor-mediated endocytosis to fusion of the viral envelope with the endosome membrane and subsequent liberation of the viral genome in the cytosol.

3. Impact of Alphaviral Infection on Regulated Cell Death

During a viral infection, RCD is generally described as a defense mechanism, induced to limit virus replication and production, to prevent infection of neighboring cells and, to some extent, to participate in immune response induction. Cell death induced by alphavirus infection has been observed and studied in several cell types infected by different alphaviruses, mainly CHIKV, SFV, SINV, VEEV, and EEV.

3.1. Apoptotic Pathways in Alphavirus-Infected Cells

Infections of baby hamster kidney (BHK), rat prostatic adenocarcinoma (AT-3), and mouse neuroblastoma (N18) cells with SINV result in clear nuclear condensation and membrane blebbing, 24 h post-infection [96]. In the same context, SFV infection of AT-3 induces apoptotic features, correlated to strain virulence. Grandgirard et al. [97] have described, in rat embryonic fibroblasts, a potential caspase-dependent Bcl-2 cleavage in SFV- or SINV-infected cells leading to cell death and viral replication, even in a context of Bcl-2 overexpression. In 293T and BHK cells, the BH3-only protein Bad seems to participate in SINV-induced cell death, through its interaction with some specific anti-apoptotic Bcl-2 proteins, while the other non-binding members also regulate cell death [98]. The dynamics of mitochondria are also highly altered during apoptosis induced by VEEV infection of human astrocytoma cells U87MG [99]. Infection rapidly induces MOMP and ROS increase, followed by perinuclear localization and fission of mitochondria, and then mitophagy. Moreover, VEEV capsid co-localizes with mitochondria and could participate in mitochondria dysregulations.

Lin et al. showed a link between SINV induced cell death, oxidative stress, NF-κB, and Bcl-2 expression. In AT-3 and N18 infected cells, NF-κB activation and cell death were indeed inhibited either by antioxidant agents or Bcl-2 overexpression, with no effect on viral entry or replication however [100]. Likewise, MRC5 human fibroblasts present a SINV persistent infection, when manganese-superoxide dismutase (Mn-SOD) is over-expressed, confirming that oxidative pathways are implicated in the effects of SINV [101]. Chikungunya virus infection has been studied in the neuroblastoma cell line SH-SY5Y by Dhanwani et al. [102]. Intrinsic apoptosis features (CYC release, CASP3 activation, PARP cleavage) are observed 24 h and 36 h post-infection. Moreover, the infection is followed, 36 h and 48 h after by an elevation of ROS, a decrease of anti-oxidant enzymes expression and glutathione (GSH) depletion.

Another cellular response to stress induced by viral infection is the unfolded protein response (UPR) of the ER due to the accumulation of newly synthetized viral proteins in the ER, leading to translation blocking and intrinsic apoptosis [103]. Endoplasmic reticulum stress response has been described in *flavivirus*-infected cells, but little is known about alphaviruses. However, SFV envelope glycoproteins, but not capsid, seem able to induce and accelerate apoptotic cell death [104], while VEEV glycoproteins induce UPR and apoptosis in primary astrocytes [105,106].

Finally, intrinsic apoptosis may be triggered by alphavirus non-structural protein activity. Indeed, SINV, VEEV, and EEEV nsp2 and nsp3 have been shown to be responsible for viral cytopathic effects, enabling persistent infections when specifically mutated [107–109]. Frolov and colleagues have described the nuclear translocation of nsp2 and global transcriptional shutoff through RNA polymerase II degradation [110–112] and a subsequent nsp3-dependent translational shutoff for arthritogenic viruses [112], while capsid would play this role for encephalitic viruses [113]. However, CHIKV nsp2 seems to inhibit UPR as well through its transcriptional shutoff activity [114] and to interfere with the IFNβ signaling pathway [115,116].

However, Sarid et al. characterized a CASP8 and TNFα/TNFR1-dependent PC-12 RCD after infection by the SINV SVNI strain (neurovirulent and cytotoxic). Indeed, they described an upregulation of TNFα expression in infected cells and a cell death inhibition following cFlip overexpression [117]. Nava and colleagues [118] have shown, in SINV-infected BHK cells and in mice, an inhibition of death after treatment with the pan-caspase inhibitor zVAD-fmk and the CASP1 and CASP8 inhibitor CrmA (a serine proteinase inhibitor from Cowpox virus). In addition, by using SFV replicon vectors or a wild-type SFV strain, Kiiver et al. have neither shown any Bcl-2 protective action against virally induced cellular protein synthesis shutdown post-infection nor cell death. Moreover, AT-3 and BHK cells did not present any CYC release after infection [119]. In addition, unlike poxvirus, that blocks CASP1 and CASP8 [120,121], or herpesviruses [122], alphaviruses have never been described as inducing necroptosis.

Thus, on the one hand, alphaviruses appear to induce apoptosis through mitochondria, oxidative, and ER stress or the transcriptional and translational shutoff induced by nsp (intrinsic pathway). On the other hand, some studies suggest the implication of death receptors and CASP8 activation, without mitochondrial involvement (extrinsic pathway). This apparent discrepancy may be explained by the different apoptotic cells that have been observed (infected or neighbor cells) and the duration of infection (few hours to days post-infection). Indeed, Joubert et al. suggest that, in a first wave, CHIKV-infected cells die through intrinsic apoptosis (CASP9 positive cells) and that, in a second time, during antiviral response, infected, and neighbor cells die through the extrinsic pathway (death receptors and CASP8 activation). These secondary pathways seem to be independent of ER and oxidative stress [123,124].

Other explanations could be found also in two recent studies, which involve two newly described pathways in SFV-induced cell death. Using 3T9 MEFs, Urban et al. [125] characterized SFV-induced RCD. In their study, they first showed that RCD was triggered by SFV replication and not only by viral entry, as previously described [126]. Secondly, SFV-induced cell death occurred in a Bak dependent MOMP, leading to apoptosome activation. Moreover, they excluded the involvement of TRAILR, Fas or TNFR1 in this process. Surprisingly, CASP8 and tBid seemed to be activated downstream of apoptosome, maybe through CASP6 [127], acting as an amplification loop. In fact, although CASP6 has long been considered as an executioner caspase based on its homology with CASP3 and CASP7, recent data suggest that CASP6 may actually be involved in RCD initiation [127,128]. Additional investigation is required to elucidate the function of CASP6 in mammalian cells. Secondly, in several SFV-infected cell types, El Maadidi et al. recently described a new mitochondrial platform, involving the innate immune factor MAV and the initiator CASP8, comparable to the death-inducing signaling complex (DISC) of the death receptors signaling. However, this complex does not involve FADD but another potential, not yet characterized adaptor. The platform is activated via the dsRNA sensors MDA5 or RIG-I and acts in parallel of the classical Bax/Bak dependent MOMP, also leading to CASP3 activation, independently of type I IFN signaling factors (IRF3, IRF7, IFNβ, PKR, etc.) and mitochondrial depolarization [129]. As MAVs platforms have been described in MAMs, it is tempting to hypothesize that the MAV/CASP8 should be localized in these subcellular structures thus interacting with other metabolic pathways (such as ROS, Ca^{2+}, lipid, autophagy) during antiviral response [130].

Hence, it appears that several alphaviruses induce regulated cell death in numerous cell types, involving mitochondria depolarization, ER stress, and CASP8 activation, maybe in a time-dependent regulation or through alternative processes. However, after nearly 30 years of study, large parts of the molecular pathway leading an infected cell to death remain to be deciphered. It would be of interest in the future studies to focus on the cell types relevant for the viral tropism (e.g., skin cells, muscle cells, neural cells) and to favor innovative cell culture technics that mimic better the natural cell characteristics (e.g., 3D culture, explants, iPS (induced Pluripotent Stem cells)).

3.2. *Inflammasome and Pyroptosis in Alphaviral Infection*

Activation of the inflammasome pathway and pyroptosis has been intensively studied for flaviviruses infection, especially for their involvement in pathogenesis [131–133], but little is known for alphaviruses. Even if inflammasome pathways seem to participate in the pathology and the response against alphaviruses, the involvement of pyroptosis has never been described. In dermal fibroblasts, CHIKV and the *flavivirus* West Nile virus (WNV) both induce IL1β production and CASP1 activation through the AIM2 inflammasome sensor, but only CHIKV replication and propagation can be controlled by CASP1 [134]. In PBMC from CHIKV-infected patients, high levels of NLRP3, IL18, and CASP1 are found [135]. Moreover, in mice, NLRP3 activation is correlated to inflammatory symptoms such as bone damage and myositis. NLRP3 inhibition leads to a reduction of the inflammatory pathology induced by CHIKV but not by WNV [135]. The alphavirus Mayaro (MAYV) induces the expression of inflammasome proteins in macrophages, and inflammatory cytokines production through the NLRP3 sensor, activated by ROS and K^+ efflux. In mice, NLRP3 is also involved in MAYV induced pathogenesis [136].

Thus, inflammasome activation has been mainly involved in global inflammatory response to alphavirus in vivo, but the molecular pathways activated in the cell remain to be described. Indeed, it is still unknown if pyroptosis may participate in inflammatory cytokines secretion during alphavirus infection, and, to our knowledge, there is no molecular study of this process.

4. Interplay between Cell Death and Alphaviral Replication and Spread in Mammals

Apoptosis appears to be a strong antiviral process. Indeed, Bcl-2 overexpression converts SINV infection from lytic to persistent in vitro [96] and in vivo [137]. Moreover, Bcl-2 seems to be able to restrict SFV replication by inhibiting early stages of infection and appears to prolong survival of productively infected cells [138].

As described above, autophagy usually blocks apoptosis, and viruses have developed strategies to take advantage of this property. The first connection between alphavirus infection and autophagy has been made by Liang and colleagues [139], when they identified Beclin-1 as a new Bcl-2-interacting protein through a yeast two-hybrids screening. Beclin-1 is a major factor of autophagy, involved in autophagosome initiation and maturation. In this study, Beclin-1 protected SINV-infected mice against fatal encephalitis, with a significantly lower viral replication rate in mice brains. The author correlated these observations with the previously observed protective role of Bcl-2 against in vivo

SINV infection [96,137]. However, years after this study, Bcl-2 was described as a Beclin-1 inhibitor, thus participating in ER stress connected autophagy regulation [140]. Another crucial protein of the autophagy pathway, Atg5, protects mouse neurons from SINV-induced cell death [141], with no apparent impact on viral replication. Moreover, the adaptor protein p62 seems to be linked to viral capsid clearance by direct interaction and target of autophagosomes, thus promoting cell survival [141]. Finally, in SFV-infected cells, autophagosomes accumulate but autophagy modulation has no effect on viral replication, and this autophagosomes accumulation seems to be due to the inhibition of their degradation rather than an induction by SFV infection [142].

In HEK293 cells, CHIKV infection induces autophagy features (LC3 positive vesicles and electron microscopy observation). In this study, autophagy has a clear pro-viral role, increasing the number of infected cells and viral RNA in the cell culture supernatant [143]. Moreover, in vitro and in vivo, CHIKV infection has also been shown to induce an autophagy flux, through ER and oxidative stress [123,124]. In these models, autophagy limits (i) extrinsic and intrinsic RCD induced by CHIKV infection, (ii) mice lethality, and (iii) viral propagation. Autophagy, as a host response to infection, limits indeed the cytopathic effects of CHIKV and regulates the pathogenesis of acute chikungunya disease. However, during late phases of in vitro infection (48 h post-infection), a switch between autophagy and apoptosis is observed and cells die. Finally, in HeLa cells, autophagy promotes CHIKV infection and inhibits cell death. Indeed, in addition to p62-dependent capsid clearing, another autophagy receptor, NDP52, interacts with nsp2, localizes near the CHIKV replication complex and restricts cell shutoff thus promoting viral replication and cell survival [144].

Hence, alphaviruses may exploit autophagy to delay cell death through (i) direct inhibition of intrinsic and extrinsic apoptosis and (ii) a limitation of viral proteins production, allowing cell survival and a longer viral replication.

However, several Old World Alphaviruses, such as CHIKV, SFV, and RRV, seem able to activate the phosphatidylinositol-3-kinase (PI3K)–AKT–mTOR pathway, involved in cell survival and autophagy inhibition. Furthermore, inhibition of this pathway has a negative effect on viral replication [145]. This apparent discrepancy with the previous observations may indicate that, more than autophagy per se, cell survival is the key process which favors viral replication.

Finally, it is of interest to note that, in some cases, apoptosis has been shown to enhance viral spread. Indeed, in their study, Krejbich-Trotot and colleagues [146] first confirmed the dual nature of the alphavirus-induced apoptosis (intrinsic and extrinsic) in HeLa and primary fibroblasts infected with CHIKV accompanied by CASP8 activation in neighbor cells. More interestingly still, inhibitors of blebbing or engulfment drastically reduced infection rates. Finally, they detected infective CHIKV in apoptotic corpses and in the macrophages which phagocyted them, leading to macrophages infection and viral production. As macrophages are refractory to CHIKV infection in vitro, this study highlights a possible role of apoptotic blebs in viral propagation. This phenomenon, called "apoptotic mimicry", is used by a large number of viruses to exploit the PS receptors present on numerous cells membranes, enhancing viral spread and limiting immune response [147].

5. Impact of Apoptosis on Virus Pathogenesis in Mammals

5.1. Alphavirus Encephalitis

The first cell death analysis was documented in vivo, using SINV, VEEV, and SFV, three alphaviruses causing encephalitis. In SINV-infected mice, the apoptotic cells were detected principally in the brain and contained viral antigens, suggesting that apoptosis was correlated to neurovirulence [148]. The in vivo mouse infection of VEEV was also associated to cell death in brain, demonstrated by TUNEL assay (DNA fragmentation) and morphological changes [149].

Comparing SINV infections with SVNI (neurovirulent and cytotoxic) or SVA (avirulent and leading to persistent infection) strains in PC-12 cells and astrocytes, revealed that SVNI induces Bax overexpression while SVA induces Bcl-2 expression [150].

Intranasally SFV-infected rats develop encephalitis, where infiltrating leucocytes and neural precursor cells undergo apoptosis while productively infected neurons present necrotic features, apparently due to the local inflammation [151].

Hence, it appears that alphaviruses pathogenicity is linked to its cytopathic effects in infected cells, at least in the case of the encephalitic group.

5.2. Alphaviral Chronic Infection: What about Cell Death?

One characteristic of the *Alphavirus* genus is the ability of some of them (CHIKV, SINV, MAYV, RRV, etc.) to induce chronic pain, such as arthritis and myalgia, which may last for years, with detectable viral genome in the organism. This persistent infection implies that some cells may be chronically infected, and in some way able to delay or block cell death. However, despite an intense immune response observed in chronic patients, damaged synovial tissues present strong apoptosis features. Chikungunya virus has been found in synovial macrophages several months after infection but joints do not seem to be the viral reservoir [152]. In addition, RRV-infected human monocyte acute leukemia MM6 cell line presents very low replication rates, without innate immune control, and apoptosis features at late stages of infection. This indicates that monocytes could be persistently infected and participate in the chronic form of RRV or CHIKV [153]. Young et al. [154] propose that dermal and muscular fibroblasts, as well as myofibers, may survive the acute CHIKV infection and harbor persistent CHIKV RNA during chronic phase of the disease. Moreover, they observe that synovial cells are not infected in large numbers in vivo and suggest that synovial cells may be infected but do not survive.

How these cells survive remains to be understood. The mechanisms involved in alphaviral persistence are mostly unknown. They may depend on infected cell type and the highly complex interplay between virus, immune response, and different RCD pathways.

6. Interplay between Cell Death and Susceptibility of Mosquito Species to Arboviruses

The interplay between arbovirus and arthropods is still poorly understood and the primary point of study concerns flaviviruses (mainly Dengue virus (DENV)) in mosquito and/or in *Drosophila*, used as a genetic model for insect immunity. Hence, very little is known concerning the impact of alphavirus infection in mosquitoes and which factors may explain the tolerance versus resistance of different mosquito strains.

It is classically admitted that alphaviruses do not induce any major pathology in their vectors. However, several lesions are observed in tissues which are critical for viral propagation and transmission. Indeed, after feeding on infected blood, cellular response in the midgut plays a decisive role in vector competence. EEEV infection of *Culiseta melanura* mosquito induces severe lesions in midgut epithelial cells and basal lamina, associated to viral spread [155]. Likewise, infection of more or less susceptible *Culex tarsalis* strains with WEEV revealed lesions and apparent necrotic cell death only in the sensitive mosquito's gut [156]. Transcriptomic analysis of *Aedes Aegypti* fed with CHIKV in blood or different buffers reveals the over-expression of matrix metallo proteinases (MMP) and other peptidases in the midgut, as well as the decrease of Collagen IV, a component of the basal lamina [157]. Intrathoracically SINV-injected *Aedes Albopictus* present colocalization of virus antigen with structural lesions and TUNEL positive cells in salivary glands [158] and midgut-associated visceral muscles [159]. Furthermore, organ-associated muscles respond differently to SINV [160]: 10 days post-infection, the virus has cleared from the midgut, is persistent in the hindgut, and unable to infect ovary associated muscle cells. High viral titers induce pathology limited to gut associated muscles and gut epithelium. Finally, in *Aedes Aegypti* mosquitoes, AeIAP1 (IAP ortholog) downregulation leads to a higher replication of SINV in the midgut, while AeDronc (CASP9 ortholog) inhibition is associated to a lower viral replication and dissemination towards salivary glands [161] (see Box 1 for RCD pathway description in insect).

Box 1. Comparative cell death pathways in mammal, *Drosophila* and *Aedes*. PM: plasma membrane; CYC: cytochrome C; IAP: Inhibitors of apoptosis; *Ae*: *Aedes*; RHG: Reaper, Hid and Grim; IMD: immune deficiency.

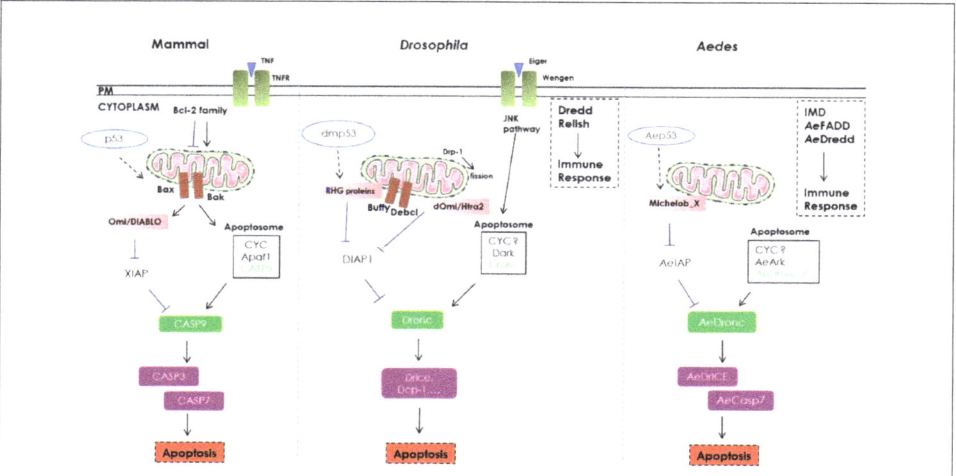

Recent knowledge concerning apoptosis in mosquitoes has been acquired through gene homology with *Drosophila melanogaster*. Apoptosis is under the control of initiator and effector caspases [162], expressed ubiquitously and synthesized as inactive procaspases. The Apaf-1-related killer (Ark) molecule [163] assembles itself into an apoptosome-like complex [164] to activate Dronc, but the role of CYC in the insect apoptosome is very controversial [165]. Two Bcl-2 orthologues have been identified in *Drosophila*: Buffy and Debcl [166], whose pro- or anti-apoptotic roles are not clear [166,167]. Finally, mitochondrial fission, through Drp1 activation, seems to be required for efficient cell death [168]. However, the pro-apoptotic activity of caspases is mainly regulated by members of insect IAP family [169,170]. IAP antagonists (dOmi/HtrA2 and RHG proteins in drosophila; Michelob_X in mosquito) are localized in the mitochondrial intermembrane space in living cells and released into the cytosol, but remain near the mitochondria, after an apoptotic stimulus [170,171], where they compete for caspase binding through their IAP-binding domain. Additionally, RHG proteins can induce DIAP1 ubiquitination and degradation [168,172]. Interestingly, RHG and Mx promoters present different response elements regulated by transcription factors, such as dmp53, activated by developmental or environmental signals, leading to cell death [173]. In insects, no clear distinction can be made between intrinsic or extrinsic apoptotic pathways. Nevertheless orthologues of TNF (Eiger) and TNFR (Wengen) have been described in *Drosophila* [174] and induce cell death through a JNK-mediated pathway, requiring apoptosome components [175]. Immune response against pathogens is triggered by the NF-κB (Relish) pathway induction through the Immune Deficiency IMD/dFADD/Dredd (CASP8) pathway [176] in *Drosophila* and IMD/AeFADD/AeDredd in *Aedes* [177].

Finally, autophagy-dependent cell death is finely controlled in insects, and its role in development has been largely studied in *Drosophila* [178,179]. To our knowledge, pyroptosis and necroptosis have not yet been described in insects.

Thus, caspase activity may be required for dissemination of SINV from the midgut to the secondary organs by participating in the remodeling of the basal lamina, as suggested in baculovirus-infected lepidopteran, where caspase and MMP activity is necessary to cross the midgut barrier [180]. However, cell death modulation in vitro, in mosquito cells, does not seem to alter alphaviral replication. In fact, recombinant SINVs, expressing Reaper (*Drosophila* IAP inhibitor) or Michelob_X (Mx, *Aedes* IAP inhibitor), induce apoptosis in infected *Aedes Albopictus* C636 cells, with no inhibitory effect on viral production in the initial phase of infection. Moreover, in these conditions, inhibition of caspase activity has no effect on viral replication neither [181]. However, recombinant SINV expressing Reaper induces cell death in vivo in *Aedes Aegypti*'s midgut, a delayed infection and propagation in the saliva [182]. More importantly, this last study also describes a rapid genetic selection of SINV variants in vivo against Reaper expression.

Hence, cell death, with tissue degradation features, seems important for alphavirus propagation in mosquito organism, with no clear effect on replication in vitro. However, as described above for mammal cells, RCD is also associated to efficient immune response against viruses and may be one of the key processes involved in mosquito resistance to virus.

How cell death is modulated in infected tolerant mosquito cells remains to be understood. Oxidative stress response may play an important role in the mosquito's response. Indeed, CHIKV infection induces upregulation of antioxidant pathways in mosquito midgut, which delays cell death [183]. During arbovirus infection, oxidative stress is actually detected in both mammal [101,123,184] and insect cells. Oxidative stress is defined by loss of homeostasis between accumulation of ROS and production of antioxidant enzymes such as superoxide dismutase (SOD), catalase (CAT) or glutathione transferase and reductase [185]. After blood feeding, the midgut is in contact with sugar, iron, heme and other components of vertebrate blood. Mosquitoes have developed protective adaptation against the damage caused by heme and iron uptake. Indeed, the heme can induce lipid peroxidation, protein degradation, and ultimately cell death. Once in the epithelial cells, these components are detoxified, and a strong antioxidant and protective response is engaged [186]. Concomitantly, pathogens present in the blood could take advantage of this antioxidant response, blocking cell death, to infect and replicate into midgut epithelial cells.

The pro-survival pathway PI3k–Akt–mTOR may also be involved in insect tolerance. Indeed, in drosophila, activation of the PI3k–Akt–mTOR pathway is associated to an increase of SINV infection, potentially through apoptosis and autophagy inhibition and a more efficient cap-dependent translation of viral genome [187]. Finally, the role of autophagy seems to be limited in CHIKV infection of mosquito cells. Even infection induces autophagy in Aag2 cells, every pharmacological modulation of autophagy (inducer or blocker) leads to a replication increase in mosquito cells [188].

Hence, the sensitivity may depend on the better resistance of midgut cells to oxidative stress induced by viral infection, leading to a delayed cell death but the involvement of autophagy in these regulations remains to be understood.

Few studies have been conducted to investigate the interplay between alphaviruses and their vector. However, recent findings in mosquito and drosophila underlie the role of p53 isoforms in cell response to oxidative stress and to DENV infection. The balance between a rapid apoptosis and a delayed, secondary necrosis may explain in part the differences between tolerant and resistant mosquitoes' strains. For more details see [177,189–196] and Figure 3 for suggested mechanisms which may be involved in alphavirus infection in sensitive versus resistant mosquitoes, extrapolated from other arboviruses.

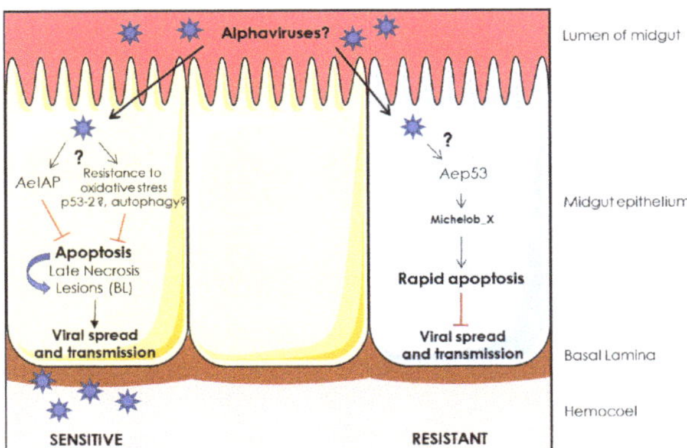

Figure 3. Possible interplay between cell death and alphavirus infection in *Aedes* mosquitoes: extrapolation from other arboviruses studies. BL: basal lamina, *Ae*IAP: *Aedes* inhibitor of apoptosis protein. See Box 1 for RCD pathway description in insect.

In addition to the virus' ability to penetrate the cells of a particular mosquito (specific receptor, lipid membrane composition), the difference between resistant and tolerant strains may also lie in the rapidity of cell response to viral infection in the first targeted tissue, i.e., the mosquito gut. This response has to be apoptotic and not necrotic to ensure mosquito resistance. A p53/Mx- [189] and caspase- [177] dependent cell death has been linked to *Aedes Aegypti* resistance to DENV infection. Phagocytosis of dead cells and apoptotic bodies also seem to be important for a virus specific immune response, at least in drosophila [190].

Sensitive mosquitoes may tolerate viral infection through IAP-dependent apoptosis inhibition, as shown for the arbovirus bluetongue virus (BTV) [191,192]. Another tolerance mechanism to DENV seems to rely on resistance to oxidative stress through CAT protection in mosquito gut [193] or p53 isoforms regulation leading to cell survival in drosophila [194] and mosquito [195,196]. Delayed apoptosis may then lead to secondary necrosis, impairing a proper innate immune response, favoring basal lamina damage and viral spread [189].

7. Conclusions

The comparative overview of cell response to alphavirus infection, in mammals and mosquitoes, underlines the complexity of cell death regulation among different species facing the same pathogen. In both, host and vector, the first cells in contact with the virus will influence the progression of the infection, through immune response and cell death.

In mammals, alphavirus-induced apoptosis is also linked to pathogenesis in the organs secondarily infected. Some discrepancies between intrinsic and extrinsic apoptotic pathways in mammals could be explained by a secondary amplification loop through CASP6 activation or by a newly described CASP8 activation platform which directly links viral RNA sensors and apoptosis. However, the precise mechanisms need to be studied further, in other cell types and for different alphaviruses.

Experimentally interfering with apoptosis in vitro does not seem to influence the viral replication rate but can lead to persistence. It is possible that autophagy may be a way for alphaviruses to delay cell death, allowing replication over a longer period of time. Alphavirus non-structural proteins 2 and 3 appear to be mostly responsible for mammal cell death, which is not the case for mosquito cells from sensitive strains. Mammals are often considered to be an "accidental host" in arboviral infections, suggesting that nsp-induced cell death may be also "accidental" and in some way deleterious for alphaviruses. Some alphaviruses have indeed evolved to be restricted to mosquitos and do not rely on transmission to mammals any longer [197–199].

How tolerant mosquito cells survive to arbovirus infection remains unknown. In addition to a strong action of insect IAPs, a higher control of oxidative stress due to the fact of infection through p53 isoforms and autophagy may be an answer, but supplementary studies are needed for alphaviruses. Moreover, in the case of arthritogenic alphaviruses in mammals, clarification of the processes leading to chronic infection and a possible survival of infected cells is needed. The differences in cell types may explain these discrepancies and further studies would help to decipher how these cells overcome cell death.

Finally, even if it is generally admitted that alphavirus infection is silent in competent mosquitoes, a certain level of tissue destruction is observed and needed in gut epithelia and salivary glands to allow viral propagation and transmission. As alphaviruses are still poorly studied to understand the link between cell death and vector competence, we can only extrapolate from recent DENV studies. Indeed, in addition to immune response, the mosquito's ability to rapidly eliminate infected epithelial cells through apoptosis, instead of a delayed cell death and secondary necrosis, may explain the difference between a resistant mosquito and a tolerant one. A better understanding and subsequent manipulation of vector tolerance could help to control arboviral propagation, as has recently been suggested [15,18].

Funding: This research received no external funding.

Acknowledgments: The authors would like to thank Aurélie Paillet for constructive proofreading of the manuscript.

Conflicts of Interest: The authors declare no conflict of interest.

References

1. Arthropod-borne and rodent-borne viral diseases. Report of a WHO Scientific Group. *World Health Organ. Tech. Rep. Ser.* **1985**, *719*, 1–116.
2. Weaver, S.C.; Barrett, A.D. Transmission cycles, host range, evolution and emergence of arboviral disease. *Nat. Rev. MicroBiol.* **2004**, *2*, 789–801. [CrossRef] [PubMed]
3. Powers, A.M.; Roehrig, J.T. Alphaviruses. *Methods Mol. Biol.* **2011**, *665*, 17–38. [CrossRef] [PubMed]
4. Suhrbier, A.; Jaffar-Bandjee, M.C.; Gasque, P. Arthritogenic alphaviruses–an overview. *Nat. Rev. Rheumatol* **2012**, *8*, 420–429. [CrossRef] [PubMed]
5. Figueiredo, M.L.; Figueiredo, L.T. Emerging alphaviruses in the Americas: Chikungunya and Mayaro. *Rev. Soc. Bras. Med. Trop* **2014**, *47*, 677–683. [CrossRef] [PubMed]
6. Baxter, V.K.; Heise, M.T. Immunopathogenesis of alphaviruses. *Adv. Virus Res.* **2020**, *107*, 315–382. [CrossRef] [PubMed]
7. Schwartz, O.; Albert, M.L. Biology and pathogenesis of chikungunya virus. *Nat. Rev. MicroBiol.* **2010**, *8*, 491–500. [CrossRef] [PubMed]
8. Konopka, J.L.; Penalva, L.O.; Thompson, J.M.; White, L.J.; Beard, C.W.; Keene, J.D.; Johnston, R.E. A two-phase innate host response to alphavirus infection identified by mRNP-tagging in vivo. *PLoS Pathog.* **2007**, *3*, e199. [CrossRef]
9. MacDonald, G.H.; Johnston, R.E. Role of dendritic cell targeting in Venezuelan equine encephalitis virus pathogenesis. *J. Virol.* **2000**, *74*, 914–922. [CrossRef]
10. Solignat, M.; Gay, B.; Higgs, S.; Briant, L.; Devaux, C. Replication cycle of chikungunya: A re-emerging arbovirus. *Virology* **2009**, *393*, 183–197. [CrossRef] [PubMed]
11. da Silveira, I.D.; Petersen, M.T.; Sylvestre, G.; Garcia, G.A.; David, M.R.; Pavan, M.G.; Maciel-de-Freitas, R. Zika Virus Infection Produces a Reduction on Aedes aegypti Lifespan but No Effects on Mosquito Fecundity and Oviposition Success. *Front. MicroBiol.* **2018**, *9*, 3011. [CrossRef]
12. Grimstad, P.R.; Ross, Q.E.; Craig, G.B., Jr. Aedes triseriatus (Diptera: Culicidae) and La Crosse virus. II. Modification of mosquito feeding behavior by virus infection. *J. Med. Entomol.* **1980**, *17*, 1–7. [CrossRef]
13. Ciota, A.T.; Styer, L.M.; Meola, M.A.; Kramer, L.D. The costs of infection and resistance as determinants of West Nile virus susceptibility in Culex mosquitoes. *BMC Ecol.* **2011**, *11*, 23. [CrossRef]
14. Oliveira, J.H.; Bahia, A.C.; Vale, P.F. How are arbovirus vectors able to tolerate infection? *Dev. Comp. Immunol.* **2020**, *103*, 103514. [CrossRef]
15. Lambrechts, L.; Saleh, M.C. Manipulating Mosquito Tolerance for Arbovirus Control. *Cell Host Microbe* **2019**, *26*, 309–313. [CrossRef]
16. Ruckert, C.; Ebel, G.D. How Do Virus-Mosquito Interactions Lead to Viral Emergence? *Trends Parasitol* **2018**, *34*, 310–321. [CrossRef]
17. Powell, J.R. An Evolutionary Perspective on Vector-Borne Diseases. *Front. Genet.* **2019**, 10. [CrossRef]
18. Clem, R.J. Arboviruses and apoptosis: The role of cell death in determining vector competence. *J. Gen. Virol.* **2016**, *97*, 1033–1036. [CrossRef]
19. Galluzzi, L.; Vitale, I.; Aaronson, S.A.; Abrams, J.M.; Adam, D.; Agostinis, P.; Alnemri, E.S.; Altucci, L.; Amelio, I.; Andrews, D.W.; et al. Molecular mechanisms of cell death: Recommendations of the Nomenclature Committee on Cell Death 2018. *Cell Death Differ.* **2018**, *25*, 486–541. [CrossRef]
20. Conradt, B. Genetic control of programmed cell death during animal development. *Annu Rev. Genet.* **2009**, *43*, 493–523. [CrossRef] [PubMed]
21. Vanden Berghe, T.; Vanlangenakker, N.; Parthoens, E.; Deckers, W.; Devos, M.; Festjens, N.; Guerin, C.J.; Brunk, U.T.; Declercq, W.; Vandenabeele, P. Necroptosis, necrosis and secondary necrosis converge on similar cellular disintegration features. *Cell Death Differ.* **2010**, *17*, 922–930. [CrossRef]
22. Kepp, O.; Galluzzi, L.; Zitvogel, L.; Kroemer, G. Pyroptosis—A cell death modality of its kind? *Eur. J. Immunol.* **2010**, *40*, 627–630. [CrossRef]

23. Xie, Y.; Hou, W.; Song, X.; Yu, Y.; Huang, J.; Sun, X.; Kang, R.; Tang, D. Ferroptosis: Process and function. *Cell Death Differ.* **2016**, *23*, 369–379. [CrossRef]
24. Tang, D.; Kang, R.; Berghe, T.V.; Vandenabeele, P.; Kroemer, G. The molecular machinery of regulated cell death. *Cell Res.* **2019**, *29*, 347–364. [CrossRef]
25. Rogers, C.; Fernandes-Alnemri, T.; Mayes, L.; Alnemri, D.; Cingolani, G.; Alnemri, E.S. Cleavage of DFNA5 by caspase-3 during apoptosis mediates progression to secondary necrotic/pyroptotic cell death. *Nat. Commun* **2017**, *8*, 14128. [CrossRef]
26. Galluzzi, L.; Kepp, O.; Kroemer, G. Mitochondrial regulation of cell death: A phylogenetically conserved control. *Microb Cell* **2016**, *3*, 101–108. [CrossRef]
27. Kalkavan, H.; Green, D.R. MOMP, cell suicide as a BCL-2 family business. *Cell Death Differ.* **2018**, *25*, 46–55. [CrossRef]
28. Wei, M.C.; Zong, W.X.; Cheng, E.H.; Lindsten, T.; Panoutsakopoulou, V.; Ross, A.J.; Roth, K.A.; MacGregor, G.R.; Thompson, C.B.; Korsmeyer, S.J. Proapoptotic BAX and BAK: A requisite gateway to mitochondrial dysfunction and death. *Science* **2001**, *292*, 727–730. [CrossRef]
29. Subburaj, Y.; Cosentino, K.; Axmann, M.; Pedrueza-Villalmanzo, E.; Hermann, E.; Bleicken, S.; Spatz, J.; Garcia-Saez, A.J. Bax monomers form dimer units in the membrane that further self-assemble into multiple oligomeric species. *Nat. Commun* **2015**, *6*, 8042. [CrossRef]
30. Ren, D.; Tu, H.C.; Kim, H.; Wang, G.X.; Bean, G.R.; Takeuchi, O.; Jeffers, J.R.; Zambetti, G.P.; Hsieh, J.J.; Cheng, E.H. BID, BIM, and PUMA are essential for activation of the BAX- and BAK-dependent cell death program. *Science* **2010**, *330*, 1390–1393. [CrossRef] [PubMed]
31. Luo, X.; Budihardjo, I.; Zou, H.; Slaughter, C.; Wang, X. Bid, a Bcl2 interacting protein, mediates cytochrome c release from mitochondria in response to activation of cell surface death receptors. *Cell* **1998**, *94*, 481–490. [CrossRef]
32. Villunger, A.; Michalak, E.M.; Coultas, L.; Mullauer, F.; Bock, G.; Ausserlechner, M.J.; Adams, J.M.; Strasser, A. p53- and drug-induced apoptotic responses mediated by BH3-only proteins puma and noxa. *Science* **2003**, *302*, 1036–1038. [CrossRef]
33. Yamada, K.; Yoshida, K. Mechanical insights into the regulation of programmed cell death by p53 via mitochondria. *Biochim Biophys Acta Mol. Cell Res.* **2019**, *1866*, 839–848. [CrossRef]
34. Czabotar, P.E.; Lessene, G.; Strasser, A.; Adams, J.M. Control of apoptosis by the BCL-2 protein family: Implications for physiology and therapy. *Nat. Rev. Mol. Cell Biol.* **2014**, *15*, 49–63. [CrossRef]
35. Oltvai, Z.N.; Milliman, C.L.; Korsmeyer, S.J. Bcl-2 heterodimerizes in vivo with a conserved homolog, Bax, that accelerates programmed cell death. *Cell* **1993**, *74*, 609–619. [CrossRef]
36. Llambi, F.; Moldoveanu, T.; Tait, S.W.; Bouchier-Hayes, L.; Temirov, J.; McCormick, L.L.; Dillon, C.P.; Green, D.R. A unified model of mammalian BCL-2 protein family interactions at the mitochondria. *Mol. Cell* **2011**, *44*, 517–531. [CrossRef] [PubMed]
37. Vervliet, T.; Parys, J.B.; Bultynck, G. Bcl-2 proteins and calcium signaling: Complexity beneath the surface. *Oncogene* **2016**, *35*, 5079–5092. [CrossRef] [PubMed]
38. Scorrano, L.; Oakes, S.A.; Opferman, J.T.; Cheng, E.H.; Sorcinelli, M.D.; Pozzan, T.; Korsmeyer, S.J. BAX and BAK regulation of endoplasmic reticulum Ca^{2+}: A control point for apoptosis. *Science* **2003**, *300*, 135–139. [CrossRef] [PubMed]
39. Chen, Z.X.; Pervaiz, S. Involvement of cytochrome c oxidase subunits Va and Vb in the regulation of cancer cell metabolism by Bcl-2. *Cell Death Differ.* **2010**, *17*, 408–420. [CrossRef]
40. Chen, Z.X.; Pervaiz, S. Bcl-2 induces pro-oxidant state by engaging mitochondrial respiration in tumor cells. *Cell Death Differ.* **2007**, *14*, 1617–1627. [CrossRef]
41. Letai, A.; Bassik, M.C.; Walensky, L.D.; Sorcinelli, M.D.; Weiler, S.; Korsmeyer, S.J. Distinct BH3 domains either sensitize or activate mitochondrial apoptosis, serving as prototype cancer therapeutics. *Cancer Cell* **2002**, *2*, 183–192. [CrossRef]
42. Vance, J.E. Phospholipid synthesis and transport in mammalian cells. *Traffic* **2015**, *16*, 1–18. [CrossRef]
43. Phillips, M.J.; Voeltz, G.K. Structure and function of ER membrane contact sites with other organelles. *Nat. Rev. Mol. Cell Biol.* **2016**, *17*, 69–82. [CrossRef]
44. Missiroli, S.; Patergnani, S.; Caroccia, N.; Pedriali, G.; Perrone, M.; Previati, M.; Wieckowski, M.R.; Giorgi, C. Mitochondria-associated membranes (MAMs) and inflammation. *Cell Death Dis.* **2018**, *9*, 329. [CrossRef]

45. Li, P.; Nijhawan, D.; Budihardjo, I.; Srinivasula, S.M.; Ahmad, M.; Alnemri, E.S.; Wang, X. Cytochrome c and dATP-dependent formation of Apaf-1/caspase-9 complex initiates an apoptotic protease cascade. *Cell* **1997**, *91*, 479–489. [CrossRef]
46. Verhagen, A.M.; Ekert, P.G.; Pakusch, M.; Silke, J.; Connolly, L.M.; Reid, G.E.; Moritz, R.L.; Simpson, R.J.; Vaux, D.L. Identification of DIABLO, a mammalian protein that promotes apoptosis by binding to and antagonizing IAP proteins. *Cell* **2000**, *102*, 43–53. [CrossRef]
47. Martins, L.M. The serine protease Omi/HtrA2: A second mammalian protein with a Reaper-like function. *Cell Death Differ.* **2002**, *9*, 699–701. [CrossRef]
48. Zamzami, N.; Marchetti, P.; Castedo, M.; Decaudin, D.; Macho, A.; Hirsch, T.; Susin, S.A.; Petit, P.X.; Mignotte, B.; Kroemer, G. Sequential reduction of mitochondrial transmembrane potential and generation of reactive oxygen species in early programmed cell death. *J. Exp. Med.* **1995**, *182*, 367–377. [CrossRef]
49. Julien, O.; Wells, J.A. Caspases and their substrates. *Cell Death Differ.* **2017**, *24*, 1380–1389. [CrossRef]
50. Nagata, S. DNA degradation in development and programmed cell death. *Annu Rev. Immunol.* **2005**, *23*, 853–875. [CrossRef]
51. Martin, S.J.; Finucane, D.M.; Amarante-Mendes, G.P.; O'Brien, G.A.; Green, D.R. Phosphatidylserine externalization during CD95-induced apoptosis of cells and cytoplasts requires ICE/CED-3 protease activity. *J. Biol. Chem.* **1996**, *271*, 28753–28756. [CrossRef]
52. Coleman, M.L.; Sahai, E.A.; Yeo, M.; Bosch, M.; Dewar, A.; Olson, M.F. Membrane blebbing during apoptosis results from caspase-mediated activation of ROCK I. *Nat. Cell Biol.* **2001**, *3*, 339–345. [CrossRef]
53. Eckelman, B.P.; Salvesen, G.S.; Scott, F.L. Human inhibitor of apoptosis proteins: Why XIAP is the black sheep of the family. *EMBO Rep.* **2006**, *7*, 988–994. [CrossRef]
54. Eckelman, B.P.; Salvesen, G.S. The human anti-apoptotic proteins cIAP1 and cIAP2 bind but do not inhibit caspases. *J. Biol. Chem.* **2006**, *281*, 3254–3260. [CrossRef]
55. Kamber Kaya, H.E.; Ditzel, M.; Meier, P.; Bergmann, A. An inhibitory mono-ubiquitylation of the Drosophila initiator caspase Dronc functions in both apoptotic and non-apoptotic pathways. *PLoS Genet.* **2017**, *13*, e1006438. [CrossRef]
56. Varfolomeev, E.; Goncharov, T.; Fedorova, A.V.; Dynek, J.N.; Zobel, K.; Deshayes, K.; Fairbrother, W.J.; Vucic, D. c-IAP1 and c-IAP2 are critical mediators of tumor necrosis factor alpha (TNFalpha)-induced NF-kappaB activation. *J. Biol. Chem.* **2008**, *283*, 24295–24299. [CrossRef]
57. Witt, A.; Vucic, D. Diverse ubiquitin linkages regulate RIP kinases-mediated inflammatory and cell death signaling. *Cell Death Differ.* **2017**, *24*, 1160–1171. [CrossRef]
58. Martins, I.; Wang, Y.; Michaud, M.; Ma, Y.; Sukkurwala, A.Q.; Shen, S.; Kepp, O.; Metivier, D.; Galluzzi, L.; Perfettini, J.L.; et al. Molecular mechanisms of ATP secretion during immunogenic cell death. *Cell Death Differ.* **2014**, *21*, 79–91. [CrossRef]
59. Huang, Q.; Li, F.; Liu, X.; Li, W.; Shi, W.; Liu, F.F.; O'Sullivan, B.; He, Z.; Peng, Y.; Tan, A.C.; et al. Caspase 3-mediated stimulation of tumor cell repopulation during cancer radiotherapy. *Nat. Med.* **2011**, *17*, 860–866. [CrossRef]
60. Mehlen, P.; Rabizadeh, S.; Snipas, S.J.; Assa-Munt, N.; Salvesen, G.S.; Bredesen, D.E. The DCC gene product induces apoptosis by a mechanism requiring receptor proteolysis. *Nature* **1998**, *395*, 801–804. [CrossRef]
61. Llambi, F.; Causeret, F.; Bloch-Gallego, E.; Mehlen, P. Netrin-1 acts as a survival factor via its receptors UNC5H and DCC. *EMBO J.* **2001**, *20*, 2715–2722. [CrossRef] [PubMed]
62. Matsunaga, E.; Tauszig-Delamasure, S.; Monnier, P.P.; Mueller, B.K.; Strittmatter, S.M.; Mehlen, P.; Chedotal, A. RGM and its receptor neogenin regulate neuronal survival. *Nat. Cell Biol.* **2004**, *6*, 749–755. [CrossRef] [PubMed]
63. Rabizadeh, S.; Oh, J.; Zhong, L.T.; Yang, J.; Bitler, C.M.; Butcher, L.L.; Bredesen, D.E. Induction of apoptosis by the low-affinity NGF receptor. *Science* **1993**, *261*, 345–348. [CrossRef] [PubMed]
64. Bordeaux, M.C.; Forcet, C.; Granger, L.; Corset, V.; Bidaud, C.; Billaud, M.; Bredesen, D.E.; Edery, P.; Mehlen, P. The RET proto-oncogene induces apoptosis: A novel mechanism for Hirschsprung disease. *EMBO J.* **2000**, *19*, 4056–4063. [CrossRef]
65. Tauszig-Delamasure, S.; Yu, L.Y.; Cabrera, J.R.; Bouzas-Rodriguez, J.; Mermet-Bouvier, C.; Guix, C.; Bordeaux, M.C.; Arumae, U.; Mehlen, P. The TrkC receptor induces apoptosis when the dependence receptor notion meets the neurotrophin paradigm. *Proc. Natl. Acad. Sci. USA* **2007**, *104*, 13361–13366. [CrossRef]

66. Wang, H.; Boussouar, A.; Mazelin, L.; Tauszig-Delamasure, S.; Sun, Y.; Goldschneider, D.; Paradisi, A.; Mehlen, P. The Proto-oncogene c-Kit Inhibits Tumor Growth by Behaving as a Dependence Receptor. *Mol. Cell* **2018**, *72*, 413–425 e415. [CrossRef]
67. Negulescu, A.M.; Mehlen, P. Dependence receptors—The dark side awakens. *FEBS J.* **2018**, *285*, 3909–3924. [CrossRef]
68. Paradisi, A.; Maisse, C.; Bernet, A.; Coissieux, M.M.; Maccarrone, M.; Scoazec, J.Y.; Mehlen, P. NF-kappaB regulates netrin-1 expression and affects the conditional tumor suppressive activity of the netrin-1 receptors. *Gastroenterology* **2008**, *135*, 1248–1257. [CrossRef]
69. Paradisi, A.; Maisse, C.; Coissieux, M.M.; Gadot, N.; Lepinasse, F.; Delloye-Bourgeois, C.; Delcros, J.G.; Svrcek, M.; Neufert, C.; Flejou, J.F.; et al. Netrin-1 up-regulation in inflammatory bowel diseases is required for colorectal cancer progression. *Proc. Natl. Acad. Sci. USA* **2009**, *106*, 17146–17151. [CrossRef]
70. Aherne, C.M.; Collins, C.B.; Eltzschig, H.K. Netrin-1 guides inflammatory cell migration to control mucosal immune responses during intestinal inflammation. *Tissue Barriers* **2013**, *1*, e24957. [CrossRef]
71. Chen, Z.; Chen, Y.; Zhou, J.; Li, Y.; Gong, C.; Wang, X. Netrin-1 reduces lung ischemia-reperfusion injury by increasing the proportion of regulatory T cells. *J. Int. Med. Res.* **2020**, *48*, 300060520926415. [CrossRef] [PubMed]
72. von Karstedt, S.; Montinaro, A.; Walczak, H. Exploring the TRAILs less travelled: TRAIL in cancer biology and therapy. *Nat. Rev. Cancer* **2017**, *17*, 352–366. [CrossRef] [PubMed]
73. Wajant, H.; Siegmund, D. TNFR1 and TNFR2 in the Control of the Life and Death Balance of Macrophages. *Front. Cell Dev. Biol.* **2019**, *7*, 91. [CrossRef] [PubMed]
74. Chinnaiyan, A.M.; O'Rourke, K.; Tewari, M.; Dixit, V.M. FADD, a novel death domain-containing protein, interacts with the death domain of Fas and initiates apoptosis. *Cell* **1995**, *81*, 505–512. [CrossRef]
75. Kischkel, F.C.; Lawrence, D.A.; Chuntharapai, A.; Schow, P.; Kim, K.J.; Ashkenazi, A. Apo2L/TRAIL-dependent recruitment of endogenous FADD and caspase-8 to death receptors 4 and 5. *Immunity* **2000**, *12*, 611–620. [CrossRef]
76. Muzio, M.; Chinnaiyan, A.M.; Kischkel, F.C.; O'Rourke, K.; Shevchenko, A.; Ni, J.; Scaffidi, C.; Bretz, J.D.; Zhang, M.; Gentz, R.; et al. FLICE, a novel FADD-homologous ICE/CED-3-like protease, is recruited to the CD95 (Fas/APO-1) death–inducing signaling complex. *Cell* **1996**, *85*, 817–827. [CrossRef]
77. Tummers, B.; Mari, L.; Guy, C.S.; Heckmann, B.L.; Rodriguez, D.A.; Ruhl, S.; Moretti, J.; Crawford, J.C.; Fitzgerald, P.; Kanneganti, T.D.; et al. Caspase-8-Dependent Inflammatory Responses Are Controlled by Its Adaptor, FADD, and Necroptosis. *Immunity* **2020**, *52*, 994–1006.e8. [CrossRef]
78. Brenner, D.; Blaser, H.; Mak, T.W. Regulation of tumour necrosis factor signalling: Live or let die. *Nat. Rev. Immunol.* **2015**, *15*, 362–374. [CrossRef]
79. Micheau, O.; Lens, S.; Gaide, O.; Alevizopoulos, K.; Tschopp, J. NF-kappaB signals induce the expression of c-FLIP. *Mol. Cell Biol.* **2001**, *21*, 5299–5305. [CrossRef]
80. Grootjans, S.; Vanden Berghe, T.; Vandenabeele, P. Initiation and execution mechanisms of necroptosis: An overview. *Cell Death Differ.* **2017**, *24*, 1184–1195. [CrossRef]
81. Li, J.; McQuade, T.; Siemer, A.B.; Napetschnig, J.; Moriwaki, K.; Hsiao, Y.S.; Damko, E.; Moquin, D.; Walz, T.; McDermott, A.; et al. The RIP1/RIP3 necrosome forms a functional amyloid signaling complex required for programmed necrosis. *Cell* **2012**, *150*, 339–350. [CrossRef]
82. Dondelinger, Y.; Declercq, W.; Montessuit, S.; Roelandt, R.; Goncalves, A.; Bruggeman, I.; Hulpiau, P.; Weber, K.; Sehon, C.A.; Marquis, R.W.; et al. MLKL compromises plasma membrane integrity by binding to phosphatidylinositol phosphates. *Cell Rep.* **2014**, *7*, 971–981. [CrossRef]
83. He, S.; Liang, Y.; Shao, F.; Wang, X. Toll-like receptors activate programmed necrosis in macrophages through a receptor-interacting kinase-3-mediated pathway. *Proc. Natl. Acad. Sci. USA* **2011**, *108*, 20054–20059. [CrossRef]
84. Upton, J.W.; Kaiser, W.J.; Mocarski, E.S. DAI/ZBP1/DLM-1 complexes with RIP3 to mediate virus-induced programmed necrosis that is targeted by murine cytomegalovirus vIRA. *Cell Host Microbe* **2012**, *11*, 290–297. [CrossRef]
85. Kaiser, W.J.; Sridharan, H.; Huang, C.; Mandal, P.; Upton, J.W.; Gough, P.J.; Sehon, C.A.; Marquis, R.W.; Bertin, J.; Mocarski, E.S. Toll-like receptor 3-mediated necrosis via TRIF, RIP3, and MLKL. *J. Biol. Chem.* **2013**, *288*, 31268–31279. [CrossRef]

86. McComb, S.; Cessford, E.; Alturki, N.A.; Joseph, J.; Shutinoski, B.; Startek, J.B.; Gamero, A.M.; Mossman, K.L.; Sad, S. Type-I interferon signaling through ISGF3 complex is required for sustained Rip3 activation and necroptosis in macrophages. *Proc. Natl. Acad. Sci. USA* **2014**, *111*, E3206–E3213. [CrossRef]
87. Man, S.M.; Karki, R.; Kanneganti, T.D. Molecular mechanisms and functions of pyroptosis, inflammatory caspases and inflammasomes in infectious diseases. *Immunol. Rev.* **2017**, *277*, 61–75. [CrossRef]
88. Franchi, L.; Eigenbrod, T.; Munoz-Planillo, R.; Nunez, G. The inflammasome: A caspase-1-activation platform that regulates immune responses and disease pathogenesis. *Nat. Immunol.* **2009**, *10*, 241–247. [CrossRef]
89. Shi, J.; Zhao, Y.; Wang, Y.; Gao, W.; Ding, J.; Li, P.; Hu, L.; Shao, F. Inflammatory caspases are innate immune receptors for intracellular LPS. *Nature* **2014**, *514*, 187–192. [CrossRef]
90. Jiang, S.; Gu, H.; Zhao, Y.; Sun, L. Teleost Gasdermin E Is Cleaved by Caspase 1, 3, and 7 and Induces Pyroptosis. *J. Immunol.* **2019**, *203*, 1369–1382. [CrossRef]
91. Yatim, N.; Cullen, S.; Albert, M.L. Dying cells actively regulate adaptive immune responses. *Nat. Rev. Immunol.* **2017**, *17*, 262–275. [CrossRef]
92. Green, D.R.; Ferguson, T.; Zitvogel, L.; Kroemer, G. Immunogenic and tolerogenic cell death. *Nat. Rev. Immunol.* **2009**, *9*, 353–363. [CrossRef]
93. Christgen, S.; Zheng, M.; Kesavardhana, S.; Karki, R.; Malireddi, R.K.S.; Banoth, B.; Place, D.E.; Briard, B.; Sharma, B.R.; Tuladhar, S.; et al. Identification of the PANoptosome: A Molecular Platform Triggering Pyroptosis, Apoptosis, and Necroptosis (PANoptosis). *Front. Cell Infect. MicroBiol.* **2020**, *10*, 237. [CrossRef]
94. Boya, P.; Gonzalez-Polo, R.A.; Casares, N.; Perfettini, J.L.; Dessen, P.; Larochette, N.; Metivier, D.; Meley, D.; Souquere, S.; Yoshimori, T.; et al. Inhibition of macroautophagy triggers apoptosis. *Mol. Cell Biol.* **2005**, *25*, 1025–1040. [CrossRef]
95. Galluzzi, L.; Baehrecke, E.H.; Ballabio, A.; Boya, P.; Bravo-San Pedro, J.M.; Cecconi, F.; Choi, A.M.; Chu, C.T.; Codogno, P.; Colombo, M.I.; et al. Molecular definitions of autophagy and related processes. *EMBO J.* **2017**, *36*, 1811–1836. [CrossRef]
96. Levine, B.; Huang, Q.; Isaacs, J.T.; Reed, J.C.; Griffin, D.E.; Hardwick, J.M. Conversion of lytic to persistent alphavirus infection by the bcl-2 cellular oncogene. *Nature* **1993**, *361*, 739–742. [CrossRef]
97. Grandgirard, D.; Studer, E.; Monney, L.; Belser, T.; Fellay, I.; Borner, C.; Michel, M.R. Alphaviruses induce apoptosis in Bcl-2-overexpressing cells: Evidence for a caspase-mediated, proteolytic inactivation of Bcl-2. *EMBO J.* **1998**, *17*, 1268–1278. [CrossRef]
98. Moriishi, K.; Koura, M.; Matsuura, Y. Induction of Bad-mediated apoptosis by Sindbis virus infection: Involvement of pro-survival members of the Bcl-2 family. *Virology* **2002**, *292*, 258–271. [CrossRef]
99. Keck, F.; Brooks-Faulconer, T.; Lark, T.; Ravishankar, P.; Bailey, C.; Salvador-Morales, C.; Narayanan, A. Altered mitochondrial dynamics as a consequence of Venezuelan Equine encephalitis virus infection. *Virulence* **2017**, *8*, 1849–1866. [CrossRef]
100. Lin, K.I.; Lee, S.H.; Narayanan, R.; Baraban, J.M.; Hardwick, J.M.; Ratan, R.R. Thiol agents and Bcl-2 identify an alphavirus-induced apoptotic pathway that requires activation of the transcription factor NF-kappa B. *J. Cell Biol.* **1995**, *131*, 1149–1161. [CrossRef]
101. Yoshinaka, Y.; Takahashi, Y.; Nakamura, S.; Katoh, I.; Takio, K.; Ikawa, Y. Induction of manganese-superoxide dismutase in MRC-5 cells persistently infected with an alphavirus, sindbis. *BioChem. Biophys Res. Commun.* **1999**, *261*, 139–143. [CrossRef]
102. Dhanwani, R.; Khan, M.; Bhaskar, A.S.; Singh, R.; Patro, I.K.; Rao, P.V.; Parida, M.M. Characterization of Chikungunya virus infection in human neuroblastoma SH-SY5Y cells: Role of apoptosis in neuronal cell death. *Virus Res.* **2012**, *163*, 563–572. [CrossRef]
103. Iranpour, M.; Moghadam, A.R.; Yazdi, M.; Ande, S.R.; Alizadeh, J.; Wiechec, E.; Lindsay, R.; Drebot, M.; Coombs, K.M.; Ghavami, S. Apoptosis, autophagy and unfolded protein response pathways in Arbovirus replication and pathogenesis. *Expert Rev. Mol. Med.* **2016**, *18*, e1. [CrossRef] [PubMed]
104. Barry, G.; Fragkoudis, R.; Ferguson, M.C.; Lulla, A.; Merits, A.; Kohl, A.; Fazakerley, J.K. Semliki forest virus-induced endoplasmic reticulum stress accelerates apoptotic death of mammalian cells. *J. Virol.* **2010**, *84*, 7369–7377. [CrossRef]
105. Dahal, B.; Lin, S.C.; Carey, B.D.; Jacobs, J.L.; Dinman, J.D.; van Hoek, M.L.; Adams, A.A.; Kehn-Hall, K. EGR1 upregulation following Venezuelan equine encephalitis virus infection is regulated by ERK and PERK pathways contributing to cell death. *Virology* **2020**, *539*, 121–128. [CrossRef] [PubMed]

106. Baer, A.; Lundberg, L.; Swales, D.; Waybright, N.; Pinkham, C.; Dinman, J.D.; Jacobs, J.L.; Kehn-Hall, K. Venezuelan Equine Encephalitis Virus Induces Apoptosis through the Unfolded Protein Response Activation of EGR1. *J. Virol.* **2016**, *90*, 3558–3572. [CrossRef] [PubMed]
107. Frolov, I.; Agapov, E.; Hoffman, T.A., Jr.; Pragai, B.M.; Lippa, M.; Schlesinger, S.; Rice, C.M. Selection of RNA replicons capable of persistent noncytopathic replication in mammalian cells. *J. Virol.* **1999**, *73*, 3854–3865. [CrossRef]
108. Petrakova, O.; Volkova, E.; Gorchakov, R.; Paessler, S.; Kinney, R.M.; Frolov, I. Noncytopathic replication of Venezuelan equine encephalitis virus and eastern equine encephalitis virus replicons in Mammalian cells. *J. Virol.* **2005**, *79*, 7597–7608. [CrossRef]
109. Perri, S.; Driver, D.A.; Gardner, J.P.; Sherrill, S.; Belli, B.A.; Dubensky, T.W., Jr.; Polo, J.M. Replicon vectors derived from Sindbis virus and Semliki forest virus that establish persistent replication in host cells. *J. Virol.* **2000**, *74*, 9802–9807. [CrossRef]
110. Garmashova, N.; Gorchakov, R.; Frolova, E.; Frolov, I. Sindbis virus nonstructural protein nsP2 is cytotoxic and inhibits cellular transcription. *J. Virol.* **2006**, *80*, 5686–5696. [CrossRef]
111. Akhrymuk, I.; Kulemzin, S.V.; Frolova, E.I. Evasion of the innate immune response: The Old World alphavirus nsP2 protein induces rapid degradation of Rpb1, a catalytic subunit of RNA polymerase II. *J. Virol.* **2012**, *86*, 7180–7191. [CrossRef] [PubMed]
112. Akhrymuk, I.; Frolov, I.; Frolova, E.I. Sindbis Virus Infection Causes Cell Death by nsP2-Induced Transcriptional Shutoff or by nsP3-Dependent Translational Shutoff. *J. Virol.* **2018**, 92. [CrossRef] [PubMed]
113. Garmashova, N.; Gorchakov, R.; Volkova, E.; Paessler, S.; Frolova, E.; Frolov, I. The Old World and New World alphaviruses use different virus-specific proteins for induction of transcriptional shutoff. *J. Virol.* **2007**, *81*, 2472–2484. [CrossRef]
114. Fros, J.J.; Major, L.D.; Scholte, F.E.M.; Gardner, J.; van Hemert, M.J.; Suhrbier, A.; Pijlman, G.P. Chikungunya virus non-structural protein 2-mediated host shut-off disables the unfolded protein response. *J. Gen. Virol.* **2015**, *96*, 580–589. [CrossRef]
115. Fros, J.J.; van der Maten, E.; Vlak, J.M.; Pijlman, G.P. The C-terminal domain of chikungunya virus nsP2 independently governs viral RNA replication, cytopathicity, and inhibition of interferon signaling. *J. Virol.* **2013**, *87*, 10394–10400. [CrossRef] [PubMed]
116. Goertz, G.P.; McNally, K.L.; Robertson, S.J.; Best, S.M.; Pijlman, G.P.; Fros, J.J. The Methyltransferase-Like Domain of Chikungunya Virus nsP2 Inhibits the Interferon Response by Promoting the Nuclear Export of STAT1. *J. Virol.* **2018**, 92. [CrossRef]
117. Sarid, R.; Ben-Moshe, T.; Kazimirsky, G.; Weisberg, S.; Appel, E.; Kobiler, D.; Lustig, S.; Brodie, C. vFLIP protects PC-12 cells from apoptosis induced by Sindbis virus: Implications for the role of TNF-alpha. *Cell Death Differ.* **2001**, *8*, 1224–1231. [CrossRef]
118. Nava, V.E.; Rosen, A.; Veliuona, M.A.; Clem, R.J.; Levine, B.; Hardwick, J.M. Sindbis virus induces apoptosis through a caspase-dependent, CrmA-sensitive pathway. *J. Virol.* **1998**, *72*, 452–459. [CrossRef]
119. Kiiver, K.; Merits, A.; Sarand, I. Novel vectors expressing anti-apoptotic protein Bcl-2 to study cell death in Semliki Forest virus-infected cells. *Virus Res.* **2008**, *131*, 54–64. [CrossRef]
120. Zhou, Q.; Snipas, S.; Orth, K.; Muzio, M.; Dixit, V.M.; Salvesen, G.S. Target protease specificity of the viral serpin CrmA. Analysis of five caspases. *J. Biol. Chem.* **1997**, *272*, 7797–7800. [CrossRef]
121. Cho, Y.S.; Challa, S.; Moquin, D.; Genga, R.; Ray, T.D.; Guildford, M.; Chan, F.K. Phosphorylation-driven assembly of the RIP1-RIP3 complex regulates programmed necrosis and virus-induced inflammation. *Cell* **2009**, *137*, 1112–1123. [CrossRef] [PubMed]
122. Nailwal, H.; Chan, F.K. Necroptosis in anti-viral inflammation. *Cell Death Differ.* **2019**, *26*, 4–13. [CrossRef] [PubMed]
123. Joubert, P.E.; Werneke, S.; de la Calle, C.; Guivel-Benhassine, F.; Giodini, A.; Peduto, L.; Levine, B.; Schwartz, O.; Lenschow, D.; Albert, M.L. Chikungunya-induced cell death is limited by ER and oxidative stress-induced autophagy. *Autophagy* **2012**, *8*, 1261–1263. [CrossRef]
124. Joubert, P.E.; Werneke, S.W.; de la Calle, C.; Guivel-Benhassine, F.; Giodini, A.; Peduto, L.; Levine, B.; Schwartz, O.; Lenschow, D.J.; Albert, M.L. Chikungunya virus-induced autophagy delays caspase-dependent cell death. *J. Exp. Med.* **2012**, *209*, 1029–1047. [CrossRef] [PubMed]
125. Urban, C.; Rheme, C.; Maerz, S.; Berg, B.; Pick, R.; Nitschke, R.; Borner, C. Apoptosis induced by Semliki Forest virus is RNA replication dependent and mediated via Bak. *Cell Death Differ.* **2008**, *15*, 1396–1407. [CrossRef]

126. Jan, J.T.; Griffin, D.E. Induction of apoptosis by Sindbis virus occurs at cell entry and does not require virus replication. *J. Virol.* **1999**, *73*, 10296–10302. [CrossRef]
127. Cowling, V.; Downward, J. Caspase-6 is the direct activator of caspase-8 in the cytochrome c-induced apoptosis pathway: Absolute requirement for removal of caspase-6 prodomain. *Cell Death Differ.* **2002**, *9*, 1046–1056. [CrossRef]
128. Zheng, M.; Karki, R.; Vogel, P.; Kanneganti, T.D. Caspase-6 Is a Key Regulator of Innate Immunity, Inflammasome Activation, and Host Defense. *Cell* **2020**. [CrossRef]
129. El Maadidi, S.; Faletti, L.; Berg, B.; Wenzl, C.; Wieland, K.; Chen, Z.J.; Maurer, U.; Borner, C. A novel mitochondrial MAVS/Caspase-8 platform links RNA virus-induced innate antiviral signaling to Bax/Bak-independent apoptosis. *J. Immunol.* **2014**, *192*, 1171–1183. [CrossRef]
130. Vazquez, C.; Horner, S.M. MAVS Coordination of Antiviral Innate Immunity. *J. Virol.* **2015**, *89*, 6974–6977. [CrossRef]
131. Pan, P.; Zhang, Q.; Liu, W.; Wang, W.; Lao, Z.; Zhang, W.; Shen, M.; Wan, P.; Xiao, F.; Liu, F.; et al. Dengue Virus M Protein Promotes NLRP3 Inflammasome Activation To Induce Vascular Leakage in Mice. *J. Virol.* **2019**, 93. [CrossRef] [PubMed]
132. Liu, T.; Tang, L.; Tang, H.; Pu, J.; Gong, S.; Fang, D.; Zhang, H.; Li, Y.P.; Zhu, X.; Wang, W.; et al. Zika Virus Infection Induces Acute Kidney Injury Through Activating NLRP3 Inflammasome Via Suppressing Bcl-2. *Front. Immunol.* **2019**, *10*, 1925. [CrossRef]
133. Wang, Z.Y.; Zhen, Z.D.; Fan, D.Y.; Qin, C.F.; Han, D.S.; Zhou, H.N.; Wang, P.G.; An, J. Axl deficiency promotes the neuroinvasion of Japanese encephalitis virus by enhancing IL-1alpha production from pyroptotic macrophages. *J. Virol.* **2020**. [CrossRef]
134. Ekchariyawat, P.; Hamel, R.; Bernard, E.; Wichit, S.; Surasombatpattana, P.; Talignani, L.; Thomas, F.; Choumet, V.; Yssel, H.; Despres, P.; et al. Inflammasome signaling pathways exert antiviral effect against Chikungunya virus in human dermal fibroblasts. *Infect. Genet. Evol.* **2015**, *32*, 401–408. [CrossRef]
135. Chen, W.; Foo, S.S.; Zaid, A.; Teng, T.S.; Herrero, L.J.; Wolf, S.; Tharmarajah, K.; Vu, L.D.; van Vreden, C.; Taylor, A.; et al. Specific inhibition of NLRP3 in chikungunya disease reveals a role for inflammasomes in alphavirus-induced inflammation. *Nat. MicroBiol.* **2017**, *2*, 1435–1445. [CrossRef]
136. de Castro-Jorge, L.A.; de Carvalho, R.V.H.; Klein, T.M.; Hiroki, C.H.; Lopes, A.H.; Guimaraes, R.M.; Fumagalli, M.J.; Floriano, V.G.; Agostinho, M.R.; Slhessarenko, R.D.; et al. The NLRP3 inflammasome is involved with the pathogenesis of Mayaro virus. *PLoS Pathog.* **2019**, *15*, e1007934. [CrossRef] [PubMed]
137. Levine, B.; Goldman, J.E.; Jiang, H.H.; Griffin, D.E.; Hardwick, J.M. Bc1-2 protects mice against fatal alphavirus encephalitis. *Proc. Natl. Acad. Sci. USA* **1996**, *93*, 4810–4815. [CrossRef]
138. Scallan, M.F.; Allsopp, T.E.; Fazakerley, J.K. bcl-2 acts early to restrict Semliki Forest virus replication and delays virus-induced programmed cell death. *J. Virol.* **1997**, *71*, 1583–1590. [CrossRef]
139. Liang, X.H.; Kleeman, L.K.; Jiang, H.H.; Gordon, G.; Goldman, J.E.; Berry, G.; Herman, B.; Levine, B. Protection against fatal Sindbis virus encephalitis by beclin, a novel Bcl-2-interacting protein. *J. Virol.* **1998**, *72*, 8586–8596. [CrossRef]
140. Pattingre, S.; Tassa, A.; Qu, X.; Garuti, R.; Liang, X.H.; Mizushima, N.; Packer, M.; Schneider, M.D.; Levine, B. Bcl-2 antiapoptotic proteins inhibit Beclin 1-dependent autophagy. *Cell* **2005**, *122*, 927–939. [CrossRef]
141. Orvedahl, A.; MacPherson, S.; Sumpter, R., Jr.; Talloczy, Z.; Zou, Z.; Levine, B. Autophagy protects against Sindbis virus infection of the central nervous system. *Cell Host Microbe* **2010**, *7*, 115–127. [CrossRef] [PubMed]
142. Eng, K.E.; Panas, M.D.; Murphy, D.; Karlsson Hedestam, G.B.; McInerney, G.M. Accumulation of autophagosomes in Semliki Forest virus-infected cells is dependent on expression of the viral glycoproteins. *J. Virol.* **2012**, *86*, 5674–5685. [CrossRef]
143. Krejbich-Trotot, P.; Gay, B.; Li-Pat-Yuen, G.; Hoarau, J.J.; Jaffar-Bandjee, M.C.; Briant, L.; Gasque, P.; Denizot, M. Chikungunya triggers an autophagic process which promotes viral replication. *Virol. J.* **2011**, *8*, 432. [CrossRef]
144. Judith, D.; Mostowy, S.; Bourai, M.; Gangneux, N.; Lelek, M.; Lucas-Hourani, M.; Cayet, N.; Jacob, Y.; Prevost, M.C.; Pierre, P.; et al. Species-specific impact of the autophagy machinery on Chikungunya virus infection. *EMBO Rep.* **2013**, *14*, 534–544. [CrossRef] [PubMed]
145. Van Huizen, E.; McInerney, G.M. Activation of the PI3K-AKT Pathway by Old World Alphaviruses. *Cells* **2020**, *9*, 970. [CrossRef] [PubMed]

146. Krejbich-Trotot, P.; Denizot, M.; Hoarau, J.J.; Jaffar-Bandjee, M.C.; Das, T.; Gasque, P. Chikungunya virus mobilizes the apoptotic machinery to invade host cell defenses. *FASEB J.* **2011**, *25*, 314–325. [CrossRef] [PubMed]
147. Amara, A.; Mercer, J. Viral apoptotic mimicry. *Nat. Rev. MicroBiol.* **2015**, *13*, 461–469. [CrossRef] [PubMed]
148. Lewis, J.; Wesselingh, S.L.; Griffin, D.E.; Hardwick, J.M. Alphavirus-induced apoptosis in mouse brains correlates with neurovirulence. *J. Virol.* **1996**, *70*, 1828–1835. [CrossRef] [PubMed]
149. Jackson, A.C.; Rossiter, J.P. Apoptotic cell death is an important cause of neuronal injury in experimental Venezuelan equine encephalitis virus infection of mice. *Acta Neuropathol.* **1997**, *93*, 349–353. [CrossRef] [PubMed]
150. Appel, E.; Katzoff, A.; Ben-Moshe, T.; Kazimirsky, G.; Kobiler, D.; Lustig, S.; Brodie, C. Differential regulation of Bcl-2 and Bax expression in cells infected with virulent and nonvirulent strains of sindbis virus. *Virology* **2000**, *276*, 238–242. [CrossRef]
151. Sammin, D.J.; Butler, D.; Atkins, G.J.; Sheahan, B.J. Cell death mechanisms in the olfactory bulb of rats infected intranasally with Semliki forest virus. *Neuropathol. Appl. NeuroBiol.* **1999**, *25*, 236–243. [CrossRef] [PubMed]
152. Hoarau, J.J.; Jaffar Bandjee, M.C.; Krejbich Trotot, P.; Das, T.; Li-Pat-Yuen, G.; Dassa, B.; Denizot, M.; Guichard, E.; Ribera, A.; Henni, T.; et al. Persistent chronic inflammation and infection by Chikungunya arthritogenic alphavirus in spite of a robust host immune response. *J. Immunol.* **2010**, *184*, 5914–5927. [CrossRef] [PubMed]
153. Krejbich-Trotot, P.; Belarbi, E.; Ralambondrainy, M.; El-Kalamouni, C.; Viranaicken, W.; Roques, P.; Despres, P.; Gadea, G. The growth of arthralgic Ross River virus is restricted in human monocytic cells. *Virus Res.* **2016**, *225*, 64–68. [CrossRef] [PubMed]
154. Young, A.R.; Locke, M.C.; Cook, L.E.; Hiller, B.E.; Zhang, R.; Hedberg, M.L.; Monte, K.J.; Veis, D.J.; Diamond, M.S.; Lenschow, D.J. Dermal and muscle fibroblasts and skeletal myofibers survive chikungunya virus infection and harbor persistent RNA. *PLoS Pathog.* **2019**, *15*, e1007993. [CrossRef] [PubMed]
155. Weaver, S.C.; Scott, T.W.; Lorenz, L.H.; Lerdthusnee, K.; Romoser, W.S. Togavirus-associated pathologic changes in the midgut of a natural mosquito vector. *J. Virol.* **1988**, *62*, 2083–2090. [CrossRef]
156. Weaver, S.C.; Lorenz, L.H.; Scott, T.W. Pathologic changes in the midgut of Culex tarsalis following infection with Western equine encephalomyelitis virus. *Am. J. Trop. Med. Hyg.* **1992**, *47*, 691–701. [CrossRef] [PubMed]
157. Dong, S.; Behura, S.K.; Franz, A.W.E. The midgut transcriptome of Aedes aegypti fed with saline or protein meals containing chikungunya virus reveals genes potentially involved in viral midgut escape. *BMC Genom.* **2017**, *18*, 382. [CrossRef]
158. Kelly, E.M.; Moon, D.C.; Bowers, D.F. Apoptosis in mosquito salivary glands: Sindbis virus-associated and tissue homeostasis. *J. Gen. Virol.* **2012**, *93*, 2419–2424. [CrossRef]
159. Bowers, D.F.; Coleman, C.G.; Brown, D.T. Sindbis virus-associated pathology in Aedes albopictus (Diptera: Culicidae). *J. Med. Entomol.* **2003**, *40*, 698–705. [CrossRef]
160. Vo, M.; Linser, P.J.; Bowers, D.F. Organ-associated muscles in Aedes albopictus (Diptera: Culicidae) respond differentially to Sindbis virus. *J. Med. Entomol.* **2010**, *47*, 215–225. [CrossRef]
161. Wang, H.; Gort, T.; Boyle, D.L.; Clem, R.J. Effects of manipulating apoptosis on Sindbis virus infection of Aedes aegypti mosquitoes. *J. Virol.* **2012**, *86*, 6546–6554. [CrossRef] [PubMed]
162. Denton, D.; Aung-Htut, M.T.; Kumar, S. Developmentally programmed cell death in Drosophila. *Biochim. Biophys. Acta* **2013**, *1833*, 3499–3506. [CrossRef] [PubMed]
163. Zhou, L.; Song, Z.; Tittel, J.; Steller, H. HAC-1, a Drosophila homolog of APAF-1 and CED-4 functions in developmental and radiation-induced apoptosis. *Mol. Cell* **1999**, *4*, 745–755. [CrossRef]
164. Yu, X.; Wang, L.; Acehan, D.; Wang, X.; Akey, C.W. Three-dimensional structure of a double apoptosome formed by the Drosophila Apaf-1 related killer. *J. Mol. Biol.* **2006**, *355*, 577–589. [CrossRef] [PubMed]
165. Dorstyn, L.; Kumar, S. A cytochrome c-free fly apoptosome. *Cell Death Differ.* **2006**, *13*, 1049–1051. [CrossRef]
166. Doumanis, J.; Dorstyn, L.; Kumar, S. Molecular determinants of the subcellular localization of the Drosophila Bcl-2 homologues DEBCL and BUFFY. *Cell Death Differ.* **2007**, *14*, 907–915. [CrossRef]
167. Igaki, T.; Miura, M. Role of Bcl-2 family members in invertebrates. *Biochim. Biophys. Acta* **2004**, *1644*, 73–81. [CrossRef]
168. Clavier, A.; Rincheval-Arnold, A.; Colin, J.; Mignotte, B.; Guenal, I. Apoptosis in Drosophila: Which role for mitochondria? *Apoptosis* **2016**, *21*, 239–251. [CrossRef]
169. Cooper, D.M.; Granville, D.J.; Lowenberger, C. The insect caspases. *Apoptosis* **2009**, *14*, 247–256. [CrossRef]

170. Liu, Q.; Clem, R.J. Defining the core apoptosis pathway in the mosquito disease vector Aedes aegypti: The roles of iap1, ark, dronc, and effector caspases. *Apoptosis* **2011**, *16*, 105–113. [CrossRef]
171. Challa, M.; Malladi, S.; Pellock, B.J.; Dresnek, D.; Varadarajan, S.; Yin, Y.W.; White, K.; Bratton, S.B. Drosophila Omi, a mitochondrial-localized IAP antagonist and proapoptotic serine protease. *EMBO J.* **2007**, *26*, 3144–3156. [CrossRef] [PubMed]
172. Chai, J.; Yan, N.; Huh, J.R.; Wu, J.W.; Li, W.; Hay, B.A.; Shi, Y. Molecular mechanism of Reaper-Grim-Hid-mediated suppression of DIAP1-dependent Dronc ubiquitination. *Nat. Struct. Biol.* **2003**, *10*, 892–898. [CrossRef] [PubMed]
173. Brodsky, M.H.; Nordstrom, W.; Tsang, G.; Kwan, E.; Rubin, G.M.; Abrams, J.M. Drosophila p53 binds a damage response element at the reaper locus. *Cell* **2000**, *101*, 103–113. [CrossRef]
174. Igaki, T.; Kanda, H.; Yamamoto-Goto, Y.; Kanuka, H.; Kuranaga, E.; Aigaki, T.; Miura, M. Eiger, a TNF superfamily ligand that triggers the Drosophila JNK pathway. *EMBO J.* **2002**, *21*, 3009–3018. [CrossRef]
175. Tafesh-Edwards, G.; Eleftherianos, I. JNK signaling in Drosophila immunity and homeostasis. *Immunol. Lett.* **2020**, *226*, 7–11. [CrossRef] [PubMed]
176. Kleino, A.; Silverman, N. The Drosophila IMD pathway in the activation of the humoral immune response. *Dev. Comp. Immunol.* **2014**, *42*, 25–35. [CrossRef]
177. Ocampo, C.B.; Caicedo, P.A.; Jaramillo, G.; Ursic Bedoya, R.; Baron, O.; Serrato, I.M.; Cooper, D.M.; Lowenberger, C. Differential expression of apoptosis related genes in selected strains of Aedes aegypti with different susceptibilities to dengue virus. *PLoS ONE* **2013**, *8*, e61187. [CrossRef] [PubMed]
178. Tettamanti, G.; Casartelli, M. Cell death during complete metamorphosis. *Philos. Trans. R. Soc. Lond. B Biol. Sci.* **2019**, *374*, 20190065. [CrossRef] [PubMed]
179. Gohel, R.; Kournoutis, A.; Petridi, S.; Nezis, I.P. Molecular mechanisms of selective autophagy in Drosophila. *Int. Rev. Cell Mol. Biol.* **2020**, *354*, 63–105. [CrossRef] [PubMed]
180. Means, J.C.; Passarelli, A.L. Viral fibroblast growth factor, matrix metalloproteases, and caspases are associated with enhancing systemic infection by baculoviruses. *Proc. Natl. Acad. Sci. USA* **2010**, *107*, 9825–9830. [CrossRef]
181. Wang, H.; Blair, C.D.; Olson, K.E.; Clem, R.J. Effects of inducing or inhibiting apoptosis on Sindbis virus replication in mosquito cells. *J. Gen. Virol.* **2008**, *89*, 2651–2661. [CrossRef] [PubMed]
182. O'Neill, K.; Olson, B.J.; Huang, N.; Unis, D.; Clem, R.J. Rapid selection against arbovirus-induced apoptosis during infection of a mosquito vector. *Proc. Natl. Acad. Sci. USA* **2015**, *112*, E1152–E1161. [CrossRef] [PubMed]
183. Tchankouo-Nguetcheu, S.; Khun, H.; Pincet, L.; Roux, P.; Bahut, M.; Huerre, M.; Guette, C.; Choumet, V. Differential protein modulation in midguts of Aedes aegypti infected with chikungunya and dengue 2 viruses. *PLoS ONE* **2010**, 5. [CrossRef] [PubMed]
184. Camini, F.C.; da Silva Caetano, C.C.; Almeida, L.T.; da Costa Guerra, J.F.; de Mello Silva, B.; de Queiroz Silva, S.; de Magalhaes, J.C.; de Brito Magalhaes, C.L. Oxidative stress in Mayaro virus infection. *Virus Res.* **2017**, *236*, 1–8. [CrossRef]
185. Felton, G.W.; Summers, C.B. Antioxidant systems in insects. *Arch. Insect BioChem. Physiol.* **1995**, *29*, 187–197. [CrossRef]
186. Whiten, S.R.; Eggleston, H.; Adelman, Z.N. Ironing out the Details: Exploring the Role of Iron and Heme in Blood-Sucking Arthropods. *Front. Physiol.* **2017**, *8*, 1134. [CrossRef]
187. Patel, R.K.; Hardy, R.W. Role for the phosphatidylinositol 3-kinase-Akt-TOR pathway during sindbis virus replication in arthropods. *J. Virol.* **2012**, *86*, 3595–3604. [CrossRef]
188. Brackney, D.E.; Correa, M.A.; Cozens, D.W. The impact of autophagy on arbovirus infection of mosquito cells. *PLoS Negl. Trop. Dis.* **2020**, *14*, e0007754. [CrossRef]
189. Liu, B.; Behura, S.K.; Clem, R.J.; Schneemann, A.; Becnel, J.; Severson, D.W.; Zhou, L. P53-mediated rapid induction of apoptosis conveys resistance to viral infection in Drosophila melanogaster. *PLoS Pathog.* **2013**, *9*, e1003137. [CrossRef]
190. Lamiable, O.; Arnold, J.; de Faria, I.; Olmo, R.P.; Bergami, F.; Meignin, C.; Hoffmann, J.A.; Marques, J.T.; Imler, J.L. Analysis of the Contribution of Hemocytes and Autophagy to Drosophila Antiviral Immunity. *J. Virol.* **2016**, *90*, 5415–5426. [CrossRef]
191. Vermaak, E.; Maree, F.F.; Theron, J. The Culicoides sonorensis inhibitor of apoptosis 1 protein protects mammalian cells from apoptosis induced by infection with African horse sickness virus and bluetongue virus. *Virus Res.* **2017**, *232*, 152–161. [CrossRef] [PubMed]

192. Li, Q.; Li, H.; Blitvich, B.J.; Zhang, J. The Aedes albopictus inhibitor of apoptosis 1 gene protects vertebrate cells from bluetongue virus-induced apoptosis. *Insect Mol. Biol.* **2007**, *16*, 93–105. [CrossRef] [PubMed]
193. Oliveira, J.H.M.; Talyuli, O.A.C.; Goncalves, R.L.S.; Paiva-Silva, G.O.; Sorgine, M.H.F.; Alvarenga, P.H.; Oliveira, P.L. Catalase protects Aedes aegypti from oxidative stress and increases midgut infection prevalence of Dengue but not Zika. *PLoS Negl. Trop. Dis.* **2017**, *11*, e0005525. [CrossRef] [PubMed]
194. Robin, M.; Issa, A.R.; Santos, C.C.; Napoletano, F.; Petitgas, C.; Chatelain, G.; Ruby, M.; Walter, L.; Birman, S.; Domingos, P.M.; et al. Drosophila p53 integrates the antagonism between autophagy and apoptosis in response to stress. *Autophagy* **2019**, *15*, 771–784. [CrossRef] [PubMed]
195. Chen, T.H.; Wu, Y.J.; Hou, J.N.; Chiang, Y.H.; Cheng, C.C.; Sifiyatun, E.; Chiu, C.H.; Wang, L.C.; Chen, W.J. A novel p53 paralogue mediates antioxidant defense of mosquito cells to survive dengue virus replication. *Virology* **2018**, *519*, 156–169. [CrossRef]
196. Chen, T.H.; Wu, Y.J.; Hou, J.N.; Chiu, C.H.; Chen, W.J. The p53 gene with emphasis on its paralogues in mosquitoes. *J. MicroBiol. Immunol. Infect.* **2017**, *50*, 747–754. [CrossRef]
197. Nasar, F.; Palacios, G.; Gorchakov, R.V.; Guzman, H.; Da Rosa, A.P.; Savji, N.; Popov, V.L.; Sherman, M.B.; Lipkin, W.I.; Tesh, R.B.; et al. Eilat virus, a unique alphavirus with host range restricted to insects by RNA replication. *Proc. Natl. Acad. Sci. USA* **2012**, *109*, 14622–14627. [CrossRef]
198. Hermanns, K.; Zirkel, F.; Kopp, A.; Marklewitz, M.; Rwego, I.B.; Estrada, A.; Gillespie, T.R.; Drosten, C.; Junglen, S. Discovery of a novel alphavirus related to Eilat virus. *J. Gen. Virol.* **2017**, *98*, 43–49. [CrossRef]
199. Elrefaey, A.M.; Abdelnabi, R.; Rosales Rosas, A.L.; Wang, L.; Basu, S.; Delang, L. Understanding the Mechanisms Underlying Host Restriction of Insect-Specific Viruses. *Viruses* **2020**, *12*, 964. [CrossRef]

Article

Importance of Endocytosis for the Biological Activity of Cedar Virus Fusion Protein

Kerstin Fischer, Martin H. Groschup and Sandra Diederich *

Institute of Novel and Emerging Diseases, Friedrich-Loeffler-Institut, Federal Research Institute for Animal Health, 17493 Greifswald—Insel Riems, Germany; Kerstin.Fischer@fli.de (K.F.); Martin.Groschup@fli.de (M.H.G.)

* Correspondence: Sandra.Diederich@fli.de; Tel.: +49-38351-7-1516

Received: 26 May 2020; Accepted: 4 September 2020; Published: 8 September 2020

Abstract: Endocytosis plays a particular role in the proteolytic activation of highly pathogenic henipaviruses Hendra (HeV) and Nipah virus (NiV) fusion (F) protein precursors. These proteins require endocytic uptake from the cell surface to be cleaved by cellular proteases within the endosomal compartment, followed by recycling to the plasma membrane for incorporation into budding virions or mediation of cell-cell fusion. This internalization largely depends on a tyrosine-based consensus motif for endocytosis present in the cytoplasmic tail of HeV and NiV F. Given the large number of tyrosine residues present in the F protein cytoplasmic domain of Cedar virus (CedV), a closely related but low pathogenic henipavirus, we aimed to investigate whether CedV F protein undergoes signal-mediated endocytosis from the cell surface controlled by tyrosine-based motifs present in its cytoplasmic tail and whether endocytosis is relevant for its biological activity. Therefore, tyrosine-based signals were mutated, and mutations were assessed for their effect on F cell surface expression, endocytosis, and biological activity. A membrane-proximal YXXΦ motif and a C-terminal di-tyrosine motif are of particular importance for cell surface expression and endocytosis rate. Furthermore, our data strongly indicate the pivotal role of endocytosis for the biological activity of the CedV F protein.

Keywords: cedar virus; henipavirus; fusion protein; endocytosis; biological activity

---

1. Introduction

Cedar virus (CedV) belongs to the *Henipavirus* genus within the *Paramyxoviridae* family and was first isolated from bat urine samples collected from an Australian *Pteropus* colony in 2012 [1]. Despite its genetic proximity to the highly pathogenic Hendra (HeV) and Nipah viruses (NiV), CedV has caused only asymptomatic infections in small animal models so far [1,2]. Therefore, research has focused on unraveling the molecular mechanisms leading to differences in the pathogenicity of these closely related viruses. One of the particularities of CedV is an impaired ability of the immunomodulatory phosphoprotein P to counteract the interferon response in cell culture [1,3]. Further differences are the receptor usage of the attachment proteins. The generally abundant expression of ephrins as cell entry receptors in numerous tissues and the high conservation among species results in a wide variety of susceptible hosts and a broad cell type tropism, which is fundamental to the zoonotic character and the pathogenesis of henipaviruses. While highly pathogenic HeV and NiV are known to utilize ephrin-B2, expressed i.e., in endothelial cells and lung tissue, and ephrin-B3, mainly found in the central nervous system, for cell entry [4–7], CedV is unable to use ephrin-B3 but rather binds to ephrin-B1, which is expressed in different tissues such as salivary glands, esophagus, and lung [8]. Therefore, recent studies have considered the distinct receptor usage of the CedV attachment protein to contribute to its reduced pathogenicity [8,9].

Besides receptor binding, the interaction of the attachment protein G with the viral fusion protein F is a prerequisite for virus entry into the host cell and virus spread. An indispensable step for the biological activity of fusion proteins and thus, viral infectivity, is the proteolytic cleavage of the precursor protein F_0 into the two subunits F_1 and F_2 [10]. Interestingly, proteolytic activation of HeV and NiV F protein differs considerably from that of other paramyxoviruses in terms of subcellular localization and protease usage. After transport along the secretory pathway, newly synthesized HeV and NiV F protein precursors require endocytosis from the cell surface to encounter the activating host cell protease and then become biologically active. Cleavage within the endosomal compartment is then followed by recycling to the cell surface before the incorporation of mature fusogenic F_1+F_2 heterodimers into newly budding virions [11–17]. Overall, both viral envelope proteins are important determinants of pathogenicity that need to act in concert to promote virus-cell membrane fusion needed for virus entry as well as cell-cell fusion resulting in syncytia formation and thus, virus spread.

While trafficking through early and late endosomes prior to fusion with cellular membranes plays a critical role in virus entry of many viruses such as influenza virus [18,19], ebolaviruses [20,21], and flaviviruses [22,23], it is dispensable during NiV entry [16]. Moreover, other viruses and their glycoproteins hijack endosomal pathways in order to support their replication in infected cells [24,25]. The viral envelope glycoprotein of human immunodeficiency virus 1 (HIV-1) for instance undergoes endocytosis during the viral replication cycle, which is hypothesized to serve as a mechanism to evade the host immune response by reducing its cell surface expression (reviewed in [26,27]). In addition, trafficking of the HIV-1 envelope glycoprotein through the endocytic recycling compartment has been recently described as an essential step for incorporation into virus particles [28]. Interestingly, endocytosis of herpesvirus glycoproteins has been discussed to play a functional role in cell–cell fusion and in the production of infectious particles by delivering the glycoproteins to the intracellular site where virus assembly takes place [25,29,30]. Noteworthy, a recent report even suggests that endocytic trafficking of HeV F protein rather than its proteolytic cleavage is a crucial step for efficient HeV virus-like particle (VLP) assembly [31].

Apart from its importance for the viral replication cycle, endocytosis represents a key process for numerous cellular functions. Characterized by the internalization of the plasma membrane and extracellular molecules from the cell surface into internal membrane compartments, endocytosis is required for many biological events such as maintaining the plasma membrane composition or transporting selected cargo molecules from the cell surface to the interior [32]. Among the different types, clathrin-mediated endocytosis (CME) is the best characterized [33,34]. CME of a transmembrane protein is a highly coordinated process that primarily involves the interaction between the cytoplasmic domain of the protein and intracellular adaptor proteins (AP) that select transmembrane cargo into clathrin-coated pits via association to the clathrin lattice. These adaptor complexes recognize specific sorting motifs within the cytoplasmic tail of the cargo protein that are usually tyrosine- or leucine-based [35–38]. One typical sorting motif in cytoplasmic tails of transmembrane proteins known to interact with specific cytosolic adaptor complex proteins is the tyrosine-based YXXΦ motif, in which X can be any amino acid and Φ stands for an amino acid with a large hydrophobic residue. Numerous studies describe YXXΦ motifs to be recognized by AP2 that selects cargo from the plasma membrane on the one hand and facilitates binding to clathrin on the other hand culminating in CME of selected transmembrane cargo [35–41]. However, adaptor complex proteins are not only known to mediate endocytic uptake from the cell surface. They have also been reported to be involved in distinct transport pathways e.g., from early endosomes to recycling or late endosomes, as well as to lysosomes by adapting to specific motifs [38,42]. Clathrin-mediated endocytic uptake of HeV and NiV F proteins strongly depends on a characteristic YXXΦ motif located in the membrane-proximal part of the cytoplasmic tail as depicted in Table 1 [11,12,43]. Consequently, disruption of this motif led to measurable effects in the biological activity of the proteins [11,12] while another di-tyrosine motif in the C-terminal part of the tail had only negligible effects on NiV F endocytosis [11].

Table 1. Boldface and underlined letters highlight (potential) endocytosis signals. Numbers above the sequence indicate the amino acid position. Cytoplasmic tails for NiV and HeV F proteins range from amino acid positions 519 to 546, cytoplasmic tail for CedV F is predicted for amino acid positions 517 to 557. Basic aa's that have been shown to be of importance for fusion protein functionality are highlighted in orange.

	519 **546**
NiV F	... EKKRNT**YSRL**EDRRVRPTSSGDL**YY**IGT
HeV F	... EKKRGN**YSRL**DDRQVRPVSNGDL**YY**IGT
	517 **557**
CedV F	... KSKHS**YKYNKF**IDDPD**YY**NDYKRERINGKASKSNNI**YY**VGD

The cytoplasmic tail of CedV F protein displays several tyrosine residues that might act as functional endocytosis motifs. In addition to a C-terminal di-tyrosine motif similarly found in HeV and NiV F protein, a second di-tyrosine motif is present. Interestingly, two degenerate motifs of the YxxΦ are obvious: A YxxN motif that has been shown to be a functional endocytosis motif for the spike protein of the Porcine Epidemic Diarrhea Virus [44] and a YxxN motif that has been described to function as an endocytosis signal in the Measles virus hemagglutinin [45]. Thus, in this study, we aimed to investigate the impact of those tyrosine-based motifs on cell surface expression, endocytosis, and biological activity of the protein. Our results suggest a signal-mediated uptake of CedV F protein and confirm the functional importance of a membrane-proximal YXXΦ and a C-terminal di-tyrosine motif for the biological activity of CedV F protein.

2. Materials and Methods

2.1. Cell Lines, Transfection

Vero76, MDCK-2 and HeLa cells (Collection of Cell Lines in Veterinary Medicine, Friedrich-Loeffler-Institut, FLI; CCLV-RIE 0228, 1061 and 0082, respectively) were maintained in Dulbecco's modified Eagle's medium (DMEM) supplemented with fetal calf serum (10%; DMEM10) and incubated at 37 °C. Vero76 and MDCK-2 cells were reverse transfected by using the Lipofectamine 3000 reagent (Invitrogen, Schwerte, Germany) following the manufacturer's protocol. HeLa cells were reverse transfected by using the METAFECTENE® transfection reagent (Biontex, Munich, Germany) according to the manufacturer's instructions.

2.2. Plasmids and Site-Directed Mutagenesis

The open reading frame (ORF) of CedV F and G (GenBank accession no. NC_025351.1) were synthesized by GeneArt (Thermo Fisher Scientific Inc., Regensburg, Germany) and subcloned into the pCAGGS expression vector. The ORF of the CedV F gene was codon-optimized for expression in human cells and HA-tagged at the C-terminus and will be designated as wild-type (wt) in the manuscript. Selected tyrosine residues within the cytoplasmic domain of CedV F protein were mutated to alanine by site-directed mutagenesis using the QuikChange Lightning Site-Directed Mutagenesis Kit (Agilent, Waldbronn, Germany) and primers designed according to manufacturer's instructions.

2.3. Generation of Polyclonal Antibodies

To obtain a polyclonal antiserum raised against the CedV F protein, a rabbit was immunized subcutaneously three times at three-week intervals with CedV F protein. The antigen was produced in 293T cells as described earlier [46]. Briefly, plasmid DNA encoding for CedV F protein was used to transfect 293T cells. At 48 h post-transfection (p.t.), the supernatant was purified through a 20% sucrose cushion at 96,000× g for 2 h at 4 °C. After washing for 30 min at 155,000× g, pellets were resuspended in Tris-sodium chloride buffer. Immunization experiments were assessed and approved by the competent authority for animal welfare issues of the Federal State of Mecklenburg-Western Pomerania, Germany (license number: LALLF 7221.3-2-042/17).

2.4. Antibody Uptake Assay

MDCK-2 cells were reverse transfected with plasmids encoding for parental and mutant CedV F proteins (1 µg/well) in a 24 well plate as mentioned above. At 24 h p.t., confluent monolayers were washed and incubated without prior fixation with the polyclonal anti-CedV F rabbit serum (1:200 in 0.35% bovine serum albumin (BSA) in PBS^{++} (PBS with 0.5 mM MgCl, 1 mM $CaCl_2$) (0.35% BSA/PBS^{++})). After incubation at 4 °C for one hour, cells were intensely washed and then either kept on ice or incubated with pre-warmed medium for 30 min at 37 °C to allow endocytic uptake. After fixing the cells with 4% paraformaldehyde (PFA), surface-bound primary antibodies were detected by an Alexa-Fluor (AF) 488-labeled secondary antibody (1:50 in 0.35% BSA/PBS^{++}; LifeTechnologies, Darmstadt, Germany) for 90 min at 4 °C. After permeabilization with methanol-acetone (1:1) for 10 min, AF 568-labelled secondary antibodies (1:500 in 0.35% BSA/PBS^{++}; LifeTechnologies, Darmstadt, Germany) were added to stain internalized primary antibodies. Images were acquired with an Eclipse Ti-S inverted microscope system and were processed with the ImageJ software version 1.45 s [47].

2.5. Metabolic Labeling

For pulse-chase analysis, MDCK-2 cells transiently expressing CedV F proteins were incubated at 24 h p.t. for 15 min with medium lacking cysteine and methionine, followed by incubation with medium containing [^{35}S]cysteine and -methionine (Hartmann Analytic, Braunschweig, Germany) at a final concentration of 100 µCi/mL for 10 min (pulse). Then, labeling medium was replaced by a serum-free nonradioactive medium, and cells were incubated at 37 °C for 2 h (chase). Afterwards, cells were washed with PBS, followed by cell lysis in radioimmunoprecipitation assay (RIPA) buffer (1% Triton X-100, 1% sodium deoxycholate, 0.1% SDS, 0.15 M NaCl, 10 mM EDTA, 50 units/mL aprotinin, and 20 mM Tris-HCl, pH 7.5). Cell lysates were centrifuged for 45 min at 20,000× g, and CedV F proteins were immunoprecipitated using a polyclonal anti-HA antibody (H6908; 1:500; Sigma, Darmstadt, Germany). Protein A-Sepharose CL-4B (Sigma, Darmstadt, Germany) suspension was added for another 45 min followed by several washes of the immune complexes with RIPA buffer. After suspension in sample buffer, proteins were separated on a 12% polyacrylamide gel under reducing conditions. Dried gels were subjected to autoradiography and analyzed with a CR35 Dark Box Image analyzer (Duerr Medical, Bietigheim-Bissingen, Germany).

2.6. Colocalization Studies

For colocalization studies of CedV F and mutants with early endosomes, antibody uptake assays were performed as described above with modifications. Briefly, at 24 h p.t., CedV F-expressing HeLa cells were incubated with the polyclonal rabbit anti-CedV F serum (1:200) for 1 h at 4 °C. After washing, cells were shifted to 37 °C for 5 or 30 min to allow endocytosis to occur. Then, surface-bound primary antibodies were blocked with a horseradish peroxidase (HRP)-conjugated secondary antibody (1:50; LifeTechnologies, Darmstadt, Germany). After fixation with 2% PFA for 20 min and permeabilization with 0.2% Triton X-100 in PBS, a mouse anti-human early endosomal antigen 1 (EEA-1) antibody (1:50; BD Biosciences, Heidelberg, Germany) was added for 1 h at 4 °C for staining of early endosomes. Internalized primary rabbit antibodies were stained with AF 488-conjugated secondary antibodies (1:500; LifeTechnologies, Darmstadt, Germany). Primary mouse antibodies bound to EEA-1 were detected with AF 568-conjugated secondary antibodies (1:500; LifeTechnologies, Darmstadt, Germany). Cell nuclei were counterstained with 4′,6-Diamidin-2-phenylindol (DAPI). Representative images were recorded with a confocal laser scanning microscope (Leica SP5) and processed with the ImageJ software version 1.45 s [47].

2.7. Surface Biotinylation and Western Blot Analysis

MDCK-2 cells were transfected with plasmid DNA encoding either CedV F or mutant CedV F protein genes. At 24 h p.t., cells were washed and incubated twice with 2 mg/mL EZ-Link® Sulfo-NHS-LC-Biotin (ThermoFisher Scientific, Waltham, MA, USA) for 20 min at 4 °C. Following cell lysis, F proteins were immunoprecipitated overnight using NeutrAvidin Agarose Resin (ThermoFisher Scientific, Waltham, MA, USA) according to the manufacturer's protocol. Proteins were separated on a 10% SDS-gel under reducing or non-reducing conditions and then transferred onto nitrocellulose. Biotinylated CedV F proteins were detected by incubation with the HA-tag specific rabbit antibody H6908 (dilution 1:2,000 in PBS-Tween (0.05%)) followed by labeling with anti-rabbit HRP-conjugated secondary antibodies (1:5.000). Under non-reducing conditions, HRP-labeled Concanavalin A (ConA; Sigma-Aldrich, Darmstadt, Germany, 1:1,000 in PBS containing 0.05% (v/v) TWEEN 20, 1 mM $CaCl_2$, 1 mM $MnCl_2$, and 1 mM $MgCl_2$ for 16 h at 20 °C) was used to stain NeutrAvidin-precipitated surface glycoproteins, which served as a loading control and as a reference to compare CedV F_0 protein intensities. Proteins were visualized using a chemiluminescent substrate (Clarity Western ECL substrate, Bio-Rad, Feldkirchen, Deutschland) and the Bio-Rad Molecular Imager Chemi Doc™ XRS+ in combination with the Image Lab software (Bio-Rad, Feldkirchen, Deutschland, Version 6.0.1).

2.8. Quantification of Endocytosis Rate by MESNA Reduction

MDCK-2 cells transiently expressing either CedV F or mutant F proteins were surface labeled with cleavable EZ-Link sulfo-NHS-SS-biotin (ThermoFisher Scientific, Waltham, MA, USA) at 4 °C. Then, wells were flooded with pre-warmed DMEM (37 °C) and shifted to 37 °C for either 5, 15, 30, or 90 min to allow endocytosis to occur. After rapid cooling to 4 °C, biotin still bound to the cell surface was cleaved by a membrane-impermeable reducing agent, 2-mercaptoethane-sulfonic acid sodium salt (MESNA; 50 mM in MESNA buffer (50 mM Tris, pH 8.5, 100 mM NaCl, 2.5 mM $CaCl_2$)) a total of three times for 20 min. One sample was kept on ice and was neither incubated at 37 °C nor cleaved with MESNA and thus served as the surface biotinylation control to determine the total amount of biotinylated protein (control). After cell lysis in RIPA buffer, CedV F and CedV mutant F proteins were immunoprecipitated using the H6908 antibody as described above and separated by SDS-PAGE under non-reducing conditions. Transfer to nitrocellulose was followed by the detection of biotinylated proteins with HRP- labeled streptavidin and a chemiluminescent substrate. Protein bands were quantified using ImageLab software (version 6.0.1). Mean internalization rates per min were calculated as the ratio of band intensities from internalized, intracellular biotinylated protein after 5, 15, 30, and 90 min and total surface-labeled F protein (control, 50%) divided by 5, 15, 30, and 90 min, respectively.

2.9. Fusion Assay

A total of 4×10^5 Vero76 cells or 3×10^5 MDCK-2 cells were co-transfected with expression plasmids encoding for CedV G (0.5 µg/well) and either F or mutant F proteins (0.5 µg/well). At 48 h p.t., cells were fixed with ethanol and stained with 1:10-diluted Giemsa staining solution. Representative images (20× magnification) were recorded with an inverted microscope.

2.10. Luciferase Reporter Gene-Based Fusion Assay

Vero76 cells were reverse transfected with expression plasmids encoding for CedV G and either F or mutant F proteins as described above and seeded in 24-well plates. In addition, pCITE Renilla (200 ng/well), a vector containing the renilla luciferase gene under the control of the T7 promoter, was co-transfected. In parallel, Vero76 cells were reverse transfected with the pCAGGS T7 polymerase vector. At 27 h p.t., Vero76 cells expressing the T7 polymerase were washed twice with EDTA-containing PBS (5 mM), detached and added to the cells pre-transfected with pCAGGS CedV F and G and pCITE Renilla to allow fusion to proceed. After 3 h, cells were lysed using Lysis Buffer (pjk, GmbH,

Kleinblittersdorf, Germany) and luciferase activity was measured by adding Renilla Glow substrate (pjk GmbH, Kleinblittersdorf, Germany) according to the manufacturer's instructions. Samples were tested in duplicate in three independent experiments. Reporter activity detected upon co-transfection of the parental CedV F with CedV G protein was set to one, which served as a reference point for fusion activity. Reporter activities measured for the fusion of mutant CedV F proteins were used to calculate their fusion activity in relation to the parental protein. Background activity of the luciferase reporter was assessed with cells transfected with pCAGGS CedV G and pCITE Renilla only overlaid with T7 polymerase expressing cells.

3. Results

Endocytosis of HeV and NiV F protein precursors represents a critical step to gain biological activity, and thus, viral infectivity. In order to investigate whether CedV F protein similarly undergoes endocytosis from the cell surface, we performed a qualitative antibody uptake assay as described previously [11]. Therefore, MDCK-2 cells were transfected with plasmid DNA encoding for the CedV F protein. At 24 h p.t. without prior fixation, cell-surface expressed CedV F protein was labeled with CedV F protein-specific antibodies. Next, cells were either kept on ice or shifted to 37 °C for 30 min to allow endocytosis to occur. Surface-bound antibodies were then detected with an AF 488-conjugated secondary antibody that was added in excess to saturate surface-bound primary antibodies. Subsequent permeabilization allowed the staining of intracellular F protein-antibody complexes with an AF 568-conjugated antibody. After incubation at 37 °C, cells expressing the CedV F protein showed both green surface staining and multiple red fluorescent intracellular vesicles indicating endocytosis of the labeled CedV F protein (Figure 1). In contrast, we observed only green fluorescent signals of F proteins in cells that were kept on ice suffering a temperature-related inhibition of endocytosis.

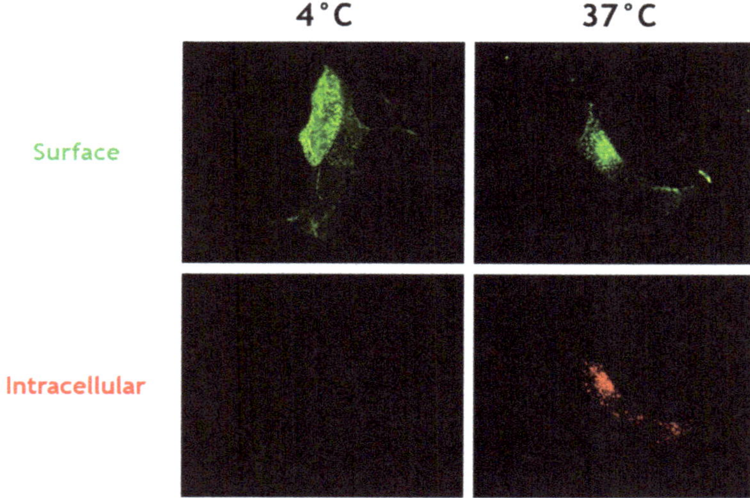

Figure 1. Endocytosis of CedV F protein in MDCK-2 cells. At 24 h p.t., CedV F-expressing MDCK-2 cells were incubated with a CedV F-specific rabbit antiserum to label surface-expressed F proteins. Then, the cells were shifted to 37 °C for 30 min to allow endocytosis to occur. AlexaFluor (AF) 488-conjugated secondary antibodies visualized surface-bound primary antibodies. After fixation and permeabilization, AF 568-conjugated secondary antibodies were used to stain internalized primary antibody-CedV F protein complexes. Magnification, ×60. n = 3.

To evaluate the functional importance of the tyrosine-based motifs within the CedV F protein cytoplasmic tail for endocytosis, we generated seven F protein mutants by tyrosine to alanine substitution (Figure 2). We constructed mutants encoding either single (e1–e5) or multiple (e3, e6) mutations as well as one mutant devoid of all putative endocytosis signals (e7) to investigate potentially additive effects of these mutated motifs (Figure 2). Apart from a classical YXXΦ motif in the membrane-proximal region (Y_{524}XXF, Figure 2), we chose to also mutate the tyrosine residue at amino acid position 522 (Y_{522}XXN, Figure 2) since a similar motif present in the cytoplasmic domain of Measles virus hemagglutinin has been described earlier to affect basolateral sorting and internalization of the protein [45].

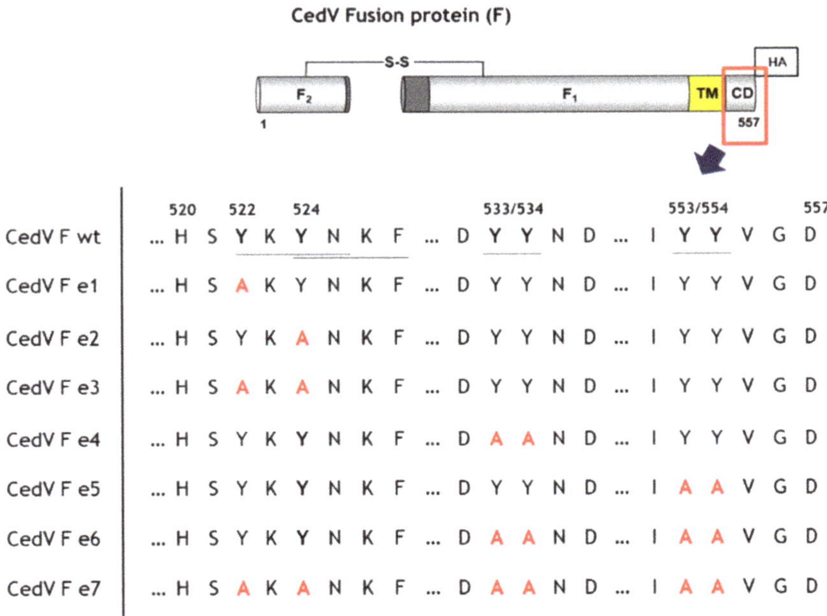

Figure 2. Schematic overview of the cytoplasmic tail mutants of CedV F protein generated in this study. Tyrosine-based putative endocytosis motifs are underlined. F_1: CedV F protein subunit F_1; F_2: CedV F protein subunit F_2; TM: transmembrane domain; CD: cytoplasmic domain; HA: HA-tag at the C-terminus of the F_1 subunit; S-S: di-sulfide bond. wt: wild-type.

To assess whether these mutations have an impact on F protein synthesis, we compared expression efficiencies using a pulse-chase analysis. At 24 h p.t., MDCK-2 cells were labeled metabolically with [35] cysteine and -methionine for 15 min. After pulse-labeling, cells were incubated for 2 h followed by immunoprecipitation and SDS-PAGE under reducing conditions. We observed similar amounts of expression for parental and mutant F proteins within 2 h demonstrating that mutations within the cytoplasmic tail do not impair protein synthesis (Figure 3). However, although mutant e7 was expressed at similar levels, proteolytic processing was much less efficient. We generated additional F protein constructs combining selected mutations of particular motifs (e8 = e3 + e4; e9 = e3 + e5). Since these mutants displayed no differences in proteolytic processing to F wt (Supplementary Figure S1), they were not included in subsequent analyses. Together, these findings suggest that proteolytic activation of the CedV F protein is markedly impaired upon the simultaneous disruption of all tyrosine-based motifs.

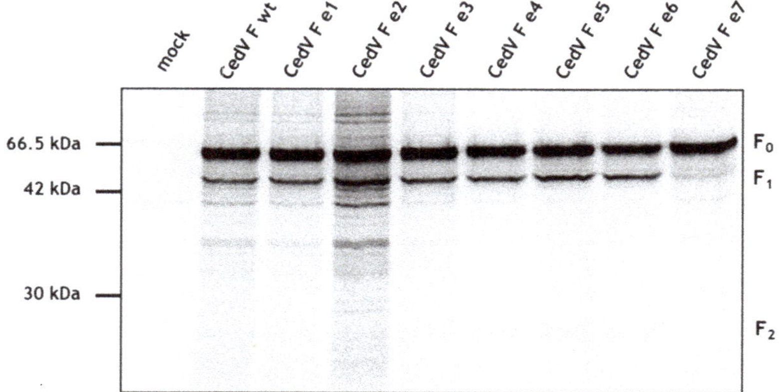

Figure 3. Effect of cytoplasmic tail mutations on CedV F protein expression and cleavage. At 24 h p.t., MDCK-2 cells expressing different CedV F proteins were metabolically labeled for 15 min (pulse) and then incubated for 2 h in serum-free nonradioactive medium (chase). After immunoprecipitation of F proteins from cell lysates and separation on a 12% SDS-gel under reducing conditions, samples were analyzed by autoradiography. n = 2; wt: wild-type.

HeV and NiV F proteins are known to be cleaved within the endosomal compartment. In analogy, blocking clathrin-mediated endocytosis by addition of chlorpromazine also inhibited proteolytic processing of CedV F protein (see Supplementary Figure S2) demonstrating that endocytosis is a prerequisite for CedV F cleavage. Thus, inefficient cleavage of CedV mutant e7 might result from an inability to undergo endocytosis. In order to study the effects of single and multiple cytoplasmic tail mutations on CedV F protein endocytosis, we performed another antibody uptake experiment as described above using all generated mutants. After a period of 30 min at 37 °C, we observed intracellular red vesicles for all mutants except mutant e7 with all tyrosine-based motifs disrupted (Figure 4a). Disruption of a single motif such as YXXN (mutant e1) or YXXF (mutant e2), the di-tyrosine motifs (mutants e4, e5) as well as the combination of several mutated tyrosine residues (mutants e3, e6) displayed no difference in comparison to the fluorescent signals of the parental CedV F protein (Figure 1). The qualitative finding of red fluorescent vesicles indicates the internalization of mutants e1 to e6 from the cell surface while endocytosis of mutant e7 was strongly impaired after 30 min of endocytosis.

To further understand the effects of single and multiple mutations on endocytosis, we next aimed to quantify the internalization of the different CedV F proteins and compare their endocytosis rate with the wt CedV F protein in a semi-quantitative biotin internalization assay. Moreover, this assay allowed us to demonstrate that the internalization of F proteins is not induced by antibody cross-linking. Briefly, we performed a surface biotinylation assay using a cleavable NHS-SS biotin derivative. Using the membrane-impermeable reducing agent MESNA, residual biotin that was not internalized from the cell surface after incubation at 37 °C for different periods can be cleaved. Cell lysis, immunoprecipitation of F proteins, and Western blot using Streptavidin-HRP were performed with subsequent detection and quantification of intracellular biotinylated F proteins. The amount of internalized protein was quantified by comparison to F-expressing cells that were neither incubated at 37 °C nor treated with MESNA, therefore, representing the total amount of biotinylated F proteins (Ctr lane, Figure 4b). Since internalization rates did not appear to be linear over time, endocytosis of biotinylated proteins is displayed additionally as a function of the time in Figure 4c. More than 80% of the parental CedV F protein was internalized after 30 min (Figure 4b,c). Similarly, mutant e1 carrying a substitution of a single tyrosine residue at aa position 522 as well as the di-tyrosine motif mutants e4, e5, and e6 displayed only a marginal decrease in internalization compared to the parental F protein.

Interestingly, mutation of the YXXΦ motif (mutant e2 and e3) clearly decreased internalization with weak signals detectable after a period of 15 to 30 min at 37 °C (Figure 4b). However, the strongest effects were observed for mutant e7. Here, biotinylated mutant e7 was only detected after 90 min of incubation at 37 °C (Figure 4b) leading to a drastically reduced internalization rate over time in contrast to the parental F protein and the other mutants (Figure 4c). An antibody uptake assay of CedV F proteins including a co-staining of endosomal marker protein early endosomal antigen-1 (EEA1) in HeLa cells confirmed the delayed uptake of mutants e2 and e3. After 5 min at 37 °C, we observed no internalization of CedV F mutants e2 and e3 and thus, no co-localization with EEA1 (Figure 4d). However, both mutants co-localized with EEA1 after 30 min. In contrast, the parental F protein showed intracellular staining and co-localization with EEA1 as early as 5 min after the shift to 37 °C (Figure 4d).

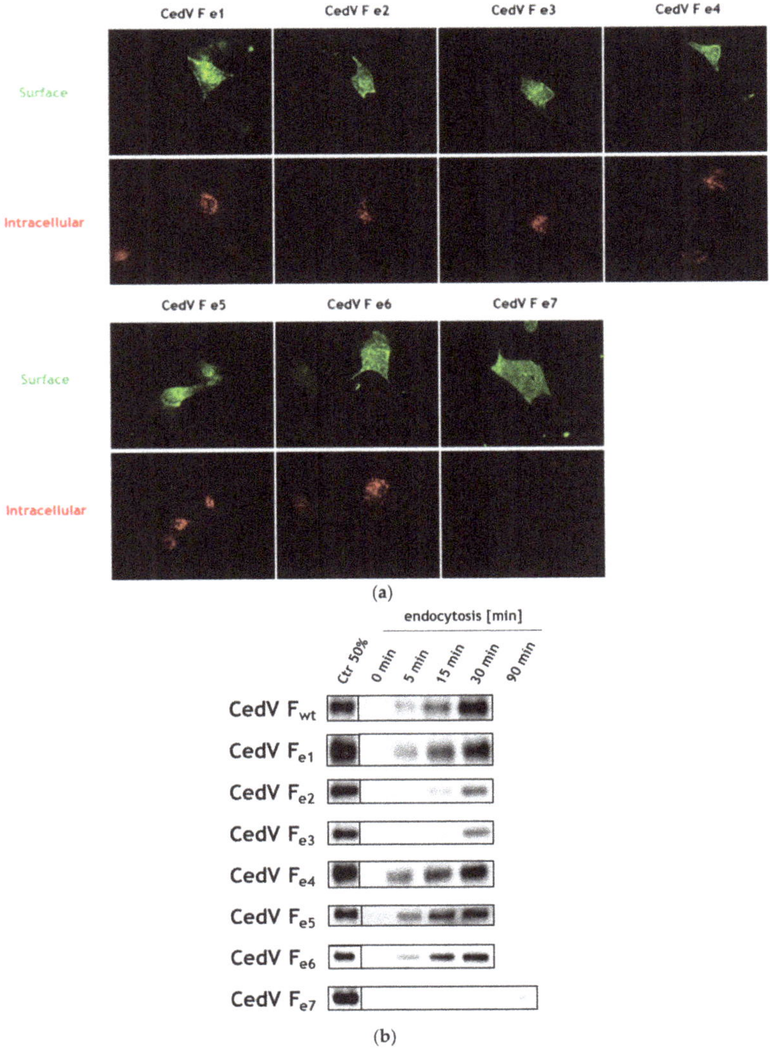

Figure 4. *Cont.*

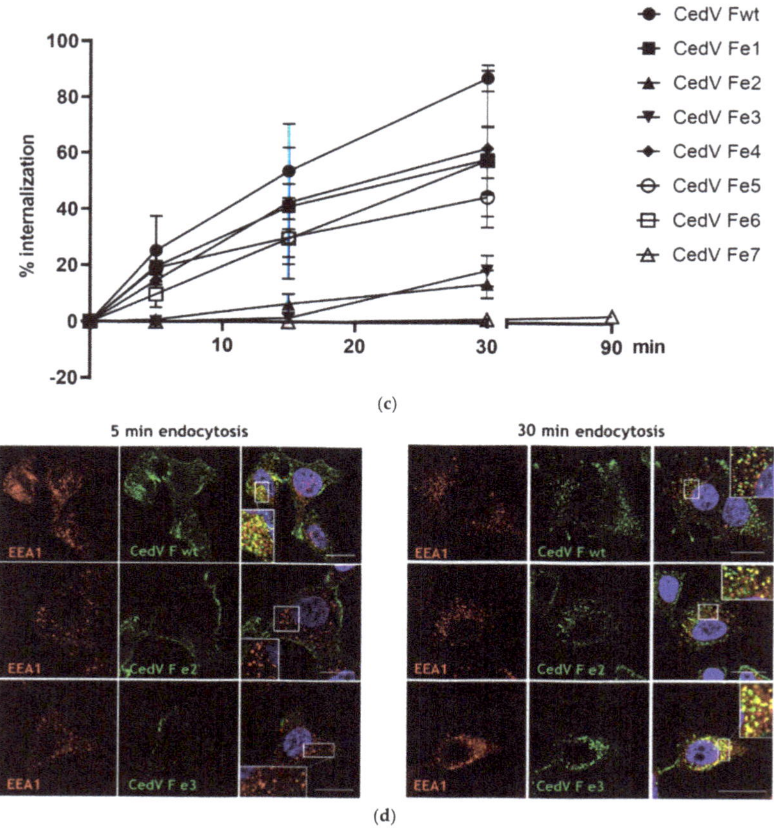

Figure 4. Endocytosis of CedV F proteins in MDCK-2 cells. (**a**) Antibody uptake assay of mutant CedV F proteins. MDCK-2 cells were transfected with plasmids encoding the indicated CedV F protein mutants. At 24 h p.t., a CedV F-specific rabbit antiserum was used to label surface-expressed F proteins at 4 °C. Then, the cells were shifted to 37 °C for 30 min to allow endocytosis to occur. AF 488-conjugated secondary antibodies visualized surface-bound primary antibodies. After fixation and permeabilization, AF 568-conjugated secondary antibodies were used to stain internalized primary antibody-CedV F protein complexes. Magnification, ×60. n = 2. (**b**) At 24 h p.t., CedV F-expressing cells were surface-labeled with cleavable NHS-SS biotin at 4 °C followed by a shift to 37 °C for the indicated times allowing endocytosis to occur. Residual biotin was then cleaved from the cell surface using MESNA. To quantify the amount of surface-biotinylated proteins that underwent endocytosis in a certain time, samples were compared to the total amount of surface-biotinylated control cells (Ctr) that were neither incubated at 37 °C nor treated with MESNA. Following cell lysis, F proteins were immunoprecipitated and samples were separated under non-reducing conditions. After transfer to nitrocellulose, biotinylated proteins were detected with peroxidase-conjugated streptavidin and chemiluminescence. The control lane represents 50% of the total amount of biotinylated F proteins (Ctr 50%). One representative blot is shown for each CedV F protein variant. (n = 3; n = 2 for mutants e1, e3, and e4). wt: wild-type (**c**) CedV F protein internalization in percentage (%) per minute. To quantify the internalization rate, the percentages of internalized F protein amounts measured in the experiment shown in Figure 4b are displayed as a function of the time of incubation at 37 °C. Error bars represent the standard error of the mean. (**d**) Intracellular localization of wt and mutant CedV F proteins in MDCK-2

cells after 5 and 30 min at 37 °C. An antibody uptake assay of CedV F proteins was performed as described above with slight modifications. After the endocytosis step, surface-bound primary antibodies were blocked using a peroxidase-labeled secondary antibody while internalized primary antibodies were detected with AF 488-conjugated rabbit-specific secondary antibodies after fixation and permeabilization. Likewise, early endosomes were visualized with a primary antibody against the early endosomal antigen-1 (EEA1) and a mouse-specific AF568-conjugated secondary antibody. Scale bars indicate 20 μm. Representative images from two independent experiments are displayed (n = 2). Inserts show magnifications of indicated areas. Magnification, ×63.

Since differences in the internalization rate might directly affect cell surface expression, we next performed a surface biotinylation assay of F-expressing MDCK-2 cells followed by immunoprecipitation of biotinylated proteins and detection of the HA-tagged proteins in Western blot. In accordance with the antibody uptake assay and the MESNA reduction, we observed that all F proteins reached the cell surface (Figure 5). However, comparing the cell surface expression of all F proteins under non-reducing conditions, the amount of F_0 differed distinctly between the parental CedV F protein and some mutant F proteins (Figure 5a). In all biotinylation assays performed mutant e1 displayed a cell surface expression similar to the parental CedV F protein, while surface expression of mutants e2–e5 was slightly increased. In contrast, the average level of cell surface expression of mutants, e6 and e7 was substantially higher. For mutant e7, we detected up to 2.5-fold more F_0 on the cell surface than for the parental F protein. Under reducing conditions (Figure 5b), both the F_0 precursor and the subunit F_1 were detected for all F proteins analyzed, indicating that all of them were proteolytically cleaved despite mutations in their cytoplasmic tails. For the parental F protein as well as for mutant e1–e6, quantification of band intensities revealed that more cleavage product F_1 than inactive precursor F_0 is found at the cell surface. However, mutant e7 rather displayed a reversed cleavage ratio, with less F_1 present on the cell surface indicating a reduced amount of cleaved F protein. Although internalization of mutants e2 and e3 was shown to be clearly delayed in the biotin internalization assay, proteolytic activation and surface expression of these mutants still seemed to reach levels comparable to the parental F protein after 24 h p.t.

Finally, to assess whether any of the observed effects influence the biological activity of the F proteins, syncytium formation was analyzed in a fusion assay. At 48 h p.t., co-expression of the parental CedV F and CedV G proteins resulted in the formation of multinucleated syncytia in MDCK2 cells (Figure 6a) and Vero76 cells (Figure 6b). Mutant CedV F e1 and e4 induced syncytium formation comparable to the parental F protein. Surprisingly, despite their reduced endocytosis rate but parental F-like cell surface expression, co-expression of mutant e2 or e3 with CedV G led to a hyperfusogenic phenotype forming syncytia that were markedly increased in number and size (Figure 6a,b). Interestingly, fusogenicity of mutant e5 and e6 displaying an increased surface expression compared to the parental F protein was slightly enhanced while fusion activity of endocytosis-deficient mutant e7 was strongly impaired in both cell lines tested (Figure 6a,b). In addition, we quantified these differences using a luciferase reporter gene-based fusion assay in Vero76 cells after 24 h p.t. The hyperfusogenic phenotype of mutant e2 and e3 displayed a 4- and 6.5-fold increase in measurable luciferase reporter activity compared to the parental F protein (Figure 6c). Co-expression of CedV G protein with mutant e5 still resulted in a 2.5-fold increase in reporter activity in Vero76 cells (Figure 6c). A marked decrease in reporter activity was confirmed for mutant e7. Since fusion activity is sensitive to the cell surface expression of both glycoproteins and to exclude that observed effects in fusogenicity were due to an altered cell-surface expression of CedV G protein, we performed a surface biotinylation assay in MDCK-2 cells co-transfected with CedV G and wt or mutant CedV F proteins. As depicted in Supplementary Figure S3, both F and G can be detected at the cell surface. Irrespective of the (mutant) F protein combination, the G expression at the cell surface seems to be similar suggesting that observed effects in fusion activity are not related to differences in CedV G cell surface expression. Taken together, these findings demonstrate that a membrane-proximal YXXΦ, as well as a C-terminal di-tyrosine motif in the CedV F protein cytoplasmic tail, are of functional relevance for endocytosis and biological activity of the protein.

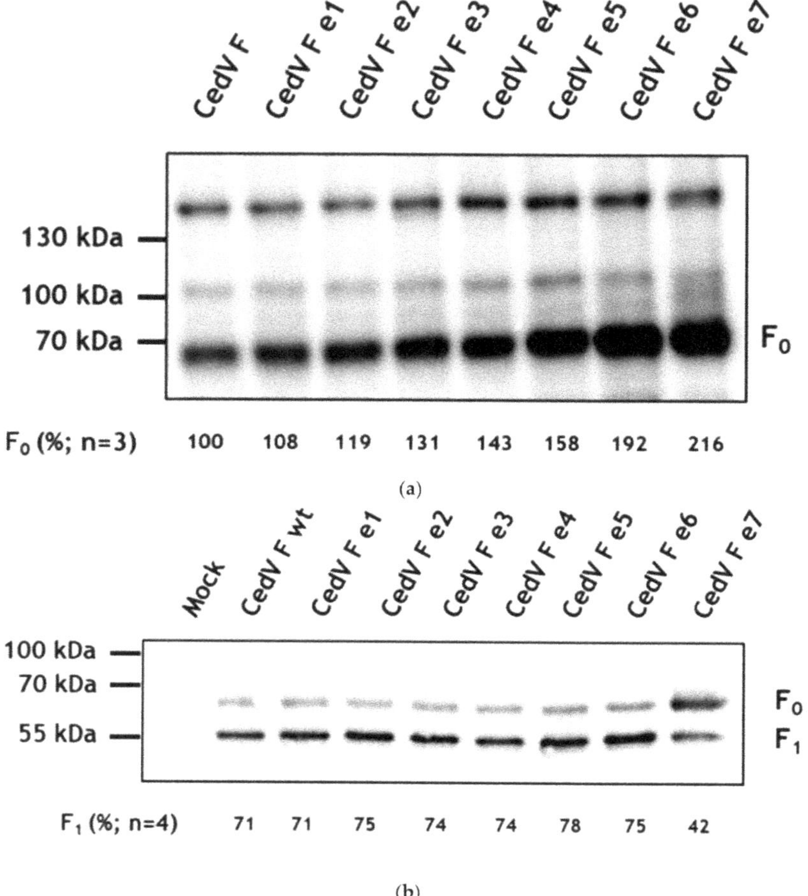

Figure 5. Cell surface expression of CedV F proteins. At 24 h p.t., MDCK-2 cells expressing F proteins were surface-labeled with biotin on ice. After cell lysis, biotinylated proteins were immunoprecipitated using NeutrAvidin beads and subjected to SDS-PAGE under non-reducing (**a**) and reducing (**b**) conditions. Precipitated F proteins were visualized using an antibody against the HA-tag (H6908), HRP-labeled secondary antibodies, and chemiluminescence. In (**a**), ConA staining is used as a loading control. Representative blots are shown from (**a**) three or (**b**) four independent experiments. The amount of F_0 and F_1 protein (in (**a**) % of CedV F_0 protein with the parental F protein set to 100% or (**b**) % of F_1 protein) is calculated as the mean of three or four independent experiments, respectively. wt: wild-type.

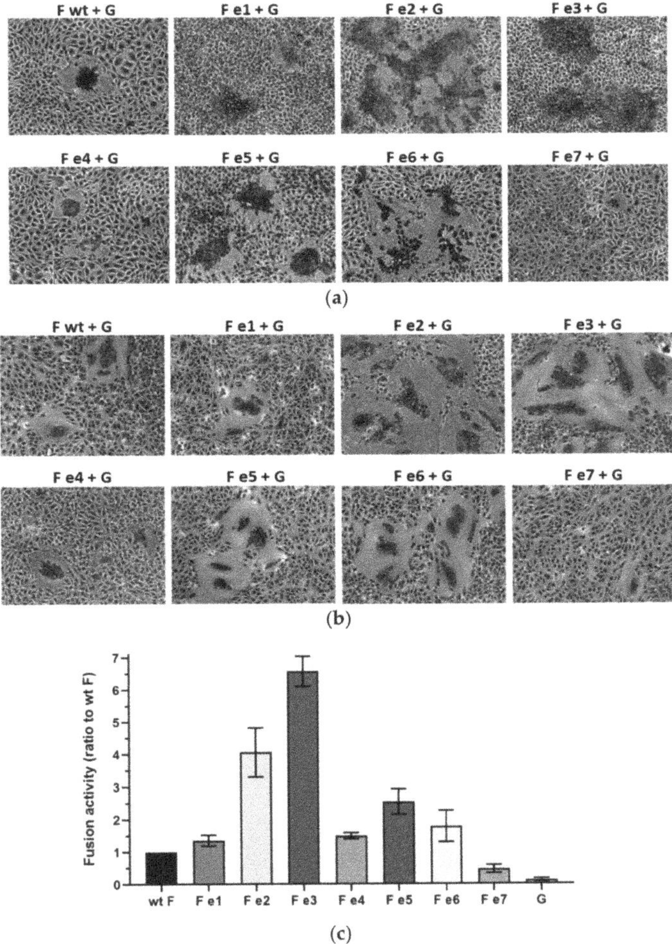

Figure 6. Effect of cytoplasmic tail mutations on CedV glycoprotein-mediated fusion activity. Syncytium formation in (**a**) MDCK-2 and (**b**) Vero76 cells co-expressing CedV F and G proteins were visualized by Giemsa staining at 48 h p.t. Magnification, ×20. n = 3; wt: wild-type (**c**) Quantitative reporter gene assay. Vero76 cells were co-transfected with plasmids encoding for the CedV glycoproteins F and G as well as a plasmid containing the luciferase gene under the control of a T7 promoter (pCITE Renilla). At 24 h p.t., Vero76 cells expressing the T7 polymerase were layered on the glycoprotein-expressing cells and incubated for 3 h at 37 °C. Then, cells were lysed, and luciferase activity measured using a luminometer. Samples were tested in duplicates in three independent experiments. Reporter activity measured for the parental CedV F protein (wt F) co-transfected with CedV G protein was set to 1 serving as a reference point for fusion activity. Bars represent the fusion activities of the different (mutant) CedV F proteins in relation to the fusion activity of the parental protein and include the standard error of the mean (SEM). Background activity of the luciferase reporter was assessed with cells transfected with pCAGGS CedV G and pCITE Renilla only, layered with T7 polymerase expressing cells. n = 3.

4. Discussion

Endocytic uptake from the cell surface plays a particular role in the replication cycle of highly pathogenic HeV and NiV [11–13]. Previous work has shown the importance of endocytosis for the maturation of F proteins to gain biological activity and thus, viral infectivity [11–13,16]. Given the fact that endocytic uptake of HeV and NiV F proteins depends on tyrosine-based motifs present in their cytoplasmic tail, the comparably high number of tyrosine residues found in the CedV F protein was of interest for this study. We first aimed to investigate whether these residues indeed mediate endocytosis of the protein. Second, we analyzed whether endocytosis is of functional relevance to the CedV F protein maturation. Taken together, our data clearly indicate that endocytic uptake of CedV F protein is signal-mediated and important for the biological activity of the protein. Furthermore, among the putative endocytosis signals, we identified a YXXΦ motif and a C-terminal di-tyrosine motif to have the strongest effects on endocytosis, cell surface expression, and fusion activity of CedV F protein.

The YXXΦ motif known to facilitate clathrin-mediated endocytosis of many transmembrane proteins is present in the membrane-proximal region of the cytoplasmic tail of both high and low pathogenic henipavirus F proteins. Mutation of the YXXΦ motif in the F protein of low pathogenic CedV resulted in a strongly decreased internalization rate for mutant e2 and e3, suggesting endocytic uptake of CedV F protein to be signal-mediated and largely dependent on this particular motif and most likely on Y524. For highly pathogenic HeV and NiV, several studies have similarly shown that an intact YXXΦ motif greatly contributes to the internalization of F proteins with a marked reduction in endocytosis rate upon motif disruption [11–13,43]. However, while the delayed endocytic uptake of NiV F protein led to a reduced fusogenicity of the mutant, the respective mutation in the HeV F protein rather resulted in an enhanced cell-to-cell fusion upon interaction with the viral attachment protein G [11,12]. Due to strong reduction but not the absence of endocytosis, the authors suggested an accumulation of recycled, fusogenic HeV F protein at the cell surface over time to account for the observed effects. For the respective CedV F mutants, we similarly found a clear delay in internalization (Figure 4b), but only moderate differences in total surface expression compared to the parental CedV F protein (Figure 5). However, we did indeed measure a strong 4- to 6.5-fold increase in the fusion activity of these two mutants in contrast to the parental CedV F protein, which cannot exclusively be explained by the only slightly increased cell surface expression. Although triggering of fusion is usually thought to primarily involve the ectodomain of the fusion protein, hyper- and hypofusogenicity were also observed in NiV F cytoplasmic tail mutants due to specific mutations in a membrane-proximal polybasic KKR motif [48]. The authors explained their findings by a mutation-dependent inside-out signaling mechanism resulting in conformational changes in the NiV F ectodomain followed by measurable effects on the fusion activity of the protein [48]. Overall, such conformational changes of the F-ectodomain due to mutations in the cytoplasmic tail can also affect the avidity of F and G interaction and thus, the coordinated processes required for cell–cell fusion [48–50]. Further, it is believed that mechanisms resulting in virus entry (viral—cellular membrane fusion) and cell–cell-fusion (fusion of neighboring cell membranes) are closely related [51]. Thus, cell–cell fusion levels often correlate to viral entry levels as shown for several Hendra and Nipah glycoprotein mutants [48,52–54]. In contrast, some hyperfusogenic G mutants with a modified O-glycosylation were described to display reduced entry levels but similar cell–cell fusion levels [55]. Also, a headless NiV G mutant readily triggered cell–cell fusion but pseudotyped NiV virions did not enter cells [56]. Hence, the underlying mechanism for the observed hyperfusogenicity of the CedV F mutants e2 and e3 and here the potential role of Y524 will have to be addressed in future studies.

Apart from a functional YXXΦ motif, the cytoplasmic tail of CedV F protein contains two additional di-tyrosine based motifs. Mutation of the di-tyrosine motif at amino acid position Y533/534A (mutant e4) showed no effect in any of the assays performed. However, mutation of the C-terminal di-tyrosine motif (Y553/554A; mutant e5 and e6) led to a detectable increase of cell surface expression with slightly enhanced fusion activity. On the one hand, this phenotype could result from a marginally reduced internalization rate. Alternatively, this di-tyrosine motif may affect the dynamics of endosomal trafficking and recycling of CedV F proteins, which could subsequently alter the availability of fusion-active F protein on the cell surface, thus having an impact on fusion activity. In a previous study, the lack of the di-tyrosine motif has been considered to delay the recycling of fusogenic NiV

F proteins to the cell surface [43]. A tail-truncated NiV F variant lacking the C-terminal di-tyrosine motif was markedly downregulated in constitutive surface expression while F protein endocytosis and endosomal cleavage remained unaffected. Consequently, the intact di-tyrosine motif was hypothesized to act as a potential cytoplasmic recycling motif affecting intracellular trafficking [43]. Interestingly, both the YXXΦ and the C-terminal di-tyrosine motif have been described to be of importance for NiV F protein trafficking and sorting in polarized microvascular endothelial cells [57], in polarized epithelial cells [58], and in polarized neuronal cells [59].

The presence of intact tyrosine-based endocytosis and/or sorting motif in the cytoplasmic tail of transmembrane proteins has been described to be important for the replication cycle, infectivity, and virulence of many viruses such as herpes-, corona- and retroviruses [24,25,29,30,44,60,61]. Apart from signal-mediated intracellular trafficking of viral proteins to the sites of viral assembly, these motifs were shown to affect cell surface expression of different viral transmembrane proteins [24,25,29,30,44,60,61]. For instance, mutating the endocytosis signal of the envelope protein of simian immunodeficiency virus led to more efficient incorporation of envelope proteins into budding virions with enhanced infectivity due to increased levels of envelope protein expressed on the cell surface [61]. In the case of the CedV F protein, mutation of the di-tyrosine motif had stronger effects on the cell surface expression than mutation of the classical YXXΦ motif. However, at this point, it is not clear whether the di-tyrosine mutants will be incorporated more efficiently into virions due to an increase in cell surface expression. Mutation of the di-tyrosine motif in the NiV F cytoplasmic tail almost completely abrogated NiV VLP budding despite the wt-like cell surface expression of the protein [62]. Additionally, the presence of an intact and functional YXXΦ endocytosis motif in the cytoplasmic tail of NiV and HeV F protein has been considered to play a critical role in endocytic trafficking and recycling, and thus, in efficient viral assembly and particle release [31,62]. Similar findings have been reported for the envelope protein of HIV-1 where recent evidence suggests that trafficking through the recycling endosome is required for efficient incorporation of viral envelope proteins into virus particles [28]. Considering the lack of an intact YXXΦ motif, the reduced endocytosis rate, and the level of cell surface expression similar to the parental F protein, it will be interesting to see whether assembly and budding of mutant CedV F e2 and e3 is impaired or rather enhanced.

The most significant effects on cell surface expression, internalization rate, and fusion activity were noted for CedV mutant e7, in which all putative tyrosine-based endocytosis motifs were mutated. Importantly, the combination of all motif mutations led to an almost abrogated fusion activity of the protein with the strongest delay in internalization, which is in accordance with previous findings for a NiV F protein mutant disrupted of all tyrosine-based motifs [11]. Though CedV F mutant e7 displayed the highest cell surface expression, the combination of all mutations led to a rather reversed cleavage ratio in contrast to the other mutants. These findings point towards a reduced proteolytic F activation, which might result from the marked decrease in the endocytosis rate. Considering what is known for the F proteins of highly pathogenic HeV and NiV that are cleaved within the endosomal compartment [12–15,17], it seems likely that the proteolytic activation of CedV F protein is quite similar in terms of subcellular localization. However, future studies are needed to unravel shared commonalities and potential differences in this process.

5. Conclusions

In conclusion, our data indicate that CedV F protein is indeed internalized from the cell surface as described for HeV and NiV F proteins. Furthermore, endocytic uptake of CedV F protein is signal-mediated and represents a key step in order to gain biological activity. Future studies should investigate proteolytic activation in more detail in order to find further commonalities and/or differences in subcellular localization and protease usage between high and low pathogenic henipaviruses. Further, tyrosine-based motifs should be analyzed for their effects on sorting and endocytic trafficking of the CedV F protein including their ability to influence virus particle assembly and their potential effect on virus-cell fusion/infectivity.

Supplementary Materials: The following are available online at http://www.mdpi.com/2073-4409/9/9/2054/s1, Figure S1: Proteolytic processing of CedV F wt and mutants e8 (e3 + e4) and e9 (e3 + e5), Figure S2: Proteolytic processing of CedV F protein is inhibited by chlorpromazine, Figure S3: Co-expression of CedV F wt, mutants and CedV G protein on the cell surface.

Author Contributions: Conceptualization, K.F. and S.D.; methodology and investigation, K.F.; analysis: K.F. and S.D.; supervision: S.D.; funding acquisition, S.D.; resources, M.H.G.; writing, K.F. and S.D. All authors have read and agreed to the published version of the manuscript.

Funding: Funding was provided by the Friedrich-Loeffler-Institut, intramural funding (S.D.) and funding as part of the VISION consortium (K.F.).

Acknowledgments: The authors wish to thank Stefanie Rößler and Carolin Rüdiger for excellent technical support and Thomas Hoenen for helpful comments and critical reading of this manuscript. The authors further acknowledge Angela Römer-Oberdörfer (pCITE) and Thomas Hoenen (pCAGGS T7) for providing the aforementioned plasmids used in this study. Many thanks to Bärbel Hammerschmidt and the animal keepers for help with the generation of polyclonal CedV F-specific serum and to Christine Luttermann for sequencing of mutants.

Conflicts of Interest: The authors declare no conflict of interest.

References

1. Marsh, G.A.; de Jong, C.; Barr, J.A.; Tachedjian, M.; Smith, C.; Middleton, D.; Yu, M.; Todd, S.; Foord, A.J.; Haring, V.; et al. Cedar Virus: A Novel Henipavirus Isolated from Australian Bats. *PLoS Pathog.* **2012**, *8*, e102836. [CrossRef]
2. Schountz, T.; Campbell, C.; Wagner, K.; Rovnak, J.; Martellaro, C.; DeBuysscher, B.L.; Feldmann, H.; Prescott, J. Differential Innate Immune Responses Elicited by Nipah Virus and Cedar Virus Correlate with Disparate In Vivo Pathogenesis in Hamsters. *Viruses* **2019**, *11*, 291. [CrossRef] [PubMed]
3. Lieu, K.G.; Marsh, G.A.; Wang, L.F.; Netter, H.J. The non-pathogenic Henipavirus Cedar paramyxovirus phosphoprotein has a compromised ability to target STAT1 and STAT2. *Antivir. Res.* **2015**, *124*, 69–76. [CrossRef] [PubMed]
4. Negrete, O.A.; Levroney, E.L.; Aguilar, H.C.; Bertolotti-Ciarlet, A.; Nazarian, R.; Tajyar, S.; Lee, B. EphrinB2 is the entry receptor for Nipah virus, an emergent deadly paramyxovirus. *Nature* **2005**, *436*, 401–405. [CrossRef]
5. Bonaparte, M.I.; Dimitrov, A.S.; Bossart, K.N.; Crameri, G.; Mungall, B.A.; Bishop, K.A.; Choudhry, V.; Dimitrov, D.S.; Wang, L.F.; Eaton, B.T.; et al. Ephrin-B2 ligand is a functional receptor for Hendra virus and Nipah virus. *Proc. Natl. Acad. Sci. USA* **2005**, *102*, 10652–10657. [CrossRef] [PubMed]
6. Negrete, O.A.; Wolf, M.C.; Aguilar, H.C.; Enterlein, S.; Wang, W.; Muhlberger, E.; Su, S.V.; Bertolotti-Ciarlet, A.; Flick, R.; Lee, B. Two key residues in ephrinB3 are critical for its use as an alternative receptor for Nipah virus. *PLoS Pathog.* **2006**, *2*, e7. [CrossRef] [PubMed]
7. Hafner, C.; Schmitz, G.; Meyer, S.; Bataille, F.; Hau, P.; Langmann, T.; Dietmaier, W.; Landthaler, M.; Vogt, T. Differential gene expression of Eph receptors and ephrins in benign human tissues and cancers. *Clin. Chem.* **2004**, *50*, 490–499. [CrossRef]
8. Pryce, R.; Azarm, K.; Rissanen, I.; Harlos, K.; Bowden, T.A.; Lee, B. A key region of molecular specificity orchestrates unique ephrin-B1 utilization by Cedar virus. *Life Sci. Alliance* **2020**, *3*. [CrossRef]
9. Laing, E.D.; Navaratnarajah, C.K.; Cheliout Da Silva, S.; Petzing, S.R.; Xu, Y.; Sterling, S.L.; Marsh, G.A.; Wang, L.F.; Amaya, M.; Nikolov, D.B.; et al. Structural and functional analyses reveal promiscuous and species specific use of ephrin receptors by Cedar virus. *Proc. Natl. Acad. Sci. USA* **2019**, *116*, 20707–20715. [CrossRef]
10. Azarm, K.D.; Lee, B. Differential Features of Fusion Activation within the Paramyxoviridae. *Viruses* **2020**, *12*, 161. [CrossRef]
11. Vogt, C.; Eickmann, M.; Diederich, S.; Moll, M.; Maisner, A. Endocytosis of the Nipah virus glycoproteins. *J. Virol.* **2005**, *79*, 3865–3872. [CrossRef] [PubMed]
12. Meulendyke, K.A.; Wurth, M.A.; McCann, R.O.; Dutch, R.E. Endocytosis plays a critical role in proteolytic processing of the Hendra virus fusion protein. *J. Virol.* **2005**, *79*, 12643–12649. [CrossRef]
13. Diederich, S.; Moll, M.; Klenk, H.D.; Maisner, A. The nipah virus fusion protein is cleaved within the endosomal compartment. *J. Biol. Chem.* **2005**, *280*, 29899–29903. [CrossRef] [PubMed]
14. Pager, C.T.; Dutch, R.E. Cathepsin L is involved in proteolytic processing of the Hendra virus fusion protein. *J. Virol.* **2005**, *79*, 12714–12720. [CrossRef] [PubMed]

15. Pager, C.T.; Craft, W.W., Jr.; Patch, J.; Dutch, R.E. A mature and fusogenic form of the Nipah virus fusion protein requires proteolytic processing by *Cathepsin* L. *Virology* **2006**, *346*, 251–257. [CrossRef]
16. Diederich, S.; Thiel, L.; Maisner, A. Role of endocytosis and cathepsin-mediated activation in Nipah virus entry. *Virology* **2008**, *375*, 391–400. [CrossRef]
17. Diederich, S.; Sauerhering, L.; Weis, M.; Altmeppen, H.; Schaschke, N.; Reinheckel, T.; Erbar, S.; Maisner, A. Activation of the Nipah virus fusion protein in MDCK cells is mediated by cathepsin B within the endosome-recycling compartment. *J. Virol.* **2012**, *86*, 3736–3745. [CrossRef]
18. Matlin, K.S.; Reggio, H.; Helenius, A.; Simons, K. Infectious entry pathway of influenza virus in a canine kidney cell line. *J. Cell Biol.* **1981**, *91*, 601–613. [CrossRef]
19. Yoshimura, A.; Ohnishi, S. Uncoating of Influenza-Virus in Endosomes. *J. Virol.* **1984**, *51*, 497–504. [CrossRef]
20. Nanbo, A.; Imai, M.; Watanabe, S.; Noda, T.; Takahashi, K.; Neumann, G.; Halfmann, P.; Kawaoka, Y. Ebolavirus Is Internalized into Host Cells via Macropinocytosis in a Viral Glycoprotein-Dependent Manner. *PLoS Pathog.* **2010**, *6*, e1001121. [CrossRef]
21. Aleksandrowicz, P.; Marzi, A.; Biedenkopf, N.; Beimforde, N.; Becker, S.; Hoenen, T.; Feldmann, H.; Schnittler, H.J. Ebola Virus Enters Host Cells by Macropinocytosis and Clathrin-Mediated Endocytosis. *J. Infect. Dis.* **2011**, *204*, S957–S967. [CrossRef] [PubMed]
22. Chu, J.J.H.; Ng, M.L. Infectious entry of West Nile virus occurs through a clathrin-mediated endocytic pathway. *J. Virol.* **2004**, *78*, 10543–10555. [CrossRef] [PubMed]
23. Blanchard, E.; Belouzard, S.; Goueslain, L.; Wakita, T.; Dubuisson, J.; Wychowski, C.; Rouille, Y. Hepatitis C virus entry depends on clathrin-mediated endocytosis. *J. Virol.* **2006**, *80*, 6964–6972. [CrossRef]
24. Marsh, M.; Pelchen-Matthews, A. Endocytosis in viral replication. *Traffic* **2000**, *1*, 525–532. [CrossRef] [PubMed]
25. Favoreel, H.W. The why's of Y-based motifs in alphaherpesvirus envelope proteins. *Virus Res.* **2006**, *117*, 202–208. [CrossRef] [PubMed]
26. Checkley, M.A.; Luttge, B.G.; Freed, E.O. HIV-1 envelope glycoprotein biosynthesis, trafficking, and incorporation. *J. Mol. Biol.* **2011**, *410*, 582–608. [CrossRef]
27. Postler, T.S.; Desrosiers, R.C. The tale of the long tail: The cytoplasmic domain of HIV-1 gp41. *J. Virol.* **2013**, *87*, 2–15. [CrossRef]
28. Kirschman, J.; Qi, M.; Ding, L.; Hammonds, J.; Dienger-Stambaugh, K.; Wang, J.J.; Lapierre, L.A.; Goldenring, J.R.; Spearman, P. HIV-1 Envelope Glycoprotein Trafficking through the Endosomal Recycling Compartment Is Required for Particle Incorporation. *J. Virol.* **2018**, *92*. [CrossRef]
29. Albecka, A.; Laine, R.F.; Janssen, A.F.J.; Kaminski, C.F.; Crump, C.M. HSV-1 Glycoproteins Are Delivered to Virus Assembly Sites Through Dynamin-Dependent Endocytosis. *Traffic* **2016**, *17*, 21–39. [CrossRef]
30. Beitia Ortiz de Zarate, I.; Cantero-Aguilar, L.; Longo, M.; Berlioz-Torrent, C.; Rozenberg, F. Contribution of endocytic motifs in the cytoplasmic tail of herpes simplex virus type 1 glycoprotein B to virus replication and cell-cell fusion. *J. Virol.* **2007**, *81*, 13889–13903. [CrossRef]
31. Cifuentes-Munoz, N.; Sun, W.; Ray, G.; Schmitt, P.T.; Webb, S.; Gibson, K.; Dutch, R.E.; Schmitt, A.P. Mutations in the Transmembrane Domain and Cytoplasmic Tail of Hendra Virus Fusion Protein Disrupt Virus-Like-Particle Assembly. *J. Virol.* **2017**, *91*. [CrossRef] [PubMed]
32. Conner, S.D.; Schmid, S.L. Regulated portals of entry into the cell. *Nature* **2003**, *422*, 37–44. [CrossRef] [PubMed]
33. Kaksonen, M.; Roux, A. Mechanisms of clathrin-mediated endocytosis. *Nat. Rev. Mol. Cell Biol.* **2018**, *19*, 313–326. [CrossRef]
34. Robinson, M.S. Forty Years of Clathrin-coated Vesicles. *Traffic* **2015**, *16*, 1210–1238. [CrossRef]
35. Bonifacino, J.S.; Traub, L.M. Signals for sorting of transmembrane proteins to endosomes and lysosomes. *Annu. Rev. Biochem.* **2003**, *72*, 395–447. [CrossRef]
36. Ohno, H.; Stewart, J.; Fournier, M.C.; Bosshart, H.; Rhee, I.; Miyatake, S.; Saito, T.; Gallusser, A.; Kirchhausen, T.; Bonifacino, J.S. Interaction of tyrosine-based sorting signals with clathrin-associated proteins. *Science* **1995**, *269*, 1872–1875. [CrossRef]
37. Reider, A.; Wendland, B. Endocytic adaptors—Social networking at the plasma membrane. *J. Cell. Sci.* **2011**, *124*, 1613–1622. [CrossRef] [PubMed]
38. Park, S.Y.; Guo, X.L. Adaptor protein complexes and intracellular transport. *Biosci. Rep.* **2014**, *34*, 381–390. [CrossRef] [PubMed]

39. Traub, L.M. Sorting it out: AP-2 and alternate clathrin adaptors in endocytic cargo selection. *J. Cell Biol.* **2003**, *163*, 203–208. [CrossRef]
40. Le Roy, C.; Wrana, J.L. Clathrin- and non-clathrin-mediated endocytic regulation of cell signalling. *Nat. Rev. Mol. Cell Biol.* **2005**, *6*, 112–126. [CrossRef]
41. Traub, L.M. Tickets to ride: Selecting cargo for clathrin-regulated internalization. *Nat. Rev. Mol. Cell Biol.* **2009**, *10*, 583–596. [CrossRef] [PubMed]
42. Ohno, H. Clathrin-associated adaptor protein complexes. *J. Cell Sci.* **2006**, *119*, 3719–3721. [CrossRef] [PubMed]
43. Weis, M.; Maisner, A. Nipah virus fusion protein: Importance of the cytoplasmic tail for endosomal trafficking and bioactivity. *Eur. J. Cell Biol.* **2015**, *94*, 316–322. [CrossRef] [PubMed]
44. Hou, Y.; Meulia, T.; Gao, X.; Saif, L.J.; Wang, Q. Deletion of both the Tyrosine-Based Endocytosis Signal and the Endoplasmic Reticulum Retrieval Signal in the Cytoplasmic Tail of Spike Protein Attenuates Porcine Epidemic Diarrhea Virus in Pigs. *J. Virol.* **2019**, *93*. [CrossRef] [PubMed]
45. Moll, M.; Klenk, H.D.; Herrler, G.; Maisner, A. A single amino acid change in the cytoplasmic domains of measles virus glycoproteins H and F alters targeting, endocytosis, and cell fusion in polarized Madin-Darby canine kidney cells. *J. Biol. Chem.* **2001**, *276*, 17887–17894. [CrossRef] [PubMed]
46. Stroh, E.; Fischer, K.; Schwaiger, T.; Sauerhering, L.; Franzke, K.; Maisner, A.; Groschup, M.H.; Blohm, U.; Diederich, S. Henipavirus-like particles induce a CD8 T cell response in C57BL/6 mice. *Vet. Microbiol.* **2019**, *237*, 108405. [CrossRef]
47. Schneider, C.A.; Rasband, W.S.; Eliceiri, K.W. NIH Image to ImageJ: 25 years of image analysis. *Nat. Methods* **2012**, *9*, 671–675. [CrossRef]
48. Aguilar, H.C.; Matreyek, K.A.; Choi, D.Y.; Filone, C.M.; Young, S.; Lee, B. Polybasic KKR motif in the cytoplasmic tail of Nipah virus fusion protein modulates membrane fusion by inside-out signaling. *J. Virol.* **2007**, *81*, 4520–4532. [CrossRef]
49. Aguilar, H.C.; Matreyek, K.A.; Filone, C.M.; Hashimi, S.T.; Levroney, E.L.; Negrete, O.A.; Bertolotti-Ciarlet, A.; Choi, D.Y.; McHardy, I.; Fulcher, J.A.; et al. N-glycans on Nipah virus fusion protein protect against neutralization but reduce membrane fusion and viral entry. *J. Virol.* **2006**, *80*, 4878–4889. [CrossRef]
50. Carter, J.R.; Pager, C.T.; Fowler, S.D.; Dutch, R.E. Role of N-linked glycosylation of the Hendra virus fusion protein. *J. Virol.* **2005**, *79*, 7922–7925. [CrossRef]
51. White, J.M.; Delos, S.E.; Brecher, M.; Schornberg, K. Structures and mechanisms of viral membrane fusion proteins: Multiple variations on a common theme. *Crit. Rev. Biochem. Mol. Biol.* **2008**, *43*, 189–219. [CrossRef] [PubMed]
52. Biering, S.B.; Huang, A.; Vu, A.T.; Robinson, L.R.; Bradel-Tretheway, B.; Choi, E.; Lee, B.; Aguilar, H.C. N-Glycans on the Nipah virus attachment glycoprotein modulate fusion and viral entry as they protect against antibody neutralization. *J. Virol.* **2012**, *86*, 11991–12002. [CrossRef] [PubMed]
53. Bradel-Tretheway, B.G.; Liu, Q.; Stone, J.A.; McInally, S.; Aguilar, H.C. Novel Functions of Hendra Virus G N-Glycans and Comparisons to Nipah Virus. *J. Virol.* **2015**, *89*, 7235–7247. [CrossRef] [PubMed]
54. Liu, Q.; Bradel-Tretheway, B.; Monreal, A.I.; Saludes, J.P.; Lu, X.; Nicola, A.V.; Aguilar, H.C. Nipah virus attachment glycoprotein stalk C-terminal region links receptor binding to fusion triggering. *J. Virol.* **2015**, *89*, 1838–1850. [CrossRef] [PubMed]
55. Stone, J.A.; Nicola, A.V.; Baum, L.G.; Aguilar, H.C. Multiple Novel Functions of Henipavirus O-glycans: The First O-glycan Functions Identified in the Paramyxovirus Family. *PLoS Pathog.* **2016**, *12*, e1005445. [CrossRef]
56. Liu, Q.; Stone, J.A.; Bradel-Tretheway, B.; Dabundo, J.; Benavides Montano, J.A.; Santos-Montanez, J.; Biering, S.B.; Nicola, A.V.; Iorio, R.M.; Lu, X.; et al. Unraveling a three-step spatiotemporal mechanism of triggering of receptor-induced Nipah virus fusion and cell entry. *PLoS Pathog.* **2013**, *9*, e1003770. [CrossRef]
57. Erbar, S.; Maisner, A. Nipah virus infection and glycoprotein targeting in endothelial cells. *Virol. J.* **2010**, *7*, 305. [CrossRef]
58. Weise, C.; Erbar, S.; Lamp, B.; Vogt, C.; Diederich, S.; Maisner, A. Tyrosine residues in the cytoplasmic domains affect sorting and fusion activity of the Nipah virus glycoproteins in polarized epithelial cells. *J. Virol.* **2010**, *84*, 7634–7641. [CrossRef]
59. Mattera, R.; Farias, G.G.; Mardones, G.A.; Bonifacino, J.S. Co-assembly of viral envelope glycoproteins regulates their polarized sorting in neurons. *PLoS Pathog.* **2014**, *10*, e1004107. [CrossRef] [PubMed]

60. Day, J.R.; Munk, C.; Guatelli, J.C. The membrane-proximal tyrosine-based sorting signal of human immunodeficiency virus type 1 gp41 is required for optimal viral infectivity. *J. Virol.* **2004**, *78*, 1069–1079. [CrossRef] [PubMed]
61. Yuste, E.; Reeves, J.D.; Doms, R.W.; Desrosiers, R.C. Modulation of Env content in virions of simian immunodeficiency virus: Correlation with cell surface expression and virion infectivity. *J. Virol.* **2004**, *78*, 6775–6785. [CrossRef] [PubMed]
62. Johnston, G.P.; Contreras, E.M.; Dabundo, J.; Henderson, B.A.; Matz, K.M.; Ortega, V.; Ramirez, A.; Park, A.; Aguilar, H.C. Cytoplasmic Motifs in the Nipah Virus Fusion Protein Modulate Virus Particle Assembly and Egress. *J. Virol.* **2017**, *91*. [CrossRef] [PubMed]

Article

The Host Factor Erlin-1 is Required for Efficient Hepatitis C Virus Infection

Christina Whitten-Bauer [1,†], Josan Chung [1,†], Andoni Gómez-Moreno [2], Pilar Gomollón-Zueco [2], Michael D. Huber [3,‡], Larry Gerace [3] and Urtzi Garaigorta [1,2,*]

1 Department of Immunology and Microbial Sciences, The Scripps Research Institute, La Jolla, CA 92037, USA; whittencr@gmail.com (C.W.-B.); chungjosan@gmail.com (J.C.)
2 Department of Molecular and Cellular Biology, Centro Nacional de Biotecnología Consejo Superior de Investigaciones Científicas (CNB-CSIC), 28049 Madrid, Spain; andoni.gomez.moreno777@gmail.com (A.G.-M.); pilar22tarazona@gmail.com (P.G.-Z.)
3 Department of Molecular Medicine, The Scripps Research Institute, La Jolla, CA 92037, USA; mhuber26.2@gmail.com (M.D.H.); lgerace@scripps.edu (L.G.)
* Correspondence: ugaraigorta@cnb.csic.es; Tel.: +34-91-585-4533
† Equal contribution.
‡ Current address: Vividion Therapeutics, San Diego, CA 92121, USA.

Received: 28 October 2019; Accepted: 29 November 2019; Published: 2 December 2019

Abstract: Development of hepatitis C virus (HCV) infection cell culture systems has permitted the identification of cellular factors that regulate the HCV life cycle. Some of these cellular factors affect steps in the viral life cycle that are tightly associated with intracellular membranes derived from the endoplasmic reticulum (ER). Here, we describe the discovery of erlin-1 protein as a cellular factor that regulates HCV infection. Erlin-1 is a cholesterol-binding protein located in detergent-resistant membranes within the ER. It is implicated in cholesterol homeostasis and the ER-associated degradation pathway. Silencing of erlin-1 protein expression by siRNA led to decreased infection efficiency characterized by reduction in intracellular RNA accumulation, HCV protein expression and virus production. Mechanistic studies revealed that erlin-1 protein is required early in the infection, downstream of cell entry and primary translation, specifically to initiate RNA replication, and later in the infection to support infectious virus production. This study identifies erlin-1 protein as an important cellular factor regulating HCV infection.

Keywords: hepatitis C virus; HCV; erlin-1; erlin-2; host factor; endoplasmic reticulum; RNA replication; protein production; virus production; lipid droplet

---

1. Introduction

Hepatitis C virus (HCV) is an enveloped virus belonging to the Flaviviridae family [1]. It is estimated that around 71 million people are chronically infected, many of whom will develop cirrhosis, end-stage liver disease and hepatocellular carcinoma [2].

The HCV genome is around 9.6 kb positive-sensed single-stranded RNA containing a unique long open reading frame (ORF) that encodes a single polyprotein of approximately 3000 amino acids [3]. The 5′ and 3′ nontranslated regions (NTR), flanking the ORF, contain essential sequences for RNA stability, translation and replication [4–6]. A highly structured internal ribosomal entry site located in the 5′ NTR drives the translation of the polyprotein that is co- and post-translationally processed by both host and viral proteases leading to the expression of three structural (core, E1 and E2) and seven non-structural proteins (p7, NS2, NS3, NS4A, NS4B, NS5A and NS5B) [7].

HCV entry into hepatocytes starts with the interaction of viral and cellular factors, present in the viral membrane, with several receptors present in the plasma membrane of the hepatocytes (reviewed in [8]). These interactions trigger the internalization of HCV via a receptor-mediated endocytosis process that is followed by the fusion of viral and cellular membranes and the release of the viral genome into the cytosol [9]. The incoming HCV RNA is then transported to intracellular membranes derived from the endoplasmic reticulum (ER) where primary translation and polyprotein processing occur. The viral replicase complex together with cellular factors is responsible for the RNA replication process, which starts by the generation of negative-sense RNA intermediates that are used as templates for the production of new positive-sense RNA molecules. Newly synthesized HCV genomes are then used as templates for translation, more rounds of RNA replication or packaging into progeny virus.

Lipid droplets (LD) have emerged as an important platform for HCV virus assembly [10], which is a highly coordinated process with several intermediate steps including: recruitment of NS5A and core proteins into LDs, encapsidation of HCV RNA into core particles and membrane envelopment of HCV RNA-containing core particles [11–15]. After budding into the ER lumen HCV exits the cell through the cellular secretory pathway. Given the tight association to the ER throughout different steps in its life cycle, it is not surprising that several ER and ER-related proteins have been identified as host factors regulating the HCV infection e.g., signal peptide peptidase [16], Sigma-1 receptor [17].

The erlin proteins, erlin-1 and erlin-2, are endoplasmic reticulum membrane lipid raft-associated proteins that belong to a larger family of proteins containing a conserved stomatin, prohibitin, flotillin, HflK/C (SPFH) domain which is proposed to organize membrane microdomains [18–20]. Both erlins are closely related as they share ~80% identity at the amino acid level [20,21] and they are evolutionarily conserved with homologous proteins found in *Caenorhabditis elegans* and *Arabidopsis thaliana* [21]. Erlin proteins are located in detergent resistant membranes (DRM) where they form high molecular weight complexes containing erlin homo- and hetero-oligomers as well as other cellular proteins [22]. Early reports described the function of erlin-1 and erlin-2 proteins in the endoplasmic reticulum associated degradation (ERAD) of inositol 1,4,5-triphosphate (IP3) receptors (IP3Rs) [23,24]. Afterwards other reports suggested that erlin-2 protein is required for the sterol-induced degradation of cholesterol biosynthetic enzyme HMG-CoA reductase [25] and for the processing of amyloid β-peptide (Aβ) precursor (APP) into Aβ by γ-secretase in the brain [26]. Besides their function in the ERAD pathway erlin proteins have been shown to regulate cholesterol homeostasis. They are cholesterol-binding proteins that interact with the sterol regulatory element binding protein (SREBP)-Scap-Insig complex restricting SREBP activation and leading to an intracellular accumulation of lipids and cholesterol [27]. More recently, Inoue and Tsai reported the first link between erlin proteins and viral infections [28]. They showed that erlin 1 and erlin 2 proteins are both required for polyomavirus SV40 infection by facilitating B12 transmembrane J-protein mobilization to specific foci in the ER, a prerequisite for the ER to cytosol transport of SV40, thus enabling the establishment of infection [28].

In view of the cellular functions and ER localization of erlin proteins, and considering the dependence of HCV on lipid metabolism and the ER for its life cycle, we decided to investigate the potential role of erlin proteins in HCV infection. In this study we describe the discovery that erlin-1 protein regulates the initiation of HCV RNA replication, the accumulation of viral proteins and therefore, the production of infectious virus, adding erlin-1 to the list of host factors required for efficient HCV infection.

2. Materials and Methods

2.1. Cells, Plasmids, Antibodies and Reagents

The origin of Huh-7 [29], Huh-7.5.1 [29], Huh-7.5.1 subclone 2 (Huh-7.5.1c2) [15] and HEK-293T [30] cells have been described previously. All cells were maintained in Dulbecco's modified Eagle's medium (DMEM) (Cellgro; Mediatech, Herndon, VA, USA) supplemented with 10% fetal calf serum (FCS) (Cellgro), 10 mM HEPES (Invitrogen, Carlsbad, CA, USA), 100 units/mL penicillin, 100 mg/mL streptomycin, and 2 mM L-glutamine (Invitrogen) in 5% CO_2 at 37 °C. The sub-genomic and full-length

JFH-1 stable replicon Huh-7 cell lines were cultured in medium supplemented with 400 or 200 μg/mL of G418, respectively, as described previously [15]. The JFH-1 genome-containing plasmid has been previously described [31]. The JFH-1 Rluc/SGR wt or JFH-1 Rluc/SGR GND plasmids carry bicistronic constructs containing the luciferase reporter gene in the first cistron and wild-type (wt) or replication-deficient (encoding a GDD-to-GND mutation in NS5B) JFH-1 sub-genomic replicon in the second cistron, respectively [32]. The rabbit polyclonal antibody for the detection of cellular erlin-2 protein was in-house generated and affinity purified against the full-length immunogen [27]. The rabbit polyclonal antibody HPA011252 against erlin-1 protein was obtained from Sigma-Aldrich (St. Louis, MO, USA). The recombinant human IgG anti-E2, the mouse monoclonal 9E10 anti-NS5A and the rabbit polyclonal MS5 anti-NS5A antibodies were kindly provided by M. Law (Scripps Research, La Jolla, CA, USA), C. M. Rice (Rockefeller University, New York, NY, USA) and M. Houghton (University of Alberta, Edmonton, AB, Canada), respectively. The monoclonal mouse antibodies against EEA1, HCV core (clone C7-50) and NS3 (clone 2E3) proteins were obtained from BD Transduction Laboratories (Franklin Lakes, NJ), Santa Cruz Biotechnology (Santa Cruz, CA, USA) and BioFront Technologies (Tallahassee, FL, USA), respectively. The HCV RNA replication inhibitor 2′-C-methyladenosine (2 mAd) was used at 10 μM final concentration and was a gift from W. Zhong (Gilead Sciences, Foster City, CA, USA). The 3-[4,5-dimethylthiazol-2-yl]-2,5-diphenyltetrazolium bromide (MTT) was purchased from Sigma Aldrich. Protease and phosphatase inhibitors were purchased from Roche (Indianapolis, IN, USA).

2.2. Silencing of Erlin Proteins by siRNA Transfection

siRNAs targeting human *ERLIN1* (siErlin 1.5: CCACAAATAGGAGCAGCAT [27]) or *ERLIN2* (siErlin 2.3: GCCTCTCCGGTACTAACAT [27]) individually, or *ERLIN1* and *ERLIN2* simultaneously (siErlin 1&2: AGAAGCAATGGCCTGGTAC [27]), and the non-targeting control siRNA (siCtrol: ACTGTCACAAGTACCTACA [24]), as well as the siRNA targeting HCV genome (siHCV: ACCTCAAAGAAAAACCAAA [17]) were all previously described. All the siRNAs were purchased from Integrated DNA Technologies (IDT, San Diego, CA, USA). Typically, Huh-7 cells were plated at a density of 1×10^5 cells per well on a 6-well plate. 24 h later cells were transfected with 20 pmol of the corresponding siRNA per well using Dharmafect 4 transfection reagent following manufacturer's instructions (GE Healthcare Dharmacon Inc, Pittsburgh, PA, USA). In brief, 20 pmol of siRNA was added into 200 μL of OptiMEM containing tube and 1 μL of Dharmafect 4 reagent was added into a second tube containing another 200 μL of OptiMEM. These mixtures were incubated for 5 min at room temperature before they were mixed together and incubated for another 20 min at room temperature. During the incubation time the culture medium of the Huh-7 cells was discarded and 1.6 mL of complete DMEM medium was added to each well. After 20 min of incubation the siRNA-Dharmafect complexes-containing medium (400 μL) was added to the corresponding wells. Fresh medium was applied after overnight incubation and the cells were assayed at the indicated times after transfection as indicated in each experiment. Cell viability was determined by evaluating cell biomass at different times after siRNA transfection by crystal violet staining and colorimetry at 570 nm, as previously described [33]. As a complementary approach, the cytotoxic effect of siRNAs was evaluated by quantifying the mitochondrial activity of siRNA-transfected cells (in MTT assays) and by determining the cellular respiration capacity. The cellular respiration capacity was calculated as the mitochondrial activity relative to the total protein content of the cells, measured by BCA in wells that were transfected in parallel.

2.3. Preparation of Viral Stocks and Infections

The original JFH-1 virus was generated by transfection of an in vitro transcribed full-length JFH-1 HCV RNA into Huh-7 cells and viral stocks were produced by inoculation of fresh Huh-7 cells at a multiplicity of infection (moi) of 0.01 as described [29]. Production of high titer cell culture adapted JFH-1 day 183 virus (D183) [34] was achieved by inoculation of highly susceptible Huh-7.5.1c2 at low moi. These D183 virus stocks were used in all low and high moi infection experiments (moi = 0.2 or 3) as described previously [15]. Intracellular infectious HCV particles were obtained by 5 freeze-thaw cycles of infected cells as described [35]. The infectivity titers present in culture supernatants and cell extracts were determined by end-point dilution and fluorescence focus forming unit (FFU) assay in Huh-7.5.1 cells, as described previously [36].

2.4. Analysis of HCV Cell Entry Using HCVpp

Retroviral particles pseudotyped with the JFH-1 E1 and E2 envelope proteins or the VSV G glycoprotein, as control, were produced in HEK293T cells [37]. For HCV entry experiments, Huh-7 cells (2×10^4 cells) were plated on 48-well plates and twenty hours later they were transfected with the indicated siRNAs as described above. Thirty-six hours later, cells were inoculated with pseudotyped retroviral particles (HCVpp or VSVpp) diluted to produce similar luciferase activity levels. Forty-eight hours later, cells were lysed and luciferase activity was detected and quantitated in a luminometer using a commercial kit (Luciferase Assay Kit; Promega, Madison, WI, USA). Luciferase results were normalized to cell density and relative infection values were calculated as a percentage of cells transfected with control siRNA. In parallel, cultures transfected with siRNAs were used: (i) to analyze knockdown efficiency by Western-blotting (WB) and (ii) to confirm the reduced susceptibility of erlin-1-deficient cells to HCV infection.

2.5. In Vitro Transcription and HCV RNA Transfection

The subgenomic JFH-1 replicon plasmid bearing a luciferase reporter gene (JFH-1 Rluc/SGR wt) and the corresponding replication-deficient construct (JFH-1 Rluc/SGR GND) have been described previously [32]. After digestion with XbaI restriction enzyme the linearized plasmids were in vitro transcribed using the T7 MEGAscript (Ambion, Austin, TX, USA) kit following the manufacturer's instructions. siRNAs targeting the indicated genes were transfected into Huh-7 cells as described above. Twenty hours later the cells were trypsinized and cells were seeded in wells of 12-well plate at a density of 1×10^5 cells per well. The following day *in vitro* transcribed subgenomic HCV RNAs were introduced into those cells by transfection using TransIT mRNA Transfection Kit (Mirus Bio LLC, Madison, WI, USA) as recommended by the manufacturer. At the indicated time points the cells were lysed. An aliquot of each lysate was used to quantitate the luciferase activity in a luminometer using a commercial kit (Promega). Another aliquot from each lysate obtained at six hours post-transfection was used to quantitate the HCV RNA levels by reverse transcription real-time quantitative PCR (RT-qPCR) as described below. The HCV RNA values were used to control for differences in the transfection efficiency between samples and were used to normalize the luciferase values. Luciferase results were normalized to cell density at each time point and the relative values were calculated as a percentage of cells transfected with control siRNA. In parallel, cultures transfected with siRNAs were used: (i) to analyze knockdown efficiency by WB and (ii) to confirm the reduced susceptibility of erlin-1-deficient cells to HCV infection.

2.6. siRNA Transfection Experiments in Acutely HCV Infected Cells

Huh-7 cells were transfected with the indicated siRNAs as described above. Thirty-six hours later, transfected cells were infected with JFH-1 D183 virus at low or high moi (0.2 or 3, respectively) as described before [15]. Five hours after HCV inoculation, cells were washed twice with PBS and medium was replaced. At the indicated times after infection supernatants were collected and extracellular infectivity was determined by end-point dilution in Huh-7.5.1 cell line. Cell extracts were harvested as indicated above and the intracellular infectivity, HCV RNA, and viral proteins were analyzed by titration, RT-qPCR and WB, respectively, as described [38]. Erlin protein down-regulation efficiency was confirmed by WB and densitometry analysis at the time post-infection indicated in each figure legend.

2.7. siRNA Transfection Experiments in Persistently Infected Cells

Huh-7 cells were infected with JFH-1 virus at a low moi = 0.01 as described [29]. Infected cells were passaged for 3 weeks before siRNAs were transfected. Three days after siRNA transfection supernatants were discarded, cells were washed twice with PBS and medium was replenished. Twenty-four hours later supernatants were collected and progeny virus released during those last twenty-four hours was determined by end-point dilution in Huh-7.5.1 cells as described above. Cell extracts were harvested and the intracellular infectivity, HCV RNA and proteins and erlin protein down-regulation were analyzed by titration, RT-qPCR and WB, respectively.

2.8. Protein Analysis by WB

Cell extracts were prepared in RIPA buffer supplemented with protease and phosphatase inhibitors and, after protein quantitation by bicinchorinic acid (BCA) assay (Thermo Scientific, Waltham, MA, USA), equivalent amounts of total protein (typically 20–25 µg) were separated in polyacrylamide-SDS gels by electrophoresis and they were transferred to Immobilon membranes (Millipore, Billerica, MA, USA) for blotting. The membranes were blocked for 1 h at room temperature with PBS-5% milk and incubated overnight at 4 °C with the primary antibodies diluted in 1% milk-0.1% Tween 20 in PBS. Primary antibodies were used at the following dilutions: EEA1 at 1:2000; erlin-1, erlin-2 and NS3 at 1:1000; NS5A at 1:500; and Core at 1:200 dilution. After washing at least 4 times for 15 min with 0.1% Tween 20 in PBS, membranes were incubated with a 1:10,000 dilution of goat-anti-rabbit or goat-anti-mouse IgG conjugated to horseradish peroxidase prepared in 1% milk-0.1% Tween 20 in PBS. After washing at least 4 more times with 0.1% Tween 20 in PBS, membranes were developed using the SuperSignal-West Pico or Femto chemiluminescence reagents (Thermo Scientific) and autoradiography. Densitometry of non-saturated films was performed using Fiji/Image J analysis software [39] and the results, shown beneath each panel, are expressed relative to control cells using EEA1 protein expression for normalization and as loading control. Two-fold serial dilutions of siCtrol samples were used to produce standards for relative quantitation.

2.9. RNA Analysis

Total RNA was extracted from cells using the guanidinium isothiocyanate extraction method [29] after adding 20 µg of glycogen (Roche) per sample as a carrier. The content of HCV RNA, *ERLIN1* and *ERLIN2* mRNAs and *GAPDH* mRNA (for normalization) in each sample was quantitated in a two-step RT-qPCR assays using the High-Capacity cDNA Reverse Transcription Kit and the Maxima SYBR Green/ROX qPCR Master Mix (2X) (Thermo Scientific). 10-fold serial dilution of plasmids containing each target sequence was used to prepare the standard curves that were used with each corresponding pair of primers: HCV (5′-TCTGCGGAACCGGTGAGTA-3′ and 5′-TCAGGCAGTACCACAAGG-3′); *ERLIN1* (5′-GGGGTTGGTGGCTGTCCTGC-3′ and 5′-TAGCCTGGTCCACTGGGGCT-3′); *ERLIN2* (5′-TGTGCACACGCTTCAAGAGGTCTA-3′ and 5′-AATGACCAGCCCAGGGGCCA-3′) and *GAPDH* (5′-GAAGGTGAAGGTCGGAGTC-3′ and 5′-GAAGATGGTGATGGGATTTC-3′). Results were normalized using *GAPDH* mRNA levels and were displayed in the figures as relative values compared to control cells.

2.10. Confocal Analysis

Huh-7 cells were transfected with siRNAs and infected at high moi (moi = 3) with JFH-1 D183 virus as described above. The day before fixation cells were seeded in glass bottom 96-well plates (Thermo Scientific) for confocal analysis. At the indicated times after virus inoculation, cells were incubated in 4% paraformaldehyde for 20 min at room temperature, washed 3 times with PBS and processed for immunofluorescence analysis and LD staining as previously described [38]. In brief, fixed cells were first incubated for 1 h at room temperature in blocking buffer (1xPBS, 10% FBS, 3% BSA, 0.3% Triton X-100), washed 3 times with PBS and then incubated for 1 h with a mixture of primary antibodies at the appropriate concentrations that were prepared in binding buffer (1xPBS, 3% BSA, 0.3% Triton X-100). The primary antibodies used for immunofluorescence were: a purified mouse monoclonal anti-core (clone C7-50) at 1:200 dilution and a rabbit polyclonal anti-NS5A (MS5) at 1:1000 dilution. After 3 washes with PBS, cells were incubated for one more hour in binding buffer containing a mixture of fluorophore-conjugated secondary antibodies at 1:1000 dilution (Invitrogen) and Hoechst dye (Invitrogen) for nuclei staining. LD staining was performed by incubating the cells for 30 min at room temperature with the LipidTox reagent (Invitrogen) at 1:500 dilution in PBS. Images of 1024 × 1024 pixels at 8-bit gray scale were acquired with a 40× objective (pixel size 0.345 µm) on a LSM 710 confocal laser scanning microscope (Zeiss, Dublin, CA, USA). All images were taken with exactly the same settings. Fiji/ImageJ software analysis package [39] was used to quantitate the fluorescence intensity signals in each image.

2.11. Statistical Analysis

GraphPad Prism v.5.0a software (GraphPad Software, San Diego, CA, USA) was used to analyze the data, prepare the graphs and perform all statistical analysis. All experimental results (quantitative RT-qPCR analysis, MTT assays, titration assays, luciferase activity assays, and quantitative fluorescence intensity analysis) were displayed in the graphs as the mean ± standard deviation (S.D.). The normal distribution of data was confirmed using Kolmogorov-Smirnov test in experiments with sample sizes n ≥ 6. Experiments where normality test could not be performed because of sample size limitations (n = 3) normal distribution of the data was assumed. The differences among means between multiple (more than two) groups were analyzed by One-way ANOVA or Two-way ANOVA followed by Dunnett's or Bonferroni's Multiple Comparison Test as indicated in each figure legend. One-way ANOVA was used for single variable analysis in Figures 1D,E, 2, 3D and 4–9, while Two-way ANOVA was used in kinetic experiments for two variable analysis in Figures 1A,C and 3B. The statistical significance was set as: * $p < 0.05$; ** $p < 0.01$; *** $p < 0.001$.

3. Results

3.1. Erlin-1 Protein Is a Host Factor Required for Efficient HCV Infection

To investigate the potential role of erlin-1 and erlin-2 proteins in the HCV lifecycle we took advantage of siRNAs that have being previously validated and used to characterize the cellular functions of erlin proteins [24,27]. We transfected Huh-7 cells with siRNAs targeting specifically erlin-1 (i.e., siErlin 1.5) or erlin-2 (i.e., siErlin 2.3) individually, or siRNAs targeting both erlin proteins simultaneously (siErlin 1&2), siRNAs targeting HCV as a positive control (siHCV) and a non-targeting siRNA as a negative control (siCtrol). siRNA transfected cells were inoculated at low multiplicity of infection (moi = 0.2) with a cell culture-adapted HCV virus (D183v) [34] and their susceptibility to infection was assessed by measuring virus production three and five days later. Analysis of the supernatants of erlin-1 down-regulated HCV-infected cells showed ~ 7–10 fold reduction of progeny virus production compared to non-targeting control siRNA-transfected cells that were infected in parallel (Figure 1A).

Remarkably, erlin-2 protein down-regulation did not reduce HCV virus production. As expected the HCV-targeting siRNA prevented the production of infectious HCV virus. Consistent with the effect on virus production, the accumulation of intracellular HCV proteins (e.g., NS3 and NS5A) was also reduced in erlin-1, but not in erlin-2, down-regulated cells (Figure 1B). Despite a small defect in the cellular proliferation rate observed in siErlin 1.5-transfected cells (Figure 1C,D), the cellular respiration capacity of the different siRNA-transfected cells was comparable to that of control cells (Figure 1E). Collectively, these results suggest that the effects observed in erlin-1 down-regulated cells (in siErlin 1.5 and siErlin 1&2) are not due to a reduction in cell viability and most probably reflect a specific requirement of erlin-1 protein for efficient HCV infection.

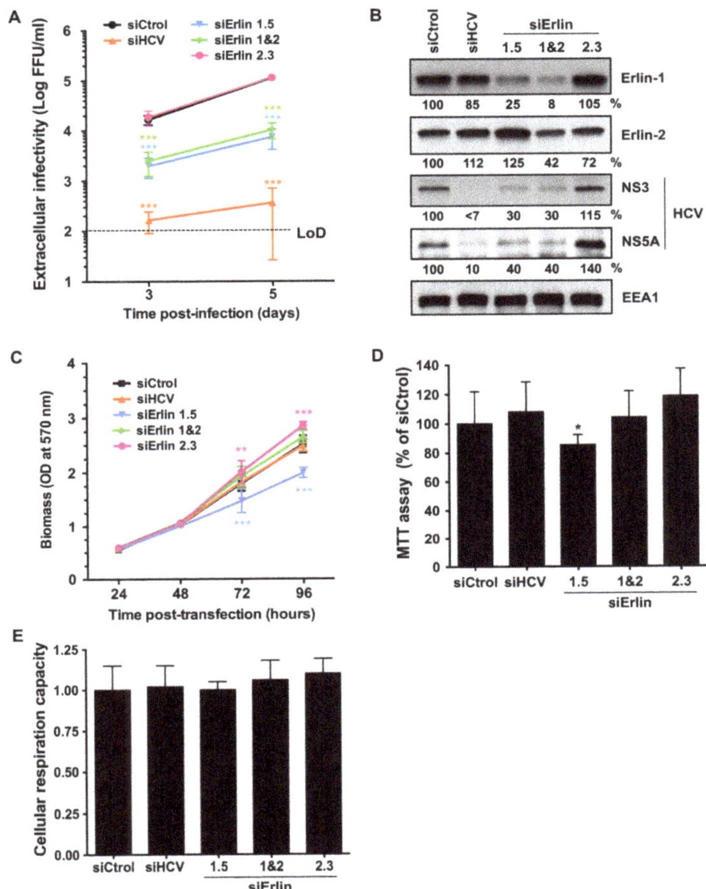

Figure 1. Erlin-1 protein down-regulation impairs efficient HCV infection. Huh-7 cells were transfected with individual siRNAs targeting erlin-1 (siErlin 1.5), erlin-2 (siErlin 2.3), both erlins simultaneously (siErlin 1&2), HCV (siHCV) or a non-targeting siRNA control (siCtrol) as indicated in Material and Methods section. Transfected cells were infected 36 h later with JFH-1 D183 virus at low multiplicity of infection (moi = 0.2). (**A**) The extracellular infectivity present in supernatants of infected cells was determined on days three and five post-inoculation by titration assay. Infectivity titers are represented as the FFUs per ml of supernatant in logarithmic scale and are displayed as the average and standard deviation (Mean; SD; n = 3). The horizontal black dotty line represents the limit of detection (LoD) of the assay. (**B**) Three days after virus inoculation cellular erlin-1, erlin-2 and EEA1 (as loading control) and viral NS3 and NS5A protein expression was determined in HCV-infected cell extracts by WB. Relative protein levels were determined by densitometry and are shown below each panel. The ratio of intensities of each protein and EEA1 in siCtrol-transfected cells was set as 100 and it was used to calculate the relative amount of each protein in each sample. Results in panels A and B are representative of two independent experiments, each one performed in triplicate. (**C**) Cells plated in 96-well plate format were transfected with siRNAs and the cell biomass was quantitated by crystal violet staining at different times post-transfection. Data are displayed as the average and standard deviation of two independent experiments with three replica wells per condition at each time point in

each experiment (Mean; SD; n = 6). (**D**) Cells plated in 96-well plate format were transfected and infected as described above and were used to perform MTT assays three days after HCV inoculation. Data are displayed as the average and standard deviation (Mean; SD; n = 7) relative to the levels in siCtrol condition, set as 100. Results are representative of two independent experiments with seven replica wells per condition in each experiment. (**E**) The cellular respiration capacity was calculated as described in Material and Methods section. Data are displayed as the average and standard deviation (Mean; SD; n = 3) relative to that of siCtrol, set as 1. Statistical significance was determined using: Two-way ANOVA followed by a Bonferroni posttest for panels A and C (** $p < 0.01$; *** $p < 0.001$), and One-way ANOVA followed by the Dunnett's Multiple Comparison Test for panels D and E (* $p < 0.05$).

3.2. Erlin-1 Protein Is Not Required for HCV Cell-Entry

The multiple cycle infection experiments described above suggested that erlin-1 protein plays an important role in the HCV lifecycle. Next, we set out to determine the step in the virus life cycle in which erlin-1 protein was required. Many aspects of HCV cell entry have been studied using HCV-pseudotyped retroviral vectors. Therefore, we used retroviral particles pseudotyped with the JFH-1 envelope glycoproteins E1 and E2 (HCVpp) or the vesicular stomatitis virus protein G (VSVpp), as control, to determine their ability to infect erlin-1 down-regulated cells. siRNA-transfected Huh-7 cells were infected with the corresponding retroviral particles and the intracellular luciferase activity was measured forty-height hours later. Down-regulation efficiency was verified in parallel cultures by WB (Figure 2A). As shown in Figure 2B, transfection of *ERLIN1*-targeting siRNAs had no effect on viral entry efficiency despite strong reduction in erlin-1 protein expression achieved in those cells. As expected *ERLIN2*- and HCV-specific siRNAs had no effect on the HCVpp entry process. These results suggest that reduced erlin-1 protein expression does not affect HCV cell entry.

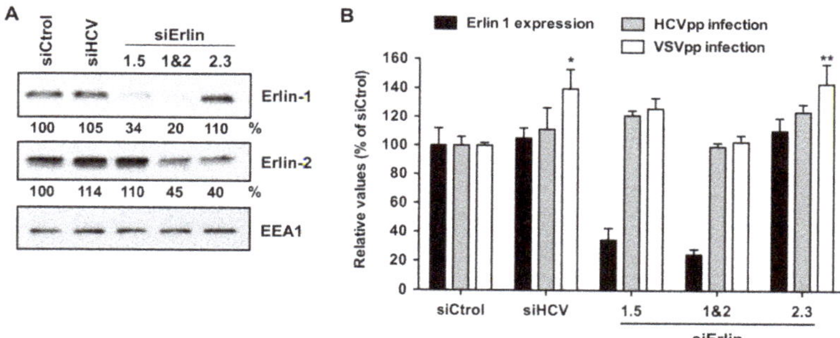

Figure 2. Erlin-1 protein down-regulation does not inhibit HCV cell-entry. Huh-7 cells were transfected with individual siRNAs as described in Material and Methods. (**A**) 36 h later one set of wells was harvested for erlin protein analysis by WB. Relative protein expression levels were determined by densitometry as described in the legend of Figure 1 and they are shown below each panel. (**B**) Another set of wells was inoculated with HCVpp (grey bars) or particles harboring the VSV-G glycoprotein (VSVpp; white bars) as control. 48 h after inoculation the intracellular luciferase activity was measured. Results are displayed as percentage of the luciferase activity present in siCtrol-transfected cell extracts for each virus particle. Data are displayed as the average and standard deviation (Mean; SD; n = 3). The luciferase values (units per well) in siCtrol-transfected cell extracts were $9 \times 10^3 \pm 5.1 \times 10^2$ and $1.2 \times 10^4 \pm 2 \times 10^2$ for HCVpp and VSVpp, respectively. These results are representative of two independent experiments, each one performed in triplicate. The average erlin-1 protein expression level of the two experiments is shown in black bars. One-way ANOVA followed by Dunnett's Multiple Comparison Test analysis of the HCVpp data did not show any statistically significant effect on the luciferase levels among different siRNAs. Instead, the entry of VSVpp was significantly higher in siHCV- and siErlin 2.3-transfected cells (* $p < 0.05$; ** $p < 0.01$).

3.3. Erlin-1 Protein Down-Regulation Impairs the Establishment of HCV RNA Replication but Does Not Affect Primary Translation or Maintenance Of Replication

To test whether erlin-1 protein regulates early steps downstream HCV entry such as the primary translation of the incoming RNA or the establishment of RNA replication, we used a highly sensitive approach based on the transfection of an in vitro transcribed JFH-1 subgenomic RNA bearing a luciferase reporter gene (JFH-1 Rluc/SGR RNA). In this system the luciferase activity measured at early time points after transfection (i.e., at six hours post-transfection) derives exclusively from the translation of transfected RNA, while the accumulation of luciferase at later time points (e.g., at 48, 72 or 96 h post-transfection) is the result of both translation and RNA replication processes. Consistent with this, luciferase activity derived from JFH-1 Rluc/SGR wt RNA was equivalent in the presence (white bars) or absence (black bars) of the HCV polymerase inhibitor 2′-C-methyladenosine (2 mAd) six hours after transfection (Figure 3A). However, it was clearly reduced in 2 mAd-treated cells at 96 h post-transfection.

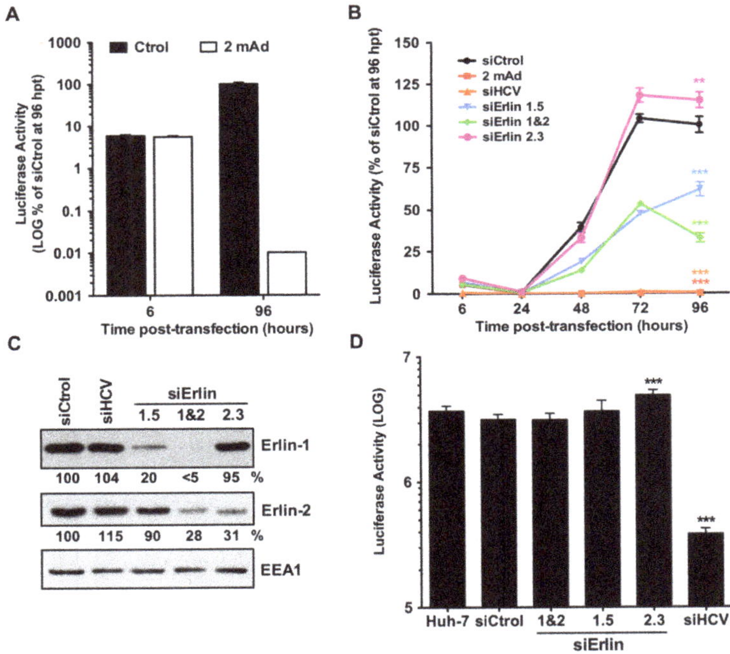

Figure 3. Erlin-1 protein down-regulation impairs the initiation of HCV RNA replication without affecting HCV IRES-dependent translation. Huh-7 cells were transfected with individual siRNAs as described in Material and Methods. 48 h later cells were transfected with in vitro transcribed bicistronic renilla luciferase (Rluc)-containing wt or replication-deficient (GND mutant) sub-genomic (SGR) HCV RNAs. As control for the inhibition of HCV RNA replication process, transfected cells were treated with the HCV polymerase inhibitor 2′-C-methyladenosine (2 mAd). (**A**) Quantitation of renilla luciferase activity derived from Rluc/SGR wt RNA in the presence (white bars) or absence (black bars) of 2 mAd at early (6 h) and late (96 h) time points. Data are displayed as the average and standard deviation (Mean; SD; n = 3) of luciferase values relative to that of control cells at 96 h post-transfection (hpt), that was set as 100. (**B**) Time course of renilla luciferase activity derived from RLuc/SGR wt RNA in siRNA-transfected cells. For each independent experiment, Rluc activity was normalized to cell density and was displayed as a percentage of that determined in siCtrol-transfected cell extracts at 96 hpt. Data shown are averages and standard deviation of two independent experiments, each one performed in

triplicate (Mean; SD; n = 6). The luciferase activity in siCtrol-transfected cell extracts at 96 h post-transfection was $1.3 \times 10^7 \pm 9.1 \times 10^5$ light units per well. To assess the specificity of Rluc activity, cells were treated with 2 mAd (red line) or were transfected with siRNAs targeting directly HCV RNA (orange line). Statistical analysis performed using Two-way ANOVA followed by Bonferroni posttest showed statistically significant differences in: 2 mAd-treated, siHCV-, siErlin 1.5- and siErlin 1&2-transfected cells at 48-, 72- and 96-hpt, and siErlin 2.3-transfected cells at 72- and 96-hpt compared to siCtrol-transfected cells (** $p < 0.01$; *** $p < 0.001$). (**C**) Erlin protein down-regulation was determined at the time of Rluc/SGR RNA transfection by WB and densitometry (shown below each panel) as described in the legend of Figure 1. (**D**) Luciferase activity of replication-deficient (GND mutant) Rluc/SGR RNA was determined at six hours post-transfection. For each independent experiment Rluc activity was normalized to cell density. Data shown are averages and standard deviation of two independent experiments, each one performed in triplicate (Mean; SD; n = 6). Statistical analysis performed using One-way ANOVA followed by Dunnett's Multiple Comparison Test showed statistically significant differences in siErlin 2.3- and siHCV-transfected cells compared to siCtrol-transfected cells (*** $p < 0.001$). No differences were observed in erlin-1 downregulated cells compared to siCtrol-transfected cells.

Luciferase activity measured at early time points (6 and 24 h post-transfection) was similar in all siRNA-transfected cell lines, suggesting that primary translation is not affected by the levels of erlin-1 or erlin-2 protein expression (Figure 3B). Interestingly, the luciferase activity in erlin-1 down-regulated cells (siErlin 1.5- and 1&2-transfected cells) was around 50% lower than that of control cells at 48 and 72 h post-transfection. Moreover, the luciferase activity measured at 96 h time point showed greater differences in siErlin 1&2-transfected cells (30% of siCtrol) than in siErlin 1.5-transfected cells (60% of siCtrol), correlating with erlin-1 down-regulation efficiencies achieved by the corresponding siRNAs (Figure 3C). These results suggest that erlin-1 protein regulates the establishment of HCV RNA replication without affecting the primary translation. Supporting this notion, luciferase activity measured six hours after the transfection of a replication deficient JFH-1 RLuc/SGR GND mutant RNA was comparable in erlin-1 down-regulated cells and control cells (Figure 3D). Altogether, these data suggest that erlin-1 protein regulates the initiation of RNA replication without affecting the translation of incoming HCV RNA.

Next, we analyzed the effect of erlin-1 protein down-regulation on the maintenance of HCV RNA replication in the stably replicating JFH-1 subgenomic replicon cells (SGR cells). SGR cells were transfected with the indicated siRNAs and three days later the intracellular erlin protein levels and HCV protein and RNA levels were analyzed by WB and RT-qPCR, respectively. As expected, transfection of the HCV-targeting siRNA produced a significant reduction of HCV proteins (Figure 4A) and RNA (Figure 4B). However, no effect was observed when siRNAs targeting erlin-1 protein were transfected despite the 50–70% reduction in erlin-1 protein expression achieved. Similar results were obtained when JFH-1 full-length replicon cells were used (data not shown). These results suggest that erlin-1 protein does not regulate the ongoing HCV RNA replication process.

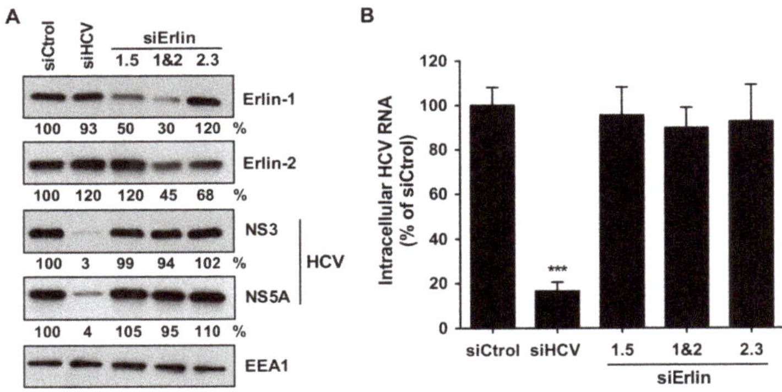

Figure 4. Erlin-1 protein down-regulation does not interfere with ongoing HCV RNA replication. JFH-1 sub-genomic replicon (SGR) bearing Huh-7 cells were transfected with siRNAs as described in Material and Methods. Three days after transfection cells were harvested for analysis. (**A**) Cellular erlin-1, erlin-2 and EEA1, and HCV NS3 and NS5A protein accumulation were determined by WB and densitometry (shown below each panel) as described in the legend of Figure 1. (**B**) The intracellular HCV RNA content was quantitated in infected cell extracts three days after siRNA transfection by RT-qPCR. Data were normalized relative to *GAPDH* mRNA levels in the same samples and are displayed as percentage of the HCV RNA present in siCtrol-transfected cells. The HCV RNA content in siCtrol-transfected cell extracts was $1.4 \times 10^6 \pm 1.2 \times 10^5$ copies per μg of total RNA. Data shown are averages of three independent experiments, each one performed in triplicate (Mean; SD; n = 9). Only the HCV RNA level in siHCV-transfected cells was significantly lower than that in siCtrol-transfected cells as determined by One-way ANOVA followed by Dunnett's Multiple Comparison Test (*** $p < 0.001$).

3.4. Erlin-1 Protein Down-Regulation Interferes with HCV Protein and Intracellular Infectious Virus Accumulation

Collectively the results from the experiments described above suggested that reduction of erlin-1 protein expression impairs HCV infection by inhibiting the establishment of RNA replication (Figure 3) without affecting HCV entry (Figure 2), HCV IRES dependent primary translation (Figure 3) or ongoing HCV RNA replication (Figure 4). Consistent with those results, quantitation of the intracellular HCV RNA levels 48 h after a high moi (moi = 3) infection showed a modest but statistically significant two to three-fold decrease in erlin-1 down-regulated cells compared to control cells (Figure 5A). Interestingly, the accumulation of intracellular (Figure 5B) and extracellular (Figure 5C) infectious virus was strongly reduced (around ten-fold) in those same cells. These results were confirmed with a different *ERLIN1*-targeting siRNA (siErlin 1.3 in Supplementary Figure S1). Similarly, the accumulation of HCV core, NS3 and NS5A proteins was reduced by four to ten-fold in erlin-1 down-regulated cells (Figure 5D). The disproportionate effect on the infectivity levels compared to the effect on the intracellular HCV RNA level suggested that erlin-1 protein is required not only for the establishment of RNA replication as shown in Figure 3, but also for a post-replication step in the virus lifecycle. In fact, the assembly rate of infectious virus particles (Figure 5E) but not the secretion rate (Figure 5F) was reduced in erlin-1 down-regulated cells compared the control cells. Supporting the latest, siRNAs that target directly the HCV genome (i.e., siHCV) inhibited HCV RNA accumulation and downstream HCV protein accumulation as well as intra- and extra-cellular infectious virus accumulation to the same extent (around 15–20 fold).

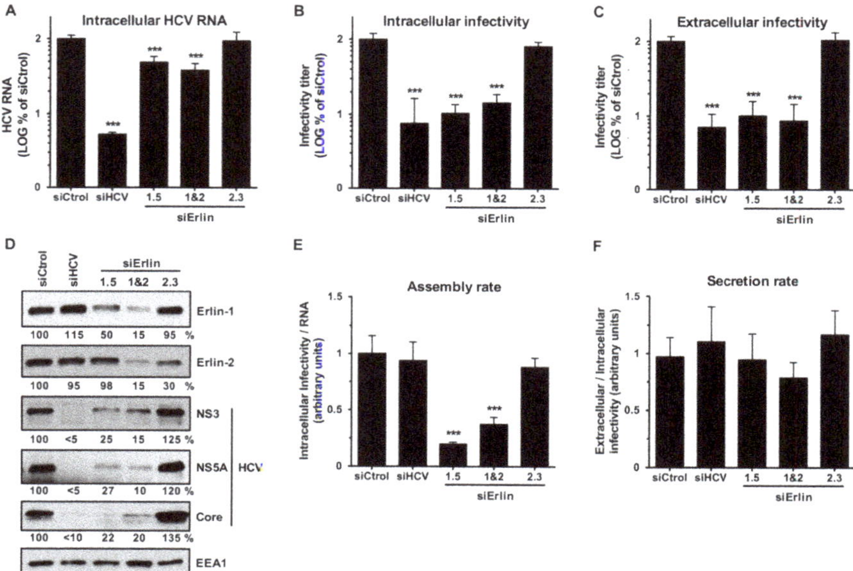

Figure 5. Erlin-1 protein down-regulation interferes with HCV in single-cycle infection experiments. Huh-7 cells were transfected with siRNAs as described in Material and Methods. 36 h later transfected cells were inoculated with JFH-1 D183 virus at high multiplicity of infection (moi = 3). 48 h after virus inoculation cellular extracts were prepared for intracellular RNA (**A**), infectivity (**B**) and protein (**D**) analysis and supernatants were collected for extracellular infectivity (**C**) analysis. Assembly (**E**) and secretion (**F**) rates were calculated using data from panels A, B and C. The RNA and infectivity results are displayed as percentage of the levels in siCtrol-transfected cells. The HCV RNA and the intracellular and extracellular infectivity levels in siCtrol-transfected cells were: $1.2 \times 10^7 \pm 1.3 \times 10^6$ HCV RNA copies per µg of total RNA, $3.5 \times 10^4 \pm 6.1 \times 10^3$ ffus per well, and $1.3 \times 10^5 \pm 2 \times 10^4$ ffus per ml of supernatant, respectively. Data shown are averages of three independent experiments, each one performed in triplicate (Mean; SD; n = 9). One-way ANOVA followed by Dunnett's Multiple Comparison Test was used to determine the statistical significance (*** $p < 0.001$).

Moreover, dose-dependent and parallel reductions in HCV parameters (i.e., intracellular HCV RNA, protein and infectious virus, and extracellular infectious virus) were observed when cells were infected in the presence of different doses of the HCV polymerase inhibitor 2 mAd (Figure 6). These results imply that the disproportionate effect in virus infectivity levels observed in erlin-1 down-regulated cells could not be explained solely by the modest reduction in intracellular HCV RNA accumulation suggesting an independent requirement for erlin-1 protein at a post RNA replication step.

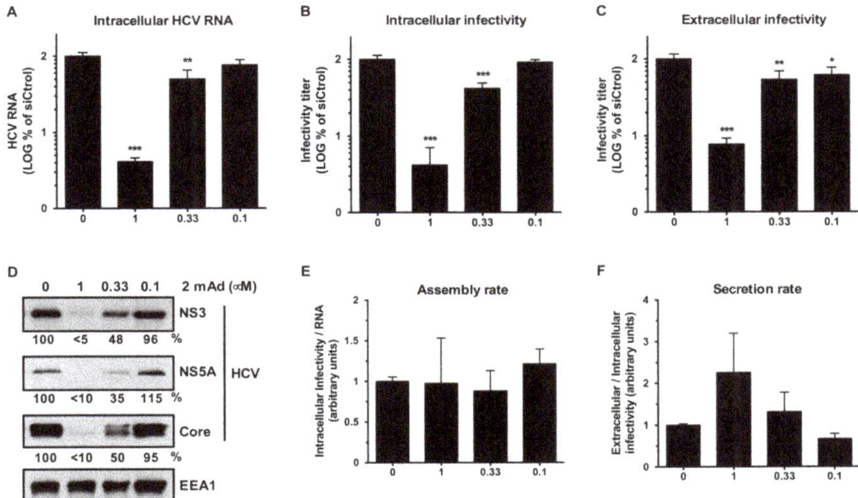

Figure 6. Dose-dependent reduction on HCV parameters upon treatment with a replication inhibitor. Huh-7 cells were inoculated with JFH-1 D183 virus at high multiplicity of infection (moi = 3) in the absence or presence of increasing amounts (0, 0.1, 0.33 and 1 µM) of the HCV polymerase inhibitor 2 mAd. 48 h later supernatants were collected, cell extracts were harvested and samples were subjected to the same analysis described in Figure 5. Note the proportional and dose response reduction of all viral parameters (RNA (**A**), infectivities (**B**,**C**) and protein (**D**)) in 2 mAd-treated cells and the absence of any significant effect in the assembly (**E**) and secretion (**F**) rates. The HCV RNA and the intracellular and extracellular infectivity levels in untreated cells were: $2 \times 10^7 \pm 2 \times 10^6$ HCV RNA copies per µg of total RNA, $2 \times 10^4 \pm 2.5 \times 10^3$ ffus per well, and $1.6 \times 10^5 \pm 2.5 \times 10^4$ ffus per ml of supernatant, respectively. Data shown are averages of two independent experiments, each one performed in duplicate (Mean; SD; n = 4). One-way ANOVA followed by Dunnett´s Multiple Comparison Test analysis was used for statistical significance analysis (* $p < 0.05$; ** $p < 0.01$; *** $p < 0.001$).

To prove unequivocally that erlin-1 protein is a rate limiting factor for a later step in the HCV virus life cycle we took advantage of the possibility of manipulating erlin-1 protein expression after the HCV infection is fully established. This approach allowed us to avoid the effect of erlin-1 protein down-regulation in the establishment of RNA replication described above. siRNAs were transfected into persistently HCV-infected cells and four days later the intracellular HCV RNA (Figure 7A) and the intracellular (Figure 7B) and extracellular (Figure 7C) infectious virus accumulation were quantitated by RT-qPCR and titration assays, respectively. While siRNAs targeting HCV genome significantly suppressed HCV RNA and infectious virus accumulation, *ERLIN1*-targeting siRNAs reduced infectious virus production by 70–80% with a modest 10–20% reduction on HCV RNA levels. This was reflected in a lower assembly rate (Figure 7E), but not in the secretion rate (Figure 7F) in erlin-1 down-regulated cells. These results were confirmed with a different *ERLIN1*-targeting siRNA (siErlin 1.3 in Supplementary Figure S2). HCV protein analysis showed stronger defect in NS3 than in NS5A and core protein accumulation in erlin-1 down-regulated cells (Figure 7D). These results strongly suggest that erlin-1 protein regulates later event(s) that lead to infectious virus production.

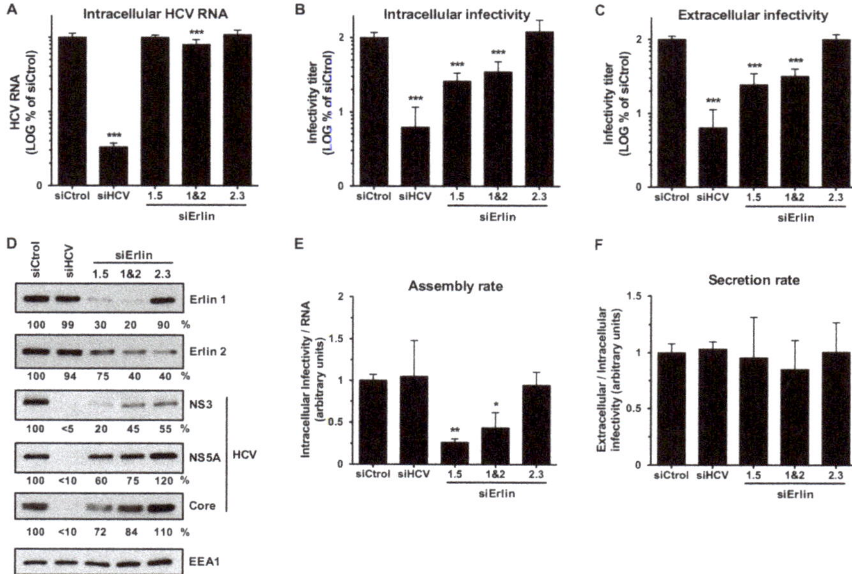

Figure 7. Erlin-1 protein down-regulation impairs infectious virus production in an ongoing HCV infection cell culture system. Persistently infected Huh-7 cells were transfected with siRNAs as described in Material and Methods. Four days after siRNA transfection cellular extracts were prepared for intracellular RNA (**A**), infectivity (**B**) and protein (**D**) analysis and supernatants were collected for extracellular infectivity (**C**) analysis. Assembly (**E**) and secretion (**F**) rates were calculated using data from panels A, B and C. The RNA and infectivity results are displayed as percentage of the levels in siCtrol-transfected cells. The HCV RNA and the intracellular and extracellular infectivity levels in siCtrol-transfected cells were: $5.8 \times 10^5 \pm 7 \times 10^4$ HCV RNA copies per μg of total RNA, $2.3 \times 10^3 \pm 2.8 \times 10^2$ ffus per well, and $1 \times 10^4 \pm 1.1 \times 10^3$ ffus per mL of supernatant, respectively. Data shown are averages of three independent experiments, each one performed in triplicate (Mean; SD; n = 9). One-way ANOVA followed by Dunnett's Multiple Comparison Test was used to determine the statistical significance (* $p < 0.05$; ** $p < 0.01$; *** $p < 0.001$).

3.5. Erlin-1 Protein Down-Regulation Increases LD Accumulation in Huh-7 Cells

We have previously described that down-regulation of erlin proteins lead to an intracellular fatty acids and cholesterol increase in Hela cells [27]. Furthermore, it is well known that HCV requires LDs for assembly of infectious virus and that HCV induces LD accumulation in infected cells [10]. Therefore, we analyzed the effect of erlin protein down-regulation on LD accumulation in HCV-infected Huh-7 cells. To do so, erlin down-regulated and control cells were infected with JFH-1 D183 virus at high moi (moi = 3) and fixed forty-eight hours later for analysis. Fixed cells were incubated with LipidTox, a reagent that stains specifically neutral lipids, and they were analyzed by confocal microscopy. As shown in Figure 8A, reduction in erlin-1 and erlin-2 protein expression produced an increase in LD accumulation compared to control cells. Quantitation of fluorescence intensity in individual cells confirmed those results and revealed a statistically significant increase in LD accumulation in cells where *ERLIN1* and *ERLIN2* were simultaneously down-regulated (i.e., siErlin 1&2) compared to that of control cells (Figure 8B). These results suggest that erlin protein down-regulation increases the intracellular accumulation of LDs not only in Hela cells, as previously described [27], but also in Huh-7 cells.

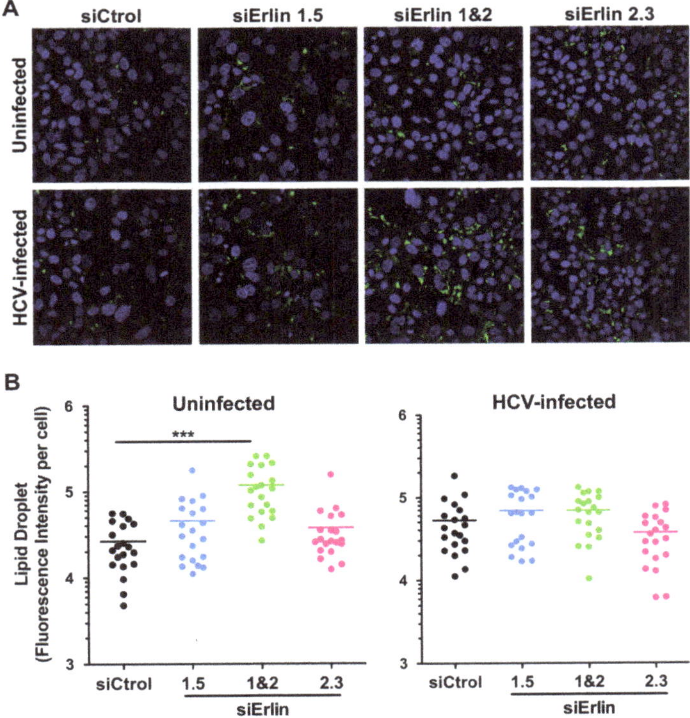

Figure 8. Erlin-1 protein down-regulation increases LD content in Huh-7 cells. siRNA-transfected Huh-7 cells were inoculated with JFH-1 D183 virus at high multiplicity of infection (moi = 3) and the intracellular LD content was analyzed 48 h after infection as described in Material and Methods. (**A**) Representative confocal image sections of intracellular LD accumulation (in green) in siRNA-transfected HCV-infected and uninfected cells. Nuclei (in blue) were counterstained with Hoechst dye. (**B**) Quantitation of the total LD fluorescence intensity signal per cell from confocal image sections. ImageJ software was used to quantitate the total LD fluorescence intensity signal in twenty individual cells of each condition. Each dot in the graph represents the LD fluorescence intensity of an individual cell and the horizontal lines show the average LD fluorescence intensity in each group of data. One-way ANOVA followed by Dunnett's Multiple Comparison Test was used to determine the statistical significance (*** $p < 0.001$).

3.6. Erlin-1 Protein Deficiency Does Not Impair HCV Core and NS5A Protein Re-Localization to LDs

Late in the infection both core and NS5A proteins are known to be associated to LDs to facilitate virus assembly [10]. Thus, we analyzed the distribution of core and NS5A proteins in *ERLIN* downregulated cells by confocal microscopy. As shown in Figure 9A, core and NS5A proteins localized to LDs similarly in both control and *ERLIN* down-regulated cells. A quantitative analysis of the localization of core and NS5A proteins surrounding LDs (Figure 9B) confirmed that the reduced infectious virus accumulation observed in *ERLIN1*-deficient cells was not due to a mislocalization of core or NS5A proteins during infection.

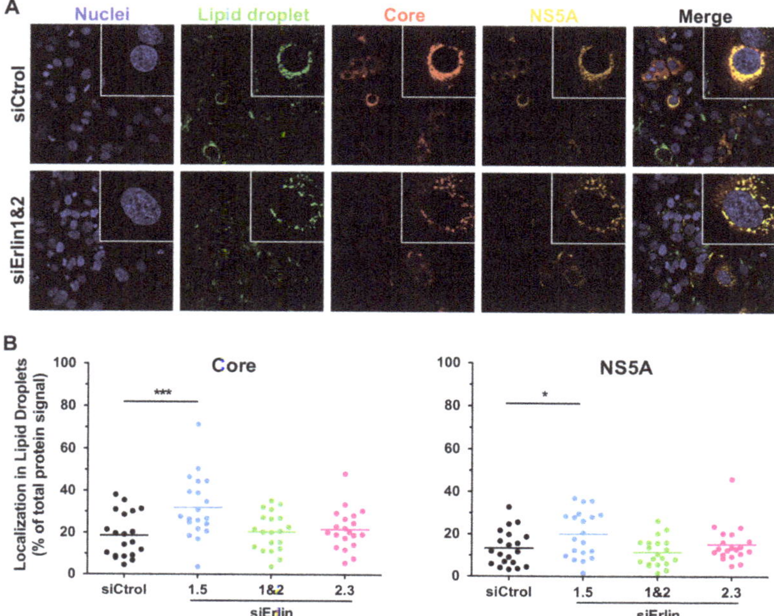

Figure 9. Erlin-1 protein down-regulation does not impair the localization of HCV core and NS5A proteins surrounding LDs. siRNA-transfected Huh-7 cells were inoculated with JFH-1 D183 virus at high multiplicity of infection (moi = 3) and the intracellular HCV protein and LD content was analyzed as described in Material and Methods. (**A**) Representative confocal images of the intracellular localization of HCV core (in red) and NS5A (in orange) proteins surrounding LDs (in green) in *ERLIN* down-regulated (siErlin 1&2) and control (siCtrol) HCV-infected cells. Nuclei (in blue) were counterstained with Hoechst dye. The panels on the right side show merged images of the four channels of each image. White boxes in the upper right side of each panel show zoom-in images of single cells for a more detailed observation. (**B**) The fluorescence intensity signal of core and NS5A proteins associated to LDs was quantitated in twenty randomly selected HCV-infected cells of each condition. Each dot in the graph represents the percentage of the fluorescence intensity signal of core or NS5A protein associated to LDs in a given cell, while the horizontal lines show the average of all data points in each group. One-way ANOVA followed by Dunnett's Multiple Comparison Test was used to determine the statistical significance (* $p < 0.05$; *** $p < 0.001$).

4. Discussion

The development of an in vitro HCV infection system in 2005 [29,40,41] allowed the identification of viral proteins and cellular factors and pathways that regulate different aspects of the HCV infection (reviewed in [8,42,43]). Using this infection model, we and others have previously identified cellular factors required for initiation of RNA replication [17,32] and HCV assembly and secretion [15,38], and cellular pathways required for the induction [44,45] and evasion [36,46] of the innate immune system. In this study, we have analyzed the consequences of erlin-1 and erlin-2 protein down-regulation on the susceptibility of Huh-7 cells to HCV infection. Our data suggest a role for erlin-1 protein early and late in the HCV life cycle.

Silencing of erlin-1 but not erlin-2 protein expression by siRNAs led to a strong reduction in the production of infectious progeny virus (Figure 1) suggesting that erlin-1 protein is required for efficient HCV infection. Single cycle infection experiments in *ERLIN1* down-regulated cells showed a modest 2–3 fold reduction in intracellular HCV RNA accumulation compared to a ~4–10 fold reduction in HCV protein accumulation and ~10-fold reduction in intracellular infectious virus accumulation (Figure 5 and Supplementary Figure S1). Further studies performed in HCV surrogate models that interrogate specific steps in the virus life cycle revealed that erlin-1 protein is required for the initiation of HCV RNA replication (Figure 3) without affecting earlier steps such as HCV cell entry (Figure 2) and HCV IRES dependent translation of the incoming RNA (Figure 3). Interestingly, *ERLIN1* silencing in subgenomic replicon cells (Figure 4) or in persistently infected cells (Figure 7) had no effect on the steady-state level of HCV RNA but revealed differences in the steady-state levels of HCV proteins. Collectively, these results suggested that erlin-1 protein plays a role in the initiation of RNA replication but it seems to be dispensable for the maintenance of RNA replication once it has been established. The formation of the first viral replication complexes and the initiation of RNA replication are thought to take place in the presence of very low levels of viral proteins. Therefore, it is conceivable that the initiation step is more likely to be dependent on host factors than later steps when the accumulation of viral factors required for RNA replication is higher and the virus has induced a more favorable environment for replication in the cell by inducing the membranous web [47]. Our results are consistent with this notion and add erlin-1 protein to the list of factors previously reported to show a differential requirement for the initiation and maintenance of the RNA replication processes [17,38,48]. Alternatively, these results might suggest the existence of a different threshold for the requirement of erlin-1 protein in the regulation of the initiation and maintenance of RNA replication i.e., higher threshold for initiation than for maintenance.

Analysis of virus accumulation in acutely (Figure 5 and Supplementary Figure S1) and persistently (Figure 7 and Supplementary Figure S2) infected *ERLIN1*-deficient cells showed a strong reduction in the production of both intracellular and extracellular infectious virus. Biophysical analysis of infectious virus particles produced in *ERLIN1*-deficient cells did not reveal any difference in terms of their specific infectivity (FFU per RNA copies) or density profiles compared to the virus produced in control cells (data not shown) suggesting that *ERLIN1*-deficient cells produce qualitatively normal but quantitatively less infectious virus than *ERLIN1*-sufficient cells.

The disproportionate reduction in virus production compared to the decrease in HCV RNA levels suggested that erlin-1 protein regulates not only early steps leading to RNA and protein accumulation, but also later steps affecting virus production. In fact, direct targeting of RNA replication by an HCV polymerase inhibitor led to a proportional decrease of HCV RNA, protein and infectious virus, suggesting that the strong inhibition of virus production observed in *ERLIN1* down-regulated cells could not be solely explained by the modest reduction in the HCV RNA levels (Figure 6). As mentioned above, the magnitude of HCV NS3, NS5A and core protein reduction was disproportionate to the magnitude of HCV RNA reduction suggesting that *ERLIN1* down-regulation affects the steady-state levels of HCV proteins. The HCV-IRES translation results indicated that the translation of HCV RNA itself is not inhibited in *ERLIN1*-deficient cells (Figure 3) suggesting that the effect is likely to be at the level of protein stability or degradation and not at the translational level. This hypothesis is very appealing since erlin proteins have been implicated in the modulation of the ERAD pathway [23–26]. In contract to the reduced HCV protein accumulation observed in erlin-1 protein deficient HCV-infected cells (Figure 5), *ERLIN1* and *ERLIN2* deficiency produces an accumulation of cellular substrates that are typically degraded through the ERAD pathway [23–26]. These opposite effects on protein homeostasis (HCV vs. cellular proteins) may suggest that cellular proteome changes specifically induced by *ERLIN1* down-regulation could indirectly be responsible, at least in part, of the effects observed in the HCV infection. Comparison of cellular proteomes of *ERLIN1*- and *ERLIN2*-deficient cells could identify cellular candidate proteins contributing to the effects observed in *ERLIN1*-deficient HCV-infected cells.

Recruitment of viral NS5A and core proteins to LDs is a prerequisite for virus assembly [10]. Indeed, interfering with viral protein trafficking to LDs reduces virus production [13,49]. However, viral protein localization into LDs is not sufficient for virus particle assembly. For instance, a single mutation in the core protein (Y136A) of J6/JFH-1 virus has been associated with an increased accumulation of core in LD, decreased colocalization with E2 protein and reduced HCV particle assembly [50].

Confocal analysis of infected cells confirmed the reduced expression of HCV NS5A and core proteins and revealed a typical localization of these proteins surrounding LDs late in the infection, suggesting that *ERLIN1*-deficiency is affecting their abundance (Figures 5 and 7) but not their recruitment to the HCV assembly factories (Figure 9). Collectively these results suggest that the reduction in HCV protein levels could be responsible for the strong inhibition of infectious virus production observed in erlin-1 down-regulated cells. Importantly, given the differential effect in HCV RNA levels and virus production, these results suggest to us the existence of different HCV protein thresholds required to maintain RNA replication and assembly processes.

Another interesting aspect of these studies is that the HCV-related effects are specific to *ERLIN1* deficiency but not to *ERLIN2* deficiency despite their high similarity and functional cellular redundancy. This difference could be interpreted as a function of protein depletion efficiency achieved by the different siRNAs. *ERLIN1*-targeting siRNAs (1.3, 1.5 and 1&2) were in general more efficient (stronger reduction and longer-lasting effect) reducing erlin-1 protein expression than *ERLIN2*-targeting siRNAs (2.3 and 1&2) in silencing erlin-2 protein. The reason for this could be in part the longer half-life of erlin-2 compared to erlin-1 protein and the relative abundance of those proteins in Huh-7 cells, with erlin-2 being ~2–3 times more abundant than erlin-1 protein (data not shown). Alternatively, our results might indeed reflect a specific function of erlin-1 protein that is not shared by erlin-2 protein and that would be the first function exclusively assigned to erlin-1 protein. Importantly, our results identify erlin-1 protein as a new positive modulator of HCV infection and together with the results described by Inoue et al. in the SV40 infection system [28] point to the need of investigating the role of erlin proteins in other viral infections, especially in those in which their life cycle is tightly associated with the ER e.g., dengue virus or zika virus infection.

Supplementary Materials: The following are available online at http://www.mdpi.com/2073-4409/8/12/1555/s1, Figure S1: Erlin-1 protein down-regulation interferes with HCV in single-cycle infection experiments, Figure S2: Erlin-1 protein down-regulation impairs infectious virus production in an ongoing HCV infection cell culture system.

Author Contributions: Conceptualization, L.G. and U.G.; methodology, C.W.-B., J.C., A.G.-M., P.G.-Z., M.D.H., and U.G.; investigation, C.W.-B., J.C., A.G.-M., P.G.-Z. and U.G.; formal analysis, C.W.-B., J.C. and U.G.; writing—original draft preparation, U.G.; writing—review and editing, A.G.-M., M.D.H., L.G. and U.G.; supervision, U.G.; funding acquisition, U.G.

Funding: This research was funded by the Spanish Ministry of Economy, Industry and Competitiveness, grant numbers RYC-2014-15805 and SAF2016-75169-R (AEI/FEDER, UE) to U.G. and A.G.-M. is supported by an FPU fellowship (FPU17/03424) from the Spanish Ministry of Education, Culture and Sports.

Acknowledgments: We are thankful to Frank Chisari (The Scripps Research Institute, La Jolla, CA, USA) for inspiration, guidance and support and in whose laboratory these experiments were begun, to Takaji Wakita (National Institute of Infectious Diseases, Tokyo, Japan) for providing JFH-1 cDNA, to François-Loïc Cosset (Ecole Normale Superieure, Lyon, France) for providing the HCVpp system, to Ralf Bartenschlager (University of Heidelberg, Heidelberg, Germany) for providing JFH-1 replicons and to Mansun Law (The Scripps Research Institute, La Jolla, USA) and Charles M. Rice (Rockefeller University, New York, USA) for providing antibodies against E2 and NS5A, respectively. We would like to acknowledge Sylvia Gutierrez Erlandsson and the Advanced Light Microscopy Facility personnel for exceptional technical assistance. We are also thankful to Frank Chisari and Pablo Gastaminza (Centro Nacional de Biotecnología, Consejo Superior de Investigaciones Científicas, Madrid, Spain) for critically reading the manuscript.

Conflicts of Interest: The authors declare no conflict of interest. The funders had no role in the design of the study; in the collection, analyses, or interpretation of data; in the writing of the manuscript, or in the decision to publish the results.

References

1. Maniloff, J. Identification and classification of viruses that have not been propagated. *Arch. Virol.* **1995**, *140*, 1515–1520. [CrossRef] [PubMed]
2. Alter, H.J.; Seeff, L.B. Recovery, persistence, and sequelae in hepatitis C virus infection: A perspective on long-term outcome. *Semin. Liver Dis.* **2000**, *20*, 17–35. [CrossRef] [PubMed]

3. Choo, Q.L.; Richman, K.H.; Han, J.H.; Berger, K.; Lee, C.; Dong, C.; Gallegos, C.; Coit, D.; Medina-Selby, R.; Barr, P.J. Genetic organization and diversity of the hepatitis C virus. *Proc. Natl. Acad. Sci. USA* **1991**, *88*, 2451–2455. [CrossRef] [PubMed]
4. Friebe, P.; Boudet, J.; Simorre, J.-P.; Bartenschlager, R. Kissing-loop interaction in the 3′ end of the hepatitis C virus genome essential for RNA replication. *J. Virol.* **2005**, *79*, 380–392. [CrossRef]
5. Friebe, P.; Lohmann, V.; Krieger, N.; Bartenschlager, R. Sequences in the 5′ nontranslated region of hepatitis C virus required for RNA replication. *J. Virol.* **2001**, *75*, 12047–12057. [CrossRef]
6. Honda, M.; Beard, M.R.; Ping, L.H.; Lemon, S.M. A phylogenetically conserved stem-loop structure at the 5′ border of the internal ribosome entry site of hepatitis C virus is required for cap-independent viral translation. *J. Virol.* **1999**, *73*, 1165–1174.
7. Penin, F.; Dubuisson, J.; Rey, F.A.; Moradpour, D.; Pawlotsky, J.-M. Structural biology of hepatitis C virus. *Hepatology* **2004**, *39*, 5–19. [CrossRef]
8. Alazard-Dany, N.; Denolly, S.; Boson, B.; Cosset, F.-L. Overview of HCV Life Cycle with a Special Focus on Current and Possible Future Antiviral Targets. *Viruses* **2019**, *11*, 30. [CrossRef]
9. Sharma, N.R.; Mateu, G.; Dreux, M.; Grakoui, A.; Cosset, F.-L.; Melikyan, G.B. Hepatitis C virus is primed by CD81 protein for low pH-dependent fusion. *J. Biol. Chem.* **2011**, *286*, 30361–30376. [CrossRef]
10. Miyanari, Y.; Atsuzawa, K.; Usuda, N.; Watashi, K.; Hishiki, T.; Zayas, M.; Bartenschlager, R.; Wakita, T.; Hijikata, M.; Shimotohno, K. The lipid droplet is an important organelle for hepatitis C virus production. *Nat. Cell Biol.* **2007**, *9*, 1089–1097. [CrossRef]
11. Zayas, M.; Long, G.; Madan, V.; Bartenschlager, R. Coordination of Hepatitis C Virus Assembly by Distinct Regulatory Regions in Nonstructural Protein 5A. *PLoS Pathog.* **2016**, *12*, e1005376. [CrossRef] [PubMed]
12. Popescu, C.-I.; Callens, N.; Trinel, D.; Roingeard, P.; Moradpour, D.; Descamps, V.; Duverlie, G.; Penin, F.; Héliot, L.; Rouille, Y.; et al. NS2 protein of hepatitis C virus interacts with structural and non-structural proteins towards virus assembly. *PLoS Pathog.* **2011**, *7*, e1001278. [CrossRef] [PubMed]
13. Boson, B.; Granio, O.; Bartenschlager, R.; Cosset, F.-L. A concerted action of hepatitis C virus p7 and nonstructural protein 2 regulates core localization at the endoplasmic reticulum and virus assembly. *PLoS Pathog.* **2011**, *7*, e1002144. [CrossRef] [PubMed]
14. Gentzsch, J.; Brohm, C.; Steinmann, E.; Friesland, M.; Menzel, N.; Vieyres, G.; Perin, P.M.; Frentzen, A.; Kaderali, L.; Pietschmann, T. Hepatitis C Virus p7 is Critical for Capsid Assembly and Envelopment. *PLoS Pathog.* **2013**, *9*, e1003355. [CrossRef]
15. Gastaminza, P.; Cheng, G.; Wieland, S.; Zhong, J.; Liao, W.; Chisari, F.V. Cellular determinants of hepatitis C virus assembly, maturation, degradation, and secretion. *J. Virol.* **2008**, *82*, 2120–2129. [CrossRef]
16. McLauchlan, J.; Lemberg, M.K.; Hope, G.; Martoglio, B. Intramembrane proteolysis promotes trafficking of hepatitis C virus core protein to lipid droplets. *EMBO J.* **2002**, *21*, 3980–3988. [CrossRef]
17. Friesland, M.; Mingorance, L.; Chung, J.; Chisari, F.V.; Gastaminza, P. Sigma-1 receptor regulates early steps of viral RNA replication at the onset of hepatitis C virus infection. *J. Virol.* **2013**, *87*, 6377–6390. [CrossRef]
18. Browman, D.T.; Hoegg, M.B.; Robbins, S.M. The SPFH domain-containing proteins: More than lipid raft markers. *Trends Cell Biol.* **2007**, *17*, 394–402. [CrossRef]
19. Langhorst, M.F.; Reuter, A.; Stuermer, C.A.O. Scaffolding microdomains and beyond: The function of reggie/flotillin proteins. *Cell. Mol. Life Sci.* **2005**, *62*, 2228–2240. [CrossRef]
20. Browman, D.T.; Resek, M.E.; Zajchowski, L.D.; Robbins, S.M. Erlin-1 and erlin-2 are novel members of the prohibitin family of proteins that define lipid-raft-like domains of the ER. *J. Cell Sci.* **2006**, *119*, 3149–3160. [CrossRef]
21. Ikegawa, S.; Isomura, M.; Koshizuka, Y.; Nakamura, Y. Cloning and characterization of a novel gene (C8orf2), a human representative of a novel gene family with homology to C. elegans C42.C1.9. *Cytogenet. Cell Genet.* **1999**, *85*, 227–231. [CrossRef] [PubMed]
22. Hoegg, M.B.; Browman, D.T.; Resek, M.E.; Robbins, S.M. Distinct regions within the erlins are required for oligomerization and association with high molecular weight complexes. *J. Biol. Chem.* **2009**, *284*, 7766–7776. [CrossRef] [PubMed]
23. Pearce, M.M.P.; Wormer, D.B.; Wilkens, S.; Wojcikiewicz, R.J.H. An endoplasmic reticulum (ER) membrane complex composed of SPFH1 and SPFH2 mediates the ER-associated degradation of inositol 1,4,5-trisphosphate receptors. *J. Biol. Chem.* **2009**, *284*, 10433–10445. [CrossRef] [PubMed]

24. Pearce, M.M.P.; Wang, Y.; Kelley, G.G.; Wojcikiewicz, R.J.H. SPFH2 mediates the endoplasmic reticulum-associated degradation of inositol 1,4,5-trisphosphate receptors and other substrates in mammalian cells. *J. Biol. Chem.* **2007**, *282*, 20104–20115. [CrossRef] [PubMed]
25. Jo, Y.; Sguigna, P.V.; DeBose-Boyd, R.A. Membrane-associated ubiquitin ligase complex containing gp78 mediates sterol-accelerated degradation of 3-hydroxy-3-methylglutaryl-coenzyme A reductase. *J. Biol. Chem.* **2011**, *286*, 15022–15031. [CrossRef]
26. Teranishi, Y.; Hur, J.-Y.; Gu, G.J.; Kihara, T.; Ishikawa, T.; Nishimura, T.; Winblad, B.; Behbahani, H.; Kamali-Moghaddam, M.; Frykman, S.; et al. Erlin-2 is associated with active γ-secretase in brain and affects amyloid β-peptide production. *Biochem. Biophys. Res. Commun.* **2012**, *424*, 476–481. [CrossRef]
27. Huber, M.D.; Vesely, P.W.; Datta, K.; Gerace, L. Erlins restrict SREBP activation in the ER and regulate cellular cholesterol homeostasis. *J. Cell Biol.* **2013**, *203*, 427–436. [CrossRef]
28. Inoue, T.; Tsai, B. Regulated Erlin-dependent release of the B12 transmembrane J-protein promotes ER membrane penetration of a non-enveloped virus. *PLoS Pathog.* **2017**, *13*, e1006439. [CrossRef]
29. Zhong, J.; Gastaminza, P.; Cheng, G.; Kapadia, S.; Kato, T.; Burton, D.R.; Wieland, S.F.; Uprichard, S.L.; Wakita, T.; Chisari, F.V. Robust hepatitis C virus infection in vitro. *Proc. Natl. Acad. Sci. USA* **2005**, *102*, 9294–9299. [CrossRef]
30. Graham, F.L.; Smiley, J.; Russell, W.C.; Nairn, R. Characteristics of a human cell line transformed by DNA from human adenovirus type 5. *J. Gen Virol.* **1977**, *36*, 59–74. [CrossRef]
31. Kato, T.; Date, T.; Miyamoto, M.; Furusaka, A.; Tokushige, K.; Mizokami, M.; Wakita, T. Efficient replication of the genotype 2a hepatitis C virus subgenomic replicon. *Gastroenterology* **2003**, *125*, 1808–1817. [CrossRef] [PubMed]
32. Dreux, M.; Gastaminza, P.; Wieland, S.F.; Chisari, F.V. The autophagy machinery is required to initiate hepatitis C virus replication. *Proc. Natl. Acad. Sci. USA* **2009**, *106*, 14046–14051. [CrossRef] [PubMed]
33. Gastaminza, P.; Whitten-Bauer, C.; Chisari, F.V. Unbiased probing of the entire hepatitis C virus life cycle identifies clinical compounds that target multiple aspects of the infection. *Proc. Natl. Acad. Sci. USA* **2010**, *107*, 291–296. [CrossRef] [PubMed]
34. Zhong, J.; Gastaminza, P.; Chung, J.; Stamataki, Z.; Isogawa, M.; Cheng, G.; Mckeating, J.A.; Chisari, F.V. Persistent hepatitis C virus infection in vitro: Coevolution of virus and host. *J. Virol.* **2006**, *80*, 11082–11093. [CrossRef]
35. Gastaminza, P.; Kapadia, S.B.; Chisari, F.V. Differential biophysical properties of infectious intracellular and secreted hepatitis C virus particles. *J. Virol.* **2006**, *80*, 11074–11081. [CrossRef]
36. Garaigorta, U.; Chisari, F.V. Hepatitis C virus blocks interferon effector function by inducing protein kinase R phosphorylation. *Cell Host Microbe* **2009**, *6*, 513–522. [CrossRef]
37. Bartosch, B.; Dubuisson, J.; Cosset, F.-L. Infectious Hepatitis C Virus Pseudo-particles Containing Functional E1–E2 Envelope Protein Complexes. *J. Exp. Med.* **2003**, *197*, 633–642. [CrossRef]
38. Garaigorta, U.; Heim, M.H.; Boyd, B.; Wieland, S.; Chisari, F.V. Hepatitis C virus (HCV) induces formation of stress granules whose proteins regulate HCV RNA replication and virus assembly and egress. *J. Virol.* **2012**, *86*, 11043–11056. [CrossRef]
39. Schindelin, J.; Arganda-Carreras, I.; Frise, E.; Kaynig, V.; Longair, M.; Pietzsch, T.; Preibisch, S.; Rueden, C.; Saalfeld, S.; Schmid, B.; et al. Fiji: An open-source platform for biological-image analysis. *Nat. Methods* **2012**, *9*, 676–682. [CrossRef]
40. Wakita, T.; Pietschmann, T.; Kato, T.; Date, T.; Miyamoto, M.; Zhao, Z.; Murthy, K.; Habermann, A.; Kräusslich, H.G.; Mizokami, M.; et al. Production of infectious hepatitis C virus in tissue culture from a cloned viral genome. *Nat. Med.* **2005**, *11*, 791–796. [CrossRef]
41. Lindenbach, B.D.; Evans, M.J.; Syder, A.J.; Wolk, B.; Tellinghuisen, T.L.; Liu, C.C.; Maruyama, T.; Hynes, R.O.; Burton, D.R.; Mckeating, J.A.; et al. Complete Replication of Hepatitis C Virus in Cell Culture. *Science* **2005**, *309*, 623–626. [CrossRef] [PubMed]
42. Alvisi, G.; Madan, V.; Bartenschlager, R. Hepatitis c virus and host cell lipids: An intimate connection. *RNA Biol.* **2014**, *8*, 258–269. [CrossRef] [PubMed]
43. Zhou, L.-Y.; Zhang, L.-L. Host restriction factors for hepatitis C virus. *World J. Gastroenterol.* **2016**, *22*, 1477–1486. [CrossRef] [PubMed]

44. Takahashi, K.; Asabe, S.; Wieland, S.; Garaigorta, U.; Gastaminza, P.; Isogawa, M.; Chisari, F.V. Plasmacytoid dendritic cells sense hepatitis C virus-infected cells, produce interferon, and inhibit infection. *Proc. Natl. Acad. Sci. USA* **2010**, *107*, 7431–7436. [CrossRef]
45. Dreux, M.; Garaigorta, U.; Boyd, B.; Décembre, E.; Chung, J.; Whitten-Bauer, C.; Wieland, S.; Chisari, F.V. Short-Range Exosomal Transfer of Viral RNA from Infected Cells to Plasmacytoid Dendritic Cells Triggers Innate Immunity. *Cell Host Microbe* **2012**, *12*, 558–570. [CrossRef]
46. Padmanabhan, P.; Garaigorta, U.; Dixit, N.M. Emergent properties of the interferon-signalling network may underlie the success of hepatitis C treatment. *Nat. Commun.* **2014**, *5*, 3872. [CrossRef]
47. Egger, D.; Wolk, B.; Gosert, R.; Bianchi, L.; Blum, H.E.; Moradpour, D.; Bienz, K. Expression of Hepatitis C Virus Proteins Induces Distinct Membrane Alterations Including a Candidate Viral Replication Complex. *J. Virol.* **2002**, *76*, 5974–5984. [CrossRef]
48. Mingorance, L.; Castro, V.; Ávila-Pérez, G.; Calvo, G.; Rodriguez, M.J.; Carrascosa, J.L.; Pérez-del-Pulgar, S.; Forns, X.; Gastaminza, P. Host phosphatidic acid phosphatase lipin1 is rate limiting for functional hepatitis C virus replicase complex formation. *PLoS Pathog.* **2018**, *14*, e1007284. [CrossRef]
49. Herker, E.; Harris, C.; Hernandez, C.; Carpentier, A.; Kaehlcke, K.; Rosenberg, A.R.; Farese, R.V.; Ott, M. Efficient hepatitis C virus particle formation requires diacylglycerol acyltransferase-1. *Nat. Med.* **2010**, *16*, 1295–1298. [CrossRef]
50. Neveu, G.; Barouch-Bentov, R.; Ziv-Av, A.; Gerber, D.; Jacob, Y.; Einav, S. Identification and targeting of an interaction between a tyrosine motif within hepatitis C virus core protein and AP2M1 essential for viral assembly. *PLoS Pathog.* **2012**, *8*, e1002845. [CrossRef]

Review

Host Calcium Channels and Pumps in Viral Infections

Xingjuan Chen [1,2], Ruiyuan Cao [2,*] and Wu Zhong [2,*]

1 Institute of Medical Research, Northwestern Polytechnical University, Xi'an 710072, China; xjchen@nwpu.edu.cn

2 National Engineering Research Center for the Emergency Drug, Beijing Institute of Pharmacology and Toxicology, Beijing 100850, China

* Correspondence: 21cc@163.com (R.C.); zhongwu@bmi.ac.cn (W.Z.)

Received: 11 December 2019; Accepted: 24 December 2019; Published: 30 December 2019

Abstract: Ca^{2+} is essential for virus entry, viral gene replication, virion maturation, and release. The alteration of host cells Ca^{2+} homeostasis is one of the strategies that viruses use to modulate host cells signal transduction mechanisms in their favor. Host calcium-permeable channels and pumps (including voltage-gated calcium channels, store-operated channels, receptor-operated channels, transient receptor potential ion channels, and Ca^{2+}-ATPase) mediate Ca^{2+} across the plasma membrane or subcellular organelles, modulating intracellular free Ca^{2+}. Therefore, these Ca^{2+} channels or pumps present important aspects of viral pathogenesis and virus–host interaction. It has been reported that viruses hijack host calcium channels or pumps, disturbing the cellular homeostatic balance of Ca^{2+}. Such a disturbance benefits virus lifecycles while inducing host cells' morbidity. Evidence has emerged that pharmacologically targeting the calcium channel or calcium release from the endoplasmic reticulum (ER) can obstruct virus lifecycles. Impeding virus-induced abnormal intracellular Ca^{2+} homeostasis is becoming a useful strategy in the development of potent antiviral drugs. In this present review, the recent identified cellular calcium channels and pumps as targets for virus attack are emphasized.

Keywords: virus; calcium channels; calcium pumps; virus–host interaction; antiviral

1. Introduction

Viruses exploit the environment of host cells to replicate, thereby inducing host cells' dysfunction. Virus–host interaction is the foundation of pathogenesis and closely associated with disease severity and incidence. The prevention and therapy of virus infections are often confounded by the high mutation rates that facilitate the viral evasion of antiviral strategies that target virally encoded proteins. Modulations of the intracellular environment have become an important strategy in antiviral drug discovery and development. In mammalian cells, Ca^{2+}, as an important second messenger, mediates the sensor input and responses output for almost all known cellular progress, such as stress responses, synaptic plasticity, immunodefenses, protein transport, and endosome formation [1,2]. It has been demonstrated that the host cell dysfunction following infection with a virus is accompanied by abnormal intracellular Ca^{2+} concentration [3]. A virus can hijack the host intracellular Ca^{2+} system to achieve successful replication via multiple routes; for instance, viral proteins directly bind to Ca^{2+} or disturb the membrane permeability for Ca^{2+} by manipulating Ca^{2+} apparatus.

The host cell plasma membrane is the first barrier against the invasion of viruses. Various Ca^{2+} channels and pumps are distributed on the cell plasma membrane. Therefore, these membrane proteins become the direct target of virus infection. Interaction between viruses and these membrane proteins is the foremost approach of viruses perturbing the host cell calcium signal system. This interaction may inhibit or stimulate calcium influx and modulate free cytosolic Ca^{2+} concentrations. After entry into the host cell, viruses stimulate or inhibit the calcium release from internal stores via an effect on calcium-permeable channels, transporters, and exchangers on organellar membranes. Then, the change in cytosolic calcium concentration may trigger further distortion of the host cell system, which benefits virus survival and replication.

This review concentrates on host cell membranes' calcium channels and pumps in viral infection. Blockers for these membrane proteins or preventing viruses from grabbing these host calcium-signaling components may lower the probability of virus stability, replication, and release, as well as infection-related host–cell apoptosis and reactive oxygen species production, neurotoxicity, and enterotoxin, making these membrane proteins potential targets for antiviral drugs.

2. Calcium Channels and Pumps in Host Ca^{2+} Homeostasis

Cellular Ca^{2+} is from two major sources: the internal Ca^{2+} store (mainly endoplasmic reticulum (ER) or sarcoplasmic reticulum (SR)) and the extracellular medium. Calcium channels on cell plasma membrane mediate the entry of Ca^{2+} from the extracellular medium. These channels are activated by specific stimuli, such as voltage-gated calcium channels (VGCCs), which are stimulated by membrane depolarization, specific receptor-operated channels (ROC), which are stimulated by external agonists, or intracellular messengers and store-operated calcium channel (SOC), which are stimulated by the depletion of internal Ca^{2+} stores. The IP_3 receptor (IP_3R) and the ryanodine receptors (RyR) are the main players in mediating the release of Ca^{2+} from the internal stores. Inositol-1,4,5-triphosphate (IP_3) activates IP_3R, triggers Ca^{2+} release from stores, and further increases IP_3R's sensitivity to Ca^{2+}. Calcium pumps (the plasma membrane Ca^{2+}-ATPase (PMCA), sarcoplasmic/endoplasmic reticulum Ca^{2+}-ATPase (SERCA)) and the Na^+/Ca^{2+} exchanger (NCX) are responsible for transporting Ca^{2+} from the cytosol to external medium or into cellular calcium stores (Figure 1). The normal function of these calcium channels and pump is important for cells to maintain intracellular Ca^{2+} homeostasis.

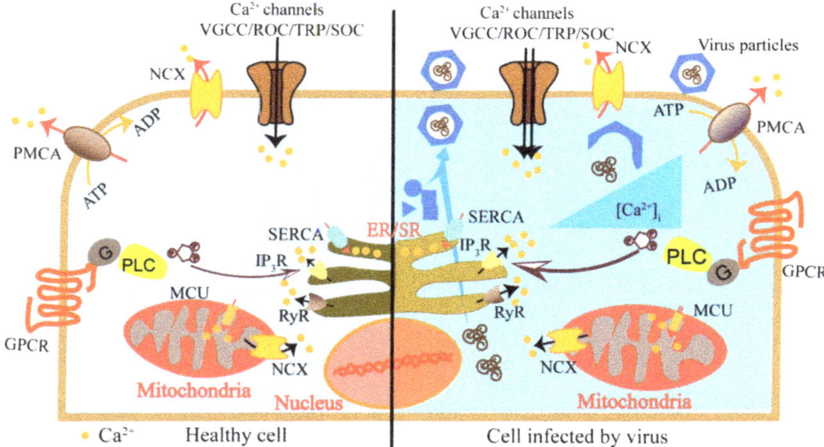

Figure 1. Schematics of host cell elevated cytosolic calcium concentration induced by a virus. Calcium channels (voltage-gated calcium channels (VGCCs), receptor-operated channels (ROC), store-operated Ca^{2+} (SOC), channels and transient receptor potential (TRP) channels) mediate the entry of Ca^{2+} from extracellular medium (black arrows). The IP_3 receptor (IP_3R) and the ryanodine receptors (RyR) on the endoplasmic reticulum (ER) mediate the release of Ca^{2+} from internal stores (black arrows). Calcium pumps (the plasma membrane Ca^{2+}-ATPase (PMCA), sarcoplasmic/endoplasmic reticulum Ca^{2+}-ATPase (SERCA)) and the Na^+/Ca^{2+} exchanger (NCX) are responsible for transporting Ca^{2+} from the cytosol to external medium or into cellular calcium stores (red arrows). Viruses utilize these calcium components to elevate cytosolic calcium concentration to activate Ca^{2+}-dependent/sensitive enzymes and transcriptional factors to promote virus replication (right panel).

These channels and pumps are activated in a flexible and precise manner to generate specific Ca^{2+} signaling, satisfying various spatiotemporal requirements. During the viral infections, host cells modulate these calcium-signaling components in response to the infection. On the other hand, viruses utilize these components to create a cellular environment that benefits their own lifecycles [4]. Viruses induce elevated cytosolic calcium concentration to activate Ca^{2+} dependent/sensitive enzymes and transcriptional factors to promote virus replication. Mitochondria could take the Ca^{2+} to generate more energy to support continuous virus replication. Moreover, regulating the calcium concentration in ER or Golgi may inhibit host proteins trafficking and promote virus protein maturation. The inhibition of host proteins frustrates host antiviral immune responses, while the promotion of virus protein maturation benefits virus propagation (Figure 1). The journey begins when a virus encounters the host cell and attaches to the cell surface. The virus particle penetrates the cytoplasm via direct membrane fusion or receptor-mediated endocytosis followed by exposing its genome to cellular machinery for replication. When the viral proteins and viral genomes are accumulated, they are assembled to form a progeny virion particle and then released [5]. During the viral lifecycles, they utilize various calcium channels and pumps to obviate the cell membrane barriers, enter the host cell, complete replication, acquire infection ability, and release.

3. Viruses Control Host Voltage-Gated Calcium Channels (VGCCs) and Two-Pore Channels (TPCs)

VGCCs are widely found in the membrane of excitable cells [6] and one of the most well studied viral targets because of the availability of specific channel blockers. There are several different types of VGCCs: L-type (Long-conducting) channels are mostly found in skeletal and smooth myocytes, bone (osteoblasts), and ventricular myocytes; N-type (Non-L or Neuronal), P/Q-type (Purkinje and Granular), and R-type (dihydropyridine- Resistant) channels display longer-lasting conductance and are expressed throughout the nervous system; T-type (Transient) channels produce short-term conductance and are found in neurons and cells that have pacemaker activity [6]. The activation of the channels results in a Ca^{2+} cascade, which is associated with numerous host physiological functions including excitation–contraction–relaxation coupling of muscles, synaptic transmission, immunoprotection, etc.

Since 1984, it has been known that verapamil, the blocker of VGCCs, inhibits influenza A virus (IAV) infection [7]. Moreover, IAV infection induces Ca^{2+} influx, and the elevated intracellular Ca^{2+} promotes endocytic uptake of IAV [8]. Until recently, the underlying mechanism was revealed by Fujioka et al. [9]. They reported that several VGCC blockers and siRNA against the $Ca_V1.2$ channel (L-type channel) inhibited H1N1 and H3N2 IAV infection in multiple cell lines. IAV attaches to the target cells by the viral hemagglutinin protein (HA), which is a type I transmembrane protein embedded in the viral envelope, binding to sialic acids [10]. Fujioka et al. showed that the virus HA binds to domain IV of $Ca_V1.2$, which contains two potential sialylated asparagine residues (N1436 and N1487) [9,11]. When $Ca_V1.2$ mutants in one or both these residues were replaced with glutamine (N1436Q, N1487Q, and N1436Q + N1487Q), the interaction between HA and the Cav1.2 was attenuated compared with the wild-type fragment. They demonstrated that a VGCC blocker, diltiazem, significantly prolonged the survival of IAV-infected mice and allowed the recovery of the survivors. Therefore, $Ca_V1.2$ may serve as a host cell surface receptor that binds IAV and is critical for IAV entry (Figure 2A). New world hemorrhagic fever arenaviruses (NWA) are another virus that is reportedly sensitive to the VGCC blocker club member [12,13]. An siRNA screen with Junín virus glycoprotein-pseudotyped viruses identified that VGCCs are involved in NWA entry. Treatment with the channel blocker gabapentin, an FDA (U.S. Food and Drug Administration)-approved analgesic that targets the channel $\alpha2\delta2$ subunit, protects against NWA infection [13]. This research work demonstrated that the interaction between a virus and VGCCs promotes virus entry at the virus-cell fusion step.

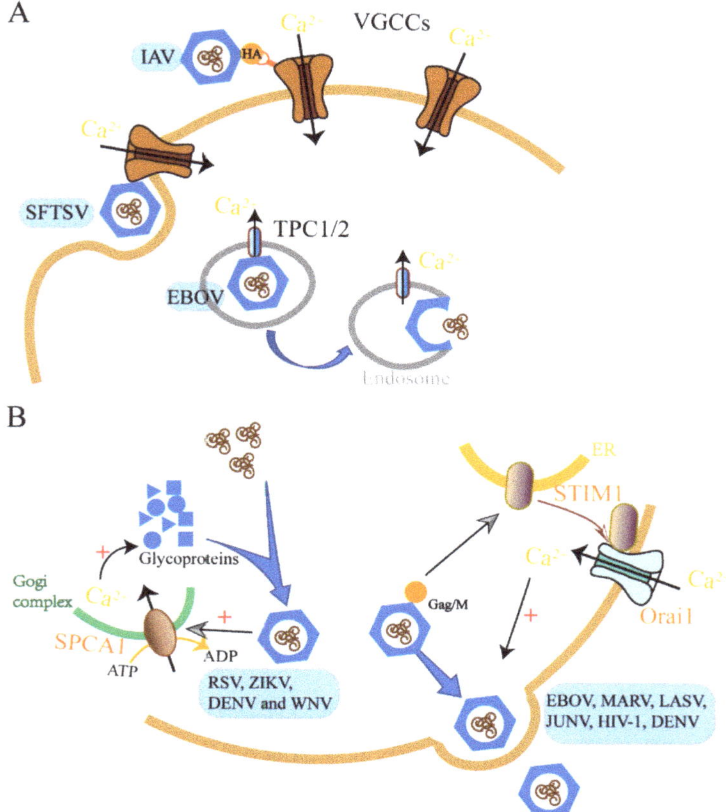

Figure 2. Examples of viruses interplaying with host calcium channels or pumps to achieve viral entry (**A**) and release (**B**). VGCCs are important for influenza A virus (IAV) and severe fever with thrombocytopenia syndrome virus (SFTSV) entry into the host cell as well as TPC1/2 for EBOV (**A**). RSV, Zika virus (ZIKA), dengue virus (DENV) and West Nile virus (WNV) hijack SPCA1 to facilitate their release as well as Ebola virus (EBOV), MARV, LASV, JUNV, HIV-1 and DENV manipulates STIM1/ORAI1 (**B**). For a complete list of definitions, see Table 1.

The effect on VGCCs is not only restricted to virus entry. A high-throughput screening of an FDA-approved drug library for inhibitors of Japanese encephalitis virus (JEV) identified five hit drugs, three of which are VGCCs blockers (manidipine, cilnidipine, and benidipine hydrochloride) [14]. Recombinant viral particles (RVPs) with the luciferase-reporting replicon were used to quantify the efficiency of JEV replication, which confirmed that these drugs inhibited JEV infection at the stage of replication. These drugs were subsequently validated for their antiviral activities against other flavivirus, such as Zika virus (ZIKV), dengue virus (DENV), and West Nile virus (WNV). Similarly, another research group screened the FDA-approved drug library and found that nifedipine and benidipine hydrochloride inhibited severe fever with thrombocytopenia syndrome virus (SFTSV) replication in vitro [12]. Moreover, the retrospective clinical investigation on SFTS patients showed that nifedipine can significantly inhibit SFTSV infection. These studies indicate that the VGCCs blockers are excellent candidates for broad-spectrum anti-virus treatment. Most of these FDA-approved VGCC blockers are clinically used to treat cardiovascular diseases. Similar to a double-edged sword, the cardiovascular effect of VGCC blockers may limit their antiviral application. Therefore, repurposing these drugs requires more analysis before clinical trials.

Further examples of virus-regulating VGCCs to service their replication can be found in HIV-1 and herpes simplex virus (HSV)-1. Two HIV-1 proteins, the membrane glycoprotein gp120 [15,16] and the transcriptical transactivator Tat [17,18], have been identified to induce the elevation of host intracellular Ca^{2+} in various cell types, including neuronal, immune, and epithelial cells, via targeting the activity of VGCCs. The detailed information about the dysregulation of the L-type calcium channel by Tat can be found in the review [16,17].

Taking together, the infections of viruses could increase the host's intracellular calcium to facilitate viral entry and replication by manipulating host VGCCs. In addition, herpes simplex virus (HSV)-1 could downregulate the VGCCs on infected neuronal cells to escape detection by host cells. T-type Ca^{2+} channels were reported as the targets of HSV-1 in sensory-like ND7-23 cells [19,20]. HSV-1 infection of differentiated ND7-23 cells causes a significant loss of T-type Ca^{2+} channels from the membrane, which depends on viral replication and protein synthesis. This downregulation of T-type Ca^{2+} channel expression may alter the ability of sensory neurons to transmit pain information. Thus, the lower expression of VGCCs may diminish the detection for viral infection by the host, which benefits virus survival and further infection.

Ebola virus (EBOV) used to be considered a VGCC blocker-sensitive virus, and several research groups independently reported that compounds blocking L-type channels (such as verapamil, nimodipine, and diltiazem) inhibited EBOV infection in vivo [21–23]. However, gabapentin, representing a fifth distinct class of L-type channel inhibitor, had no effect even at high concentrations. It has been shown that verapamil, nimodipine, and diltiazem also inhibit the activity of two-pore channels (TPCs) [24]. TPCs are intracellular voltage-gated and receptor-operated calcium permeable channels, playing an integral role in membrane trafficking pathways [25,26]. Mouse embryonic fibroblasts (MEFs) lacking TPC1 or TPC2 expression ($Tpcn1^{-/-}$, $Tpcn2^{-/-}$) resisted EBOV infection [27]. It turns out that the target of EBOV may not be classical L-type calcium channels but rather endosomal calcium channels termed TPCs (Figure 2A). The calcium channels inhibitors prevented virus–endosome membrane fusion and virus capsid releasing into the cell cytoplasm, which is a late entry step. EBOV acts on TPCs, which control the movement of endosomes containing virus particles, and thereby facilitate its intracellular trafficking [27,28]. The channel inhibitor, tetrandrine, significantly enhanced the survival of mice challenged with mouse-adapted EBOV without any detectable side effects. This indicates that tetrandrine is highly effective against EBOV disease in mice.

4. Store-Operated Calcium (SOC) Channel in Viral Assembly and Egress

SOC channels are the major Ca^{2+} entry pathways in non-excitable cells. The protein Orai1 on the plasma membrane and STIM1 (stromal interaction molecule) on ER are the molecular identities that are responsible for SOC channels activation. The depletion of ER Ca^{2+} stores promotes STIM1 proteins aggregation and interaction with Orai1 to open the channel, mediating Ca^{2+} entry [29,30].

Most enveloped viruses are released extracellularly via exocytosis, also called budding, as an analogy of buds in plants [31,32]. The budding process of the enveloped viruses is triggered by a peptide motif (termed late (L) domains), which was discovered in the Gag polyproteins of retroviruses and M (matrix) proteins of rhabdoviruses [33]. These L domains interact with cellular proteins to promote the formation of virus vesicles that bud away from the cytoplasm [32,34]. Research works established that this essential final step of related viruses depends on the host Ca^{2+} signal mediated by SOC channels (STIM1/Orai1).

The research work done on four distinct hemorrhagic fever viruses (Ebolavirus (EBOV), Marburgvirus (MARV), Lassa Virus (LASV), and Junín Virus (JUNV)) demonstrated that EBOV, MARV VP40, and JUNV Z proteins trigger host cell Ca^{2+} signals by activating the ER Ca^{2+} sensing protein STIM1 and the plasma membrane ORAI1 channel. The STIM1/Orai1-mediated Ca^{2+} signal is critical for EBOV and related viruses budding from host cells [35]. The suppression of STIM1 expression and genetic inactivation or the pharmacological blockade of ORAI1 inhibits infections of EBOV, MARV, and JUNV in cultured cells (Figure 2B). Obviously, the matrix proteins or live virus activates STIM1 and the ORAI channel. Similarly to hemorrhagic fever viruses, the HIV-1 matrix protein Gag, directing HIV-1 budding, and mediating VLP (Virus-like Particle) formation also exhibit dependence on Ca^{2+} regulation [36,37]. Therefore, further study is needed to validate the role for STIM1 and the ORAI channel in HIV- 1 budding.

Infections of other enveloped RNA viruses that buds in similar mechanisms may also be inhibited by STIM1 and ORAI1 inhibitors. Indeed, SOC channel antagonists significantly reduced DENV yield [38]. When the human hepatic HepG2 and Huh-7 cells are infected by DENV, STIM1 and ORAI1 were shown to be co-localized in infected cells, indicating activation of the SOC channel [38,39]. Therefore, DENV infection alters cell Ca^{2+} homeostasis possibly via promoting the interaction between STIM1 and ORAI1.

It is possible that the viral proteins trigger the Ca^{2+} depletion in ER or prevent ER refilled with Ca^{2+} to maintain resting ER Ca^{2+} levels [40–42]. Alternately, these viral proteins may directly modify STIM1 and uncouple the activation of STIM1 from Ca^{2+} levels [35,43]. Thus, STIM1/ORAI might represent a conserved target to regulating the budding of enveloped RNA viruses and possibly DNA viruses that rely on similar host cellular proteins for budding. The example is that the hepatitis B virus X (HBx) protein upregulates the activity of the STIM1–ORAI1 channel complex [44]. The mechanisms by which these viruses activate STIM1 and ORAI1 represent novel therapeutic targets for controlling budding.

Calcium influx through the SOC channel also contributes to the elevated cytosolic calcium concentration induced by rotavirus (RV) infection. It has been well established that the dramatically increased cellular Ca^{2+} is the hallmark of rotavirus (RV) infection [45,46]. Rotavirus nonstructural protein 4 (NSP4) is an ER-localized viroporin that functionally depletes ER calcium. Thus, in rotavirus-infected cells, STIM1 is constitutively active and colocalizes with the ORAI1 channel. The knockdown of STIM1 or the pharmacological inhibition of SOC channels significantly reduced rotavirus yield, indicating that the SOC channel plays a critical role in the RV lifecycle [47,48].

5. Host Transient Receptor Potential (TRP) Channels and Receptor-Operated Calcium (ROC) Channels

5.1. TRP Channels

The TRP channel is a non-selective cation channel predominately permeable for Ca^{2+} [49,50]. It is divided into six sub-families according to their amino acid sequence: TRP canonical or classical (TRPC), TRP vanilloid (TRPV), TRP melastatin (TRPM), TRP ankyrin (TRPA), TRP polycystin (TRPP), and TRP mucolipin (TRPML) [50]. They are ubiquitously expressed in different tissues and cell types and are a key player in the regulation of intracellular calcium. Nevertheless, reports about virus control TRP channels are not numerous.

TRPV4 mediates intracellular Ca^{2+} signals in response to several stimuli, including hypotonic cell swelling, mechanical forces, moderating heat, and UVB radiation [51]. When cells are exposed to the Zika virus or the purified viral envelope protein, TRPV4 mediates Ca^{2+} influx and drives the nuclear translocation of DEAD-box RNA helicase (DDX3X) [52]. DDX3X is an ATP-dependent RNA helicase from the DEAD-box helicase family and is involved in multiple stages of the RNA metabolism, from transcription to translation [53,54]. Moreover, diverse RNA viruses hijack DDX3X to multiply efficiently. Targeting TRPV4 reduces the infectivity of dengue, hepatitis C, and Zika viruses [52]. Overall, the research work demonstrated the role of TRPV4 in the regulation of DDX3X-dependent control of the RNA metabolism and viral infectivity [52].

The TRPV1, TRPA1, and TRPM8 channels are directly activated by chemical, thermal, and mechanical stimuli. They are potentially associated with respiratory virus-induced airway hypersensitivity [55]. The infection of respiratory viruses, such as respiratory syncytial virus (RSV), measles virus (MV), and human rhinovirus (HRV), was reported to increase the expression of these channels in sensory neurons and human bronchial epithelium cells [56,57]. The increase in TRPA1 and

TRPV1 levels can be mediated by soluble factors induced by infection, whereas TRPM8 requires virus replication [56,57]. These reports may explain the possible mechanism by which respiratory viruses induce cough. Alternatively, the up-regulated TRP channels may be utilized by the respiratory viruses to create a favorable calcium environment. Further investigation is required to determine the possible pathway by which this happens.

5.2. N-Methyl-D-Aspartate (NMDA) Receptors

NMDArs and IP_3Rs are the two receptor-operated calcium (ROC) channels reported to be involved in virus–host interactions. NMDAr, which is activated by the endogenously synthesized excitant amino acid, glutamate, is widely expressed throughout the mammalian central nervous system and is particularly enriched in the cerebral cortex and hippocampus.

The Zika virus infection is a "neurodegenerative" disease. Costa and colleagues showed that blockage of the NMDAr channel activity with FDA-approved memantine or other antagonists prevents neuronal cell death and microgliosis induced by Zika in vitro and in vivo, without affecting the ability of Zika to replicate in the host [58,59]. It seems that NMDAr mainly contributes to Zika, triggering the neuronal cell death progress. Anyway, the blockade of NMDAr may be a viable treatment for patients at risk for Zika infection-induced neurodegeneration [59]. NMDAr blockade also significantly abrogated neuronal cell death and inflammatory response triggered by JEV infection [60]. Therefore, NMDAr are probably common attack targets of flavivirus, inducing host neuron cell death.

5.3. IP_3 Receptors

The engagement of receptors or agonist binding to the cell surface receptors activate phospholipase C (PLC), which hydrolyses phosphatidylinositol-4,5-biphosphate (PIP_2) to produce IP_3. The activation of IP_3R triggers Ca^{2+} release from stores and further increases IP_3R's sensitivity to Ca^{2+}. Targeting IP_3Rs is more universal among different virus types. In turn, this process generates a more profound effect on host cell physiology, including metabolic stress, neurotoxicity, and enterotoxicity. Modifying IP_3 and directly interfering with IP_3R are the main ways in which viruses affect calcium release through IP_3R. For example, human cytomegalovirus (HCMV) UL37x1 protein interacts with the host P2Y2 purinergic receptor to increase intracellular Ca^{2+} levels via the PLC–IP_3 pathway, and this activity is important to viral replication [61]. HIV-1 Gp120 and Tat upregulate intracellular IP_3 [37,62], while HIV Nef [63,64] and p12(I) human T-cell lymphotropic virus type 1 (HTLV-1) [65] activate IP_3R directly as an agonist. Glycoprotein E of HSV redistributes the density of IP_3Rs within infected cells [66]. Some virus proteins activate IP_3R through both ways. Enterotoxin NSP4 of RV and HCMV UL37x1-encoded protein increases the basal permeability of the ER as IP_3 diffuses and binds to IP_3Rs to stimulate Ca^{2+} release. These virus proteins deplete ER Ca^{2+} storage during early stages of viral infection to increase the replication ability of viruses. On the other hand, the depletion of ER calcium storage triggers the calcium influx through SOCs (STIM1/ORAI1).

6. Calcium Pumps

The calcium pumps (Ca^{2+} ATPase) and the Na^+–Ca^{2+} exchanger are the main regulators of intracellular Ca^{2+} concentrations [67]. Three Ca^{2+} ATPase types (pumps) have been described in animal cells located in the membranes of the sarcoplasmic reticulum (SERCA pump), in those of the Golgi network (Secretory Pathway Ca^{2+} ATPase, SPCA pump), and in the plasma membrane (PMCA pump) [68]. At the end of a stimulus, these pumps participate in returning the cell to its resting state by the decrease in the cytosolic Ca^{2+} concentration. For example, SERCA transports Ca^{2+} from the cytoplasm into the lumen of the ER. PMCA pumps function to remove Ca^{2+} from the cell. They serve as the main regulators of intracellular and extracellular Ca^{2+} concentrations. All three calcium pumps are reported to be involved in different viral lifecycles.

Maturation is the last assembly step of a virus particle. In this step, the virus acquires the infectivity, which is essential for its lifecycle. The results obtained from a genome-wide knockout screen indicated that SPCA1 in the Golgi network is critical for human respiratory syncytial virus (RSV) infection [69–71]. Further study demonstrated that similar to RSV, other viruses from the Paramyxoviridae, Flaviviridae, and Togaviridae families also failed to spread in SPCA1-deficient cells. Studies on this mechanism revealed that Ca^{2+} pumped into Golgi by SPCA1 is the trigger to produce normal functional viral glycoproteins that are essential for virus spread. Therefore, SPCA1 may be a promising target for therapeutic intervention against a diverse set of viruses (Figure 2B). However, in this study, herpes simplex 1 (HSV-1), vesicular stomatitis (VSV), bunyamwera (BUNV), and LCMV were unaffected by the loss of SPCA1. That is probably because the maturation of these viral glycoproteins requires different triggers.

Cui et al. [72] screened the natural product library and found that cyclopiazonic acid, an inhibitor of SERCA, was shown to have activity in the low micromolar range against RSV strains at the step of virus genome replication. Further study found that another SERCA inhibitor BHQ had a similar effect [72]. It is probable that SERCA inhibitors prevent cytosolic calcium returning to the ER from the cytosol via SERCA, resulting in an increase in intracellular calcium concentration. A continuous higher concentration of intracellular calcium may impair viral genome replication and/or transcription, thereby reducing virus yield.

7. Conclusions

The journey of viral infections is also a procedure during which a virus grabs the host intracellular calcium signaling system and homeostasis to benefit its own lifecycle. The disturbance of the host Ca^{2+} system by virus may suppress T-cell responsiveness, antiapoptotic, and other protentional functions. In this review, we summarized the calcium permeable channels and calcium pumps located on host cell membranes that may be potential therapeutic targets for viral infections (Table 1). However, the calcium channel family is continuously being refreshed with the discovery of new members, and the effect of viruses on the host intracellular calcium environment is not often achieved through a single pathway; it is more complicated. Both upstream messengers and downstream Ca^{2+} binding proteins are targets for pharmacological intervention. Thus, explicating the entire picture of all aspects of the calcium–virus interplay at the molecular level will be a challenge in the future.

The prevention of and therapeutic efforts against virus infections are often challenged by the high mutation rates of viral proteins. However, those host calcium apparatus proteins required by viral replication and transmission are highly conserved, essentially immutable, and thus are potentially an Achilles' heel of virus infection. As described above, almost all the calcium permeable channels or calcium pumps are possible attack targets of viruses and customized host intracellular calcium signaling via effects on these channel proteins. Targeting those host cellular proteins required by virus infection may be one treatment strategy with therapeutic potential, which may avoid or delay the development of resistance. The identified calcium-permeable proteins discussed above offer suitable target proteins for inhibiting virus infection. VGCCs and NMDA receptors are two of most well studied targets because of the availability of specific channel antagonists. Moreover, some of those compounds are FDA-approved drugs with clinical application. For example, verapamil, amlodipine, verapamil, and diltiazem, which are widely used to treat cardiovascular disease, have shown inhibition for various virus infection. Memantine, which is an FDA-approved Alzheimer's drug, reduces neurological complications associated with Zika virus infection. Obviously, this direction depends on the identification of selective and potent small molecule modulators of calcium channels. Currently, aside from some blockers for VGCCs and the NMDA receptor, almost no other channel modulators can be used in clinics, although important progress has been made in recent years toward high quality and widely available channel modulators. Furthermore, modulators should be generated with properties that make them suitable for in vivo use. This review outlines promising progress made with the potentiation for a host cell calcium channels modulator to become an antiviral drug. We believe that modulators for host calcium channels are the pots of gold for antiviral drug development.

Table 1. Calcium channels/pumps utilized by a virus.

Cellular Targets	Virus	Consequences [Ref.]
VGCCs	IAV	$Ca_V1.2$ serves as a host cell surface receptor that binds IAV and is critical for IAV entry [9].
	SFTSV	Benidipine hydrochloride, VGCC blocker, inhibits SFTSV infection via impairing virus internalization and genome replication [12].
	NWV	Virus binds to VGCCs and promotes virus entry at the virus–cell fusion step [13].
	Flavivirus (JEV, ZIKV, DENV, and WNV	VGCCs blockers inhibit flavivirus (JEV, ZIKV, DENV and WNV) infection at the stage of replication [14].
	HIV-1	Tat/gp120 overactivate VGCCs [15–17].
	HSV-1	HSV-1 downregulates the $Ca_V3.2$ channel and diminishes the detection of viral infection by host [19,20].
TPCs	EBOV	Facilitates virus–endosome membrane fusion and releases of virus capsid into the cell cytoplasm [27].
STIM1/ORAI1	EBOV, MARV, LASV, JUNV, HIV-1, DENV, and HBV	Promote virion assembly and budding [35–38,44].
TRPV4	ZIKV	Activation of TRPV4- releases DDX3X and promote the viral RNA metabolism [52].
NMDAr	ZIKV, JEV	NMDAr contributes to ZIKA by triggering the neuronal cell death progress [58,60].
	HIV-1	Increases Ca^{2+} influx [62].
IP_3R	HIV-1, HSV, HRV, and HCMV	These viral proteins deplete ER Ca^{2+} store during early stages of viral infection to increase the replication ability of viruses [37,62–64,66].
SPCA1	RSV, ZIKV, DENV and WNV	Trigger to produce functional viral glycoproteins that are essential for virus spread [70]

Abbreviations: Influenza A virus (IAV), Severe fever with thrombocytopenia syndrome virus (SFTSV), New world hemorrhagic fever arenaviruses (NWV), Japanese encephalitis virus (JEV), Zika virus (ZIKV), Dengue virus (DENV), West Nile virus (WNV), Herpes simplex virus (HSV)-1, Ebolavirus (EBOV), Marburgvirus (MARV), Lassa Virus (LASV), Junín Virus (JUNV), Hepatitis B virus (HBV), Respiratory syncytial virus (RSV), Human rhinovirus (HRV), and Human cytomegalovirus (HCMV). Voltage-gated calcium channels (VGCCs), Two-pore channels (TPCs), Stromal interaction molecule 1 (STIM1), Transient receptor potential vanilloid 4 (TRPV4), N-methyl-D-aspartate receptor (NMDAR), IP_3 receptor (IP_3R), Secretory pathway Ca^{2+}-ATPase (SPCA1). DEAD-box RNA helicase (DDX3X), Endoplasmic reticulum (ER).

Author Contributions: X.C. and W.Z. conceived this article; X.C. and R.C. wrote the original draft and prepared figures; R.C. and W.Z. edited and reviewed the manuscript. All authors have read and agreed to the published version of the manuscript.

Funding: This work was funded by the National Natural Science Foundation of China, grant numbers 81773631 and 81900402), National Science and Technology Major Projects, grant number 2018ZX09711003 and Fundamental Research Funds for the Central Universities grant number G2019KY0501.

Conflicts of Interest: The authors declare no conflict of interest.

References

1. Martinez de Victoria, E. Calcium, essential for health. *Nutr. Hosp.* **2016**, *33* (Suppl. S4), 341.
2. Berridge, M.J.; Bootman, M.D.; Lipp, P. Calcium—A life and death signal. *Nature* **1998**, *395*, 645–648. [CrossRef] [PubMed]
3. Olivier, M. Modulation of host cell intracellular Ca^{2+}. *Parasitol. Today* **1996**, *12*, 145–150. [CrossRef]
4. Clark, K.B.; Eisenstein, E.M. Targeting host store-operated Ca^{2+} release to attenuate viral infections. *Curr. Top. Med. Chem.* **2013**, *13*, 1916–1932. [CrossRef] [PubMed]
5. Gonzales-van Horn, S.R.; Sarnow, P. Making the Mark: The Role of Adenosine Modifications in the Life Cycle of RNA Viruses. *Cell Host Microbe* **2017**, *21*, 661–669. [CrossRef] [PubMed]
6. Atlas, D. Voltage-gated calcium channels function as Ca^{2+}-activated signaling receptors. *Trends Biochem. Sci.* **2014**, *39*, 45–52. [CrossRef] [PubMed]
7. Nugent, K.M.; Shanley, J.D. Verapamil inhibits influenza A virus replication. *Arch. Virol.* **1984**, *81*, 163–170. [CrossRef]
8. Fujioka, Y.; Tsuda, M.; Nanbo, A.; Hattori, T.; Sasaki, J.; Sasaki, T.; Miyazaki, T.; Ohba, Y. A Ca^{2+}-dependent signalling circuit regulates influenza a virus internalization and infection. *Nat. Commun.* **2013**, *4*, 2763. [CrossRef]
9. Fujioka, Y.; Nishide, S.; Ose, T.; Suzuki, T.; Kato, I.; Fukuhara, H.; Fujioka, M.; Horiuchi, K.; Satoh, A.O.; Nepal, P. A Sialylated Voltage-Dependent Ca^{2+} Channel Binds Hemagglutinin and Mediates Influenza A Virus Entry into Mammalian Cells. *Cell Host Microbe* **2018**, *23*, 809–818. [CrossRef]
10. Chandrasekaran, A.; Srinivasan, A.; Raman, R.; Viswanathan, K.; Raguram, S.; Tumpey, T.M.; Sasisekharan, V.; Sasisekharan, R. Glycan topology determines human adaptation of avian H5N1 virus hemagglutinin. *Nat. Biotechnol.* **2008**, *26*, 107–113. [CrossRef]
11. Lazniewska, J.; Weiss, N. Glycosylation of voltage-gated calcium channels in health and disease. *Biochim. Biophys. Acta Biomembr.* **2017**, *1859*, 662–668. [CrossRef] [PubMed]
12. Li, H.; Zhang, L.K.; Li, S.F.; Zhang, S.F.; Wan, W.W.; Zhang, Y.L.; Xin, Q.L.; Dai, K.; Hu, Y.Y.; Wang, Z.B. Calcium channel blockers reduce severe fever with thrombocytopenia syndrome virus (SFTSV) related fatality. *Cell Res.* **2019**, *29*, 739–753. [CrossRef] [PubMed]
13. Lavanya, M.; Cuevas, C.D.; Thomas, M.; Cherry, S.; Ross, S.R. siRNA screen for genes that affect Junin virus entry uncovers voltage-gated calcium channels as a therapeutic target. *Sci. Transl. Med.* **2013**, *5*, 204ra131. [CrossRef] [PubMed]
14. Wang, S.; Liu, Y.; Guo, J.; Wang, P.; Zhang, L.; Xiao, G.; Wang, W. Screening of FDA-Approved Drugs for Inhibitors of Japanese Encephalitis Virus Infection. *J. Virol.* **2017**, *91*, e01055-17. [CrossRef]
15. Dreyer, E.B.; Kaiser, P.K.; Offermann, J.T.; Lipton, S.A. HIV-1 coat protein neurotoxicity prevented by calcium channel antagonists. *Science* **1990**, *248*, 364–367. [CrossRef]
16. Haughey, N.J.; Mattson, M.P. Calcium dysregulation and neuronal apoptosis by the HIV-1 proteins Tat and gp120. *J. Acquir. Immune. Defic. Syndr.* **2002**, *31* (Suppl. S2), S55–S61. [CrossRef]
17. Hu, X.T. HIV-1 Tat-Mediated Calcium Dysregulation and Neuronal Dysfunction in Vulnerable Brain Regions. *Curr. Drug Targets* **2016**, *17*, 4–14. [CrossRef]
18. Kruman, I.I.; Nath, A.; Mattson, M.P. HIV-1 protein Tat induces apoptosis of hippocampal neurons by a mechanism involving caspase activation, calcium overload, and oxidative stress. *Exp. Neurol.* **1998**, *154*, 276–288. [CrossRef]
19. Zhang, Q.; Hsia, S.C.; Martin-Caraballo, M. Regulation of T-type Ca^{2+} channel expression by herpes simplex virus-1 infection in sensory-like ND7 cells. *J. Neurovirol.* **2017**, *23*, 657–670. [CrossRef]

20. Zhang, Q.; Hsia, S.C.; Martin-Caraballo, M. Regulation of T-type Ca^{2+} channel expression by interleukin-6 in sensory-like ND7/23 cells post-herpes simplex virus (HSV-1) infection. *J. Neurochem.* **2019**, *151*, 238–254. [CrossRef]
21. Kolokoltsov, A.A.; Saeed, M.F.; Freiberg, A.N.; Holbrook, M.R.; Davey, R.A. Identification of novel cellular targets for therapeutic intervention against Ebola virus infection by siRNA screening. *Drug Dev. Res.* **2009**, *70*, 255–265. [CrossRef] [PubMed]
22. Gehring, G.; Rohrmann, K.; Atenchong, N.; Mittler, E.; Becker, S.; Dahlmann, F.; Pohlmann, S.; Vondran, F.W.; David, S.; Manns, M.P. The clinically approved drugs amiodarone, dronedarone and verapamil inhibit filovirus cell entry. *J. Antimicrob. Chemother.* **2014**, *69*, 2123–2131. [CrossRef] [PubMed]
23. Madrid, P.B.; Chopra, S.; Manger, I.D.; Gilfillan, L.; Keepers, T.R.; Shurtleff, A.C.; Green, C.E.; Iyer, L.V.; Dilks, H.H.; Davey, R.A. A systematic screen of FDA-approved drugs for inhibitors of biological threat agents. *PLoS ONE* **2013**, *8*, e60579. [CrossRef] [PubMed]
24. Calcraft, P.J.; Ruas, M.; Pan, Z.; Cheng, X.; Arredouani, A.; Hao, X.; Tang, J.; Rietdorf, K.; Teboul, L.; Chuang, K.T. NAADP mobilizes calcium from acidic organelles through two-pore channels. *Nature* **2009**, *459*, 596–600. [CrossRef] [PubMed]
25. Zhu, M.X.; Ma, J.; Parrington, J.; Calcraft, P.J.; Galione, A.; Evans, A.M. Calcium signaling via two-pore channels: Local or global, that is the question. *Am. J. Physiol. Cell Physiol.* **2010**, *298*, C430–C441. [CrossRef] [PubMed]
26. Morgan, A.J. Ca^{2+} dialogue between acidic vesicles and ER. *Biochem. Soc. Trans.* **2016**, *44*, 546–553. [CrossRef] [PubMed]
27. Sakurai, Y.; Kolokoltsov, A.A.; Chen, C.C.; Tidwell, M.W.; Bauta, W.E.; Klugbauer, N.; Grimm, C.; Wahl-Schott, C.; Biel, M.; Davey, R.A. Two-pore channels control Ebola virus host cell entry and are drug targets for disease treatment. *Science* **2015**, *347*, 995–998. [CrossRef]
28. Kintzer, A.F.; Stroud, R.M. Structure, inhibition and regulation of two-pore channel TPC1 from Arabidopsis thaliana. *Nature* **2016**, *531*, 258–262. [CrossRef]
29. Hogan, P.G.; Lewis, R.S.; Rao, A. Molecular basis of calcium signaling in lymphocytes: STIM and ORAI. *Annu. Rev. Immunol.* **2010**, *28*, 491–533. [CrossRef]
30. Nwokonko, R.M.; Cai, X.; Loktionova, N.A.; Wang, Y.; Zhou, Y.; Gill, D.L. The STIM-Orai Pathway: Conformational Coupling between STIM and Orai in the Activation of Store-Operated Ca^{2+} Entry. *Adv. Exp. Med. Biol.* **2017**, *993*, 83–98.
31. Sepulveda, C.S.; Garcia, C.C.; Damonte, E.B. Determining the Virus Life-Cycle Stage Blocked by an Antiviral. *Methods Mol. Biol.* **2018**, *1604*, 371–392. [PubMed]
32. Hartlieb, B.; Weissenhorn, W. Filovirus assembly and budding. *Virology* **2006**, *344*, 64–70. [CrossRef] [PubMed]
33. Chen, B.J.; Lamb, R.A. Mechanisms for enveloped virus budding: Can some viruses do without an ESCRT? *Virology* **2008**, *372*, 221–232. [CrossRef] [PubMed]
34. Licata, J.M.; Simpson-Holley, M.; Wright, N.T.; Han, Z.; Paragas, J.; Harty, R.N. Overlapping motifs (PTAP and PPEY) within the Ebola virus VP40 protein function independently as late budding domains: Involvement of host proteins TSG101 and VPS-4. *J. Virol.* **2003**, *77*, 1812–1819. [CrossRef]
35. Han, Z.; Madara, J.J.; Herbert, A.; Prugar, L.I.; Ruthel, G.; Lu, J.; Liu, Y.; Liu, W.; Liu, X.; Wrobel, J.E. Calcium Regulation of Hemorrhagic Fever Virus Budding: Mechanistic Implications for Host-Oriented Therapeutic Intervention. *PLoS Pathog.* **2015**, *11*, e1005220. [CrossRef]
36. Ehrlich, L.S.; Carter, C.A. HIV Assembly and Budding: Ca^{2+} Signaling and Non-ESCRT Proteins Set the Stage. *Mol. Biol. Int.* **2012**, *2012*, 851670. [CrossRef]
37. Ehrlich, L.S.; Medina, G.N.; Carter, C.A. ESCRT machinery potentiates HIV-1 utilization of the PI (4,5) P (2)-PLC-IP3R-Ca^{2+} signaling cascade. *J. Mol. Biol.* **2011**, *413*, 347–358. [CrossRef]
38. Dionicio, C.L.; Pena, F.; Constantino-Jonapa, L.A.; Vazquez, C.; Yocupicio-Monroy, M.; Rosales, R.; Zambrano, J.L.; Ruiz, M.C.; Del Angel, R.M.; Ludert, J.E. Dengue virus induced changes in Ca^{2+} homeostasis in human hepatic cells that favor the viral replicative cycle. *Virus Res.* **2018**, *245*, 17–28. [CrossRef]
39. Cheshenko, N.; Liu, W.; Satlin, L.M.; Herold, B.C. Multiple receptor interactions trigger release of membrane and intracellular calcium stores critical for herpes simplex virus entry. *Mol. Biol. Cell* **2007**, *18*, 3119–3130. [CrossRef]

40. Robinson, L.C.; Marchant, J.S. Enhanced Ca^{2+} leak from ER Ca^{2+} stores induced by hepatitis C NS5A protein. *Biochem. Biophys. Res. Commun.* **2008**, *368*, 593–599. [CrossRef]
41. Zhadina, M.; Bieniasz, P.D. Functional interchangeability of late domains, late domain cofactors and ubiquitin in viral budding. *PLoS Pathog.* **2010**, *6*, e1001153. [CrossRef] [PubMed]
42. Ruiz, M.C.; Aristimuno, O.C.; Diaz, Y.; Pena, F.; Chemello, M.E.; Rojas, H.; Ludert, J.E.; Michelangeli, F. Intracellular disassembly of infectious rotavirus particles by depletion of Ca^{2+} sequestered in the endoplasmic reticulum at the end of virus cycle. *Virus Res.* **2007**, *130*, 140–150. [CrossRef] [PubMed]
43. Freedman, B.D.; Harty, R.N. Calcium and filoviruses: A budding relationship. *Future Microbiol.* **2016**, *11*, 713–715. [CrossRef] [PubMed]
44. Yao, J.H.; Liu, Z.J.; Yi, J.H.; Wang, J.; Liu, Y.N. Hepatitis B Virus X Protein Upregulates Intracellular Calcium Signaling by Binding C-terminal of Orail Protein. *Curr. Med. Sci.* **2018**, *38*, 26–34. [CrossRef]
45. Michelangeli, F.; Ruiz, M.C.; del Castillo, J.R.; Ludert, J.E.; Liprandi, F. Effect of rotavirus infection on intracellular calcium homeostasis in cultured cells. *Virology* **1991**, *181*, 520–527. [CrossRef]
46. Pham, T.; Perry, J.L.; Dosey, T.L.; Delcour, A.H.; Hyser, J.M. The Rotavirus NSP4 Viroporin Domain is a Calcium-conducting Ion Channel. *Sci. Rep.* **2017**, *7*, 43487. [CrossRef]
47. Hyser, J.M.; Utama, B.; Crawford, S.E.; Broughman, J.R.; Estes, M.K. Activation of the endoplasmic reticulum calcium sensor STIM1 and store-operated calcium entry by rotavirus requires NSP4 viroporin activity. *J. Virol.* **2013**, *87*, 13579–13588. [CrossRef]
48. Chang-Graham, A.L.; Perry, J.L.; Strtak, A.C.; Ramachandran, N.K.; Criglar, J.M.; Philip, A.A.; Patton, J.T.; Estes, M.K.; Hyser, J.M. Rotavirus Calcium Dysregulation Manifests as Dynamic Calcium Signaling in the Cytoplasm and Endoplasmic Reticulum. *Sci. Rep.* **2019**, *9*, 10822. [CrossRef]
49. Li, J.; Zhang, X.; Song, X.; Liu, R.; Zhang, J.; Li, Z. The structure of TRPC ion channels. *Cell Calcium.* **2019**, *80*, 25–28. [CrossRef]
50. Hicks, G.A. TRP channels as therapeutic targets: Hot property, or time to cool down? *Neurogastroenterol. Motil.* **2006**, *18*, 590–594. [CrossRef]
51. Garcia-Elias, A.; Mrkonjic, S.; Jung, C.; Pardo-Pastor, C.; Vicente, R.; Valverde, M.A. The TRPV4 channel. *Handb. Exp. Pharm.* **2014**, *222*, 293–319.
52. Donate-Macian, P.; Jungfleisch, J.; Perez-Vilaro, G.; Rubio-Moscardo, F.; Peralvarez-Marin, A.; Diez, J.; Valverde, M.A. The TRPV4 channel links calcium influx to DDX3X activity and viral infectivity. *Nat. Commun.* **2018**, *9*, 2307. [CrossRef] [PubMed]
53. Ariumi, Y. Multiple functions of DDX3 RNA helicase in gene regulation, tumorigenesis, and viral infection. *Front. Genet.* **2014**, *5*, 423. [CrossRef] [PubMed]
54. Valiente-Echeverria, F.; Hermoso, M.A.; Soto-Rifo, R. RNA helicase DDX3: At the crossroad of viral replication and antiviral immunity. *Rev. Med. Virol.* **2015**, *25*, 286–299. [CrossRef]
55. Fujita, T.; Liu, Y.; Higashitsuji, H.; Itoh, K.; Shibasaki, K.; Fujita, J.; Nishiyama, H. Involvement of TRPV3 and TRPM8 ion channel proteins in induction of mammalian cold-inducible proteins. *Biochem. Biophys. Res. Commun.* **2018**, *495*, 935–940. [CrossRef]
56. Omar, S.; Clarke, R.; Abdullah, H.; Brady, C.; Corry, J.; Winter, H.; Touzelet, O.; Power, U.F.; Lundy, F.; McGarvey, L.P. Respiratory virus infection up-regulates TRPV1, TRPA1 and ASICS3 receptors on airway cells. *PLoS ONE* **2017**, *12*, e0171681. [CrossRef]
57. Abdullah, H.; Heaney, L.G.; Cosby, S.L.; McGarvey, L.P. Rhinovirus upregulates transient receptor potential channels in a human neuronal cell line: Implications for respiratory virus-induced cough reflex sensitivity. *Thorax* **2014**, *69*, 46–54. [CrossRef]
58. Costa, V.V.; Del Sarto, J.L.; Rocha, R.F.; Silva, F.R.; Doria, J.G.; Olmo, I.G.; Marques, R.E.; Queiroz-Junior, C.M.; Foureaux, G.; Araujo, J.M.S. N-Methyl-d-Aspartate (NMDA) Receptor Blockade Prevents Neuronal Death Induced by Zika Virus Infection. *MBio* **2017**, *8*, e00543-17. [CrossRef]
59. Sirohi, D.; Kuhn, R.J. Can an FDA-Approved Alzheimer's Drug Be Repurposed for Alleviating Neuronal Symptoms of Zika Virus? *MBio* **2017**, *8*, e00916-17. [CrossRef]
60. Chen, Z.; Wang, X.; Ashraf, U.; Zheng, B.; Ye, J.; Zhou, D.; Zhang, H.; Song, Y.; Chen, H.; Zhao, S. Activation of neuronal N-methyl-D-aspartate receptor plays a pivotal role in Japanese encephalitis virus-induced neuronal cell damage. *J. Neuroinflammation* **2018**, *15*, 238. [CrossRef]

61. Chen, S.; Shenk, T.; Nogalski, M.T. P2Y2 purinergic receptor modulates virus yield, calcium homeostasis, and cell motility in human cytomegalovirus-infected cells. *Proc. Natl. Acad. Sci. USA* **2019**, *116*, 18971–18982. [CrossRef] [PubMed]
62. Ehrlich, L.S.; Medina, G.N.; Photiadis, S.; Whittredge, P.B.; Watanabe, S.; Taraska, J.W.; Carter, C.A. Tsg101 regulates PI (4,5) P2/Ca^{2+} signaling for HIV-1 Gag assembly. *Front. Microbiol.* **2014**, *5*, 234. [CrossRef] [PubMed]
63. Ehrlich, L.S.; Medina, G.N.; Khan, M.B.; Powell, M.D.; Mikoshiba, K.; Carter, C.A. Activation of the inositol (1,4,5)-triphosphate calcium gate receptor is required for HIV-1 Gag release. *J. Virol.* **2010**, *84*, 6438–6451. [CrossRef] [PubMed]
64. Manninen, A.; Saksela, K. HIV-1 Nef interacts with inositol trisphosphate receptor to activate calcium signaling in T cells. *J. Exp. Med.* **2002**, *195*, 1023–1032. [CrossRef]
65. Ding, W.; Albrecht, B.; Kelley, R.E.; Muthusamy, N.; Kim, S.J.; Altschuld, R.A.; Lairmore, M.D. Human T-cell lymphotropic virus type 1 p12 (I) expression increases cytoplasmic calcium to enhance the activation of nuclear factor of activated T cells. *J. Virol.* **2002**, *76*, 10374–10382. [CrossRef]
66. Cheshenko, N.; Del Rosario, B.; Woda, C.; Marcellino, D.; Satlin, L.M.; Herold, B.C. Herpes simplex virus triggers activation of calcium-signaling pathways. *J. Cell Biol.* **2003**, *163*, 283–293. [CrossRef]
67. Strehler, E.E.; Zacharias, D.A. Role of alternative splicing in generating isoform diversity among plasma membrane calcium pumps. *Physiol. Rev.* **2001**, *81*, 21–50. [CrossRef]
68. Brini, M.; Cali, T.; Ottolini, D.; Carafoli, E. Calcium pumps: Why so many? *Compr. Physiol.* **2012**, *2*, 1045–1060.
69. Griffiths, C.; Drews, S.J.; Marchant, D.J. Respiratory Syncytial Virus: Infection, Detection, and New Options for Prevention and Treatment. *Clin. Microbiol. Rev.* **2017**, *30*, 277–319. [CrossRef]
70. Hoffmann, H.H.; Schneider, W.M.; Blomen, V.A.; Scull, M.A.; Hovnanian, A.; Brummelkamp, T.R.; Rice, C.M. Diverse Viruses Require the Calcium Transporter SPCA1 for Maturation and Spread. *Cell Host Microbe* **2017**, *22*, 460–470. [CrossRef]
71. Cervantes-Ortiz, S.L.; Zamorano Cuervo, N.; Grandvaux, N. Respiratory Syncytial Virus and Cellular Stress Responses: Impact on Replication and Physiopathology. *Viruses* **2016**, *8*, 124. [CrossRef] [PubMed]
72. Cui, R.; Wang, Y.; Wang, L.; Li, G.; Lan, K.; Altmeyer, R.; Zou, G. Cyclopiazonic acid, an inhibitor of calcium-dependent ATPases with antiviral activity against human respiratory syncytial virus. *Antivir. Res.* **2016**, *132*, 38–45. [CrossRef] [PubMed]

Review

On the Host Side of the Hepatitis E Virus Life Cycle

Noémie Oechslin, Darius Moradpour and Jérôme Gouttenoire *

Division of Gastroenterology and Hepatology, Lausanne University Hospital and University of Lausanne, CH-1011 Lausanne, Switzerland; Noemie.Oechslin@unil.ch (N.O.); Darius.Moradpour@chuv.ch (D.M.)
* Correspondence: Jerome.Gouttenoire@chuv.ch

Received: 28 April 2020; Accepted: 21 May 2020; Published: 22 May 2020

Abstract: Hepatitis E virus (HEV) infection is one of the most common causes of acute hepatitis in the world. HEV is an enterically transmitted positive-strand RNA virus found as a non-enveloped particle in bile as well as stool and as a quasi-enveloped particle in blood. Current understanding of the molecular mechanisms and host factors involved in productive HEV infection is incomplete, but recently developed model systems have facilitated rapid progress in this area. Here, we provide an overview of the HEV life cycle with a focus on the host factors required for viral entry, RNA replication, assembly and release. Further developments of HEV model systems and novel technologies should yield a broader picture in the future.

Keywords: HEV; host factor; particle production; viral replication; virus entry

---

1. Introduction

Hepatitis E virus (HEV) has been identified as a cause of the waterborne hepatitis outbreaks in the early 1980s [1,2]. The viral genome was cloned and sequenced in 1990, allowing the development of serological tests to study its epidemiology [3,4]. The virus has been classified in the *Hepeviridae* family, and most human pathogenic strains belong to species Orthohepevirus A [5]. Members of this species can be classified into 8 genotypes (gt): gt 1 and 2 are restricted to humans and are transmitted via the fecal-oral route, mainly through contaminated drinking water. Gt 3 and 4 cause zoonotic infections and the transmission occurs mainly via the consumption of un(der)cooked pork, wild boar or deer meat. Gt 5 and 6 are found in wild boar, and gt 7 as well as 8 infect dromedary and Bactrian camels, respectively. Gt 7 has been identified in an immunosuppressed patient after consumption of camel milk and meat [6] (reviewed in [7]) but no transmission to humans has thus far been reported for gt 5, 6 and 8. More recently, rabbit HEV (closely related to gt 3 within species Orthohepevirus A) and rat HEV (belonging to species Orthohepevirus C) have also been found to infect humans [8–11].

HEV is a small, non-enveloped, icosahedral virus with a diameter of 27–34 nm [2]. It contains a 7.2 kb single-stranded, positive-sense RNA genome which possesses a m7G cap at its 5′ and a poly-A tail at its 3′ end (Figure 1A). The HEV genome harbors 3 open reading frames (ORF). ORF1 encodes the viral replicase, ORF2 the capsid and ORF3 a small protein involved in virion secretion via its potential ion channel activity [12]. The first contact between HEV and host cells occurs through interaction with as yet poorly characterized entry factor(s). After endocytosis, the viral genome is released into the cytoplasm and the host translational machinery produces the ORF1 replicase, which drives viral RNA replication (Figure 1B). During this step, two RNA species are produced from a negative-strand RNA intermediate: a full-length genomic RNA and a subgenomic RNA of 2.2 kb [13,14]. Translation of the subgenomic RNA yields the ORF2 and ORF3 proteins. Later steps of the HEV life cycle include viral assembly and release of newly produced virions. Very similar to hepatitis A virus (HAV), another hepatotropic positive-strand RNA virus, HEV is found as a 'quasi-enveloped' virion (eHEV) wrapped in exosomal membranes in blood and as a naked particle in bile and feces (reviewed in [15]) (Figure 1B).

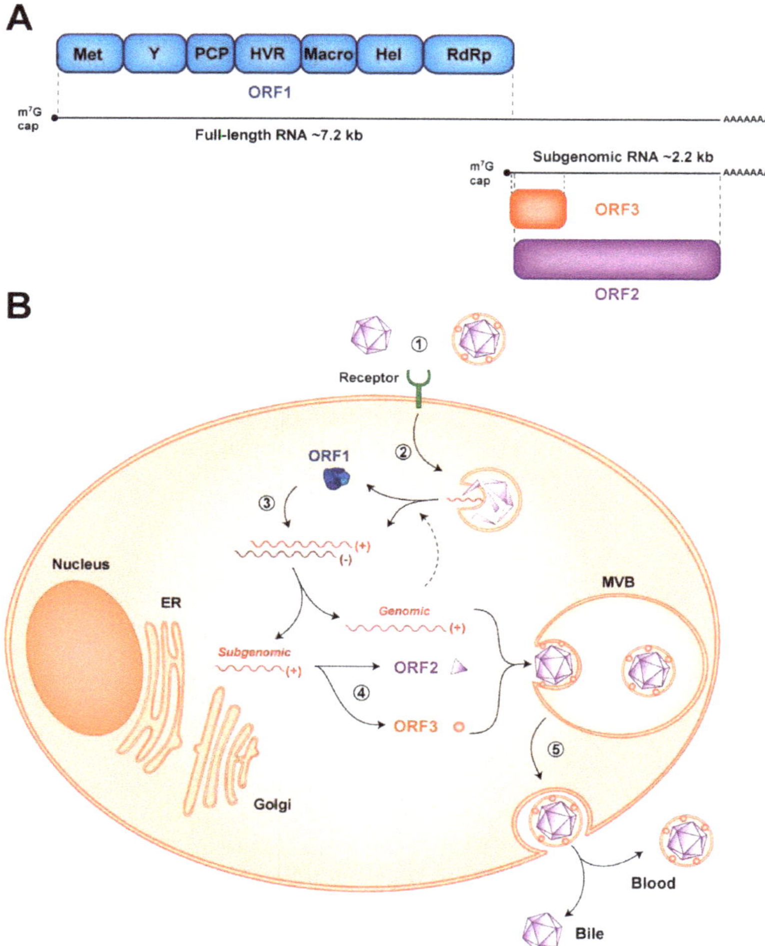

Figure 1. Genome organization and life cycle of hepatitis E virus (HEV). (**A**) The 7.2 kb positive-strand RNA genome has a 5′ 7-methylguanylate cap (m^7G cap) and a 3′ polyadenylated tail (poly-A). It harbors 3 open reading frames (ORFs). ORF1 encodes a replicase of about 190 kDa comprising different functional domains, including a methyltransferase (Met), an RNA helicase (Hel) and an RNA-dependent RNA polymerase (RdRp), as well as less well-characterized domains, such as the Y domain, a putative papain-like cysteine protease (PCP), a hypervariable region (HVR) and the Macro domain. ORF2 and ORF3 encode the viral capsid and a small protein involved in virus secretion respectively, which are translated from a 2.2 kb subgenomic RNA generated during viral replication. (**B**) The HEV life cycle can be dissected into the following steps: (1) viral entry by as yet unidentified receptor(s), (2) endocytosis and release of the viral positive-strand RNA genome (+) into the cytosol, (3) translation of the ORF1 protein to allow replication of the full-length and generation of the subgenomic RNA through a negative-strand RNA intermediate (-), (4) translation of the subgenomic RNA to produce the ORF2 and ORF3 proteins and (5) genome packaging, virion assembly and release of the virus into the bloodstream and the bile from the basolateral and apical sides, respectively. ER, endoplasmic reticulum; MVB, multivesicular body.

As obligate intracellular pathogens, viruses have developed strategies to hijack and manipulate host cell pathways in order to ensure productive infection. Moreover, RNA viruses, especially those with relatively limited genome size and coding capacity, such as HEV, are particularly dependent on the host cell machinery. Because the tools to study HEV have been limited until recently, only little is known about the host factors involved in the various steps of the viral life cycle. Studies performed in heterologous settings, such as the yeast two hybrid system, identified cellular factors interacting with HEV proteins [16,17]. However, most of these candidates remain to be validated and further studied using infectious cell culture systems, in vivo models and liver biopsies from patients with hepatitis E. In this review, we shall focus on host factors whose involvement in the viral life cycle has been validated in HEV infection settings.

2. HEV Entry

As HEV is present in a non-enveloped ("naked") and a quasi-enveloped form (eHEV), the entry pathway of the virus may differ for these two forms. Our knowledge of HEV entry remains scarce but studies using virus-like particles (VLP) as a model system have highlighted possible host factors involved in the initial attachment to the cell and virus internalization, including the 78-kDa glucose-regulated protein (GRP78), ATP synthase subunit β (ATPB5) and asialoglycoprotein receptor (ASGPR) [18–20]. Notably, non-enveloped HEV was shown to interact with heparan sulfate proteoglycans (HSPG), likely syndecans [21,22], which are expressed on the surface of many cell types. Hence, treatment of susceptible hepatoma cell lines with heparinase I considerably reduced VLP binding as well as HEV infection [21] (Figure 2). Of note, HSPG are known to mediate cell attachment of several enveloped and nonenveloped viruses, including, among others, herpes simplex virus, hepatitis C virus (HCV), norovirus and human immunodeficiency virus [23–27].

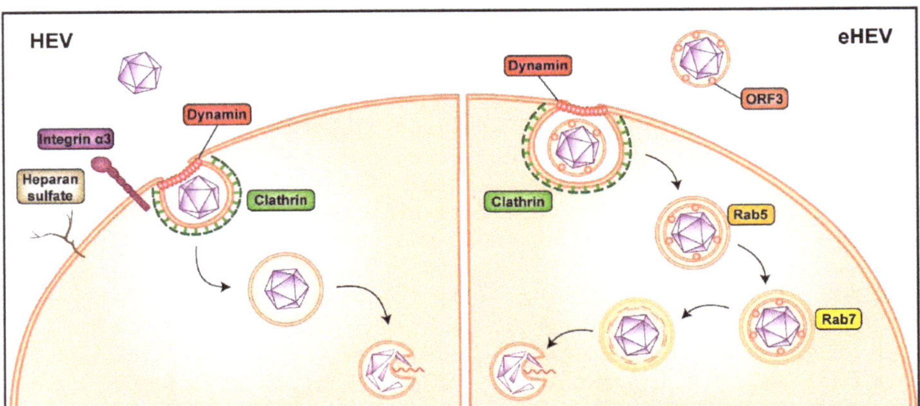

Figure 2. Non-enveloped versus quasi-enveloped hepatitis E virus entry. Non-enveloped hepatitis E virus (HEV) is believed to first bind to heparan sulfate proteoglycans and integrin α3 at the cell surface (left panel). The virus is then internalized via clathrin- and dynamin 2-dependent endocytosis, followed by release of the viral genome into the cytoplasm by an as yet unknown mechanism, possibly involving a conformational change of the capsid. The cofactor(s) and receptor(s) allowing entry of quasi-enveloped HEV (eHEV) are unknown (right panel). Internalization requires clathrin- and dynamin 2-dependent endocytosis as well as trafficking through Rab5- (early) as well as Rab7-positive (late) endosomes and eventually lysosomes to allow release of the viral genome into the cytoplasm, likely by a process similar to that of naked HEV.

A recent microarray analysis comparing gene expression in permissive versus non-permissive cells identified integrin α3 as an entry factor for HEV [28]. Integrins belong to a family of transmembrane proteins localized at the cell surface which can bind to extracellular matrix (ECM) as well as cell surface and intracellular ligands. These interactions trigger a myriad of intracellular signals modulating cell behavior through effects on actin microfilaments, thereby connecting the extracellular space and the cytoskeleton (reviewed in [29]). Integrins were shown to function as receptors for Kaposi's sarcoma-associated herpesvirus [30] and, interestingly, also for HAV [31]. In the case of HEV, overexpression of integrin α3 in non-permissive cells allowed non-enveloped HEV infection only and, conversely, knockout of integrin α3 gene in permissive cells prevented entry of non-enveloped HEV but not of eHEV [28]. Together with the physical interaction between integrin α3 and HEV, the data strongly suggests that this ECM receptor is an entry factor for the non-enveloped viral particle (Figure 2). Integrins are expressed in a broad array of tissues, including in the intestine [26,32], where HEV is known to replicate and which may represent the initial site of infection [33,34]. However, in some organs, such as the liver, integrin α3 expression is rather low [35–37], raising the possibility that integrin α3 may act as a cofactor for HEV entry rather than as the key receptor. Further studies are needed to clarify the entry pathway used by non-enveloped HEV.

Internalization of non-enveloped HEV is believed to occur through clathrin- and dynamin 2-dependent endocytosis [22,38,39] (Figure 2). In this context, membrane cholesterol was also shown to be important for HEV entry, as treatment of cells with cholesterol sequestering agents significantly reduced VLP uptake [39]. Following endocytosis, the HEV genome needs to be uncoated to be released into the cytoplasm. It is likely that the capsid protein itself plays a crucial role in this poorly understood process, possibly by undergoing conformational changes induced by interaction with a host protein, as it is seen in other non-enveloped viruses, such as human papilloma virus, murine polyomavirus and poliovirus [40–42].

Since the viral capsid is not exposed at the surface of eHEV (see Section 4), the quasi-enveloped particle may use different, as yet unknown entry factor(s). However, as for the naked virus, eHEV is internalized through clathrin- and dynamin 2-dependent endocytosis [38,39]. Unlike non-enveloped HEV, eHEV entry is dependent on the small GTPases, Rab5 and Rab7, both of which are involved in endosomal trafficking. In fact, depletion of one or the other has been shown to considerably reduce virus infection [22] (Figure 2). In addition, eHEV entry depends on endosomal acidification [22], indicating that transition of the endosome to a more acidic cell compartment is necessary (Figure 2). Lysosomal lipid degradation appears to be a required step in eHEV entry. Indeed, depletion of the late endosomal and lysosomal Niemann-Pick disease type C1 protein, involved in cholesterol extraction, significantly reduced eHEV infection [22]. Moreover, treatment with an inhibitor of lysosomal acid lipase, responsible for lipid degradation, resulted in a dose-dependent reduction of eHEV cell entry [22].

These observations suggest that the quasi-envelope of eHEV is removed in endolysosomes. The viral capsid may subsequently interact with an as yet unidentified host factor and undergo the conformational changes required for genome release into the cytoplasm. A similar mechanism was recently proposed for HAV [31]. Since both quasi-enveloped and non-enveloped HEV are internalized in vesicles belonging to the endosomal pathway, it is plausible that they use a common host factor to allow uncoating and release of the genome into the cytoplasm (detailed in [43]).

The ORF3 protein is present within eHEV and interacts with the capsid [44]. However, its role in eHEV entry, uncoating and genome release remains to be explored.

Taken together, some HEV entry factors and pathways have been identified, but key elements are still missing. The capacity of HEV to infect a broad array of cell types [45–51], possibly related to the diverse extrahepatic manifestations of hepatitis E [52–55], strongly suggests that the entry receptor(s) is (are) not specific to hepatocytes but ubiquitously expressed.

3. Viral RNA Replication

Upon release of the viral positive-strand RNA genome into the cytoplasm, the host translational machinery, including ribosomal subunits and elongation factors, seeds on the 5′ untranslated region to start translation of the ORF1 protein. Among the different components, the eukaryotic translation initiation factor 4F (eIF4F) complex has been identified as being important for HEV replication. This complex is known to be involved in the cap-dependent translation and replication of several viruses (reviewed in [56]). In this context, an RNA interference-based loss-of-function study showed that components of the eIF4F complex are required for efficient HEV replication, while known negative regulatory factors of this pathway limit viral RNA synthesis [57]. Along these lines, silvestrol, a natural compound inhibiting part of the eIF4F complex, efficiently inhibits HEV replication in vitro, and in a mouse model in vivo, ORF2 protein production and virus spread [58,59]. Interestingly, silvestrol was reported to also inhibit corona-, picorna-, alpha-, flavi- and filo-viruses and, therefore, displays broad antiviral activity [60–63].

ORF1 encodes the functional domains required for viral RNA synthesis, including an RNA-dependent RNA polymerase (RdRp) located at the C-terminal end of the polyprotein. In addition to the RdRp, the replicase encoded by ORF1 comprises an RNA helicase, a methyltransferase as well as less well-characterized domains, including the Macro and Y domains, a hypervariable region and a putative papain-like cysteine protease (PCP) (Figure 1A). Of note, the PCP has been associated with deubiquitination and the Macro domain with deribosylation activities likely involved in the posttranslational modifications of host proteins [64,65]. Positive-strand RNA viruses commonly process their polyproteins into individual functional proteins by viral, and in some instances, cellular proteases. In the case of HEV, however, it is unsettled whether, and if so, by which mechanism, the ORF1 protein is processed. While some studies reported processing of the ORF1 polyprotein, including by the PCP or the cellular proteases thrombin and factor Xa [66], others have found that the major form of ORF1 protein is unprocessed (reviewed in [53,67]). Further studies in complete cell culture systems and using more sensitive techniques for the detection of ORF1 protein are required [68].

As for all positive-strand RNA viruses, replication includes synthesis of a complementary negative-strand RNA which serves as a template for the production of positive-strand genomic and, in the case of HEV, an additional subgenomic RNA. The key enzyme responsible for these steps is the RdRp. The mechanisms regulating the production of these RNA species from the negative-strand RNA are still unknown but imply cis-acting elements within the genome [69]. Regulation of transcription may also involve host factor(s), such as heterogeneous nuclear ribonucleoproteins (hnRNP) [70,71]. hnRNPs play a role in nuclear RNA metabolism and have been reported to re-localize to the cytoplasm of HEV-infected cells [71], similarly to what had been observed for infection with other RNA viruses [72–76], suggesting a potential role of hnRNPs in genomic and subgenomic HEV RNA transcription.

Replication of positive-strand RNA viruses takes place in membrane-associated replication complexes composed of viral proteins, the replicating viral RNA, rearranged cellular membranes and other host factors [77]. The subcellular site of HEV RNA replication has not been identified to date but recent advances such as the development of tagged functional HEV genomes may facilitate progress in this area [68]. Insertion of a hemagglutinin epitope tag in the ORF1 polyprotein allowed us to visualize the HEV replicase together with viral RNA in cytoplasmic dot-like structures, likely indicating the site of active replication. ORF1 protein was found to colocalize best with exosomal markers as well as with the ORF2 and ORF3 proteins, suggesting that HEV RNA replication takes place in close proximity to virion assembly sites [68]. Virus-induced membrane rearrangements may serve several purposes, including the physical support and organization of the replication complex, the compartmentalization and local concentration of viral and host factors required for RNA replication, tethering of the viral RNA during unwinding and coordination of its translation, replication and packaging, provision of lipid constituents important for replication, and physical protection from host antiviral defenses (reviewed in [77,78]). Of note, HCV as well as picorna- and coronaviruses require the guanine nucleotide-exchange factor Golgi brefeldin A resistance factor 1 (GBF1) for the induction of membrane alterations making up their viral replication complexes [79–83]. Interestingly, similar observations have been made for HEV [84]. However, no colocalization of GBF1 with HEV ORF1 and ORF2 proteins or relocalization of GBF1 upon HEV infection have been observed. One may thus hypothesize that GBF1 is not recruited to HEV replication sites and is not involved in replication complex formation but rather plays an indirect role in HEV replication [84], as also proposed for HCV [79,80] and mouse hepatitis virus [85].

4. Virion Assembly and Infectious Particle Release

Virion assembly consists of the packaging of genomic viral RNA in the capsid. While the subcellular site of HEV assembly has not been identified yet, it is likely tightly connected to replication complexes (see above). The mechanisms driving HEV assembly are poorly understood but early observations showed that RNA and the ORF2 protein can spontaneously assemble into virus-like particles in insect cells [86] (reviewed in [87]). These findings argue in favor of a self-assembly process involving a limited number of host factors.

The ORF2 protein exists in several forms, of which a non-glycosylated form is involved in virion formation [88–91]. It is currently debated whether this form, harboring a truncated and non-functional signal peptide, results from translation at an alternative start site [90] or from proteolytic cleavage, as suggested by mass spectrometry analyses [89]. The non-glycosylated ORF2 protein may represent the major intracellular form detected in the cytosol as well as the cell nucleus [89,91,92]. While the role of nuclear ORF2 protein is unknown, its nucleocytoplasmic shuttling likely involves the host nuclear import/export machineries. In contrast to the non-glycosylated form packaging the viral genome, the glycosylated form of ORF2 protein is rapidly released extracellularly through the secretory pathway [88–91,93,94]. This implies recognition of its signal peptide by the translocon, translocation into the ER lumen, cleavage by signal peptidase, followed by sialylation as well as *N*- and *O*-glycosylation by yet to be defined glycosyltransferases in the ER and Golgi [89], and secretion. Glycosylated ORF2 protein is present in at least two forms, of which the smaller may result from cleavage by furin-like proteases at an RRR motif [91]. Overall, these studies suggest that ORF2 protein likely has more than one function in genome packaging and point toward the implication of different host factors.

HEV assembly involves the non-glycosylated ORF2 protein and the viral RNA but does not require the ORF3 protein, as genomes harboring a mutated ORF3 start codon yield infectious particles; however, these are not secreted from the cell [95,96]. Infectious virus is believed to be released as quasi-enveloped particles from both the basolateral and apical sides of hepatocytes, facing the liver sinusoids and the bile canaliculi, respectively [33,97,98]. The pseudo-envelope of virions secreted into the bile is likely delipidated by bile acids and/or pancreatic enzymes (Figure 3). Interestingly, HEV is preferentially secreted from the apical side, explaining the high viral load detected in feces. Based on these observations, different host factors may be involved in virus secretion from the apical and basolateral sides of hepatocytes. Moreover, HEV ORF3 protein is essential for the secretion of infectious particles, possibly by connecting the capsid with the host factors required for egress. Phosphorylation of ORF3 protein at a serine residue (Ser 70 and Ser 71 in gt 3 and gt 1, respectively) may trigger the interaction with assembled non-glycosylated ORF2 protein [44]. Based on sequence information, phosphorylation may involve the p34cdc2 kinase and mitogen-activated protein kinase [99]. Furthermore, ORF3 protein recruits Tsg101, a member of the endosomal sorting complex required for transport-I (ESCRT-I), via a highly conserved PSAP motif [100–102]. The presence of this motif in the *C*-terminal region of the ORF3 protein is required for virus release [101]. It is known that proline-rich motifs P(S/T)AP and PPXY (X being any amino acid), which are also called "late domains", are essential for budding of enveloped viruses such as human immunodeficiency virus (HIV)-1 or Ebola [103,104]. The ESCRT machinery is composed of 4 complexes, ESCRT-0 to ESCRT-III, which are involved in membrane remodeling, leading to budding reactions or membrane involution. These complexes can further recruit accessory proteins, such as vacuolar protein sorting 4 (Vps4) and apoptosis-linked gene-2 interacting protein X (Alix), both interacting with the ESCRT-III complex, to close newly formed vesicles (reviewed in [105]). Further confirming the requirement for the ESCRT machinery, hepatocyte growth factor-regulated tyrosine kinase substrate (Hrs), a member of ESCRT-0, and Vps4 were shown to be required for HEV particle release [102,106]. In addition, the envelope wrapping the virion has been shown to be derived from exosomes, small vesicles generated from multivesicular bodies (MVB) by the ESCRT pathway [102,106]. These MVB, which fuse with the plasma membrane to release their content into the extracellular milieu, require the Rab27a protein, which has been shown to colocalize with ORF3 protein [102,106] (Figure 3).

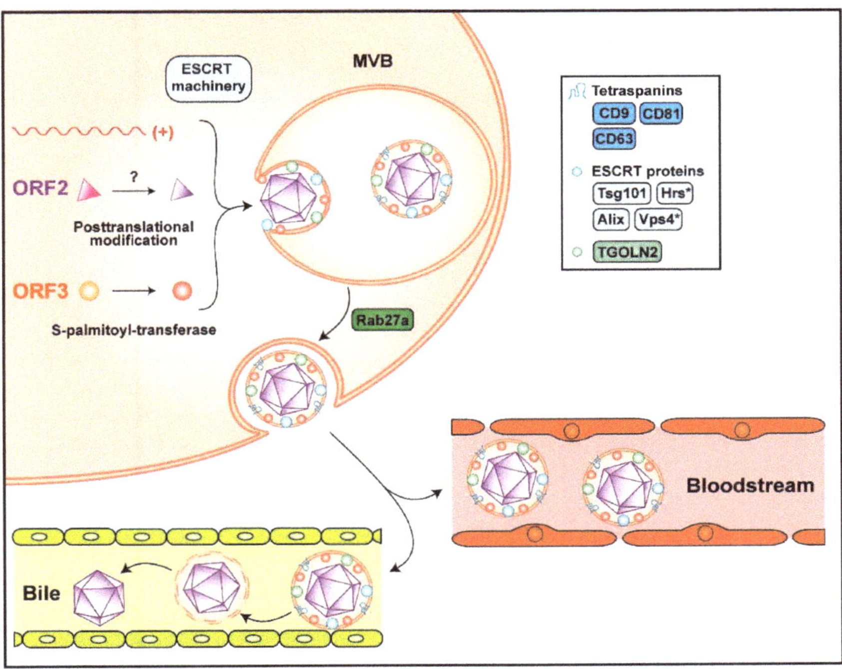

Figure 3. Assembly and release of infectious hepatitis E virus. Packaging of the viral genome into the capsid is believed to occur by spontaneous self-assembly of the non-glycosylated ORF2 protein. The non-glycosylated ORF2 protein may undergo post-translational modification by yet unknown enzyme(s). Formation of the quasi-enveloped particle involves phosphorylation and palmitoylation of the ORF3 protein and the ESCRT machinery, of which the components Tsg101, Hrs, Vps4 and Alix were shown to be required (see text for abbreviations). The virion is wrapped in an exosomal membrane harboring the tetraspanins CD9, CD63 and CD81, as well as the trans-Golgi network protein 2 (TGOLN2), Alix and Tsg101. Release of quasi-enveloped HEV (eHEV) involves Rab27a-dependent trafficking of multivesicular bodies (MVB) and fusion with the plasma membrane. Secreted particles remain associated with the lipid membrane in the bloodstream while they are delipidated in the bile. Asterisks indicate host factors that were not found on the quasi-envelope of eHEV.

Current evidence indicates that none of the viral proteins are present on the surface of eHEV but identified ORF3 protein beneath the quasi-envelope [107–109]. Palmitoylation at *N*-terminal cysteine residues mediates membrane association as well as stability of the ORF3 protein and is required for virus secretion [110]. A membrane topology where ORF3 protein is located entirely on the cytosolic side, corresponding to the exosome lumen, has therefore been proposed [110] and is further supported by the interaction of its PSAP motif with Tsg101. Of note, palmitoylation is a reversible protein modification taking place in the cytosol, which increases the hydrophobicity of a protein and contributes to its membrane association (reviewed in [111]). This implies the requirement for one or more as yet unidentified *S*-palmitoyl-transferase(s) of the host cell [110].

Given the origin of the quasi-envelope, host proteins, and more specifically, exosomal proteins, can be present on eHEV. Indeed, particles released from infected cells and positive for ORF2 and ORF3 proteins were shown to harbor classical exosomal markers, including the tetraspanins CD81, CD63 and CD9, as well as the ESCRT components Alix and Tsg101 [106,108,112]. Interestingly, trans-Golgi network protein 2 (TGOLN2) has also been detected on eHEV [107] (Figure 3). This observation and the fact that TGOLN2 is located in the cytoplasm confirms that quasi-enveloped particles are produced within the cell and not at the plasma membrane.

The quasi-enveloped nature of virions circulating in blood may provide several advantages to HEV, including protection from neutralizing antibodies. However, the presence of eHEV in blood may also favor virus dissemination to organs other than the liver, as exosomes containing viral genetic material were shown to lead to productive infection by other viruses and to modulate cellular responses (reviewed in [113]). Hence, it is plausible that the production of quasi-enveloped virions has additional functions beyond the spread of HEV within the liver.

5. Conclusions and Perspectives

Although HEV research is a rapidly growing field, our current knowledge of the virus life cycle is still limited by important gaps. We lack key information on virus entry, including the cellular receptor(s), and on the uncoating of the viral RNA. Moreover, future efforts should concentrate on the molecular mechanisms and subcellular compartments involved in RNA replication and assembly. While virus release remains one of the best studied steps of the viral life cycle, many aspects need to be clarified, in particular the contribution of host factors and the composition and role of quasi-enveloped particles. Considering the ability of HEV to replicate in different tissues as well as to infect a wide range of animals, one may hypothesize that HEV host dependency is not very selective, facilitating the crossing of species barriers.

Studies on the HEV life cycle currently rely on the use of in vitro model systems, which have certain limitations, especially with respect to host factors present in differentiated hepatocytes. As an example, a stimulatory effect of cyclophilin inhibitors on HEV replication was reported initially in hepatoma cells [114]. However, in stem cell-derived hepatocyte-like cells, this observation was confirmed only with a cell culture-adapted infectious clone, but not with natural HEV isolates [115]. Future improvements of in vitro models should include the use of natural HEV isolates together with primary and stem cell-derived hepatocyte-like cells, polarized cell culture models, as well as ex vivo and in vivo infection model systems to confirm in vitro findings. Ultimately, key findings will have to be validated in liver specimens from patients with hepatitis E.

Obtaining a broader and unbiased view of the host factors involved in the HEV life cycle will likely depend on novel technologies, such as clustered regularly interspaced short palindromic repeats (CRISPR)/Cas9-based genome-wide screening and proteomic proximity labeling approaches. CRISPR/Cas9-based screens have advanced the understanding of, among others, virus entry [116,117] and replication [118,119], and have also facilitated the identification of new antiviral targets [120]. Proximity labeling has been successfully used, for example, to characterize the microenvironment of coronavirus replication complexes [121]. In the future, the combination of improved HEV model systems and of novel technologies should improve our knowledge of the host factors required for productive HEV infection.

Author Contributions: N.O. and J.G. conceived the article; N.O. and J.G. wrote the original draft and prepared figures; D.M. edited and reviewed the manuscript. All authors have read and agreed to the published version of the manuscript.

Funding: This work was funded by the Swiss National Science Foundation (31003A_179424 to D.M. and CRSK-3_190706 to J.G.), the Novartis Foundation (18C140 to D.M.) and the Gilead Sciences International Research Scholars Program in Liver Disease (Award 2015 to J.G.).

Conflicts of Interest: The authors declare no conflict of interest.

References

1. Khuroo, M.S. Study of an epidemic of non-A, non-B hepatitis. Possibility of another human hepatitis virus distinct from post-transfusion non-A, non-B type. *Am. J. Med.* **1980**, *68*, 818–824. [CrossRef]
2. Balayan, M.S.; Andjaparidze, A.G.; Savinskaya, S.S.; Ketiladze, E.S.; Braginsky, D.M.; Savinov, A.P.; Poleschuk, V.F. Evidence for a virus in non-A, non-B hepatitis transmitted via the fecal-oral route. *Intervirology* **1983**, *20*, 23–31.
3. Reyes, G.R.; Purdy, M.A.; Kim, J.P.; Luk, K.C.; Young, L.M.; Fry, K.E.; Bradley, D.W. Isolation of a cDNA from the virus responsible for enterically transmitted non-A, non-B hepatitis. *Science* **1990**, *247*, 1335–1339. [CrossRef] [PubMed]
4. Tam, A.W.; Smith, M.M.; Guerra, M.E.; Huang, C.C.; Bradley, D.W.; Fry, K.E.; Reyes, G.R. Hepatitis E virus (HEV): Molecular cloning and sequencing of the full-length viral genome. *Virology* **1991**, *185*, 120–131. [CrossRef]
5. Smith, D.B.; Simmonds, P.; Jameel, S.; Emerson, S.U.; Harrison, T.J.; Meng, X.J.; Okamoto, H.; Van der Poel, W.H.; Purdy, M.A. Consensus proposals for classification of the family Hepeviridae. *J. Gen. Virol.* **2015**, *96*, 1191–1192. [CrossRef] [PubMed]
6. Lee, G.H.; Tan, B.H.; Teo, E.C.; Lim, S.G.; Dan, Y.Y.; Wee, A.; Aw, P.P.; Zhu, Y.; Hibberd, M.L.; Tan, C.K.; et al. Chronic infection with camelid hepatitis E virus in a liver transplant recipient who regularly consumes camel meat and milk. *Gastroenterology* **2016**, *150*, 355.e353–357.e353. [CrossRef]
7. Nimgaonkar, I.; Ding, Q.; Schwartz, R.E.; Ploss, A. Hepatitis E virus: Advances and challenges. *Nat. Rev. Gastroenterol. Hepatol.* **2018**, *15*, 96–110. [CrossRef]
8. Abravanel, F.; Lhomme, S.; El Costa, H.; Schvartz, B.; Peron, J.M.; Kamar, N.; Izopet, J. Rabbit hepatitis E virus infections in humans, France. *Emerg. Infect. Dis.* **2017**, *23*, 1191–1193. [CrossRef]
9. Sahli, R.; Fraga, M.; Semela, D.; Moradpour, D.; Gouttenoire, J. Rabbit HEV in immunosuppressed patients with hepatitis E acquired in Switzerland. *J. Hepatol.* **2019**, *70*, 1023–1025. [CrossRef]
10. Sridhar, S.; Yip, C.C.Y.; Wu, S.; Cai, J.; Zhang, A.J.; Leung, K.H.; Chung, T.W.H.; Chan, J.F.W.; Chan, W.M.; Teng, J.L.L.; et al. Rat hepatitis E virus as cause of persistent hepatitis after liver transplant. *Emerg. Infect. Dis.* **2018**, *24*, 2241–2250. [CrossRef]
11. Sridhar, S.; Yip, C.C.; Wu, S.; Chew, N.F.; Leung, K.H.; Chan, J.F.; Zhao, P.S.; Chan, W.M.; Poon, R.W.; Tsoi, H.W.; et al. Transmission of rat hepatitis E virus infection to humans in Hong Kong: A clinical and epidemiological analysis. *Hepatology* **2020**. [CrossRef] [PubMed]
12. Ding, Q.; Heller, B.; Capuccino, J.M.; Song, B.; Nimgaonkar, I.; Hrebikova, G.; Contreras, J.E.; Ploss, A. Hepatitis E virus ORF3 is a functional ion channel required for release of infectious particles. *Proc. Natl. Acad. Sci. USA* **2017**, *114*, 1147–1152. [CrossRef] [PubMed]
13. Graff, J.; Torian, U.; Nguyen, H.; Emerson, S.U. A bicistronic subgenomic mRNA encodes both the ORF2 and ORF3 proteins of hepatitis E virus. *J. Virol.* **2006**, *80*, 5919–5926. [CrossRef] [PubMed]
14. Ichiyama, K.; Yamada, K.; Tanaka, T.; Nagashima, S.; Jirintai; Takahashi, M.; Okamoto, H. Determination of the 5′-terminal sequence of subgenomic RNA of hepatitis E virus strains in cultured cells. *Arch. Virol.* **2009**, *154*, 1945–1951. [CrossRef]
15. Feng, Z.; Hensley, L.; McKnight, K.L.; Hu, F.; Madden, V.; Ping, L.; Jeong, S.H.; Walker, C.; Lanford, R.E.; Lemon, S.M. A pathogenic picornavirus acquires an envelope by hijacking cellular membranes. *Nature* **2013**, *496*, 367–371. [CrossRef]
16. Geng, Y.; Yang, J.; Huang, W.; Harrison, T.J.; Zhou, Y.; Wen, Z.; Wang, Y. Virus host protein interaction network analysis reveals that the HEV ORF3 protein may interrupt the blood coagulation process. *PLoS ONE* **2013**, *8*, e56320. [CrossRef]
17. Subramani, C.; Nair, V.P.; Anang, S.; Mandal, S.D.; Pareek, M.; Kaushik, N.; Srivastava, A.; Saha, S.; Shalimar; Nayak, B.; et al. Host-virus protein interaction network reveals the involvement of multiple host processes in the life cycle of hepatitis E virus. *mSystems* **2018**, *3*, e00135-17. [CrossRef]
18. Yu, H.; Li, S.; Yang, C.; Wei, M.; Song, C.; Zheng, Z.; Gu, Y.; Du, H.; Zhang, J.; Xia, N. Homology model and potential virus-capsid binding site of a putative HEV receptor Grp78. *J. Mol. Model.* **2011**, *17*, 987–995. [CrossRef]
19. Ahmed, Z.; Holla, P.; Ahmad, I.; Jameel, S. The ATP synthase subunit β (ATP5B) is an entry factor for the hepatitis E virus. *bioRxiv* **2016**. [CrossRef]

20. Zhang, L.; Tian, Y.; Wen, Z.; Zhang, F.; Qi, Y.; Huang, W.; Zhang, H.; Wang, Y. Asialoglycoprotein receptor facilitates infection of PLC/PRF/5 cells by HEV through interaction with ORF2. *J. Med. Virol.* **2016**, *88*, 2186–2195. [CrossRef]
21. Kalia, M.; Chandra, V.; Rahman, S.A.; Sehgal, D.; Jameel, S. Heparan sulfate proteoglycans are required for cellular binding of the hepatitis E virus ORF2 capsid protein and for viral infection. *J. Virol.* **2009**, *83*, 12714–12724. [CrossRef] [PubMed]
22. Yin, X.; Ambardekar, C.; Lu, Y.; Feng, Z. Distinct entry mechanisms for nonenveloped and quasi-enveloped hepatitis E viruses. *J. Virol.* **2016**, *90*, 4232–4242. [CrossRef] [PubMed]
23. WuDunn, D.; Spear, P.G. Initial interaction of herpes simplex virus with cells is binding to heparan sulfate. *J. Virol.* **1989**, *63*, 52–58. [CrossRef] [PubMed]
24. Shieh, M.T.; WuDunn, D.; Montgomery, R.I.; Esko, J.D.; Spear, P.G. Cell surface receptors for herpes simplex virus are heparan sulfate proteoglycans. *J. Cell Biol.* **1992**, *116*, 1273–1281. [CrossRef] [PubMed]
25. Mondor, I.; Ugolini, S.; Sattentau, Q.J. Human immunodeficiency virus type 1 attachment to HeLa CD4 cells is CD4 independent and gp120 dependent and requires cell surface heparans. *J. Virol.* **1998**, *72*, 3623–3634. [CrossRef] [PubMed]
26. Barth, H.; Schafer, C.; Adah, M.I.; Zhang, F.; Linhardt, R.J.; Toyoda, H.; Kinoshita-Toyoda, A.; Toida, T.; Van Kuppevelt, T.H.; Depla, E.; et al. Cellular binding of hepatitis C virus envelope glycoprotein E2 requires cell surface heparan sulfate. *J. Biol. Chem.* **2003**, *278*, 41003–41012. [CrossRef] [PubMed]
27. Tamura, M.; Natori, K.; Kobayashi, M.; Miyamura, T.; Takeda, N. Genogroup II noroviruses efficiently bind to heparan sulfate proteoglycan associated with the cellular membrane. *J. Virol.* **2004**, *78*, 3817–3826. [CrossRef]
28. Shiota, T.; Li, T.C.; Nishimura, Y.; Yoshizaki, S.; Sugiyama, R.; Shimojima, M.; Saijo, M.; Shimizu, H.; Suzuki, R.; Wakita, T.; et al. Integrin alpha3 is involved in non-enveloped hepatitis E virus infection. *Virology* **2019**, *536*, 119–124. [CrossRef]
29. Takada, Y.; Ye, X.; Simon, S. The integrins. *Genome Biol.* **2007**, *8*, 215. [CrossRef]
30. Akula, S.M.; Pramod, N.P.; Wang, F.Z.; Chandran, B. Integrin alpha3beta1 (CD 49c/29) is a cellular receptor for Kaposi's sarcoma-associated herpesvirus (KSHV/HHV-8) entry into the target cells. *Cell* **2002**, *108*, 407–419. [CrossRef]
31. Rivera-Serrano, E.E.; González-López, O.; Das, A.; Lemon, S.M. Cellular entry and uncoating of naked and quasi-enveloped human hepatoviruses. *Elife* **2019**, *8*, e43983. [CrossRef] [PubMed]
32. Choy, M.Y.; Richman, P.I.; Horton, M.A.; MacDonald, T.T. Expression of the VLA family of integrins in human intestine. *J. Pathol.* **1990**, *160*, 35–40. [CrossRef] [PubMed]
33. Marion, O.; Lhomme, S.; Nayrac, M.; Dubois, M.; Pucelle, M.; Requena, M.; Migueres, M.; Abravanel, F.; Peron, J.M.; Carrere, N.; et al. Hepatitis E virus replication in human intestinal cells. *Gut* **2020**, *69*, 901–910. [CrossRef] [PubMed]
34. Oechslin, N.; Moradpour, D.; Gouttenoire, J. Hepatitis E virus finds its path through the gut. *Gut* **2020**, *69*, 796–798. [CrossRef]
35. de Melker, A.A.; Sterk, L.M.; Delwel, G.O.; Fles, D.L.; Daams, H.; Weening, J.J.; Sonnenberg, A. The A and B variants of the alpha 3 integrin subunit: Tissue distribution and functional characterization. *Lab. Invest.* **1997**, *76*, 547–563.
36. Volpes, R.; van den Oord, J.J.; Desmet, V.J. Distribution of the VLA family of integrins in normal and pathological human liver tissue. *Gastroenterology* **1991**, *101*, 200–206. [CrossRef]
37. Volpes, R.; van den Oord, J.J.; Desmet, V.J. Integrins as differential cell lineage markers of primary liver tumors. *Am. J. Pathol.* **1993**, *142*, 1483–1492.
38. Kapur, N.; Thakral, D.; Durgapal, H.; Panda, S.K. Hepatitis E virus enters liver cells through receptor-dependent clathrin-mediated endocytosis. *J. Viral Hepat.* **2012**, *19*, 436–448. [CrossRef]
39. Holla, P.; Ahmad, I.; Ahmed, Z.; Jameel, S. Hepatitis E virus enters liver cells through a dynamin-2, clathrin and membrane cholesterol-dependent pathway. *Traffic* **2015**, *16*, 398–416. [CrossRef]
40. Bubeck, D.; Filman, D.J.; Cheng, N.; Steven, A.C.; Hogle, J.M.; Belnap, D.M. The structure of the poliovirus 135S cell entry intermediate at 10-angstrom resolution reveals the location of an externalized polypeptide that binds to membranes. *J. Virol.* **2005**, *79*, 7745–7755. [CrossRef]
41. Cavaldesi, M.; Caruso, M.; Sthandier, O.; Amati, P.; Garcia, M.I. Conformational changes of murine polyomavirus capsid proteins induced by sialic acid binding. *J. Biol. Chem.* **2004**, *279*, 41573–41579. [CrossRef] [PubMed]

42. Sapp, M.; Bienkowska-Haba, M. Viral entry mechanisms: human papillomavirus and a long journey from extracellular matrix to the nucleus. *FEBS J.* **2009**, *276*, 7206–7216. [CrossRef] [PubMed]
43. Yin, X.; Feng, Z. Hepatitis E virus entry. *Viruses* **2019**, *11*, 883. [CrossRef]
44. Tyagi, S.; Korkaya, H.; Zafrullah, M.; Jameel, S.; Lal, S.K. The phosphorylated form of the ORF3 protein of hepatitis E virus interacts with its non-glycosylated form of the major capsid protein, ORF2. *J. Biol. Chem.* **2002**, *277*, 22759–22767. [CrossRef] [PubMed]
45. Drave, S.A.; Debing, Y.; Walter, S.; Todt, D.; Engelmann, M.; Friesland, M.; Wedemeyer, H.; Neyts, J.; Behrendt, P.; Steinmann, E. Extra-hepatic replication and infection of hepatitis E virus in neuronal-derived cells. *J. Viral Hepat.* **2016**, *23*, 512–521. [CrossRef]
46. Fu, R.M.; Decker, C.C.; Dao Thi, V.L. Cell culture models for hepatitis E virus. *Viruses* **2019**, *11*, 608. [CrossRef]
47. Gouilly, J.; Chen, Q.; Siewiera, J.; Cartron, G.; Levy, C.; Dubois, M.; Al-Daccak, R.; Izopet, J.; Jabrane-Ferrat, N.; El Costa, H. Genotype specific pathogenicity of hepatitis E virus at the human maternal-fetal interface. *Nat. Commun.* **2018**, *9*, 4748. [CrossRef]
48. Knegendorf, L.; Drave, S.A.; Dao Thi, V.L.; Debing, Y.; Brown, R.J.P.; Vondran, F.W.R.; Resner, K.; Friesland, M.; Khera, T.; Engelmann, M.; et al. Hepatitis E virus replication and interferon responses in human placental cells. *Hepatol. Commun.* **2018**, *2*, 173–187. [CrossRef]
49. Meister, T.L.; Bruening, J.; Todt, D.; Steinmann, E. Cell culture systems for the study of hepatitis E virus. *Antiviral Res.* **2019**, *163*, 34–49. [CrossRef]
50. Shukla, P.; Nguyen, H.T.; Torian, U.; Engle, R.E.; Faulk, K.; Dalton, H.R.; Bendall, R.P.; Keane, F.E.; Purcell, R.H.; Emerson, S.U. Cross-species infections of cultured cells by hepatitis E virus and discovery of an infectious virus-host recombinant. *Proc/ Natl. Acad. Sci. USA* **2011**, *108*, 2438–2443. [CrossRef]
51. Zhou, X.; Huang, F.; Xu, L.; Lin, Z.; de Vrij, F.M.S.; Ayo-Martin, A.C.; van der Kroeg, M.; Zhao, M.; Yin, Y.; Wang, W.; et al. Hepatitis E virus infects neurons and brains. *J. Infect. Dis.* **2017**, *215*, 1197–1206. [CrossRef] [PubMed]
52. Kamar, N.; Bendall, R.; Legrand-Abravanel, F.; Xia, N.S.; Ijaz, S.; Izopet, J.; Dalton, H.R. Hepatitis E. *Lancet* **2012**, *379*, 2477–2488. [CrossRef]
53. Debing, Y.; Moradpour, D.; Neyts, J.; Gouttenoire, J. Update on hepatitis E virology: Implications for clinical practice. *J. Hepatol.* **2016**, *65*, 200–212. [CrossRef] [PubMed]
54. Geng, Y.; Zhao, C.; Huang, W.; Harrison, T.J.; Zhang, H.; Geng, K.; Wang, Y. Detection and assessment of infectivity of hepatitis E virus in urine. *J. Hepatol.* **2016**, *64*, 37–43. [CrossRef] [PubMed]
55. Kamar, N.; Abravanel, F.; Lhomme, S.; Rostaing, L.; Izopet, J. Hepatitis E virus: Chronic infection, extra-hepatic manifestations, and treatment. *Clin. Res. Hepatol. Gastroenterol.* **2015**, *39*, 20–27. [CrossRef] [PubMed]
56. Montero, H.; Perez-Gil, G.; Sampieri, C.L. Eukaryotic initiation factor 4A (eIF4A) during viral infections. *Virus Genes* **2019**, *55*, 267–273. [CrossRef]
57. Zhou, X.; Xu, L.; Wang, Y.; Wang, W.; Sprengers, D.; Metselaar, H.J.; Peppelenbosch, M.P.; Pan, Q. Requirement of the eukaryotic translation initiation factor 4F complex in hepatitis E virus replication. *Antiviral Res.* **2015**, *124*, 11–19. [CrossRef]
58. Glitscher, M.; Himmelsbach, K.; Woytinek, K.; Johne, R.; Reuter, A.; Spiric, J.; Schwaben, L.; Grunweller, A.; Hildt, E. Inhibition of hepatitis E virus spread by the natural compound silvestrol. *Viruses* **2018**, *10*, 301. [CrossRef]
59. Todt, D.; Moeller, N.; Praditya, D.; Kinast, V.; Friesland, M.; Engelmann, M.; Verhoye, L.; Sayed, I.M.; Behrendt, P.; Dao Thi, V.L.; et al. The natural compound silvestrol inhibits hepatitis E virus (HEV) replication in vitro and in vivo. *Antiviral Res.* **2018**, *157*, 151–158. [CrossRef]
60. Biedenkopf, N.; Lange-Grunweller, K.; Schulte, F.W.; Weisser, A.; Muller, C.; Becker, D.; Becker, S.; Hartmann, R.K.; Grunweller, A. The natural compound silvestrol is a potent inhibitor of Ebola virus replication. *Antiviral Res.* **2017**, *137*, 76–81. [CrossRef]
61. Elgner, F.; Sabino, C.; Basic, M.; Ploen, D.; Grunweller, A.; Hildt, E. Inhibition of Zika virus replication by silvestrol. *Viruses* **2018**, *10*, 149. [CrossRef] [PubMed]
62. Henss, L.; Scholz, T.; Grunweller, A.; Schnierle, B.S. Silvestrol inhibits Chikungunya virus replication. *Viruses* **2018**, *10*, 592. [CrossRef] [PubMed]
63. Muller, C.; Schulte, F.W.; Lange-Grunweller, K.; Obermann, W.; Madhugiri, R.; Pleschka, S.; Ziebuhr, J.; Hartmann, R.K.; Grunweller, A. Broad-spectrum antiviral activity of the eIF4A inhibitor silvestrol against corona- and picornaviruses. *Antiviral Res.* **2018**, *150*, 123–129. [CrossRef] [PubMed]

64. Karpe, Y.A.; Lole, K.S. Deubiquitination activity associated with hepatitis E virus putative papain-like cysteine protease. *J. Gen. Virol.* **2011**, *92*, 2088–2092. [CrossRef]
65. Li, C.; Debing, Y.; Jankevicius, G.; Neyts, J.; Ahel, I.; Coutard, B.; Canard, B. Viral macro domains reverse protein ADP-ribosylation. *J. Virol.* **2016**, *90*, 8478–8486. [CrossRef]
66. Kanade, G.D.; Pingale, K.D.; Karpe, Y.A. Activities of thrombin and factor Xa are essential for replication of hepatitis E virus and are possibly implicated in ORF1 polyprotein processing. *J. Virol.* **2018**, *92*, e01853-17. [CrossRef]
67. LeDesma, R.; Nimgaonkar, I.; Ploss, A. Hepatitis E virus replication. *Viruses* **2019**, *11*, 719. [CrossRef]
68. Szkolnicka, D.; Pollan, A.; Da Silva, N.; Oechslin, N.; Gouttenoire, J.; Moradpour, D. Recombinant hepatitis E viruses harboring tags in the ORF1 protein. *J. Virol.* **2019**, *93*, e00459-19. [CrossRef]
69. Ding, Q.; Nimgaonkar, I.; Archer, N.F.; Bram, Y.; Heller, B.; Schwartz, R.E.; Ploss, A. Identification of the intragenomic promoter controlling hepatitis E virus subgenomic RNA transcription. *mBio* **2018**, *9*, e00769-18. [CrossRef]
70. Kanade, G.D.; Pingale, K.D.; Karpe, Y.A. Protein interactions network of hepatitis E virus RNA and polymerase with host proteins. *Front. Microbiol.* **2019**, *10*, 2501. [CrossRef]
71. Pingale, K.D.; Kanade, G.D.; Karpe, Y.A. Heterogeneous nuclear ribonucleoproteins participate in hepatitis E virus (HEV) replication. *J. Mol. Biol.* **2020**, *432*, 2369–2387. [CrossRef] [PubMed]
72. Burnham, A.J.; Gong, L.; Hardy, R.W. Heterogeneous nuclear ribonuclear protein K interacts with Sindbis virus nonstructural proteins and viral subgenomic mRNA. *Virology* **2007**, *367*, 212–221. [CrossRef] [PubMed]
73. Pettit Kneller, E.L.; Connor, J.H.; Lyles, D.S. hnRNPs relocalize to the cytoplasm following infection with vesicular stomatitis virus. *J. Virol.* **2009**, *83*, 770–780. [CrossRef] [PubMed]
74. Bourai, M.; Lucas-Hourani, M.; Gad, H.H.; Drosten, C.; Jacob, Y.; Tafforeau, L.; Cassonnet, P.; Jones, L.M.; Judith, D.; Couderc, T.; et al. Mapping of Chikungunya virus interactions with host proteins identified nsP2 as a highly connected viral component. *J. Virol.* **2012**, *86*, 3121–3134. [CrossRef] [PubMed]
75. Brunetti, J.E.; Scolaro, L.A.; Castilla, V. The heterogeneous nuclear ribonucleoprotein K (hnRNP K) is a host factor required for dengue virus and Junin virus multiplication. *Virus Res.* **2015**, *203*, 84–91. [CrossRef] [PubMed]
76. Poenisch, M.; Metz, P.; Blankenburg, H.; Ruggieri, A.; Lee, J.Y.; Rupp, D.; Rebhan, I.; Diederich, K.; Kaderali, L.; Domingues, F.S.; et al. Identification of hnRNPK as regulator of hepatitis C virus particle production. *PLoS Pathog.* **2015**, *11*, e1004573. [CrossRef]
77. Paul, D.; Madan, V.; Bartenschlager, R. Hepatitis C virus RNA replication and assembly: Living on the fat of the land. *Cell Host Microbe* **2014**, *16*, 569–579. [CrossRef]
78. Moradpour, D.; Penin, F.; Rice, C.M. Replication of hepatitis C virus. *Nature Rev. Microbiol.* **2007**, *5*, 453–463. [CrossRef]
79. Farhat, R.; Seron, K.; Ferlin, J.; Feneant, L.; Belouzard, S.; Goueslain, L.; Jackson, C.L.; Dubuisson, J.; Rouillé, Y. Identification of class II ADP-ribosylation factors as cellular factors required for hepatitis C virus replication. *Cell. Microbiol.* **2016**, *18*, 1121–1133. [CrossRef]
80. Goueslain, L.; Alsaleh, K.; Horellou, P.; Roingeard, P.; Descamps, V.; Duverlie, G.; Ciczora, Y.; Wychowski, C.; Dubuisson, J.; Rouille, Y. Identification of GBF1 as a cellular factor required for hepatitis C virus RNA replication. *J. Virol.* **2010**, *84*, 773–787. [CrossRef]
81. Belov, G.A.; Feng, Q.; Nikovics, K.; Jackson, C.L.; Ehrenfeld, E. A critical role of a cellular membrane traffic protein in poliovirus RNA replication. *PLoS Pathog.* **2008**, *4*, e1000216. [CrossRef] [PubMed]
82. Lanke, K.H.; van der Schaar, H.M.; Belov, G.A.; Feng, Q.; Duijsings, D.; Jackson, C.L.; Ehrenfeld, E.; van Kuppeveld, F.J. GBF1, a guanine nucleotide exchange factor for Arf, is crucial for coxsackievirus B3 RNA replication. *J. Virol.* **2009**, *83*, 11940–11949. [CrossRef] [PubMed]
83. Wang, J.; Du, J.; Jin, Q. Class I ADP-ribosylation factors are involved in enterovirus 71 replication. *PLoS ONE* **2014**, *9*, e99768. [CrossRef] [PubMed]
84. Farhat, R.; Ankavay, M.; Lebsir, N.; Gouttenoire, J.; Jackson, C.L.; Wychowski, C.; Moradpour, D.; Dubuisson, J.; Rouille, Y.; Cocquerel, L. Identification of GBF1 as a cellular factor required for hepatitis E virus RNA replication. *Cell. Microbiol.* **2018**, *20*, e12804. [CrossRef]
85. Verheije, M.H.; Raaben, M.; Mari, M.; Te Lintelo, E.G.; Reggiori, F.; van Kuppeveld, F.J.; Rottier, P.J.; de Haan, C.A. Mouse hepatitis coronavirus RNA replication depends on GBF1-mediated ARF1 activation. *PLoS Pathog.* **2008**, *4*, e1000088. [CrossRef]

86. Xing, L.; Li, T.C.; Mayazaki, N.; Simon, M.N.; Wall, J.S.; Moore, M.; Wang, C.Y.; Takeda, N.; Wakita, T.; Miyamura, T.; et al. Structure of hepatitis E virion-sized particle reveals an RNA-dependent viral assembly pathway. *J. Biol. Chem.* **2010**, *285*, 33175–33183. [CrossRef]
87. Mori, Y.; Matsuura, Y. Structure of hepatitis E viral particle. *Virus Res.* **2011**, *161*, 59–64. [CrossRef]
88. Graff, J.; Zhou, Y.H.; Torian, U.; Nguyen, H.; St Claire, M.; Yu, C.; Purcell, R.H.; Emerson, S.U. Mutations within potential glycosylation sites in the capsid protein of hepatitis E virus prevent the formation of infectious virus particles. *J. Virol.* **2008**, *82*, 1185–1194. [CrossRef]
89. Montpellier, C.; Wychowski, C.; Sayed, I.M.; Meunier, J.C.; Saliou, J.M.; Ankavay, M.; Bull, A.; Pillez, A.; Abravanel, F.; Helle, F.; et al. Hepatitis E virus lifecycle and identification of 3 forms of the ORF2 capsid protein. *Gastroenterology* **2018**, *154*, 211.e218–223.e218. [CrossRef]
90. Yin, X.; Ying, D.; Lhomme, S.; Tang, Z.; Walker, C.M.; Xia, N.; Zheng, Z.; Feng, Z. Origin, antigenicity, and function of a secreted form of ORF2 in hepatitis E virus infection. *Proc. Natl. Acad. Sci. USA* **2018**, *115*, 4773–4778. [CrossRef]
91. Ankavay, M.; Montpellier, C.; Sayed, I.M.; Saliou, J.M.; Wychowski, C.; Saas, L.; Duvet, S.; Aliouat-Denis, C.M.; Farhat, R.; de Masson d'Autume, V.; et al. New insights into the ORF2 capsid protein, a key player of the hepatitis E virus lifecycle. *Sci. Rep.* **2019**, *9*, 6243. [CrossRef] [PubMed]
92. Lenggenhager, D.; Gouttenoire, J.; Malehmir, M.; Bawohl, M.; Honcharova-Biletska, H.; Kreutzer, S.; Semela, D.; Neuweiler, J.; Hurlimann, S.; Aepli, P.; et al. Visualization of hepatitis E virus RNA and proteins in the human liver. *J. Hepatol.* **2017**, *67*, 471–479. [CrossRef] [PubMed]
93. Jameel, S.; Zafrullah, M.; Ozdener, M.H.; Panda, S.K. Expression in animal cells and characterization of the hepatitis E virus structural proteins. *J. Virol.* **1996**, *70*, 207–216. [CrossRef] [PubMed]
94. Zafrullah, M.; Ozdener, M.H.; Kumar, R.; Panda, S.K.; Jameel, S. Mutational analysis of glycosylation, membrane translocation, and cell surface expression of the hepatitis E virus ORF2 protein. *J. Virol.* **1999**, *73*, 4074–4082. [CrossRef]
95. Emerson, S.U.; Nguyen, H.; Torian, U.; Purcell, R.H. ORF3 protein of hepatitis E virus is not required for replication, virion assembly, or infection of hepatoma cells in vitro. *J. Virol.* **2006**, *80*, 10457–10464. [CrossRef]
96. Yamada, K.; Takahashi, M.; Hoshino, Y.; Takahashi, H.; Ichiyama, K.; Nagashima, S.; Tanaka, T.; Okamoto, H. ORF3 protein of hepatitis E virus is essential for virion release from infected cells. *J. Gen. Virol.* **2009**, *90*, 1880–1891. [CrossRef]
97. Capelli, N.; Marion, O.; Dubois, M.; Allart, S.; Bertrand-Michel, J.; Lhomme, S.; Abravanel, F.; Izopet, J.; Chapuy-Regaud, S. Vectorial release of hepatitis E virus in polarized human hepatocytes. *J. Virol.* **2019**, *93*. [CrossRef]
98. Dao Thi, V.L.; Wu, X.; Belote, R.L.; Andreo, U.; Takacs, C.N.; Fernandez, J.P.; Vale-Silva, L.A.; Prallet, S.; Decker, C.C.; Fu, R.M.; et al. Stem cell-derived polarized hepatocytes. *Nat. Commun.* **2020**, *11*, 1677. [CrossRef]
99. Zafrullah, M.; Ozdener, M.H.; Panda, S.K.; Jameel, S. The ORF3 protein of hepatitis E virus is a phosphoprotein that associates with the cytoskeleton. *J. Virol.* **1997**, *71*, 9045–9053. [CrossRef]
100. Surjit, M.; Oberoi, R.; Kumar, R.; Lal, S.K. Enhanced alpha1 microglobulin secretion from hepatitis E virus ORF3-expressing human hepatoma cells is mediated by the tumor susceptibility gene 101. *J. Biol. Chem.* **2006**, *281*, 8135–8142. [CrossRef]
101. Nagashima, S.; Takahashi, M.; Jirintai; Tanaka, T.; Yamada, K.; Nishizawa, T.; Okamoto, H. A PSAP motif in the ORF3 protein of hepatitis E virus is necessary for virion release from infected cells. *J. Gen. Virol.* **2011**, *92*, 269–278. [CrossRef] [PubMed]
102. Nagashima, S.; Takahashi, M.; Jirintai, S.; Tanaka, T.; Nishizawa, T.; Yasuda, J.; Okamoto, H. Tumour susceptibility gene 101 and the vacuolar protein sorting pathway are required for the release of hepatitis E virions. *J. Gen. Virol.* **2011**, *92*, 2838–2848. [CrossRef] [PubMed]
103. Garrus, J.E.; von Schwedler, U.K.; Pornillos, O.W.; Morham, S.G.; Zavitz, K.H.; Wang, H.E.; Wettstein, D.A.; Stray, K.M.; Cote, M.; Rich, R.L.; et al. Tsg101 and the vacuolar protein sorting pathway are essential for HIV-1 budding. *Cell* **2001**, *107*, 55–65. [CrossRef]
104. Martin-Serrano, J.; Zang, T.; Bieniasz, P.D. HIV-1 and Ebola virus encode small peptide motifs that recruit Tsg101 to sites of particle assembly to facilitate egress. *Nat. Med.* **2001**, *7*, 1313–1319. [CrossRef] [PubMed]
105. Vietri, M.; Radulovic, M.; Stenmark, H. The many functions of ESCRTs. *Nat. Rev. Mol. Cell Biol.* **2020**, *21*, 25–42. [CrossRef] [PubMed]

106. Nagashima, S.; Jirintai, S.; Takahashi, M.; Kobayashi, T.; Tanggis; Nishizawa, T.; Kouki, T.; Yashiro, T.; Okamoto, H. Hepatitis E virus egress depends on the exosomal pathway, with secretory exosomes derived from multivesicular bodies. *J. Gen. Virol.* **2014**, *95*, 2166–2175. [CrossRef]
107. Nagashima, S.; Takahashi, M.; Jirintai, S.; Tanggis; Kobayashi, T.; Nishizawa, T.; Okamoto, H. The membrane on the surface of hepatitis E virus particles is derived from the intracellular membrane and contains trans-Golgi network protein 2. *Arch. Virol.* **2014**, *159*, 979–991. [CrossRef]
108. Nagashima, S.; Takahashi, M.; Kobayashi, T.; Nishizawa, T.; Nishiyama, T.; Primadharsini, P.P.; Okamoto, H. Characterization of the quasi-enveloped hepatitis E virus particles released by the cellular exosomal pathway. *J. Virol.* **2017**, *91*, e00822-17. [CrossRef]
109. Qi, Y.; Zhang, F.; Zhang, L.; Harrison, T.J.; Huang, W.; Zhao, C.; Kong, W.; Jiang, C.; Wang, Y. Hepatitis E virus produced from cell culture has a lipid envelope. *PLoS ONE* **2015**, *10*, e0132503. [CrossRef]
110. Gouttenoire, J.; Pollan, A.; Abrami, L.; Oechslin, N.; Mauron, J.; Matter, M.; Oppliger, J.; Szkolnicka, D.; Dao Thi, V.L.; van der Goot, F.G.; et al. Palmitoylation mediates membrane association of hepatitis E virus ORF3 protein and is required for infectious particle secretion. *PLoS Pathog.* **2018**, *14*, e1007471. [CrossRef]
111. Blaskovic, S.; Blanc, M.; van der Goot, F.G. What does S-palmitoylation do to membrane proteins? *FEBS J.* **2013**, *280*, 2766–2774. [CrossRef] [PubMed]
112. Chapuy-Regaud, S.; Dubois, M.; Plisson-Chastang, C.; Bonnefois, T.; Lhomme, S.; Bertrand-Michel, J.; You, B.; Simoneau, S.; Gleizes, P.E.; Flan, B.; et al. Characterization of the lipid envelope of exosome encapsulated HEV particles protected from the immune response. *Biochimie* **2017**, *141*, 70–79. [CrossRef] [PubMed]
113. Chahar, H.S.; Bao, X.; Casola, A. Exosomes and their role in the life cycle and pathogenesis of RNA viruses. *Viruses* **2015**, *7*, 3204–3225. [CrossRef] [PubMed]
114. Wang, Y.; Zhou, X.; Debing, Y.; Chen, K.; Van Der Laan, L.J.; Neyts, J.; Janssen, H.L.; Metselaar, H.J.; Peppelenbosch, M.P.; Pan, Q. Calcineurin inhibitors stimulate and mycophenolic acid inhibits replication of hepatitis E virus. *Gastroenterology* **2014**, *146*, 1775–1783. [CrossRef]
115. Wu, X.; Dao Thi, V.L.; Liu, P.; Takacs, C.N.; Xiang, K.; Andrus, L.; Gouttenoire, J.; Moradpour, D.; Rice, C.M. Pan-genotype hepatitis E virus replication in stem cell-derived hepatocellular systems. *Gastroenterology* **2018**, *154*, 663.e7–674.e7. [CrossRef]
116. Meertens, L.; Hafirassou, M.L.; Couderc, T.; Bonnet-Madin, L.; Kril, V.; Kummerer, B.M.; Labeau, A.; Brugier, A.; Simon-Loriere, E.; Burlaud-Gaillard, J.; et al. FHL1 is a major host factor for Chikungunya virus infection. *Nature* **2019**, *574*, 259–263. [CrossRef]
117. Zhang, R.; Kim, A.S.; Fox, J.M.; Nair, S.; Basore, K.; Klimstra, W.B.; Rimkunas, R.; Fong, R.H.; Lin, H.; Poddar, S.; et al. Mxra8 is a receptor for multiple arthritogenic alphaviruses. *Nature* **2018**, *557*, 570–574. [CrossRef]
118. Han, J.; Perez, J.T.; Chen, C.; Li, Y.; Benitez, A.; Kandasamy, M.; Lee, Y.; Andrade, J.; tenOever, B.; Manicassamy, B. Genome-wide CRISPR/Cas9 screen identifies host factors essential for influenza virus replication. *Cell Rep.* **2018**, *23*, 596–607. [CrossRef]
119. Richardson, R.B.; Ohlson, M.B.; Eitson, J.L.; Kumar, A.; McDougal, M.B.; Boys, I.N.; Mar, K.B.; De La Cruz-Rivera, P.C.; Douglas, C.; Konopka, G.; et al. A CRISPR screen identifies IFI6 as an ER-resident interferon effector that blocks flavivirus replication. *Nat. Microbiol.* **2018**, *3*, 1214–1223. [CrossRef]
120. Flint, M.; Chatterjee, P.; Lin, D.L.; McMullan, L.K.; Shrivastava-Ranjan, P.; Bergeron, E.; Lo, M.K.; Welch, S.R.; Nichol, S.T.; Tai, A.W.; et al. A genome-wide CRISPR screen identifies N-acetylglucosamine-1-phosphate transferase as a potential antiviral target for Ebola virus. *Nat. Commun.* **2019**, *10*, 285. [CrossRef]
121. V'Kovski, P.; Gerber, M.; Kelly, J.; Pfaender, S.; Ebert, N.; Braga Lagache, S.; Simillion, C.; Portmann, J.; Stalder, H.; Gaschen, V.; et al. Determination of host proteins composing the microenvironment of coronavirus replicase complexes by proximity-labeling. *Elife* **2019**, *8*, e42037. [CrossRef] [PubMed]

 *cells*

Review

Environmental Restrictions: A New Concept Governing HIV-1 Spread Emerging from Integrated Experimental-Computational Analysis of Tissue-Like 3D Cultures

Samy Sid Ahmed [1], Nils Bundgaard [2], Frederik Graw [2,*] and Oliver T. Fackler [1,*]

1 Department of Infectious Diseases, Integrative Virology, University Hospital Heidelberg, 69120 Heidelberg, Germany; Samy.SidAhmed@med.uni-heidelberg.de

2 BioQuant – Center for Quantitative Biology, Heidelberg University, 69120 Heidelberg, Germany; nils.bundgaard@bioquant.uni-heidelberg.de

* Correspondence: frederik.graw@bioquant.uni-heidelberg.de (F.G.); oliver.fackler@med.uni-heidelberg.de (O.T.F.); Tel.: +49-(0)6221-54 51309 (F.G.); +49-(0)6221-561322 (O.T.F.); Fax: +49-(0)6221-54 14427 (F.G.); +49-(0)6221-565003 (O.T.F.)

Received: 24 March 2020; Accepted: 28 April 2020; Published: 30 April 2020

Abstract: HIV-1 can use cell-free and cell-associated transmission modes to infect new target cells, but how the virus spreads in the infected host remains to be determined. We recently established 3D collagen cultures to study HIV-1 spread in tissue-like environments and applied iterative cycles of experimentation and computation to develop a first in silico model to describe the dynamics of HIV-1 spread in complex tissue. These analyses (i) revealed that 3D collagen environments restrict cell-free HIV-1 infection but promote cell-associated virus transmission and (ii) defined that cell densities in tissue dictate the efficacy of these transmission modes for virus spread. In this review, we discuss, in the context of the current literature, the implications of this study for our understanding of HIV-1 spread in vivo, which aspects of in vivo physiology this integrated experimental–computational analysis takes into account, and how it can be further improved experimentally and in silico.

Keywords: HIV-1 spread; cell-free infection; cell–cell transmission; 3D cultures; mathematical modeling; environmental restriction

1. Introduction

As obligate intracellular parasites, the replication of viruses depends on the infection of host cells that support the viral life cycle and the production of viral progeny. In order to establish virus replication in a new host, the virus has to efficiently spread following the initial infection at the portal of entry. The production of infectious progeny and infection of new target cells represents the central mechanism for virus spread. In principle, this can be achieved by the release of virus particles into the extracellular space, which can encounter and infect new target cells (cell-free infection) (Figure 1a). In addition, viruses can be transferred from infected donor cells to uninfected target cells via close physical contact between the cells (cell–cell transmission) (Figure 1b–d). Cell-associated modes of virus transmission include the short-distance transmission of cell-free virus at cell–cell contacts (Figure 1d), the transport of virus particles along or within cell protrusions connecting donor and target cells (Figure 1b,c), as well as cell–cell fusion [1,2], and are generally considered more efficient than cell-free infections. While cell-associated modes of virus transmission have been less explored than cell-free infection, evidence for the use of this transmission mode is steadily increasing and has been documented, e.g., for Vaccinia virus [3], Hepatitis C virus [4], Herpes Simplex virus [5], Epstein–Barr Virus [6] Dengue Virus [7], and the pathogenic human retroviruses Human Immunodeficiency Virus type

1 (HIV-1) and Human T-cell Lymphotropic Virus type 1 (HTLV-1) [8–12]. Most of these viruses are known to be able to spread by cell-free and cell-associated modes of transmission, but some viruses, such as HTLV-I, specialize in cell–cell transmission and appear to exclusively rely on this transfer mode, as cell-free infectious virus can seldom be isolated [12]. For viruses using both cell-free and cell-associated transmission, the relative contribution of each transmission mode to overall spread is difficult to assess, and the pathophysiological relevance of cell-associated transmission often remains unclear. It is therefore not surprising that traditional concepts in virology have focused on cell-free infections, which is still reflected in the majority of experimental studies conducted.

HIV-1 is an example of a virus for which the modes of transmission are particularly well studied. Initially assumed to spread exclusively via cell-free virus, early studies indicated that infected cells are a much better inoculum to drive virus spread in a new culture than cell-free virus [13]. The demonstration that constant agitation of infected $CD4^+$ T cells or physical separation of infected from uninfected cells by transwells disrupts the formation of cell–cell contacts as well as efficient virus spread then suggested that, in fact, cell-associated modes of transmission are essential for efficient HIV-1 spread in $CD4^+$ T-cell cultures [14,15]. A large series of imaging-based studies has now established that in addition to infection with cell-free virions, HIV-1 can efficiently spread via cell-cell contacts. Although probably relying on a slightly divergent mechanism, HIV-1 cell–cell transmission is observed between $CD4^+$ T cells, for the transfer from dendritic cells to $CD4^+$ T cells, between $CD4^+$ T cells and macrophages, and between myeloid cells [16]. The cell–cell contacts involved are referred to as virological synapses (VSs) between productively infected donor and target cells, or as infectious synapses when donor cells such as dendritic cells store virus for transmission without being productively infected [17–22]. This contact-dependent transmission mode has been found to be much more efficient than cell-free virus uptake, with an estimated 10-fold to 18,000-fold higher efficiency in mediating viral spread [14,23–25]. Despite this overwhelming evidence that HIV-1 efficiently uses cell-associated modes of transmission in experimental HIV-1 infection, the technical barriers to studying this aspect of HIV-1 biology in vivo limit the generation of evidence for the use of this transmission mode in the infected host. Only a few studies have reported visualization of cell–cell contacts reminiscent of a VS in vivo [26–28], or established the motility of cells loaded with infectious HIV-1 as essential for efficient HIV-1 spread in infected humanized mice [27,29]. By which transmission mode HIV-1 spreads in vivo, and how this might depend on the specific tissue environment, thus remains to be established.

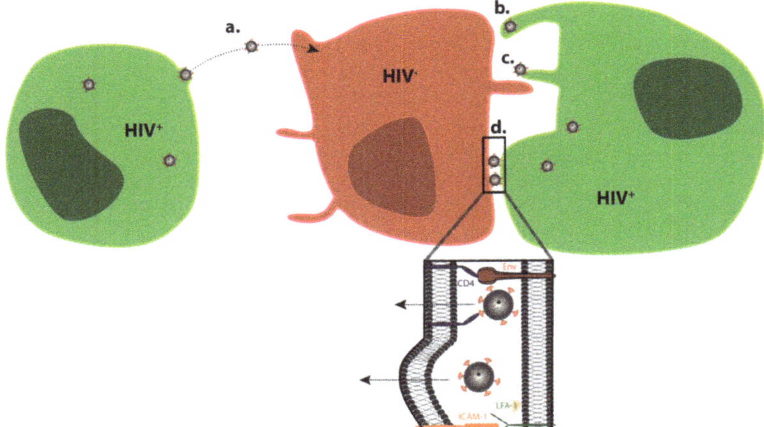

Figure 1. Transmission modes of HIV-1. Viral particles infect target cells via cell-free (**a**) or cell-associated (**b**-**d**) modes of transmission. (**a**) Viral particles bud at the surface of infected donor cells, mature, diffuse, and infect non-adjacent target cells. (**b**,**c**) Virions can bud at the tip (**b**) and surf along (**c**) filopodia to enter in adjacent target cells. In addition, infected and non-infected cells establish close contact, forming a virological synapse (**d**). Whether HIV-1 enters the target cell via fusion at the plasma membrane or following prior internalization [30,31] remains a matter of debate, and may depend on the nature of the target cell (reviewed in Reference [32]).

2. 3D Culture Systems

Questions such as the relevance of individual virus transmission modes for viral spread and disease progression would be best addressed in in vivo infection models, as these reflect the complex tissue organization and pathogen–host interactions of infected patients [33]. In the case of HIV-1 infection, various humanized mouse models that mirror a range of but not all pathological events of AIDS are available for such studies [34]. However, experimental parameters are very difficult to control in these complex in vivo systems, and the number of experiments that can be conducted is limited by logistic, financial, and/or ethical concerns. In turn, simple monotypic, two-dimensional (2D) cultures do not reflect the complex architecture and cell heterogeneity of HIV-1 target tissue and, even for suspension cells, such as $CD4^+$ T cells, quickly organize into a cell monolayer with dense cell packing. In vivo, the physical distance between donor and target cells often is the main barrier for efficient virus transmission, be it via diffusion of cell-free particles or the motility of the cell towards forming cell–cell contacts for virus transmission. Thus, densely packed 2D cultures are not suitable to study such processes.

As these constraints apply to all areas of biology, major efforts are being undertaken to establish and exploit experimental 3D systems that faithfully reflect specific aspects of tissue organization and physiology. One major area of development involves establishing organ or organoid cultures, either by culturing original tissues obtained from surgery or by reconstituting organ-like tissue ex vivo by differentiation from induced pluripotent stem cells [35,36]. In the case of HIV-1, such organotypic cell systems include cultures of human tonsil tissue as surrogates for lymph node tissues or mucosa explants as mimics of sites of virus transmission during primary HIV-1 infection [33,37]. While these models are of great value for advancing our understanding of pathological mechanisms, they do not offer tight control over critical experimental parameters such as cell density and composition, are subject to significant donor-to-donor variability, and their availability is limited. These limitations have precluded their use for defining the relative contribution of individual virus transmission modes and generated the need for the development of alternative experimental approaches. In, for example, immunology, significant efforts have been made to establish synthetic 3D systems to complement our portfolio of experimental systems with approaches that offer tight experimental control and direct access to visualization and genetic modification of individual cells. Particularly for lymphocytes and dendritic cells, 3D matrices made of collagen have emerged as the gold standard for such studies [38–40]. Collagen has the advantage of being a major constituent of extracellular matrix, supporting cell viability for up to several weeks, and allowing the easy removal of cells by collagenase digestion. Moreover, the good optical properties of the matrix provide access to live cell microscopy and the architecture of the matrix can be modified easily by adjusting polymerization conditions (collagen concentration, polymerization temperature, and medium used) [41–43]. The bovine and rat tail collagen classically used in 3D culture systems share high amino-acid similarities with the human type I collagen protein (91.1% identity for the alpha I chains and 87.4% for the alpha 2 chains), suggesting that they likely mirror the properties of human collagen. Consistent with this idea, $CD4^+$ cells embedded in bovine and rat collagen matrices adopt migrating behaviors reminiscent of in vivo situations [44]. The xenogeneic effects associated with culturing human cells in this type of culture thus seem to be minimal with respect to cell migration, although species differences in molecular regulation between different collagens cannot be entirely excluded. These aspects allow quantitative insight to be gained into, for example, $CD4^+$ T-cell function, such as their interaction and communication with DCs, or the ability to incorporate specific small RNAs into exosomes that match their behavior in vivo [45,46]. We therefore embarked on an attempt to reconstitute key aspects of HIV-1 spread in target tissue using 3D collagen cultures [47].

3. New Insights from Combining the Study of HIV-1 Spread in 3D Collagen Cultures with Computation

In our recent proof-of-concept study [47], 3D cultures used to study HIV-1 spread included exclusively lymphocytes, considering HIV-1-infected and uninfected $CD4^+$ T cells. Using a reporter virus expressing a sortable cell surface tag enabled the generation of pure HIV-1-infected donor cell populations that were mixed with uninfected target cells at a defined ratio prior to being embedded in 3D collagen. Cells were viable over several weeks in this system and cultures allowed the quantification of viral titers, quantification of expansion or depletion of specific cell populations, and the determination of cell motility and cell–cell contact formation/duration over time. Finally, the system allows for the use of collagen types with different density, with rat and bovine collagen resulting in denser or looser meshworks, respectively [48]. Recording the dynamics of HIV-1 spread and $CD4^+$ T-cell depletion in parallel suspension and 3D cultures immediately revealed that the accurate interpretation of this plethora of quantitative kinetic data required computational approaches. Moreover, addressing the potential relationship between cell motility and virus spread, as well as disentangling the relative contribution of cell-free vs. cell-associated virus transmission to the infection dynamics, was impossible to assess experimentally but required the development of customized computational models. Our analysis relied on several iterative cycles between experimental quantification and computation and resulted in an Integrative method to Study Pathogen spread by Experiment and Computation within Tissue-like 3D cultures(INSPECT-3D) workflow depicted in Figure 2.

Together, the development and first application of INSPECT-3D revealed that HIV-1 spread is delayed relative to suspension cultures when cells are placed in a dense matrix but, after an initial phase characterized by low virus replication, follows kinetics that are similar to those seen in suspension cultures. In contrast, low-density collagen matrices did not support a marked spread of HIV-1. Several findings allowed these surprising observations to be explained. First, 3D collagen, irrespective of its density, potently suppresses the infectivity of cell-free HIV-1 particles (approx. 20-fold). Moreover, virion diffusion rates are too slow to efficiently deliver virus particles to new target cells in the spacing of 3D collagen (i.e., diffusion of virions to the nineteen nearest neighboring cells requires ~22.6 h, in comparison to half-life HIV-1 particle infectivity of $t_{1/2} = 17.9$ h). Together, these results revealed that 3D tissue-like environments potently restrict cell-free HIV-1 infection (see discussion of environmental restriction below). Analysis of the experimental data by mathematical modeling allowed us to estimate the relative contribution of cell-free and cell-associated HIV transmission under these various conditions. Hereby, consistent with the experimental observations, we found that HIV-1 spread in 3D collagen predominantly relies on cell–cell transmission (~78% (73–100%) and ~63% (55.8–100%) of all infections for loose and dense collagen, respectively). In contrast, in suspension cultures, there was no dominance of cell–cell transmission found and its contribution ranged widely between 0 and 100% (Table 1). Viral spread could thus be explained by the exclusive use of cell-free or cell–cell transmission. This revealed that in suspension cultures, no selection pressures for any of the transmission modes exist. This flexibility is consistent with the observation that physical separation of cells in suspension by agitation or transwells, which create conditions in which (i) cell-free infection depends on long-range diffusion of virus particles which compromises their infectivity and (ii) cell–cell contact is limited, markedly reduces virus spread [14,15]. Moreover, our analyses revealed that transmission dynamics substantially differ between suspension cultures and collagen environments.

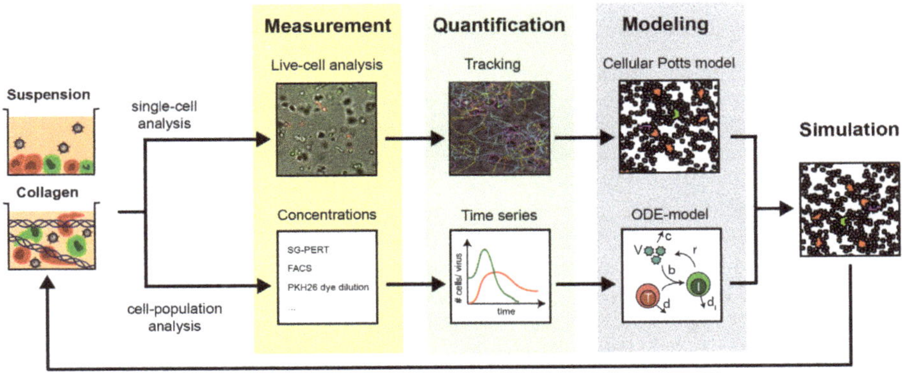

Figure 2. Schematic of the INSPECT-3D workflow that combined single-cell and cell-population measurements by mathematical modeling to reveal the dynamics of infection spread. Infected and non-infected Peripheral Blood Mononuclear Cells (PBMCs) were cultured in suspension or collagen cultures at a 1:20 ratio. In suspension cultures, cells rapidly settled at the bottom of the culture dish to form dense 2D monolayers. In collagen, cells were spatially separated from another, migrated along the 3D scaffold, and eventually made contact with other cells. Measurements from live-cell imaging of single cells and bulk dynamics allowed quantification of individual cell motilities and interactions, as well as long-term kinetics of viral load and cell concentrations. Appropriate mathematical model systems that either rely on spatially resolved cellular Potts models describing single-cell behavior or systems of ordinary differential equations (ODE) enabled us to disentangle processes of cell motility, cellular turnover, and viral transmission and replication. Theses analyses led to an integrative computational model for HIV-1 spread that provides a mechanistic and quantitative understanding of viral transmission within these multicellular systems. Recurrent experimental validation of model predictions by in silico manipulation of experimental conditions was used to refine the model system and reveal key processes governing the dynamics of HIV-1 spread.

Since 3D collagen suppressed cell-free HIV-1 infection and virus spread depended on cell-associated virus transmission, we investigated this scenario in more detail and found that in dense 3D collagen, contacts of extended duration between donor and target cells were more frequent in comparison to in loose collagen. These findings suggested that the dense 3D environment specifically promotes cell-associated HIV-1 transmission, while the low virus replication rates observed in loose collagen stem from the combination of suppression of cell-free infectivity by the 3D matrix and the lack of support for cell-associated virus transmission due to a lack of long-lasting cell–cell contacts. Importantly, the computational model was able to predict the extent to which cell densities had to be increased experimentally to overcome these barriers to HIV-1 spread in loose 3D collagen, suggesting that adopting a 2D monolayer type of configuration could override adverse effects of a 3D matrix on HIV-1 spread. These findings also imply that in a dense matrix, in which cells are not unphysiologically densely packed, cell motility is a key parameter governing the efficacy of HIV-1 spread, i.e., allowing transport of the infection to other areas as cell-free infection is impaired. Quantitative assessments of the relationship of HIV-1 spread in specific target organs will hence require the individual reconstitution of cell and matrix density as well as of the spacing between donor and target cells.

Correlating cell–cell contact duration and productive HIV-1 transmission also allowed us to predict the minimal donor–target cell contact time required for productive infection of new target cells, which was estimated to be in the range of ~25 min. Low-resolution inspection of productive donor–target cell contacts in dense 3D collagen revealed a strikingly close and intimate organization of these contacts, and it will be interesting to dissect how this morphology translates into elevated rates of virus transmission and how dense, but not loose, 3D environments induce these events.

4. How Well Do Experimental and Computational INSPECT-3D Data Match In Vivo?

Although ex vivo 3D collagen cultures allow an improved experimental assessment of physiologically relevant cellular behaviour, they still only provide a reduced representation of relevant tissue conditions and do not completely mirror specific target tissues. However, comparison of the kinetics inferred for viral transmission and cellular turnover dynamics within these culture systems can help to assess their relevance for understanding the spread of HIV-1 in vivo. Mathematical modeling has long been used as an essential tool to dissect the multifactorial complexity of viral transmission and replication dynamics (reviewed in References [49,50]). Standard models of viral dynamics describing the turnover of uninfected and infected cells and the progression of the viral load by systems of ordinary differential equations have been applied to various experimental and clinical data assessing the dynamics of HIV-1 viral spread [51–53]. In particular, these models allowed the quantification of rates describing the kinetics of viral production (ρ) or the death of infected cells (δ) based on patient data. Using HIV-1 plasma RNA levels as an indicator for viral spread in patients, several studies estimated an average production of ~10^{10} HIV-1 virions in total per day, with estimates of the death rate of infected cells between $\delta \approx 0.46–1.40$ day^{-1}, corresponding to average half-lives of infected cells between $t_{1/2}$ = 0.5 and 1.5 days ($t_{1/2}$ = log2/ δ) [51–54]. Studies based on standard in vitro culture systems have produced estimates that are at the upper end of the half-lives determined for infected cells in vivo, with estimates of the corresponding death rates being in the order of $\delta = 0.5 +/- 0.1$ day^{-1} [55]. The estimates on the death rate of infected cells assessed via the INSPECT-3D workflow are within the same range, with a slight tendency towards higher death rates within 3D collagen environments compared to 2D suspension cultures (δ = 0.46–0.56 day^{-1} in dense and loose collagen vs. δ = 0.40–0.44 day^{-1} in suspension; ranges based on 95% confidence intervals of estimates, Table 1). Although the differences are minor, this might indicate that environmental factors, in combination with additional clearance mechanisms such as immune responses, could also contribute to the shorter half-lives of infected cells observed in vivo. While mathematical analyses of HIV-1 viral load kinetics in patients usually allow an appropriate quantification of viral and infected cell decay dynamics, as they work under the assumption that Highly Active Antiretroviral Therapy (HAART) effectively blocks novel infections [50,52], assessing the kinetics of viral transmission and spread in vivo is much more difficult. This requires an approximation of the available number of target cells, which might vary dependent on the time-point of infection and the different compartments of viral replication [56]. In vitro systems, with their associated complete knowledge of the initial experimental conditions, provide an advantage for assessing these dynamics, but have to account for the general lack of potential physiological relevance. As cell-free and cell-associated transmission modes are synergistic, unraveling their relative contributions to viral spread in vitro usually requires the impairment of at least one of them [4,55,57–59]. Iwami et al. [55] set out to assess the transmission rates for HIV-1 using shaken suspension cultures where cell–cell transmission was considered to be impaired [14]. Based on their data, they estimated that cell-associated viral transmission contributed to approximately 60% of viral infections. Without the assumption of efficiently blocking one transmission mode, we found similar contributions of cell-associated transmission modes to viral spread within 3D collagen (with only ~22% (0%, 27%) and ~37% (0%, 44.2%) of infections within loose and dense collagen, respectively, due to cell-free transmission). Thus, there was a clear difference in this contribution compared to spread within 2D suspension cultures. The simultaneous analysis of all three environments revealed a ~4–7-fold higher probability of cells becoming infected by cell–cell transmission within dense and loose collagen compared to suspension. The impaired contribution of cell-free transmission within 3D collagen compared to suspension cultures is further indicated by a ~50% reduction in viral production rates, and a measured relative infectivity of virions that was reduced to ~14% within collagen (β_f, Table 1). In addition to previous reports that blocking $CD4^+$ T-cell exit from lymph nodes restricted HIV-1 spread [27], these findings obtained for viral spread within 3D collagen environments support the argument that cell–cell transmission is the main contributor to HIV-1 spread in vivo. The importance of cell-contact-dependent transmission modes in tissue-like conditions is further indicated by the increased fraction of $CD4^+$ T cells predicted to be initially refractory to infection in collagen (loose ~50%; dense ~30%) compared to suspension (~17%; fractions at 2.5 days post infection). This could point towards cells that are inaccessible for contact-dependent transmission modes due to physical barriers, as they are blocked by collagen within certain areas of the culture.

Table 1. Estimates for HIV-1 infection dynamics obtained from the analysis of experimental infection in cell lines or primary cells and in vivo patient data. Obtained estimates vary over several orders of magnitude due to different mathematical models and data types used. Please note that viral production rates vary in terms of units due to different quantification methods used. Numbers marked with (*) indicate estimates obtained for Simian Immunodeficiency Virus (SIV) infection.

Parameter	Description	Unit	Cell Lines	Primary Cells [47]			In Vivo
				2D Suspension	3D Collagen		
					Loose	Dense	
ρ	Viral production rate	$\times 10^4$ day^{-1}	2.61 (1.55, 3.70) (RNA copies) [60]	1.02 (0.80, 1.38) (RT cell^{-1})	0.48 (0.37, 0.66) (RT cell^{-1})	0.43 (0.32, 0.61) (RT cell^{-1})	5.0 * (1.3, 12.0) (virion cell^{-1}) [61]
							0.07–0.34 [56,62]
ρI	Total viral production rate	$\times 10^{10}$ virions day^{-1}					1.03 ± 1.17 [52,53]
δ_I	Death rate of infected cells	day^{-1}	0.50 ± 0.10 [55]	0.42 (0.40, 0.44)	0.48 (0.46, 0.50)	0.52 (0.51, 0.56)	0.48–1.36 ± 0.16 [53,63]
							0.7 ± 0.25 [51]
							1.0 ± 0.3 [54]
							0.88 * ± 0.40 [61]
β_f	Cell-free infection rate	$\times 10^{-5}$ day^{-1}	0.42 ± 0.14 [55] (p24)	2.3 (0, 3.0) (RT)	0.14 × suspension	0.14 × suspension	
β_c	Cell–cell infection rate	$\times 10^{-5}$ (cell × day)$^{-1}$	0.11 ± 0.03 [55]	10^{-6} (0, 7.0)	4.3 (3.6, 5.4)	1.7 (1.4, 2.6)	
P_f	Predicted percentage of infections by cell-free transmission	%	57% ± 7% [55]	99.6% (0.0%, 100%)	22.0% (0.0%, 27.0%)	37% (0.0%, 44.2%)	

Combining live-cell microscopy data with a spatially detailed cellular Potts model in our INSPECT3D workflow also allowed us to estimate a minimal contact duration between infected and uninfected cells required for productive infection of ~25 min. Importantly, this matches very well the time required for the transfer of fluorescent viral material across the VS ex vivo [17,20], indicating that the events observed by these imaging approaches typically result in the productive infection of the target cell.

Based on the relative comparison of the viral kinetics between 2D suspension and 3D culture systems, our study indicated the need to account for physiological conditions when aiming to quantitatively understand viral replication and transmission dynamics in vivo. Arguably, the dense collagen conditions we used [47] might come close to mucosal tissue without reflecting the cellular composition of this tissue environment. Extending the existing culture systems by incorporating additional cell types and molecular factors will be needed to approximate specific tissue conditions. Other relevant physiological parameters that will have to be considered to be implemented to advance the experimental 3D model further include dynamic tissue perfusion, which might impact HIV-1 spread, e.g., by affecting the dynamics of cell-free infections or the frequency and/or stability of cell–cell interactions. It will be interesting to assess how these additional factors might impact the quantification of the viral processes obtained so far.

5. Requirements and Limitations of Current Computational Approaches to Simulating HIV-1 Spread in Tissue

The majority of previous studies combining mathematical models and experimental data have relied on model systems based on ordinary differential equations that use time-resolved measurements of cell or viral concentrations [50]. These analyses have increased our knowledge of various aspects of HIV-1 life-cycle kinetics and transmission dynamics, and are still the method of choice for gathering most types of experimental and clinical data. However, a more detailed analysis, and thus computational representation, of single-cell dynamics within tissues is needed to understand the dynamics of HIV-1 spread within multicellular systems.

Population dynamic models have their limitations when it comes to analyzing cell-based transmission modes, as their ability to describe single-cell transmission dynamics appropriately is, by definition, hampered. Several extensions have been developed that adjust standard models for viral dynamics to account for cell-contact requirements dependent on the tissue environment [64,65]. Nevertheless, the need to clearly quantify the kinetics of cell-associated viral transmission modes, and to determine the contributions of cell-free and cell-associated transmission, as well as local immunity to viral spread, provides novel challenges for computational analysis.

Individual-cell-based models that describe single-cell behavior have been used previously to simulate and analyze viral spread [66]. However, these model systems were mainly used for qualitative assessment of dynamics, e.g., for the spread of HIV-1 [67,68], as appropriate data and tools for model parameterization remained limited. Both limitations have now been overcome. Advanced imaging and visualization techniques used in vitro and in vivo [33,69,70] have increased our ability to observe and quantify cellular behavior and infection processes at a single-cell level. In addition, improved parameter inference tools and modeling systems have enabled us to quantitatively adapt individual-cell-based models to these novel types of data [71,72]. However, several challenges remain to appropriately analyze HIV-1 spread within different tissue environments using mathematical and computational models (Figure 3). A major challenge for computational approaches that analyze the spatiotemporal dynamics of HIV-1 spread is the appropriate representation of cell motility and the underlying tissue architecture, e.g., as represented by the collagen matrix within 3D ex vivo culture systems. Determining the influence of the environment on cell and infection dynamics requires a detailed description of these matrices and networks that structure the tissue or culture system. Previous attempts to describe the topology of the fibroblastic reticular network within lymph nodes in order to analyze T-cell motility and activation dynamics have shown the difficulty of such tasks [73–76]. High-resolution images of collagen matrices and fibroblastic reticular networks help to improve our understanding of the structures shaping and influencing tissue architecture and cell motility [48]. Nevertheless, the identification of appropriate quantities for network description, such as the pore size, fiber length, and connectivity, and how they relate to network density [42,43,77–79], is also required in order to provide a reliable computational representation of 3D tissue environments that accounts for the different resolutions of fibers and cells [73,76,80–82]. The challenges of describing cell motility based on live-cell microscopy data, and possible ways to refine these analyses, have been already discussed elsewhere [83]. Considering relevant physiological tissue conditions will also have to account for various cell types, such as T cells, dendritic cells, and macrophages, and requires a detailed characterization of their complex morphology and dynamics in 3D. In particular, this also applies to the appropriate representation of the diffusion dynamics of soluble factors, such as virions or chemokines, that are important for cell–cell communication, infection spread, or cell movement. Being able to measure gradients and concentrations of these factors, especially their release dynamics from single cells, would help to mimic their dynamics and to determine the influence of these factors on viral spread and the efficacy of the counteracting immune response [84].

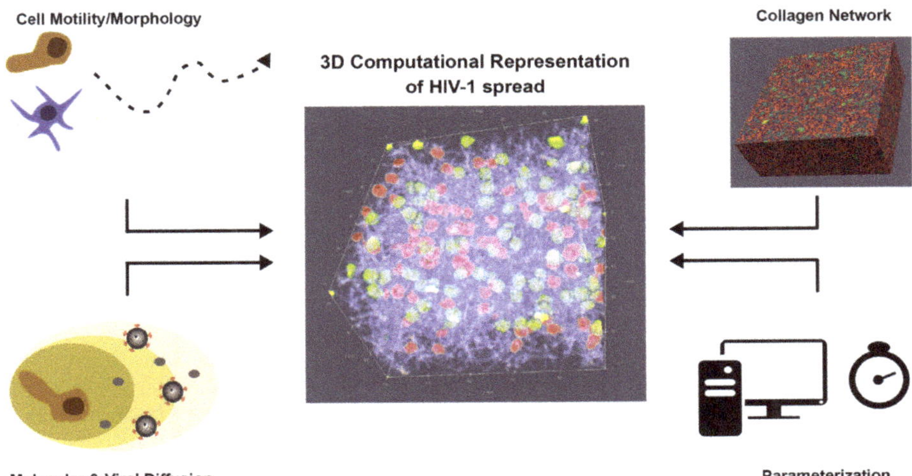

Figure 3. 3D computational representation of HIV-1 spread: Appropriate representations to infer the processes that govern the dynamics of infection require a detailed description and quantification of individual cell motility and morphology, the underlying tissue structure, and the diffusion dynamics of soluble components (chemokines, virions, etc.) within these environments. Data-driven parameterization and subsequent simulation of these 3D model systems requires computationally efficient methods. The sketch shows a computational representation of 100 HIV-1-infected (green) and 100 uninfected (red) cells within a 3D collagen matrix (blue fibers).

Finally, a major challenge for the computational modeling of 3D tissue conditions on a single-cell level is the increased level of detail required for its description, and especially its parameterization. The higher complexity of the model systems and the adaptation of their stochastic dynamics to heterogeneous experimental data requires the use of increased computational resources, including high-performance computing clusters and sophisticated parameter inference tools. Improved parallelization and advanced parameter estimation methods, such as approximated Bayesian computing [72], could help to reduce computational run time costs. In addition, simulation areas in terms of the size and number of cells considered have to be chosen with care in order to balance the need for sufficient tissue volumes that account for stochastic variability while simultaneously ensuring computational efficiency for the simulation of long-term infection dynamics. Thus, novel advances in imaging techniques and computational methods will improve our ability to conduct data-driven computational modeling of HIV-1 spread within multicellular systems, which will be essential to disentangling and understanding the processes that shape these dynamics within physiologically relevant conditions.

6. Environmental Restriction to Cell-Free HIV-1 Infection

A central finding of the studies on HIV-1 spread in 3D collagen is that residing within such a matrix reduces the infectivity of cell-free virus particles approx. 20-fold [47]. Since single-particle tracking has revealed that particles frequently undergo short and transient interactions with the collagen matrix, this negative impact on particle infectivity likely represents the consequence of physical stress imposed on virions in the context of these interactions. It is noteworthy that marked effects of the tissue environment on virus infectivity have been reported. One example is the interaction of viruses with bile acids, which can induce fusogenicity (e.g., norovirus [85]) or restrict infection (cytomegalovirus [86]; Hepatitis Delta virus [87]). Similarly, seminal fluids can impact virion infectivity via peptide-mediated enhancement of virus infectivity (e.g., retroviruses including HIV-1 [88], Ebola virus [89]) or by affecting particle aggregation [90,91]. Since the infectivity reduction in 3D collagen results from negative physical impact by the surrounding matrix and thus represents a direct effect of tissue architecture rather than the consequence of compounds released in the extracellular space, we propose the novel term "environmental restriction" for this phenomenon. Considering that this environmental restriction seems to represent a built-in antiviral activity of the 3D collagen matrix, this is conceptually similar to intrinsic immune barriers such as host cell restriction factors, and could thus be viewed as a novel element of innate immunity (Figure 4). In contrast to cell-autonomous mechanisms exerted by, for example, restriction factors, this activity would reflect a tissue-autonomous restriction. Since the restriction of virus spread by extracellular tissue architecture has only recently begun to become apparent, we can only speculate about the effector function mediating the restriction. Such environmental restrictions could include direct effects on the pathogen, such as the impairment of cell-free infectivity [92] (see below), but could also impact the host cells (e.g., via the promotion of longer-lasting cell contacts with architecture optimized for virus transmission). It can also be envisioned that, for example, transcriptional and epigenetic changes resulting from mechanosensing of the cells in the 3D environment would affect expression and activity of pro- and antiviral host cell factors. Assessing the effect of tissue-like 3D environments on the permissivity of various HIV-1 target cells for infection and their ability to sense the infection will be an important area of future research. Cell-associated HIV-1 transmission is currently considered to be beneficial for the virus, as it allows for more polarized and therefore efficient short-term transfer of particles to target cells, allows host cell restriction factors such as tetherin to be partially overcome, and avoids the recognition of certain neutralizing antibodies [25,93–95]. In addition, we propose that cell-associated HIV-1 transmission evolved from the need to bypass the environmental restriction against cell-free virion infectivity exerted by HIV-1 target tissue.

Following the initial description of the environmental restrictions on cell-free HIV-1 infectivity in 3D collagen cultures, it will be important to dissect the mechanism by which tissue-like 3D matrices impair the infectivity of cell-free virions. Intuitively, physical interactions with the matrix could enhance shedding of the Env glycoprotein. However, overall Env levels in virions were unaffected by interactions of virus particles with the 3D matrix. Alternatively, the 3D matrix could affect virion infectivity by alteration of their aggregation state [90]. However, single-particle tracking of HIV-1 particles in 3D matrices did not reveal virion aggregates in either suspension or 3D collagen, suggesting that HIV-1 particles do not form aggregates under these experimental conditions. Finally, the reduced particle infectivity in 3D collagen may also reflect adaptation of the producer cells to the 3D environment; however, the reduction of virion infectivity in 3D collagen was similar when particles were produced within the matrix or embedded in a cell-free matrix, indicating that the matrix exerts direct effects on HIV-1 virions. Future studies will be thus required to dissect whether the 3D matrix affects HIV-1 infection at the level of particle entry into target cells, e.g., by affecting the conformation of Env glycoproteins on virions, or at subsequent post entry steps (Figure 5). Assessing how conserved this effect may be among viruses other than HIV-1 will be helpful in dissecting the underlying mechanisms. Another key question is whether this barrier is indeed in place in HIV-1 target tissue in infected individuals. Analyzing this question is complicated by the fact that HIV-1 particles derived from patient samples represent a pool of particles of heterogeneous and undefined origin, and that potential effects of environmental tissue restrictions could not easily be distinguished from effects due to soluble factors present in bodily fluids. Organotypic cultures will likely be instrumental for addressing this question.

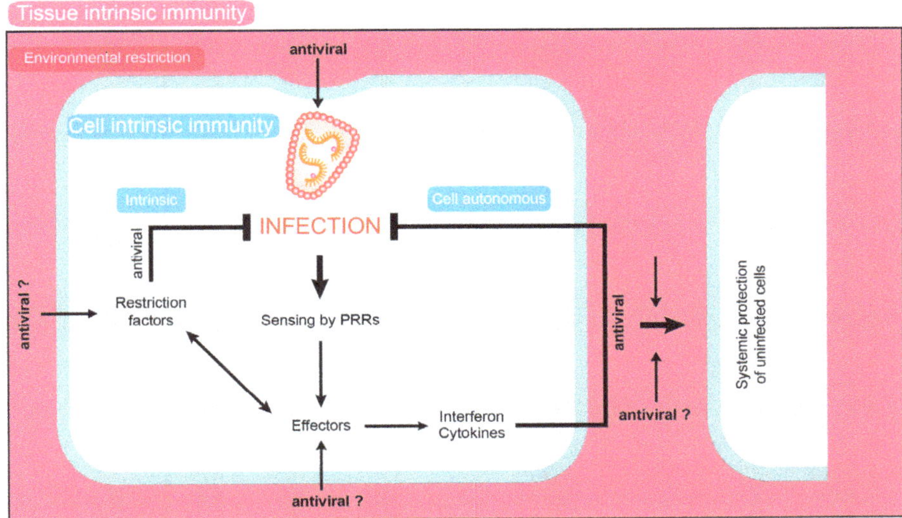

Figure 4. Schematic representation of the environmental restriction concept. Human cells bear cell-intrinsic mechanisms to limit virus spread, such as intrinsic immunity by restriction factors and cell-autonomous immunity by recognition of viral components by pattern-recognition receptors (PRRs). Both pathways can lead to the induction of antiviral effectors such as interferons and cytokines, which can impair virus replication in both the infected and bystander cells. In addition, we propose that the tissue environment can provide tissue-intrinsic immunity by exerting environmental restrictions on virus replication. Such activities could be exerted via direct effects on virus particles, but we also hypothesize that tissue interactions impact the expression and activity of cell-autonomous immune mechanisms.

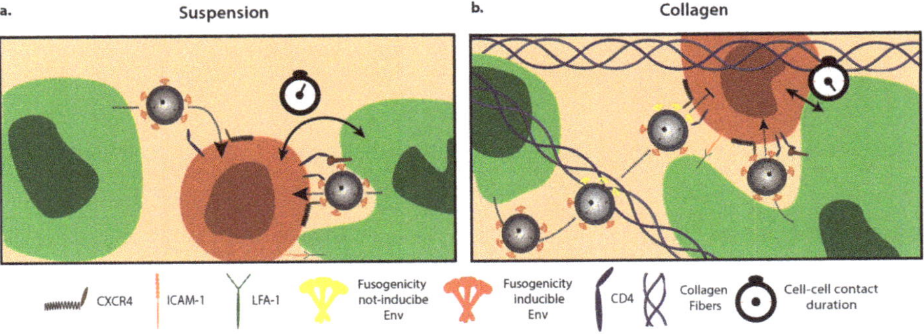

Figure 5. HIV-1 spread in suspension and 3D collagen cultures. (**a**) In suspension cultures, virions can be transmitted to uninfected target cells via both cell-free and cell-associated modes of transmission. (**b**) In collagen cultures, cell-associated transmission is the main driving force for HIV-1 spread. This reflects on one hand that the infectivity of cell-free particles is impaired, which coincides with frequent transient interactions with collagen fibres. These interactions could impair the ability of Env proteins to adopt a fusogenic conformation in response to interaction with the HIV-1 receptor/coreceptor complex on target cells, and thus inhibit virus entry. Moreover, cell-associated HIV-1 transmission may benefit from alterations in the architecture of contacts between donor and target cells, which are more stable and involve larger areas of membrane compared to cells in suspension.

7. Conclusions and Future Perspectives

The efficacy of virus spread in vivo is governed by a complex array of parameters that can only be disentangled by the combined application of synthetic and organoid 3D models and computational analyses. Our proof-of-concept study on HIV-1 spread in 3D collagen cultures revealed a significant impact of this tissue-like environment on modes and efficacy of HIV-1 spread, and suggests the existence of tissue-intrinsic mechanisms that suppress viral replication (environmental restriction). With the constant development of physiological ex vivo 3D culture systems and improved computational analyses, this approach is likely to continue to reveal more relevant details about the mechanisms of virus spread in vivo. An important aspect of such future efforts will be to decipher the contribution of antibody-mediated neutralization of HIV-1 during cell–cell spread (reviewed in Reference [96]) by extending the mathematical model and experimental validations within the INSPECT-3D workflow. Moreover, exosomes that resemble retroviral particles can also play a role in viral spread by affecting susceptibility to infection or the life-span of bystander cells [97–99]. However, the impact of tissue environments on these activities is unexplored. Finally, the INSPECT-3D workflow can directly be used to investigate the spread of other T-cell-tropic viruses, such as HTLV-1, for which the concept of cell-associated transmission via a VS was first established [10], and for which infection is thought to almost exclusively depend on cell–cell transmission, even in suspension cultures [100,101].

Author Contributions: O.T.F. and F.G. conceptualized the review. S.S.A., F.G. and O.T.F. wrote the manuscript. N.B. provided important intellectual content and contributed to the figures. All authors have read and agreed to the published version of the manuscript.

Funding: This work was funded by the Deutsche Forschungsgemeinschaft (German research foundation, DFG—Project number 240245660—SFB1129 (project 8) to O.T.F and by the Center for Modeling and Simulation in the Biosciences (BIOMS) to F.G. F.G. is member of the IWR and additionally supported by a Fellowship from the Chica and Heinz Schaller Foundation.

Conflicts of Interest: The authors declare to have no conflict of interest.

References

1. Zhong, P.; Agosto, L.M.; Munro, J.B.; Mothes, W. Cell-to-cell transmission of viruses. *Curr. Opin. Virol.* **2013**, *3*, 44–50. [CrossRef] [PubMed]
2. Sattentau, Q. Avoiding the void: Cell-to-cell spread of human viruses. *Nat. Rev. Microbiol.* **2008**, *6*, 815–826. [CrossRef] [PubMed]
3. Roberts, K.L.; Smith, G.L. Vaccinia virus morphogenesis and dissemination. *Trends Microbiol.* **2008**, *16*, 472–479. [CrossRef] [PubMed]
4. Graw, F.; Martin, D.N.; Perelson, A.S.; Uprichard, S.L.; Dahari, H. Quantification of Hepatitis C Virus Cell-to-Cell Spread Using a Stochastic Modeling Approach. *J. Virol.* **2015**, *89*, 6551–6561. [CrossRef] [PubMed]
5. Polcicova, K.; Goldsmith, K.; Rainish, B.L.; Wisner, T.W.; Johnson, D.C. The extracellular domain of herpes simplex virus gE is indispensable for efficient cell-to-cell spread: Evidence for gE/gI receptors. *J. Virol.* **2005**, *79*, 11990–12001. [CrossRef] [PubMed]
6. Imai, S.; Nishikawa, J.; Takada, K. Cell-to-cell contact as an efficient mode of Epstein-Barr virus infection of diverse human epithelial cells. *J. Virol.* **1998**, *72*, 4371–4378. [CrossRef] [PubMed]
7. Yang, C.F.; Tu, C.H.; Lo, Y.P.; Cheng, C.C.; Chen, W.J. Involvement of Tetraspanin C189 in Cell-to-Cell Spreading of the Dengue Virus in C6/36 Cells. *PLoS Negl. Trop. Dis.* **2015**, *9*, e0003885. [CrossRef]
8. Jolly, C.; Kashefi, K.; Hollinshead, M.; Sattentau, Q.J. HIV-1 cell to cell transfer across an Env-induced, actin-dependent synapse. *J. Exp. Med.* **2004**, *199*, 283–293. [CrossRef]
9. Bracq, L.; Xie, M.; Benichou, S.; Bouchet, J. Mechanisms for Cell-to-Cell Transmission of HIV-1. *Front. Immunol.* **2018**, *9*, 260. [CrossRef]
10. Igakura, T.; Stinchcombe, J.C.; Goon, P.K.; Taylor, G.P.; Weber, J.N.; Griffiths, G.M.; Tanaka, Y.; Osame, M.; Bangham, C.R. Spread of HTLV-I between lymphocytes by virus-induced polarization of the cytoskeleton. *Science* **2003**, *299*, 1713–1716. [CrossRef]

11. Pais-Correia, A.M.; Sachse, M.; Guadagnini, S.; Robbiati, V.; Lasserre, R.; Gessain, A.; Gout, O.; Alcover, A.; Thoulouze, M.I. Biofilm-like extracellular viral assemblies mediate HTLV-1 cell-to-cell transmission at virological synapses. *Nat. Med.* **2010**, *16*, 83–89. [CrossRef] [PubMed]
12. Gross, C.; Thoma-Kress, A.K. Molecular Mechanisms of HTLV-1 Cell-to-Cell Transmission. *Viruses* **2016**, *8*, 74. [CrossRef] [PubMed]
13. Dimitrov, D.S.; Willey, R.L.; Sato, H.; Chang, L.J.; Blumenthal, R.; Martin, M.A. Quantitation of human immunodeficiency virus type 1 infection kinetics. *J. Virol.* **1993**, *67*, 2182–2190. [CrossRef] [PubMed]
14. Sourisseau, M.; Sol-Foulon, N.; Porrot, F.; Blanchet, F.; Schwartz, O. Inefficient human immunodeficiency virus replication in mobile lymphocytes. *J. Virol.* **2007**, *81*, 1000–1012. [CrossRef]
15. Orlandi, C.; Flinko, R.; Lewis, G.K. A new cell line for high throughput HIV-specific antibody-dependent cellular cytotoxicity (ADCC) and cell-to-cell virus transmission studies. *J. Immunol. Methods* **2016**, *433*, 51–58. [CrossRef]
16. Xie, M.; Leroy, H.; Mascarau, R.; Woottum, M.; Dupont, M.; Ciccone, C.; Schmitt, A.; Raynaud-Messina, B.; Verollet, C.; Bouchet, J.; et al. Cell-to-Cell Spreading of HIV-1 in Myeloid Target Cells Escapes SAMHD1 Restriction. *Mbio* **2019**, *10*. [CrossRef]
17. Hubner, W.; McNerney, G.P.; Chen, P.; Dale, B.M.; Gordon, R.E.; Chuang, F.Y.; Li, X.D.; Asmuth, D.M.; Huser, T.; Chen, B.K. Quantitative 3D video microscopy of HIV transfer across T cell virological synapses. *Science* **2009**, *323*, 1743–1747. [CrossRef]
18. Sowinski, S.; Jolly, C.; Berninghausen, O.; Purbhoo, M.A.; Chauveau, A.; Kohler, K.; Oddos, S.; Eissmann, P.; Brodsky, F.M.; Hopkins, C.; et al. Membrane nanotubes physically connect T cells over long distances presenting a novel route for HIV-1 transmission. *Nat. Cell Biol.* **2008**, *10*, 211–219. [CrossRef]
19. Felts, R.L.; Narayan, K.; Estes, J.D.; Shi, D.; Trubey, C.M.; Fu, J.; Hartnell, L.M.; Ruthel, G.T.; Schneider, D.K.; Nagashima, K.; et al. 3D visualization of HIV transfer at the virological synapse between dendritic cells and T cells. *Proc. Natl. Acad. Sci. USA* **2010**. [CrossRef]
20. Real, F.; Sennepin, A.; Ganor, Y.; Schmitt, A.; Bomsel, M. Live Imaging of HIV-1 Transfer across T Cell Virological Synapse to Epithelial Cells that Promotes Stromal Macrophage Infection. *Cell Rep.* **2018**, *23*, 1794–1805. [CrossRef]
21. McDonald, D.; Wu, L.; Bohks, S.M.; KewalRamani, V.N.; Unutmaz, D.; Hope, T.J. Recruitment of HIV and its receptors to dendritic cell-T cell junctions. *Science* **2003**, *300*, 1295–1297. [CrossRef] [PubMed]
22. Earl, L.A.; Lifson, J.D.; Subramaniam, S. Catching HIV 'in the act' with 3D electron microscopy. *Trends Microbiol.* **2013**, *21*, 397–404. [CrossRef] [PubMed]
23. Chen, P.; Hubner, W.; Spinelli, M.A.; Chen, B.K. Predominant mode of human immunodeficiency virus transfer between T cells is mediated by sustained Env-dependent neutralization-resistant virological synapses. *J. Virol.* **2007**, *81*, 12582–12595. [CrossRef] [PubMed]
24. Martin, N.; Welsch, S.; Jolly, C.; Briggs, J.A.; Vaux, D.; Sattentau, Q.J. Virological synapse-mediated spread of human immunodeficiency virus type 1 between T cells is sensitive to entry inhibition. *J. Virol.* **2010**, *84*, 3516–3527. [CrossRef]
25. Kolodkin-Gal, D.; Hulot, S.L.; Korioth-Schmitz, B.; Gombos, R.B.; Zheng, Y.; Owuor, J.; Lifton, M.A.; Ayeni, C.; Najarian, R.M.; Yeh, W.W.; et al. Efficiency of cell-free and cell-associated virus in mucosal transmission of human immunodeficiency virus type 1 and simian immunodeficiency virus. *J. Virol.* **2013**, *87*, 13589–13597. [CrossRef]
26. Sewald, X.; Gonzalez, D.G.; Haberman, A.M.; Mothes, W. In vivo imaging of virological synapses. *Nat. Commun.* **2012**, *3*, 1320. [CrossRef]
27. Murooka, T.T.; Deruaz, M.; Marangoni, F.; Vrbanac, V.D.; Seung, E.; von Andrian, U.H.; Tager, A.M.; Luster, A.D.; Mempel, T.R. HIV-infected T cells are migratory vehicles for viral dissemination. *Nature* **2012**, *490*, 283–287. [CrossRef]
28. Ladinsky, M.S.; Khamaikawin, W.; Jung, Y.; Lin, S.; Lam, J.; An, D.S.; Bjorkman, P.J.; Kieffer, C. Mechanisms of virus dissemination in bone marrow of HIV-1-infected humanized BLT mice. *ELife* **2019**, *8*. [CrossRef]
29. Sewald, X.; Ladinsky, M.S.; Uchil, P.D.; Beloor, J.; Pi, R.; Herrmann, C.; Motamedi, N.; Murooka, T.T.; Brehm, M.A.; Greiner, D.L.; et al. Retroviruses use CD169-mediated trans-infection of permissive lymphocytes to establish infection. *Science* **2015**, *350*, 563–567. [CrossRef]
30. Miyauchi, K.; Kim, Y.; Latinovic, O.; Morozov, V.; Melikyan, G.B. HIV enters cells via endocytosis and dynamin-dependent fusion with endosomes. *Cell* **2009**, *137*, 433–444. [CrossRef]

31. Dale, B.M.; McNerney, G.P.; Thompson, D.L.; Hubner, W.; de Los Reyes, K.; Chuang, F.Y.; Huser, T.; Chen, B.K. Cell-to-cell transfer of HIV-1 via virological synapses leads to endosomal virion maturation that activates viral membrane fusion. *Cell Host Microbe* **2011**, *10*, 551–562. [CrossRef] [PubMed]
32. Melikyan, G.B. HIV entry: A game of hide-and-fuse? *Curr. Opin. Virol.* **2014**, *4*, 1–7. [CrossRef] [PubMed]
33. Fackler, O.T.; Murooka, T.T.; Imle, A.; Mempel, T.R. Adding new dimensions: Towards an integrative understanding of HIV-1 spread. *Nat. Rev. Microbiol.* **2014**, *12*, 563–574. [CrossRef] [PubMed]
34. Marsden, M.D.; Zack, J.A. Humanized Mouse Models for Human Immunodeficiency Virus Infection. *Annu. Rev. Virol.* **2017**, *4*, 393–412. [CrossRef] [PubMed]
35. Rossi, G.; Manfrin, A.; Lutolf, M.P. Progress and potential in organoid research. *Nat. Rev. Genet.* **2018**, *19*, 671–687. [CrossRef] [PubMed]
36. Bar-Ephraim, Y.E.; Kretzschmar, K.; Clevers, H. Organoids in immunological research. *Nat. Rev. Immunol.* **2019**. [CrossRef]
37. Merbah, M.; Introini, A.; Fitzgerald, W.; Grivel, J.C.; Lisco, A.; Vanpouille, C.; Margolis, L. Cervico-vaginal tissue ex vivo as a model to study early events in HIV-1 infection. *Am. J. Reprod. Immunol.* **2011**, *65*, 268–278. [CrossRef]
38. Sixt, M.; Lammermann, T. In vitro analysis of chemotactic leukocyte migration in 3D environments. *Methods Mol. Biol.* **2011**, *769*, 149–165. [CrossRef]
39. Wolf, K.; Alexander, S.; Schacht, V.; Coussens, L.M.; von Andrian, U.H.; van Rheenen, J.; Deryugina, E.; Friedl, P. Collagen-based cell migration models in vitro and in vivo. *Semin. Cell Dev. Biol.* **2009**, *20*, 931–941. [CrossRef]
40. Lammermann, T.; Bader, B.L.; Monkley, S.J.; Worbs, T.; Wedlich-Soldner, R.; Hirsch, K.; Keller, M.; Forster, R.; Critchley, D.R.; Fassler, R.; et al. Rapid leukocyte migration by integrin-independent flowing and squeezing. *Nature* **2008**, *453*, 51–55. [CrossRef]
41. Miron-Mendoza, M.; Seemann, J.; Grinnell, F. The differential regulation of cell motile activity through matrix stiffness and porosity in three dimensional collagen matrices. *Biomaterials* **2010**, *31*, 6425–6435. [CrossRef] [PubMed]
42. Olivares, V.; Condor, M.; Del Amo, C.; Asin, J.; Borau, C.; Garcia-Aznar, J.M. Image-based Characterization of 3D Collagen Networks and the Effect of Embedded Cells. *Microsc. Microanal.* **2019**, *25*, 971–981. [CrossRef] [PubMed]
43. Jansen, K.A.; Licup, A.J.; Sharma, A.; Rens, R.; MacKintosh, F.C.; Koenderink, G.H. The Role of Network Architecture in Collagen Mechanics. *Biophys. J.* **2018**, *114*, 2665–2678. [CrossRef] [PubMed]
44. Friedl, P.; Brocker, E.B. The biology of cell locomotion within three-dimensional extracellular matrix. *Cell. Mol. Life Sci.* **2000**, *57*, 41–64. [CrossRef]
45. Thippabhotla, S.; Zhong, C.; He, M. 3D cell culture stimulates the secretion of in vivo like extracellular vesicles. *Sci. Rep.* **2019**, *9*, 13012. [CrossRef]
46. Abu-Shah, E.; Demetriou, P.; Balint, S.; Mayya, V.; Kutuzov, M.A.; Dushek, O.; Dustin, M.L. A tissue-like platform for studying engineered quiescent human T-cells' interactions with dendritic cells. *ELife* **2019**, *8*. [CrossRef]
47. Imle, A.; Kumberger, P.; Schnellbacher, N.D.; Fehr, J.; Carrillo-Bustamante, P.; Ales, J.; Schmidt, P.; Ritter, C.; Godinez, W.J.; Muller, B.; et al. Experimental and computational analyses reveal that environmental restrictions shape HIV-1 spread in 3D cultures. *Nat. Commun.* **2019**, *10*, 2144. [CrossRef]
48. Wolf, K.; Te Lindert, M.; Krause, M.; Alexander, S.; Te Riet, J.; Willis, A.L.; Hoffman, R.M.; Figdor, C.G.; Weiss, S.J.; Friedl, P. Physical limits of cell migration: Control by ECM space and nuclear deformation and tuning by proteolysis and traction force. *J. Cell Biol.* **2013**, *201*, 1069–1084. [CrossRef]
49. Graw, F.; Perelson, A.S. Modeling Viral Spread. *Annu. Rev. Virol.* **2016**, *3*, 555–572. [CrossRef]
50. Perelson, A.S. Modelling viral and immune system dynamics. *Nat. Rev. Immunol.* **2002**, *2*, 28–36. [CrossRef]
51. Perelson, A.S.; Neumann, A.U.; Markowitz, M.; Leonard, J.M.; Ho, D.D. HIV-1 dynamics in vivo: Virion clearance rate, infected cell life-span, and viral generation time. *Science* **1996**, *271*, 1582–1586. [CrossRef] [PubMed]
52. Ramratnam, B.; Bonhoeffer, S.; Binley, J.; Hurley, A.; Zhang, L.; Mittler, J.E.; Markowitz, M.; Moore, J.P.; Perelson, A.S.; Ho, D.D. Rapid production and clearance of HIV-1 and hepatitis C virus assessed by large volume plasma apheresis. *Lancet* **1999**, *354*, 1782–1785. [CrossRef]

53. Perelson, A.S.; Essunger, P.; Cao, Y.; Vesanen, M.; Hurley, A.; Saksela, K.; Markowitz, M.; Ho, D.D. Decay characteristics of HIV-1-infected compartments during combination therapy. *Nature* **1997**, *387*, 188–191. [CrossRef] [PubMed]
54. Markowitz, M.; Louie, M.; Hurley, A.; Sun, E.; Di Mascio, M.; Perelson, A.S.; Ho, D.D. A novel antiviral intervention results in more accurate assessment of human immunodeficiency virus type 1 replication dynamics and T-cell decay in vivo. *J. Virol.* **2003**, *77*, 5037–5038. [CrossRef] [PubMed]
55. Iwami, S.; Takeuchi, J.S.; Nakaoka, S.; Mammano, F.; Clavel, F.; Inaba, H.; Kobayashi, T.; Misawa, N.; Aihara, K.; Koyanagi, Y.; et al. Cell-to-cell infection by HIV contributes over half of virus infection. *ELife* **2015**, *4*. [CrossRef]
56. De Boer, R.J.; Ribeiro, R.M.; Perelson, A.S. Current estimates for HIV-1 production imply rapid viral clearance in lymphoid tissues. *PLoS Comput. Biol.* **2010**, *6*, e1000906. [CrossRef]
57. Timpe, J.M.; Stamataki, Z.; Jennings, A.; Hu, K.; Farquhar, M.J.; Harris, H.J.; Schwarz, A.; Desombere, I.; Roels, G.L.; Balfe, P.; et al. Hepatitis C virus cell-cell transmission in hepatoma cells in the presence of neutralizing antibodies. *Hepatology* **2008**, *47*, 17–24. [CrossRef]
58. Barretto, N.; Sainz, B., Jr.; Hussain, S.; Uprichard, S.L. Determining the involvement and therapeutic implications of host cellular factors in hepatitis C virus cell-to-cell spread. *J. Virol.* **2014**, *88*, 5050–5061. [CrossRef]
59. Barretto, N.; Uprichard, S.L. Hepatitis C virus Cell-to-cell Spread Assay. *Bio Protoc.* **2014**, *4*. [CrossRef]
60. Iwami, S.; Holder, B.P.; Beauchemin, C.A.; Morita, S.; Tada, T.; Sato, K.; Igarashi, T.; Miura, T. Quantification system for the viral dynamics of a highly pathogenic simian/human immunodeficiency virus based on an in vitro experiment and a mathematical model. *Retrovirology* **2012**, *9*, 18. [CrossRef]
61. Chen, H.Y.; Di Mascio, M.; Perelson, A.S.; Ho, D.D.; Zhang, L. Determination of virus burst size in vivo using a single-cycle SIV in rhesus macaques. *Proc. Natl. Acad. Sci. USA* **2007**, *104*, 19079–19084. [CrossRef] [PubMed]
62. Reilly, C.; Wietgrefe, S.; Sedgewick, G.; Haase, A. Determination of simian immunodeficiency virus production by infected activated and resting cells. *AIDS (London, England)* **2007**, *21*, 163–168. [CrossRef] [PubMed]
63. Althaus, C.L.; De Vos, A.S.; De Boer, R.J. Reassessing the human immunodeficiency virus type 1 life cycle through age-structured modeling: Life span of infected cells, viral generation time, and basic reproductive number, R0. *J. Virol.* **2009**, *83*, 7659–7667. [CrossRef] [PubMed]
64. Kumberger, P.; Durso-Cain, K.; Uprichard, S.L.; Dahari, H.; Graw, F. Accounting for Space-Quantification of Cell-To-Cell Transmission Kinetics Using Virus Dynamics Models. *Viruses* **2018**, *10*, 200. [CrossRef] [PubMed]
65. Pilyugin, S.S.; Antia, R. Modeling immune responses with handling time. *Bull. Math. Biol.* **2000**, *62*, 869–890. [CrossRef] [PubMed]
66. Bauer, A.L.; Beauchemin, C.A.; Perelson, A.S. Agent-based modeling of host-pathogen systems: The successes and challenges. *Infor. Sci.* **2009**, *179*, 1379–1389. [CrossRef]
67. Strain, M.C.; Richman, D.D.; Wong, J.K.; Levine, H. Spatiotemporal dynamics of HIV propagation. *J. Theor. Biol.* **2002**, *218*, 85–96. [CrossRef]
68. dos Santos, R.M.Z.; Coutinho, S. Dynamics of HIV Infection: A cellular automata approach. *Phys. Rev. Lett.* **2001**, *87*, 168102. [CrossRef]
69. Coombes, J.L.; Robey, E.A. Dynamic imaging of host-pathogen interactions in vivo. *Nat. Rev. Immunol.* **2010**, *10*, 353–364. [CrossRef]
70. Germain, R.N.; Miller, M.J.; Dustin, M.L.; Nussenzweig, M.C. Dynamic imaging of the immune system: Progress, pitfalls and promise. *Nat. Rev. Immunol.* **2006**, *6*, 497–507. [CrossRef]
71. Starruss, J.; de Back, W.; Brusch, L.; Deutsch, A. Morpheus: A user-friendly modeling environment for multiscale and multicellular systems biology. *Bioinformatics* **2014**, *30*, 1331–1332. [CrossRef] [PubMed]
72. Klinger, E.; Rickert, D.; Hasenauer, J. pyABC: Distributed, likelihood-free inference. *Bioinformatics* **2018**, *34*, 3591–3593. [CrossRef] [PubMed]
73. Novkovic, M.; Onder, L.; Cheng, H.W.; Bocharov, G.; Ludewig, B. Integrative Computational Modeling of the Lymph Node Stromal Cell Landscape. *Front. Immunol.* **2018**, *9*, 2428. [CrossRef] [PubMed]
74. Novkovic, M.; Onder, L.; Bocharov, G.; Ludewig, B. Graph Theory-Based Analysis of the Lymph Node Fibroblastic Reticular Cell Network. *Meth. Mol. Biol.* **2017**, *1591*, 43–57. [CrossRef]

75. Novkovic, M.; Onder, L.; Cupovic, J.; Abe, J.; Bomze, D.; Cremasco, V.; Scandella, E.; Stein, J.V.; Bocharov, G.; Turley, S.J.; et al. Topological Small-World Organization of the Fibroblastic Reticular Cell Network Determines Lymph Node Functionality. *PLoS Biol.* **2016**, *14*, e1002515. [CrossRef]
76. Novkovic, M.; Onder, L.; Bocharov, G.; Ludewig, B. Topological Structure and Robustness of the Lymph Node Conduit System. *Cell Rep.* **2020**, *30*, 893–904. [CrossRef]
77. Licup, A.J.; Munster, S.; Sharma, A.; Sheinman, M.; Jawerth, L.M.; Fabry, B.; Weitz, D.A.; MacKintosh, F.C. Stress controls the mechanics of collagen networks. *Proc. Natl. Acad. Sci. USA* **2015**, *112*, 9573–9578. [CrossRef]
78. Lindstrom, S.B.; Vader, D.A.; Kulachenko, A.; Weitz, D.A. Biopolymer network geometries: Characterization, regeneration, and elastic properties. *Phys. Rev. E.* **2010**, *82*, 051905. [CrossRef]
79. Stein, A.M.; Vader, D.A.; Jawerth, L.M.; Weitz, D.A.; Sander, L.M. An algorithm for extracting the network geometry of three-dimensional collagen gels. *J. Microsc.* **2008**, *232*, 463–475. [CrossRef]
80. Harjanto, D.; Zaman, M.H. Modeling extracellular matrix reorganization in 3D environments. *PLoS ONE* **2013**, *8*, e52509. [CrossRef]
81. Graw, F.; Balagopal, A.; Kandathil, A.J.; Ray, S.C.; Thomas, D.L.; Ribeiro, R.M.; Perelson, A.S. Inferring viral dynamics in chronically HCV infected patients from the spatial distribution of infected hepatocytes. *PLoS Computat. Biol.* **2014**, *10*, e1003934. [CrossRef]
82. Beltman, J.B.; Henrickson, S.E.; von Andrian, U.H.; de Boer, R.J.; Maree, A.F. Towards estimating the true duration of dendritic cell interactions with T cells. *J. Immunol. Methods* **2009**, *347*, 54–69. [CrossRef] [PubMed]
83. Beltman, J.B.; Maree, A.F.; de Boer, R.J. Analysing immune cell migration. *Nat. Rev. Immunol.* **2009**, *9*, 789–798. [CrossRef] [PubMed]
84. Howat, T.J.; Barreca, C.; O'Hare, P.; Gog, J.R.; Grenfell, B.T. Modelling dynamics of the type I interferon response to in vitro viral infection. *J. R. Soc. Interface* **2006**, *3*, 699–709. [CrossRef] [PubMed]
85. Nelson, C.A.; Wilen, C.B.; Dai, Y.N.; Orchard, R.C.; Kim, A.S.; Stegeman, R.A.; Hsieh, L.L.; Smith, T.J.; Virgin, H.W.; Fremont, D.H. Structural basis for murine norovirus engagement of bile acids and the CD300lf receptor. *Proc. Natl. Acad. Sci. USA* **2018**, *115*, E9201–E9210. [CrossRef]
86. Schupp, A.K.; Trilling, M.; Rattay, S.; Le-Trilling, V.T.K.; Haselow, K.; Stindt, J.; Zimmermann, A.; Haussinger, D.; Hengel, H.; Graf, D. Bile Acids Act as Soluble Host Restriction Factors Limiting Cytomegalovirus Replication in Hepatocytes. *J. Virol.* **2016**, *90*, 6686–6698. [CrossRef] [PubMed]
87. Veloso Alves Pereira, I.; Buchmann, B.; Sandmann, L.; Sprinzl, K.; Schlaphoff, V.; Dohner, K.; Vondran, F.; Sarrazin, C.; Manns, M.P.; Pinto Marques Souza de Oliveira, C.; et al. Primary biliary acids inhibit hepatitis D virus (HDV) entry into human hepatoma cells expressing the sodium-taurocholate cotransporting polypeptide (NTCP). *PLoS ONE* **2015**, *10*, e0117152. [CrossRef] [PubMed]
88. Munch, J.; Rucker, E.; Standker, L.; Adermann, K.; Goffinet, C.; Schindler, M.; Wildum, S.; Chinnadurai, R.; Rajan, D.; Specht, A.; et al. Semen-derived amyloid fibrils drastically enhance HIV infection. *Cell* **2007**, *131*, 1059–1071. [CrossRef]
89. Bart, S.M.; Cohen, C.; Dye, J.M.; Shorter, J.; Bates, P. Enhancement of Ebola virus infection by seminal amyloid fibrils. *Proc. Natl. Acad. Sci. USA* **2018**, *115*, 7410–7415. [CrossRef]
90. Sanjuan, R. Collective Infectious Units in Viruses. *Trends Microbiol.* **2017**, *25*, 402–412. [CrossRef]
91. Andreu-Moreno, I.; Sanjuan, R. Collective Infection of Cells by Viral Aggregates Promotes Early Viral Proliferation and Reveals a Cellular-L0evel Allee Effect. *Curr. Biol.* **2018**, *28*, 3212–3219. [CrossRef] [PubMed]
92. Zotova, A.; Atemasova, A.; Pichugin, A.; Filatov, A.; Mazurov, D. Distinct Requirements for HIV-1 Accessory Proteins during Cell Coculture and Cell-Free Infection. *Viruses* **2019**, *11*, 390. [CrossRef] [PubMed]
93. Malbec, M.; Porrot, F.; Rua, R.; Horwitz, J.; Klein, F.; Halper-Stromberg, A.; Scheid, J.F.; Eden, C.; Mouquet, H.; Nussenzweig, M.C.; et al. Broadly neutralizing antibodies that inhibit HIV-1 cell to cell transmission. *J. Exp. Med.* **2013**, *210*, 2813–2821. [CrossRef] [PubMed]
94. Zhong, P.; Agosto, L.M.; Ilinskaya, A.; Dorjbal, B.; Truong, R.; Derse, D.; Uchil, P.D.; Heidecker, G.; Mothes, W. Cell-to-Cell Transmission Can Overcome Multiple Donor and Target Cell Barriers Imposed on Cell-Free HIV. *PLoS ONE* **2013**, *8*, e53138. [CrossRef]
95. Jolly, C.; Booth, N.J.; Neil, S.J. Cell-cell spread of human immunodeficiency virus type 1 overcomes tetherin/BST-2-mediated restriction in T cells. *J. Virol.* **2010**, *84*, 12185–12199. [CrossRef]
96. Dufloo, J.; Bruel, T.; Schwartz, O. HIV-1 cell-to-cell transmission and broadly neutralizing antibodies. *Retrovirology* **2018**, *15*, 51. [CrossRef]

97. Chen, P.; Chen, B.K.; Mosoian, A.; Hays, T.; Ross, M.J.; Klotman, P.E.; Klotman, M.E. Virological synapses allow HIV-1 uptake and gene expression in renal tubular epithelial cells. *J. Am. Soc. Nephrol.* **2011**, *22*, 496–507. [CrossRef]
98. Lee, J.H.; Schierer, S.; Blume, K.; Dindorf, J.; Wittki, S.; Xiang, W.; Ostalecki, C.; Koliha, N.; Wild, S.; Schuler, G.; et al. HIV-Nef and ADAM17-Containing Plasma Extracellular Vesicles Induce and Correlate with Immune Pathogenesis in Chronic HIV Infection. *EBioMedicine* **2016**, *6*, 103–113. [CrossRef]
99. Ostalecki, C.; Wittki, S.; Lee, J.H.; Geist, M.M.; Tibroni, N.; Harrer, T.; Schuler, G.; Fackler, O.T.; Baur, A.S. HIV Nef- and Notch1-dependent Endocytosis of ADAM17 Induces Vesicular TNF Secretion in Chronic HIV Infection. *EBioMedicine* **2016**, *13*, 294–304. [CrossRef]
100. Mazurov, D.; Ilinskaya, A.; Heidecker, G.; Lloyd, P.; Derse, D. Quantitative comparison of HTLV-1 and HIV-1 cell-to-cell infection with new replication dependent vectors. *PLoS Pathog.* **2010**, *6*, e1000788. [CrossRef]
101. Shunaeva, A.; Potashnikova, D.; Pichugin, A.; Mishina, A.; Filatov, A.; Nikolaitchik, O.; Hu, W.S.; Mazurov, D. Improvement of HIV-1 and Human T Cell Lymphotropic Virus Type 1 Replication-Dependent Vectors via Optimization of Reporter Gene Reconstitution and Modification with Intronic Short Hairpin RNA. *J. Virol.* **2015**, *89*, 10591–10601. [CrossRef] [PubMed]

Article

Teratogenic Rubella Virus Alters the Endodermal Differentiation Capacity of Human Induced Pluripotent Stem Cells

Nicole C. Bilz [1,†], Edith Willscher [2,†], Hans Binder [2], Janik Böhnke [1,‡], Megan L. Stanifer [3], Denise Hübner [1], Steeve Boulant [3,4], Uwe G. Liebert [1] and Claudia Claus [1,*]

1 Institute of Virology, University of Leipzig, 04103 Leipzig, Germany
2 Interdisciplinary Center for Bioinformatics, University of Leipzig, 04107 Leipzig, Germany
3 Schaller Research Group at CellNetworks, Department of Infectious Diseases, Virology, Heidelberg University Hospital, 69120 Heidelberg, Germany
4 Research Group "Cellular Polarity and Viral Infection" (F140), German Cancer Research Center (DKFZ), 69120 Heidelberg, Germany
* Correspondence: claudia.claus@medizin.uni-leipzig.de; Tel.: +49-341-9714321
† These authors contributed equally.
‡ Current address: Institute for Biomedical Engineering, Department of Cell Biology, RWTH Aachen University, Medical School, 52074 Aachen, Germany.

Received: 14 July 2019; Accepted: 7 August 2019; Published: 10 August 2019

Abstract: The study of congenital virus infections in humans requires suitable ex vivo platforms for the species-specific events during embryonal development. A prominent example for these infections is rubella virus (RV) which most commonly leads to defects in ear, heart, and eye development. We applied teratogenic RV to human induced pluripotent stem cells (iPSCs) followed by differentiation into cells of the three embryonic lineages (ecto-, meso-, and endoderm) as a cell culture model for blastocyst- and gastrulation-like stages. In the presence of RV, lineage-specific differentiation markers were expressed, indicating that lineage identity was maintained. However, portrait analysis of the transcriptomic expression signatures of all samples revealed that mock- and RV-infected endodermal cells were less related to each other than their ecto- and mesodermal counterparts. Markers for definitive endoderm were increased during RV infection. Profound alterations of the epigenetic landscape including the expression level of components of the chromatin remodeling complexes and an induction of type III interferons were found, especially after endodermal differentiation of RV-infected iPSCs. Moreover, the eye field transcription factors RAX and SIX3 and components of the gene set vasculogenesis were identified as dysregulated transcripts. Although iPSC morphology was maintained, the formation of embryoid bodies as three-dimensional cell aggregates and as such cellular adhesion capacity was impaired during RV infection. The correlation of the molecular alterations induced by RV during differentiation of iPSCs with the clinical signs of congenital rubella syndrome suggests mechanisms of viral impairment of human development.

Keywords: ectoderm; mesoderm; human development; embryogenesis; interferon response; interferon-induced genes; self-organizing map (SOM) data portrayal; epigenetic signature; embryoid body; TGF-β and Wnt/β-catenin pathway

---

1. Introduction

The enveloped, single stranded (positive-sense) RNA virus rubella virus (RV) of the genus *Rubivirus* within the family *Togaviridae* is one of the few viruses that can cause an intrauterine infection. How these viruses are transmitted vertically from the infected mother to the fetus and how they impact human development is only partially resolved. In the case of the very efficient teratogen

RV, the human-specific symptoms are categorized as congenital rubella syndrome (CRS) with the classical triad of clinical symptoms being sensorineural deafness, congenital heart disease (including cardiovascular and vascular anomalies), and cataracts [1,2]. Heart defects in CRS may comprise ventricular/atrial septal defects, patent ductus arteriosus, and patent foramen ovale. In congenital rubella, ocular (ophthalmic) pathologies include cataract, microphthalmia, glaucoma, and pigmentary retinopathy [1,2]. Furthermore, in tissue samples from three fatal CRS cases RV was detected in cardiac and adventitia (aorta and pulmonary artery) fibroblasts in association with vascular lesions [3]. The risk for the development of congenital defects is especially prevalent during maternal rubella until gestational week 11 and 12 [4–6]. Thus, intrauterine RV infection is only of concern during the first trimester. While congenital malformations are common, premature delivery and stillbirths are not markedly increased after intrauterine RV infection [1].

There are a number of ethical constraints associated with the study of human embryogenesis and congenital malformations, especially as early implantation stages of human embryos are inaccessible [7]. With embryonic stem cells (ESCs) and induced pluripotent stem cells (iPSCs), as the two types of human pluripotent stem cells (PSCs), these novel ex vivo cell culture platforms allow for the analysis of human embryonic germ layer segregation and as well as for developmental toxicity testing [8]. As a cell culture model, they represent a blastocyst-like stage, which can be extended to gastrulation-like stages through their differentiation into derivatives of the embryonic germ layers (ectoderm, mesoderm and endoderm). Additionally, their suitability as a developmental model has been demonstrated for cardiac commitment during development [9] as the heart is the first organ to develop and cardiac cell fate decisions occur very early. Furthermore, cultivation of ESCs in combination with suitable 3D matrices or together with trophoblast cells enables the formation of blastoids, gastruloids, and even embryoids (or embryo-like entities) as culture dish models for human embryogenesis [7,10].

PSCs and PSC-based differentiation models, especially the mouse (m) ESC test, are already validated for testing of teratogenic and embryotoxic substances such as thalidomide (brand name Contergan®), [11,12]. However, their potential for the study of infections during pregnancy is just at the beginning of evaluation [13,14]. In line with the limited number of viruses that can cause perinatal infection, iPSCs possess intrinsic mechanisms that restrict virus infections. In addition, compared to differentiated somatic cells, iPSCs have a higher expression level of a distinct set of interferon (IFN)-induced genes [14]. This appears to counterbalance the absence of a type I IFN response in iPSCs as an essential component of antiviral innate immunity [15].

Teratogenic RV can be maintained in iPSCs over several passages followed by directed differentiation into embryonic germ layer cells [13], highlighting iPSCs as a promising model for the very early mechanisms involved in rubella embryopathy. As a follow-up to this study we aimed at the identification of RV-induced molecular alterations in these cells before and after initiation of directed differentiation through transcriptomics. The most profound effects associated with RV infection were detected in endodermal cells derived from RV-infected iPSCs. Markers for definitive endoderm were upregulated, which occurred in association with profound epigenetic changes, an upregulation of factors involved in vasculogenesis, and reduced activity of the TGF-β signaling pathway. Additionally, ectodermal cells revealed an altered expression profile of essential transcription factors for eye field development during RV infection. Thus, the study of RV infection on iPSCs and derived lineages provides insights into viral alterations of early developmental pathways and as such into congenital diseases in general.

2. Materials and Methods

2.1. Cell Lines and Cultivation

Vero (green monkey kidney epithelial cell line, ATCC CCL-81) and A549 (human lung carcinoma epithelial cells, ATCC, LGC Standards GmbH, Wesel, Germany) were cultured in Dulbecco's modified Eagle's medium (DMEM; Thermo Fisher Scientific, Darmstadt, Germany) with high glucose, GlutaMAX,

10% fetal calf serum (FCS) and 100 U/mL penicillin–streptomycin. If not otherwise indicated, the vector-free human episomal A18945 iPS cell line (alias TMOi001-A), (Thermo Fisher Scientific) was maintained in mTeSR™1 medium (StemCell Technologies, Cologne, Germany) with 10 µg/mL gentamycin on Matrigel™ (BD Biosciences, dispensed in DMEM/F-12)-coated culture plates with daily medium change. They were passaged enzymatically at a ratio of 1:6 to 1:10 every 3 to 5 days with collagenase type IV (Thermo Fisher Scientific) in DMEM-F12 with the addition of 10 µM Y-27632 ROCK inhibitor.

2.2. Directed and Undirected Differentiation of iPSCs

Directed differentiation was performed as an endpoint differentiation assay through the STEMdiff™ trilineage differentiation kit (StemCell Technologies). The differentiation protocol was performed according to the manufacturer's instructions and required cultivation of A18945 iPSCs in mTeSR™1 medium. Single cells, as obtained after treatment with Accutase (Merck/Sigma-Aldrich Chemie GmbH, Taufkirchen, Germany), were plated on Matrigel. Every 24 h medium change of the respective STEMdiff™ trilineage differentiation medium for ectoderm, mesoderm, and endoderm was performed. Samples were collected after 5 days (mesoderm and endoderm) and 7 days (ectoderm) of cultivation. Undirected differentiation was initiated 24 h after collagenase-passaging of iPSC cultures at a ratio of 1:4 through application of undirected differentiation medium (DMEM-F12, 1x MEM-NEAA, 0.2 mM L-glutamine, 20% FBS, 0.11 mM β-mercaptoethanol, and 100 U/mL penicillin) followed by further cultivation for 5 days.

2.3. Embryoid Body Formation

EB formation as based on a previous publication [16] and (http://www.biolamina.com/media.ashx/instructions-bl010.pdf) was carried out in suspension culture and single cell suspensions were obtained after Accutase (Sigma-Aldrich) treatment. A total of 1×10^6 cells was seeded in 200 µL of EB culture medium (DMEM-F12, 20% KnockOut™ Serum Replacement (Thermo Fisher Scientific), 1× MEM-NEAA, 0.2 mM L-glutamine, 0.11 mM β-mercaptoethanol and 1 mg/mL Gentamicin) medium into one well of a nontreated conical 96-well plate and centrifuged at 600× *g* for 5 min. After cultivation for 2 days the EBs were transferred according to the protocol to a low attachment flat-bottom six-well plate and medium was changed every third day.

2.4. Virus Infection and Interferon Assays

The supernatant of infected Vero cells was collected and cleared from cellular debris by centrifugation at 350× *g* for 10 min at 4 °C and filtration through a 0.45 µm syringe filter. Thereafter ultracentrifugation with a 20% sucrose cushion (*w/v* in PBS) was performed for 2 h at 25,000 rpm and 4 °C. The obtained pellets were resuspended in mTeSR1. Viral titers were determined by standard plaque assay. As described previously [13], iPSC cultures with a 40–50% confluency were acutely infected with 7.5×10^5 plaque forming units (PFU) of RV per well of a 24-well plate. This corresponds approximately to an MOI of 20. The applied MOI can only be estimated as iPSCs were passaged enzymatically in clumps. The inoculum was replaced with fresh mTesR1 medium after 2 h of incubation [13]. After 4 to 5 days of cultivation, RV-infected iPSCs were passaged.

For exogenous (or paracrine) IFN treatment human recombinant IFN lambda 1 (IL-29, #300-02L) and 2 (IL28A, #300-02K), were purchased from Peprotech (Hamburg, Germany), and 3 (IL-28B, #CS26) from Novoprotein (Novoprotein, PELOBIOTECH GmbH, Planegg/Martinsried, Germany). The Accuri C6 flow cytometer (BD Bioscience, Heidelberg, Germany) was used for IFN measurement by the LEGENDplex human type 1/2/3 IFN panel (BioLegend, San Diego, CA, USA). The double-stranded (ds) RNA analogue polyinosinic-polycytidylic acid (poly I:C; Santa Cruz Biotechnology, Heidelberg, Germany) was added either directly to the cell culture or transfected at a concentration of 1 µg using Lipofectamine 2000 (Thermo Fisher Scientific) as transfection reagent.

2.5. Calcein Live Cell Staining

For live cell staining, EBs were incubated with mTeSR1 plus calcein FM (Sigma-Aldrich) at 1 μM. After an incubation period for 30 min, EBs were washed twice with PBS and analyzed on an inverted fluorescence microscope.

2.6. RNA Isolation

Total RNA was extracted from mock- and RV-infected cells by Trizol reagent (Thermo Fisher Scientific). The purification was performed with the Direct-zol RNA kit (Zymo Research, Freiburg, Germany) according to manufacturer's instructions. The integrity of the RNA samples was confirmed through analysis on a fragment analyzer (Advanced Analytical). Only samples with a RIN (RNA integrity number as a means of quality assessment) equal to 7 or greater were subjected to further analysis.

2.7. Microarray Gene Expression Analysis and SOM Portrayal

Isolated RNA was processed and hybridized to Illumina HT-12 v4 Expression BeadChips (Illumina, San Diego, CA, USA) and measured on the Illumina HiScan. Raw intensity data of 47,323 gene probes was extracted by Illumina GenomeStudio and subsequently background corrected, transformed into $\log_{10}$-scale, quantile normalized, and centralized to obtain gene expression estimates. Two independent samples per condition and cell type were processed.

Expression data were then further processed using self-organizing map (SOM) machine learning. The method distributes the gene-centered expression values among 2500 microclusters called meta-genes, which were arranged in a two-dimensional 50 × 50 lattice and colored in maroon-to-blue for high-to-low meta-gene expression values. These mosaic images visualize the transcriptome patterns of each individual sample and therefore can be understood as their molecular portraits exhibiting clusters of coexpressed genes in the samples studied [17]. Mean portraits over replicates were calculated by averaging the meta-gene landscapes of replicated samples while difference portraits between different cell types were obtained by subtracting the respective metagene values to highlight differentially expressed genes. Clusters of coexpressed genes were identified by selecting so-called 'spot-areas' in the SOM portraits using overexpression criteria as described previously [17]. For functional interpretation of the expression-modules, we applied gene set enrichment analysis using the gene set Z-score (GSZ), [17]. Enrichment of functional gene sets in the spot cluster was calculated by applying Fisher's exact test. We considered gene sets related to biological processes (BP) of the gene ontology (GO) classification, standard literature sets [17,18], and literature sets curated by our group. Downstream analysis methods were described previously [17,19] and are implemented in the R-package 'oposSOM' used for analysis [20].

Pathway activity was analyzed based on pathway topologies and gene expression data using the pathway signal flow method as implemented in oposSOM [21].

2.8. Quantitative Real-Time PCR Analysis of Viral and Cellular RNA

For determination of the mRNA expression level of selected cellular genes, 1.2 μg of total RNA were reverse transcribed with Oligo(dT)$_{18}$ primer and AMV reverse transcriptase (Promega, Mannheim, Germany) at 42 °C for 1 h. This was followed by an incubation step at 70 °C for 10 min. The carousel-based LightCycler 2.0 (Roche, Mannheim, Germany) was used for quantitative real-time PCR (qRT-PCR) experiments. These experiments included a 1:5 dilution of the respective cDNA samples together with 1 μg BSA and the *GoTaq® qPCR master mix* (Promega). Supplement Table S1 lists oligonucleotides and probes targeting viral p90 gene that were used for quantification of viral RNA as described [22]. Two different approaches for relative expression analysis were pursued. For direct comparison of one sample type after mock- and RV-infection, comparative delta delta Ct (ΔΔCt) was used. For comparison of gene expression levels among different cell types within a large data set, a modified version of the comparative delta delta Ct (ΔΔCt) method was used. The normalized relative quantity (NRQ) values were derived from qbase+ software (Biogazelle, Zulte, Belgium) which are based on the mean expression values of all samples and replicates within a given data set [23].

2.9. Immunofluorescence

For assessment of viral proteins, immunofluorescence was carried out as described [13]. Briefly, cells were fixed with 2% (*w*/*v*) paraformaldehyde in PBS and permeabilized with 0.1 Triton X-100 followed by incubation with mAb anti-E1 from Viral Antigens (Viral Antigens Incorporation, Memphis, TN, USA) at a 1:200 dilution as primary antibody.

2.10. Statistical Analysis

All statistical calculations were done with Graph Pad Prism software (GraphPad Software, Inc., La Jolla, CA, USA). Asterisks (* $p < 0.05$, ** $p < 0.01$, *** $p < 0.001$, **** $p < 0.0001$) highlight the level of significance in diagrams which include data as means ± standard deviation (SD). For comparison of normalized mRNA expression levels in RV-infected samples with the corresponding mock controls, a paired Student's *t* test (consistent ratios of paired values) was applied. Statistical analysis for different samples was based on one-way ANOVA followed by Bonferroni's multiple comparison test.

3. Results

3.1. In the Presence of RV, iPSCs Maintain Pluripotent Properties and Lineage Identity after Initiation of Differentiation

Specification to one of the three germ lineages is the first critical step in directing differentiation to downstream cellular phenotypes. Therefore, directed as well as undirected differentiation was induced in RV-infected iPSCs which were subcultured for two to five passages. Passaging of infected iPSCs results in a homogenous level of infection within iPSC cultures without affecting the protein expression level of the pluripotency marker OCT4 [13]. During passaging of RV-infected iPSCs, replication occurred at a rather constant rate as assessed by viral titer and E1 protein expression rate [13]. Furthermore, passaging allows for adaptation of RV to iPSCs and excludes any possible effects of the differentiation process itself on the otherwise acute infection with RV (Figure 1A). Undirected differentiation is spontaneous and was thus induced to assess whether RV, without a specific differentiation stimulus, directs a gene expression profile different from the mock-infected population. Directed differentiation of RV-infected iPSCs into ecto-, endo-, or mesodermal cells (thereafter referred to as RV-infected) was initiated with the STEMdiff trilineage differentiation kit as an endpoint differentiation approach to determine which of the early cell fate decision pathways could be affected by RV.

RV establishes a noncytopathic infection of iPSCs with a homogenous distribution of infected cells within the respective colony (Figure 1B), [13]. Differentiation into all three embryonic germ layers supported RV replication at a comparable rate (Figure 1C [i,ii]), [13]. As a next step, we generated an expression heatmap of selected marker genes (based on microarray whole transcriptome data) for assessment of pluripotency and lineage identity (Figure 1D). In agreement with the maintenance of OCT4 (octamer-binding transcription factor 4, also known as POU5F1) expression in RV-infected iPSCs [13], high expression of pluripotency markers CDH1 and OCT4 was noted. Their expression was maintained to some degree in endodermal cells, which is in agreement with the conditions of the STEMdiff trilineage differentiation kit. The same applies to the expression of the pluripotency marker SOX2 (SRY (Sex Determining Region Y)-Box 2) in ectodermal cells. The expression profile of lineage-specific markers confirmed ectodermal (PAX6 (Paired Box 6), DLK1 (Delta-Like 1 Homolog), and FABP7 (Fatty Acid Binding Protein 7)), mesodermal (HAND1 (Heart and Neural Crest-Derived Transcript 1), CDX2 (Caudal Type Homeobox 2), APLNR (Apelin Receptor)) and endodermal (LEFTY1 (Left-Right Determination Factor 1), EOMES (Eomesodermin), NODAL (Nodal Growth Differentiation

Factor)) identity after initiation of directed differentiation in mock- and RV-infected iPSCs (Figure 1D). Additionally, some overlap between lineages, especially between mesoderm and endoderm, was noted for RV-infected samples (Figure 1D). Transcriptomic data was confirmed by RT-qPCR of pluripotency marker OCT4 and of selected markers for ectoderm (PAX6), mesoderm (HAND1), and endoderm (NODAL) lineages followed by relative quantification by qbase+ method (Figure 1E). Figure 1D and E indicate that during undirected differentiation, especially mesodermal markers were expressed, which occurred at a comparable level between mock- and RV-infected cells. Among the lineage-specific markers, the expression level of HAND1 was significantly downregulated in mesodermal cells after RV infection. Additionally, RV infection did not alter stemness-related expression signatures as indicated by the transcriptomic activity of the GO gene set telomere maintenance (Figure 1F). Telomere maintenance is active in stem cells, but gets deactivated in differentiated somatic cells [24].

In summary, comparable to the mock-control, RV maintained the pluripotent properties of iPSCs and enabled initiation of differentiation into embryonic germ layer cells as indicated by expression of essential germ layer markers.

3.2. *High-Resolution Transcriptomic Maps Reveal Modules of Coregulated Genes Promoted by RV Infection during Endodermal Differentiation*

As we found out that RV infection did not affect unspecific differentiation of iPSCs and enabled their lineage-specific differentiation, we wanted to focus on the effect of RV on lineage identity. The self-organizing map (SOM) transcriptome data portrayal provides a high-resolution visualization of the transcriptome landscape of each cell system studied in terms of a quadratic mosaic image and decomposes into clusters of coregulated genes. They are represented as colored spot-like areas where red and blue colors code activated and deactivated gene clusters, respectively. These transcriptomic portraits were then used to evaluate the mutual relatedness between the cell systems by means of a phylogenetic similarity tree (Figure 2A). The tree structure results from the fact that common and different spot patterns in the portraits reflect mutual similarities and differences of the activated cellular programs which enable judging the effect of RV-infection on the different lineages (see the portraits in Figure 2A). For an overview, we generated a spot-summary map in Figure 2B which shows the activated spots observed in any of the samples together with their functional context as extracted by means of gene set enrichment analysis of the genes in each of the spot-clusters of coexpressed genes (see also Supplement Table S1). In total, we identified five relevant spots labeled with capital letters A–E. Each of the spots is characterized by a specific expression profile (Supplement Figures S1 and S2) which, in turn, shows close similarities with the expression profiles of distinct gene sets (shown as 'barcode' plots in Figure 2C).

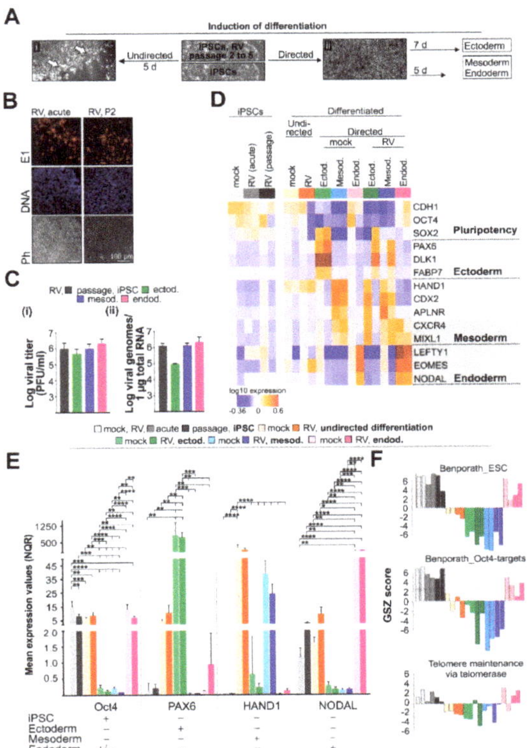

Figure 1. The identity of iPSCs and derived lineages was maintained in the presence of rubella virus (RV). (**A**) Overview of the methods applied to initiate differentiation in mock- and RV-infected iPSCs. (**A**) [**i**] Undirected differentiation in the respective induction medium started at the rim region of iPSC colonies (indicated by white arrows) and extended to their center over time of incubation. (**A**) [**ii**] Additionally, directed differentiation into the primary germ layers ectoderm, mesoderm, and endoderm was induced with the STEMdiff differentiation kit. (**B**) Immunofluorescence analysis with anti-E1 antibody (shown in red) was performed to monitor distribution of RV-positive cells within iPSC colonies. Nuclei are shown in blue. Ph, phase contrast (**C**) To assess RV replication in iPSCs and derived lineages, (**i**) virus progeny, and (**ii**) the amount of genomic viral RNAs was determined by standard plaque assay (n = 11 for passaged iPSCs, otherwise n = 3) and TaqMan-based reverse transcription-quantitative PCR (passaged iPSCs and ectodermal cells n = 3, mesodermal cells n = 7, endodermal cells n = 4), respectively. (**D**) Expression heatmap of selected marker genes of pluripotency and lineage identity (based on microarray whole transcriptome data) in mock- and RV-infected iPSCs and iPSC-derived lineages. Shading indicates overlap in the expression of some of the marker genes between iPSCs and iPSC-derived lineage cells, respectively. (**E**) The expression of indicated target genes in mock- and Wb-12-infected iPSCs and derived cells after initiation of undirected and directed differentiation was determined by real-time quantitative PCR (RT-qPCR) and analyzed by qbase+ software. For normalization, chromosome 1 open reading frame 43 (C1orf43) and hypoxanthine guanine phosphoribosyl transferase 1 (HPRT1) were used. Relative gene expression was calculated as normalized relative quantity (NRQ) and given as means ± SD (n = 3 to 5). As a reference, lineage-specific expression levels were assigned based on literature data and transcriptomics data by Stemcell Technologies for the Trilineage differentiation kit. (**F**) Analysis of stemness-related expression signatures was based on the GO gene set telomere maintenance. An embryonal stem cell signature and a gene set collecting OCT4 targets were both taken from [25].

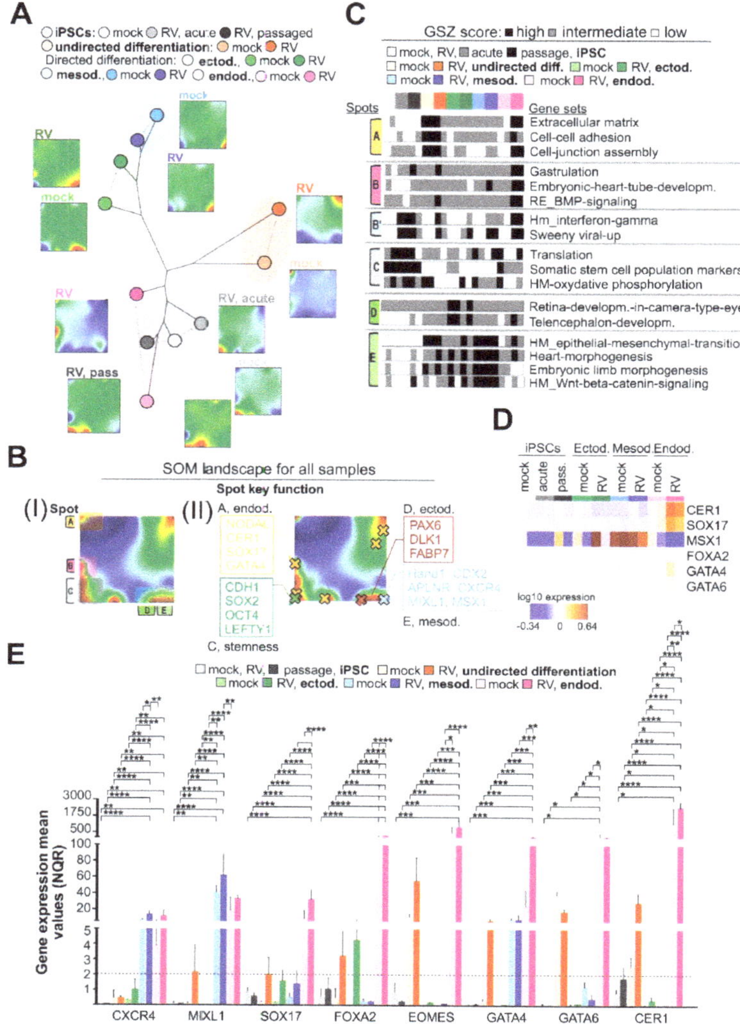

Figure 2. RV supports expression of markers for definite endoderm. (**A**) Similarity tree of the gene expression portraits of the cell systems studied. Both, mock- and RV-infected mesodermal and ectodermal cells share relatively high mutual similarity of their transcriptomes. In contrast, mock- and RV-infected endodermal cells form a separate branch that is closer to the mock- and RV-infected iPSCs. (**B**) Spot summary map (**I**) provides an overview of activated cellular programs and their functional context, which is depicted in more detail in (**II**). (**C**) Barplot representation of the expression profiles of gene sets related to different functions. Their genes were enriched in the spots that were identified in the SOM portraits. They are thus indicated accordingly (see also the overview map in part B of the Figure and Table S2). (**D**) Selected marker genes for definite endoderm are illustrated in the heatmap of the transcriptome of mock- and RV-infected iPSCs and iPSC-derived lineages. (**E**) The expression of marker genes for definite endoderm was determined by RT-qPCR in mock- and Wb-12-infected iPSCs and iPSC-derived lineages by qbase+ software. For normalization, C1orf43 and HPRT1 were used. Relative gene expression was calculated as normalized relative quantity (NRQ) and given as means ± SD (n = 3 to 5). See also Figures S1 and S2.

Mapping of the position of genes into the SOM enables us to deduce their expression profile in the cell systems studied according to the gene's location in or near the spots and the respective spot profiles (Figure 2B and Supplement Figure S1). Importantly, key genes of ectoderm and mesoderm development are found in spots D and E, respectively, and are confirmed to be upregulated in the ectoderm and mesoderm cells. On the other hand, genes related to gastrulation and heart-tube development were enriched in spot B and found to be activated in RV-infected endoderm cells. This occurred together with genes that are activated upon interferon response (see next subsection). Stemness key genes locate in spot C together with part of the developmental genes of the endoderm (Supplement Figure S1). Difference portraits clearly indicate that spot B associates with RV infection (Figure 2B), which contains the genes NODAL, CER1, SOX17, and GATA4, reflecting their activation upon RV infection in endodermal cells. In addition, we are able to show that mesoderm and ectoderm cells share similar expression patterns characterized by upregulated spots D and E and downregulated stemness genes (spot C), which, in contrast, are upregulated in endoderm cells. This is in agreement with the expression level of pluripotency genes in the endodermal lineage which was higher than in the remaining two embryonal germ layers (Figure 1D,E). As a consequence, mesoderm and ectoderm cells occupy neighboring positions at one end of the similarity tree, while endoderm cells are found at the opposite end. These findings highlight two important aspects: (I) there is a close similarity of transcriptional patterns among ecto-, mesoderm, and iPSC samples without and with RV infection; and (II) the endodermal lineage is an exception, where RV infection induces a notable shift away from the corresponding mock sample.

To further elucidate this specific effect of RV, we analyzed the transcriptomic data for markers for definitive endoderm (CXCR4 (C-X-C Motif Chemokine Receptor 4), MIXL1, SOX17, FOXA2 (Forkhead Box A2), EOMES, GATA6, CER1 (Cerberus, DAN family BMP Antagonist), and LEFTY, [26]. The expression heatmap shown in Figure 2D indicates an increased expression level of CER1 and SOX17 after RV infection. Figure 2E shows qPCR analysis followed by relative quantification by qbase+ method and highlights that the definite endoderm markers CXCR4, MIXL1, SOX17, GATA4/6, and CER1 were significantly higher expressed in RV-infected endodermal cells as compared to their mock-infected counterparts. While RV infection did not induce these markers above a cut-off of a two-fold increase during mesodermal and ectodermal differentiation, EOMES, GATA4, and CER1 were specifically induced by RV during undirected differentiation.

In conclusion, similarity analysis supports the hypothesis that ectodermal and mesodermal lineage identity was maintained after infection with RV, while endodermal cells derived from RV-infected iPSCs were enriched in markers for definitive endoderm.

3.3. RV Infection Activates IFN Type III Response Pathways on iPSCs and Derived Lineages

Difference portraits (Figure 3A) indicated that RV-infection specifically upregulated genes in spot B, which were associated with "IFN response" characteristics. This is supported by the profiles of gene expression signatures of viral infections such as by influenza virus and pneumonia that is accompanied by interferon activation. These genes were consistently upregulated in RV-infected samples with the largest observed effect in endodermal cells (Supplement Figure S3), [27–29]. Spot B highlighted in the SOM landscapes contained genes involved in IFN and viral response mechanisms (Figure 3A, a list of genes is given in Figure 3B). This is further emphasized by the expression heatmap shown in Figure 3C. Whereas genes involved in IFN-sensing, including the type III IFN receptor IFNLR1 (IFN lambda receptor 1), were not altered in their expression level, the IFN-signaling components STAT1 (signal transducer and activator of transcription 1) and IRF9 (interferon regulatory factor 9) appeared to be slightly upregulated at their mRNA expression level after infection with RV, especially in endodermal cells. The highest level of upregulation was found for IFN-stimulated genes (ISGs), notably for MX1 (MX dynamin like GTPase 1), IFITM2 (interferon induced transmembrane protein 2), and ISG15 (interferon-stimulated gene 15). Mapping of these genes into SOM space further underlines these findings: The IFN-signaling genes and the ISGs accumulate in and around spots B and D, respectively, while IFN-sensing genes are located outside these spot regions (Figure 3B). The increase in the expression level of selected marker genes of the IFN pathway in the presence of RV was confirmed by RT-qPCR (Figure 3D). Compared to RV-infected iPSCs and ecto- and mesodermal cells, the expression level of IRF9 and STAT1 and selected ISGs (IFITM1/2, IFIT1, and ISG15) was significantly higher in RV-infected endodermal cells. The highest increase in mRNA expression after RV infection was noted for the ISGs IFIT1 and ISG15. Therefore, we determined whether this gene expression pattern was indeed associated with IFN generation during RV infection through quantification of type I (α and β), type II (γ), and type III (λ1 and λ2/3) IFNs by the LEGENDplex assay from the supernatants of RV-infected cells (Figure 3E). In iPSCs as well as iPSC-derived lineage cells, RV infection induced secretion of type III IFNs, namely IFN λ2/3 (Figure 3E). As a positive control, the synthetic dsRNA analog poly I:C was used, which was either transfected into iPSCs or added directly to the supernatant. Either application of poly I:C did not lead to secretion of any type I, II, or III interferons (Figure 3E). The activation of type III IFNs by RV was also confirmed at the mRNA level by RT-qPCR (Figure 3F). Compared to the mock-infected control, RV induced a significant increase in the mRNA expression of IFN λ2/3 in endodermal cells (Figure 3F). Thereafter, we addressed the discrepancy between IFN λ2/3 protein (Figure 3E) and mRNA level (Figure 3F). Gene set analysis revealed that mRNAs associated with the KH type-splicing regulatory protein (KHSRP) were specifically enriched in endodermal cells (Figure 3G). KHSRP is involved in post-transcriptional regulation of mRNA expression, including IFN λ3 [30]. This could explain the discrepancy between IFN λ2/3 protein and mRNA expression level.

To address the influence of the type III IFNs secreted during RV to the cell culture supernatant on the gene expression landscape of iPSCs, type III IFNs were added exogenously for two weeks of cultivation during daily medium change of iPSCs. The zoom-in similarity tree shown in Figure 3H highlights the relatedness between the expression portraits of mock- and RV-infected iPSCs as well as iPSCs after application of type III IFNs. The gene expression profile of passaged RV-infected cells shifted away from iPSCs after exogenous IFN type III application, but closer to the mock-infected cells, suggesting an adaptation of RV to iPSCs.

In conclusion, in iPSCs and iPSC-derived embryonic lineages, RV infection induced a type III IFN response together with activation of ISGs, notably MX1 and ISG15. This activation appeared to be specifically profound in endodermal cells.

3.4. *RV Infection Is Associated with Chromatin Remodeling*

Alterations of gene expression patterns during development are governed by epigenetic mechanisms in cooperation with regulation via transcription factor networks [32,33]. Particularly, we found that gene signatures of epigenetic impact, such as targets of the polycomb repressive complex 2 (PRC2), of H3K27me3, and of bivalently (H3K4me3 and H3K27me3) marked gene promoters, have an almost antagonistic expression profile as compared to the stemness signatures (Figure 4A, in comparison to Figure 1F).

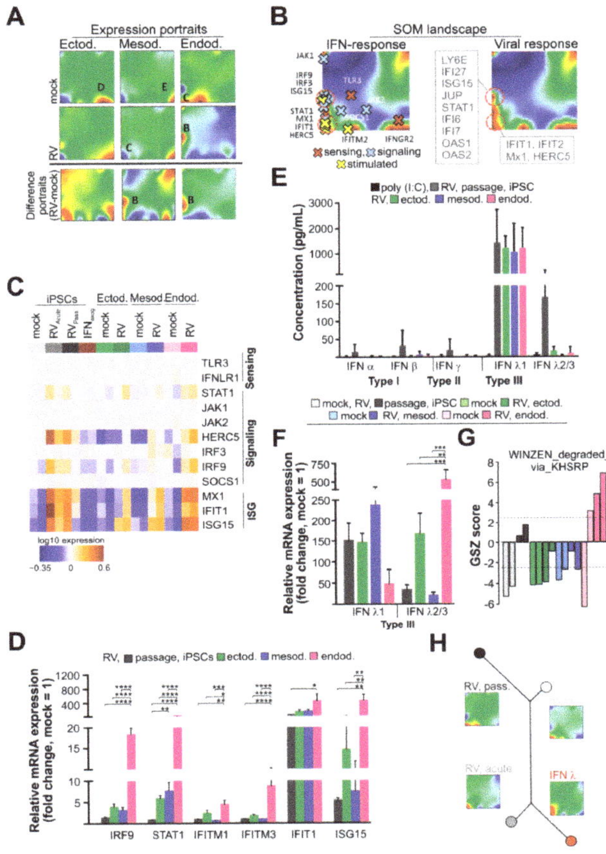

Figure 3. RV infection activates an IFN response in iPSCs and derived lineages. (**A**) Expression portraits of the embryonal germ layers before and after RV infection and the respective difference portraits reveal characteristic spot patterns, where spots C, D, and E are specific for endoderm, ectoderm, and mesoderm cells, respectively. Spot B appears after RV infection mainly in endodermal cells. (**B**) IFN response genes with signaling and stimulated functions in the IFN response pathways accumulate in spots B and C and were upregulated predominantly in iPS and endodermal cells after RV infection as also indicated by the IFN and viral response gene signature profiles. (**C**) Expression heatmap of selected marker genes involved in IFN-sensing and -signaling in mock- and RV-infected iPSCs and iPSC-derived lineages. Interferon-stimulated genes (ISGs) that were identified by the SOM analysis shown in (**A**) are included. (**D**) The mRNA expression level of the IFN-signaling components IRF9 and STAT1 and indicated ISGs was verified by RT-qPCR analysis. Data are given as means ± SD (n = 3 and n = 5 for RV-infected mesoderm, IRF9 and STAT1). (**E**) The IFN profile for RV-infected iPSCs and derived lineages was determined by the LEGENDplex IFN panel for undiluted supernatants collected after five (iPSCs and mesodermal cells) and seven days (endodermal and ectodermal) of cultivation. (**F**) The mRNA expression level of type III IFNs was verified by qPCR analysis and given as means ± SD (n = 3). (**G**) Mean expression of a gene set (gene set Z-score, GSZ) that is controlled by KSRP, which appears to keep inflammatory gene expression within defined limits [31]. (**H**) Similarity tree of the gene expression portraits of mock- and RV-infected iPSCs in comparison to iPSCs after cultivation in the presence of exogenous type III IFNs. (**D**,**F**) For normalization of qRT-PCR data in the $2^{-\Delta\Delta Ct}$ method, the HPRT1 gene was used. See also Figure S3.

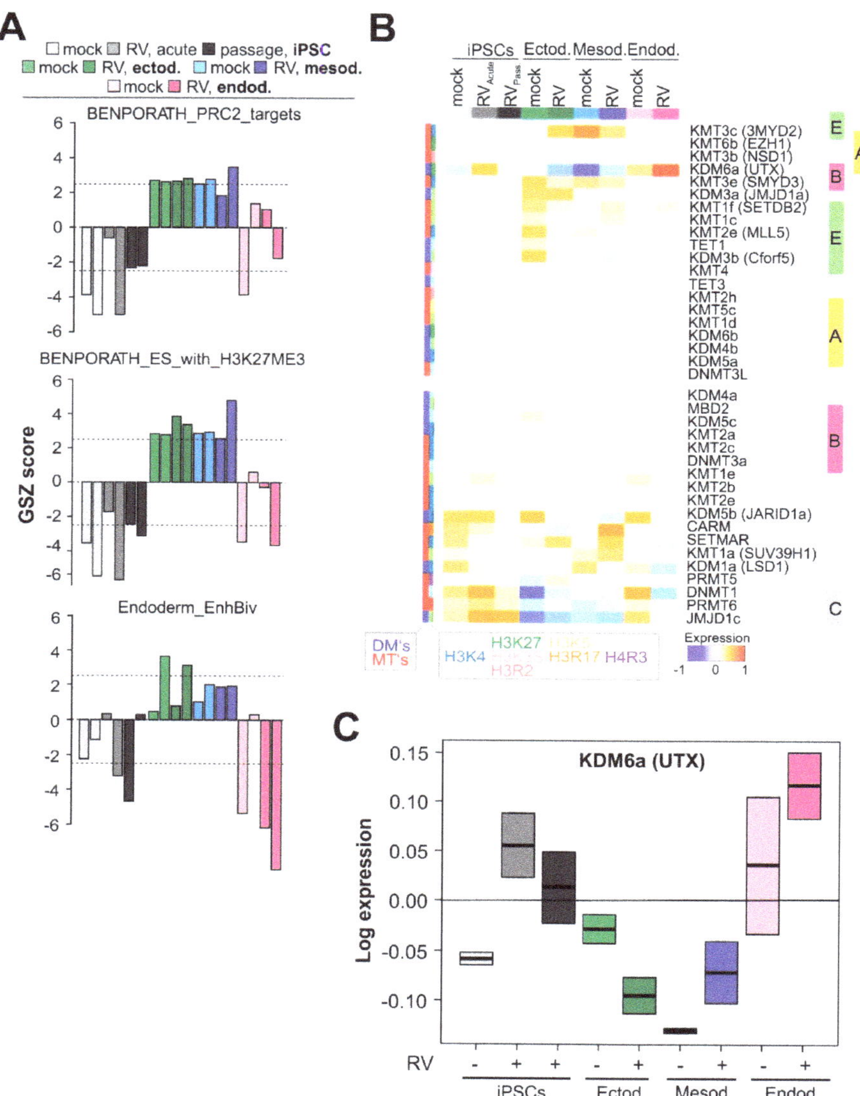

Figure 4. RV infection is accompanied by an altered expression of components of the SWI/SNF and NURF chromatin remodeling complexes. (**A**) Mean GSZ expression signatures of stemness-related transcriptional programs which act via epigenetic programming (PRC2 targets, repressive and bivalent chromatin marks). (**B**) Expression heatmap of chromatin modifying enzymes in the cell systems studied. The gene expression data of methyltransferases (MTs) and demethylases (DMs) of DNA cytosines, histone lysine and arginine side chains were assigned to the expression spot-cluster A–E according to their expression profiles. This suggests their involvement in the regulation of chromatin structure as writers and erasers of methylation marks at histone lysine and arginine side chains and affecting DNA methylation. Notably, a very strong variability was observed for KDM6a (alias UTX), a constituent of the SWI/SNF ATP-dependent chromatin remodeling machinery. (**C**) Expression of KDM6a directly relates to stemness programs (Figure 1F) and inversely relates to programs repressing stemness functions (part A of the Figure). See also Figure S4.

Next, we focused on chromatin remodeling which is essential for lineage segregation [9,34]. The analysis of transcription factor networks that act in regions of euchromatin with transcriptionally active genes and regulatory elements revealed that genes in repressed and bivalent states of endoderm progenitors become more rigorously deactivated in RV-infected endoderm cells than genes from these states of mesoderm progenitors in RV-infected mesoderm cells (Supplement Figure S4A). This suggests that RV infection specifically dedifferentiates endoderm cells by suppressing developmental suppressors. We then studied enzymes affecting methylation of DNA and of arginine and lysine side chains of histones such as H3K4, H3K9, H3K27, and H3K36, with potential impact on chromatin structure. The profiles of the DNA methylation maintenance methyltransferase DNMT1, of PRMT6, a methyltransferase of histone arginine side chains, and of JMJD1c, a H3K9 demethylase, and KDM5b demethylating H3K4me3 correlate with the 'stemness' spot cluster C upregulated in iPS and endodermal cells.

Notably, the gene encoding the H3K4 demethylase KDM6a (alias UTX) was markedly upregulated in RV-infected iPSCs and, especially, endoderm-derived cells (Figure 4B). KDM6a is a constituent of the SWI/SNF ATP-dependent chromatin remodeling machinery. Figure 4C highlights that in comparison to iPSCs and ecto- and mesodermal cells, the upregulation of KDM6a was highest in endodermal cells. The alterations of its expression suggest its role in chromatin remodeling after RV infection described above. This motivated us to estimate the expression patterns of other genes encoding components of the SWI/SNF and of the NURF chromatin remodeling complexes [9] by mapping them into the SOM (Supplement Figure S4B). We found that, indeed, Smarcc2, Smarcd3, and Btpf were all upregulated in endoderm-derived cells after RV infection, which further supports the assumption that SWI/SNF and possibly also NURF contribute to chromatin remodeling during RV infection. In conclusion, expression changes of different sets of genes involved in epigenetic regulation and of constituents of the ATP-dependent chromatin remodeling complexes such as KDM6a-UTX were detected in association with RV-infection, especially during endodermal differentiation.

3.5. *RV Infection Impairs Aggregation of iPSCs into Embryoid Bodies*

The progression of embryogenesis does not only involve the activation of developmental pathways, but also requires cell–cell interactions based on adhesive forces [35]. The relevance of these observations for RV-infected iPSCs was emphasized by gene ontology analysis regarding focal adhesion and regulation of cell adhesion (Figure 5A). Transcriptomic analysis revealed that THY-1 (also known as cluster of differentiation (CD) 90) was among the targets affected by RV (Figure 5B). The relevance of THY-1 (CD90) for cellular adhesion capacity was highlighted for CD90 negative carcinoma, which compared to their CD90 positive counterparts lack the ability to form spheres [36]. Accordingly, we have addressed whether RV alters the spontaneous aggregation capacity of iPSCs into 3D aggregates called embryoid bodies (EBs). EB formation relies on cell–cell adhesive interactions [35]. Compared to the mock control, EBs generated from RV-infected iPSCs were reduced in diameter and of irregular shape (Figure 5C). Furthermore, during cultivation they lost stability and small-sized debris was generated (Figure 5C). In contrast to RV-infected iPSCs, the mock-infected controls generated viable EBs as indicated by staining with calcein performed after two weeks of cultivation (Figure 5D). In conclusion, RV infection impaired the adhesion capacity of iPSCs as shown by their reduced ability to assemble into EBs. This suggests an impaired cell–cell interaction capacity during lineage segregation.

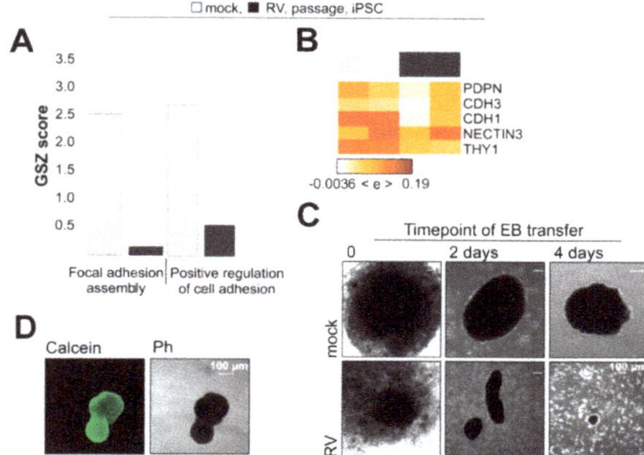

Figure 5. RV impairs the cellular adhesion capacity of iPSCs. (**A**) Gene signatures related to focal adhesion in mock- and RV-infected iPSCs. (**B**) Expression heatmap of selected marker genes involved in cellular adhesion. (**C**) Assessment of 3D stability through embryoid body (EB) formation. Shown are images before and after cultivation in suspension. (**D**) To verify viability, EBs were stained with calcein after cultivation for two weeks.

3.6. RV Infection Specifically Affects Developmental Pathways during Endodermal Differentiation

For assessment of the effect of RV on global cellular signaling networks, we focused on two important signaling pathways, namely transforming growth factor β (TGF-β) and Wnt/β-catenin (Wnt), (Figure 6A,B, respectively). The TGF-β signaling pathway is involved in cell growth and differentiation during embryogenesis [37]. The Wnt signaling pathway regulates the interaction between cellular pathways involved in primary germ layer formation and is required for mesodermal differentiation from pluripotent stem cells [38]. A more detailed view of the TGF-β and Wnt signaling pathways is provided in Supplementary Figures S5 and S6, respectively. As expected, the highest TGF-β signaling pathway activity was observed in endodermal cells, while the Wnt signaling pathway was most active in mesodermal cells (Supplement Figures S5 and S6). Thus, these two lineages were depicted in Figure 6A,B, respectively, to highlight the effect of RV on their activity in comparison to the respective controls. Within the TGF-β signaling pathway, mock-infected endodermal cells show high cell cycle activity induced by CDKN2B and its downstream interaction partners, which became deactivated during RV infection (Figure 6A and Supplement Figure S5). Additionally, RV infection in endodermal cells was specifically accompanied by a strong activation of NODAL, an essential component of the TGF-β signaling pathway (Figure 6A and Supplement Figure S5). During RV infection, the transcriptional activity of Wnt signaling pathway was reduced in mesodermal cells (Figure 6B), whereas for ectodermal and endodermal cells, almost no alteration in its activity was detected (Supplement Figure S6).

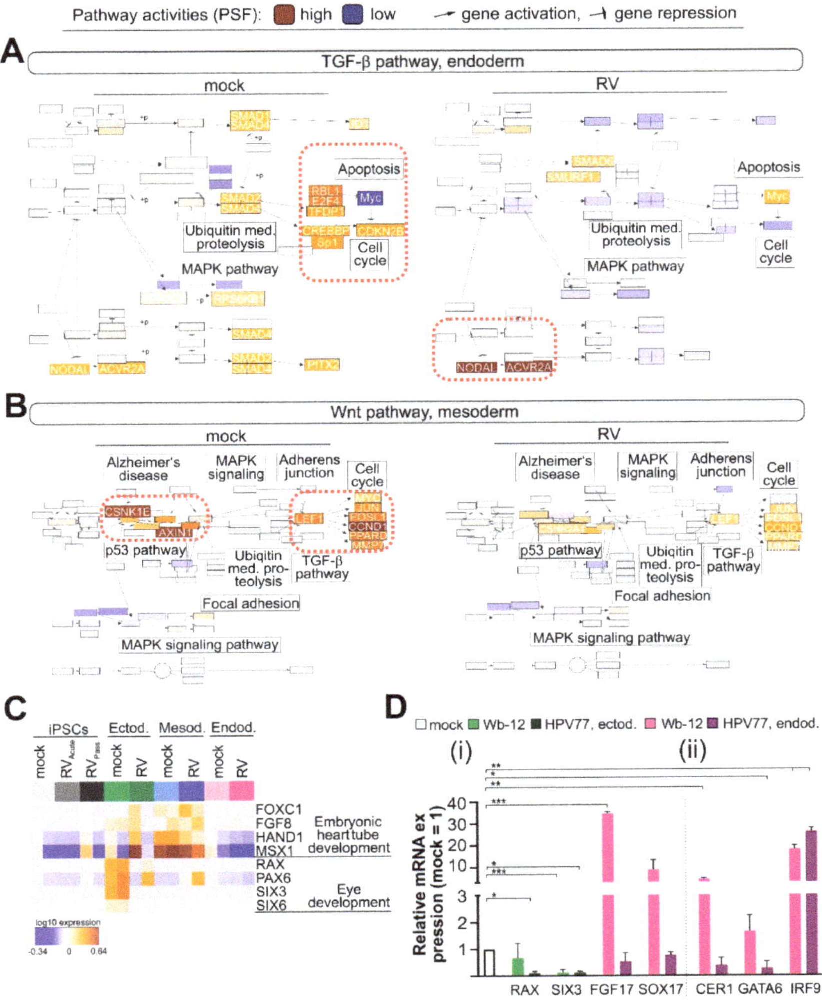

Figure 6. RV exerts lineage-specific effects on developmental signaling pathways and alters expression of transcription and growth factors. Pathway signal flow (PSF) activity plot of (**A**) the TGF-β signaling pathway in endodermal cells (highlighted are genes with higher (NODAL, ACVR2A, Myc) and lower (RBL1 and E2F4) activity after RV infection) and (**B**) of the Wnt signaling pathway in mesodermal cells (highlighted are genes with lower (the CSNK1E/AXIN1E and the LEF1/CCND1 axis) activity after RV infection). The calculation of the activity of the nodes was based on the PSF algorithm using the respective gene expression values and the wirings between the nodes [21]. (**C**) Selected genes within pathways that were specifically affected by RV infection are illustrated in the heatmap of the transcriptome of mock- and RV-infected iPSCs and iPSC-derived lineages. (**D**) For qRT-PCR expression analysis, the $2^{-\Delta\Delta Ct}$ method based on normalization to HPRT1 gene was used. Values are given as means ± SD (n = 4 for Wb-12-infected ectoderm, n = 2 for HPV77-infected ectoderm, otherwise n = 3). See also Figures S5–S7.

As congenital rubella leads to defects in heart and eye development, we analyzed the impact of RV infection on the underlying molecular pathways. For members of the gene annotation embryonic heart tube development and the gene set heart morphogenesis only a slight effect of RV infection was noted (Supplement Figure S7), suggesting the involvement of other factors. In mesodermal cells, expression of HAND1, which is involved in embryonic heart tube development, was reduced after RV infection as compared to the mock controls (Figures 6C and 1E). The ectoderm gives rise to components of the eye. At the molecular level, RV infection of ectodermal cells impaired the gene set eye development (Figure 6C), (Supplement Figure S7). Specifically, SIX3 and SIX6, as key transcription factors for mammalian eye development [39,40], were reduced in their expression level (Figure 6C). Among others, SIX3, together with RAX, initiates transcription of genes required for lens placode formation [39].

To further assess developmental pathways with relevance for the teratogenic outcome of RV infection, we determined the mRNA expression of RAX and SIX3 (as important factors for eye development) besides FGF17 (Fibroblast Growth Factor 17) and SOX17 (as contributing factors for endodermal differentiation and cardiovascular development) by RT-qPCR in ectodermal and endodermal cells, respectively (Figure 6D). Here, the vaccine strain HPV77 was used in addition to Wb-12 strain. Attenuated vaccine strains such as HPV77 are not teratogenic as revealed after immunization of unknowingly pregnant women [4]. Thus, any alteration at the molecular level that is present during wild-type Wb-12, but not HPV77 infection, emphasizes its possible contribution to congenital rubella. In comparison to the mock control, a similar reduction in the expression of RAX and SIX3 was detected in ectodermal cells after infection with both RV strains. However, a different picture emerged for FGF17 and SOX17. Figure 6D [i] highlights an increase in the expression of FGF17 and SOX17, which was significant for FGF17 compared to the mock control, but only for endodermal cells derived from Wb-12-infected iPSCs, not for endodermal cells derived from HPV77-infected iPSCs. Moreover, the increase in the expression of the definitive endoderm markers CER and GATA6 after infection with Wb-12 as shown in Figure 2E was not detected after infection with HPV77 (Figure 6D [ii]). However, both RV strains induced a significant increase of the IFN-signaling component IRF9 (Figure 6D [ii]). The endoderm plays an essential role in the crosstalk between the lineages and contributes to the epithelial lining of many organs, including the vascular network. Accordingly, the gene set vasculogenesis, but not angiogenesis, was affected by RV infection (Supplement Figure S7). This emphasizes our notion on the correlation between the impact of RV infection on endodermal cells and congenital rubella.

In summary, specific signatures including the TGF-β signaling pathway were affected by RV infection, but in a lineage-specific manner. In ectodermal cells, RV infection significantly reduced expression of SIX3 as key transcription factors for eye field development. Only for the clinical isolate Wb-12, but not for the vaccine strain HPV77, was an impact on the growth factor FGF17 and the endodermal transcription factor SOX17 noted.

Figure 7 summarizes the findings of this study in correlation to the main CRS symptoms. The noncytolytic course of infection of RV during directed differentiation is in agreement with its persistence in multiple organs and tissues during congenital rubella. We have not identified any indication at the molecular level that could contribute to the defects in ear development during congenital rubella. However, sensorineural deafness is often a late-onset symptom and could be associated with pathological alterations in the brain of the infected infants [3].

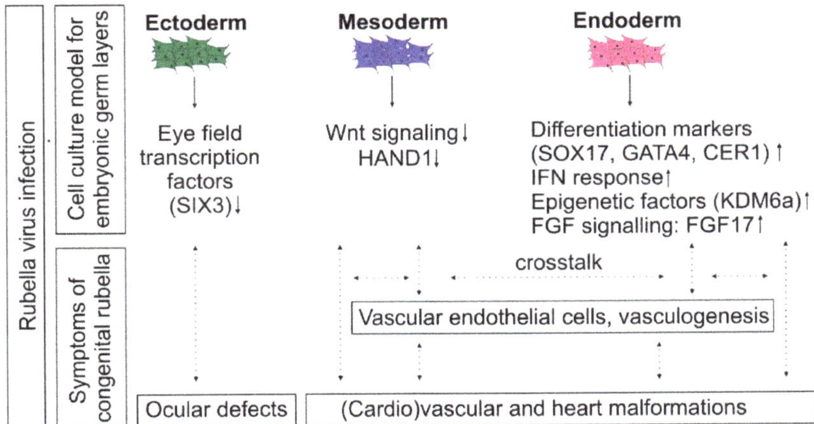

Figure 7. Graphical summary of the identified molecular alterations induced by RV during directed differentiation of iPSCs into the three embryonic germ layer cells. The data was set in a possible relation to the prevailing symptoms of congenital rubella embryopathy. Especially endodermal cells were characterized by profound alterations in their gene expression landscape, including the expression of markers for definitive endoderm and epigenetic factors. This could impair the crosstalk between endodermal and mesodermal cells during differentiation.

4. Discussion

Knowledge on developmental signaling networks is an essential prerequisite to understand congenital abnormalities, either caused by pathogenic, hereditary or environmental risk factors. Models for developmental toxicity testing range from iPSCs to iPSC-derived EBs and three-dimensional organoids. They have different properties regarding high-throughput screening capacity and relevance for in vivo developmental processes [41]. Their proper assessment requires compounds or pathogens with well-known symptoms arising from embryotoxic or teratogenic alterations during embryonal development. Here, we used RV to correlate clinical observations for congenital rubella syndrome with its impact on the differentiation capacity of iPSCs. Although iPSC-based cell culture models reflect only transient stages during human embryogenesis, they allow us to recapitulate essential developmental pathways that are otherwise inaccessible [42].

Among human pathogens, RV is rather exceptional in its ability to replicate noncytopathically in iPSCs, which in general represent a rather restrictive environment to most viral infections [43]. The protection of human development from a pathogenic insult involves several mechanisms, including transcriptional silencing of viruses in pluripotent stem cells [44] and an intrinsic high expression level of IFN-induced genes [14]. This includes interferon-induced transmembrane protein 1 (IFITM1) and its capacity to restrict the potentially harmful reactivation of human endogenous retroviruses [43]. Otherwise, the antiviral innate immune response in iPSCs is rather refractive [15]. The constitutive overexpression of an active IRF7 as a master regulator of the type I IFN system revealed the harmful effects an activated type I IFN response would have on the expression of pluripotency and lineage specific genes, especially of endodermal cells [45]. In contrast to the engineered type I IFN response in iPSCs through overexpression of IRF7, no morphological changes were noted after infection of iPSCs with RV [13]. However, in agreement with the study on the effect of type I IFNS on differentiation capacity of iPSCs [45], the impact of RV on directed differentiation was most profound during endodermal differentiation. The differences in the signaling cascades of type I and III IFNs [46] might explain the milder effects noted after RV-associated type III IFN activation as compared to the severe effects of an engineered type I IFN response [45]. Our data complements a recent study on the impact of Influenza A virus (IAV) on the pluripotency and proteome of hiPSCs [47]. Whereas, in contrast to RV,

IAV reduces the pluripotency of iPSCs, both virus infections induce ISG15 and IFN λ1 [47], highlighting this observation as an innate immune mechanism that is already developed in iPSCs. Further studies need to address whether the impact of RV infection on endodermal differentiation is correlated with the activation of the type III IFN signaling pathway and how this affects the course of infection of RV in iPSCs. In ectodermal cells, RV infection was associated with the downregulation of SIX3, an essential transcription factor for early eye development [48]. Together with SIX6, SIX3 suppresses Wnt signaling, which could contribute to the slight activation of this essential developmental signaling pathway in ectodermal cells derived from RV-infected iPSCs [40]. Their functional importance during retinal development and eye field specification was recently shown by the use of iPSC-derived retinal organoids [39]. Our study complements a previous study on the gene expression profile of fetal (HUVEC originating from umbilical cord veins) and adult (HSaVEC derived from the saphenous vein) endothelial cells which revealed a specific enrichment of 18 downregulated genes within the GO terms "sensory organ development", "eye development", and "ear development" [49].

Among the embryonic germ layers, especially differentiation to definite endoderm appeared to be affected by RV infection. In addition to its role in formation of organs of the digestive tract, the interaction of endodermal cells with precardiac mesoderm drives specification and differentiation of cardiac myocytes and cells of endocardial endothelium [50]. This is supported by studies on the contribution of signals from endodermal cells and the interactive crosstalk between the endoderm and mesoderm to differentiation of ESCs to a cardiomyogenic lineage [9]. RV infection does not only target the endoderm, but also signals that facilitate this interactive crosstalk. This includes Cerberus as a bone morphogenetic protein (BMP) antagonist [51]. The secretion of Cerberus from endodermal cells initiates differentiation of the neighboring tissue, namely the overlying cardiac mesoderm [51,52]. Furthermore, the analysis of endoderm-depleted frog and avian embryos revealed that the endoderm contributes to vasculogenesis and vascular tube formation [53]. Thus, as summarized in Figure 7, the molecular events identified in RV-infected endodermal cells could contribute to cardiovascular defects during congenital rubella [2].

Besides the mere expression level of essential components of developmental pathways, post-translational histone modifications are involved in the regulation of gene expression during development. The balance between H3K4me as an active and H3K27 as an inactive state histone modification directs the switch between active and inactive pathways during differentiation [54]. The activity of the KDM6A (UTX) demethylase was especially upregulated in endodermal cells during RV infection. KDM6A demethylase activity was reported to counteract DNA damage response and cell death induction in differentiating ESCs [55], which could also apply to RV-infected endodermal cells.

RV infection was associated with an upregulation of definitive endoderm-enriched transcription factors, including GATA4, EOMES, and SOX17 [56]. In a context- and dose-dependent manner, the transcription factor EOMES directs cardiac development as well as endoderm specification [57]. Whereas SOX17-null mice revealed a downregulation of several genes involved in heart development [58], the ectopic overexpression of SOX17 during hematopoiesis impaired survival of early hematopoietic precursors due to induction of apoptosis [59]. This indicates that normal embryonal development, especially cardiac specification, requires fine-tuned expression of several factors [60], which appears to be affected by RV infection.

The characterization of teratogens such as RV on iPSC-based models is an essential requirement to emphasize their suitability for the assessment of embryotoxicants and to identify relevant parameters to increase their predictive power. Congenital heart malformations are not only caused by pathogens such as RV, they are the most common among human developmental defects identified for human births. iPSC-based models enable valuable insights into human development and processes that might disturb its normal progression, which will broaden our diagnostic and treatment options for congenital defects. Further studies are needed to correlate the identified transcriptional changes with functional consequences for pathways directing embryonal development.

Supplementary Materials: The following are available online at http://www.mdpi.com/2073-4409/8/8/870/s1, Table S1: Related to description of quantitative real-time PCR analysis, Table S2: Related to Figure 2C. Spot cluster characteristics, Figure S1: Mean expression profiles of 'spot'-clusters of genes which were denoted with capital letters A–E., Figure S2: Gene set enrichment for pathway analysis of mock- and RV-infected iPSCs and iPSC-derived lineages., Figure S3: Characterization of the IFN-response gene signature in RV-infected iPSCs and iPSC-derived lineages., Figure S4: Characterization of gene expression signatures related to epigenetic regulation., Figure S5: Pathway signal flow (PSF) activity plot of the TGF-beta signaling pathway in ecto-, meso-, and endodermal cells derived from mock- and RV-infected iPSCs., Figure S6: Pathway signal flow (PSF) activity plot of the Wnt signaling pathway in ecto-, meso-, and endodermal cells derived from mock- and RV-infected iPSCs., Figure S7: Gene set expression signatures of developmental programs in mock- and RV-infected iPSCs and iPSC-derived lineages.

Author Contributions: C.C. supervision of the study; funding acquisition, C.C., H.B. wrote the first draft of the manuscript, all authors read and revised the final manuscript, N.C.B., J.B., D.H. performed experiments, E.W. and H.B. performed mathematical modelling and processing and analysis of transcriptomic data. M.L.S., S.B. and U.G.L. provided resources.

Funding: This work was supported by DFG grant CL 459/3-1 to C.C. We acknowledge support from the German Research Foundation (DFG) and Leipzig University within the program of Open Access Publishing.

Acknowledgments: For provision of Wb-12 strain the authors want to thank B. Weißbrich (University of Wuerzburg, Germany). We want to thank Knut Krohn and Petra Süptitz from the core unit DNA technologies, IZKF Leipzig, Medical Faculty of the University of Leipzig, Leipzig, Germany for RNA quality assessment and technical support of the microarray experiments. The authors also want to thank Sandra Bergs for technical support.

Conflicts of Interest: The authors declare no conflict of interest.

References

1. Dudgeon, J.A. Congenital rubella. *J. Pediatr.* **1975**, *87*, 1078–1086. [CrossRef]
2. Duszak, R.S. Congenital rubella syndrome—Major review. *Optometry* **2009**, *80*, 36–43. [CrossRef] [PubMed]
3. Lazar, M.; Perelygina, L.; Martines, R.; Greer, P.; Paddock, C.D.; Peltecu, G.; Lupulescu, E.; Icenogle, J.; Zaki, S.R. Immunolocalization and distribution of rubella antigen in fatal congenital rubella syndrome. *EBioMedicine* **2016**, *3*, 86–92. [CrossRef] [PubMed]
4. Freij, B.J.; South, M.A.; Sever, J.L. Maternal rubella and the congenital rubella syndrome. *Clin. Perinatol.* **1988**, *15*, 247–257. [CrossRef]
5. Bouthry, E.; Picone, O.; Hamdi, G.; Grangeot-Keros, L.; Ayoubi, J.M.; Vauloup-Fellous, C. Rubella and pregnancy: Diagnosis, management and outcomes. *Prenat. Diagn.* **2014**, *34*, 1246–1253. [CrossRef] [PubMed]
6. Enders, G.; Nickerl-Pacher, U.; Miller, E.; Cradock-Watson, J.E. Outcome of confirmed periconceptional maternal rubella. *Lancet* **1988**, *1*, 1445–1447. [CrossRef]
7. Rossant, J.; Tam, P.P.L. Exploring early human embryo development. *Science* **2018**, *360*, 1075–1076. [CrossRef] [PubMed]
8. Kugler, J.; Huhse, B.; Tralau, T.; Luch, A. Embryonic stem cells and the next generation of developmental toxicity testing. *Expert Opin. Drug Metab. Toxicol.* **2017**, *13*, 833–841. [CrossRef]
9. Van Vliet, P.; Wu, S.M.; Zaffran, S.; Puceat, M. Early cardiac development: A view from stem cells to embryos. *Cardiovasc. Res.* **2012**, *96*, 352–362. [CrossRef]
10. Simunovic, M.; Brivanlou, A.H. Embryoids, organoids and gastruloids: New approaches to understanding embryogenesis. *Development* **2017**, *144*, 976–985. [CrossRef]
11. Meganathan, K.; Jagtap, S.; Wagh, V.; Winkler, J.; Gaspar, J.A.; Hildebrand, D.; Trusch, M.; Lehmann, K.; Hescheler, J.; Schluter, H.; et al. Identification of thalidomide-specific transcriptomics and proteomics signatures during differentiation of human embryonic stem cells. *PLoS ONE* **2012**, *7*, e44228. [CrossRef]
12. Luz, A.L.; Tokar, E.J. Pluripotent stem cells in developmental toxicity testing: A review of methodological advances. *Toxicol. Sci.* **2018**, *165*, 31–39. [CrossRef]
13. Hubner, D.; Jahn, K.; Pinkert, S.; Bohnke, J.; Jung, M.; Fechner, H.; Rujescu, D.; Liebert, U.G.; Claus, C. Infection of ipsc lines with miscarriage-associated coxsackievirus and measles virus and teratogenic rubella virus as a model for viral impairment of early human embryogenesis. *ACS Infect. Dis.* **2017**, *3*, 886–897. [CrossRef]
14. Wu, X.; Dao Thi, V.L.; Huang, Y.; Billerbeck, E.; Saha, D.; Hoffmann, H.H.; Wang, Y.; Silva, L.A.V.; Sarbanes, S.; Sun, T.; et al. Intrinsic immunity shapes viral resistance of stem cells. *Cell* **2018**, *172*, 423–438. [CrossRef]

15. Hong, X.X.; Carmichael, G.G. Innate immunity in pluripotent human cells: Attenuated response to interferon-beta. *J. Biol. Chem.* **2013**, *288*, 16196–16205. [CrossRef]
16. Dziedzicka, D.; Markouli, C.; Barbe, L.; Spits, C.; Sermon, K.; Geens, M. A high proliferation rate is critical for reproducible and standardized embryoid body formation from laminin-521-based human pluripotent stem cell cultures. *Stem Cell Rev.* **2016**, *12*, 721–730. [CrossRef]
17. Wirth, H.; Loffler, M.; von Bergen, M.; Binder, H. Expression cartography of human tissues using self organizing maps. *BMC Bioinform.* **2011**, *12*, 306. [CrossRef]
18. Subramanian, A.; Tamayo, P.; Mootha, V.K.; Mukherjee, S.; Ebert, B.L.; Gillette, M.A.; Paulovich, A.; Pomeroy, S.L.; Golub, T.R.; Lander, E.S.; et al. Gene set enrichment analysis: A knowledge-based approach for interpreting genome-wide expression profiles. *Proc. Natl. Acad. Sci. USA* **2005**, *102*, 15545–15550. [CrossRef]
19. Wirth, H.; von Bergen, M.; Murugaiyan, J.; Rosler, U.; Stokowy, T.; Binder, H. Maldi-typing of infectious algae of the genus prototheca using som portraits. *J. Microbiol. Methods* **2012**, *88*, 83–97. [CrossRef]
20. Loffler-Wirth, H.; Kalcher, M.; Binder, H. Opossom: R-package for high-dimensional portraying of genome-wide expression landscapes on bioconductor. *Bioinformatics* **2015**, *31*, 3225–3227. [CrossRef]
21. Nersisyan, L.; Löffler-Wirth, H.; Arakelyan, A.; Binder, H. Gene set- and pathway- centered knowledge discovery assigns transcriptional activation patterns in brain, blood, and colon cancer: A bioinformatics perspective. *Int. J. Knowl. Discov. Bioinform.* **2014**, *4*, 46–69. [CrossRef]
22. Claus, C.; Bergs, S.; Emmrich, N.C.; Hubschen, J.M.; Mankertz, A.; Liebert, U.G. A sensitive one-step taqman amplification approach for detection of rubella virus clade i and ii genotypes in clinical samples. *Arch. Virol.* **2017**, *162*, 477–486. [CrossRef]
23. Hellemans, J.; Mortier, G.; De Paepe, A.; Speleman, F.; Vandesompele, J. Qbase relative quantification framework and software for management and automated analysis of real-time quantitative pcr data. *Genome Biol.* **2007**, *8*, R19. [CrossRef]
24. Wang, H.; Zhang, K.; Liu, Y.; Fu, Y.; Gao, S.; Gong, P.; Wang, H.; Zhou, Z.; Zeng, M.; Wu, Z.; et al. Telomere heterogeneity linked to metabolism and pluripotency state revealed by simultaneous analysis of telomere length and rna-seq in the same human embryonic stem cell. *BMC Biol.* **2017**, *15*, 114. [CrossRef]
25. Ben-Porath, I.; Thomson, M.W.; Carey, V.J.; Ge, R.; Bell, G.W.; Regev, A.; Weinberg, R.A. An embryonic stem cell-like gene expression signature in poorly differentiated aggressive human tumors. *Nat. Genet.* **2008**, *40*, 499–507. [CrossRef]
26. Chu, L.F.; Leng, N.; Zhang, J.; Hou, Z.; Mamott, D.; Vereide, D.T.; Choi, J.; Kendziorski, C.; Stewart, R.; Thomson, J.A. Single-cell rna-seq reveals novel regulators of human embryonic stem cell differentiation to definitive endoderm. *Genome Biol.* **2016**, *17*, 173. [CrossRef]
27. Hopp, L.; Loeffler-Wirth, H.; Nersisyan, L.; Arakelyan, A.; Binder, H. Footprints of sepsis framed within community acquired pneumonia in the blood transcriptome. *Front Immunol* **2018**, *9*, 1620. [CrossRef]
28. Andres-Terre, M.; McGuire, H.M.; Pouliot, Y.; Bongen, E.; Sweeney, T.E.; Tato, C.M.; Khatri, P. Integrated, multi-cohort analysis identifies conserved transcriptional signatures across multiple respiratory viruses. *Immunity* **2015**, *43*, 1199–1211. [CrossRef]
29. Sweeney, T.E.; Wong, H.R.; Khatri, P. Robust classification of bacterial and viral infections via integrated host gene expression diagnostics. *Sci. Transl. Med.* **2016**, *8*, 346ra91. [CrossRef]
30. Schmidtke, L.; Schrick, K.; Saurin, S.; Kafer, R.; Gather, F.; Weinmann-Menke, J.; Kleinert, H.; Pautz, A. The kh-type splicing regulatory protein (ksrp) regulates type iii interferon expression post-transcriptionally. *Biochem. J.* **2019**, *476*, 333–352. [CrossRef]
31. Winzen, R.; Thakur, B.K.; Dittrich-Breiholz, O.; Shah, M.; Redich, N.; Dhamija, S.; Kracht, M.; Holtmann, H. Functional analysis of ksrp interaction with the au-rich element of interleukin-8 and identification of inflammatory mrna targets. *Mol. Cell. Biol.* **2007**, *27*, 8388–8400. [CrossRef]
32. Dambacher, S.; Hahn, M.; Schotta, G. Epigenetic regulation of development by histone lysine methylation. *Heredity* **2010**, *105*, 24–37. [CrossRef]
33. Thalheim, T.; Hopp, L.; Binder, H.; Aust, G.; Galle, J. On the cooperation between epigenetics and transcription factor networks in the specification of tissue stem cells. *Epigenomes* **2018**, *2*, 20. [CrossRef]
34. Grandy, R.A.; Whitfield, T.W.; Wu, H.; Fitzgerald, M.P.; VanOudenhove, J.J.; Zaidi, S.K.; Montecino, M.A.; Lian, J.B.; van Wijnen, A.J.; Stein, J.L.; et al. Genome-wide studies reveal that h3k4me3 modification in bivalent genes is dynamically regulated during the pluripotent cell cycle and stabilized upon differentiation. *Mol. Cell. Biol.* **2016**, *36*, 615–627. [CrossRef]

35. Bratt-Leal, A.M.; Carpenedo, R.L.; McDevitt, T.C. Engineering the embryoid body microenvironment to direct embryonic stem cell differentiation. *Biotechnol. Prog.* **2009**, *25*, 43–51. [CrossRef]
36. Zhang, K.T.; Che, S.Y.; Su, Z.; Zheng, S.Y.; Zhang, H.Y.; Yang, S.L.; Li, W.D.; Liu, J.P. Cd90 promotes cell migration, viability and sphere-forming ability of hepatocellular carcinoma cells. *Int. J. Mol. Med.* **2018**, *41*, 946–954. [CrossRef]
37. Liu, C.; Peng, G.; Jing, N. Tgf-beta signaling pathway in early mouse development and embryonic stem cells. *Acta Biochim. Biophys. Sin.* **2018**, *50*, 68–73. [CrossRef]
38. Lindsley, R.C.; Gill, J.G.; Kyba, M.; Murphy, T.L.; Murphy, K.M. Canonical wnt signaling is required for development of embryonic stem cell-derived mesoderm. *Development* **2006**, *133*, 3787–3796. [CrossRef]
39. Weed, L.S.; Mills, J.A. Strategies for retinal cell generation from human pluripotent stem cells. *Stem Cell Investig.* **2017**, *4*, 65. [CrossRef]
40. Diacou, R.; Zhao, Y.; Zheng, D.; Cvekl, A.; Liu, W. Six3 and six6 are jointly required for the maintenance of multipotent retinal progenitors through both positive and negative regulation. *Cell Rep.* **2018**, *25*, 2510–2523. [CrossRef]
41. Worley, K.E.; Rico-Varela, J.; Ho, D.; Wan, L.Q. Teratogen screening with human pluripotent stem cells. *Integr. Biol.* **2018**, *10*, 491–501. [CrossRef]
42. Rathjen, J. The states of pluripotency: Pluripotent lineage development in the embryo and in the dish. *ISRN Stem Cells* **2014**, *2014*, 19. [CrossRef]
43. Fu, Y.; Zhou, Z.; Wang, H.; Gong, P.; Guo, R.; Wang, J.; Lu, X.; Qi, F.; Liu, L. Ifitm1 suppresses expression of human endogenous retroviruses in human embryonic stem cells. *FEBS Open Bio* **2017**, *7*, 1102–1110. [CrossRef]
44. Wolf, D.; Goff, S.P. Embryonic stem cells use zfp809 to silence retroviral dnas. *Nature* **2009**, *458*, 1201–1204. [CrossRef]
45. Eggenberger, J.; Blanco-Melo, D.; Panis, M.; Brennand, K.J.; tenOever, B.R. Type i interferon response impairs differentiation potential of pluripotent stem cells. *Proc. Natl. Acad. Sci. USA* **2019**, *116*, 1384–1393. [CrossRef]
46. Pervolaraki, K.; Talemi, S.R.; Albrecht, D.; Bormann, F.; Bamford, C.; Mendoza, J.L.; Garcia, K.C.; McLauchlan, J.; Hofer, T.; Stanifer, M.L.; et al. Differential induction of interferon stimulated genes between type i and type iii interferons is independent of interferon receptor abundance. *PLoS Pathog.* **2018**, *14*, e1007420. [CrossRef]
47. Zahedi-Amiri, A.; Sequiera, G.L.; Dhingra, S.; Coombs, K.M. Influenza a virus-triggered autophagy decreases the pluripotency of human-induced pluripotent stem cells. *Cell Death Dis.* **2019**, *10*, 337. [CrossRef]
48. Heavner, W.; Pevny, L. Eye development and retinogenesis. *Cold Spring Harb. Perspect. Biol.* **2012**, *4*, a008391. [CrossRef]
49. Geyer, H.; Bauer, M.; Neumann, J.; Ludde, A.; Rennert, P.; Friedrich, N.; Claus, C.; Perelygina, L.; Mankertz, A. Gene expression profiling of rubella virus infected primary endothelial cells of fetal and adult origin. *Virol. J.* **2016**, *13*, 21. [CrossRef]
50. Lough, J.; Sugi, Y. Endoderm and heart development. *Dev. Dyn.* **2000**, *217*, 327–342. [CrossRef]
51. Mulloy, B.; Rider, C.C. The bone morphogenetic proteins and their antagonists. *Vitam. Horm.* **2015**, *99*, 63–90.
52. Foley, A.C.; Korol, O.; Timmer, A.M.; Mercola, M. Multiple functions of cerberus cooperate to induce heart downstream of nodal. *Dev. Biol.* **2007**, *303*, 57–65. [CrossRef]
53. Vokes, S.A.; Krieg, P.A. Endoderm is required for vascular endothelial tube formation, but not for angioblast specification. *Development* **2002**, *129*, 775–785.
54. Mikkelsen, T.S.; Ku, M.; Jaffe, D.B.; Issac, B.; Lieberman, E.; Giannoukos, G.; Alvarez, P.; Brockman, W.; Kim, T.K.; Koche, R.P.; et al. Genome-wide maps of chromatin state in pluripotent and lineage-committed cells. *Nature* **2007**, *448*, 553–560. [CrossRef]
55. Hofstetter, C.; Kampka, J.M.; Huppertz, S.; Weber, H.; Schlosser, A.; Muller, A.M.; Becker, M. Inhibition of kdm6 activity during murine esc differentiation induces DNA damage. *J. Cell Sci.* **2016**, *129*, 788–803. [CrossRef]
56. Fisher, J.B.; Pulakanti, K.; Rao, S.; Duncan, S.A. Gata6 is essential for endoderm formation from human pluripotent stem cells. *Biol Open* **2017**, *6*, 1084–1095. [CrossRef]
57. Pfeiffer, M.J.; Quaranta, R.; Piccini, I.; Fell, J.; Rao, J.; Ropke, A.; Seebohm, G.; Greber, B. Cardiogenic programming of human pluripotent stem cells by dose-controlled activation of eomes. *Nat. Commun.* **2018**, *9*, 440. [CrossRef]

58. Pfister, S.; Jones, V.J.; Power, M.; Truisi, G.L.; Khoo, P.L.; Steiner, K.A.; Kanai-Azuma, M.; Kanai, Y.; Tam, P.P.; Loebel, D.A. Sox17-dependent gene expression and early heart and gut development in sox17-deficient mouse embryos. *Int. J. Dev. Biol.* **2011**, *55*, 45–58. [CrossRef]
59. Serrano, A.G.; Gandillet, A.; Pearson, S.; Lacaud, G.; Kouskoff, V. Contrasting effects of sox17- and sox18-sustained expression at the onset of blood specification. *Blood* **2010**, *115*, 3895–3898. [CrossRef]
60. George, R.M.; Firulli, A.B. Hand factors in cardiac development. *Anat. Rec.* **2019**, *302*, 101–107. [CrossRef]

Review

Oncolytic Virus Encoding a Master Pro-Inflammatory Cytokine Interleukin 12 in Cancer Immunotherapy

Hong-My Nguyen †, Kirsten Guz-Montgomery † and Dipongkor Saha *

Department of Immunotherapeutics and Biotechnology, Texas Tech University Health Sciences Center School of Pharmacy, Abilene, TX 79601, USA; My.Nguyen@ttuhsc.edu (H.-M.N.); Kirsten.Montgomery@ttuhsc.edu (K.G.-M.)

* Correspondence: dipongkor.saha@ttuhsc.edu; Tel.: +1-325-696-0583

† These authors contributed equally to this work.

Received: 21 December 2019; Accepted: 29 January 2020; Published: 10 February 2020

Abstract: Oncolytic viruses (OVs) are genetically modified or naturally occurring viruses, which preferentially replicate in and kill cancer cells while sparing healthy cells, and induce anti-tumor immunity. OV-induced tumor immunity can be enhanced through viral expression of anti-tumor cytokines such as interleukin 12 (IL-12). IL-12 is a potent anti-cancer agent that promotes T-helper 1 (Th1) differentiation, facilitates T-cell-mediated killing of cancer cells, and inhibits tumor angiogenesis. Despite success in preclinical models, systemic IL-12 therapy is associated with significant toxicity in humans. Therefore, to utilize the therapeutic potential of IL-12 in OV-based cancer therapy, 25 different IL-12 expressing OVs (OV-IL12s) have been genetically engineered for local IL-12 production and tested preclinically in various cancer models. Among OV-IL12s, oncolytic herpes simplex virus encoding IL-12 (OHSV-IL12) is the furthest along in the clinic. IL-12 expression locally in the tumors avoids systemic toxicity while inducing an efficient anti-tumor immunity and synergizes with anti-angiogenic drugs or immunomodulators without compromising safety. Despite the rapidly rising interest, there are no current reviews on OV-IL12s that exploit their potential efficacy and safety to translate into human subjects. In this article, we will discuss safety, tumor-specificity, and anti-tumor immune/anti-angiogenic effects of OHSV-IL12 as mono- and combination-therapies. In addition to OHSV-IL12 viruses, we will also review other IL-12-expressing OVs and their application in cancer therapy.

Keywords: cancer immunotherapy; oncolytic virus; herpes simplex virus; immune checkpoint inhibitor; angiogenesis inhibitor

---

1. Introduction

Interleukin 12 (IL-12) is a powerful master regulator of both innate and adaptive anti-tumor immune responses. As a heterodimeric cytokine, it produces multifaceted anti-tumor effects [1,2], including stimulation of growth and cytotoxic activity of natural killer (NK) cells and T cells (both $CD4^+$ and $CD8^+$) [1,3–5], induction of differentiation of $CD4^+$ T cells towards Th1 phenotype [6,7], increased production of IFN-γ from NK and T cells [1,8,9], and inhibition of tumor angiogenesis [1,10]. Despite encouraging success in preclinical studies [4], the early stages of IL-12 clinical trials did not meet expectations. Severe adverse events were first reported on 15 out of 17 patients in a phase II clinical trial following intravenous IL-12 administration, and the trial was immediately terminated by the FDA following two cases of death [11,12]. Although success was observed in cutaneous T cell lymphoma variants [13,14], AIDS-related Kaposi sarcoma [15] and non–Hodgkin's lymphoma [16], severity of side effects outweighed effectiveness of IL-12 based therapies in the vast majority of oncology clinical trials [17]. In an effort to optimize efficacy and enhance the safety profile, alternative approaches are being studied to localize IL-12 expression at the tumor microenvironment.

Recent studies show that systemic toxicity of IL-12 is limited when expressed by oncolytic viruses (OVs) locally in the tumors [18–20] and in the brains of non-human primates [21]. OVs are a distinct class of anti-cancer agents with unique mechanisms of action: (i) selectively replicating in and killing cancer cells (i.e., oncolysis) without harming healthy cells or tissues [22–24], and (ii) exposing viral/tumor antigens, which promote a cascade of anti-tumor innate and adaptive immune responses (i.e., in situ vaccine effects) [25,26]. The OV-induced vaccine effects can be further enhanced through viral expression of anti-tumor cytokines such as IL-12 [18–20], as illustrated in Figure 1. Cancer immunotherapy involving OVs is an emerging and increasingly examined therapeutic approach for the treatment of cancer [27,28]. Among OVs, oncolytic herpes simplex virus (OHSV) is the furthest along in the clinic and approved by the FDA for the treatment of advanced melanoma [29]. To utilize the therapeutic potential of IL-12, there are several OHSVs encoding IL-12 which have been genetically engineered and tested in various cancers (Table 1). In addition to OHSVs, several different OVs such as adenoviruses, measles virus, maraba virus, Newcastle disease virus, Semliki forest virus, vesicular stomatitis virus and Sindbis virus are also being engineered to express IL-12 (Table 2). Our literature research found that 25 different types of OV-IL12s are either being or have recently been explored (see Tables 1 and 2). Despite this rapidly rising interest, there are no reviews on IL-12 expressing viruses that exploit their potential efficacy and safety to translate into human subjects. This review presents the most current data on this topic and provides a basic understanding of OV-IL12 as a promising treatment approach in cancer immunotherapy, which ultimately could support continued research in the future. More specifically, in this review, we will discuss safety, tumor-specificity, and anti-tumor/anti-vascular effects of OHSV-IL12 as monotherapy or combination therapy. In addition to OHSV-IL12 viruses, we will also review other IL-12 expressing OVs and their application in cancer therapy.

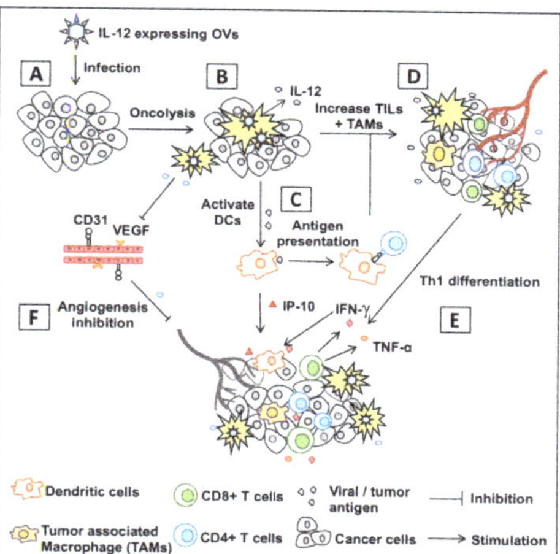

Figure 1. Graphic presentation of mechanism of action of oncolytic virus encoding IL-12. (**A**) Infection of tumor with oncolytic virus encoding IL-12 (OV-IL12). (**B**) OV-IL12 replicates in and kills cancer cells (i.e., oncolysis) and releases IL-12 in the tumor microenvironment. (**C**) Neoantigens from lysed cancer cells activate and recruit dendritic cells (DCs) into the tumor microenvironment. DCs process neoantigens, travel to nearest lymphoid organs, and present the antigen to T cells ($CD4^+$ and $CD8^+$ T cells). (**D**,**E**) T cells migrate to the site of infection (referred as tumor-infiltrating T cells or TILs), differentiate into Th1 cells, produce anti-tumor cytokines and kill cancer cells. (**F**) IL-12-induced production of IFN-γ and interferon inducible protein 10 (IP-10) produces anti-angiogenetic effect through reduction of tumoral vascular endothelial growth factor (VEGF) and $CD31^+$ tumor endothelial cells.

Table 1. List of OHSV-IL12s and their efficacy in pre-clinical cancer models.

Virus	Genomic Modification	Cancer Model	RoA	Dose (pfu)	Efficacy	Ref.
G47Δ-mIL12	ΔICP6, ΔΔICP34.5, ΔICP47, ○LacZ, ○mIL-12	Intracranial 005 GSC (Glioblastoma)	I.T.	5×10^5	Inhibited intracranial tumor growth and extended survival. Promoted IL-12 expression, stimulated IFN-γ production, upregulated IP-10, and inhibited VEGF. Polarized T_H1 response and inhibited T-regs function.	[19,30]
T-mfIL12	ΔICP6, ΔΔICP34.5, ΔICP47, ○mIL-12	Intracerebral Neuro2a (neuroblastma)	I.V.	5×10^6	Prolonged survival (Mock vs. T-mfIL12, $p < 0.05$), although not statistically significant versus T-01 treatments.	[31]
NV1042	ΔICP0, ΔICP4, ΔICP34.5, ΔUL56, ΔICP47, Us11Δ, Us10Δ, UL56 (duplicated), ○mIL-12	Subcutaneous SCC VII (Squamous Cell Carcinoma)	I.T.	1×10^7	Reduced tumor volume and improved survival (3 doses of 2×10^7 pfu). 57% of mice from NV1042 group rejected subsequent SCC re-challenge in the contralateral flank compared with 14% in NV1023 or NV1034 group.	[32]
			I.V.	5×10^7	NV1042 treatment resulted in 100% survival, in contrast to 70% of NV1023 and 0% of PBS.	[33]
M002	ΔΔICP34.5, ○mIL-12	Intracranial X21415 (Pediatric embryonal tumor); D456 (pediatric glioblastoma); GBM-12 and UAB106 (adult glioblastoma)	I.T.	1×10^7	M002 significantly prolonged survival in mice bearing all 4 types of tumor compared to saline. No difference in survival was observed compared with G207, excluding X21415 with high levels of nectin-1	[34]
		Intracranial SCK (brain metastasized breast cancer)	I.T.	1.5×10^7	Single injection of M002 extended the survival of treated animals more effectively than a non-cytokine control virus.	[35]
		Xenograft SK-N-AS and SK-N-BE (2) (human neuroblastoma); subcutaneous Neuro-2a (murine neuroblastoma)	I.T.	1×10^7	Significant decrease in tumor growth were observed in both SK-N-AS and SK-N-BE (2) cell lines. Extended median survival compared to the parent R3659.	[36]
		HuH6 (human hepatoblastoma; G401 (human malignant rhabdoid kidney tumor); SK-NEP-1 (renal Ewing sarcoma)	I.T.	1×10^7	M002 significantly reduced tumor volume and increased survival over those treated with vehicle alone in all three different xenograft models.	[37]
R-115	Virulent with retargeted HER-2, ○mIL-12	pLV-HER2-nectin-puro	I.P.	1×10^8 to 2×10^9	Induced greater local and systemic anti-tumor immunity and durable response than unarmed R-LM113 in both early and late schedule. All mice that survived from primary tumor challenge were protected from the distant challenge tumor and subsequent re-challenge. Increased number of CD8+ and CD141+ cells, PD-L1+ tumor cells, and Treg with a decrease in the number of CD11b+ cells. Enhanced Th1 polarization and increased expression of IFN-γ, IL-2, Granzyme B, T-bet and TNFα and tumor infiltrating lymphocytes	[38]
		Orthotopic mHGG^{pdgf}-hHER2 (glioblastoma)	I.T.	Low dose: 2×10^6 High dose: 1×10^8	27% of mice treated with R-115 ($n = 6$, 4 low-dose arm and 2 high-dose arm) alive 100 days after the virus treatment versus all mice died within 48 days. Increased infiltrating CD4+ and CD8+, and expression of IFN-γ	[39]

Table 1. *Cont.*

Virus	Genomic Modification	Cancer Model	RoA	Dose (pfu)	Efficacy	Ref.
vHSV-IL-12	ΔICP6, ΔΔICP34.5, ∘mIL-12	Subcutaneous Neuro2a (neuroblastoma)	I.T.	1×10^4	Significantly reduced tumor growth versus vHSV-null and other cytokine armed groups.	[40]
T2850	ΔIR 15,091bp, ∘mIL-12	Subcutaneous A20 (Murine B Lymphoma), MC38 (colon adenocarcinoma), MFC (Murine Forestomach Carcinoma)	I.T.	1×10^7	Reduced tumor volume compared to IL-12 unarmed parental group. IFN-γ level was markedly increased in the tumor bed and sera of mice infected with both T2850 and T3855 by day 4.	[41]
T3855	ΔIR 15,091bp, ∘mIL-12, ∘mPD-1	Subcutaneous B16 (melanoma)		5×10^6, 1×10^7, 3×10^7		
T3011	ΔIR15,091bp, ∘hIL-12, ∘hPD-1	Subcutaneous B16 (melanoma)	I.T.	5×10^6, 1×10^7 or 3×10^7	Reduced tumor volume as compared with control group.	[41]

Δ—deletion, ∘ insertion, RoA—Route of administration, I.T—intratumorally, IP—intraperitoneally, pfu—plaque forming unit, ref.—reference, VEGF—vascular endothelial growth factor.

Table 2. List of IL-12 expressing oncolytic viruses (other than OHSVs) and their efficacy in pre-clinical cancer models.

Virus	Strain	Cancer Model	RoA	Dose	Efficacy	Ref
Adenovirus	Ad5-yCD/ mutTKSR39 rep-mIL12	Subcutaneous TRAMP (C2 prostate adenocarcinoma)	I.T.	5×10^8 pfu	Improved local and metastatic tumor control. Increased NK and CTL cytolytic activities. Significantly increased survival, levels of IL-12 and IFN-γ in serum and tumor.	[42,43]
	Ad-TD-IL-12, Ad-TD-nsIL-12	Subcutaneous HPD1NR (pancreatic cancer)	I.T.	1×10^9 pfu	100% tumor eradication and survival of both IL-12 modified Adenovirus treated animals. Ad-TD-IL-12, but not Ad-TD-nsIL-12 resulted in a significant increase in $CD3^+CD4^-CD8^+$ populations in the spleen. Level of splenic IFN-γ, IP-10 and lymph node IFN-γ were lower in Ad-TD-nsIL-12 compared to Ad-TD-IL-12 treated hamster.	[44]
	Ad-ΔB7/IL12/GMCSF	Subcutaneous B16-F10 (melanoma)	I.T.	5×10^7 pfu	Primary tumor growth was better controlled in Ad-ΔB7/IL12/GMCSF and Ad-ΔB7/IL12 compared to Ad-ΔB7GMCSF or PBS. Increased tumor infiltrating $CD86^+$ APCs and enhanced CD4+ and CD8+ T cell-mediated Th1 antitumor immune response.Reduced VEGF expression in the tumor treated with oncolytic Ad co-expressing IL-12 and GM-CSF or IL-12 alone. IFN-γ, TNF-α and IL-6 were higher in Ad-ΔB7/IL12/GMCSF and Ad-ΔB7/IL12 compared to Ad-ΔB7GMCSF or PBS.	[45]
	RdB/IL-12/ IL-18	Subcutaneous B16-F10 (melanoma)	I.T.	1×10^8 pfu	95% and 99% tumor growth inhibition was observed in treatment with RdB/IL-12 and RdB/IL-12/IL-18, respectively. Increased Th1/Th2 cytokine ratio and increased tumor infiltration of $CD4^+$ T, $CD8^+$ T and NK cells. Promoted differentiation of T cells expressing IL-12Rβ2 or IL-18Rα.	[46]
	RdB/IL12/ DCN	Orthotopic 4T1 (Triple negative breast cancer)	I.T.	2×10^{10} VP	Both of the IL-12-expressing oncolytic Ads showed similar tumor growth inhibition up to day 9 after initial treatment. RdB/IL12/DCN increased upregulation of IFN-γ, TNF-α, infiltrating cytotoxic lymphocytes, downregulation of TGF-β expression and T-regs compared to RdB/IL12 and RdB/DCN.	[47]
	YKL-IL12/B7	Subcutaneous B16-F10 (melanoma)	I.T.	5×10^8 pfu	Tumor growth was suppressed in both YKL-IL12 and YKL-IL12/B7 treated mice vs PBS. YKL-IL12- or YKL-IL12/B7-treated mice produced a significantly greater level of IFN-γ, infiltrating APCs, CD4+, CD8+ compared with PBS.	[48]
	Ad-ΔB7/ IL-12/4-1BBL	Subcutaneous B16-F10 (melanoma)	I.T.	5×10^9 VP	100% of mice in the Ad-ΔB7/IL-12/4-1BBL group survived >30 days after initial viral injection compared with 20% of that in virus expressing either IL-12 or 4-1BBL. Mice treated with Ad-ΔB7/IL-12 or Ad-ΔB7/IL-12/4-1BBL had greater amount of tumor infiltrating CD4+ and CD8+ compared to Ad-ΔB7/4-1BBL and Ad-ΔB7.	[49]

Table 2. *Cont.*

Virus	Strain	Cancer Model	RoA	Dose	Efficacy	Ref
Measles virus	MeVac FmIL-12	Subcutaneous MC38ce (colon carcinoma)	I.T.	5×10^5–1× 10^6 ciu	Tumor remissions in 90% of animals. Driven polarization of Th1-associated immune response and increased tumor infiltrating CD8⁺ T cells. Increased IFN-γ and TNF-α, and polarization of Th1-associated immune response. Co-expression of IL-12 and IL-15 showed synergistic effect.	[20,50]
Maraba Virus	MG1-IL12-ICV	CT26 and B16F10 peritoneal carcinomatosis	I.P	Seeding dose 5×10^5, then 1×10^4 on day 3	Reduced tumor burden and improved mouse survival. Activated and matured DCs to secrete IP-10, and activated and recruited NK cells. Increased production of IFN-γ	[51]
Newcastle disease virus	rClone30–IL 12	Orthotopic H22 (hepatocarcinoma)	I.T.	1×10^7 pfu	Reduced tumor volume and improved percentage of survival. Increased IFN-γ and IP-10. Co-expression of IL-12 and IL-2 showed synergistic effect.	[52]
Semliki Forest virus	rSFV/IL12	Subcutaneous B16 (melanoma)	I.T.	10^7 IU	Single injection with SFV-IL12 resulted in significant tumor regression. 2 days after injection, IFN-γ production increased with inhibition of tumor vascularization. Splenic IP-10 and MIG expression was increased.	[53]
	SFV/IL12	Subcutaneous P815 (mastocytoma)	I.T.	10^6 IU	Significantly delayed P815 tumor growth. 40–53% of mice exhibited complete tumor regressions. Induced high levels of IFN-γ production in draining lymph nodes.	[54]
	SFV/IL12	Subcutaneous MC38 (colon adenocarcinoma)	I.T.	10^8 particles	Reduced tumor volume and improved percentage of survival. Increased tumor-specific CD8+ T lymphocytes. Enhanced the expression of CD11c, CD8α, CD40, and CD86 of tumor-infiltrating M-MDSCs in the presence of an intact endogenous IFN-I system.	[55]
	SFV-VLP-	Syngeneic RG2 (rat glioma)	I.T.	5×10^7 (low-dose) or 5×10^8 (high-dose)	Reduction in tumor volume (70%—low dose; 87%—high dose)	[56]
Vesicular stomatitis virus (VSV)	VSV-IL12	Orthotopic SCC VII (squamous cell Carcinoma)	I.T.	MOI 0.01	Significant reduction in tumor volume, and prolonged survival.	[57]
Sindbis virus	Sin/IL12	Orthotopic ES-2 cells (ovarian clear cell Carcinoma)	I.P	10^7 pfu	Reduced tumor growth and improved survival. Activated and matured DCs, activated and recruited NK cells. Increased production of IFN-γ.	[58]

Δ—deletion, RoA—Route of administration, I.T—intratumorally, IP—intraperitoneally, pfu—plaque forming unit, ref. —reference, MOI—multiplicity of infection, VP—viral particle, IU—infectious units.

2. Genetically Engineered IL-12 Expressing OHSVs and Their Therapeutic Efficacy

2.1. IL-12 Insertion Does Not Compromise Safety and Tumor Specificity of Genetically Modified OHSV-IL12 Viruses

Herpes simplex virus type 1 (HSV-1) has a large 152 kb genome, and therefore, deletion or mutation of various genes (Figure 2) does not fundamentally alter its functional properties (such as infectivity, viral replication, etc.), but rather confers tumor cell-specific replication, with no or reduced toxicity (e.g., neuropathogenicity) [59]. For example, G47Δ-IL12 (an IL-12 expressing OHSV) has three genomic modifications that endow its tumor-specificity and safety: ICP6 inactivation, γ-(ICP)34.5 and ICP47 deletion, and IL-12 insertion [30,60]. ICP6, encodes for viral ribonucleotide reductase, controls nucleotide metabolism and helps HSV to replicate in healthy or non-dividing cells that are inherently lacking sufficient nucleotide pools [59,61] (Figure 3A). ICP6 inactivation by fusion of LacZ does not hamper DNA synthesis in cancer cells (Figure 3B) [59,62] but the lack of ribonucleotide reductase results in no nucleotide metabolism and no viral DNA replication in healthy cells (Figure 3C) [59,61–63]. Therefore, mutation in the ICP6 gene makes viral infection and replication tumor-specific and thereby increases safety. Similar to ICP6 inactivation, deletion of γ-34.5 also increases safety and cancer selectivity [22,59,64]. Healthy cells have various anti-viral defense mechanisms. For example, protein kinase R (PKR) phosphorylates eukaryotic translation initiation factor eIF2α, which shuts down synthesis of foreign proteins or viral antigens (Figure 4A) [65,66]. HSV with intact γ-34.5 overturns anti-viral defense in healthy cells through γ-34.5-mediated dephosphorylation of eIF2α and helps in viral protein synthesis and viral replication in healthy cells (Figure 4B) [67,68]. Therefore, γ-34.5 deletion results in no eIF2α dephosphorylation, and thereby, no protein synthesis and viral replication (Figure 4C) [67,68]. However, in cancer cells, γ-34.5-deleted HSV can freely replicate, since cancer cells usually have defects in anti-viral pathways such as PKR-eIF2α pathway (Figure 4D). ICP47 downregulates MHC class I presentation through inhibition of transporter-associated protein (TAP) channel [69,70] and prevents detection of OHSV-infected cancer cells by virus-specific $CD8^+$ cells [71]. Thus, deletion of ICP47 enhances MHC class I expression and immune response against virus-infected tumor cells and the host's anti-tumor immunity (innate and adaptive) [59,60,72]. ICP47 deletion also complements the loss of γ-34.5 through immediate early (IE) expression of unique short sequence US11 under the control of ICP47-IE promoter [59,73]. US11 is a true late gene that binds with PKR, preventing it from phosphorylating eIF2a [59,74,75]. IE expression of US11 keeps eIF2α dephosphorylated and helps viral protein translation and synthesis [59,74,75]. Finally, in order to improve efficacy, the master anti-tumor cytokine IL-12 has been inserted into the ICP6 region of OHSV G47Δ to create G47Δ-IL12 [30]. Viral expression of IL-12 does not compromise its tumor specificity and safety, but rather significantly improves its anti-tumor properties [19,26,30,76]. T-mfIL12 (another IL-12 expressing OHSV) has the same backbone as G47Δ but with mIL-12 inserted in the ICP6 deletion region [77]. In vivo safety data showed that intravenous application of T-mfIL12 (5×10^6 pfu) in mice bearing subcutaneous, intracerebral or intravenously disseminated tumors is as safe as non-IL12 expressing OHSV [31].

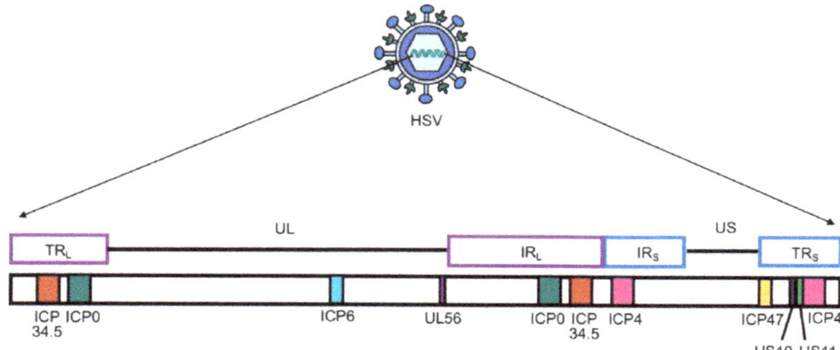

Figure 2. Schematic presentation of herpes simplex virus (HSV) genome with unique long (UL) and short (US) sequences. TR$_{L\ or\ S}$—terminal repeat long or short; IR$_{L\ or\ S}$—internal repeat long or short. Only genes that are modified and/or deleted during construction of OHSV-IL12 are presented. ICP, infected cell protein.

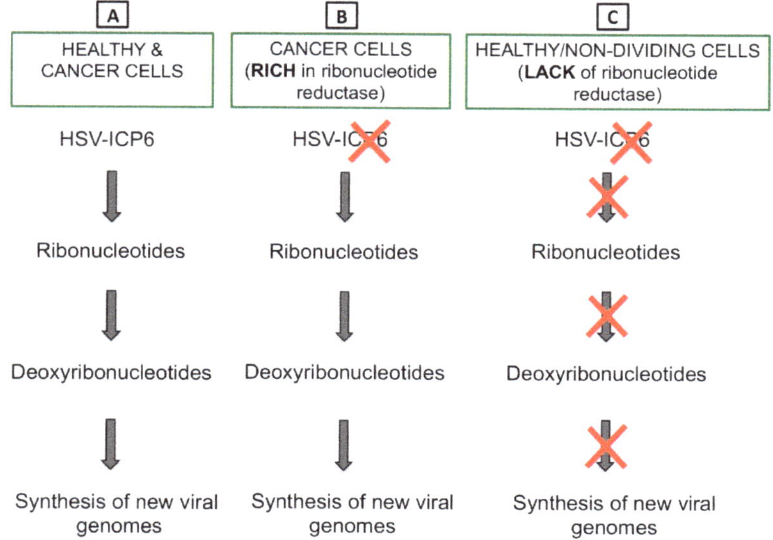

Figure 3. ICP6 inactivation drives tumor-specific replication of OHSV-IL12. (**A**) ICP6 encodes for large subunit of ribonucleotide reductase, which controls nucleotide metabolism and helps HSV to replicate in normal or healthy host cells that are inherently lacking or have insufficient nucleotide pools. (**B**) Cancer cells are rich in ribonucleotide reductase, thus HSV with an inactivated ICP6 does not hamper DNA synthesis in cancer cells. (**C**) Healthy or non-dividing cells lack ribonucleotide reductase. Thus, infection of healthy cells with an ICP6-inactivated HSV leads to no nucleotide metabolism and no viral DNA replication.

Similar to OHSV G47Δ-IL12, a HSV-1/2 recombinant vaccine strain encoding IL-12 is constructed (designated NV1042), which has also multiple deletions/mutations: (i) deletion of one copy of ICP0, ICP4, ICP34.5, and one copy of *UL56* at the $U_{L/S}$ junction, (ii) insertion of *Esherichia coli LacZ* gene under the control of the α47 promoter at the α47 locus, (iii) deletion of ICP47, and (iv) insertion of mIL-12 under the control of a hybrid a4-TK (thymidine kinase) promoter [32,59,78,79]. ICP0 is an important immediate early (IE) protein in switching viral lytic and latent phases that affects defense mechanisms of the host by blocking nuclear factor kappa B (NF-κB)-mediated transcription of immunomodulatory cytokines, inhibiting interferon regulatory factor 3 (IRF3) translocation to the nucleus, inhibiting gamma-interferon inducible protein 16 (IFI16), and degrading mature dendritic cell (DC) markers (CD83) [24,80]. After translocating to the host's nucleus, ICP0 modulates different overlapping cellular pathways to regulate intrinsic and innate antiviral defense mechanism of host cells, allowing the virus to replicate and persist [80,81]. ICP4 blocks apoptosis and positively regulates many other genes in the HSV-1 genome necessary for viral growth [82]. Function of UL56 has not been fully studied but is thought to be involved in neuro-invasiveness of HSV-1 [78]. Therefore, removal of ICP0, ICP4, ICP34.5 and UL56 attenuates virulence and ensures selective viral replication in cancer. In vivo experiment shows no toxicity after intravenous administration of NV1042 (5×10^7 pfu), as demonstrated by lack of cytopathic effects in vital organs (such as lung, brain, spleen, liver, and pancreas) during three months follow up [33]. However, its safety and tumor-selective replication is still a major concern especially for the treatment of tumors located in the central nervous system, since it has 1 intact copy of γ-34.5 (responsible for neuropathogenicity) and intact ribonucleotide reductase ICP6.

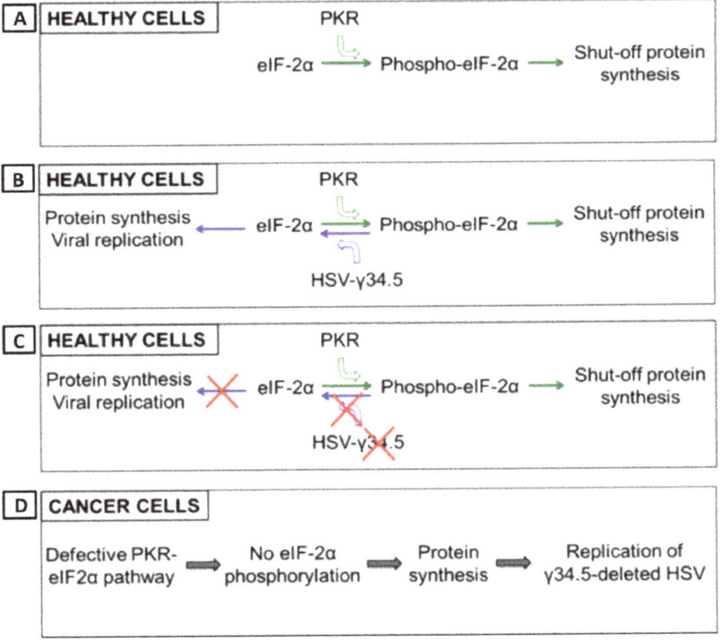

Figure 4. γ-34.5 deletion enhances safety and tumor-specificity of OHSV-IL12. (**A**) Healthy cells have inherent anti-viral defense mechanisms, such as protein kinase R (PKR). PKR phosphorylates translation initiation factor eIF2α, which shuts down synthesis of foreign proteins or viral antigens. (**B**) OHSV with an intact γ-34.5 overturns anti-viral defense in healthy cells through γ-34.5-mediated dephosphorylation of eIF2α and helps in viral protein synthesis/viral replication in healthy cells, leading to development of disease. (**C**) γ-34.5 deletion results in no eIF2α dephosphorylation in normal or healthy cells, and thereby, no protein synthesis and viral replication. (**D**) Cancer cells usually have defective PKR-eIF2α pathway, thus no inhibition of foreign protein synthesis. Therefore, γ-34.5-deleted OHSV can freely replicate in cancer cells.

The OHSV M002 and M032 have deletion of both copies of γ-34.5, with murine and human IL-12 cDNA (p35 and p40 subunits, connected by an IRES), respectively, inserted into each of the γ-34.5 deleted regions [83–86]. M002 has been reported to be safe with no significant toxicity seen after intracerebral inoculation into mice or HSV-sensitive primate Aotus nancymae, despite long-term persistence of viral DNA [87]. M032, with demonstrated safety in non-human primates [21], is now in clinical trial in patients with recurrent glioblastoma (GBM) (see clinical section) [88].

Introducing multiple mutations or deletions in the OHSV genome to confer safety and cancer selectivity may lead to over-attenuation or undermine replication efficiency in cancer cells as opposed to its wild-type or lowly mutated/deleted HSV counterparts [38]. To address this issue, a recent next-generation retargeted IL-12-expressing OHSV known as R-115 has been developed. This OHSV contains no major mutation or deletion and expresses mouse IL-12 under a CMV promoter [38,89]. IL-12-armed R-115 is a derivative of R-LM113 [90]. R-LM113 is a recombinant human epidermal growth factor receptor 2 (HER2) retargeted OHSV with no IL-12 expression, and is successfully engineered by deleting amino acid residues 6 to 38 and by moving the site of single-chain antibody insertion in front of the nectin 1 interacting surface (i.e., at residue 39) [90]. Because of retargeting, it enters and spreads from cancer cell to cell solely via HER2 receptors, and has lost the ability to enter cells through natural glycoprotein D (gD) receptors, herpes virus entry mediator (HVEM) and nectin 1 [90]. Safety profile of R-115 is evaluated in immunocompetent (wt-C57BL/6) model and HER2-transgenic/tolerant counterparts. Mice receiving R-LM113 or R-115 resist very high intraperitoneal OHSV dose of 2×10^9 PFU, which is a lethal dose for wild-type HSV that kills 83% animals [38]. In addition, 4 consecutive intratumoral injections of R-115 at 3–4 days interval shows no viral DNA in vital organs (blood, brain, heart, kidney, liver, brain and spleen) [38]. This indicates that IL12-armed R-115 is safe in mice. However, HER2 specificity makes R-115 applicable only in HER2-expressing tumors, such as mammary tumors, and merely suitable for the treatment of other lethal cancers, such as glioblastoma or other non-HER2-expressing tumors [90]. Recently, another IL-12-expressing OHSV MH1006 (ICP47 deletion and human IL-12 insertion) has been developed and found to be safe in an immunocompetent subcutaneous model of neuroblastoma [91]. However, MH1006 has an intact neurovirulence gene γ-34.5 and an intact ribonucleotide reductase ICP6, thus raising safety concerns in the brain [91]. In Table 1, we have listed all IL-12-expressing OHSVs with their genomic modifications and preclinical applications.

2.2. *IL-12 Expressing OHSVs Produce Superior Anti-Tumor Immunity/Efficacy than Non-IL12 OHSVs*

In an orthotopic, immunocompetent intracranial glioblastoma (GBM) stem-like cell model (005 GSC model), OHSV G47Δ-IL12 therapy provides superior results compared to OHSV G47Δ treatment alone. For example, G47Δ-IL12 causes significant extension of survival of mice compared with either G47Δ-empty (i.e., an OHSV with no IL-12 transgene expression; $p < 0.005$) or mock treatment ($p < 0.001$, >50% increase in median survival), with 10% of mice surviving long term [30]. Anti-tumor efficacy of G47Δ-IL12 is associated with a significant reduction of tumor cells (i.e., GFP^+ 005 GBM stem-like cells or GSCs) and a robust immune alteration in the tumor, following single intra-tumoral virus injection [19]. The immune alteration mainly includes, but is not limited to, increased tumor infiltration of $CD3^+$ and $CD8^+$ T cells, reduction of immunosuppressive $FoxP3^+$ regulatory T cells, and an increased ratio of $CD8^+$ T cells/regulatory T cells ($CD4^+FoxP3^+$) (a hall mark of clinical efficacy) (Figure 5) [19,30]. The role of T cells in therapeutic efficacy is further investigated in athymic nude mice (i.e., T cell-deficient mice) bearing orthotopic brain tumors [30]. In the absence of T cells, G47Δ-IL12 treatment is unable to significantly enhance survival over G47Δ-empty treatment, indicating a critical role of T cells in the IL12-mediated anti-tumor activity [30]. While replicating in vivo in the tumor, G47Δ-IL12 treatment causes increased local production of IL-12 in the tumor, which is accompanied by a marked release of downstream Th1 mediator interferon gamma (IFN-γ) in the tumors and, to a lesser extent, in the blood [30]. IL-12/IFN-γ promotes differentiation of T cells towards Th1 phenotype [92], which further produces IFN-γ and anti-tumor immune effects, as opposed to Th2 type T cells [9,93]. Similarly, T-bet^+ Th1 type cells are increased in the tumor following intra-tumoral G47Δ-IL12 treatment [19,26], though it has not been determined whether this increase is directly associated with OHSV-IL12-mediated IFN-γ production in the tumor. G47Δ-IL12 treatment promotes polarization of macrophages from pro-tumoral M2 towards anti-tumoral M1 (e.g., increased expression of iNOS$^+$ and pSTAT1$^+$ cells) without affecting total tumor-associated macrophage (TAM) population (Figure 5) [19], possibly because of IL-12 induced M1-polarizing IFN-γ expression [30]. G47Δ-IL12-mediated anti-cancer immune responses, i.e., in situ vaccine effect opens the door to combination treatment strategies involving other cancer immunotherapy drugs. In orthotopic malignant peripheral nerve sheath tumor (MPNST) models, a single intra-tumoral injection of G47Δ in sciatic nerve tumors, derived from human MPNST stem-like cells in athymic mice or mouse MPNST cells in immunocompetent mice, significantly inhibits tumor growth and prolongs survival, as compared to mock treatment [94]. Local IL-12 expression (i.e., G47Δ-IL12) further significantly improves the efficacy of G47Δ in an immunocompetent orthotopic MPNST model, indicating that IL-12 expression induces anti-MPNST immune responses and improves overall efficacy [94]. These studies support the application of G47Δ-IL12 in combination immunotherapies for MPNST tumors.

Similar to anti-tumor efficacy with G47Δ-IL12, NV1042 (i.e., another OHSV with IL-12 expression) treatment results in a striking reduction in squamous cell carcinoma (SCC) tumor volume compared with the tumors treated with NV1023 (i.e., OHSV lacking IL-12 expression) and NV1034 (i.e., OHSV lacking IL-12, but with GM-CSF expression) [32]. Fifty-seven percent of mice treated with NV1042 reject subsequent SCC re-challenge in the contralateral flank, indicating strong global anti-cancer immune response, as opposed to 14% mice treated with NV1023 or NV1034 [32]. Besides local application, NV1042 was intravenously administered for the treatment of spontaneous primary and metastatic prostate cancer in the transgenic TRAMP mice. Systemic IL12-expressing NV1042 was significantly more efficacious than non-IL12 expressing OHSV NV1023 in reducing the frequency of prostate cancer development and lung metastases [95]. NV1042 DNA was detected in primary and metastatic tumors at 2 weeks after the final systemic virus injection but not in liver or blood [95]. Similarly, anti-cancer efficacy of intravenously delivered NV1042 was also observed in disseminated pulmonary SCC. Compared to PBS and parental NV1023, the group treated with IL-12 expressing NV1042 completely showed no sign of pulmonary nodules at day 12. In a low tumor burden model, NV1042 treatment resulted in 100% survival, in contrast to 70% in NV1023-treated group and 0% in PBS-treated group [33]. Depletion of CD4$^+$ and CD8$^+$ T cells reduces anti-cancer efficacy of IL-12 expressing NV1042, which is similar to anti-cancer effects of non-IL12-expressing NV1023, indicating IL-12 expression plays an important role in enhancing oncolytic efficacy through immune modulation [33].

M002 treatment resulted in prolonged survival in both pediatric and adult intracranial patient-derived tumor xenograft models [34]. The better survival benefit is associated with OHSV receptor nectin-1 expression in tumor cells, which is usually higher in pediatric brain tumors than in adult GBMs [34]. In an immunocompetent breast cancer metastasis model, IL-12-armed M002 treatment significantly improved survival of mice over its parental unarmed OHSV R3659 (no IL-12 expression) [35], indicating IL-12 played a critical role for anti-tumor efficacy. In a syngeneic neuroblastoma model, single intracranial injection of M002 produced a minimal survival benefit over untreated mice [85], indicating the need for IL-12 in immunocompetent models. Similarly, mice bearing intracranial neuroblastoma treated intramuscularly (IM) with M002-infected irradiated neuroblastoma cells did not show any survival advantage over mice treated with non-infected irradiated tumor cells. However, a prime-boost vaccine strategy, such as IM injection of M002-infected irradiated tumor cells seven days prior to tumor implantation and seven days post-tumor implantation, produced sustained anti-tumor T-cell responses and significant survival advantage, as opposed to irradiated control tumor cells [85]. Because an important control group is missing in this experimental setup (i.e., unarmed OHSV-infected irradiated tumor cells), it is not clear whether this anti-cancer vaccine effect in neuroblastoma was due to OHSV, local IL-12 expression, or both. In syngeneic sarcoma models, M002 did not produce any survival benefit compared to its parental virus R3659 (no IL-12 expression) [96], despite M002 inducing a significant anti-tumor immune effect over R3659 treatment, such as an increased percentage of intra-tumoral CD8$^+$ T cells and activated monocytes, a decreased percentage of myeloid-derived suppressor cells (MDSCs), and increased CD8:MDSC and CD8:T regulatory cell ratios [96]. In recently performed pilot experiments in an ovarian cancer metastatic model, systemic intraperitoneal application of M002 resulted in a robust tumor-antigen specific CD8$^+$ T cell response in the peritoneal cavity and the omentum [97], which are the primary sites of ovarian cancer metastasis [98]. Because of the tumor-specific immunity, M002 treatment was more successful in controlling ovarian cancer metastasis and produced a significantly longer overall survival than mock treatment [97]. Whether the anti-tumor efficacy is minimal or better, local IL-12 expression (M002) creates a more favorable immune-active tumor microenvironment than unarmed OHSV, which makes tumors more responsive to other forms of immunotherapies, such as immune checkpoint blockades.

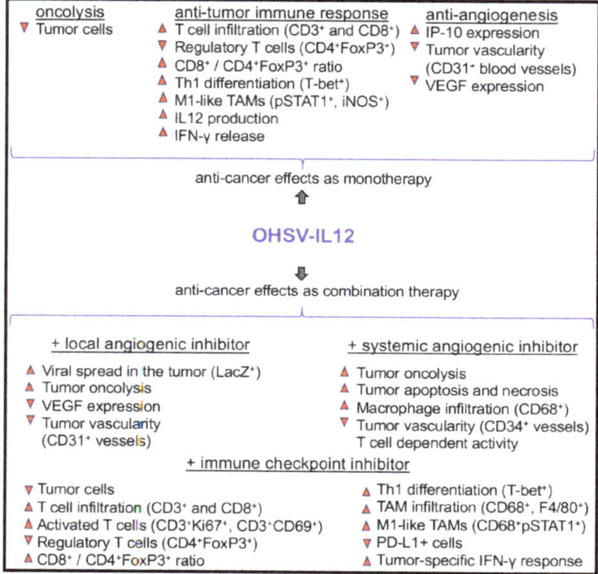

Figure 5. Anti-tumor effects of OHSV-IL12 treatment as mono- and combination-therapies. OHSV-IL12 treatment as monotherapy leads to three distinct anti-cancer effects: 1. Oncolysis, leading to reduction of cancer cells; 2. Induction of anti-tumor immunity, which is characterized by increased intratumoral infiltration of T cells, reduction of regulatory T cells, increased T effector/regulatory T cell ratio, enhanced Th1 differentiation, increased polarization of macrophages toward anti-tumoral M1-phenotype, and increased production of IL-12 and IFN-γ; and 3. Inhibition of tumor angiogenesis as demonstrated by reduced CD31$^+$ blood vessels and increased expression of vascular endothelial growth factor (VEGF) and interferon inducible protein 10 (IP-10). Because of these three aforementioned unique anti-cancer potentials, OHSV-IL12 was tested in combination with local or systemic antiangiogenic inhibitors and immune checkpoint blockade. The combination of OHSV-IL12 + local angiogenic inhibitor produces anti-tumor effects by increasing intratumoral virus spread (as determined by X-gal staining for viral LacZ expression) and oncolysis, and by reducing CD31$^+$ tumor vascularity and VEGF expression. The anti-tumor effects of the OHSV-IL12 + systemic angiogenic inhibitor are characterized by increased lysis of cancer cells and macrophage (CD68$^+$) infiltration into tumors, increased apoptosis and necrosis in the tumor microenvironment, reduced tumor vascularity, and T cell dependent anti-tumor activity. OHSV-IL12 + immune checkpoint inhibitor produces robust and multifaceted anti-cancer activities, which include: oncolysis, increased infiltration of T cells and activated T cells into tumors, reduction of immunosuppressive regulatory T cells, increased effector (CD8$^+$)/regulatory T cell (CD4$^+$FoxP3$^+$) ratio, Th1 differentiation, tumor-associated macrophage (TAM) infiltration and macrophage polarization towards M1-type, reduction of immune checkpoint PD-L1 positive cells, and induction of tumor-specific IFN-γ response. Upward and downward triangles indicate 'increase' and 'decrease', respectively.

IL-12-armed R115 was superior in inducing local and systemic anti-tumor immunity and durable response over unarmed R-LM113 in both early and late schedules [38]. All mice that survived the primary tumor were protected from the distant tumor challenge and subsequent re-challenge. Treatment with R115 drove Th1 polarization, increased immunomodulatory cytokines such as IFN-γ, IL-2, Granzyme B, T-bet and TNF-α, and tumor infiltrating lymphocytes [38]. Tumor microenvironment of R115 group showed an increase in number of CD8$^+$ and CD141$^+$ cells, PD-L1$^+$ tumor cells, and FoxP3$^+$ T regulatory cells with a decrease in the number of CD11b$^+$ cells [38]. In another study, a single R-115 injection in established tumors resulted in complete tumor eradication in about 30% of animals [39]. The treatment also induced a significant improvement in the overall median survival time of mice and a resistance to recurrence from the same neoplasia [39]. Interestingly, treatment with R-115 increases the number of CD4$^+$ and CD8$^+$ T cells infiltrating into the tumor microenvironment, while the vast majority of CD4$^+$ and CD8$^+$ T cells in the R-LM113 treatment group accumulated at the edge of the tumors, indicating the effects of IL-12 [39].

2.3. Anti-Tumor Anti-Vascular Effects of OHSV-IL12

IL-12 does not only enhances the anti-tumor immune effects of virotherapy, but it also suppresses the development of new blood vessels, a process termed angiogenesis [99], making it an anti-angiogenic cytokine. IL-12 elicits its anti-angiogenic effects through release of IFN-γ, which activates IFN-inducible protein 10 [IP-10 or CXC chemokine ligand (CXCL) 10], a chemokine that mediates chemotaxis of lymphocytes and angiostatic effects [10,17,100]. It has been demonstrated that IL-12-armed OHSV produces significantly higher level of anti-angiogenic effects (i.e., reduction of CD31$^+$ blood vessels in the tumors) through IFN-γ/IP-10 pathway in an immunocompetent brain tumor model (Figure 5) [30], compared to non-IL12 OHSV. IL12-armed OHSV treatment also causes a reduction in vascular endothelial growth factor (VEGF) expression, another likely contributor in tumor angiogenesis (Figure 5) [30].

In a model of prostate cancer in transgenic TRAMP mice, treatment with IL-12 armed NV1042 significantly reduces expression of CD31$^+$ vascularity compared to either NV1023 or mock treated tumors [95]. Anti-angiogenic property of NV1042 is confirmed by another study in a SCC model. Intratumoral delivery of NV1042 results in release of a high level of IL-12, as well as other secondary angiogenic mediators such as IFN-γ, monokine induced by gamma interferon (MIG), and IP-10. In contrast, IL-12 unarmed parental NV1023 treatment shows no increase in IL-12 expression and lower level of secondary angiogenic mediators [101]. These studies indicate that IL-12 gene transfer could significantly enhance unique anti-tumor and anti-angiogenic effects of virotherapy. These anti-angiogenic features allow OHSV-IL12 to be tested with other local or systemic angiogenesis inhibitors for an improved therapeutic outcome.

2.4. Inhibition of Tumor Angiogenesis Enhances Anti-Tumor Potential of OHSV-IL12 Treatment

Angiogenesis is one of the hallmarks of cancer. It plays a key role in cancer progression [102–110] and anti-angiogenic therapy has been an interesting target to control tumor growth [108,111,112]. Efforts to disrupt the vascular supply and starve the tumor from nutrients and oxygen have resulted in 11 anti-VEGF drugs approved for certain advanced cancers, either alone or in combination with chemotherapy or other targeted therapies. Unfortunately, this success has had only limited impact on overall survival of cancer patients, and rarely resulted in durable responses. Bevacizumab (Avastin), an FDA approved anti-angiogenic drug (anti-VEGF), did not show significant improvement in overall survival [112–114]. Therefore, other anti-angiogenic agents and combinatorial strategies are being tested to target complex tumor microenvironment.

Because OHSV G47Δ-IL12 does not only induce anti-tumor immunity but also produce anti-angiogenic activities [30], it is hypothesized that anti-tumor effects of G47Δ-IL12 treatment would synergize with anti-vascular drugs. Axitinib (AG-013736) is an FDA approved, orally administered potent small molecule tyrosine kinase inhibitor (TKI), which inhibits VEGF receptor

(VEGFR) 1-3, platelet-derived growth factor receptor beta (PDGFR-β) and receptor tyrosine kinase c-KIT (CD117) [115], and shows promising anti-vascular and anti-tumor activity in a variety of advanced stage cancers, including GBM [116–118]. In addition to anti-vascular effects, it also induces anti-tumor immune effects [119,120]. Therefore, anti-vascular/immune axitinib was tested in combination with anti-angiogenic/immune stimulatory G47Δ-IL12 in highly angiogenic patient and mouse GSC-derived GBM models [76]. This combination significantly extends survival in both models and involves multifaceted anti-tumor activities including: direct oncolysis of tumor cells, extensive tumor apoptosis and necrosis, increased macrophage infiltration to the tumor, greatly reduced tumor vascularity (i.e., $CD34^+$ blood vessels) and inhibition of angiogenic PDGFR/ERK pathway in patient GSC-derived GBM model, and T cell dependent activity in mouse GSC-derived GBM model (Figure 5) [76]. Since the anti-tumor efficacy of the dual combination therapy (G47Δ-IL12+axitinib) was T cell dependent, it is hypothesized that ICI (i.e., anti-PD-1 or anti-CTLA4) will improve the therapeutic outcome of G47Δ-IL12+axitinib dual combination. Interestingly, ICI did not improve anti-tumor effects of axitinib or G47Δ-IL12+axitinib combination therapy. This is in sharp contrast with the findings from other investigators, since they observed synergistic anti-tumor immune effects following axitinib + ICI combination therapy in preclinical syngeneic tumor models [121]. The underlying mechanism(s) behind the failure of ICI combination therapy in orthotopic brain tumor model is not clear and warrants further investigation. It is speculated that reduced vascular permeability after axitinib therapy may inhibit extravasation of T cells into the tumor [122]. Indeed, axitinib treatment significantly reduces T cell ($CD3^+$ and $CD8^+$) infiltration into brain tumors [76]. Because both axitinib and OHSV are already in clinical trials for brain tumors as monotherapy with limited efficacy, dual combination therapy (OHSV-IL12+axitinib) that shows anti-tumor efficacy in both immune deficient and immune competent orthotopic brain tumor models has translational relevance [76]. Because systemic anti-angiogenic therapy (i.e., axitinib) is often associated with renal toxicities [123,124], G47Δ-IL12 was tested in combination with a local OHSV expression of angiostatin (OHSV G47Δ-angio), an anti-angiogenic polypeptide, in hypervascular human GBM models [125]. The combination of two OHSVs (G47Δ-IL12+G47Δ-angio) significantly prolongs survival compare to each armed OHSV alone, which is associated with increased viral spread and reduced expression of VEGF and $CD31^+$ blood vessels in the tumor (Figure 5) [125]. This study supports further engineering of OHSV to express both IL-12 and angiostatin locally in the tumor. Use of one virus rather than two is practical in the context of future FDA approval. Similar to G47Δ-IL12 and anti-angiogenesis studies, the combination of another IL-12 expressing OHSV NV1042 with the anti-cancer chemotherapy drug vinblastine results in significant reduction of tumor burden in athymic mice bearing subcutaneous CWR22 prostate tumors, which is most likely associated with diminishing the number of $CD31^+$ endothelial cells [126].

2.5. Immune Checkpoint Inhibition Enhances OHSV-IL12 Treatment-Induced Anti-Tumor Immunity

Though anti-tumor effects of OHSV-IL12 therapy is multifaceted, virotherapy alone does not improve significant survival in cancer [19,30,76,127]. For example, G47Δ-IL12 monotherapy shows limited efficacy in preclinical immune competent models of prostate and malignant peripheral nerve sheath tumors [94]. Since OHSV-IL12 treatment changes immune phenotypes of the tumor microenvironment, it is tested in combination with other forms of immunotherapies (e.g., ICIs) to obtain a better therapeutic response [19,76]. ICIs, such as cytotoxic T lymphocyte antigen 4 (CTLA-4) and programmed death 1 (PD-1) suppress T cell-mediated anti-tumor immune responses (Figure 6), leading to tumor progression [128]. ICI antibodies (i.e., anti-CTLA-4 or anti-PD-1) are effective in unleashing tumor-induced immunosuppression and activating effector immune cells (Figure 6) [129]. Since OHSV-IL12 induces robust anti-tumor immunity [19], it is hypothesized that OHSV-IL12 will synergize with ICI antibodies and will improve the anti-cancer efficacy of G47Δ-IL12. Indeed, in 005 GSC-derived orthotopic brain tumor models, dual combination modestly extends survival compared to ICI antibody alone or G47Δ-IL12 therapy alone [19]. The modest anti-tumor efficacy of the dual combination is not due to the inability of ICI antibodies to cross the blood-brain barrier, since ICI antibodies were detected in the tumor [19]. Because CTLA-4 and PD-1 regulate anti-tumor immunity via distinct and non-redundant immune evasion mechanisms [130–132], it is hypothesized that combinatorial blockade of two immune inhibitory pathways will produce enhanced anti-tumor immune effects and will synergize with the anti-tumor efficacy of G47Δ-IL12 (i.e., triple combination therapy: G47Δ-IL12+anti-PD-1+anti-CTLA-4). Indeed, triple combination therapy leads to a significant percentage (89%) of long-term survivors (i.e., survived six months post-tumor implantation) [19]. These survivors remain protected following lethal tumor re-challenge in the contralateral hemisphere, surviving another three months until the experiment was terminated, which indicates development of long-term memory protection [19]. These unprecedented findings were reproduced in a second aggressive immune competent CT-2A GBM model [19]. The survival efficacy of the triple combination therapies is associated with a significant but complex immune alteration in the tumor microenvironment, as opposed to mock, single or dual combination therapies, which includes: i) increase tumor infiltration of T cells ($CD3^+$ and $CD8^+$); ii) increase number of proliferating T cells ($CD3^+Ki67^+$); iii) increase activated T cells ($CD8^+CD69^+$); iv) reduce regulatory T cells ($FoxP3^+$); v) increase T effector ($CD8^+$)/regulatory T cell ($CD4^+FoxP3^+$) ratio; vi) increase TAMs ($CD68^+$, $F4/80^+$); vii) skew TAMs towards M1-phenotypes ($iNOS^+$, $pSTAT1^+$, $CD68^+pSTAT1^+$); viii) increase Th1 differentiation ($T\text{-}bet^+$); ix) reduce PD-$L1^+$ cells; x) increase tumor-cell specific IFN-γ response; and (xi) reduce tumor cells (Figure 5) [19,26]. Depletion and inhibition of immune cell subtypes (i.e., $CD4^+$ cell depletion by anti-CD4, $CD8^+$ cell depletion by anti-CD8, peripheral macrophage depletion by liposomal clodronate, or CSF-1R inhibition by brain penetrant drug BLZ945 to target TAMs) confirms the necessity of $CD4^+$ cells, $CD8^+$ cells, and macrophages in the therapeutic efficacy, with $CD4^+$ cells playing the critical role [19]. It remains to be determined how immune cells, especially $CD4^+$ cells and M1-like macrophage polarization contribute to therapeutic efficacy. IL-12 appears to be critical for this exceptional anti-tumor efficacy, since another triple combination therapy involving the base G47Δ (without IL-12 expression) plus two systemic ICI antibodies results in only 13% long-term survivors [25], as opposed to 89% from G47Δ-IL12 + anti-PD-1 + anti-CTLA-4 combination [19].

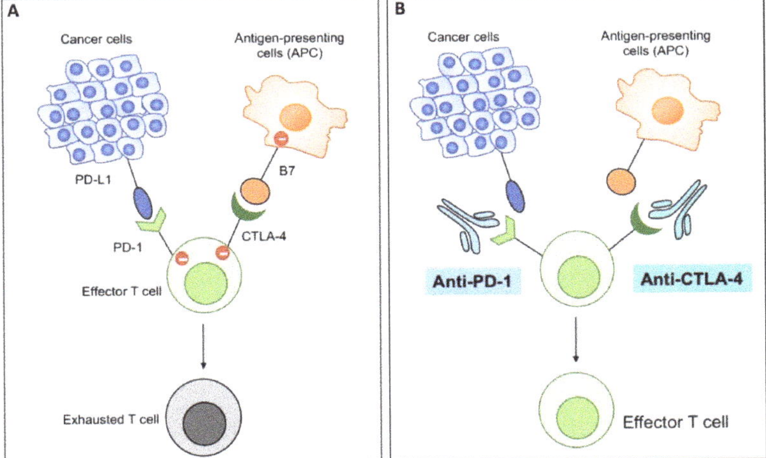

Figure 6. Graphic presentation of mechanism of action of immune checkpoint inhibitors (ICIs). (**A**) T cells express immune checkpoints programmed death 1 (PD-1) and cytotoxic T lymphocyte antigen 4 (CTLA-4), which interact with their corresponding ligands, i.e., programmed death ligand 1 (PD-L1) on cancer cells and B7 molecules on antigen presenting cells (APC), respectively. PD-1:PD-L1 and CTLA-4:B7 interactions send negative signals to immune cells, leading to exhaustion of T cells and no effector activity. (**B**) Antibodies to PD-1 (anti-PD-1) and CTLA-4 (anti-CTLA4) block those interactions and unleash anti-tumor immunity by enhancing activity of effector T cells.

3. Anti-Cancer Potential of other OVs Encoding IL-12

As IL-12 is a master anti-tumor cytokine and has distinct multifaceted anti-cancer properties [1–3,6,8,10], IL-12 expression by OVs is not limited to OHSVs. Several other OVs encoding IL-12 have been developed (Table 2), including adenoviruses [18,42,46–49,133–138], measles virus [20,50], maraba virus [51], Newcastle disease virus [52], Semliki forest virus [53–56,139–141], vesicular stomatitis virus [57,142], and Sindbis virus [58]. In these above-mentioned studies, IL-12 is expressed either alone [18,20,51,57,137–139] or co-expressed alongside with GM-CSF [134–136], pericellular matrix proteoglycan decorin [47], tumor necrosis factor-related apoptosis-inducing ligand (TRAIL) [133], IL-2 proinflammatory cytokine [52], IFN-γ inducing factor IL-18 [46], T cell co-stimulatory ligand 4-1BBL [49], CD8$^+$ co-receptor for CD28 and CTLA-4 [48], or suicide genes [42]. The anti-tumor efficacy of these engineered OVs are tested in various mouse or hamster pre-clinical cancer models and produce superior anti-tumor immunity either alone [18,20,42,46–48,51,52,57,133,135,137] or in combination with other immunotherapeutic agents such as dendritic cell (DC) vaccine [49,134,136], ICI anti-PD-1 and anti-PD-L1 [139], or cytokine-induced killer cells [138].

The tumor-targeting oncolytic adenovirus (Ad-TD) delivers either wild-type IL-12 or non-secreting IL-12 (nsIL-12) directly to pancreatic tumor cells (Table 2) [44]. In a Syrian hamster tumor model, treatment with both Ad-TD-IL12 and Ad-TD-nsIL-12 results in 100% tumor eradication and animal survival, and treatment with Ad-TD-IL12 in particular produces a significant increase in populations of CD3$^+$CD4$^-$CD8$^+$ in the spleen [44]. In addition, treatment with Ad-TD-nsIL-12 results in lower levels of lymph node IFN-γ, and splenic IFN-γ and IP-10 production [44]. Another oncolytic adenovirus expressing IL-12 (RdB/IL-12) inhibits tumor growth in murine melanoma lines by 95%, while adenovirus expressing both IL-12 and IL-18 (RdB/IL-12/IL-18) inhibits growth by 99% [46]. RdB/IL-12/IL-18 also increases the cytokine ratio of Th1/Th2, increases tumor infiltration of CD4$^+$ T, CD8$^+$ T and NK cells, and promotes differentiation of T cells expressing IL-12Rβ2 or IL-18Rα [46]. In another engineered oncolytic adenovirus (Ad-ΔB7/IL-12/4-1BBL), co-expression of both IL-12 and the cytokine 4-1BB ligand (4-1BBL) produces significantly higher survival in mice, with 100% of mice surviving more than 30 days after viral injection [49]. This is considerable when in comparison to the 20% survival rate of mice treated with virus expressing only IL-12 or 4-1BBL. In this study, mice treated with either Ad-ΔB7/IL-12 or Ad-ΔB7/IL-12/4-1BBL have a higher amount of tumor infiltrating CD4$^+$ and CD8$^+$ cells in comparison to treatment with Ad-ΔB7/4-1BBL or Ad-ΔB7 [49].

An oncolytic measles virus encoding an IL-12 fusion protein (MeVac FmIL-12) as a single agent produces potent anti-tumor immune effects in an immunocompetent colon cancer model with 90% complete remission [20]. This robust anti-cancer efficacy is dependent on T cells, particularly CD8$^+$ cells, and is associated with activation of early NK cells and effector T cells, and upregulation of effector anti-tumor cytokines IFN-γ and TNF-α [20]. The findings are similar to what we observed in GBM with OHSV G47Δ-IL12 virus [19,30]. Although it was not examined whether the long-term survivors in MeVac FmIL-12 group developed any memory protection, the potent anti-tumor immune efficacy indicates that MeVac FmIL-12 is also an attractive candidate in cancer therapy [20]. Similar to the MeVac FmIL-12 virus, IL12-expressing oncolytic maraba virus (designated MG1-IL12-ICV) treatment also leads to complete eradication of peritoneal carcinomatosis, which is associated with significant recruitment of NK cells in the tumor microenvironment [51]. The lentogenic Newcastle disease virus Clone30 strain was generated to express IL-12 (designated rClone30-IL-12) that displays improved survival and reduced tumor volume [52]. In this case, co-expression of two cytokines (IL-12 and IL-2) produced synergistic effects on treatment, with the rClone30-IL-12-IL-2 virus inducing the greatest release of IFN-γ and IP-10 [52].

The Semliki Forest virus (SFV) has a broad host range and is suicidal, causing apoptosis in infected cells, which makes it a promising viral vector [53]. In a murine tumor model using B16 cells, a single intratumoral injection of SFV expressing IL-12 (SFV-IL12) results in significant tumor regression seven days after injection. This follows a distinct inhibition of tumor vascularization and an increase in IFN-γ production at two days post-injection [53]. Furthermore, treatment with SFV-IL12 results in an increase in expression of splenic IP-10 and monokine induced by interferon-γ (MIG) [53]. Similar results are obtained on the murine colon adenocarcinoma cell line, MC38, with SFV-IL12 treatment resulting in increased tumor-specific CD8$^+$ T lymphocytes, reduced tumor volume, and improved survival [55]. Of particular notice, treatment of MC38 cells with SFV-IL12 results in tumor-infiltrating monocytic myeloid-derived suppressor cells (M-MDSCs) displaying increased expression of CD11c, CD8α, CD40, and CD86 in the presence of an intact, endogenous host type-I interferon (IFN-I) system [55].

Vesicular stomatitis virus (VSV) is another example of a viral vector [57]. Oncolytic VSV carrying IL-12 (rVSV-IL12) succeeds in effectively reducing tumor volume and prolonging survival in both human and murine SCC tumors, with 40% of mice treated with rVSV-IL12 surviving past 100 days post-injection, in comparison to mice treated with the fusogenic OV, rVSV-F, surviving only to 60 days post-injection ($p < 0.0001$) [57]. Similarly, treatment with the Sindbis viral vector carrying IL-12 (Sin/IL12) increases production of IFN-γ, and results in reduced tumor growth and improved survival in ovarian clear cell carcinoma [58]. Treatment with Sin/IL12 also modulates the regulatory functions of NK cells, increasing activation and recruitment of the cells [58]. These and afore-mentioned studies clearly suggest that IL-12 is a useful anti-cancer agent for oncolytic immunovirotherapy, and boosts anti-cancer immune properties of OVs.

4. Clinical Perspectives

Regardless of cancer types, IL-12 expressing OVs have been tested and found effective against various cancers (such as glioma, neuroblastoma, squamous cell carcinoma, metastatic breast cancer, hepatoblastoma, sarcoma, kidney cancer, lymphoma, prostate cancer, pancreatic cancer, colon cancer, ovarian cancer, melanoma, etc.) (Tables 1 and 2). This clearly suggests that IL-12 expressing OVs are an attractive therapeutic candidate for clinical translation against any forms of cancer. In general, when translating results from bench to bed side, safety remains the most concerning aspect, along with dose, route of administration, viral pharmacokinetics and resistance mechanism of host cells [143]. Recent FDA approval of OHSV T-VEC in 2015 has heightened the field of oncolytic virus-based immunotherapy. The FDA approved OHSV expresses GM-CSF instead of IL-12. The safety and efficacy of T-VEC in immune-privileged organs, such as brain, has not been extensively elucidated. Moreover, T-VEC has not demonstrated durable responses in a majority of advanced melanoma patients [29], especially those with visceral metastases [144,145], raising questions about its possible long-term efficacy in patients with difficult-to-treat metastatic cancers.

Safety and anti-tumor efficacy of OHSV-IL12 as monotherapy or combination therapy has been demonstrated in preclinical tumor models [19,30,76]. Currently, G47Δ expressing human IL-12 is under development for clinical use. Safety of M002 has been established in the brain after intracerebral administration to non-human primates [21]. M032 is now in a phase I clinical trial (NCT02062827) in patients with recurrent/progressive GBM, anaplastic astrocytoma, or GBM [88]. Similar to OHSV-IL12, an adenovirus expressing human IL-12 is also under clinical trial investigation as monotherapy in prostate cancer (NCT02555397, NCT00406939), pancreatic cancer (NCT03281382), breast cancer (NCT00849459) and melanoma (NCT01397708). In addition, Ad-RTS-hIL-12 (another adenovirus encoding for IL-12) is under clinical trial evaluation in combination with veledimex (an oral activator ligand to promote release of IL-12 by an OV) in pediatric brain tumor (NCT03330197) and adult glioblastoma or malignant glioma (NCT02026271). Triple combination of Ad-RTS-hIL-12, veledimex and nivolumab (anti-PD-1) are also in active status in glioblastoma patients (NCT03636477). In our opinion, since OV-IL12 treatment induces prominent anti-tumor immunity including increased expression of PD-L1 in the tumor microenvironment, OV-IL12 therapy may improve the response rate to anti-PD-L1 treatment, especially in cancer patients who inherently lack PD-L1 expression and/or previously unresponsive to anti-PD-L1 treatment. Thus, combination studies involving OV-IL12 and ICI warrant clinical investigation and could be an attractive treatment strategy for cancer patients.

5. Conclusions and Future Directions

Genomic manipulation and understanding of pathogenicity have made OVs an attractive candidate for cancer therapy. Among OVs, OHSV is FDA approved for cancer treatment and is the furthest along in the clinic [29,144–146]. Moreover, the availability of antiviral drugs, such as acyclovir, makes OHSV a safer anti-cancer candidate over other OVs [59,147]. However, we all have recognized that OHSV expression of IL-12 and its application as a monotherapy does not provide a desired therapeutic outcome, as demonstrated in several preclinical cancer models [30,84,86,148,149]. It requires synergistic or additive combination approach with other anti-cancer therapies for an improved therapeutic outcome [19,25,76,148,150,151]. Similar to OHSV-IL12, a combination immunotherapeutic approach is also required to enhance anti-tumor efficacy of other OVs encoding IL-12 [49,134,136,138,139].

Although ICI-based cancer immunotherapy is rapidly evolving in the field of oncology, the combination therapy involving single or multiple ICIs is often associated with significant toxicity in humans [152–156]. Moreover, ICI immunotherapy has not been successful in devastating cancer types such as GBM (CheckMate-143) [157,158] or triple-negative breast cancer (KEYNOTE-119) [159]. In contrast, OV expressing single or multiple immune stimulator does not cause toxicity when expressed locally in the tumor by the virus [21,52,160–163], even in an immune-privileged brain [21], but rather induces robust anti-tumor immunity. Thus, developing appropriately designed and stronger version of viral vectors expressing multiple immune stimulators alongside the master anti-tumor cytokine IL-12 may induce superior local anti-tumor immune responses, while reducing the need for multiple systemic ICI or other systemic drugs, and eventually thwarting the current limitation of systemic cancer immunotherapy.

Besides considering construction of new viral vectors, another issue that needs to be addressed before utilizing the full potential of OVs is limited viral spread in tumors [19] due to presence of anti-viral genes or up-regulation of anti-viral factors following OV treatment [164]. Restricted viral replication and spread in the tumor may result in reduced tumor oncolysis with limited in situ vaccine effect. Identification, followed by inhibition of anti-viral factor(s) will provide tools to develop new OV-based immunotherapeutic strategies to enhance viral replication and spread in the tumor, and to induce potent anti-tumor immunity [164]. Designing a better rationale OV-based combination treatment strategy without compromising safety will be the key for clinical success.

Author Contributions: Conceptualization, original draft preparation, review and editing, funding acquisition, D.S.; H.-M.N. and K.G.-M. contributed equally in this manuscript, prepared the figures and wrote the manuscript. All authors have read and agreed to the published version of the manuscript.

Funding: D.S. is supported by startup funds from Dodge Jones Foundation-Abilene and TTUHSC-School of Pharmacy. This work received no external funding.

Conflicts of Interest: The authors declare no conflict of interest. The funders had no role in the writing of the manuscript or in the decision to publish the manuscript.

References

1. Berraondo, P.; Etxeberria, I.; Ponz-Sarvise, M.; Melero, I. Revisiting Interleukin-12 as a Cancer Immunotherapy Agent. *Clin. Cancer Res.* **2018**, *24*, 2716–2718. [CrossRef] [PubMed]
2. Lu, X. Impact of IL-12 in Cancer. *Curr. Cancer Drug Targets* **2017**, *17*, 682–697. [CrossRef] [PubMed]
3. Zeh, H.J., 3rd; Hurd, S.; Storkus, W.J.; Lotze, M.T. Interleukin-12 promotes the proliferation and cytolytic maturation of immune effectors: Implications for the immunotherapy of cancer. *J. Immunother. Emphasis Tumor Immunol.* **1993**, *14*, 155–161. [CrossRef] [PubMed]
4. Lasek, W.; Zagozdzon, R.; Jakobisiak, M. Interleukin 12: Still a promising candidate for tumor immunotherapy? *Cancer Immunol. Immunother.* **2014**, *63*, 419–435. [CrossRef]
5. Lehmann, D.; Spanholtz, J.; Sturtzel, C.; Tordoir, M.; Schlechta, B.; Groenewegen, D.; Hofer, E. IL-12 directs further maturation of ex vivo differentiated NK cells with improved therapeutic potential. *PLoS ONE* **2014**, *9*, e87131. [CrossRef]

6. Trinchieri, G.; Wysocka, M.; D'Andrea, A.; Rengaraju, M.; Aste-Amezaga, M.; Kubin, M.; Valiante, N.M.; Chehimi, J. Natural killer cell stimulatory factor (NKSF) or interleukin-12 is a key regulator of immune response and inflammation. *Prog. Growth Factor Res.* **1992**, *4*, 355–368. [CrossRef]
7. Trinchieri, G. Interleukin-12 and the regulation of innate resistance and adaptive immunity. *Nat. Rev. Immunol.* **2003**, *3*, 133–146. [CrossRef]
8. Otani, T.; Nakamura, S.; Toki, M.; Motoda, R.; Kurimoto, M.; Orita, K. Identification of IFN-gamma-producing cells in IL-12/IL-18-treated mice. *Cell Immunol.* **1999**, *198*, 111–119. [CrossRef]
9. Tugues, S.; Burkhard, S.H.; Ohs, I.; Vrohlings, M.; Nussbaum, K.; Vom Berg, J.; Kulig, P.; Becher, B. New insights into IL-12-mediated tumor suppression. *Cell Death Differ.* **2015**, *22*, 237–246. [CrossRef]
10. Angiolillo, A.L.; Sgadari, C.; Tosato, G. A role for the interferon-inducible protein 10 in inhibition of angiogenesis by interleukin-12. *Ann. N. Y. Acad. Sci.* **1996**, *795*, 158–167. [CrossRef]
11. Lamont, A.G.; Adorini, L. IL-12: A key cytokine in immune regulation. *Immunol. Today* **1996**, *17*, 214–217. [CrossRef]
12. Jenks, S. After initial setback, IL-12 regaining popularity. *J. Natl. Cancer Inst.* **1996**, *88*, 576–577. [CrossRef]
13. Duvic, M.; Sherman, M.L.; Wood, G.S.; Kuzel, T.M.; Olsen, E.; Foss, F.; Laliberte, R.J.; Ryan, J.L.; Zonno, K.; Rook, A.H. A phase II open-label study of recombinant human interleukin-12 in patients with stage IA, IB, or IIA mycosis fungoides. *J. Am. Acad. Dermatol.* **2006**, *55*, 807–813. [CrossRef] [PubMed]
14. Rook, A.H.; Wood, G.S.; Yoo, E.K.; Elenitsas, R.; Kao, D.M.; Sherman, M.L.; Witmer, W.K.; Rockwell, K.A.; Shane, R.B.; Lessin, S.R.; et al. Interleukin-12 therapy of cutaneous T-cell lymphoma induces lesion regression and cytotoxic T-cell responses. *Blood* **1999**, *94*, 902–908. [CrossRef] [PubMed]
15. Little, R.F.; Pluda, J.M.; Wyvill, K.M.; Rodriguez-Chavez, I.R.; Tosato, G.; Catanzaro, A.T.; Steinberg, S.M.; Yarchoan, R. Activity of subcutaneous interleukin-12 in AIDS-related Kaposi sarcoma. *Blood* **2006**, *107*, 4650–4657. [CrossRef]
16. Younes, A.; Pro, B.; Robertson, M.J.; Flinn, I.W.; Romaguera, J.E.; Hagemeister, F.; Dang, N.H.; Fiumara, P.; Loyer, E.M.; Cabanillas, F.F.; et al. Phase II clinical trial of interleukin-12 in patients with relapsed and refractory non-Hodgkin's lymphoma and Hodgkin's disease. *Clin. Cancer Res.* **2004**, *10*, 5432–5438. [CrossRef] [PubMed]
17. Del Vecchio, M.; Bajetta, E.; Canova, S.; Lotze, M.T.; Wesa, A.; Parmiani, G.; Anichini, A. Interleukin-12: Biological properties and clinical application. *Clin. Cancer Res.* **2007**, *13*, 4677–4685. [CrossRef]
18. Poutou, J.; Bunuales, M.; Gonzalez-Aparicio, M.; Garcia-Aragoncillo, E.; Quetglas, J.I.; Casado, R.; Bravo-Perez, C.; Alzuguren, P.; Hernandez-Alcoceba, R. Safety and antitumor effect of oncolytic and helper-dependent adenoviruses expressing interleukin-12 variants in a hamster pancreatic cancer model. *Gene Ther.* **2015**, *22*, 696–706. [CrossRef]
19. Saha, D.; Martuza, R.L.; Rabkin, S.D. Macrophage Polarization Contributes to Glioblastoma Eradication by Combination Immunovirotherapy and Immune Checkpoint Blockade. *Cancer Cell* **2017**, *32*, 253–267.e255. [CrossRef]
20. Veinalde, R.; Grossardt, C.; Hartmann, L.; Bourgeois-Daigneault, M.C.; Bell, J.C.; Jager, D.; von Kalle, C.; Ungerechts, G.; Engeland, C.E. Oncolytic measles virus encoding interleukin-12 mediates potent antitumor effects through T cell activation. *Oncoimmunology* **2017**, *6*, e1285992. [CrossRef]
21. Roth, J.C.; Cassady, K.A.; Cody, J.J.; Parker, J.N.; Price, K.H.; Coleman, J.M.; Peggins, J.O.; Noker, P.E.; Powers, N.W.; Grimes, S.D.; et al. Evaluation of the safety and biodistribution of M032, an attenuated herpes simplex virus type 1 expressing hIL-12, after intracerebral administration to aotus nonhuman primates. *Hum. Gene Ther. Clin Dev.* **2014**, *25*, 16–27. [CrossRef]
22. Saha, D.; Ahmed, S.S.; Rabkin, S.D. Exploring the Antitumor Effect of Virus in Malignant Glioma. *Drugs Future* **2015**, *40*, 739–749. [CrossRef]
23. Saha, D.; Martuza, R.L.; Curry, W.T. Viral oncolysis of glioblastoma. In *Neurotropic viral infections*, 2nd ed.; Reiss, C.S., Ed.; Springer: New York, NY, USA, 2016; Volume 2, pp. 481–517.
24. Saha, D.; Wakimoto, H.; Rabkin, S.D. Oncolytic herpes simplex virus interactions with the host immune system. *Curr. Opin. Virol.* **2016**, *21*, 26–34. [CrossRef]
25. Saha, D.; Martuza, R.L.; Rabkin, S.D. Oncolytic herpes simplex virus immunovirotherapy in combination with immune checkpoint blockade to treat glioblastoma. *Immunotherapy* **2018**, *10*, 779–786. [CrossRef] [PubMed]

26. Saha, D.; Martuza, R.L.; Rabkin, S.D. Curing glioblastoma: Oncolytic HSV-IL12 and checkpoint blockade. *Oncoscience* **2017**, *4*, 67–69. [CrossRef] [PubMed]
27. Kaufman, H.L.; Bommareddy, P.K. Two roads for oncolytic immunotherapy development. *J. Immunother. Cancer* **2019**, *7*, 26. [CrossRef]
28. Raja, J.; Ludwig, J.M.; Gettinger, S.N.; Schalper, K.A.; Kim, H.S. Oncolytic virus immunotherapy: Future prospects for oncology. *J. Immunother. Cancer* **2018**, *6*, 140. [CrossRef]
29. Andtbacka, R.H.; Kaufman, H.L.; Collichio, F.; Amatruda, T.; Senzer, N.; Chesney, J.; Delman, K.A.; Spitler, L.E.; Puzanov, I.; Agarwala, S.S.; et al. Talimogene Laherparepvec Improves Durable Response Rate in Patients With Advanced Melanoma. *J. Clin. Oncol.* **2015**, *33*, 2780–2788. [CrossRef]
30. Cheema, T.A.; Wakimoto, H.; Fecci, P.E.; Ning, J.; Kuroda, T.; Jeyaretna, D.S.; Martuza, R.L.; Rabkin, S.D. Multifaceted oncolytic virus therapy for glioblastoma in an immunocompetent cancer stem cell model. *Proc. Natl. Acad. Sci. USA* **2013**, *110*, 12006–12011. [CrossRef]
31. Guan, Y.; Ino, Y.; Fukuhara, H.; Todo, T. Antitumor Efficacy of Intravenous Administration of Oncolytic Herpes Simplex Virus Expressing Interleukin 12. *Mol. Ther.* **2006**, *13*, S108. [CrossRef]
32. Wong, R.J.; Patel, S.G.; Kim, S.; DeMatteo, R.P.; Malhotra, S.; Bennett, J.J.; St-Louis, M.; Shah, J.P.; Johnson, P.A.; Fong, Y. Cytokine gene transfer enhances herpes oncolytic therapy in murine squamous cell carcinoma. *Hum. Gene. Ther.* **2001**, *12*, 253–265. [CrossRef]
33. Wong, R.J.; Chan, M.K.; Yu, Z.; Kim, T.H.; Bhargava, A.; Stiles, B.M.; Horsburgh, B.C.; Shah, J.P.; Ghossein, R.A.; Singh, B.; et al. Effective intravenous therapy of murine pulmonary metastases with an oncolytic herpes virus expressing interleukin 12. *Clin. Cancer Res.* **2004**, *10*, 251–259. [CrossRef] [PubMed]
34. Friedman, G.K.; Bernstock, J.D.; Chen, D.; Nan, L.; Moore, B.P.; Kelly, V.M.; Youngblood, S.L.; Langford, C.P.; Han, X.; Ring, E.K.; et al. Enhanced Sensitivity of Patient-Derived Pediatric High-Grade Brain Tumor Xenografts to Oncolytic HSV-1 Virotherapy Correlates with Nectin-1 Expression. *Sci. Rep.* **2018**, *8*, 13930. [CrossRef] [PubMed]
35. Cody, J.J.; Scaturro, P.; Cantor, A.B.; Yancey Gillespie, G.; Parker, J.N.; Markert, J.M. Preclinical evaluation of oncolytic deltagamma(1)34.5 herpes simplex virus expressing interleukin-12 for therapy of breast cancer brain metastases. *Int. J. Breast Cancer* **2012**, *2012*, 628697. [CrossRef] [PubMed]
36. Gillory, L.A.; Megison, M.L.; Stewart, J.E.; Mroczek-Musulman, E.; Nabers, H.C.; Waters, A.M.; Kelly, V.; Coleman, J.M.; Markert, J.M.; Gillespie, G.Y.; et al. Preclinical evaluation of engineered oncolytic herpes simplex virus for the treatment of neuroblastoma. *PLoS ONE* **2013**, *8*, e77753. [CrossRef] [PubMed]
37. Megison, M.L.; Gillory, L.A.; Stewart, J.E.; Nabers, H.C.; Mroczek-Musulman, E.; Waters, A.M.; Coleman, J.M.; Kelly, V.; Markert, J.M.; Gillespie, G.Y.; et al. Preclinical evaluation of engineered oncolytic herpes simplex virus for the treatment of pediatric solid tumors. *PLoS ONE* **2014**, *9*, e86843. [CrossRef] [PubMed]
38. Leoni, V.; Vannini, A.; Gatta, V.; Rambaldi, J.; Sanapo, M.; Barboni, C.; Zaghini, A.; Nanni, P.; Lollini, P.L.; Casiraghi, C.; et al. A fully-virulent retargeted oncolytic HSV armed with IL-12 elicits local immunity and vaccine therapy towards distant tumors. *PLoS Pathog* **2018**, *14*, e1007209. [CrossRef]
39. Alessandrini, F.; Menotti, L.; Avitabile, E.; Appolloni, I.; Ceresa, D.; Marubbi, D.; Campadelli-Fiume, G.; Malatesta, P. Eradication of glioblastoma by immuno-virotherapy with a retargeted oncolytic HSV in a preclinical model. *Oncogene* **2019**, *38*, 4467–4479. [CrossRef]
40. Ino, Y.; Saeki, Y.; Fukuhara, H.; Todo, T. Triple combination of oncolytic herpes simplex virus-1 vectors armed with interleukin-12, interleukin-18, or soluble B7-1 results in enhanced antitumor efficacy. *Clin. Cancer Res.* **2006**, *12*, 643–652. [CrossRef]
41. Yan, R.; Zhou, X.; Chen, X.; Liu, X.; Ma, J.; Wang, L.; Liu, Z.; Zhan, B.; Chen, H.; Wang, J.; et al. Enhancement of Oncolytic Activity of oHSV Expressing IL-12 and Anti PD-1 Antibody by Concurrent Administration of Exosomes Carrying CTLA-4 miRNA. *Immunotherapy* **2019**, *5*, 1–10. [CrossRef]
42. Freytag, S.O.; Barton, K.N.; Zhang, Y. Efficacy of oncolytic adenovirus expressing suicide genes and interleukin-12 in preclinical model of prostate cancer. *Gene Ther.* **2013**, *20*, 1131–1139. [CrossRef] [PubMed]
43. Freytag, S.O.; Zhang, Y.; Siddiqui, F. Preclinical toxicology of oncolytic adenovirus-mediated cytotoxic and interleukin-12 gene therapy for prostate cancer. *Mol. Ther. Oncolytics.* **2015**, *2*. [CrossRef] [PubMed]
44. Wang, P.; Li, X.; Wang, J.; Gao, D.; Li, Y.; Li, H.; Chu, Y.; Zhang, Z.; Liu, H.; Jiang, G.; et al. Re-designing Interleukin-12 to enhance its safety and potential as an anti-tumor immunotherapeutic agent. *Nat. Commun.* **2017**, *8*, 1395. [CrossRef] [PubMed]

45. Choi, K.J.; Zhang, S.N.; Choi, I.K.; Kim, J.S.; Yun, C.O. Strengthening of antitumor immune memory and prevention of thymic atrophy mediated by adenovirus expressing IL-12 and GM-CSF. *Gene Ther.* **2012**, *19*, 711–723. [CrossRef]
46. Choi, I.K.; Lee, J.S.; Zhang, S.N.; Park, J.; Sonn, C.H.; Lee, K.M.; Yun, C.O. Oncolytic adenovirus co-expressing IL-12 and IL-18 improves tumor-specific immunity via differentiation of T cells expressing IL-12Rbeta2 or IL-18Ralpha. *Gene Ther.* **2011**, *18*, 898–909. [CrossRef]
47. Oh, E.; Choi, I.K.; Hong, J.; Yun, C.O. Oncolytic adenovirus coexpressing interleukin-12 and decorin overcomes Treg-mediated immunosuppression inducing potent antitumor effects in a weakly immunogenic tumor model. *Oncotarget* **2017**, *8*, 4730–4746. [CrossRef]
48. Lee, Y.S.; Kim, J.H.; Choi, K.J.; Choi, I.K.; Kim, H.; Cho, S.; Cho, B.C.; Yun, C.O. Enhanced antitumor effect of oncolytic adenovirus expressing interleukin-12 and B7-1 in an immunocompetent murine model. *Clin. Cancer Res.* **2006**, *12*, 5859–5868. [CrossRef]
49. Huang, J.H.; Zhang, S.N.; Choi, K.J.; Choi, I.K.; Kim, J.H.; Lee, M.G.; Kim, H.; Yun, C.O. Therapeutic and tumor-specific immunity induced by combination of dendritic cells and oncolytic adenovirus expressing IL-12 and 4-1BBL. *Mol. Ther.* **2010**, *18*, 264–274. [CrossRef]
50. Backhaus, P.S.; Veinalde, R.; Hartmann, L.; Dunder, J.E.; Jeworowski, L.M.; Albert, J.; Hoyler, B.; Poth, T.; Jager, D.; Ungerechts, G.; et al. Immunological Effects and Viral Gene Expression Determine the Efficacy of Oncolytic Measles Vaccines Encoding IL-12 or IL-15 Agonists. *Viruses* **2019**, *11*, 914. [CrossRef]
51. Alkayyal, A.A.; Tai, L.H.; Kennedy, M.A.; de Souza, C.T.; Zhang, J.; Lefebvre, C.; Sahi, S.; Ananth, A.A.; Mahmoud, A.B.; Makrigiannis, A.P.; et al. NK-Cell Recruitment Is Necessary for Eradication of Peritoneal Carcinomatosis with an IL12-Expressing Maraba Virus Cellular Vaccine. *Cancer Immunol. Res.* **2017**, *5*, 211–221. [CrossRef]
52. Ren, G.; Tian, G.; Liu, Y.; He, J.; Gao, X.; Yu, Y.; Liu, X.; Zhang, X.; Sun, T.; Liu, S.; et al. Recombinant Newcastle Disease Virus Encoding IL-12 and/or IL-2 as Potential Candidate for Hepatoma Carcinoma Therapy. *Technol. Cancer Res. Treat.* **2016**, *15*, NP83-94. [CrossRef] [PubMed]
53. Asselin-Paturel, C.; Lassau, N.; Guinebretiere, J.M.; Zhang, J.; Gay, F.; Bex, F.; Hallez, S.; Leclere, J.; Peronneau, P.; Mami-Chouaib, F.; et al. Transfer of the murine interleukin-12 gene in vivo by a Semliki Forest virus vector induces B16 tumor regression through inhibition of tumor blood vessel formation monitored by Doppler ultrasonography. *Gene Ther.* **1999**, *6*, 606–615. [CrossRef] [PubMed]
54. Colmenero, P.; Chen, M.; Castanos-Velez, E.; Liljestrom, P.; Jondal, M. Immunotherapy with recombinant SFV-replicons expressing the P815A tumor antigen or IL-12 induces tumor regression. *Int. J. Cancer* **2002**, *98*, 554–560. [CrossRef]
55. Melero, I.; Quetglas, J.I.; Reboredo, M.; Dubrot, J.; Rodriguez-Madoz, J.R.; Mancheno, U.; Casales, E.; Riezu-Boj, J.I.; Ruiz-Guillen, M.; Ochoa, M.C.; et al. Strict requirement for vector-induced type I interferon in efficacious antitumor responses to virally encoded IL12. *Cancer Res.* **2015**, *75*, 497–507. [CrossRef]
56. Roche, F.P.; Sheahan, B.J.; O'Mara, S.M.; Atkins, G.J. Semliki Forest virus-mediated gene therapy of the RG2 rat glioma. *Neuropathol. Appl. Neurobiol.* **2010**, *36*, 648–660. [CrossRef]
57. Shin, E.J.; Wanna, G.B.; Choi, B.; Aguila, D., 3rd; Ebert, O.; Genden, E.M.; Woo, S.L. Interleukin-12 expression enhances vesicular stomatitis virus oncolytic therapy in murine squamous cell carcinoma. *Laryngoscope* **2007**, *117*, 210–214. [CrossRef]
58. Granot, T.; Venticinque, L.; Tseng, J.C.; Meruelo, D. Activation of cytotoxic and regulatory functions of NK cells by Sindbis viral vectors. *PLoS ONE* **2011**, *6*, e20598. [CrossRef]
59. Peters, C.; Rabkin, S.D. Designing Herpes Viruses as Oncolytics. *Mol Ther Oncolytics* **2015**, *2*. [CrossRef]
60. Todo, T.; Martuza, R.L.; Rabkin, S.D.; Johnson, P.A. Oncolytic herpes simplex virus vector with enhanced MHC class I presentation and tumor cell killing. *Proc. Natl. Acad. Sci. USA* **2001**, *98*, 6396–6401. [CrossRef]
61. Goldstein, D.J.; Weller, S.K. Factor(s) present in herpes simplex virus type 1-infected cells can compensate for the loss of the large subunit of the viral ribonucleotide reductase: characterization of an ICP6 deletion mutant. *Virology* **1988**, *166*, 41–51. [CrossRef]
62. Aghi, M.; Visted, T.; Depinho, R.A.; Chiocca, E.A. Oncolytic herpes virus with defective ICP6 specifically replicates in quiescent cells with homozygous genetic mutations in p16. *Oncogene* **2008**, *27*, 4249–4254. [CrossRef] [PubMed]

63. Cameron, J.M.; McDougall, I.; Marsden, H.S.; Preston, V.G.; Ryan, D.M.; Subak-Sharpe, J.H. Ribonucleotide reductase encoded by herpes simplex virus is a determinant of the pathogenicity of the virus in mice and a valid antiviral target. *J. Gen. Virol.* **1988**, *69*, 2607–2612. [CrossRef] [PubMed]
64. Markert, J.M.; Razdan, S.N.; Kuo, H.C.; Cantor, A.; Knoll, A.; Karrasch, M.; Nabors, L.B.; Markiewicz, M.; Agee, B.S.; Coleman, J.M.; et al. A phase 1 trial of oncolytic HSV-1, G207, given in combination with radiation for recurrent GBM demonstrates safety and radiographic responses. *Mol. Ther.* **2014**, *22*, 1048–1055. [CrossRef] [PubMed]
65. Pasieka, T.J.; Baas, T.; Carter, V.S.; Proll, S.C.; Katze, M.G.; Leib, D.A. Functional genomic analysis of herpes simplex virus type 1 counteraction of the host innate response. *J. Virol.* **2006**, *80*, 7600–7612. [CrossRef] [PubMed]
66. Wylie, K.M.; Schrimpf, J.E.; Morrison, L.A. Increased eIF2alpha phosphorylation attenuates replication of herpes simplex virus 2 vhs mutants in mouse embryonic fibroblasts and correlates with reduced accumulation of the PKR antagonist ICP34.5. *J. Virol.* **2009**, *83*, 9151–9162. [CrossRef]
67. He, B.; Gross, M.; Roizman, B. The gamma(1)34.5 protein of herpes simplex virus 1 complexes with protein phosphatase 1alpha to dephosphorylate the alpha subunit of the eukaryotic translation initiation factor 2 and preclude the shutoff of protein synthesis by double-stranded RNA-activated protein kinase. *Proc. Natl. Acad. Sci. USA* **1997**, *94*, 843–848.
68. Li, Y.; Zhang, C.; Chen, X.; Yu, J.; Wang, Y.; Yang, Y.; Du, M.; Jin, H.; Ma, Y.; He, B.; et al. ICP34.5 protein of herpes simplex virus facilitates the initiation of protein translation by bridging eukaryotic initiation factor 2alpha (eIF2alpha) and protein phosphatase 1. *J. Biol. Chem.* **2011**, *286*, 24785–24792. [CrossRef]
69. Orr, M.T.; Edelmann, K.H.; Vieira, J.; Corey, L.; Raulet, D.H.; Wilson, C.B. Inhibition of MHC class I is a virulence factor in herpes simplex virus infection of mice. *PLoS Pathog* **2005**, *1*, e7. [CrossRef]
70. Eggensperger, S.; Tampe, R. The transporter associated with antigen processing: A key player in adaptive immunity. *Biol. Chem.* **2015**, *396*, 1059–1072. [CrossRef]
71. Goldsmith, K.; Chen, W.; Johnson, D.C.; Hendricks, R.L. Infected cell protein (ICP)47 enhances herpes simplex virus neurovirulence by blocking the CD8+ T cell response. *J. Exp. Med.* **1998**, *187*, 341–348. [CrossRef]
72. Liu, B.L.; Robinson, M.; Han, Z.Q.; Branston, R.H.; English, C.; Reay, P.; McGrath, Y.; Thomas, S.K.; Thornton, M.; Bullock, P.; et al. ICP34.5 deleted herpes simplex virus with enhanced oncolytic, immune stimulating, and anti-tumour properties. *Gene Ther.* **2003**, *10*, 292–303. [CrossRef] [PubMed]
73. Wollmann, G.; Ozduman, K.; van den Pol, A.N. Oncolytic virus therapy for glioblastoma multiforme: concepts and candidates. *Cancer J.* **2012**, *18*, 69–81. [CrossRef] [PubMed]
74. Mulvey, M.; Poppers, J.; Sternberg, D.; Mohr, I. Regulation of eIF2alpha phosphorylation by different functions that act during discrete phases in the herpes simplex virus type 1 life cycle. *J. Virol.* **2003**, *77*, 10917–10928. [CrossRef] [PubMed]
75. Cassady, K.A.; Gross, M.; Roizman, B. The herpes simplex virus US11 protein effectively compensates for the gamma1(34.5) gene if present before activation of protein kinase R by precluding its phosphorylation and that of the alpha subunit of eukaryotic translation initiation factor 2. *J. Virol.* **1998**, *72*, 8620–8626. [CrossRef] [PubMed]
76. Saha, D.; Wakimoto, H.; Peters, C.W.; Antoszczyk, S.J.; Rabkin, S.D.; Martuza, R.L. Combinatorial effects of VEGFR kinase inhibitor axitinib and oncolytic virotherapy in mouse and human glioblastoma stem-like cell models. *Clin. Cancer Res.* **2018**. [CrossRef] [PubMed]
77. Todo, T. "Armed" oncolytic herpes simplex viruses for brain tumor therapy. *Cell Adh. Migr.* **2008**, *2*, 208–213. [CrossRef]
78. Kelly, K.J.; Wong, J.; Fong, Y. Herpes simplex virus NV1020 as a novel and promising therapy for hepatic malignancy. *Expert Opin. Investig. Drugs* **2008**, *17*, 1105–1113. [CrossRef]
79. Bennett, J.J.; Malhotra, S.; Wong, R.J.; Delman, K.; Zager, J.; St-Louis, M.; Johnson, P.; Fong, Y. Interleukin 12 secretion enhances antitumor efficacy of oncolytic herpes simplex viral therapy for colorectal cancer. *Ann. Surg.* **2001**, *233*, 819–826. [CrossRef]
80. Smith, M.C.; Boutell, C.; Davido, D.J. HSV-1 ICP0: Paving the way for viral replication. *Future Virol.* **2011**, *6*, 421–429. [CrossRef]
81. Lanfranca, M.P.; Mostafa, H.H.; Davido, D.J. HSV-1 ICP0: An E3 Ubiquitin Ligase That Counteracts Host Intrinsic and Innate Immunity. *Cells* **2014**, *3*, 438–454. [CrossRef]

82. Jacobs, A.; Breakefield, X.O.; Fraefel, C. HSV-1-based vectors for gene therapy of neurological diseases and brain tumors: Part I. HSV-1 structure, replication and pathogenesis. *Neoplasia* **1999**, *1*, 387–401. [CrossRef] [PubMed]
83. Parker, J.N.; Pfister, L.A.; Quenelle, D.; Gillespie, G.Y.; Markert, J.M.; Kern, E.R.; Whitley, R.J. Genetically engineered herpes simplex viruses that express IL-12 or GM-CSF as vaccine candidates. *Vaccine* **2006**, *24*, 1644–1652. [CrossRef] [PubMed]
84. Parker, J.N.; Gillespie, G.Y.; Love, C.E.; Randall, S.; Whitley, R.J.; Markert, J.M. Engineered herpes simplex virus expressing IL-12 in the treatment of experimental murine brain tumors. *Proc. Natl. Acad. Sci. USA* **2000**, *97*, 2208–2213. [CrossRef] [PubMed]
85. Bauer, D.F.; Pereboeva, L.; Gillespie, G.Y.; Cloud, G.A.; Elzafarany, O.; Langford, C.; Markert, J.M.; Lamb, L.S., Jr. Effect of HSV-IL12 Loaded Tumor Cell-Based Vaccination in a Mouse Model of High-Grade Neuroblastoma. *J. Immunol. Res.* **2016**, *2016*, 2568125. [CrossRef] [PubMed]
86. Hellums, E.K.; Markert, J.M.; Parker, J.N.; He, B.; Perbal, B.; Roizman, B.; Whitley, R.J.; Langford, C.P.; Bharara, S.; Gillespie, G.Y. Increased efficacy of an interleukin-12-secreting herpes simplex virus in a syngeneic intracranial murine glioma model. *Neuro. Oncol.* **2005**, *7*, 213–224. [CrossRef] [PubMed]
87. Markert, J.M.; Cody, J.J.; Parker, J.N.; Coleman, J.M.; Price, K.H.; Kern, E.R.; Quenelle, D.C.; Lakeman, A.D.; Schoeb, T.R.; Palmer, C.A.; et al. Preclinical evaluation of a genetically engineered herpes simplex virus expressing interleukin-12. *J. Virol.* **2012**, *86*, 5304–5313. [CrossRef]
88. Patel, D.M.; Foreman, P.M.; Nabors, L.B.; Riley, K.O.; Gillespie, G.Y.; Markert, J.M. Design of a Phase I Clinical Trial to Evaluate M032, a Genetically Engineered HSV-1 Expressing IL-12, in Patients with Recurrent/Progressive Glioblastoma Multiforme, Anaplastic Astrocytoma, or Gliosarcoma. *Hum. Gene Ther. Clin. Dev.* **2016**, *27*, 69–78. [CrossRef]
89. Menotti, L.; Avitabile, E.; Gatta, V.; Malatesta, P.; Petrovic, B.; Campadelli-Fiume, G. HSV as A Platform for the Generation of Retargeted, Armed, and Reporter-Expressing Oncolytic Viruses. *Viruses* **2018**, *10*, 352. [CrossRef]
90. Menotti, L.; Cerretani, A.; Hengel, H.; Campadelli-Fiume, G. Construction of a fully retargeted herpes simplex virus 1 recombinant capable of entering cells solely via human epidermal growth factor receptor 2. *J. Virol.* **2008**, *82*, 10153–10161. [CrossRef]
91. Liu, X.J.; Wang, X.Y.; Guo, J.X.; Zhu, H.J.; Zhang, C.R.; Ma, Z.H. Oncolytic property of HSV-1 recombinant viruses carrying the human IL-12. *Zhonghua Yi Xue Za Zhi* **2017**, *97*, 2135–2140. [CrossRef]
92. Athie-Morales, V.; Smits, H.H.; Cantrell, D.A.; Hilkens, C.M. Sustained IL-12 signaling is required for Th1 development. *J. Immunol.* **2004**, *172*, 61–69. [CrossRef] [PubMed]
93. Grivennikov, S.I.; Greten, F.R.; Karin, M. Immunity, inflammation, and cancer. *Cell* **2010**, *140*, 883–899. [CrossRef] [PubMed]
94. Antoszczyk, S.; Spyra, M.; Mautner, V.F.; Kurtz, A.; Stemmer-Rachamimov, A.O.; Martuza, R.L.; Rabkin, S.D. Treatment of orthotopic malignant peripheral nerve sheath tumors with oncolytic herpes simplex virus. *Neuro. Oncol.* **2014**, *16*, 1057–1066. [CrossRef] [PubMed]
95. Varghese, S.; Rabkin, S.D.; Nielsen, G.P.; MacGarvey, U.; Liu, R.; Martuza, R.L. Systemic therapy of spontaneous prostate cancer in transgenic mice with oncolytic herpes simplex viruses. *Cancer Res.* **2007**, *67*, 9371–9379. [CrossRef]
96. Ring, E.K.; Li, R.; Moore, B.P.; Nan, L.; Kelly, V.M.; Han, X.; Beierle, E.A.; Markert, J.M.; Leavenworth, J.W.; Gillespie, G.Y.; et al. Newly Characterized Murine Undifferentiated Sarcoma Models Sensitive to Virotherapy with Oncolytic HSV-1 M002. *Mol. Ther. Oncolytics* **2017**, *7*, 27–36. [CrossRef]
97. Thomas, E.D.; Meza-Perez, S.; Bevis, K.S.; Randall, T.D.; Gillespie, G.Y.; Langford, C.; Alvarez, R.D. IL-12 Expressing oncolytic herpes simplex virus promotes anti-tumor activity and immunologic control of metastatic ovarian cancer in mice. *J. Ovarian Res.* **2016**, *9*, 70. [CrossRef]
98. Krist, L.F.; Kerremans, M.; Broekhuis-Fluitsma, D.M.; Eestermans, I.L.; Meyer, S.; Beelen, R.H. Milky spots in the greater omentum are predominant sites of local tumour cell proliferation and accumulation in the peritoneal cavity. *Cancer Immunol. Immunother.* **1998**, *47*, 205–212. [CrossRef]
99. Sorensen, E.W.; Gerber, S.A.; Frelinger, J.G.; Lord, E.M. IL-12 suppresses vascular endothelial growth factor receptor 3 expression on tumor vessels by two distinct IFN-gamma-dependent mechanisms. *J. Immunol.* **2010**, *184*, 1858–1866. [CrossRef]

100. Cicchelero, L.; Denies, S.; Haers, H.; Vanderperren, K.; Stock, E.; Van Brantegem, L.; de Rooster, H.; Sanders, N.N. Intratumoural interleukin 12 gene therapy stimulates the immune system and decreases angiogenesis in dogs with spontaneous cancer. *Vet. Comp. Oncol.* **2017**, *15*, 1187–1205. [CrossRef]
101. Wong, R.J.; Chan, M.K.; Yu, Z.; Ghossein, R.A.; Ngai, I.; Adusumilli, P.S.; Stiles, B.M.; Shah, J.P.; Singh, B.; Fong, Y. Angiogenesis inhibition by an oncolytic herpes virus expressing interleukin 12. *Clin. Cancer Res.* **2004**, *10*, 4509–4516. [CrossRef]
102. Das, S.; Marsden, P.A. Angiogenesis in glioblastoma. *N Engl. J. Med.* **2013**, *369*, 1561–1563. [CrossRef]
103. Gatson, N.N.; Chiocca, E.A.; Kaur, B. Anti-angiogenic gene therapy in the treatment of malignant gliomas. *Neurosci. Lett.* **2012**, *527*, 62–70. [CrossRef]
104. Hardee, M.E.; Zagzag, D. Mechanisms of glioma-associated neovascularization. *Am. J. Pathol.* **2012**, *181*, 1126–1141. [CrossRef] [PubMed]
105. Jain, R.K.; di Tomaso, E.; Duda, D.G.; Loeffler, J.S.; Sorensen, A.G.; Batchelor, T.T. Angiogenesis in brain tumours. *Nat. Rev. Neurosci.* **2007**, *8*, 610–622. [CrossRef] [PubMed]
106. Zirlik, K.; Duyster, J. Anti-Angiogenics: Current Situation and Future Perspectives. *Oncol. Res. Treat.* **2018**, *41*, 166–171. [CrossRef] [PubMed]
107. Ramjiawan, R.R.; Griffioen, A.W.; Duda, D.G. Anti-angiogenesis for cancer revisited: Is there a role for combinations with immunotherapy? *Angiogenesis* **2017**, *20*, 185–204. [CrossRef] [PubMed]
108. Rajabi, M.; Mousa, S.A. The Role of Angiogenesis in Cancer Treatment. *Biomedicines* **2017**, *5*, 34. [CrossRef]
109. Melegh, Z.; Oltean, S. Targeting Angiogenesis in Prostate Cancer. *Int. J. Mol. Sci.* **2019**, *20*, 2676. [CrossRef]
110. van Moorselaar, R.J.; Voest, E.E. Angiogenesis in prostate cancer: Its role in disease progression and possible therapeutic approaches. *Mol. Cell Endocrinol.* **2002**, *197*, 239–250. [CrossRef]
111. Batchelor, T.T.; Reardon, D.A.; de Groot, J.F.; Wick, W.; Weller, M. Antiangiogenic therapy for glioblastoma: Current status and future prospects. *Clin. Cancer Res.* **2014**, *20*, 5612–5619. [CrossRef]
112. Wick, W.; Chinot, O.L.; Bendszus, M.; Mason, W.; Henriksson, R.; Saran, F.; Nishikawa, R.; Revil, C.; Kerloeguen, Y.; Cloughesy, T. Evaluation of pseudoprogression rates and tumor progression patterns in a phase III trial of bevacizumab plus radiotherapy/temozolomide for newly diagnosed glioblastoma. *Neuro. Oncol.* **2016**, *18*, 1434–1441. [CrossRef] [PubMed]
113. Lombardi, G.; Pambuku, A.; Bellu, L.; Farina, M.; Della Puppa, A.; Denaro, L.; Zagonel, V. Effectiveness of antiangiogenic drugs in glioblastoma patients: A systematic review and meta-analysis of randomized clinical trials. *Crit. Rev. Oncol. Hematol.* **2017**, *111*, 94–102. [CrossRef] [PubMed]
114. Kelly, W.K.; Halabi, S.; Carducci, M.; George, D.; Mahoney, J.F.; Stadler, W.M.; Morris, M.; Kantoff, P.; Monk, J.P.; Kaplan, E.; et al. Randomized, double-blind, placebo-controlled phase III trial comparing docetaxel and prednisone with or without bevacizumab in men with metastatic castration-resistant prostate cancer: CALGB 90401. *J. Clin. Oncol.* **2012**, *30*, 1534–1540. [CrossRef]
115. Hu-Lowe, D.D.; Zou, H.Y.; Grazzini, M.L.; Hallin, M.E.; Wickman, G.R.; Amundson, K.; Chen, J.H.; Rewolinski, D.A.; Yamazaki, S.; Wu, E.Y.; et al. Nonclinical antiangiogenesis and antitumor activities of axitinib (AG-013736), an oral, potent, and selective inhibitor of vascular endothelial growth factor receptor tyrosine kinases 1, 2, 3. *Clin. Cancer Res.* **2008**, *14*, 7272–7283. [CrossRef] [PubMed]
116. Ho, A.L.; Dunn, L.; Sherman, E.J.; Fury, M.G.; Baxi, S.S.; Chandramohan, R.; Dogan, S.; Morris, L.G.; Cullen, G.D.; Haque, S.; et al. A phase II study of axitinib (AG-013736) in patients with incurable adenoid cystic carcinoma. *Ann. Oncol.* **2016**, *27*, 1902–1908. [CrossRef] [PubMed]
117. McNamara, M.G.; Le, L.W.; Horgan, A.M.; Aspinall, A.; Burak, K.W.; Dhani, N.; Chen, E.; Sinaei, M.; Lo, G.; Kim, T.K.; et al. A phase II trial of second-line axitinib following prior antiangiogenic therapy in advanced hepatocellular carcinoma. *Cancer* **2015**, *121*, 1620–1627. [CrossRef] [PubMed]
118. Schiller, J.H.; Larson, T.; Ou, S.H.; Limentani, S.; Sandler, A.; Vokes, E.; Kim, S.; Liau, K.; Bycott, P.; Olszanski, A.J.; et al. Efficacy and safety of axitinib in patients with advanced non-small-cell lung cancer: Results from a phase II study. *J. Clin. Oncol.* **2009**, *27*, 3836–3841. [CrossRef]
119. Du Four, S.; Maenhout, S.K.; Benteyn, D.; De Keersmaecker, B.; Duerinck, J.; Thielemans, K.; Neyns, B.; Aerts, J.L. Disease progression in recurrent glioblastoma patients treated with the VEGFR inhibitor axitinib is associated with increased regulatory T cell numbers and T cell exhaustion. *Cancer Immunol. Immunother.* **2016**, *65*, 727–740. [CrossRef]

120. Du Four, S.; Maenhout, S.K.; De Pierre, K.; Renmans, D.; Niclou, S.P.; Thielemans, K.; Neyns, B.; Aerts, J.L. Axitinib increases the infiltration of immune cells and reduces the suppressive capacity of monocytic MDSCs in an intracranial mouse melanoma model. *Oncoimmunology* **2015**, *4*, e998107. [CrossRef]
121. Laubli, H.; Muller, P.; D'Amico, L.; Buchi, M.; Kashyap, A.S.; Zippelius, A. The multi-receptor inhibitor axitinib reverses tumor-induced immunosuppression and potentiates treatment with immune-modulatory antibodies in preclinical murine models. *Cancer Immunol. Immunother.* **2018**, *67*, 815–824. [CrossRef]
122. Wilmes, L.J.; Pallavicini, M.G.; Fleming, L.M.; Gibbs, J.; Wang, D.; Li, K.L.; Partridge, S.C.; Henry, R.G.; Shalinsky, D.R.; Hu-Lowe, D.; et al. AG-013736, a novel inhibitor of VEGF receptor tyrosine kinases, inhibits breast cancer growth and decreases vascular permeability as detected by dynamic contrast-enhanced magnetic resonance imaging. *Magn Reson Imaging* **2007**, *25*, 319–327. [CrossRef] [PubMed]
123. Chen, Y.; Rini, B.I.; Motzer, R.J.; Dutcher, J.P.; Rixe, O.; Wilding, G.; Stadler, W.M.; Tarazi, J.; Garrett, M.; Pithavala, Y.K. Effect of Renal Impairment on the Pharmacokinetics and Safety of Axitinib. *Target Oncol.* **2016**, *11*, 229–234. [CrossRef]
124. Gross-Goupil, M.; Francois, L.; Quivy, A.; Ravaud, A. Axitinib: A review of its safety and efficacy in the treatment of adults with advanced renal cell carcinoma. *Clin. Med. Insights Oncol.* **2013**, *7*, 269–277. [CrossRef] [PubMed]
125. Zhang, W.; Fulci, G.; Wakimoto, H.; Cheema, T.A.; Buhrman, J.S.; Jeyaretna, D.S.; Stemmer Rachamimov, A.O.; Rabkin, S.D.; Martuza, R.L. Combination of oncolytic herpes simplex viruses armed with angiostatin and IL-12 enhances antitumor efficacy in human glioblastoma models. *Neoplasia* **2013**, *15*, 591–599. [CrossRef] [PubMed]
126. Passer, B.J.; Cheema, T.; Wu, S.; Wu, C.L.; Rabkin, S.D.; Martuza, R.L. Combination of vinblastine and oncolytic herpes simplex virus vector expressing IL-12 therapy increases antitumor and antiangiogenic effects in prostate cancer models. *Cancer Gene Ther.* **2013**, *20*, 17–24. [CrossRef]
127. Chen, C.Y.; Hutzen, B.; Wedekind, M.F.; Cripe, T.P. Oncolytic virus and PD-1/PD-L1 blockade combination therapy. *Oncolytic Virother* **2018**, *7*, 65–77. [CrossRef]
128. Huang, J.; Liu, F.; Liu, Z.; Tang, H.; Wu, H.; Gong, Q.; Chen, J. Immune Checkpoint in Glioblastoma: Promising and Challenging. *Front. Pharmacol.* **2017**, *8*, 242. [CrossRef]
129. Sharma, P.; Allison, J.P. Immune checkpoint targeting in cancer therapy: Toward combination strategies with curative potential. *Cell* **2015**, *161*, 205–214. [CrossRef]
130. Twyman-Saint Victor, C.; Rech, A.J.; Maity, A.; Rengan, R.; Pauken, K.E.; Stelekati, E.; Benci, J.L.; Xu, B.; Dada, H.; Odorizzi, P.M.; et al. Radiation and dual checkpoint blockade activate non-redundant immune mechanisms in cancer. *Nature* **2015**, *520*, 373–377. [CrossRef]
131. Curran, M.A.; Montalvo, W.; Yagita, H.; Allison, J.P. PD-1 and CTLA-4 combination blockade expands infiltrating T cells and reduces regulatory T and myeloid cells within B16 melanoma tumors. *Proc. Natl. Acad. Sci. USA* **2010**, *107*, 4275–4280. [CrossRef]
132. Topalian, S.L.; Drake, C.G.; Pardoll, D.M. Immune checkpoint blockade: A common denominator approach to cancer therapy. *Cancer Cell* **2015**, *27*, 450–461. [CrossRef]
133. El-Shemi, A.G.; Ashshi, A.M.; Na, Y.; Li, Y.; Basalamah, M.; Al-Allaf, F.A.; Oh, E.; Jung, B.K.; Yun, C.O. Combined therapy with oncolytic adenoviruses encoding TRAIL and IL-12 genes markedly suppressed human hepatocellular carcinoma both in vitro and in an orthotopic transplanted mouse model. *J. Exp. Clin. Cancer Res.* **2016**, *35*, 74. [CrossRef]
134. Zhang, S.N.; Choi, I.K.; Huang, J.H.; Yoo, J.Y.; Choi, K.J.; Yun, C.O. Optimizing DC vaccination by combination with oncolytic adenovirus coexpressing IL-12 and GM-CSF. *Mol. Ther.* **2011**, *19*, 1558–1568. [CrossRef] [PubMed]
135. Kim, W.; Seong, J.; Oh, H.J.; Koom, W.S.; Choi, K.J.; Yun, C.O. A novel combination treatment of armed oncolytic adenovirus expressing IL-12 and GM-CSF with radiotherapy in murine hepatocarcinoma. *J. Radiat. Res.* **2011**, *52*, 646–654. [CrossRef] [PubMed]
136. Oh, E.; Oh, J.E.; Hong, J.; Chung, Y.; Lee, Y.; Park, K.D.; Kim, S.; Yun, C.O. Optimized biodegradable polymeric reservoir-mediated local and sustained co-delivery of dendritic cells and oncolytic adenovirus co-expressing IL-12 and GM-CSF for cancer immunotherapy. *J. Control. Release* **2017**, *259*, 115–127. [CrossRef] [PubMed]

137. Bortolanza, S.; Bunuales, M.; Otano, I.; Gonzalez-Aseguinolaza, G.; Ortiz-de-Solorzano, C.; Perez, D.; Prieto, J.; Hernandez-Alcoceba, R. Treatment of pancreatic cancer with an oncolytic adenovirus expressing interleukin-12 in Syrian hamsters. *Mol. Ther.* **2009**, *17*, 614–622. [CrossRef]
138. Yang, Z.; Zhang, Q.; Xu, K.; Shan, J.; Shen, J.; Liu, L.; Xu, Y.; Xia, F.; Bie, P.; Zhang, X.; et al. Combined therapy with cytokine-induced killer cells and oncolytic adenovirus expressing IL-12 induce enhanced antitumor activity in liver tumor model. *PLoS ONE* **2012**, *7*, e44802. [CrossRef]
139. Quetglas, J.I.; Labiano, S.; Aznar, M.A.; Bolanos, E.; Azpilikueta, A.; Rodriguez, I.; Casales, E.; Sanchez-Paulete, A.R.; Segura, V.; Smerdou, C.; et al. Virotherapy with a Semliki Forest Virus-Based Vector Encoding IL12 Synergizes with PD-1/PD-L1 Blockade. *Cancer Immunol. Res.* **2015**, *3*, 449–454. [CrossRef]
140. Schirmacher, V.; Forg, P.; Dalemans, W.; Chlichlia, K.; Zeng, Y.; Fournier, P.; von Hoegen, P. Intra-pinna anti-tumor vaccination with self-replicating infectious RNA or with DNA encoding a model tumor antigen and a cytokine. *Gene Ther.* **2000**, *7*, 1137–1147. [CrossRef]
141. Quetglas, J.I.; Dubrot, J.; Bezunartea, J.; Sanmamed, M.F.; Hervas-Stubbs, S.; Smerdou, C.; Melero, I. Immunotherapeutic synergy between anti-CD137 mAb and intratumoral administration of a cytopathic Semliki Forest virus encoding IL-12. *Mol. Ther.* **2012**, *20*, 1664–1675. [CrossRef]
142. Klas, S.D.; Robison, C.S.; Whitt, M.A.; Miller, M.A. Adjuvanticity of an IL-12 fusion protein expressed by recombinant deltaG-vesicular stomatitis virus. *Cell Immunol.* **2002**, *218*, 59–73. [CrossRef]
143. Lawler, S.E.; Speranza, M.C.; Cho, C.F.; Chiocca, E.A. Oncolytic Viruses in Cancer Treatment: A Review. *JAMA Oncol.* **2017**, *3*, 841–849. [CrossRef] [PubMed]
144. Senzer, N.N.; Kaufman, H.L.; Amatruda, T.; Nemunaitis, M.; Reid, T.; Daniels, G.; Gonzalez, R.; Glaspy, J.; Whitman, E.; Harrington, K.; et al. Phase II clinical trial of a granulocyte-macrophage colony-stimulating factor-encoding, second-generation oncolytic herpesvirus in patients with unresectable metastatic melanoma. *J. Clin. Oncol* **2009**, *27*, 5763–5771. [CrossRef]
145. Johnson, D.B.; Puzanov, I.; Kelley, M.C. Talimogene laherparepvec (T-VEC) for the treatment of advanced melanoma. *Immunotherapy* **2015**, *7*, 611–619. [CrossRef]
146. Bommareddy, P.K.; Patel, A.; Hossain, S.; Kaufman, H.L. Talimogene Laherparepvec (T-VEC) and Other Oncolytic Viruses for the Treatment of Melanoma. *Am. J. Clin. Dermatol* **2017**, *18*, 1–15. [CrossRef] [PubMed]
147. Vere Hodge, R.A.; Field, H.J. Antiviral agents for herpes simplex virus. *Adv. Pharmacol.* **2013**, *67*, 1–38. [CrossRef] [PubMed]
148. Parker, J.N.; Meleth, S.; Hughes, K.B.; Gillespie, G.Y.; Whitley, R.J.; Markert, J.M. Enhanced inhibition of syngeneic murine tumors by combinatorial therapy with genetically engineered HSV-1 expressing CCL2 and IL-12. *Cancer Gene Ther.* **2005**, *12*, 359–368. [CrossRef]
149. Ghouse, S.M.; Bommareddy, P.K.; Nguyen, H.M.; Guz-Montgomery, K.; Saha, D. Oncolytic herpes simplex virus encoding IL12 controls triple-negative breast cancer growth and metastasis in CD8-dependent manner. In Proceedings of the International Oncolytic Virus Conference, Rochester, MN, USA, 9–12 October 2019.
150. Esaki, S.; Nigim, F.; Moon, E.; Luk, S.; Kiyokawa, J.; Curry, W., Jr.; Cahill, D.P.; Chi, A.S.; Iafrate, A.J.; Martuza, R.L.; et al. Blockade of transforming growth factor-beta signaling enhances oncolytic herpes simplex virus efficacy in patient-derived recurrent glioblastoma models. *Int. J. Cancer* **2017**, *141*, 2348–2358. [CrossRef]
151. Ning, J.; Wakimoto, H.; Peters, C.; Martuza, R.L.; Rabkin, S.D. Rad51 Degradation: Role in Oncolytic Virus-Poly(ADP-Ribose) Polymerase Inhibitor Combination Therapy in Glioblastoma. *J. Natl. Cancer Inst.* **2017**, *109*, 1–13. [CrossRef]
152. Johnson, D.B.; Chandra, S.; Sosman, J.A. Immune Checkpoint Inhibitor Toxicity in 2018. *JAMA* **2018**, *320*, 1702–1703. [CrossRef]
153. Lyon, A.R.; Yousaf, N.; Battisti, N.M.L.; Moslehi, J.; Larkin, J. Immune checkpoint inhibitors and cardiovascular toxicity. *Lancet Oncol.* **2018**, *19*, e447–e458. [CrossRef]
154. Pai, C.S.; Simons, D.M.; Lu, X.; Evans, M.; Wei, J.; Wang, Y.H.; Chen, M.; Huang, J.; Park, C.; Chang, A.; et al. Tumor-conditional anti-CTLA4 uncouples antitumor efficacy from immunotherapy-related toxicity. *J. Clin. Investig.* **2019**, *129*, 349–363. [CrossRef] [PubMed]
155. Palmieri, D.J.; Carlino, M.S. Immune Checkpoint Inhibitor Toxicity. *Curr. Oncol Rep.* **2018**, *20*, 72. [CrossRef] [PubMed]

156. Rota, E.; Varese, P.; Agosti, S.; Celli, L.; Ghiglione, E.; Pappalardo, I.; Zaccone, G.; Paglia, A.; Morelli, N. Concomitant myasthenia gravis, myositis, myocarditis and polyneuropathy, induced by immune-checkpoint inhibitors: A life-threatening continuum of neuromuscular and cardiac toxicity. *eNeurologicalSci* **2019**, *14*, 4–5. [CrossRef]
157. Filley, A.C.; Henriquez, M.; Dey, M. Recurrent glioma clinical trial, CheckMate-143: The game is not over yet. *Oncotarget* **2017**, *8*, 91779–91794. [CrossRef]
158. Omuro, A.; Vlahovic, G.; Lim, M.; Sahebjam, S.; Baehring, J.; Cloughesy, T.; Voloschin, A.; Ramkissoon, S.H.; Ligon, K.L.; Latek, R.; et al. Nivolumab with or without ipilimumab in patients with recurrent glioblastoma: Results from exploratory phase I cohorts of CheckMate 143. *Neuro. Oncol.* **2018**, *20*, 674–686. [CrossRef]
159. Emens, L.A. Breast Cancer Immunotherapy: Facts and Hopes. *Clin. Cancer Res.* **2018**, *24*, 511–520. [CrossRef]
160. Conry, R.M.; Westbrook, B.; McKee, S.; Norwood, T.G. Talimogene laherparepvec: First in class oncolytic virotherapy. *Hum. Vaccin. Immunother.* **2018**, *14*, 839–846. [CrossRef]
161. Siurala, M.; Havunen, R.; Saha, D.; Lumen, D.; Airaksinen, A.J.; Tahtinen, S.; Cervera-Carrascon, V.; Bramante, S.; Parviainen, S.; Vaha-Koskela, M.; et al. Adenoviral Delivery of Tumor Necrosis Factor-alpha and Interleukin-2 Enables Successful Adoptive Cell Therapy of Immunosuppressive Melanoma. *Mol. Ther.* **2016**, *24*, 1435–1443. [CrossRef]
162. Tahtinen, S.; Blattner, C.; Vaha-Koskela, M.; Saha, D.; Siurala, M.; Parviainen, S.; Utikal, J.; Kanerva, A.; Umansky, V.; Hemminki, A. T-Cell Therapy Enabling Adenoviruses Coding for IL2 and TNFalpha Induce Systemic Immunomodulation in Mice With Spontaneous Melanoma. *J. Immunother.* **2016**, *39*, 343–354. [CrossRef]
163. Martinet, O.; Divino, C.M.; Zang, Y.; Gan, Y.; Mandeli, J.; Thung, S.; Pan, P.Y.; Chen, S.H. T cell activation with systemic agonistic antibody versus local 4-1BB ligand gene delivery combined with interleukin-12 eradicate liver metastases of breast cancer. *Gene Ther.* **2002**, *9*, 786–792. [CrossRef]
164. Kurokawa, C.; Iankov, I.D.; Anderson, S.K.; Aderca, I.; Leontovich, A.A.; Maurer, M.J.; Oberg, A.L.; Schroeder, M.A.; Giannini, C.; Greiner, S.M.; et al. Constitutive Interferon Pathway Activation in Tumors as an Efficacy Determinant Following Oncolytic Virotherapy. *J. Natl. Cancer Inst.* **2018**, *110*, 1123–1132. [CrossRef]

Communication

Development of Feline Ileum- and Colon-Derived Organoids and Their Potential Use to Support Feline Coronavirus Infection

Gergely Tekes [1,*,†], Rosina Ehmann [2], Steeve Boulant [3,4] and Megan L. Stanifer [5,*]

1 Institute of Virology, Justus Liebig University Giessen, 35390 Giessen, Germany
2 Bundeswehr Institute of Microbiology, 80937 Munich, Germany; RosinaEhmann@bundeswehr.org
3 Department of Infectious Diseases, Virology, Heidelberg University Hospital, 69120 Heidelberg, Germany; s.boulant@dkfz.de
4 Research Group "Cellular Polarity and Viral Infection", German Cancer Research Center (DKFZ), 69120 Heidelberg, Germany
5 Department of Infectious Diseases, Molecular Virology, Heidelberg University Hospital, 69120 Heidelberg, Germany
* Correspondence: gergely.tekes@web.de (G.T.); m.stanifer@dkfz.de (M.L.S.); Tel.: +49-(0)-6221567858 (M.L.S.)
† Current address: Elanco Animal Health, Germany.

Received: 14 August 2020; Accepted: 10 September 2020; Published: 12 September 2020

Abstract: Feline coronaviruses (FCoVs) infect both wild and domestic cat populations world-wide. FCoVs present as two main biotypes: the mild feline enteric coronavirus (FECV) and the fatal feline infectious peritonitis virus (FIPV). FIPV develops through mutations from FECV during a persistence infection. So far, the molecular mechanism of FECV-persistence and contributing factors for FIPV development may not be studied, since field FECV isolates do not grow in available cell culture models. In this work, we aimed at establishing feline ileum and colon organoids that allow the propagation of field FECVs. We have determined the best methods to isolate, culture and passage feline ileum and colon organoids. Importantly, we have demonstrated using GFP-expressing recombinant field FECV that colon organoids are able to support infection of FECV, which were unable to infect traditional feline cell culture models. These organoids in combination with recombinant FECVs can now open the door to unravel the molecular mechanisms by which FECV can persist in the gut for a longer period of time and how transition to FIPV is achieved.

Keywords: feline coronavirus; feline enteric coronavirus; FECV; feline infectious peritonitis virus; FIPV; feline intestinal organoids

1. Introduction

The diverse family of the *Coronaviridae* causes infections in a wide range of mammals, birds and humans. Feline coronavirus (FCoV) is a highly prevalent member of the *Coronaviridae* family and is found in both domestic and wild cat populations worldwide [1]. FCoVs occur in two different biotypes: feline enteric coronavirus (FECV) and feline infectious peritonitis virus (FIPV). FECV infections commonly manifest as mild or asymptomatic infections of the feline enteric tract. Infections are often persistent and display intermittent shedding of virus over long periods of time which greatly contributes to the high seropositivity levels found in domestic cats. In single household cats, 20–60% of cats display signs of exposure to the virus and up to 90% of cats in multi-cat populations are seropositive [2–4].

FIPV emerges through mutations from the harmless FECV and can lead to a fatal clinical condition known as feline infectious peritonitis (FIP) [5–9]. Although the molecular pathogenesis of FIP is poorly understood [10–12], promising therapeutical approaches have recently been described [13–18]. It is important to note that both biotypes exist in two serotypes [19–21]. Serotype II FCoVs are the results of recombination between a serotype I FCoV and a closely related canine coronavirus (CCoV) [22–25] and can easily be grown in vitro. In sharp contrast, the more relevant and prevalent serotype I FCoVs cannot be propagated in cell culture. Accordingly, serotype II viruses were often used in the past to gain insight into FCoV biology instead of serotype I FCoVs. To elucidate the molecular pathogenesis of FIP, cell culture-adapted serotype I FIPV laboratory strains were obtained over time [26]. However, these viruses proved to be unsuitable to study the pathogenesis of FIP due to the loss of pathogenicity via cell culture adaptation [1,9,26]. The first reverse genetic system that enabled genetic manipulation of the entire FCoV genome was described by Tekes et al. (2008) for serotype I FIPV laboratory strain Black using a vaccinia virus vector [26–28]. However, animal experiments showed that like many other laboratory strains, serotype I FIPV Black lost its capability to induce FIP [26]. On the contrary, another commonly used serotype II FIPV laboratory strain, 79-1146, [26,29,30] is much more pathogenic and thus does not appropriately resemble most of the field strains either. Due to the lack of suitable in vitro systems for field serotype I FECVs, it is critical to establish a suitable in vitro system that enables the growth of serotype I FECV. This culture system for serotype I FECV field viruses would not only provide insight into the molecular mechanism by which FECVs persist in the gut for a longer period of time but it might also contribute to the understanding of how FIPV can evolve from FECV during a harmless persistent infection.

Over the past years, organoids have been employed as an in vitro system to support the growth of several human viruses that were unable to be cultured using standard cell culture methods [31,32]. Organoids are derived from either induced pluripotent stem cells (iPSCs) or from tissue-derived stem cells, which are grown and differentiated as three-dimensional structures that closely recapitulate the cellular composition and functions of their originating organ [33]. Tissue-derived organoids rely on the ability to isolate stem cell containing crypts. These crypts are then grown in the presence of differentiation factors (Wnt3a, R-Spondin, Noggin and EGF), allowing them to grow into three-dimensional mini-gut organoids [33]. As these complex cultures more closely resemble the multi-cell types found in their natural tissue counterparts, they often contain factors, which are required for the replication and propagation of viruses that are missing in standard cell cultures. To determine if these model systems could be used to support the relevant serotype I FECV growth, we established a cat intestinal organoid culture system and show that it is capable of supporting infection with GFP-expressing recombinant serotype I FECV generated by reverse genetics. This model will now open the doors to study the molecular mechanism of serotype I FECV-persistence in its natural enteric environment.

2. Materials and Methods

2.1. Viruses and Cell Lines

Serotype I recFECV-GFP and recFECV-S_{79}-GFP were produced in vitro using the reverse genetic system for FCoV field strains described previously [34]. Recombinant virus stocks of recFECV-S_{79}-GFP were titrated by plaque assay on routinely used felis catus whole fetus (FCWF) cells [26–28,35,36]. Virus stocks of recFECV-GFP which cannot be cultivated in standard cell culture systems were quantified by comparative Western blot analysis of the FCoV M protein together with recFECV-S_{79}-GFP [34]. The cell line FCWF was provided by the diagnostic laboratory at the Justus Liebig University Giessen and maintained in culture media (DMEM with 1× penicillin/streptomycin (Thermo, Waltham, MA, USA) and 10% FBS (Biochrom, Cambridge, UK)). According to our experience with propagation of FCoV laboratory strains, the FCWF cells were used at a confluency of approximately 90% for the infection with recFECV-S_{79}-GFP.

2.2. Animals

Handling of the animals used for enteric tissue donation was performed according to the guidelines of the Hungarian legislation on animal protection. The protocol was approved by the Pest

Megyei Kormany-hivatal, Budapest (assurance numbers PE/EA/2441-6/2016 and TMF/657-12/2016). The animals were euthanized according to the designate protocol. Female specific-pathogen-free (SPF) cats were raised and housed in pathogen free conditions for laboratory use. These were not cats taken from outside veterinary practices. The SPF cats were euthanized at an age of 25 weeks and tissue samples from the gut were collected, 6 donors were used for this study. The cats were not tested prior to isolation specifically for FCoV. Animals had access to feed ad libitum prior to euthanization and tissue collection.

2.3. Chemicals and Solutions

Conditioned media containing Wnt3a, R-Spondin and Noggin was produced from the L-WRN cell line (ATCC CRL-3276) as per manufacturer's instructions. A 293T cell line which produces only R-Spondin was a kind gift from Calvin Kuo (Standford University) and conditioned media was made as previously described [37]. All organoid media were made from a base of advanced DMEM/F12 (Thermo, Waltham, MA, USA) which contained 1% penicillin/streptomycin (Thermo), 2 mM GlutaMAX (Thermo) and 10 mM HEPES (Thermo) and is referred to as Ad DMEM/F12++. All other organoid media components are found in Table 1.

Table 1. Human and mouse media compositions tested for their ability to support feline ileum and colon organoid growth. Final concentrations and manufactures of each component are listed.

Reagent	Company	Final Concentration	Human Media	Mouse Intestine Media	Mouse Colon Media
L-WRN (Wnt3s, R-Spondin, Noggin containing conditioned media)	Made in the lab	50%	X		X
Ad DMEM/F12++	Thermo	50% (Human and mouse colon) 90% for mouse intestine	X	X	X
B27	Thermo	1X	X	X	X
Nicotinamide	Sigma (Munich, Germany)	10 mM	X		
N-acetylcysteine	Sigma	1 mM	X	X	X
A-83-01	Tocris (Bristol, UK)	500 nM	X		
SB202190	Sigma	500 nM	X		
Leu-Gastrin	Sigma	10 nM	X		
Mouse recombinant	Thermo	50 ng/mL	X	X	X
R-Spondin conditioned media	Made in the lab	10%		X	
Mouse recombinant Noggin	Peprotech (Rocky Hill, NJ, USA)	100 ng/mL		X	
Y-27632	Sigma	10 µM	X	X	X
Matrigel, Growth factor reduced (GFR), phenol free	Corning (Corning, NY, USA)	100%	X	X	X

2.4. Isolation of Feline Intestinal Cells

Ileum and colon sections (10 cm each) were harvested from 6 donors and stored in cold transport buffer (1× phosphate buffered saline (PBS), 50 ng/mL gentamicin (Thermo), 1% penicillin/streptomycin (Thermo), 1% fetal bovine serum (FBS) and 250 µg/mL Fungizone (Thermo)) until the time of isolation. The tissue was cut in half and washed 3 × 10 min with shaking in cold PBS to ensure that all fecal material had been removed. Crypts containing stem cells were isolated within 16 h of sacrificing the animals by first cutting the tissue into smaller 1 cm pieces and then either adding 2 mM EDTA to tissue sample for 1 h at 4 °C or 20 mL of Gentle Cell Dissociation Reagent (Stem Cell Technologies, Vancouver, BC, Canada) for 1 h at room temp. Tissue sections were transferred to a clean tube and 10 mL of cold PBS + 1% BSA was added to tube. Tubes were shaken to release the crypt fraction. Fractions enriched in crypts were filtered with 70 µm filters and the process was repeated to collect four to five fractions for each tissue. The fractions were observed under a light microscope and those containing the highest number of crypts were pooled and spun at 500× *g* for 5 min at 4 °C. The supernatant was removed, and crypts were washed 1× with cold DMEM/F12 (Thermo) and spun at 500× *g* for 5 min at 4 °C. The media was removed, and crypts were re-suspended in 100% Matrigel, plated in 50 µL drops in 24-well non-tissue culture treated plates (Corning) and following polymerization of the Matrigel, 500 µL organoid media was added to each well and was replaced every 48 h. Following isolation of the crypts and seeding into organoid media, the size of the organoids was monitored over time at day 3, 6, 9, 12, and 15 days by observing them under bright field microscopy with a Nikon Eclipse Ti-S microscope. Their size was measured with a 10× objective using the Nikon NIS software.

2.5. RNA Isolation, cDNA, and qPCR

RNA was harvested from cells using NucleoSpin RNA extraction kit (Macherey-Nagel, Dueren, Germany) as per manufacturer's instructions. cDNA was made using iSCRIPT reverse transcriptase (BioRad, Hercules, CA, USA) from 250 ng of total RNA as per manufacturer's instructions. Quantitative-PCR was performed using iTaq SYBR green (BioRad) as per manufacturer's instructions, GAPDH was used as normalizing gene. Pre-designed feline specific KiCqStart SYBR Green pre-designed primers were purchased from Sigma-Aldrich (Table 2).

Table 2. Feline specific primers used to control cell population in tissue and organoids.

Gene	Cell Type	Gene ID	Sequence ID
GAPDH	House keeping	493876	NM_001009307
LGR5	Stem cell	101080720	XM_003989046
SMOC2	Stem cell	101082409	XM_003986725
MUC2	Goblet cell	101096605	XM_003993797
SI	Enterocyte	100144605	NM_001123332
SYP	Enteroendocrine	101084343	XM_004000526
LYZ	Paneth cell	100127109	XM_003989032

2.6. Passaging of Feline Mini-Gut Organoids

Ileum and colon mini-gut organoids were monitored under a light microscope and were passaged when centers became dark and filled with dead cells. For passaging, media was removed and Matrigel was dissolved in cold PBS. Organoids were spun at 500× *g* for 5 min at 4 °C and PBS was removed. Subsequently three different methods were used for passaging:

i. Mechanical passaging: 1 mL of cold PBS was used to re-suspend the organoids. Using a 27-gauge needle on a 1 mL syringe, organoids were broken down by passing the solution up and down 10 times through the needle. Organoids were then spun at 500× *g* for 5 min at 4 °C. The supernatant was removed, and the crypts were re-suspended in 100% Matrigel, plated in 50 µL drops in 24-well non-tissue culture treated plates (Corning) and following polymerization of the Matrigel, organoid media 500 µL was added to each well.
ii. Trypsin-based passaging: 1 mL of cold PBS was used to re-suspend the organoids. Organoids were then spun at 500× *g* for 5 min at 4 °C. Organoids were washed a second time in cold PBS by resuspending the pellet in 1 mL of PBS and spinning at 500× *g* for 5 min at 4 °C.

The supernatant was removed and organoids were incubated in 0.05% Trypsin-EDTA (Gibco) for 5 min at 37 °C. Trypsin digestion was stopped with the addition of serum containing media and samples were spun at 500× *g* for 5 min at 4 °C. Organoids were washed a second time in cold PBS by resuspending the pellet in 1 mL of PBS and spinning at 500× *g* for 5 min at 4 °C. The supernatant was removed, and the crypts were re-suspended in 100% Matrigel, plated in 50 μL drops in 24-well non-tissue culture treated plates (Corning) and following polymerization of the Matrigel, 500 μL organoid media was added to each well.

iii. Gentle Cell Dissociation Reagent method: Media was removed and Gentle Cell Dissociation Reagent (Stem cell technologies) was added to the organoid containing pellet and incubated for 10 min at room temp. Organoids were spun at 500× *g* for 5 min at 4 °C and the supernatant was removed. Organoids were washed in DMEM/F12 and then spun at 500× *g* for 5 min at 4 °C. The supernatant was removed and the crypts were re-suspended in 100% Matrigel, plated in 50 μL drops in 24-well non-tissue culture treated plates (Corning) and following polymerization of the Matrigel, 500 μL organoid media was added to each well.

2.7. Infection of Cell Culture with Recombinant Viruses

Approximately 90% confluent monolayers of FCWF cells were washed with serum free DMEM culture media. Cells were inoculated with recFECV-GFP and recFECV-S_{79}-GFP at a multiplicity of infection (MOI) of 0.01 or serum free culture media for the mock control. The infection was incubated for one hour and subsequently the inoculum was replaced by culture media containing FBS. The formation of plaques and the GFP signal was monitored 48 hours post-infection (hpi).

2.8. Infection of Organoids

Organoids were removed from Matrigel by adding cold PBS for 5 min, liquefied Matrigel and organoids were separated by centrifugation (500× *g*, 5 min). Supernatant and Matrigel were removed and organoids were resuspended in media and gently disrupted with a 27 G needle to allow virus to access both the apical and basolateral sides of the organoids. Following disruption, organoid media containing 10^4 pfu of FCoV was added to the organoids and allowed to incubate for 6 h in suspension. Following the 6-h incubation, Matrigel was added back to the cultures and organoids were observed over a three-day period.

2.9. Statistical Methods

Statistics were calculated by Prizm using an unpaired *t*-test.

3. Results and Discussion

Currently, infectious disease research of feline intestinal pathogens has been hampered by the lack of cell-based systems allowing for the screening of viruses and antiviral compounds. To fill this gap, we developed a method to isolate and propagate ileum and colon mini-gut organoids from felines. Six SPF cats were euthanized and the ileum and colon were collected. Tissue sections were washed thoroughly to remove all fecal material and were stored at 4 °C in a transport buffer containing PBS as well as antibiotics and antifungals. Organoids were prepared within 16 h of tissue harvesting. To determine the best method to isolate crypts containing stem cells from feline intestinal tissues, samples were split in two parts and two commonly used approaches (EDTA-based dissociation and Gentle Cell Dissociation Reagent™) for harvesting human and murine intestinal crypts [38,39] were compared side by side. Feline ileum and colon samples were washed thoroughly to remove all mucus or remaining contaminants and crypts containing stem cells were then isolated following incubation in EDTA or Gentle Cell Dissociation Reagent™. Microscopic evaluation of the crypts prior to seeding showed no large difference in the quantity or quality of crypts between the two isolation methods (data not shown). Equal numbers of crypts were seeded into Matrigel and the number and size of organoids was followed over 12 days. Unlike human and murine crypts, which form organoids with 16–24 h [38–40], the feline crypts took 72 h to seal and form cyst like organoids (Figure 1A). Following initial cyst formation, the ileum and colon organoids continued to grow over the twelve-day period (Figure 1A). To determine which isolation method produced the best organoids, the number and size of organoids was followed over time. Results show that organoids isolated using the EDTA approach produced a greater number of organoids and the organoids were larger in size compared to the Gentle Cell Dissociation Reagent™ method (Figure 1B,C). The increase in the number of organoids between day 3 and 6 does not illustrate a growth in organoid number but is due to the fact that at day 3 the crypts are too small to be effectively counted. This number of organoids then remained constant over the rest of the period as they continued to grow larger (Figure 1B,C), which was consistent with murine and human organoid generation [38,39].

To determine if the organoids displayed similar cell types as the natural feline intestine, tissue and organoids samples from three donors were lysed and the relative expression of each cell type was quantified by q-RT-PCR. As the natural intestine is made of stem cells, absorptive enterocytes and secretory cells we evaluated known markers from each of these populations. As there are no markers to analyze feline organoids, we used the best-known markers for murine and human, hypothesizing that they will constitute a marker for feline tissue. As a marker of stem cells, we used LGR5 and SMOC2. LGR5 is the historical marker described by the Clevers lab, however while it constitutes an excellent marker in tissue it is known to be suboptimal for organoids [38,39,41]. On the contrary, SMOC2 represents an excellent marker for both tissue and tissue-derived organoids. As a marker for Goblet we used mucin 2 (Muc2), for enterocytes we used sucrose isomaltase (SI), for enteroendocrine cells we used synaptophysin (SYP) and for Paneth cells we used lysozyme (LYZ). Results showed that ileum from feline tissue samples displays markers for stem cells (LGR5, SMOC2), Goblet cells (Muc2), enterocytes (SI), enteroendocrine cells (SYP) and Paneth cells (LYZ) (Figure 2A). Similarly, feline colon tissues expressed similar amounts of markers for all cell types except Paneth cells which are not present in colon tissue. The ratios found in feline tissues are similar to those observed in humans [41].

Importantly, organoids derived from feline ileum expressed the same cell type-specific markers as their tissue counterpart (Figure 2B). The relative expressions of the different markers were slightly different compared to the expression in tissue (Figure 2A,B). Ileum organoids displayed a higher relative expression of the stem cells marker (SMOC2) and a lower relative expression of the enterocyte maker (SI) compared to normal feline tissue. This suggests that there are more stem cells than enterocytes. The increase in stem cell number is expected as organoids are cultured under high Wnt3a conditions which favor stem cell numbers. Similarly, colon organoids also displayed a high amount of stem cell markers and their cellular composition looked similar to those of ileum organoids (Figure 2B). These differences in expression of the different cell type specific markers are not specific to the feline ileum and colon organoids but this is also observed in both murine and human intestinal organoids [40,42]. This shows that although organoids are extremely close to the originating tissue, the fact that they are ex vivo mini-organs causes subtle differences in their expression profile and differentiation pathways. All together these data show that we have developed a novel protocol to isolate and generate intestinal organoids from feline intestinal tissue and that these organoids closely resemble their tissue counterpart.

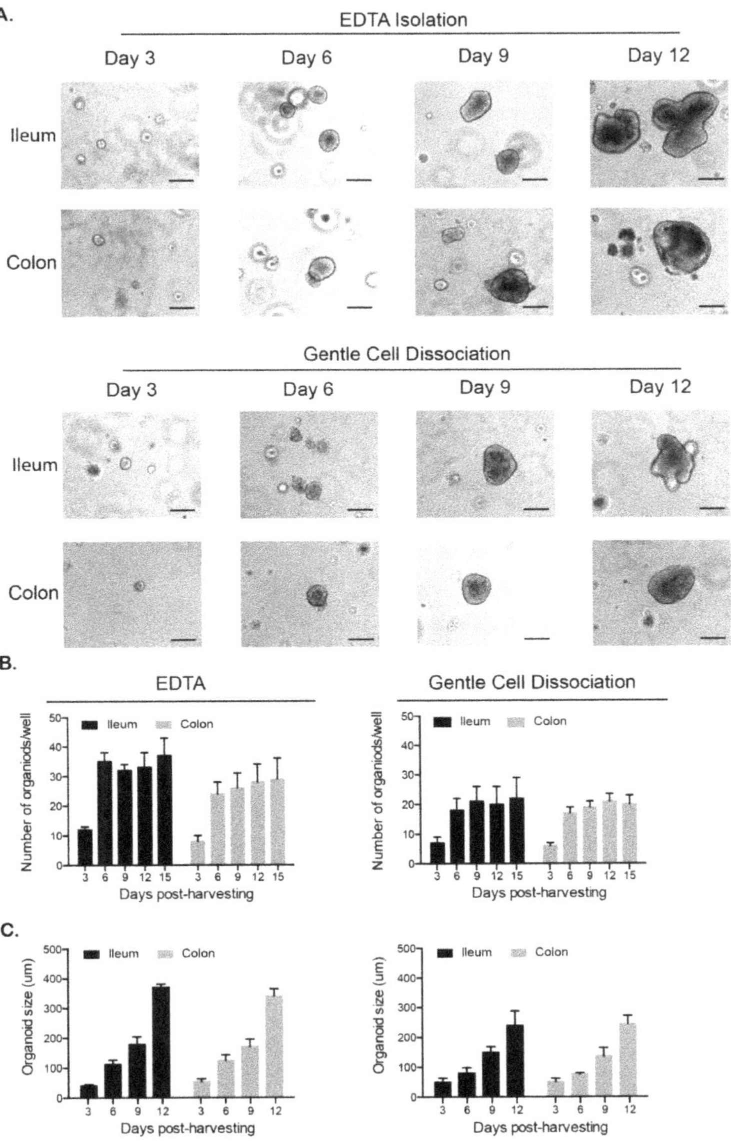

Figure 1. Ileum and colon organoids can be established from primary feline tissue. (**A–C**). The 1 cm^2 sections of feline ileum and colon tissue were incubated with EDTA or Gentle Cell Dissociation Reagent to allow for the isolation of intestinal crypts. Isolated crypts were resuspended in Matrigel and followed over a 12-day period. (**A**). Bright field images of ileum and colon organoids. Representative images are shown (n = 6 donors). Scale bar = 100 µm. (**B**). The number of organoids from EDTA and Gentle cell Dissociation Reagent isolation were counted over the indicated time course. Error bars represent standard deviation (n = 6 donors). (**C**). Following EDTA and Gentle cell Dissociation Reagent isolation, organoids were imaged in three-day intervals using a Nikon Eclipse Ti-S. Their size was measured using the Nikon NIS software. Error bars represent standard deviation (n = 6 donors).

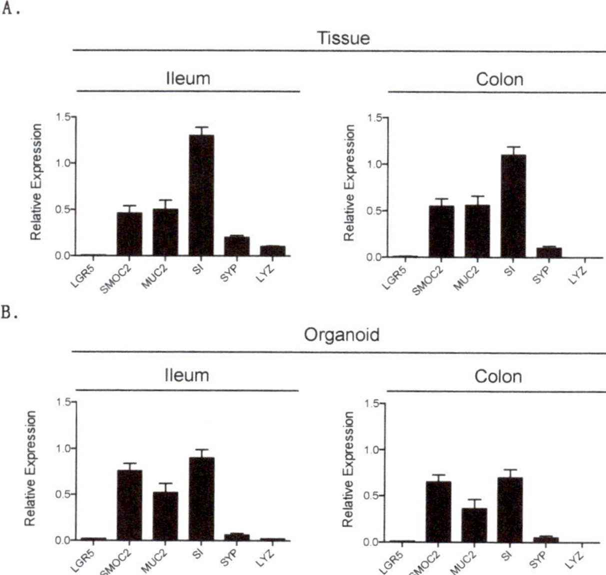

Figure 2. Feline organoids and originating tissue contain similar cell types. (**A**). The relative expression of intestinal cell type specific markers was evaluated in feline ileum and colon tissue by q-RT-PCR. (**B**). Same as A except using feline ileum- and colon-derived organoids. Results are mean +/− s.d. and are expressed as a relative expression to the house-keeping gene GAPDH. n = 3 donors and q-RT-PCR was done as technical triplicate.

3.1. *Passaging and Maintenance of Feline Intestinal or Ganoids*

As the ileum and colon organoids continued to grow and retain a similar identity to the natural tissue, we wanted to determine whether they could be passaged and maintained in culture. Trypsinization, mechanical disruption and passaging using Gentle Cell Dissociation Reagent™ were tested to establish the method that best supported the continued maintenance of feline ileal and colon organoids. For all methods, organoids were removed from the Matrigel and either incubated with low concentrations of trypsin, mechanical disrupted using a 27 G needle, or were incubated with Gentle Cell Dissociation Reagent™ (see materials and methods for full details). Following disruption, organoids were centrifuged to separate out organoid structures from dead cells. The organoids were then seeded into Matrigel and the formation of new organoids was followed for five days. One day post-passaging many small organoids could be seen in all conditions for both the ileum and colon organoids (Figure 3A). Organoids which were passaged by trypsinization were small and stressed, as shown by their loss of tight borders (dark rim at the periphery of the organoids under phase microscope) and the presence of many dissociated cells. On the contrary, organoids that were obtained using the mechanical disruption and Gentle Cell Dissociation Reagent™ methods were larger and their borders look more discrete (Figure 3A). Similar to the original isolation, many organoids were too small to be counted on the first day after passaging for all methods (Figure 3B). Observations five days post passaging showed that using mechanical disruption for passaging both ileum and colon organoids leads to a greater number of organoids that are larger in comparison to organoids that were passaged using trypsin (Figure 3B,C).

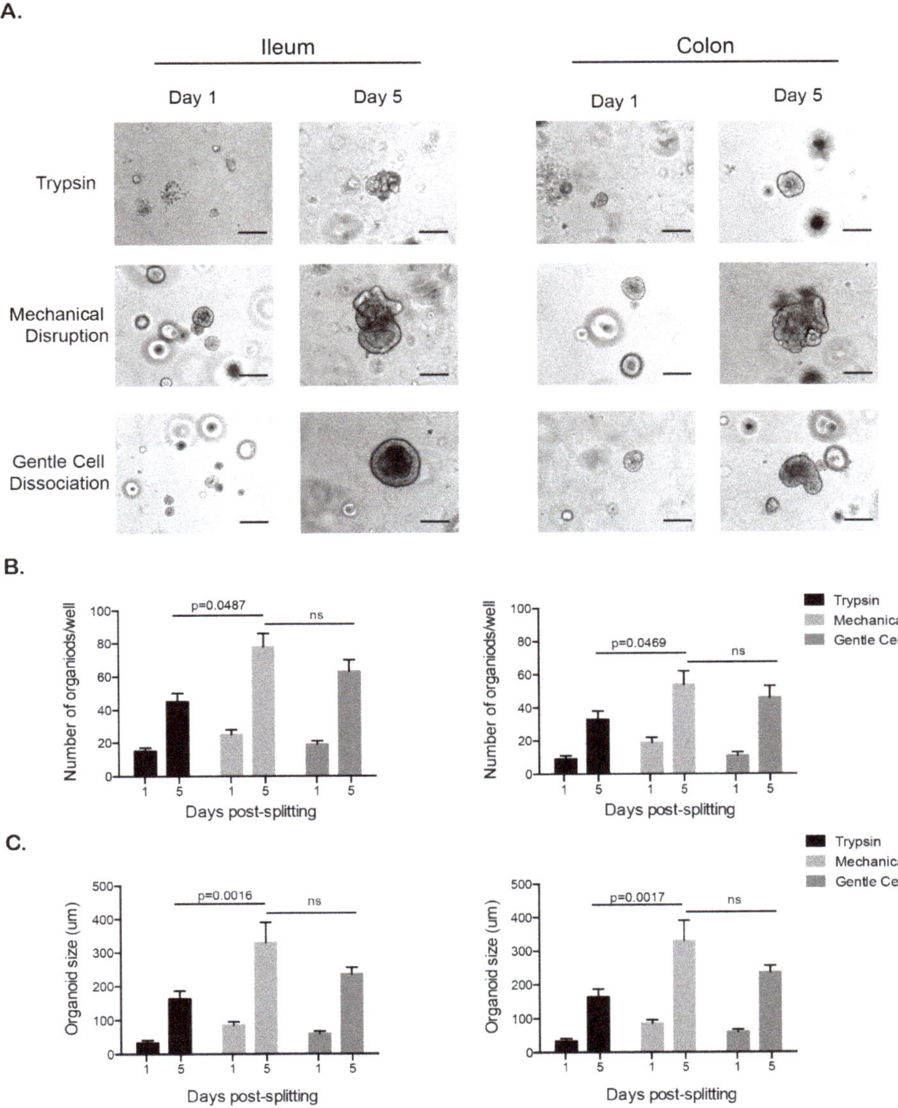

Figure 3. Mechanical disruption is the preferred method for passaging feline organoids. (**A**–**C**). Feline ileum- and colon-derived organoids were grown for 10 days post-harvesting prior to passaging. Three methods of passaging were used, and new organoids were followed over five days. (**A**). Bright field images of ileum and colon organoids. Representative images are shown. N = 3 donors and each donor were followed overtime as a triplicate series. Scale bar = 100 µm (**B**). The number of organoids from each passing method was counted over the indicated time course. Results are mean +/− s.d. (n = 3 donors). (**C**). The size of the organoids was determined using a Nikon Eclipse Ti-S. Results are mean +/− s.d. (n = 3 donors).

Upon isolation of feline ileum and colon crypts, organoids were maintained in a media based upon one that is commonly used to support the growth of human organoids [39,43,44]. To determine if additional media compositions would support or enhance feline intestinal organoid growth, we tested three different formulations normally used to support human intestinal and colon organoids, murine small intestine or murine colon (see material and methods) [37–40,45]. Feline ileum and colon organoids were passaged using mechanical disruption and were maintained in each of the media conditions. Ten days post passaging, the number of organoids and the size of the organoids were counted for each condition. Results show that the human media condition supported a greater number and a larger size of both ileum and colon organoids (Table 3). Additionally, to determine how many passages each media type could support, organoids were split using mechanical dissociation and followed over time. The mouse intestine media did not support long term growth of either ileum or colon organoids and both types of organoids stopped growing and died within two passages in this media (Table 3). The mouse colon media supported longer passaging than its intestine counterpart, however this was still limited to a few passages. The human media was the only one that supported longer term passaging. However, unlike human and mouse organoids [38–40], both the feline ileum and colon organoids became arrested after 14–15 passages (Table 3). This was reproduced with several animals (n = 6) and was consistent between animals suggesting that an additional media component may be required to support culturing to the extent that human and murine organoids can reach. Importantly, it seems that Wnt3A is a critical component required for the maintenance of feline organoids as the mouse intestine media, which lacks Wnt3A, supported the fewest number of passages. Overall, we could see that feline intestinal and colon organoids could be maintained using mechanical disruption and Wnt3A containing media. In recent years, intestine- and colon-derived organoids have been generated from both large farm animals (bovine and porcine) [46] and domestic animals (canine) [47,48]. Bovine, porcine and canine intestinal and colon organoids have been shown to require Wnt3a for their generation and growth [46–48]. Additionally, while many of these studies did not compare different passaging techniques, they often used mechanical disruption as a common method to passage and maintain the organoids [46,47].

Table 3. Optimization of feline ileum and colon culturing conditions. The number and size of organoids were measured 10 days post-splitting for every other passage.

	Organoid	Avg # of Organoids/Well	Avg. Size of Organoids	No of Passages
Human Media	Colon	83 +/− 11	384.2 +/− 31.5	15
Mouse Intestine Media	Colon	54 +/− 6	312.5 +/− 23.6	2
Mouse Colon Media	Colon	68 +/− 9	359.4 +/− 29.1	6
Human Media	Ileum	68 +/− 9	346.2 +/− 24.8	14
Mouse Intestine Media	Ileum	49 +/− 6	297.6 +/− 21.3	2
Mouse Colon Media	Ileum	53 +/− 7	318.9 +/− 35.4	4

3.2. Infection of Feline Intestinal Organoids

Currently, the molecular mechanism by which FECVs can persist in the gut could not been studied due to the lack of cell culture models supporting the replication of FECV field viruses. To evaluate if the feline intestinal and colon organoid system could support FECV infection, two different FCoV viruses (recFECV-GFP and recFECV-S79-GFP) were constructed for this study (Figure 4A). recFECV-GFP is a recombinant FCoV with the entire genome sequence of a serotype FECV field strain while recFECV-S79-GFP contains the genomic backbone of the same serotype I FECV field strains with the spike protein gene of cell-culture-adapted serotype II laboratory strain 79–1146. To determine the ability of recFECV-GFP and recFECV-S79-GFP to replicate in cell culture, these recombinant viruses were first tested on FCWF cells. Inoculation of FCWF cells with recFECV-GFP, did not lead to any formation of cytopathic effect (CPE) or fluorescence as standard cell culture systems do not support the growth of serotype I FCoV field strains (Figure 4B), while as expected, recFECV-S79-GFP growth was supported in FCWF cells demonstrated by their ability to form a characteristic plaque phenotype accompanied by green fluorescence (Figure 4B).

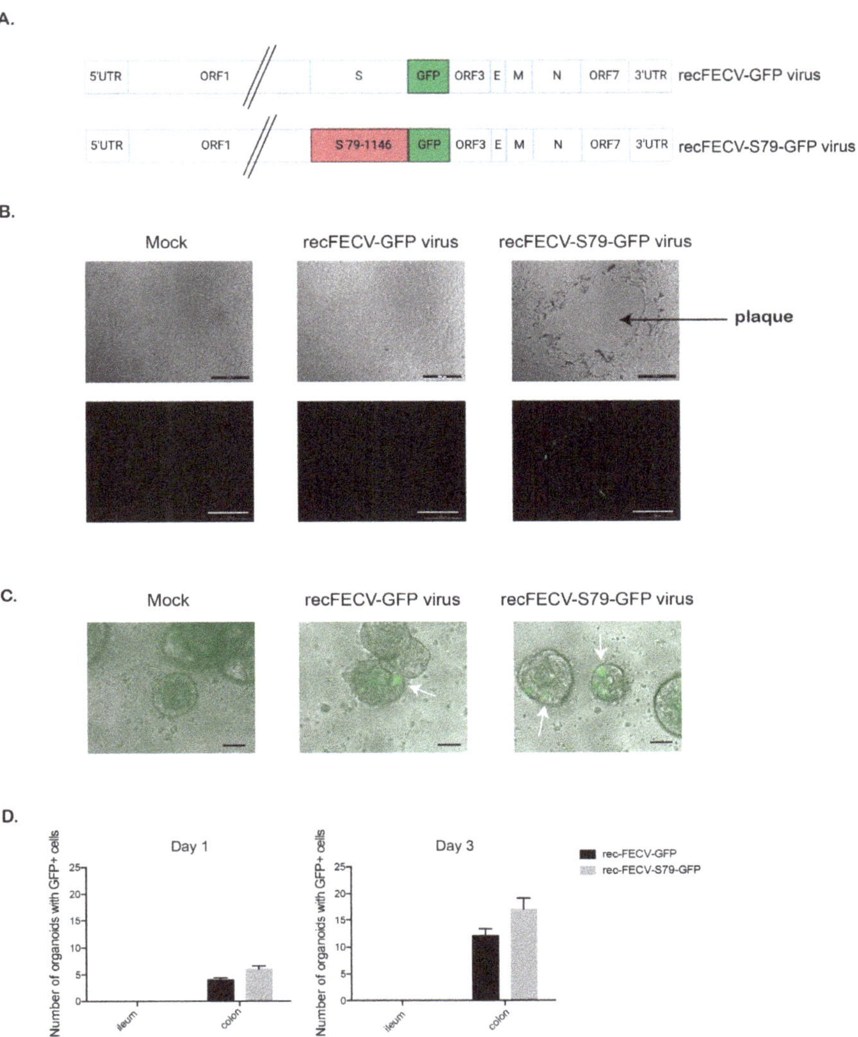

Figure 4. Feline colon organoids support infection of feline coronavirus (FCOV). (**A**). Diagram depicting the newly generated feline enteric coronavirus (FECV) recombinant viruses. (**B**). Felis catus whole fetus (FCWF) feline cells were infected with rec-FECV-GFP or recFECV-S79-GFP viruses and their ability to express GFP and produce plaques was monitored 48 h post-infection. Mock samples indicate media alone conditions. Representative images are shown. Scale bar is 200 µm. (**C**). Feline colon organoids infected with rec-FECV-GFP or recFECV-S79-GFP for 72 h. Representative images shown. White arrow heads indicate GFP positive cells. Scale bar = 100 µm (**D**). Feline ileum and colon organoids (passage 3) were infected with rec-FECV-GFP or recFECV-S79-GFP viruses. Mock samples indicate media alone conditions. Infections were monitored 1 and 3 days post-infection by wide field microscopy. Results are mean +/− s.d. n = 3 donors for each infection.

Previously, Desmarets et al. (2013) [49] established permanent feline intestinal epithelial cell cultures of ileocyte and colonocyte origin that were shown to sustain propagation of FECV field strains in vitro. However, such valuable cell lines are not broadly available to the scientific community. To test if the established feline ileum and colon organoids could be used to support feline coronavirus infection, five days post-passaging, organoids were removed from the Matrigel and incubated with feline coronavirus recFECV-GFP and recFECV-S79-GFP and followed over 72 h. Unexpectedly, no fluorescence was obtained upon infection of ileum organoids with either recFECV-GFP or recFECV-S79-GFP (Figure 4D). The lack of fluorescence of ileum organoids upon infection with both viruses is somewhat surprising, since ileum has been described as a tissue supporting FCoV-growth in vivo [50]. Whether the lack of green fluorescence of ileum organoids is due to (i) a very low viral replication rate and GFP expression which is below the detection limit or (ii) the complete lack of virus infection, cannot be ruled out in the current experimental setup. For example, luciferase-expressing recFECVs might be more suitable tools to further investigate these hypotheses. The low/no infection of ileum organoids with the recombinant viruses may reflect the limitation of ileum organoids to study the biology of serotype I FECVs in vitro. In contrast, both recombinant viruses were found to infect feline colon organoids (Figure 4C,D). One day post-infection, GFP positive cells could be detected and the number of infected cells increased over three days for both viruses (Figure 4C,D). As opposed to the FCWF cells, our feline colon organoids supported infection of recFECV-GFP. Since the number of GFP positive cells increased over time, this data strongly suggests that colon organoids support initial infection, production of de novo virus and spreading of serotype I FECV field strain-infection.

In summary, we propose that feline colon-derived organoids represent a primary intestinal cell model supporting serotype I field stain FECV infection. These cultures now open the doors for researchers to unravel the molecular mechanisms leading to FECV persistence and possibly FECV-FIPV conversion. The here described protocol provides the community with a step-by-step approach to generate feline colon-derived organoids which provides a solution for the lack of easily available culture models. Furthermore, as primary intestinal epithelium cells are often more immune-responsive than their immortalize counterparts, as such, the here described organoid model will allow the community to better study host/pathogen interactions [37] and immune response [40,51] against FCoV.

Author Contributions: M.L.S. designed the methods, prepared the organoids, and analyzed the results, R.E. constructed the GFP expressing virus. G.T. and S.B. contributed to the concept of the study and critical discussions. All authors have read and agreed to the published version of the manuscript.

Funding: S.B. received financial support from the Deutsche Forschungsgemeinschaft (DFG): Project number 240245660 (SFB1129), 415089553 (Heisenberg), 278001972 (TRR186) and 272983813 (TRR179). M.L.S. was supported by DFG project number 416072091. The work of G.T. and R.E. was supported by the DFG, SFB 1021, project B01.

Conflicts of Interest: The authors declare no conflict of interest

References

1. Pedersen, N.C. A review of feline infectious peritonitis virus infection: 1963–2008. *J. Feline Med. Surg.* **2009**, *11*, 225–258. [CrossRef] [PubMed]
2. Addie, D.D.; Schaap, I.A.T.; Nicolson, L.; Jarrett, O. Persistence and transmission of natural type I feline coronavirus infection. *J. Gen. Virol.* **2003**, *84*, 2735–2744. [CrossRef] [PubMed]
3. Pedersen, N.C. Serologic studies of naturally occurring feline infectious peritonitis. *Am. J. Vet. Res.* **1976**, *37*, 1449–1453.
4. Sparkes, A.H.; Gruffydd-Jones, T.J.; Harbour, D.A. Feline coronavirus antibodies in UK cats. *Vet. Rec.* **1992**, *131*, 223–224. [CrossRef]
5. Kipar, A.; May, H.; Menger, S.; Weber, M.; Leukert, W.; Reinacher, M. Morphologic features and development of granulomatous vasculitis in feline infectious peritonitis. *Vet. Pathol.* **2005**, *42*, 321–330. [CrossRef] [PubMed]
6. Weiss, R.C.; Scott, F.W. Pathogenesis of feline infetious peritonitis: Pathologic changes and immunofluorescence. *Am. J. Vet. Res.* **1981**, *42*, 2036–2048.

7. Weiss, R.C.; Scott, F.W. Pathogenesis of feline infectious peritonitis: Nature and development of viremia. *Am. J. Vet. Res.* **1981**, *42*, 382–390.
8. Pedersen, N.C. Morphologic and physical characteristics of feline infectious peritonitis virus and its growth in autochthonous peritoneal cell cultures. *Am. J. Vet. Res.* **1976**, *37*, 567–572.
9. Tekes, G.; Thiel, H.J. Feline Coronaviruses: Pathogenesis of Feline Infectious Peritonitis. *Adv. Virus Res.* **2016**, *96*, 193–218. [CrossRef]
10. Pedersen, N.C.; Liu, H.; Gandolfi, B.; Lyons, L.A. The influence of age and genetics on natural resistance to experimentally induced feline infectious peritonitis. *Vet. Immunol. Immunopathol.* **2014**, *162*, 33–40. [CrossRef]
11. Pedersen, N.C. An update on feline infectious peritonitis: Diagnostics and therapeutics. *Vet. J.* **2014**, *201*, 133–141. [CrossRef] [PubMed]
12. Pedersen, N.C. An update on feline infectious peritonitis: Virology and immunopathogenesis. *Vet. J.* **2014**, *201*, 123–132. [CrossRef] [PubMed]
13. Kim, Y.; Liu, H.; Galasiti Kankanamalage, A.C.; Weerasekara, S.; Hua, D.H.; Groutas, W.C.; Chang, K.O.; Pedersen, N.C. Reversal of the Progression of Fatal Coronavirus Infection in Cats by a Broad-Spectrum Coronavirus Protease Inhibitor. *PLoS Pathog.* **2016**, *12*, e1005531. [CrossRef]
14. Dickinson, P.J.; Bannasch, M.; Thomasy, S.M.; Murthy, V.D.; Vernau, K.M.; Liepnieks, M.; Montgomery, E.; Knickelbein, K.E.; Murphy, B.; Pedersen, N.C. Antiviral treatment using the adenosine nucleoside analogue GS-441524 in cats with clinically diagnosed neurological feline infectious peritonitis. *J. Vet. Intern. Med.* **2020**, *34*, 1587–1593. [CrossRef] [PubMed]
15. Perera, K.D.; Rathnayake, A.D.; Liu, H.; Pedersen, N.C.; Groutas, W.C.; Chang, K.O.; Kim, Y. Characterization of amino acid substitutions in feline coronavirus 3C-like protease from a cat with feline infectious peritonitis treated with a protease inhibitor. *Vet. Microbiol.* **2019**, *237*, 108398. [CrossRef] [PubMed]
16. Pedersen, N.C.; Perron, M.; Bannasch, M.; Montgomery, E.; Murakami, E.; Liepnieks, M.; Liu, H. Efficacy and safety of the nucleoside analog GS-441524 for treatment of cats with naturally occurring feline infectious peritonitis. *J. Feline Med. Surg.* **2019**, *21*, 271–281. [CrossRef]
17. Murphy, B.G.; Perron, M.; Murakami, E.; Bauer, K.; Park, Y.; Eckstrand, C.; Liepnieks, M.; Pedersen, N.C. The nucleoside analog GS-441524 strongly inhibits feline infectious peritonitis (FIP) virus in tissue culture and experimental cat infection studies. *Vet. Microbiol.* **2018**, *219*, 226–233. [CrossRef]
18. Pedersen, N.C.; Kim, Y.; Liu, H.; Galasiti Kankanamalage, A.C.; Eckstrand, C.; Groutas, W.C.; Bannasch, M.; Meadows, J.M.; Chang, K.O. Efficacy of a 3C-like protease inhibitor in treating various forms of acquired feline infectious peritonitis. *J. Feline Med. Surg.* **2018**, *20*, 378–392. [CrossRef]
19. Hohdatsu, T.; Sasamoto, T.; Okada, S.; Koyama, H. Antigenic analysis of feline coronaviruses with monoclonal antibodies (MAbs): Preparation of MAbs which discriminate between FIPV strain 79-1146 and FECV strain 79-1683. *Vet. Microbiol.* **1991**, *28*, 13–24. [CrossRef]
20. Hohdatsu, T.; Okada, S.; Koyama, H. Characterization of monoclonal antibodies against feline infectious peritonitis virus type II and antigenic relationship between feline, porcine, and canine coronaviruses. *Arch. Virol.* **1991**, *117*, 85–95. [CrossRef]
21. Pedersen, N.C.; Black, J.W.; Boyle, J.F.; Evermann, J.F.; McKeirnan, A.J.; Ott, R.L. Pathogenic differences between various feline coronavirus isolates. *Adv. Exp. Med. Biol.* **1984**, *173*, 365–380. [CrossRef] [PubMed]
22. Decaro, N.; Buonavoglia, C. An update on canine coronaviruses: Viral evolution and pathobiology. *Vet. Microbiol.* **2008**, *132*, 221–234. [CrossRef] [PubMed]
23. Herrewegh, A.A.; Smeenk, I.; Horzinek, M.C.; Rottier, P.J.; de Groot, R.J. Feline coronavirus type II strains 79-1683 and 79-1146 originate from a double recombination between feline coronavirus type I and canine coronavirus. *J. Virol.* **1998**, *72*, 4508–4514. [CrossRef] [PubMed]
24. Lin, C.N.; Chang, R.Y.; Su, B.L.; Chueh, L.L. Full genome analysis of a novel type II feline coronavirus NTU156. *Virus Genes* **2013**, *46*, 316–322. [CrossRef]
25. Terada, Y.; Matsui, N.; Noguchi, K.; Kuwata, R.; Shimoda, H.; Soma, T.; Mochizuki, M.; Maeda, K. Emergence of pathogenic coronaviruses in cats by homologous recombination between feline and canine coronaviruses. *PLoS ONE* **2014**, *9*, e106534. [CrossRef]
26. Tekes, G.; Spies, D.; Bank-Wolf, B.; Thiel, V.; Thiel, H.J. A reverse genetics approach to study feline infectious peritonitis. *J. Virol.* **2012**, *86*, 6994–6998. [CrossRef]

27. Tekes, G.; Hofmann-Lehmann, R.; Stallkamp, I.; Thiel, V.; Thiel, H.J. Genome organization and reverse genetic analysis of a type I feline coronavirus. *J. Virol.* **2008**, *82*, 1851–1859. [CrossRef]
28. Tekes, G.; Hofmann-Lehmann, R.; Bank-Wolf, B.; Maier, R.; Thiel, H.J.; Thiel, V. Chimeric feline coronaviruses that encode type II spike protein on type I genetic background display accelerated viral growth and altered receptor usage. *J. Virol.* **2010**, *84*, 1326–1333. [CrossRef]
29. Thiel, V.; Thiel, H.J.; Tekes, G. Tackling feline infectious peritonitis via reverse genetics. *Bioengineered* **2014**, *5*, 396–400. [CrossRef]
30. Haijema, B.J.; Volders, H.; Rottier, P.J. Switching species tropism: An effective way to manipulate the feline coronavirus genome. *J. Virol.* **2003**, *77*, 4528–4538. [CrossRef]
31. Ettayebi, K.; Crawford, S.E.; Murakami, K.; Broughman, J.R.; Karandikar, U.; Tenge, V.R.; Neill, F.H.; Blutt, S.E.; Zeng, X.L.; Qu, L.; et al. Replication of human noroviruses in stem cell-derived human enteroids. *Science* **2016**, *353*, 1387–1393. [CrossRef] [PubMed]
32. Pyrc, K.; Sims, A.C.; Dijkman, R.; Jebbink, M.; Long, C.; Deming, D.; Donaldson, E.; Vabret, A.; Baric, R.; van der Hoek, L.; et al. Culturing the unculturable: Human coronavirus HKU1 infects, replicates, and produces progeny virions in human ciliated airway epithelial cell cultures. *J. Virol.* **2010**, *84*, 11255–11263. [CrossRef]
33. Sato, T.; Clevers, H. Growing self-organizing mini-guts from a single intestinal stem cell: Mechanism and applications. *Science* **2013**, *340*, 1190–1194. [CrossRef] [PubMed]
34. Ehmann, R.; Kristen-Burmann, C.; Bank-Wolf, B.; Konig, M.; Herden, C.; Hain, T.; Thiel, H.J.; Ziebuhr, J.; Tekes, G. Reverse Genetics for Type I Feline Coronavirus Field Isolate to Study the Molecular Pathogenesis of Feline Infectious Peritonitis. *MBio* **2018**, *9*. [CrossRef] [PubMed]
35. Mettelman, R.C.; O'Brien, A.; Whittaker, G.R.; Baker, S.C. Generating and evaluating type I interferon receptor-deficient and feline TMPRSS2-expressing cells for propagating serotype I feline infectious peritonitis virus. *Virology* **2019**, *537*, 226–236. [CrossRef] [PubMed]
36. O'Brien, A.; Mettelman, R.C.; Volk, A.; André, N.M.; Whittaker, G.R.; Baker, S.C. Characterizing replication kinetics and plaque production of type I feline infectious peritonitis virus in three feline cell lines. *Virology* **2018**, *525*, 1–9. [CrossRef] [PubMed]
37. Stanifer, M.L.; Mukenhirn, M.; Muenchau, S.; Pervolaraki, K.; Kanaya, T.; Albrecht, D.; Odendall, C.; Hielscher, T.; Haucke, V.; Kagan, J.C.; et al. Asymmetric distribution of TLR3 leads to a polarized immune response in human intestinal epithelial cells. *Nat. Microbiol.* **2020**, *5*, 181–191. [CrossRef]
38. Sato, T.; Vries, R.G.; Snippert, H.J.; van de Wetering, M.; Barker, N.; Stange, D.E.; van Es, J.H.; Abo, A.; Kujala, P.; Peters, P.J.; et al. Single Lgr5 stem cells build crypt-villus structures in vitro without a mesenchymal niche. *Nature* **2009**, *459*, 262–265. [CrossRef]
39. Sato, T.; Stange, D.E.; Ferrante, M.; Vries, R.G.; Van Es, J.H.; Van den Brink, S.; Van Houdt, W.J.; Pronk, A.; Van Gorp, J.; Siersema, P.D.; et al. Long-term expansion of epithelial organoids from human colon, adenoma, adenocarcinoma, and Barrett's epithelium. *Gastroenterology* **2011**, *141*, 1762–1772. [CrossRef]
40. Pervolaraki, K.; Stanifer, M.L.; Munchau, S.; Renn, L.A.; Albrecht, D.; Kurzhals, S.; Senis, E.; Grimm, D.; Schroder-Braunstein, J.; Rabin, R.L.; et al. Type I and Type III Interferons Display Different Dependency on Mitogen-Activated Protein Kinases to Mount an Antiviral State in the Human Gut. *Front. Immunol.* **2017**, *8*, 459. [CrossRef]
41. Triana, S.; Stanifer, M.L.; Shahraz, M.; Mukenhirn, M.; Kee, C.; Ordoñez-Rueda, D.; Paulsen, M.; Benes, V.; Boulant, S.; Alexandrov, T. Single-cell transcriptomics reveals immune response of intestinal cell types to viral infection. *bioRxiv Preprint Serv. Biol.* 2020. Available online: https://doi.org/10.1101/2020.08.19.255893 (accessed on 19 August 2020).
42. Fujii, M.; Matano, M.; Toshimitsu, K.; Takano, A.; Mikami, Y.; Nishikori, S.; Sugimoto, S.; Sato, T. Human Intestinal Organoids Maintain Self-Renewal Capacity and Cellular Diversity in Niche-Inspired Culture Condition. *Cell Stem Cell* **2018**, *23*, 787–793. [CrossRef] [PubMed]
43. Koo, B.K.; Stange, D.E.; Sato, T.; Karthaus, W.; Farin, H.F.; Huch, M.; van Es, J.H.; Clevers, H. Controlled gene expression in primary Lgr5 organoid cultures. *Nat. Methods* **2011**, *9*, 81–83. [CrossRef]
44. Jung, P.; Sato, T.; Merlos-Suarez, A.; Barriga, F.M.; Iglesias, M.; Rossell, D.; Auer, H.; Gallardo, M.; Blasco, M.A.; Sancho, E.; et al. Isolation and in vitro expansion of human colonic stem cells. *Nat. Med.* **2011**, *17*, 1225–1227. [CrossRef] [PubMed]

45. Stanifer, M.L.; Kee, C.; Cortese, M.; Zumaran, C.M.; Triana, S.; Mukenhirn, M.; Kraeusslich, H.G.; Alexandrov, T.; Bartenschlager, R.; Boulant, S. Critical Role of Type III Interferon in Controlling SARS-CoV-2 Infection in Human Intestinal Epithelial Cells. *Cell Rep.* **2020**, *32*, 107863. [CrossRef] [PubMed]
46. Derricott, H.; Luu, L.; Fong, W.Y.; Hartley, C.S.; Johnston, L.J.; Armstrong, S.D.; Randle, N.; Duckworth, C.A.; Campbell, B.J.; Wastling, J.M.; et al. Developing a 3D intestinal epithelium model for livestock species. *Cell Tissue Res.* **2019**, *375*, 409–424. [CrossRef]
47. Kramer, N.; Pratscher, B.; Meneses, A.M.C.; Tschulenk, W.; Walter, I.; Swoboda, A.; Kruitwagen, H.S.; Schneeberger, K.; Penning, L.C.; Spee, B.; et al. Generation of Differentiating and Long-Living Intestinal Organoids Reflecting the Cellular Diversity of Canine Intestine. *Cells* **2020**, *9*, 822. [CrossRef]
48. Chandra, L.; Borcherding, D.C.; Kingsbury, D.; Atherly, T.; Ambrosini, Y.M.; Bourgois-Mochel, A.; Yuan, W.; Kimber, M.; Qi, Y.; Wang, Q.; et al. Derivation of adult canine intestinal organoids for translational research in gastroenterology. *BMC Biol.* **2019**, *17*, 33. [CrossRef]
49. Desmarets, L.M.; Theuns, S.; Olyslaegers, D.A.; Dedeurwaerder, A.; Vermeulen, B.L.; Roukaerts, I.D.; Nauwynck, H.J. Establishment of feline intestinal epithelial cell cultures for the propagation and study of feline enteric coronaviruses. *Vet. Res.* **2013**, *44*, 71. [CrossRef]
50. Kipar, A.; Meli, M.L.; Baptiste, K.E.; Bowker, L.J.; Lutz, H. Sites of feline coronavirus persistence in healthy cats. *J. Gen. Virol.* **2010**, *91*, 1698–1707. [CrossRef]
51. Pervolaraki, K.; Rastgou Talemi, S.; Albrecht, D.; Bormann, F.; Bamford, C.; Mendoza, J.L.; Garcia, K.C.; McLauchlan, J.; Hofer, T.; Stanifer, M.L.; et al. Differential induction of interferon stimulated genes between type I and type III interferons is independent of interferon receptor abundance. *PLoS Pathog.* **2018**, *14*, e1007420. [CrossRef]

Article

Astrocyte Infection during Rabies Encephalitis Depends on the Virus Strain and Infection Route as Demonstrated by Novel Quantitative 3D Analysis of Cell Tropism

Madlin Potratz [1], Luca Zaeck [1], Michael Christen [1], Verena te Kamp [2], Antonia Klein [1], Tobias Nolden [3], Conrad M. Freuling [1], Thomas Müller [1] and Stefan Finke [1,*]

[1] Friedrich-Loeffler-Institut (FLI), Federal Research Institute for Animal Health, Institute of Molecular Virology and Cell Biology, 17493 Greifswald-Insel Riems, Germany; Madlin.Potratz@fli.de (M.P.); Luca.Zaeck@fli.de (L.Z.); m.christen1994@gmail.com (M.C.); Antonia.Klein@fli.de (A.K.); Conrad.Freuling@fli.de (C.M.F.); Thomas.Mueller@fli.de (T.M.)

[2] Thescon GmbH, 48653 Coesfeld, Germany; verena_tekamp@gmx.de

[3] ViraTherapeutics GmbH, 6020 Innsbruck, Austria; tobias.nolden@boehringer-ingelheim.com

* Correspondence: stefan.finke@fli.de; Tel.: +49-38351-71283

Received: 14 January 2020; Accepted: 5 February 2020; Published: 11 February 2020

Abstract: Although conventional immunohistochemistry for neurotropic rabies virus (RABV) usually shows high preference for neurons, non-neuronal cells are also potential targets, and abortive astrocyte infection is considered a main trigger of innate immunity in the CNS. While in vitro studies indicated differences between field and less virulent lab-adapted RABVs, a systematic, quantitative comparison of astrocyte tropism in vivo is lacking. Here, solvent-based tissue clearing was used to measure RABV cell tropism in infected brains. Immunofluorescence analysis of 1 mm-thick tissue slices enabled 3D-segmentation and quantification of astrocyte and neuron infection frequencies. Comparison of three highly virulent field virus clones from fox, dog, and raccoon with three lab-adapted strains revealed remarkable differences in the ability to infect astrocytes in vivo. While all viruses and infection routes led to neuron infection frequencies between 7–19%, striking differences appeared for astrocytes. Whereas astrocyte infection by field viruses was detected independent of the inoculation route (8–27%), only one lab-adapted strain infected astrocytes route-dependently [0% after intramuscular (i.m.) and 13% after intracerebral (i.c.) inoculation]. Two lab-adapted vaccine viruses lacked astrocyte infection altogether (0%, i.c. and i.m.). This suggests a model in which the ability to establish productive astrocyte infection in vivo functionally distinguishes field and attenuated lab RABV strains.

Keywords: rabies; uDISCO; 3D imaging; rabies pathogenicity; astrocyte infection

---

1. Introduction

The rabies virus (RABV) is a highly neurotropic virus, which inevitably causes lethal disease in mammals after onset of neurological signs [1]. As a non-segmented, single-stranded RNA virus of negative RNA polarity, RABV belongs to the *Rhabdoviridae* family in the order *Mononegavirales* [2]. With nucleoprotein N, phosphoprotein P, matrix protein M, glycoprotein G, and the large polymerase L, the 12 kb genome of RABV encodes five virus proteins, all of which are essential for virus replication and spread [3]. In addition to essential roles of the virus proteins in genome replication and virus assembly, multiple accessory functions of the RABV proteins have been identified. RABV pathogenicity has mainly been attributed to a potent interference with the innate immune system by N, P, and M [4–10], and neuronal survival regulation by G [11–14]. Most pathogenicity studies, however, were performed on already attenuated virus backbones. Thus, differences in their ability to cause disease between highly virulent field virus isolates and lab-adapted, less pathogenic RABV strains are poorly understood. Moreover, it is unclear how molecular differences identified in virulent and attenuated viruses affect virus replication and spread in the infected animal and how the complex virus–host interplay eventually results in either disease or an abortive infection.

In vivo, after infection of neurons, RABV spreads trans-synaptically from infected to connected neurons [15]. Retrograde axonal transport of RABV over long distances [16,17] along microtubules [18,19] is a key step in RABV neuroinvasion and is essential for infection of the central nervous system (CNS) through the peripheral nervous system. Co-internalization together with the neuronal p75NTR (tumor necrosis factor receptor superfamily member 16; TNFRSF16) receptor, subsequent retrograde axonal transport of RABV particles in endocytic vesicles, and post-replicative anterograde axonal transport of newly formed RABV have been visualized by live virus particle tracking in sensory neurons [20,21], emphasizing the capacity of hijacking neuron-specific machineries for long distance transport to synaptic membranes. However, internalization and axonal transport of lab-adapted viruses [20,21] together with the use of vaccine virus vectors for trans-synaptic tracing [22,23] demonstrate that the general capacity of axonal transport and trans-synaptic spread cannot explain mechanistic differences between highly virulent RABV and more attenuated lab strains. Differences between RABV lab strains in the efficiency of trans-synaptic spread [24] indicate that the efficacy of the involved processes, more than the capability itself, may contribute to RABV spread in vivo.

With nAChR (nicotinic acetylcholine receptor), NCAM (neuronal cell adhesion molecule), p75NTR, and mGluR2 (metabotropic glutamate receptor subtype 2) supporting RABV entry [25–28], several RABV receptors have been discussed. However, none of these receptors are essential for CNS infection by RABV, and a broad panel of non-neuronal cell types can be infected in vitro [29], indicating that cell tropism of RABV is not restricted to neurons by receptor specificity.

Most in vivo studies report a strict neurotropic infection. Directly after exposure by bite, however, muscle cells are infected (reviewed in [30]), and infection of non-neuronal cells in the CNS can occur [29,31,32]. Use of recombinant Cre recombinase-expressing RABV led to the identification of abortively infected glial cells in infected mouse brains, strongly suggesting infection of and virus elimination from these cells by a potent type I interferon response [33]. Accordingly, abortive infection of non-neuronal cells and induction of innate immune responses may play an important role in the infection process itself and in regulating downstream adaptive immune pathways. Indeed, a model based on in vitro-infected astrocytes suggests that, in contrast to wild-type RABV, attenuated RABV activates inflammatory responses in astrocytes through increased double-strand RNA (dsRNA) synthesis and recognition by retinoic acid-inducible gene I (RIG-I)/melanoma differentiation-associated protein 5 (MDA5) [34]. Highly virulent field RABV isolates are able to evade or at least delay host immune reactions [35], which may allow virus replication to reach pathogenic levels, whereas early innate immune induction via astrocytes or other glial cells by attenuated viruses does not. Nevertheless, all productive RABV infections eventually cause rabies as a disease, which is always associated with an encephalitis. However, differences appear to exist as field viruses cause less tissue damage and neuroinflammation than attenuated lab RABV [34,36,37].

To compare the cell tropism of highly virulent field and lab-adapted RABVs in the CNS, we employed the novel immunostaining-compatible tissue clearing technique uDISCO (ultimate 3D imaging of solvent-cleared organs) [38–40] for detection and quantification of RABV infection in neurons and astrocytes in solvent-cleared brain tissue. After confocal laser scan acquisition of large confocal z-stacks in thick tissue slices, three-dimensional (3D) reconstructions were performed to visualize the cellular context of RABV infection. Frequencies of RABV infection in neurons and astrocytes were determined after intramuscular (i.m.) and intracerebral (i.c.) inoculation, leading to novel insights regarding the ability to establish RABV infection in non-neuronal astrocytes, its correlation with RABV virulence, and its dependence on the infection route. These data strongly support a model in which, contrary to in vitro conditions where all viruses are able to infect non-neuronal astrocytes, there are differences between field and lab-adapted viruses in their ability to replicate to detectable levels in astrocytes in vivo. The infection of astrocytes with virulent field RABV and the accumulation of interferon antagonistic virus proteins may block rapid innate immune induction, thereby contributing to immune evasion, even in the context of glial cell infection.

2. Materials and Methods

2.1. Cell Lines, Primary Brain Cell Preparation, and Cultivation

Mouse neuroblastoma cells (Na42/13) were used for virus amplification. All cells were provided by the Collection of Cell Lines in Veterinary Medicine (CCLV), Friedrich-Loeffler-Institut, Insel Riems, Germany. Primary neurons and astrocytes were prepared from 1-day-old neonatal Sprague Dawley rats (P0–P1) [41]. The rats were decapitated, the heads were disinfected in 70% ethanol, the brains were removed, and the hippocampi were separated and mechanically minced. The hippocampal cells were transferred into ice-cold Hank´s Balanced Salt Solution (HBSS) and stored on ice. After preparation, the brain cells were dissociated by adding 0.25% trypsin + EDTA (Gibco; Thermo Fisher Scientific, Darmstadt, Germany) and DNase I (Applichem, Darmstadt, Germany) and incubated for 15 min at 37 °C. Afterwards, further dissociation in Neurobasal-A Medium (NB-A, Gibco; Thermo Fisher Scientific) was performed by using a glass Pasteur pipette with a fire-polished tip. The cells were purified with an Optiprep^{TM} gradient (with concentrations from bottom to the top: 17.3%, 12.4%, 9.9%, and 7.4%) and centrifuged for 15 min at 800× g without brake. Next, the cells were washed with NB-A medium and counted with the LUNA-II Automated Cell Counter (Logos Biosystems, Villeneuve d'Ascq, France). The neuronal cells were seeded on coverslips in a 24-well plate and cultured for two weeks at 37 °C and 5% CO_2 in serum-free NB-A supplemented with 2% B27 (Gibco; Thermo Fisher Scientific), 1% GlutaMAX (Gibco; Thermo Fisher Scientific), 1% Penicillin-Streptomycin (stock: 10,000 U/mL penicillin and 10 mg/mL streptomycin; Sigma-Aldrich, Taufkirchen, Germany), and 0.2% Gentamicin (stock: 50 mg/mL; Gibco; Thermo Fisher Scientific) until they were used for infection experiments.

2.2. Viruses

RABV isolates used in this study comprised five recombinant virus clones and one non-recombinant lab virus. The recombinant field virus clones rRABV Dog and rRABV Fox have been described before [42]. SAD L16 is a recombinant virus clone of attenuated vaccine virus SAD B19 live vaccine virus [43]. The Evelyn Rokitnicki Abelseth (ERA) strain [44] is a progenitor virus strain of live vaccine virus SAD B19 [45] and was obtained from the FLI virus archive (FLI ID N 12829).

A cDNA full length clone (pRABV Rac) from a raccoon RABV isolate (Alabama, USA 1991; FLI archive ID N 13205) [46] was generated by full length RT-PCR amplification of the 12 kb cDNA genome with the primer pair 5′-TCGATCCCGGGTCACGCTTAACAACAAAA-3′/ 5′-TAATACACCTGCCCATGCCGACCCACGCTTAACAAAAAAACAA-3′. After PCR amplification of a 2.7 kb vector DNA fragment from pCMV HaHd ampR Br 322 ori with the primers 5′-TCTGTTTGCTTGATGGTTTTTTTGTCTTTGTTGTTTTTTGTTAAGCGTGGGTCGGCATGGC ATCTCCAC-3′ and 5′-TTTTTGTAGATGATACTGTCTACTTCTTCTCTGATTTTGTTGTTAAGCGTGA CCCGGGACTCCGGGTTTCGTC-3′, the fragments were combined by linear to linear recombination, and the resultant recombinant rRABV Rac virus was rescued and amplified according to previously described protocols [42]. The sequence of the full-length cDNA clone was deposited at GenBank (accession no. MN862283).

A CVS-11 (Challenge Virus Street-11; sequence accession no. LT839616) cDNA full length clone (pCVS-11) was generated by RT-PCR amplification of 7.5kb and 3.7 kb cDNA fragments from CVS-11 RNA with the primers pairs 5′-AGTTTCAGACGTCTCAGTC-3′/5′-CTAGTAGGGATGATCTAGATC-3′ and 5′-GACTGAGACGTCTGAAACT-3′/5′ CATTGCAGATAGGATAGAG-3′, respectively. The resultant fragments were assembled in combination with EcoRI-linearized 3.4 kb vector plasmid pCVS-11-termini by Hot Fusion reaction [47]. Plasmid pCVS-11-termini was generated by PCR amplification of a 2.7 kb DNA fragment from pCMV HaHd ampR Br 322 ori [42] with the primers 5′-GACCCGGGACTCCGGGTTTC-3′ and 5′-GGGTCGGCATGGCATCTCCA-3′, and Hot Fusion assembly with a synthetic DNA fragment containing the CVS-11 (accession no. LT839616) regions from nucleotide positions 1–580 (3′-genome end) and 11759–11927 (5′-genome end) separated by a non-viral EcoRI restriction site. The full-length cDNA clone was 99.99% identical to CVS-11 GenBank sequence no. LT839616 (1 nt mismatch in the N gene at nucleotide position 1263, which was already present in #LT839616 as a minor SNP (single nucleotide polymorphism), resulting in the amino acid exchange E398V). Recombinant rCVS-11 was rescued in Na42/13 neuroblastoma cells by co-transfection of pCVS-11, pCAGGS-based expression plasmids for RABV N, P, and L [48], and a pCAGGS-T7Pol vector comprising a codon-optimized bacteriophage T7 RNA polymerase gene. Three to six days after transfection, the supernatants were transferred to new Na42/13 cells. Two days after the transfer, infectious virus was identified by N and G protein-specific indirect immunofluorescence.

rRABV Dog, rRABV Fox, rRABV Rac, and rCVS-11 were amplified on Na42/13 cells. BSR T7/5 cells [49] were used for amplification of SAD L16 and ERA viruses. Infectious virus titers in cell culture supernatants were determined by end point dilution and titration on Na42/13 cells.

2.3. Antibodies

To verify the presence of RABV, a polyclonal rabbit serum against recombinant RABV P protein (P160-5, immunofluorescence 1:5000, uDISCO 1:3000) and a polyclonal goat serum against RABV N (goat anti-RV N, immunofluorescence 1:4000), which have been described previously [40,50], were used. The polyclonal chicken anti-glial fibrillary acidic protein (GFAP) antibody (Thermo Fisher Scientific, Darmstadt, Germany; #PA1-10004, uDISCO 1:1500), the polyclonal rabbit anti-GFAP (Dako, #Z0334, immunofluorescence 1:500), the polyclonal guinea pig anti-NeuN antibody (Synaptic Systems, Goettingen, Germany; #266004, uDISCO 1:800), and the rabbit anti-MAP2 antibody (Abcam, Cambridge, UK; #ab32454, immunofluorescence 1:250) were purchased from their respective suppliers.

2.4. Infection of in Vitro Cell Cultures and Immunofluorescence Staining

Two-week-old primary rat hippocampal cell cultures were infected with 1×10^3 infectious units of rRABV Dog, rRABV Fox, rCVS-11, and SAD L16, and cultivated for 24 h at 37 °C and 5% CO_2. Indirect immunofluorescence was performed by standard techniques after fixation with 4% paraformaldehyde (PFA) in phosphate-buffered saline (PBS) for 30 min and 15 min permeabilization with 0.5% Triton X-100 in PBS. Afterwards, samples were blocked with 0.025% skim milk powder in PBS for 15 min. Immunostainings were executed by 1.5 h incubation with primary antibodies, three wash steps with PBS, followed by 1 h incubation with secondary antibodies and additional Hoechst33342 (1 µg/mL) for staining of the nuclear chromatin. Specimens were mounted on coverslips and were analyzed by confocal laser scanning microscopy.

2.5. Brain Samples and Mouse Infections

Three- to four-week-old BALB/c mice (Charles-River, Germany) were infected with rRABV Rac, rCVS-11, ERA, and SAD L16 viruses using two different inoculation routes and two different viral doses, essentially as described before [51]. Two groups of six animals each were anesthetized and infected i.m. with 10^2 or 10^5 $TCID_{50}$/30 µL, and an additional group of three mice was infected i.c. with 10^2 $TCID_{50}$/30 µL. The weight and the clinical score, ranging from zero to four, of all mice were observed for 21 days post infection (dpi). When reaching a clinical score of two or three (ruffled fur, slowed movement, weight loss >15%), the animals were anaesthetized with isoflurane and euthanized through cervical dislocation. Samples were taken, fixed with 4% paraformaldehyde (PFA) for one week, and stored for further processing. All remaining animals were euthanized at 21 dpi. Mouse experimental studies on the characterization of lyssaviruses were evaluated by the responsible animal care, use, and ethics committee of the State Office for Agriculture, Food Safety, and Fishery in Mecklenburg-Western Pomerania (LALFF M-V) and gained approval with permissions 7221.3–2–001/18.

2.6. Archived Mouse Brains Infected with rRABV Dog and rRABV Fox

To minimize animal experiments, archived PFA-fixed brains from previous pathogenicity trials [42] were used for the analysis of rRABV Dog and rRABV Fox virus infections in mice. Similar to the mouse experiments with rRABV Rac, rCVS-11, ERA, and SAD L16 described above, mice were inoculated via the i.c. or the i.m. route.

2.7. Ultimate 3D Imaging of Solvent-Cleared Organs (uDISCO)

The clearing of brain tissue slices was performed as described previously [40] in modification of earlier publications [38,39]. Briefly, the PFA-fixed tissues were sectioned into 1 mm-thick slices using a vibratome (Leica VT1200S; Leica Biosystems, Wetzlar, Germany). All subsequent incubations steps were performed with gentle oscillation. To increase antibody diffusion and reduce tissue autofluorescence [38], the sections were pretreated with increasing concentrations of methanol (20%, 40%, 60%, 80%, and twice in 100%; dilutions with distilled water, incubation for 1 h each) and bleached by overnight incubation at 4 °C with 5% H_2O_2 in 100% methanol. After removal of the bleaching solution, the samples were rehydrated with decreasing concentrations of methanol (80%, 60%, 40%, and 20%; dilutions with distilled water, incubation for 1 h each) and a subsequent wash with PBS for 1 h. To permeabilize the samples, they were washed twice for 1 h each with 0.2% Triton X-100 in PBS and subsequently incubated for 48 h at 37 °C in 0.2% Triton X-100/20% DMSO/0.3 M glycine in PBS. The samples were then blocked by incubation with 0.2% Triton X-100/10% DMSO/6% donkey serum in PBS for 48 h at 37 °C. Primary antibodies were diluted in 3% donkey serum/5% DMSO in PTwH (0.2% Tween-20 in PBS with 10 μg/mL heparin), and incubation of samples was performed at 37 °C for 5 days. The antibody solution was refreshed after 2.5 days. The samples were washed with PTwH by exchanging the solution four times during the course of the day and subsequently incubated overnight in PTwH. Incubation with secondary antibodies was performed in 3% donkey serum in PTwH for 5 days at 37 °C, refreshing the secondary antibody solution once after 2.5 days. Subsequent washing was performed as described above following primary antibody incubation.

For tissue clearing, the samples were dehydrated with a series of *tert*-butanol (TBA) solutions (30%, 50%, 70%, 80%, 90%, and 96%; dilutions with distilled water, incubation for 2 h each), leaving 96% TBA on overnight. Following further dehydration in 100% TBA for 2 h, the samples were cleared in BABB-D15 [39] [1:2 mixture of benzyl alcohol (BA) and benzyl benzoate (BB), which is mixed with diphenyl ether (DPE) at a ratio of 15:1 and supplemented with 0.4 vol% DL-α-tocopherol] until they were optically transparent (2–6 h).

For confocal laser scanning microscopy, the samples were mounted in 3D-printed imaging chambers (printer: Ultimaker 2 + [Ultimaker, Utrecht, Netherlands], material: co-polyester, nozzle: 0.25 mm, layer height: 0.06 mm, wall thickness: 0.88 mm, wall count: 4, infill: 100%, no support structure; the corresponding .STL file is provided in the Supplementary Materials of Zaeck et al. [40]).

2.8. Confocal Laser Scanning Microscopy and Image Processing

The immunofluorescent staining of primary brain cells and infected tissues was visualized with a confocal laser scanning microscope (Leica DMI 6000 TCS SP5; Leica Microsystems, Wetzlar, Germany) equipped with a long free working distance 40× water immersion objective (NA = 1.1; Leica, #15506360). For image processing, a Dell Precision 7920 workstation was used (CPU: Intel Xeon Gold 5118, GPU: Nvidia Quadro P5000, RAM: 128 GB 2666 MHz DDR4, SSD: 2 TB; Dell, Frankfurt am Main, Germany).

For quantification of infected cells in thick tissue sections, the image was split into individual channels using Fiji, an ImageJ (v1.52h) distribution package [52]. A bleach correction was performed (simple ratio; background intensity: 5.0), and brightness and contrast were adjusted for each channel. Objects were identified and counted with the 3D Objects Counter plugin [53]. The resulting objects map was overlaid with the RABV P channel to quantify infected objects. For each sample, at least six regions were imaged and analyzed. The 3D projections in Figure 2d,e were generated with Icy [54]. All other maximum z- and 3D projections, including the Supplementary Videos S1–S4, were generated with Fiji.

2.9. Statistical Analysis

Statistical significance was determined using two-way ANOVA followed by Tukey's multiple comparison test using GraphPad Prism 7.05.

3. Results

3.1. Astrocytes Are Readily Infected by both Field Viruses and Lab Strains in Mixed Primary Brain Cell Cultures

To investigate whether field and lab-adapted RABVs differ in their ability to infect primary neurons and non-neuronal astrocytes, hippocampal brain cells were prepared from neonatal rats and were cultivated as mixed cultures containing neurons and glial cells. After 13 days of cultivation, the cultures were infected with 10^3 infectious units of the recombinant field viruses rRABV Dog and rRABV Fox, and the lab-adapted strains rCVS-11 and SAD L16. After 24 h of infection, the cells were fixed and analyzed by confocal laser scanning microscopy using indirect immunofluorescence stainings against RABV nucleoprotein N, neuron marker MAP2 (microtubule-associated protein 2) and astrocyte marker GFAP (glial fibrillary acidic protein). All viruses infected both MAP2-positive neurons and GFAP-positive astrocytes, as demonstrated by the formation of RABV-positive cytoplasmic inclusion bodies (Figure 1). These data indicated that there was no obvious difference between the tested field and the lab-adapted viruses in their ability to infect neurons and non-neuronal astrocytes.

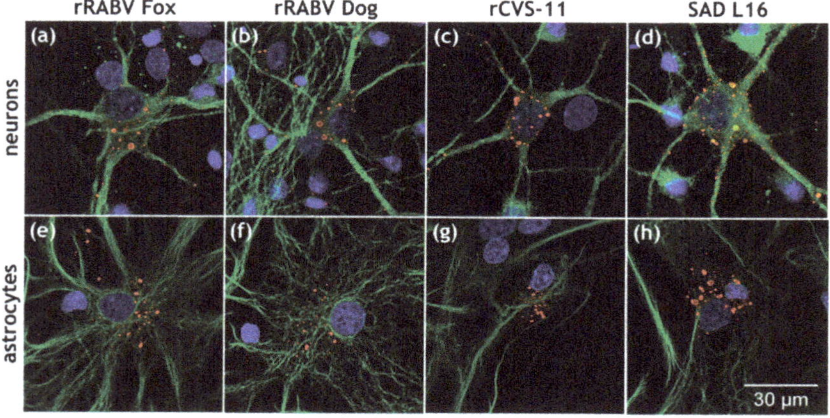

Figure 1. In vitro infection of primary hippocampus neurons and astrocytes by field and lab rabies virus (RABV). (**a–d**) Indirect immunofluorescence detection of RABV nucleoprotein N (red) and microtubule-associated protein 2 (MAP2) (green). Nuclei were counterstained with Hoechst33342 (blue). (**e–f**) Indirect immunofluorescence detection of RABV nucleoprotein N (red) and glial fibrillary acidic protein (GFAP) (green). Nuclei were counterstained with Hoechst33342 (blue). Negative controls for RABV detection are provided in Supplementary Figure S1.

Although all four viruses were able to infect GFAP-positive cells in the hippocampal cell cultures, the majority of the infected cells were MAP2-positive neurons. To quantitatively compare astrocyte infection between viruses, the number of RABV N-positive cells was counted for an area of 25 mm^2, and the percentage of GFAP-positive and -negative RABV-infected cells was determined (Table 1). All four viruses resulted in infection of GFAP-positive cells at levels ranging from 4.9% to 6.7%. Although a higher percentage of GFAP-positive cells could not be excluded because a high number of cells could not be classified into either positive or negative for GFAP, these results indicated that all four viruses could readily establish an infection in cultivated astrocytes and did not substantially differ in their ability to do so.

Table 1. Infection of GFAP- and MAP2-expressing cells in hippocampus cell cultures. RABV-infected cells were counted after indirect immunofluorescence detection of RABV N protein and subdivided in cells that were either negative (−) or positive (+) for GFAP/MAP2. Because of overlaying positive and negative cells in GFAP and MAP2 stainings, a clear assignment was not always possible. Such cells were classified as uncertain. CVS: Challenge Virus Street.

Virus	GFAP (−)		GFAP (+)		Uncertain	
	Count	%	Count	%	Count	%
SAD L16	256	48.3	30	5.7	244	46
rCVS-11	58	64.4	6	6.7	26	28.9
rRABV Dog	252	46.8	29	5.4	258	47.9
rRABV Fox	1002	55.2	89	4.9	712	39.5
Virus	**MAP2 (−)**		**MAP2 (+)**		**Uncertain**	
	Count	%	Count	%	Count	%
SAD L16	18	3.5	387	74.2	116	22.3
rCVS-11	1	1.0	44	43.2	57	55.9
rRABV Dog	27	5.3	347	73.3	109	21.4
rRABV Fox	190	7.3	2096	80.3	324	12.4

3.2. Field RABV Infects Neurons and Astrocytes In Vivo as Demonstrated by High-Resolution 3D Analysis of Infected Brain Tissue

In order to test whether the primary cell culture infection experiments are comparable to the more complex in vivo situation, a sample preparation and imaging pipeline was established, which allows three-dimensional, high-resolution confocal laser scanning image acquisition from RABV-infected tissue for 3D reconstruction and quantification of infected cells. To this end, 1 mm-thick brain slices (Figure 2a) from clinically diseased rRABV Fox field virus-infected mice (i.m. infection route) were immunostained for RABV P, GFAP, and NeuN, optically cleared, and imaged. Acquisition of z-stacks by confocal laser scanning microscopy resulted in z-volumes of 400 μm × 400 μm × 50–100 μm (x, y, and z) from different areas of the infected brain. RABV-infected neurons of variable morphologies were detected by presence of NeuN and RABV P protein accumulation in neuronal cell bodies and neurites (Figure 2b–e; details in Supplementary Figure S2). Astrocytes were detected by filamentous GFAP-positive structures (Figure 2b–g).

Notably, besides accumulation of RABV P in NeuN-positive neurons, P also accumulated at GFAP-positive filaments (Figure 2b,c,f,g; white arrows), indicating a robust infection of astrocytes by field virus rRABV Fox. Evaluation of brain areas with differing localization patterns of the highlighted cellular subpopulations (Figure 2b,c) revealed that astrocyte infections were not restricted to particular regions of the brain but appeared in all imaged neuron layers or brain areas.

Because of the complex 3D morphology of both neurons (long axons in Figure 2b,c) and astrocytes (filamentous structure of GFAP signals in Figure 2f,g), 3D reconstruction of the imaged tissues (Figure 2d,e,g; Supplementary Videos S1 and S2) was performed in order to achieve reliable visualization and quantification of cells in downstream analyses.

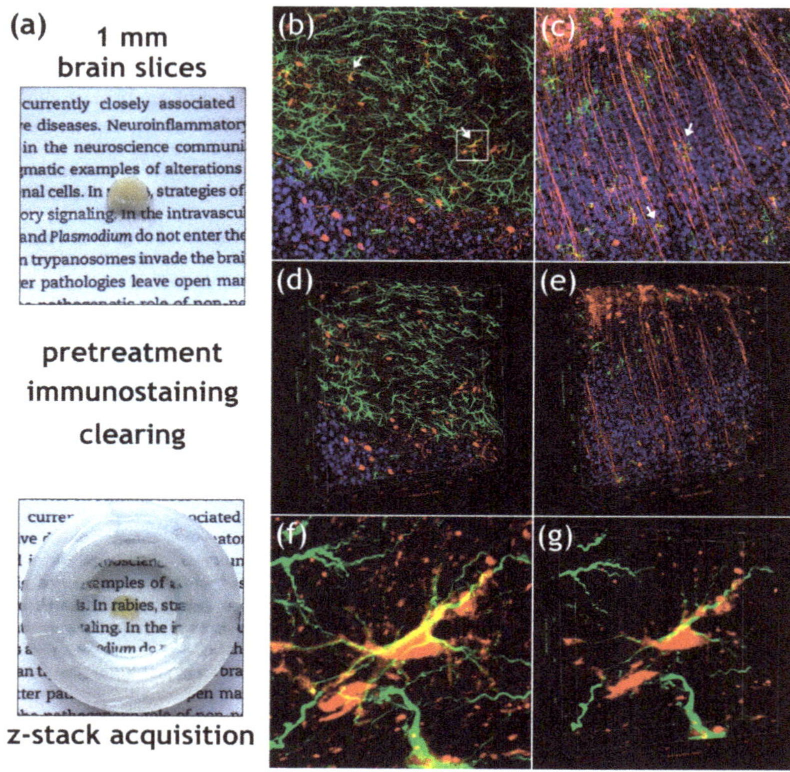

Figure 2. 3D immunofluorescence imaging of field RABV-infected neurons and astrocytes in a mouse brain. (**a**) Workflow for 3D immunofluorescence imaging, including vibratome sectioning into 1 mm slices, pretreatment, immunostaining, and subsequent optical clearing with organic solvents. Confocal imaging and acquisition of cleared tissue slices was done in custom-made imaging containers (see lower image). (**b**,**c**) Maximum z- and (**d**–**e**) 3D projections of z-stack (x, y, z = 400 μm, 400 μm, 59 μm for D and 400 μm, 400 μm, 103 μm for E) after indirect immunofluorescence for RABV phosphoprotein P (red), GFAP (green), and NeuN (blue). White arrows in Figure 2b indicate RABV P and GFAP-positive astrocytes. (**f**) Maximum projection of detail from Figure 2b (see white box) with GFAP-positive cell (green) and associated RABV P fluorescence (red). (**g**) 3D projection of detail view from Figure 2f.

3.3. *Similar Levels of Field Virus Infection in both Neurons and Astrocytes in the Infected Mouse Brain*

Whereas immunofluorescence analysis in Figure 2 clearly demonstrated infection of astrocytes and neurons, they also revealed that only a fraction of both cell types was positive for RABV P. To quantify the ratio of infected and non-infected neurons and astrocytes, 3D object segmentation and counting was performed for NeuN- (Figure 3a,b) or GFAP (Figure 3d,e)-positive cells. The object map was merged with their respective RABV fluorescence (Figure 3c,f; Supplementary Video S3), and the number of RABV-positive neurons and astrocytes was determined by manual counting of RABV P- and cell marker-positive cells. For the z-stack shown in Figures 2b and 3, total numbers of 762 neurons and 272 astrocytes were counted (Table 2, region 2), of which 16 and 18 were RABV positive, respectively.

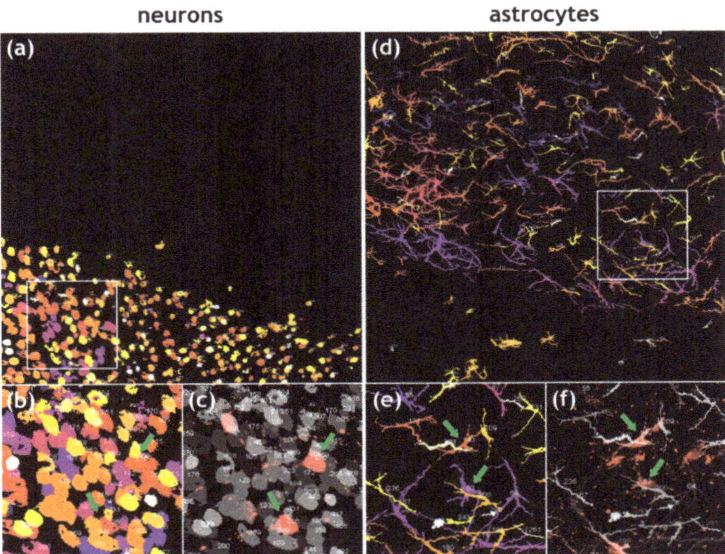

Figure 3. Quantification of RABV-infected neurons and astrocytes. (**a**) Maximum z-projection of objects map for NeuN-positive neurons generated from a confocal z-stack (see Figure 2b). Individual neurons in the z-stack were identified by NeuN-specific fluorescence and converted to objects. Numbers indicate individual cell counts (for improved legibility of the individual numbers, refer to the enlarged details in Supplementary Video S3). n = 762 neurons in a volume of 400 µm × 400 µm × 59 µm. The object colors indicate different z-positions (darker colors in the back and brighter colors in the front). (**b**) Detail of area indicated by white box in Figure 3a. Green arrows indicate RABV-positive neuron cell bodies. (**c**) Overlay of objects map (greyscale) with RABV P signals allows identification and counting of rRABV Fox-infected neurons. (**d**) Maximum z-projection of objects map for GFAP-positive astrocytes generated from a confocal z-stack (see Figure 2b). Individual astrocytes in the z-stack were identified by GFAP-specific fluorescence and converted to objects. Numbers indicate individual cell counts. n = 272 astrocytes in a volume of 400 µm × 400 µm × 59 µm. The object colors indicate different z-positions (darker colors in the back and brighter colors in the front). (**e**) Detail of area indicated by white box in Figure 3d. Green arrows indicate RABV-positive astrocytes. (**f**) Overlay of objects map (greyscale) with RABV P signals allows identification and counting of rRABV Fox-infected astrocytes.

Table 2. Quantification of rRABV Fox-infected neurons and astrocytes. Results of the analysis of six different areas of one infected mouse brain. The quantification pipeline of region 2 is depicted in Figure 3. Neurons and astrocytes were segmented, automatically counted, and overlaid with the respective RABV P signals. Infected cells were counted manually.

	Neurons					Astrocytes				
Region	**Counted**	**Infected**	**%**	**Mean**	**SD**	**Counted**	**Infected**	**%**	**Mean**	**SD**
1	485	21	4.3			362	11	3.0		
2	762	16	2.1			272	18	6.6		
3	448	13	2.9			106	18	17.0		
4	2518	132	5.2	3.9	1.4	488	9	1.8	7.0	4.9
5	3029	91	3.0			273	20	7.3		
6	3847	234	6.1			541	32	5.9		
Σ	11098	507				2042	108			

Analysis of six z-stacks from different RABV-infected brain areas of the same sample led to the detection of 11,089 neurons and 2042 astrocytes, of which 3.9% and 7.0% were RABV positive, respectively (Table 2). The fraction of RABV-positive cells ranged from 2.1% to 6.1% (SD: ± 1.4) for neurons and 1.8% to 17.0% for astrocytes (SD: ± 4.9) (Table 2). These data indicated that rRABV Fox infects astrocytes and neurons in the mouse brain to comparable levels.

Observation of three different rRABV Fox-infected mice revealed that the infection level for astrocytes differed between single animals from 1.2% to 15.6% (SD: ± 6.7) (Supplementary Table S1). Nevertheless, in spite of some variance between individual mice and/or between different analyzed brain regions, detection of infected non-neuronal astrocytes in all animals at surprisingly high levels indicates that astrocyte infection by RABV has been underestimated thus far.

3.4. Astrocyte Infection by RABV Depends on the Type of Virus (Field vs. Lab-Adapted).

To compare the astrocyte tropism of field and lab-adapted viruses after i.m. inoculation, brains of mice infected with three different field viruses (rRABV Fox, rRABV Dog, and rRABV Rac) and two lab-adapted viruses (rCVS-11 and ERA) were analyzed. SAD L16 was excluded from these analyses since it is not able to induce clinical signs after i.m. inoculation (Supplementary Figure S3d).

Whereas rRABV Fox and rRABV Dog caused disease even after infection with a very low virus dose [42], infections with rRABV Rac, rCVS-11, and ERA did not show any clinical signs at a dose of 10^2 $TCID_{50}$ (Supplementary Figure S3). High dose infection with the ERA strain (10^5 $TCID_{50}$) led to 100% disease development (Supplementary Figure S3c), whereas high dose infections with rRABV Rac and rCVS-11 only caused disease in 50% and 16.7% of the infected mice, respectively (Supplementary Figure S3a,b). Accordingly, available tissue samples for the latter two viruses were limited to two and one infected brain. Imaging and quantification of at least six confocal z-stacks from different RABV-infected regions per infected brain was performed.

RABV infections were readily detectable in the brain for all five viruses (Figure 4; Supplementary Video S4). Whereas RABV P antigen accumulated at GFAP-positive structures in rRABV Fox, rRABV Dog, and rRABV Rac-infected brains (Figure 4a,c,e), similar accumulations were not observed in rCVS-11 or ERA-infected brains, in which only infection of NeuN-positive neurons was detected (Figure 4b,d).

Infections with rRABV Fox, rRABV Dog, rRABV Rac, rCVS-11, and ERA led to a mean of 8.3%, 7.3%, 18.9%, 6.9%, and 15.1% RABV-positive neurons (Figure 5; Supplementary Table S1), respectively, demonstrating a comparable level of neuron infection by all five viruses in clinically diseased mice. However, significant differences were observed for the astrocyte infections. Whereas rRABV Fox, rRABV Dog, and rRABV Rac infected 7.6% (SD: ± 6.7), 10.1% (SD: ± 7.7), and 16.5% (SD: ± 15.0) of the astrocytes, no RABV-positive astrocytes were detected in rCVS-11 and ERA-infected samples (Figure 5; Supplementary Table S1). These data indicated that, in contrast to the tested field viruses, the lab-adapted RABVs rCVS-11 and ERA did not infect astrocytes to detectable levels in vivo after i.m. inoculation.

3.5. Confirmation of the Specific Astrocyte Tropism of Field RABV, ERA, and SAD L16 after i.c. Inoculation and Route-Dependent Astrocyte Infection by rCVS-11

To test whether the inoculation route affects astrocyte infection and whether the highly attenuated SAD L16 virus is comparable to the SAD vaccine progenitor strain ERA, brain samples from two animals i.c.-infected with rRABV Fox, rRABV Dog, rCVS-11, or SAD L16 and one infected with the ERA strain were analyzed.

With astrocyte infections at frequencies of 10.9% (SD: ± 10.3), 11.6% (SD: ± 5.6), and 27.2% (SD: ± 12.8) (Figure 6; Supplementary Table S2), rRABV Fox, rRABV Dog, and rRABV Rac led to robust astrocyte infection via the i.c. route and thus confirmed their ability to establish astrocyte infection in vivo (Figure 7a,c,e). Lack of detectable astrocyte infection for the ERA and SAD L16 further

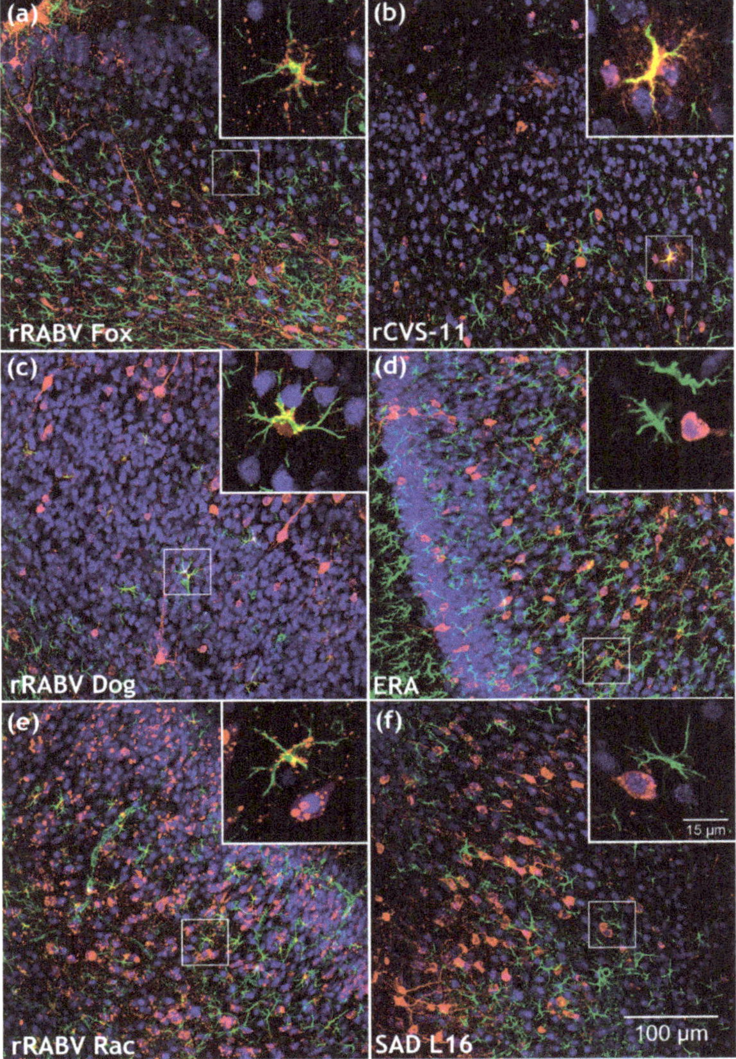

Figure 7. Comparison of field and lab RABV-infected brains after i.c. infection. (**a–f**) Maximum z-projections of z-stacks [x, y = 400 µm, 400 µm (**a–f**); z = 21 µm (**a**, rRABV Fox), 48 µm (**b**, rCVS-11), 74 µm (**c**, rRABV Dog), 98 µm (**d**, ERA), 75 µm (**e**, rRABV Rac) and 67 µm (**f**, SAD L16)] after indirect immunofluorescence for RABV phosphoprotein P (red), GFAP (green), and NeuN (blue). Insets in Figure 7a,b,c,e show RABV P accumulation (red) at GFAP-positive cells (green). To improve visualization of the maximum z-projections, some z-stacks were reduced in thickness. Insets in Figure 7d,f show NeuN- (blue) and RABV P (red)-positive neurons. For the individual channels of the detail images, see Supplementary Figure S5.

Notably, in contrast to the i.m. inoculation route and to the other i.c.-inoculated lab-adapted RABVs SAD L16 and ERA, 13.4% (SD: ± 10.4) of the astrocytes were RABV-positive in rCVS-11-infected animals, indicating that, in the case of rCVS-11, the infection route has a substantial influence on astrocyte infection in the brain. However, the ratio of rCVS-11-infected neurons (8.7%; SD: ± 4.0) remained comparable to i.m. infections.

4. Discussion

By use of a novel immunofluorescence-compatible technique for 3D immunofluorescence imaging of solvent-cleared brain tissue slices [38,39] adapted to the imaging of RABV infections in brain tissue [40], we investigated the cell tropism of six different RABV isolates and lab-adapted strains. Both neurons and astrocytes display heterogeneous morphologies with pronounced three-dimensional projections. Compared to conventional thin sections or even in silico 3D reconstructions of serial thin sections, optical slicing of 1 mm-thick tissue samples during image acquisition allowed fast and seamless 3D reconstruction of immunostained tissues and high-resolution dissection of the detected antigens [40]. For the first time, this allowed systematic analysis and comparison of the tropism of the different viruses in large 3D volumes of infected mouse brains. One example for the superiority of the employed technique is the unambiguous detection of rCVS-11-infected astrocytes after i.c. inoculation (Figures 6 and 7), whereas conventional thin layer immunohistochemistry analyses led to the assumption that infection of glial cells by CVS does not occur, even after i.c. inoculation [55].

Importantly, more than 10^4 neurons and 10^3 astrocytes (Supplementary Table S1 and S2) were quantitatively investigated for RABV infection in their authentic environment per virus and inoculation route. This provided reliable datasets about the frequencies of RABV detection in the two cellular subpopulations in vivo and statistically significant differences in astrocyte infections, which depended on the degree of virulence and the inoculation route of the tested RABVs (Figures 5 and 6).

Because of the clonal and the defined nucleotide sequences, the recombinant viruses rRABV Fox and rRABV Dog [42], rRABV Rac (this work), SAD L16 [43], and rCVS-11 (this work) were selected for the analyses (full genome nucleotide and G protein amino acid sequence comparisons are provided in Supplementary Tables S3 and S4, and Supplementary Figure S6). Since SAD L16 was directly derived from the live vaccine SAD B19 vaccine strain [56], which is equally apathogenic after intramuscular inoculation, the non-recombinant SAD B19 progenitor vaccine virus strain ERA was included to allow for the investigation of attenuated live vaccine virus astrocyte tropism after peripheral i.m. inoculation. Compared to SAD B19, the ERA strain has a less intense cell culture passage history [45] and is still pathogenic in mice after peripheral inoculation [57]. Indeed, whereas SAD L16 was not able to cause disease after i.m. inoculation, ERA was even more pathogenic than rCVS-11 at the high i.m. inoculation dose of 10^5 $TCID_{50}$ with 100% and 16.7% of mice developing clinical signs, respectively (Supplementary Figure S3).

Both the progenitor field virus isolates and the respective recombinant virus clones rRABV Fox and rRABV Dog have been shown to share comparable pathogenicity with efficient disease development after i.m. infection of mice [42]. For the two aforementioned field virus isolates as well as for the raccoon RABV isolate used here for the generation of rRABV Rac, virulence in raccoons has been demonstrated [46]. Notably, although the raccoon virus isolate was highly virulent in its natural reservoir host, the recombinant clone rRABV Rac generated here was less virulent in mice, showing a pathogenicity of 50% only at the high i.m. inoculation dose (Supplementary Figure S3).

In striking contrast to rRABV Fox and rRABV Dog, the lab virus clone rCVS-11 was less efficient via the i.m. route but still able to cause disease in one out of six mice at a high virus dose of 10^5 infectious units per mouse (Supplementary Figure S3). This perfectly matched previous studies, where the non-recombinant progenitor CVS-11 strain caused disease in only 16.7% of the mice at high i.m. dose inoculation, and, as observed here (Supplementary Figure S3), was apathogenic at low dose i.m. infection [51].

In contrast to the in vivo results, in vitro infection of mixed rat hippocampal neuron/astrocyte cultures with rRABV Fox, rRABV Dog, SAD L16, and rCVS-11 led to comparable levels of astrocyte infections, as demonstrated by GFAP-positive, RABV-infected cells (Figure 1) in a range from 4.9% to 6.7% for all viruses (Table 1). These data were in accordance with the general susceptibility of cultivated astrocytes for RABV infection repeatedly shown for field and lab-adapted viruses [29,34], with preferential replication in neurons [32]. Whereas Tsiang et al. reported 90–99% of the primary astrocytes free of RABV antigen, our results indicate that there is less variation within the tested viruses, with virus antigen detection in about 5% of the cultivated astrocytes independent of whether they were of field virus or lab strain origin (Table 1). Overall, comparison of the in vitro and the in vivo data showed that a general susceptibility of a primary CNS cell subpopulation in vitro (Figure 1) does not necessarily reflect the situation in vivo (Figures 2, 4 and 6).

Reasons for this could be altered antiviral response profiles in the more disordered cell culture conditions, different replication and spreading kinetics of the viruses in vivo, or differences in the non-synaptic release of infected neurons to allow infection of non-synaptically connected CNS cells. Whereas no information is available about the latter thus far, it is known that astrocytes are abortively infected by a chimeric SAD L16 virus expressing CVS-11 glycoprotein in mouse brains [33]. This supports the idea that replication of rCVS-11, ERA, and SAD L16 was similarly blocked in astrocytes by potent innate immune responses after infection via the i.m. or the i.c. inoculation routes, respectively. However, since neither viral genome copies nor virus mRNA levels were measured here, it remains to be clarified whether replication kinetics of the lab strains indeed differ from those of the field strains in the infected brains.

Nevertheless, it is conceivable that a more efficient replication of rRABV Fox, rRABV Dog, and rRABV Rac, and therefore the accumulation of the major interferon antagonist phosphoprotein P [8,9] in astrocytes, led to inhibition of antiviral responses. Indeed, lab-attenuated RABV has been shown to differ from wild-type RABV by the induction of an increased type I interferon (IFN) production and expression of inflammatory cytokines via the mitochondrial antiviral-signaling protein (MAVS) pathway [34]. Accordingly, the immunofluorescence detection of RABV P for the identification of infected cells (Figures 2, 4 and 6) confirmed the presence of abundant levels of the major interferon antagonist in the field virus-infected astrocytes. It is highly unlikely that the lack of P detection in the astrocytes of lab RABV-infected animals was due to a major difference in P gene expression of field viruses, since P was readily detectable in lab RABV-infected neurons.

Besides supporting virus replication in astrocytes by decreasing release of inflammatory cytokines, robust field virus replication in these cells may result in a less pronounced or delayed general antiviral response to field viruses. This would be in accordance with the observed immune escape of field viruses in infected animals [35] and may represent a major immunological difference to lab-adapted strains. However, further experimentation will be needed to test this hypothesis.

Notably, and in contrast to the ERA strain not being detected in astrocytes after i.c. inoculation, the astrocyte infection by rCVS-11 was at a frequency of 13.4% after i.c. inoculation (Figure 6 and Supplementary Table S2) and also at higher levels than observed for the field viruses rRABV Fox and rRABV Dog after both i.c. and i.m. infection. This revealed that the ability for rCVS-11 to establish an infection in these cell types depended on the infection route. Since infection from the periphery relies on trans-synaptic spread of the viruses and the virus may not become visible for immune and non-neuronal target cells, astrocyte infection could represent a late phase phenotype of brain infection, where abundant infection of neurons and non-synaptic release of virus particles may facilitate astrocyte infection. Consequently, astrocyte-mediated immune reactions would be delayed and virus elimination prior to disease onset not possible. Compared to the field viruses, slower virus replication and/or spreading kinetics of rCVS-11 could lead to lower levels or the absence of detectable astrocyte infection after i.m. inoculation (Figure 5), although the virus is principally able to infect these cell types (Figure 7b). Indeed, the efficacy of trans-synaptic retrograde spread can differ between a highly neurotropic CVS-24 virus variant and an SAD L16-like vaccine virus [24]. On a higher level, this may also distinguish highly virulent field viruses such as rRABV Fox and rRABV Dog from rCVS-11. Furthermore, it is conceivable that i.c. inoculations represent a shortcut to brain infection with simultaneous and multiple infection of neurons and astrocytes in a non-transsynaptic manner. Higher numbers of infected cells at the beginning of CNS infection compared to trans-synaptic invasion after i.m. infection may lead to faster virus spread and infection of multiple regions of the brain with abundant late phase virus release and astrocyte infection, similar to that speculated above for i.m. infections with highly virulent field viruses.

Indeed, astrocyte activation by other viruses have been shown to occur earlier after i.c. than after peripheral infections [58]. Both virus and cell response kinetics may differ between the two infection routes. Increased induction of neuronal cell death after i.c. RABV infection compared to no detectable apoptosis in i.m.-infected animals [59] further indicates qualitative differences in host reaction to the virus. The special role of astrocytes in the CNS as a main source of IFN-β expression and virus control through TLR (Toll-like receptor) and RLR (RIG-I-like receptor) activation pathways [33,60,61] in combination with the infection route-dependent differences described here, as observed for rCVS-11, may contribute to such differences in apoptosis and other host reaction patterns. However, further studies must clarify whether different infectious route-dependent and -independent virus kinetics can determine astrocyte tropism in vivo and how this affects downstream host reactions. Since differences in innate immune induction through dsRNA between field and attenuated viruses were demonstrated [34], the innate immune induction potential of rRABV Fox, rRABV Dog, rRABV Rac, rCVS-11, ERA, and SAD L16 in astrocytes has to be investigated in order to assess whether the route dependency of rCVS-11 is also affected by different levels of innate immune induction.

Most likely, lack of astrocyte infection by SAD L16 and ERA was the outcome of strong virus inhibition and elimination, as abortive astrocyte infection by a comparable virus has previously been suggested [33]. Whether less antiviral response induction may allow the more virulent rCVS-11 or the highly virulent field viruses to overcome a threshold of virus replication and antagonist expression—and thus may support further replication—will be addressed in future studies. Even though the underlying mechanisms cannot be clarified here, comparable results for SAD L16 and the still pathogenic ERA strain strongly suggest that a general block of detectable astrocyte replication of these vaccine strains represents a major difference to the other tested viruses.

Only by using novel 3D immunofluorescence techniques, we were able to provide comprehensive analyses of the cell tropism of highly virulent field RABVs and less virulent or attenuated lab strains. In particular, high-resolution 3D images and quantitative downstream analysis provided novel insights in the infection processes at the clinical phase of this deadly disease. Although independent of detectable astrocyte infection, symptoms and lethal progression of the disease occur once the virus efficiently spreads in the brain after infection with all six tested viruses (Figure 7 and Supplementary Figure S3). However, different virus kinetics and astrocyte-related innate immune reactions may affect the progression kinetics, immune pathogenicity, and further spread of the virus to peripheral salivary glands. The latter may represent a key issue in terms of field virus transmission and maintenance in host populations.

Whereas these aspects must be addressed in separate trials, this study provides a novel and quantitative basis for a new, dynamic view on RABV host interactions in vivo on the cellular level. Also, this approach showcases the potential of immunostaining-compatible tissue clearing/3D imaging techniques to specifically investigate virus–host interactions at high-resolution on cellular and subcellular levels. While this study focused on standard cellular markers for neurons and astrocytes, future approaches including immune and host pathway markers will pave the way for direct high-resolution imaging-based analysis of infection processes in complex and morphologically preserved tissues.

Supplementary Materials: The following are available online at http://www.mdpi.com/2073-4409/9/2/412/s1, Captions_SupplementaryFiles: Captions and legends to supplementary figures, tables, and videos. Figure S1: Immunofluorescence of non-infected astrocytes and neurons in vitro. Figure S2: Details of rRABV Fox-infected, NeuN-positive neurons. Figure S3: Kaplan-Meyer survival plots for rRABV Rac, rCVS-11, ERA, and SAD L16. Figure S4: Details of field virus (rRABV Fox, rRABV Dog, and rRABV Rac) and lab RABV (rCVS-11 and ERA) infections after i.m. inoculation. Figure S5: Details of field virus (rRABV Fox, rRABV Dog, and rRABV Rac) and lab RABV (rCVS-11, ERA, and SAD L16) infections after i.c. inoculation. Figure S6. Amino acid alignment of glycoprotein G of rRABV Rac, rCVS-11, SAD L16, rRABV Fox, and rRABV Dog. Table S1: Quantification of RABV-infected neurons and astrocytes after i.m. infection. Table S2: Quantification of RABV-infected neurons and astrocytes after i.c. infection. Table S3. Nucleotide sequence alignment of full genome sequences of rRABV Rac, rCVS-11, SAD L16, rRABV Dog, and rRABV Fox. Table S4. Alignment of G protein amino acid sequence of rRABV Rac, rCVS-11, SAD L16, rRABV Dog, and rRABV Fox. Video S1: 3D projections of rRABV Fox-infected astrocytes and neurons in two different areas of a mouse brain. Video S2: 3D projection of an rRABV Fox-infected astrocyte in a mouse brain. Video S3: 3D projection of the quantification of RABV-infected neurons and astrocytes. Video S4: 3D projections of field and lab RABV-infected brains after i.m. infection with rRABV Fox, rRABV Dog, rRABV Rac, rCVS-11, and ERA, and SAD L16 after i.c. infection.

Author Contributions: Conceptualization, S.F., L.Z., T.M. and C.M.F.; methodology, L.Z., V.t.K., T.N. and M.P.; investigation, M.P., L.Z., M.C. and A.K.; writing—original draft preparation, M.P., L.Z. and S.F.; writing—review and editing, M.P., L.Z., M.C., V.t.K., A.K., T.N., C.M.F., T.M. and S.F.; visualization, M.P. and L.Z.; supervision, S.F., T.M. and C.M.F.; project administration, S.F.; funding acquisition, S.F. All authors have read and agreed to the published version of the manuscript.

Funding: This work was funded by an intramural collaborative research grant on Lyssaviruses at the Friedrich-Loeffler-Institute (Ri-0372) to S.F. and C.M.F. L.Z. was supported by the Federal Excellence Initiative of Mecklenburg Western Pomerania and European Social Fund (ESF) Grant KoInfekt (ESF/14-BM-A55-0002/16).

Acknowledgments: We thank Dietlind Kretzschmar and Angela Hillner for technical assistance.

Conflicts of Interest: The authors declare no conflict of interest.

References

1. Fooks, A.R.; Cliquet, F.; Finke, S.; Freuling, C.; Hemachudha, T.; Mani, R.S.; Müller, T.; Nadin-Davis, S.; Picard-Meyer, E.; Wilde, H.; et al. Rabies. *Nat. Rev. Dis. Primers* **2017**, *3*, 17091. [CrossRef]
2. Walker, P.J.; Blasdell, K.R.; Calisher, C.H.; Dietzgen, R.G.; Kondo, H.; Kurath, G.; Longdon, B.; Stone, D.M.; Tesh, R.B.; Tordo, N.; et al. ICTV Virus Taxonomy Profile: Rhabdoviridae. *J. Gen. Virol.* **2018**, *99*, 447–448. [CrossRef] [PubMed]
3. Finke, S.; Conzelmann, K.-K. Replication strategies of rabies virus. *Virus Res.* **2005**, *111*, 120–131. [CrossRef] [PubMed]
4. Zhang, G.; Wang, H.; Mahmood, F.; Fu, Z.F. Rabies virus glycoprotein is an important determinant for the induction of innate immune responses and the pathogenic mechanisms. *Vet. Microbiol.* **2013**, *162*, 601–613. [CrossRef] [PubMed]
5. Besson, B.; Sonthonnax, F.; Duchateau, M.; Ben Khalifa, Y.; Larrous, F.; Eun, H.; Hourdel, V.; Matondo, M.; Chamot-Rooke, J.; Grailhe, R.; et al. Regulation of NF-κB by the p105-ABIN2-TPL2 complex and RelAp43 during rabies virus infection. *PLoS Pathog.* **2017**, *13*, e1006697. [CrossRef] [PubMed]
6. Masatani, T.; Ito, N.; Shimizu, K.; Ito, Y.; Nakagawa, K.; Sawaki, Y.; Koyama, H.; Sugiyama, M. Rabies virus nucleoprotein functions to evade activation of the RIG-I-mediated antiviral response. *J. Virol.* **2010**, *84*, 4002–4012. [CrossRef] [PubMed]
7. Ben Khalifa, Y.; Luco, S.; Besson, B.; Sonthonnax, F.; Archambaud, M.; Grimes, J.M.; Larrous, F.; Bourhy, H. The matrix protein of rabies virus binds to RelAp43 to modulate NF-κB-dependent gene expression related to innate immunity. *Sci. Rep.* **2016**, *6*, 39420. [CrossRef]
8. Brzózka, K.; Finke, S.; Conzelmann, K.-K. Identification of the rabies virus alpha/beta interferon antagonist: Phosphoprotein P interferes with phosphorylation of interferon regulatory factor 3. *J. Virol.* **2005**, *79*, 7673–7681. [CrossRef]
9. Brzózka, K.; Finke, S.; Conzelmann, K.-K. Inhibition of interferon signaling by rabies virus phosphoprotein P: Activation-dependent binding of STAT1 and STAT2. *J. Virol.* **2006**, *80*, 2675–2683. [CrossRef]
10. Ito, N.; Moseley, G.W.; Blondel, D.; Shimizu, K.; Rowe, C.L.; Ito, Y.; Masatani, T.; Nakagawa, K.; Jans, D.A.; Sugiyama, M. Role of interferon antagonist activity of rabies virus phosphoprotein in viral pathogenicity. *J. Virol.* **2010**, *84*, 6699–6710. [CrossRef]
11. Faber, M.; Pulmanausahakul, R.; Hodawadekar, S.S.; Spitsin, S.; McGettigan, J.P.; Schnell, M.J.; Dietzschold, B. Overexpression of the rabies virus glycoprotein results in enhancement of apoptosis and antiviral immune response. *J. Virol.* **2002**, *76*, 3374–3381. [CrossRef] [PubMed]
12. Jackson, A.C.; Rasalingam, P.; Weli, S.C. Comparative pathogenesis of recombinant rabies vaccine strain SAD-L16 and SAD-D29 with replacement of Arg333 in the glycoprotein after peripheral inoculation of neonatal mice: Less neurovirulent strain is a stronger inducer of neuronal apoptosis. *Acta Neuropathol.* **2006**, *111*, 372–378. [CrossRef] [PubMed]

13. Morimoto, K.; Hooper, D.C.; Spitsin, S.; Koprowski, H.; Dietzschold, B. Pathogenicity of different rabies virus variants inversely correlates with apoptosis and rabies virus glycoprotein expression in infected primary neuron cultures. *J. Virol.* **1999**, *73*, 510–518. [CrossRef] [PubMed]
14. Sarmento, L.; Li, X.-q.; Howerth, E.; Jackson, A.C.; Fu, Z.F. Glycoprotein-mediated induction of apoptosis limits the spread of attenuated rabies viruses in the central nervous system of mice. *J. Neurovirol.* **2005**, *11*, 571–581. [CrossRef] [PubMed]
15. Dietzschold, B.; Li, J.; Faber, M.; Schnell, M. Concepts in the pathogenesis of rabies. *Future Virol.* **2008**, *3*, 481–490. [CrossRef] [PubMed]
16. Tsiang, H.; Lycke, E.; Ceccaldi, P.E.; Ermine, A.; Hirardot, X. The anterograde transport of rabies virus in rat sensory dorsal root ganglia neurons. *J. Gen. Virol.* **1989**, *70*, 2075–2085. [CrossRef]
17. Astic, L.; Saucier, D.; Coulon, P.; Lafay, F.; Flamand, A. The CVS strain of rabies virus as transneuronal tracer in the olfactory system of mice. *Brain Res.* **1993**, *619*, 146–156. [CrossRef]
18. Lycke, E.; Tsiang, H. Rabies virus infection of cultured rat sensory neurons. *J. Virol.* **1987**, *61*, 2733–2741. [CrossRef]
19. Ceccaldi, P.E.; Gillet, J.P.; Tsiang, H. Inhibition of the transport of rabies virus in the central nervous system. *J. Neuropathol. Exp. Neurol.* **1989**, *48*, 620–630. [CrossRef]
20. Gluska, S.; Zahavi, E.E.; Chein, M.; Gradus, T.; Bauer, A.; Finke, S.; Perlson, E. Rabies Virus Hijacks and accelerates the p75NTR retrograde axonal transport machinery. *PLoS Pathog.* **2014**, *10*, e1004348. [CrossRef]
21. Bauer, A.; Nolden, T.; Schröter, J.; Römer-Oberdörfer, A.; Gluska, S.; Perlson, E.; Finke, S. Anterograde glycoprotein-dependent transport of newly generated rabies virus in dorsal root ganglion neurons. *J. Virol.* **2014**, *88*, 14172–14183. [CrossRef] [PubMed]
22. Wickersham, I.R.; Finke, S.; Conzelmann, K.-K.; Callaway, E.M. Retrograde neuronal tracing with a deletion-mutant rabies virus. *Nat. Methods* **2007**, *4*, 47–49. [CrossRef] [PubMed]
23. Wickersham, I.R.; Lyon, D.C.; Barnard, R.J.O.; Mori, T.; Finke, S.; Conzelmann, K.-K.; Young, J.A.T.; Callaway, E.M. Monosynaptic restriction of transsynaptic tracing from single, genetically targeted neurons. *Neuron* **2007**, *53*, 639–647. [CrossRef] [PubMed]
24. Reardon, T.R.; Murray, A.J.; Turi, G.F.; Wirblich, C.; Croce, K.R.; Schnell, M.J.; Jessell, T.M.; Losonczy, A. Rabies Virus CVS-N2c(ΔG) Strain Enhances Retrograde Synaptic Transfer and Neuronal Viability. *Neuron* **2016**, *89*, 711–724. [CrossRef] [PubMed]
25. Thoulouze, M.I.; Lafage, M.; Schachner, M.; Hartmann, U.; Cremer, H.; Lafon, M. The neural cell adhesion molecule is a receptor for rabies virus. *J. Virol.* **1998**, *72*, 7181–7190. [CrossRef] [PubMed]
26. Tuffereau, C.; Bénéjean, J.; Blondel, D.; Kieffer, B.; Flamand, A. Low-affinity nerve-growth factor receptor (P75NTR) can serve as a receptor for rabies virus. *EMBO J.* **1998**, *17*, 7250–7259. [CrossRef]
27. Lentz, T.; Burrage, T.; Smith, A.; Crick, J.; Tignor, G. Is the acetylcholine receptor a rabies virus receptor? *Science* **1982**, *215*, 182–184. [CrossRef]
28. Wang, J.; Wang, Z.; Liu, R.; Shuai, L.; Wang, X.; Luo, J.; Wang, C.; Chen, W.; Wang, X.; Ge, J.; et al. Metabotropic glutamate receptor subtype 2 is a cellular receptor for rabies virus. *PLoS Pathog.* **2018**, *14*, e1007189. [CrossRef]
29. Ray, N.B.; Power, C.; Lynch, W.P.; Ewalt, L.C.; Lodmell, D.L. Rabies viruses infect primary cultures of murine, feline, and human microglia and astrocytes. *Arch. Virol.* **1997**, *142*, 1011–1019. [CrossRef]
30. Davis, B.M.; Rall, G.F.; Schnell, M.J. Everything You Always Wanted to Know About Rabies Virus (But Were Afraid to Ask). *Annu. Rev. Virol.* **2015**, *2*, 451–471. [CrossRef]
31. Jackson, A.C.; Phelan, C.C.; Rossiter, J.P. Infection of Bergmann glia in the cerebellum of a skunk experimentally infected with street rabies virus. *Can. J. Vet. Res.* **2000**, *64*, 226–228. [PubMed]
32. Tsiang, H.; Koulakoff, A.; Bizzini, B.; Berwald-Netter, Y. Neurotropism of rabies virus. An in vitro study. *J. Neuropathol. Exp. Neurol.* **1983**, *42*, 439–452. [CrossRef] [PubMed]
33. Pfefferkorn, C.; Kallfass, C.; Lienenklaus, S.; Spanier, J.; Kalinke, U.; Rieder, M.; Conzelmann, K.-K.; Michiels, T.; Staeheli, P. Abortively Infected Astrocytes Appear to Represent the Main Source of Interferon Beta in the Virus-Infected Brain. *J. Virol.* **2016**, *90*, 2031–2038. [CrossRef] [PubMed]
34. Tian, B.; Zhou, M.; Yang, Y.; Yu, L.; Luo, Z.; Tian, D.; Wang, K.; Cui, M.; Chen, H.; Fu, Z.F.; et al. Lab-Attenuated Rabies Virus Causes Abortive Infection and Induces Cytokine Expression in Astrocytes by Activating Mitochondrial Antiviral-Signaling Protein Signaling Pathway. *Front. Immunol.* **2018**, *8*, 2011. [CrossRef] [PubMed]

35. Ito, N.; Moseley, G.W.; Sugiyama, M. The importance of immune evasion in the pathogenesis of rabies virus. *J. Vet. Med. Sci.* **2016**, *78*, 1089–1098. [CrossRef]
36. Suja, M.S.; Mahadevan, A.; Madhusudana, S.N.; Shankar, S.K. Role of apoptosis in rabies viral encephalitis: A comparative study in mice, canine, and human brain with a review of literature. *Patholog. Res. Int.* **2011**, *2011*, 374286. [CrossRef]
37. Wang, Z.W.; Sarmento, L.; Wang, Y.; Li, X.-q.; Dhingra, V.; Tseggai, T.; Jiang, B.; Fu, Z.F. Attenuated Rabies Virus Activates, while Pathogenic Rabies Virus Evades, the Host Innate Immune Responses in the Central Nervous System. *J. Virol.* **2005**, *79*, 12554–12565. [CrossRef]
38. Renier, N.; Wu, Z.; Simon, D.J.; Yang, J.; Ariel, P.; Tessier-Lavigne, M. iDISCO: A simple, rapid method to immunolabel large tissue samples for volume imaging. *Cell* **2014**, *159*, 896–910. [CrossRef]
39. Pan, C.; Cai, R.; Quacquarelli, F.P.; Ghasemigharagoz, A.; Lourbopoulos, A.; Matryba, P.; Plesnila, N.; Dichgans, M.; Hellal, F.; Ertürk, A. Shrinkage-mediated imaging of entire organs and organisms using uDISCO. *Nat. Methods* **2016**, *13*, 859–867. [CrossRef]
40. Zaeck, L.; Potratz, M.; Freuling, C.M.; Müller, T.; Finke, S. High-Resolution 3D Imaging of Rabies Virus Infection in Solvent-Cleared Brain Tissue. *J. Vis. Exp.* **2019**, *30*. [CrossRef]
41. Brewer, G.J.; Torricelli, J.R. Isolation and culture of adult neurons and neurospheres. *Nat. Protoc.* **2007**, *2*, 1490–1498. [CrossRef] [PubMed]
42. Nolden, T.; Pfaff, F.; Nemitz, S.; Freuling, C.M.; Höper, D.; Müller, T.; Finke, S. Reverse genetics in high throughput: Rapid generation of complete negative strand RNA virus cDNA clones and recombinant viruses thereof. *Sci. Rep.* **2016**, *6*, 23887. [CrossRef] [PubMed]
43. Schnell, M.J.; Mebatsion, T.; Conzelmann, K.K. Infectious rabies viruses from cloned cDNA. *EMBO J.* **1994**, 4195–4203. [CrossRef]
44. Abelseth, M.K. An attenuated rabies vaccine for domestic animals produced in tissue culture. *Can. Vet. J.* **1964**, *5*, 279–286.
45. Höper, D.; Freuling, C.M.; Müller, T.; Hanke, D.; von Messling, V.; Duchow, K.; Beer, M.; Mettenleiter, T.C. High definition viral vaccine strain identity and stability testing using full-genome population data–The next generation of vaccine quality control. *Vaccine* **2015**, *33*, 5829–5837. [CrossRef]
46. Vos, A.; Nolden, T.; Habla, C.; Finke, S.; Freuling, C.M.; Teifke, J.; Müller, T. Raccoons (Procyon lotor) in Germany as potential reservoir species for Lyssaviruses. *Eur J. Wildl Res.* **2013**, *59*, 637–643. [CrossRef]
47. Fu, C.; Donovan, W.P.; Shikapwashya-Hasser, O.; Ye, X.; Cole, R.H. Hot Fusion: An efficient method to clone multiple DNA fragments as well as inverted repeats without ligase. *PLoS ONE* **2014**, *9*, e115318. [CrossRef]
48. Finke, S.; Granzow, H.; Hurst, J.; Pollin, R.; Mettenleiter, T.C. Intergenotypic replacement of lyssavirus matrix proteins demonstrates the role of lyssavirus M proteins in intracellular virus accumulation. *J. Virol.* **2010**, *84*, 1816–1827. [CrossRef]
49. Buchholz, U.J.; Finke, S.; Conzelmann, K.K. Generation of bovine respiratory syncytial virus (BRSV) from cDNA: BRSV NS2 is not essential for virus replication in tissue culture, and the human RSV leader region acts as a functional BRSV genome promoter. *J. Virol.* **1999**, *73*, 251–259. [CrossRef]
50. Orbanz, J.; Finke, S. Generation of recombinant European bat lyssavirus type 1 and inter-genotypic compatibility of lyssavirus genotype 1 and 5 antigenome promoters. *Arch. Virol.* **2010**, *155*, 1631–1641. [CrossRef]
51. Eggerbauer, E.; Pfaff, F.; Finke, S.; Höper, D.; Beer, M.; Mettenleiter, T.C.; Nolden, T.; Teifke, J.-P.; Müller, T.; Freuling, C.M. Comparative analysis of European bat lyssavirus 1 pathogenicity in the mouse model. *PLoS Negl. Trop. Dis.* **2017**, *11*, e0005668. [CrossRef] [PubMed]
52. Schindelin, J.; Arganda-Carreras, I.; Frise, E.; Kaynig, V.; Longair, M.; Pietzsch, T.; Preibisch, S.; Rueden, C.; Saalfeld, S.; Schmid, B.; et al. Fiji: An open-source platform for biological-image analysis. *Nat. Methods* **2012**, *9*, 676–682. [CrossRef] [PubMed]
53. Bolte, S.; Cordelières, F.P. A guided tour into subcellular colocalization analysis in light microscopy. *J. Microsc.* **2006**, *224*, 213–232. [CrossRef] [PubMed]
54. de Chaumont, F.; Dallongeville, S.; Chenouard, N.; Hervé, N.; Pop, S.; Provoost, T.; Meas-Yedid, V.; Pankajakshan, P.; Lecomte, T.; Le Montagner, Y.; et al. Icy: An open bioimage informatics platform for extended reproducible research. *Nat. Methods* **2012**, *9*, 690–696. [CrossRef]
55. Jackson, A.C.; Reimer, D.L. Pathogenesis of experimental rabies in mice: An immunohistochemical study. *Acta Neuropathol.* **1989**, *78*, 159–165. [CrossRef]

56. Conzelmann, K.K.; Cox, J.H.; Schneider, L.G.; Thiel, H.J. Molecular cloning and complete nucleotide sequence of the attenuated rabies virus SAD B19. *Virology* **1990**, *175*, 485–499. [CrossRef]
57. Weiland, F.; Cox, J.H.; Meyer, S.; Dahme, E.; Reddehase, M.J. Rabies Virus Neuritic Paralysis: Immunopathogenesis of Nonfatal Paralytic Rabies. *J. Virol.* **1992**, 5096–5099. [CrossRef]
58. Zlotnik, I. The reaction of astrocytes to acute virus infections of the central nervous system. *Br. J. Exp. Pathol* **1968**, *49*, 555–564.
59. Jackson, A.C. *Rabies. Scientific Basis of the Disease and Its Management*, 3rd ed.; Elsevier Science: San Diego, CA, USA, 2013; ISBN 9780123965479.
60. Detje, C.N.; Lienenklaus, S.; Chhatbar, C.; Spanier, J.; Prajeeth, C.K.; Soldner, C.; Tovey, M.G.; Schlüter, D.; Weiss, S.; Stangel, M.; et al. Upon intranasal vesicular stomatitis virus infection, astrocytes in the olfactory bulb are important interferon Beta producers that protect from lethal encephalitis. *J. Virol.* **2015**, *89*, 2731–2738. [CrossRef]
61. Kallfass, C.; Ackerman, A.; Lienenklaus, S.; Weiss, S.; Heimrich, B.; Staeheli, P. Visualizing production of beta interferon by astrocytes and microglia in brain of La Crosse virus-infected mice. *J. Virol.* **2012**, *86*, 11223–11230. [CrossRef]

MDPI
St. Alban-Anlage 66
4052 Basel
Switzerland
Tel. +41 61 683 77 34
Fax +41 61 302 89 18
www.mdpi.com

Cells Editorial Office
E-mail: cells@mdpi.com
www.mdpi.com/journal/cells

www.ingramcontent.com/pod-product-compliance
Lightning Source LLC
LaVergne TN
LVHW071615170726
843515LV00009B/2402

* 9 7 8 3 0 3 6 5 0 1 4 6 8 *